10th International Seminar on Industrial Engineering and Management 2017

Sustainable Development in Industry and Management

IOP Conference Series: Materials Science and Engineering
Volume 277

Tanjung Pandan, Indonesia
7 - 9 November 2017

Editor:

Rahmi Maulidya

ISBN: 978-1-5108-5495-6
ISSN: 1757-8981

Printed from e-media with permission by:

Curran Associates, Inc.
57 Morehouse Lane
Red Hook, NY 12571

Some format issues inherent in the e-media version may also appear in this print version.

Printed by Curran Associates, Inc. (2018)

For permission requests, please contact the Institute of Physics
at the address below.

Institute of Physics
Dirac House, Temple Back
Bristol BS1 6BE UK

Phone: 44 1 17 929 7481
Fax: 44 1 17 920 0979

techtracking@iop.org

Additional copies of this publication are available from:

Curran Associates, Inc.
57 Morehouse Lane
Red Hook, NY 12571 USA
Phone: 845-758-0400
Fax: 845-758-2633
Email: curran@proceedings.com
Web: www.proceedings.com

Table of contents

Volume 277

10th International Seminar on Industrial Engineering and Management "Sustainable Development In Industry and Management"

7-9 November 2017, Tanjung Pandan - Belitung, Indonesia

Accepted papers received: 23 November 2017
Published online: 20 December 2017

Preface

Papers

Decision Support System and Artificial Intelligence (DSS)

Ergonomics

Production System

Quality Engineering & Management

Supply Chain Management

Entrepreneurship / Technopreneurship

IOP Conf. Series: Materials Science and Engineering **277** (2017) 011001 doi:10.1088/1757-899X/277/1/011001

PREFACE

This issue is published in line with the Tenth International Seminar on Industrial Engineering and Management (10th ISIEM). The articles cover a broad spectrum of topics including Quality Engineering Management, Decision Support System and Artificial Intelligent, Ergonomics, Supply Chain Management, Production System, Operation Research, and Industrial Management. The 87 articles provide an overview of critical research issues reflecting on past achievements and future challenges. Those papers were selected from 77 out of 87 abstracts. This statistics shows the high competition to get published on this proceeding.

This issue and the seminar become special as more delegates come and join us from various countries as well as universities compared to previous years. We host 82 delegates both from abroad and local. The 10th ISIEM is hosted by seven universities, i.e. Esa Unggul University, Trisakti University, Tarumanagara University, Al Azhar Indonesia University, Atma Jaya Catholic University of Indonesia, Pasundan University and Pancasila University.

Special thanks go to Prof. Matteo Mario Savino from University of Sannio, Italy; Dr. Nutthapol Asadathorn, Executive Director of TRR Sugar Refinery Group, Thailand; and Dr. Syuhaida Ismail from University of Technology Malaysia, for their contribution as keynote speakers. We are also grateful to all reviewers, for their commitment, effort and dedication in undertaking the task of reviewing all of the abstracts and full papers. Reviewing a large number of submissions in a relatively short time frame is always challenging. Without their help and dedication, it would not be possible to produce this proceeding on time. I highly appreciate all members of ISIEM committees (advisory, steering, and organizing committees) for their mutual efforts and invaluable contribution for the success of the seminar.

Have a pleasant seminar

Dr. Iphov Kumala Sriwana
Chair

IOP Conf. Series: Materials Science and Engineering **277** (2017) 011002 doi:10.1088/1757-899X/277/1/011002

The 10th ISIEM
Belitung, November 7-9, 2017

COMMITTEE

- Chair **Dr. Iphov Kumala Sriwana, S.T., M.Si.**
  (Esa Unggul University)
- Vice Chair **Vivi Triyanti, S.T., M.Sc.**
  (Atma Jaya Catholic University of Indonesia)
- Treasurer **Dr. Winnie Septiani, ST, M.Si.**
  (Trisakti University)
- Secretariat **Endro Wahyono**
  (Tarumanagara University)
- Paper coordinator:
  Nunung Nurhasanah, S.T., M.Si. (Al Azhar Indonesia University)
- Accommodation:
  Vivi Triyanti, S.T., M.Sc. (Atma Jaya Catholic University of Indonesia)
- Event coordinator:
  Stefani Prima Dias Kristiana, S.T.,M.Sc. (Atma Jaya Catholic University of Indonesia)
- Publication coordinator:
  Rahmi Maulidya, S.T., M.T. (Trisakti University)
- Fundraiser:
  Dr. Rina Fitriana, S.T., M.M. (Trisakti University)
  Ir. Rini Prasetyani, M.T. (Pancasila University)
- Indexing:
  Dr. Wisnu Sakti Dewobroto, M.Sc. (Trisakti University)
  Ir. Rini Prasetyani, M.T. (Pancasila University)
- Website Administrator:
  Ir. Wahyukaton, M.T. (Pasundan University)
  Wawan Tripiawan, S.T., M.T. (Telkom University)
- City Tour coordinator:
  Dr. Ir. Nofi Erni, M.M. (Esa Unggul University)
- Seminar coordinator:
  Dr. Lamto Widodo, S.T., M.T. (Tarumanagara University)
- Documentation:
  Ir. Wahyukaton, M.T. (Pasundan University)

Editor

- Chief Editor: Rahmi Maulidya, S.T., M.T.
- Member:
  - Dr. Wisnu Sakti Dewobroto, M.Sc.
  - Dr. Rina Fitriana, S.T., M.M.
  - Asrul Harun Ismail, S.T., M.T., PMP.
  - Bambang Cahyadi, S.T., M.T.
  - Gama Harta Nugraha NR., S.T., M.Sc.
  - Stefani Prima Dias Kristiana, S.T.,M.Sc.
  - Endro Wahyono

IOP Conf. Series: Materials Science and Engineering **277** (2017) 011003 doi:10.1088/1757-899X/277/1/011003

REVIEWER

- Chief Reviewer:
  - Nunung Nurhasanah, S.T., M.Si.
    (Production System, Al Azhar Indonesia University)
- Member:
  - Dr. Ir. Syarif Hidayat, M.Eng.Sc, M.M.
    (Supply Chain Management, Al Azhar Indonesia University)
  - Prof. Parwadi Moengin, Ph.D.
    (Optimization, Trisakti University)
  - Dr. Ir. Hj. Tjutju Tarliah Dimyati, MSIE.
    (Production System, Pasundan University)
  - Dr. Ir. Hj. Arumsari, M.Sc.
    (Quality, Pasundan University)
  - Dr. Lamto Widodo, S.T., M.T.
    (Ergonomics, Tarumanagara University)
  - Vivi Triyanti, S.T., M.Sc.
    (Decision Support System, Atma Jaya Catholic University of Indonesia)
  - Dr. Syuhaida Ismail
    (Supply Chain Management, Malaysia)
  - Dr. Wisnu Sakti Dewobroto, M.Sc.
    (Industrial Management, Trisakti University)
  - Ronald Sukwadi, S.T., M.M., Ph.D.
    (Quality, Atma Jaya Catholic University of Indonesia)
  - Dr. Ir. Yogi Yogaswara, M.T.
    (Optimization and System Modeling, Pasundan University)
  - Ir. Wahyu Katon, M.T.
    (Ergonomics, Pasundan University

IOP Conf. Series: Materials Science and Engineering **277** (2017) 011004 doi:10.1088/1757-899X/277/1/011004

Peer review statement

All papers published in this volume of *IOP Conference Series: Materials Science and Engineering* have been peer reviewed through processes administered by the proceedings Editors. Reviews were conducted by expert referees to the professional and scientific standards expected of a proceedings journal published by IOP Publishing.

Characteristic and factors of competitive maritime industry clusters in Indonesia

N Marlyana[1,2], A E Tontowi[1] and H A Yuniarto[1]

[1] Department of Mechanical and Industrial Engineering, Faculty of Engineering, Universitas Gadjah Mada, Jl. Grafika No.2 Kampus UGM, Yogyakarta, Indonesia, 55281
[2] Department of Industrial Engineering, Faculty of Industrial Technology, Universitas Islam Sultan Agung, Jl. Raya Kaligawe KM. 4, Semarang, Indonesia, 50221
novi@unissula.ac.id

Abstract. Indonesia is situated in the strategic position between two oceans therefore is identified as a maritime state. The fact opens big opportunity to build a competitive maritime industry. However, potential factors to boost the competitive maritime industry still need to be explored. The objective of this paper is then to determine the main characteristics and potential factors of competitive maritime industry cluster. Qualitative analysis based on literature review has been carried out in two aspects. First, benchmarking analysis conducted to distinguish the most relevant factors of maritime clusters in several countries in Europe (Norway, Spain, South West of England) and Asia (China, South Korea, Malaysia). Seven key dimensions are used for this benchmarking. Secondly, the competitiveness of maritime clusters in Indonesia was diagnosed through a reconceptualization of Porter's Diamond model. There were four interlinked of advanced factors in and between companies within clusters, which can be influenced in a pro-active way by government.
Keywords: maritime industry, clusters, competitive, Indonesia

1. Introduction

A cluster is a geographically proximate group of interconnected companies and associated institutions in a particular field, linked by commonalities and complementarities [1]. Maritime clusters can be defined as a network of firm, research, development and innovation units and training organizations, sometimes supported by national or local authorities, which cooperate with the aim of technology innovation and of increasing maritime industry's performance [2]. Maritime industries in broadest terms include all enterprises engaged in the business of designing, constructing, manufacturing, acquiring, operating, supplying, repairing and/or maintaining vessels, or component parts thereof: of managing and/or operating shipping lines, stevedoring and customs brokerage services, shipyards, dry docks, marine railways, marine repair shops, shipping and freight forwarding services and similar enterprises.

President Joko Widodo suggested the creation of a Global Maritime Axis (or Fulcrum) with Indonesia playing a key role as a maritime power [3]. It rests on two respects: geographical situation placing Indonesia as a maritime country and Indonesia's strategic situation prioritizing maritime economic development. Indonesia's strategic location between two continents, Asia and Australia, and between two oceans, the Indian Ocean and the Pacific Ocean, contributes to excellence, at the same time high dependency on such fields as marine and marine economy for national economic development [4].

The major challenge which national (Indonesia) industries are encountering at present is the low industrial competitiveness in international markets [5]. Porter defines international industrial competitiveness as the ability of certain industry of a country, under free international trade conditions (on conditions of non-tariff barriers to trade), to provide more products with higher productivity to consumers (including productive consumption) in international markets compared to that of other countries, as well as to continuously yield more profits [6].

One of factors leading to the Indonesia's low industrial competitiveness is structural weakness of industrial sector. The other factors include weak correlation between industries, either between upstream industries and downstream industries, or between large scale industries and small and medium-scale industries, and underdevelopment of mutually supportive industrial clusters [5]. Porter in [7] points out that the problems with which Indonesia deals in marine and fishery sectors involve weak innovation and low level of cluster-based development.

The article will discuss the characteristics and factors to develop Indonesia's maritime industries with competitiveness through qualitative analysis based on literature review. The first step is analyzing the benchmarking to differentiate the most relevant aspects of maritime industrial cluster among the following countries: Norway, Spain, South West of England, China, South Korea, and Malaysia. The second step is diagnosing the competitiveness of the maritime clusters through the conceptualization of Porter's Diamond to determine the characteristics and factors of the competitiveness of Indonesia's maritime industries.

2. Methods

Qualitative analysis based on literature review has been carried out in two things. Firstly, benchmarking analysis was intended to distinguish the most relevant factors of maritime clusters in several countries in Europe (Norway, Spain, South West of England) and Asia (China, South Korea, Malaysia). Figure as shown below displays the map of those countries. The datas from World Development Indicators in www.worldbank.org explained that Norway (NOR) have the biggest % of territorial waters (60.7%) among the other countries which benchmarked in this paper. Indonesia have 5.8% of territorial waters.

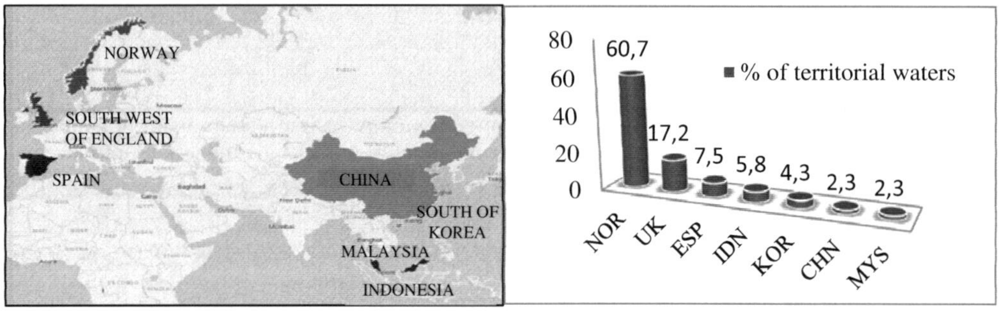

(a) (b)

Figure 1. (a) The map and (b) % of territorial waters[a] of the countries (www.worldbank.org)

[a] % of territorial waters areas are areas of intertidal or subtidal terrain--and overlying water and associated flora and fauna and historical and cultural features--that have been reserved by law or other effective means to protect part or all of the enclosed environment

Benchmarking analysis carried out by 7 (seven) driven factors of the cluster. This mentioned by [8]. The driving factors and decisive points for successes clusters are:

- Geographical concentration— is a key driver of the appearance of the clusters. Industry in a concentrated area will be able to increase efficiency, productivity, and innovation, as a result of the ease of access to knowledge, ideas and skills.
- Specialization—referred to as specialization in core activities that carried out have same markets or processes objectives among the actors who participate in the cluster
- The cluster actors— clusters is seen as a building which consists of several industry, not the relationship between a single industry with other industries as subcontractors or clients.

- Cluster dynamics and linkages: This element related to competition and cooperation relationship among the actors in the cluster. Competition conducted to produce improvement and benefit. Cooperation carried out in terms of using resources and services together to complement each other.
- Critical mass—the critical mass intend that cluster achieve inner dynamics. Critical mass is used to make cluster withstand exogenous shocks or other types of pressures, such as the company's losses.
- The cluster lifecycle—each cluster involves several stages that are not the same and the rate of their evolution is also different. Understanding the patterns of certain cluster characteristics is the logic inherent to the way of developing clusters.
- Innovation—combining changes in technical, commercial and / or organization in a broad sense.

Secondly, the competitiveness of maritime clusters in Indonesia was diagnosed through a reconceptualization of Porter's Diamond model. To understand the dynamics of maritime clusters, a conceptual model is proposed, which highlights the main forces driving a maritime cluster as well as its underlying mechanism. A reconceptualised model for the case of maritime clusters [9], inspired by the Porter Diamond, is shown in Figure 2.

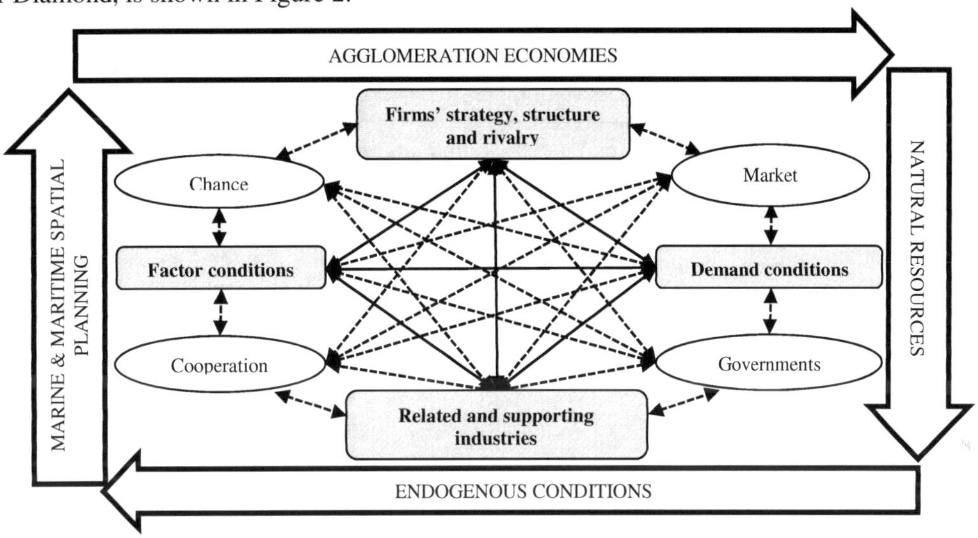

Figure 2. Reconceptualization of Porter's Diamond Model: the case of Maritime Clusters [9]

This paper describes new methodology techniques for analyzing the competitiveness of maritime clusters, using a modification of the Porter Diamond Model. There were four interlinked of advanced factors in and between companies within clusters, which can be influenced in a pro-active way by government. Porter's Competitive Advantage of Nations introduces his diamond model of competitiveness through four broad drivers that shape the environment in which firms and regions compete for business:

- Factor conditions, which include the skills, resources, technology, and infrastructure necessary to create competition in a given industry or cluster;
- Demand conditions, which include the nature of local and overseas demand for industry products and services;
- Related and supporting industries, where the presence or absence of suppliers and distributors in support of industry sectors or clusters will determine competitiveness;
- Firm strategy, structure, and rivalry, which relate to conditions in a nation governing how companies are created, organized, and managed and the nature of domestic rivalry.

3. Result and Discussion

3.1. Benchmarking Analysis

Benchmarking analysis carried out by 7 (seven) driven factors of the cluster in several countries in Europe (Norway, Spain, South West of England) and Asia (China, South Korea, Malaysia). Table 1 (a - g) shown as below:

Table 1 (a – g). Key Cluster Element. Sources: [2], [9], [10], [11], [12], [13], [14], [15]

(a)

Countries	1st Cluster key elements : Geographical concentration
Norway	The maritime cluster in Norway is not based in one region of the country. It is concentrated in different regions along the coastline.
Spain	Focus on one of the seven major seaports located in Spain, is the Port of Bilbao, located in Basque Country. This port is a driver for the development of other auxiliary sectors.
South West of England	The largest concentration of resource based industries, operations and shipping sub-sector, and equipment and service provider sub-sector is in the Devon sub-region. The largest concentration of design and construction businesses is located in Cornwall, closely followed by Devon.
China	China's largest shipbuilding cluster is located in the Yangtze River Delta region. The Yangtze River is the longest river in China. Another cluster of shipbuilding facilities is located in the area extending from the Yellow River (China's second longest river) to the Heilong River in northern China.
South Korea	The Korean shipbuilding industry is heavily concentrated in the south-eastern region of Korea and plays a very important role in the country's economy. The major cluster established in the south-eastern region is composed of four major city-regions, centred on Busan, Ulsan, Changwon and Geoje.
Malaysia	In the shipping services category; the moderately high concentration group is located in Labuan, Penang, Pahang, and Sabah region. For the category of ship building industry, high representation group is situated in Kuala Lumpur, Perak, Johor, Negeri Sembilan and Sarawak region. Ports and terminals category, the high concentration of this industry is clearly distributed in eight states, i.e. Melaka, Penang, Negeri Sembilan, Terengganu, Kedah, Pahang, Perak and Sarawak region.

(b)

Countries	2nd Cluster key elements : Specialisation
Norway	The shipping companies are the most central actors in the maritime industry. They are strongly related to most of the industries within the sector. The classification services, shipping consultants, and shipbuilding have many strong and medium strong links to the rest of the maritime sector.
Spain	The maritime cluster of the Basque Country comprises two anchor areas: ports and shipbuilding. The Port of Bilbao is included within the service sector linked to the internationalization of industrial companies, both in his role as an importer and exporter.
South West of England	Maritime cluster in South West focused in Port activities, including: coastal shipping, short sea shipping, waterborne transport and tourist travelling (e.g cruise liners, visiting pleasure craft).
China	Marine industries are classified as primary (marine agriculture), secondary (all others), and tertiary (marine services). Marine primary industries are include marine fisheries, aquaculture, and tideland farming. Marine secondary industries are offshore oil industry, ocean mining, marine shipbuilding, etc. Marine tertiary industries are include marine transportation, marine tourism, marine information service, etc.
South Korea	Maritime industry in South Korea classified in: marine transportation, harbor, fishery and marine products, and shipbuilding and other marine sectors.
Malaysia	Malaysian maritime industry cluster comprising three main sectors: shipping, ship building, ports and terminals.

(c)

Countries	3rd Cluster key elements : Multiple actors
Norway	The maritime industry in Norway constitutes a complete cluster, composed of three main groups; shipping, maritime services and ship industry. These three main groups surrounded by facilitating associations, educational & research institutions and political bodies.

Spain	The Basque Maritime Forum (BMF) is made up of a heterogeneous group of members, representing different sectors, such as: Institutions, Technology Centres, Associations in the Maritime Sector, Financial entities, Basque State University, Companies, Chambers of Commerce, Port Authorities, Museums.
South West of England	There are three official organisations which deal with marine affairs. They are: Marine South West, Marine Science and Technology and the Advanced Composites Manufacturing Centre, and University of Plymouth.
China	The China shipbuilding industry cluster composed of: associations, financial institution, educational & research institutions, other industries and governance.
South Korea	The shipbuilding industry cluster in South Korea are including: large shipbuilding company, R&D institution, university, and government.
Malaysia	The maritime industry cluster in Malaysia are including: shipping agents, forwarders, governments, associations, universities and R&D institutions

(d)

Countries	4th Cluster key elements : Competition and cooperation
Norway	Half of the Norwegian companies cooperate with other firms around R&D and the supplier-buyer cooperation to improve innovation is particularly strong. The relationships between the industries in the same sector are strong and there are strong relationships between what is traditionally labelled as shipping and the shipping supply industry.
Spain	BMF organises workshops and work groups to foster cooperation and knowledge sharing among marine industry actors and they provide them with tools to tackle issues that wouldn't be possible to deal with in an individual basis. One of the main future objectives are the creation of the network of cooperative research centres and centres of excellence research.
South West of England	The Marine South West networks are a powerful local force for the maritime industry, offering a collective voice to organisations with an interest in the sector. Members include local business personnel and related firms, who meet regularly to share best practice and develop collective knowledge.
China	The Chinese government generally supports the shipbuilding industry by exempting it from tariffs applicable to imports for key components necessary for the production of some kinds of high-tech ships, and by providing incentives for investment in R&D and innovation. The four most common types of cooperation between Chinese and foreign companies are joint ventures, partnerships, Chinese owned companies and Chinese holding companies.
South Korea	Local dynamic relationships are also found among small and medium-sized companies supplying inputs to the shipbuilding companies of the cluster. Among the former, a few large shipbuilders, with local government's assistance, have developed industrial parks close to the shipbuilders to provide cheap land with easy access.
Malaysia	Integration between maritime sectors and supply chain is crucial especially in ship building sector. The support industries shall be promoted in Malaysia by the government and finance institutions such as incentives and tax breaks.

(e)

Countries	5th Cluster key elements : Critical mass
Norway	The Norwegian maritime cluster represents the entire maritime value chain, from ship design to shipping, as well as all relevant ancillary industries. Thus, it can be argued that the Norwegian maritime cluster is one of the most complete maritime clusters in the world.
Spain	The Basque maritime companies are classed into these lines of activity: shipbuilding, fishing and merchant vessel operators and auxiliary industry (equipment manufacturers and subcontractors. The main products exported by the Basque maritime sector, some of whom are first-order global benchmarks, are FORAN Software design, equipment such as motors, pumps, tuna vessels, suction dredgers, offshore vessels, etc.
South West of England	Port cities in South West contribute 45% of the regional total business. The estimated workforces in maritime field makes up 56.7% of the overall estimate of employees. This indicates that each South West port provides a hub to develop a small maritime cluster.
China	Chinese shipyards delivered just 0.9% of all ships in 1985. The proportion increased to 4.7% in 2000 (based on dwt).

South Korea	Korea possesses territorial waters under its jurisdiction five times the size of its land, and produces 3 million tons of fish per annum, while 99.7% of its trade cargo is transported by sea. Its maritime industry has achieved phenomenal growth with international competitiveness in shipbuilding and port industry, and its fisheries sectors ranking around the 10th largest in the world.
Malaysia	The shipbuilding industry has been identified by the Third Industrial Master Plan 2006-2020 (IMP3) as a strategic industry. The IMP 3 focuses on the development of ship building and ship repairing activities to increase the capacity to build and maintain Malaysian ships.

(f)

Countries	6th Cluster key elements : Cluster life cycle
Norway	For three decades the Norwegian continental shelf has been an attractive location due to large and profitable oil fields. The recent decline in investments indicates that the Norwegian sector may be losing its attractiveness. This reveals that many Norwegian offshore suppliers, although they are technologically advanced, lack international competitiveness.
Spain	The maritime cluster association is integrated within the cluster program of the Basque Government (since 1997). Industrial reconversion of the large shipyards in the 1980s and of the small & medium shipyards in the 1990s. The current situation is becoming worse and new challenges are arising when it comes to the future of shipbuilding. The global crisis that started in 2008 has accentuated the economic recession that the shipbuilding sector was suffering due to the introduction in the offer of countries like South Korea and China with an increasing strength and a rapid demand of the ships.
South West of England	In the 1990s, the strong bilateral relationship that involves maritime sectors emerged different. As the delivery port; delivery by ship and marine equipment; maritime legal and financial services the City of London. At the beginning of the 21st century a more multi-lateral approach began to emerge, in recognizing the value of different maritime sectors working together toward better outcomes. In recent years, the integration of marine environmental law and the policy has been introduced as a new approach to the problem of fragmentation.
China	China's marine industry began to develop in the 1950s, with fisheries, transportation, shipping and salt as its commercial focus. China has actively developed its marine resources and made marine development a national economic strategy. China have a strategy to implement ocean development in 2002, the National Marine Economy Development Plan in 2003, the 11th Five-Year Plan (2006–2010), a strategy to develop China's marine industry in 2007, the National Marine Development Plan in 2008, the 12th Five-Year Plan (2011–2015), and improvement to develop marine resources to become a maritime power in 2012.
South Korea	The geographic concentration of firms through the life cycle of the cluster, there has been a strong tendency of geographic concentration of the cluster in the past, but recently the cluster disperses and even internationalizes. The Korean shipbuilding cluster continues to expand its linkages with other countries to acquire advanced skills and know-how, not simply cheap.
Malaysia	For centuries, the maritime areas have always been important to Malaysia and economic activities have blossomed based on the maritime sector which have contributed significantly to Malaysia's economic growth. The challenges to sustain economic growth for the Malaysian economy are clearly laid out in the Seventh Malaysian Plan.

(g)

Countries	7th Cluster key elements : Innovation
Norway	The Norwegian maritime industry is the source of many important innovations in for example ship design, navigation, and advanced equipment. The innovation level is, however, quite high. This seems to imply that Norwegian companies get higher returns from their R&D investments than companies in the other countries do.
Spain	The Basque government cluster program put forward policies to support research. These included the creation of generic technological centres that carry out research in several areas and, specifically, the establishment of Azti-Tecnalia, a technological centre specialized in marine technologies. Specialized education has also been supported both by the regional government and the EU. The Basque government also subsidizes private R&D projects.
South West of England	The application of information and communications technology (ICT) required, in order to achieve the optimisation of the region's transport network, since the utilisation of ICT may

	improve the overall business performance. This may also provide potentially significant spin-off benefits for the ICT sector in the region.
China	China has built up a Marine Economic Accounting System, the Industrial classification for Ocean industries and their related activities, and the Gross Ocean Product Accounting. In order to accelerate its technical capability, China has been importing advanced production methods and key equipment including complete production lines, as well as using foreign sourced hardware and software for computer-aided design and computer- aided manufacturing (CAD/CAM).
South Korea	The Korea Electronics and Telecommunications Research Institute (ETRI), a major government-supported research centre maintaining international levels of technology specializing in semiconductors, telecommunications and information technology, is another example of an institution that is active in advancing shipbuilding technology, especially in relation to electronic devices.
Malaysia	The best fast track option is emulating the world best shipbuilders through joint venture or technical packages or setup satellite suppliers for ship equipment in Malaysia. Through this strategy, the Malaysian entity is able to follow the production system established by ship building nations to build the ships efficiently.

3.2. Reconceptualization of Porter's Diamond model

The modification of the Porter Diamond Model is used to analyze the competitiveness of maritime clusters in Indonesia. The result of the analysis consists of:

- Factor conditions: Indonesia's marine area covers 70% of the total area of the archipelago [16], thus it will be in a superior position to utilize maritime assets [17]. The state policy establish Indonesia as the World Maritime Axis Country [3]. Utilization of shipbuilding industry capacity is under 50% [16].
- Demand conditions: Domestic demands improve the maritime infrastructure to support the transportation of goods and improve economic development across the archipelago [3].
- Related and supporting industries: The structure of national industry in Indonesia is dominated by Small and Medium Industry (SME) [18,19].
- Firm strategy, structure, and rivalry: The cluster of shipping industry in Indonesia has long existed although not all value chains have been filled by domestic industry, as in the upstream industry, there are still many shipbuilding industries using imported raw materials, such as machinery, instrument equipment, electrical equipment, etc. [16].
- Government: The development of strategic sector in maritime transportation is on shipbuilding industry. Establishment of shipbuilding industry clusters based on Presidential Regulation of the Republic of Indonesia (Perpres RI) No 28/2008 on National Industrial Policy and Ministerial Decree of Industry of the Republic of Indonesia (Permenperin RI) No 124/M-IND/PER/10/2009 on the Development of Shipbuilding Industry Cluster [16].
- Cooperation: To accelarate the growth of shipbuilding industries, the cluster of shipbuilding industry may be needed to be developed in some regions. This cluster consists of core industries, supporting industries and related industries [16].
- Chance: the maritime connectivity in future becomes very essential for the economy, trade, food and energy security in the region, mainly when interconnectivity within ASEAN is also crucial for creating a political, economic, and social community. Global economic competition of the 21st century will be about securing resources and markets but competition in the next decade will shift to the oceans [3]. The deployment of the shipbuilding industry is concentrated in West Area of Indonesia. However, East Area of Indonesia has the potential to be developed as it has abundant maritime potential [16].
- Market: Indonesia has a strategic position between two oceans and has the formulation of a maritime state identity, its will expand the opportunities to build a modern maritime industry and for maritime security [3].

4. Conclusion

To conclude, we present a set of important characteristic and factors that can create a framework for the success and sustainable competitive maritime industry clusters in Indonesia. Indonesia has large marine areas such as Norway. The recent potential maritime industry developed in Indonesia is the shipbuilding industry as in the Norway, Spain, South West of England, China, South Korea and Malaysia. Involvement of multiple actors began to be built like in other countries. Competition and cooperation can occur especially between domestic industries and foreign industries as is the case in all countries. And competition in the next decade will shift to the oceans. Government policies exist to promote cluster growth. While innovation should continue to be developed, especially related to shipbuilding technology.

5. References

[1] Porter M E 2000 Location, competition, and economic development: local clusters in a global economy *Econ Dev Q* **14**:15–34 doi:10.1177/089124240001400105

[2] Chang Y 2011 Maritime clusters : What can be learnt from the South West of England *Ocean Coast Manag* **54**:488–94 doi:10.1016/j.ocecoaman.2011.03.005

[3] Pattiradjawane R L and Soebagjo N 2015 Global maritime axis : Indonesia, China, and a new approach to Southeast Asian Regional resilience *Int J China Stud* **6**:175–85

[4] Kusumastanto T 2014 Arah strategi pembangunan Indonesia sebagai negara maritim *www.researchgate.net* 1–30

[5] Kementerian Perindustrian RI 2015 Peta panduan (road map) pengembangan kompetensi inti industri daerah *www.kemenperin.go.id*

[6] Li C 2009 A study on the definition of industrial international competitiveness *2008 Int Semin Bus Inf Manag* **1**:121–4 doi:10.1109/ISBIM.2008.53

[7] Yusuf M and Torbjorn T 2013 Improving Indonesia's competitiveness: innovation, value chains and cluster-bases for realising the huge potential of marine and fisheries *Int J Organ Innov* **6**:128–37.

[8] Andersson T, Serger S S, Sorvik J and Hansson E W 2004 *The Cluster Policies Whitebook*

[9] Monteiro P, Noronha T D and Neto P 2013 A differentiation framework for maritime clusters: comparisons across Europe *Sustainability* **5**:4076–105 doi:10.3390/su5094076

[10] Tsai Y C 2011 The shipbuilding industry in China *OECD J Gen Pap* 2010:37–69

[11] Shin D H and Hassink R 2011 Cluster life cycles: the case of the shipbuilding industry cluster in South Korea *Reg Stud* **45**:1387–402 doi:10.1080/00343404.2011.579594

[12] Yan X, Yan L, Yao X L and Liao M 2015 The marine industrial competitiveness of blue economic regions in China *Mar Policy* **62**:153–60 doi:10.1016/j.marpol.2015.09.015

[13] Kwak S J, Yoo S H and Chang J I 2005 The role of the maritime industry in the Korean national economy: an input-output analysis *Mar Policy* **29**:371–83 doi:10.1016/j.marpol.2004.06.004

[14] Song W L, He G S and McIlgorm A 2013 From behind the Great Wall: The development of statistics on the marine economy in China *Mar Policy* **39**:120–7 doi:10.1016/j.marpol.2012.09.006

[15] Othman M R, Bruce G J and Hamid S A 2011 The strength of Malaysian maritime cluster: the development of maritime policy *Ocean Coast Manag* **54**:557–68 doi:10.1016/j.ocecoaman.2011.02.004

[16] Habibie S, Gumelar M D and Sitorus R 2015 Pengembangan klaster industri perkapalan untuk meningkatkan daya saing industri perkapalan nasional *MPI* **9**:67–76

[17] Evers H D and Karim A 2011 The maritime potential of ASEAN economies *J Curr Southeast Asia Aff* **30**:117–24

[18] Tambunan T 2005 Promoting small and medium enterprises with a clustering approach : a policy experience from Indonesia *J Small Bus Manag* **43**:138–54

[19] Tambunan TTH 2011 Development of small and medium enterprises in a developing country *J Enterprising Communities People Places Glob Econ* **5**:68–82 doi:10.1108/17506201111119626

Warehouses information system design and development

R A Darajatun[1] and Sukanta[2]

[1]Industrial Engineering Departement, Singaperbangsa Karawang University, Indonesia
[1]dosen@rizkidarajatun.org, [2]sukanta.tsm@gmail.com

Abstract. Materials/goods handling industry is fundamental for companies to ensure the smooth running of their warehouses. Efficiency and organization within every aspect of the business is essential in order to gain a competitive advantage. The purpose of this research is design and development of Kanban of inventory storage and delivery system. Application aims to facilitate inventory stock checks to be more efficient and effective. Users easily input finished goods from production department, warehouse, customer, and also suppliers. Master data designed as complete as possible to be prepared applications used in a variety of process logistic warehouse variations. The author uses Java programming language to develop the application, which is used for building Java Web applications, while the database used is MySQL. System development methodology that I use is the Waterfall methodology. Waterfall methodology has several stages of the Analysis, System Design, Implementation, Integration, Operation and Maintenance. In the process of collecting data the author uses the method of observation, interviews, and literature.
Keywords: warehouse management system

1. Introduction

In today's products/goods handling industry, it is fundamental for companies to ensure the smooth running of their warehouses. Efficiency and organization within every aspect of the business is essential in order to gain a competitive advantage.

Disorganized warehouse spaces can cause unnecessary labor costs and the incorrect use of storage systems and racking arrangements result in many companies finding their warehouse shelves full, with no space to receive new inventory. When inventory location is not organized and easily available, pickers will take longer to find items that need to be shipped. This can ultimately lead to a backup in labor. Forklift operators often make multiple trips around the warehouse searching for slots in the storage racking for new inventory. This results in them slotting the pallet of inventory wherever they find an empty space. As the pallet family and size is not carefully considered in its allocation, you will eventually find a large amount of inventory slotted in an unorganized manner, with no room to reorganize due to limited warehouse space.

A cloud-based inventory management app that designed and developed inventory management, control, and tracking apps to make inventory data entry, search, and visualization possible from any location and at any time using web, smartphones and tablets. A smart option to save money with bring your own device (BYOD) barcode scanning and/or inventory, Scan to Spreadsheet is useful for inventory, cycle count, marketing event scanning, and other times when scanning to create a list is necessary.

2. Methods

The concept of the method in making this application is to compare between existing conditions with the expectations of the users and the general knowledge of the various literature.

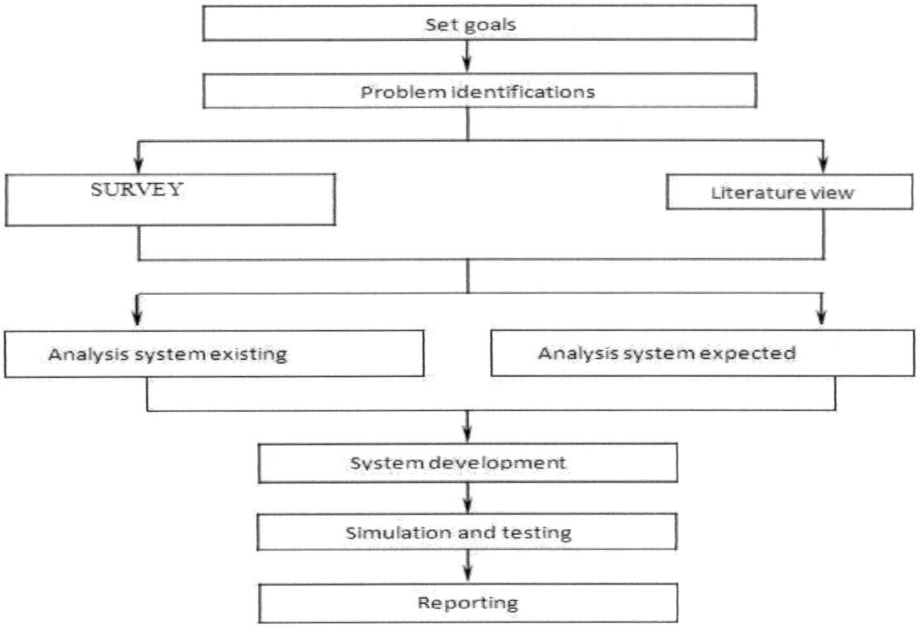

Figure 1. Methods

3. System analysis, design and development

System analysis is the translation of a complete information system into several parts of its components in order to be able to identify and evaluate various problems or obstacles that arise in the current system so that later can be done countermeasures, improvements and also development system.

3.1. System analysis

3.1.1. Current versus expected system

in this section the authors analyze the problems that exist and identify the expectations of all parties, whether internal or external users.

Table 1. Current versus expected system

Item	Current/Existing	Next/Hope
Assambly and FG Handover	Data Kanban recording by manual/visual (handwriting)	Kanban be created by barcode system.
Storage/Rack/Area	There is no system to monitor the position of goods	FG Warehouse is divided into several locations. Goods are actually monitored.
Delivery Export	Control and checking is done manually, and has not been monitored actual arrival of goods	Conducted cross check with barcode scan for items in box, pallet, packling list and case mark

IOP Conf. Series: Materials Science and Engineering **277** (2017) 012002 doi:10.1088/1757-899X/277/1/012002

Table 1. Current versus expected system (cont.)

Item	Current/Existing	Next/Hope
Control Truck	Monitor truck has been done over the phone	Control with GPS, and reporting when goods arrive at customer by App Android based
Delivery preparation process	KANBAN, DIS and DO are visually processed	Checking using Barcode scanner
Input Data	manually data input and not integrated yet	
Monitoring Stock, goods location,	all items can't be monitored in an integrated system	Monitoring by dashboard infromation system

3.1.2. Flow process

The current process can be seen in the flow diagram below.

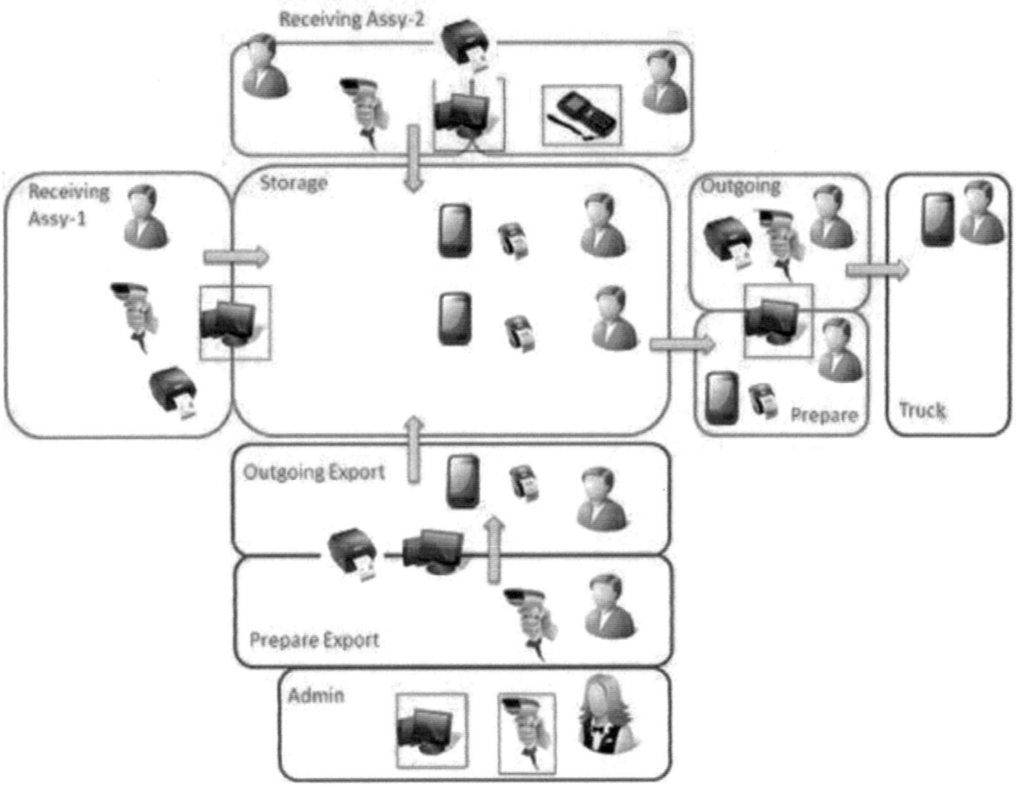

Figure 2. Warehouse expected flow

3.2. System design

3.2.1. Key features

Key features of this application are:

- Real-time inventory synching to create, disperse, manage, and track
- Barcode scanning with mobile devices (Android)
- User-friendly interfaces with easy navigation

- Applicable for any type warehouse manufacture companies (study case in three companies)

3.2.2. Technology
Technologies that are used in making this application include using Google Map API provided by Google. Details can be seen in the following figure;

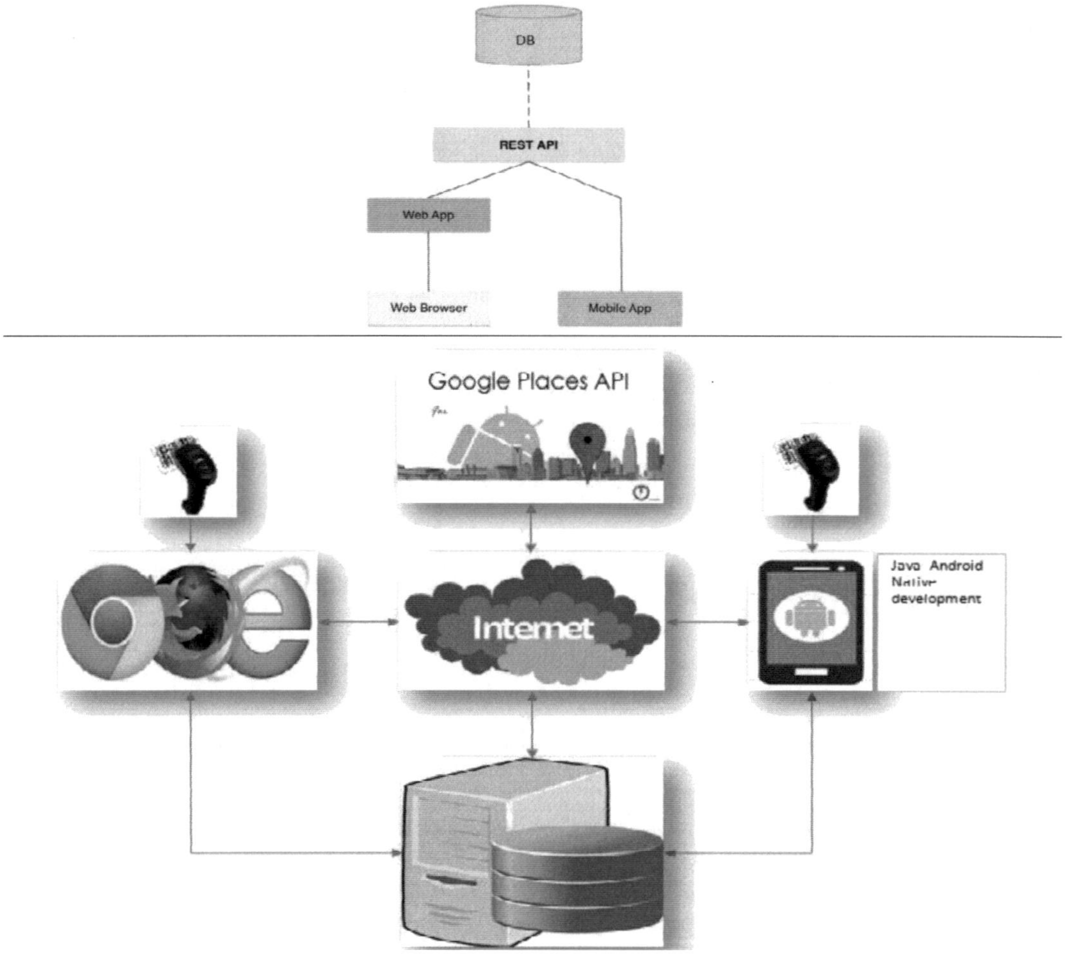

Figure 3. Technology and concept system

3.2.3. Modules
The application consists of several modules. each module. the distribution of modules by function and its users. The following is a list of modules of the created application.

Table 2. List of modules 1

Area	Process	Detail
Receiving	Input production result from assy line	Check model. Lot Number, and Quantity, give barcode each box, and seal plastic cover
		Print barcode for 1 pallet
		Make receiving Document (BSTB)

IOP Conf. Series: Materials Science and Engineering **277** (2017) 012002 doi:10.1088/1757-899X/277/1/012002

Table 2. List of modules 1 (cont.)

Area	Process	Detail
Storage	Place Pallet into warehouse storage Area	Warehouse have small section (A1, A2, Etc..)
		PIC Storage input Storage location on system by scan on area barcode
		PIC Storage able to check stock quantity, location, and FIFO
		Non pallet (partial) part handling
Pre Delivery	PIC pick up from storage area, repacking into delivery trolley	PIC pre delivery check stock location, Lot Number
		Scan delivery instruction sheet (DIS), prepare by admin, to check delivery schedule.
		Repacking from pallet into delivery trolley, and scan each box and print trolley barcode, based on DIS

Table 2. List of modules 2

Area	Process	Detail
Delivery Local Customer	Loading to Truck base on Delivery Note (DN) and monitoring truck movement	PIC prepare trolley base on DN, scan DN Barcode, and Trolley Barcode for confirmation
		Input data for trucking, vendor, plate number, driver, destination
		Hand over GPS device to driver
		Tracking Truck position
		Hand over part from Trucking driver to customer, report, take a picture
Delivery Export	PIC pick up from storage area, repacking into delivery box	PIC pre delivery check stock location, Lot Number
		Scan delivery instruction sheet (DIS), prepare by admin, to check delivery schedule.
		Repacking from pallet into carton box, and scan each box and print export pallet barcode, based on DIS
		Scan export pallet barcode and packing list barcode to re confirm
Stock Control	Admin and Storage PIC able to control and monitoring stock in warehouse	Warehouse member able to see stock level in LED Monitor, real time, data shown on graphic
		Data shown stock level quantity, each model, Lot Number
		Warehouse member can see stock position, rack position
		Warehouse member can see how long will extent for remaining stock, comparing with delivery plan. Production plan
Import Part	Input production result from Assy line	Check model. Lot Number, and Quantity, give barcode each pallet/boxes
		Record for Invoice, Packing List document from origin country
		Make receiving Document (BSTB)

IOP Conf. Series: Materials Science and Engineering **277** (2017) 012002 doi:10.1088/1757-899X/277/1/012002

Table 2. List of modules 2 (cont.)

Area	Process	Detail
In transit	Part production on KID Plant-1 delivery to KID Plant for storage, and vice versa	Part production from Plant-1, but delivery to Plant-2 for storage, and delivery to customer, part stock in Plant-2, and vice versa
Other Account	Combine Part	Part more than 1 part number, but delivery to customer only 1 part number
	Import/Export data to other system from Customer or internal (PRONES)	To prevent additional job, can use data from other system/platform
	Disposal	Part dispose/scrap if not use/NG
	Picture/Data Storage	Warehouse member can storage data (PDF,JPG format) for traceability and paper less
	Temporary Warehouse	If part pending, rework, repair, stock will be separate from OK part, until judgment OK/NG from quality. Classification
	Initial Delivery/First Delivery	Information for initial delivery, due to new model, design change
	Part Borrow	Data if part borrow by other department
	Return /Claim Part	Data if part returned from customer/claimed

Table 2. List of modules 3

Area	Process	Detail
	Transfer Stock	Change stock from part number A to part number B
	BOM	Bill of Material, Part Database
Other	Daily Sales	Report detail all PO monthly
Planning and Monitoring	Monitoring Daily	Daily report for stock level, delivery schedule, etc.
	Delivery Plan	Input delivery plan, weekly, monthly, 3 month
	Traceability	Print out data base on Model, Lot Number, delivery date, etc.
	Production plan	Input production plan weekly, monthly
Packing control	BOM Packing	List of packaging, model, size
	In-Out Packing	Input data from incoming supplier/customer and Out data base on actual delivery quantity
	Delivery Tag	Print delivery tag from system, based on part number
	Packaging Stock Report	Report for packaging stock

3.3. System development

The author uses Java programming language to develop the application, which is used for building Java Web applications, while the database used is MySQL. System development methodology that I use is the Waterfall methodology. Waterfall methodology has several stages of the Analysis, System Design, Implementation, Integration, Operation and Maintenance.

IOP Conf. Series: Materials Science and Engineering **277** (2017) 012002 doi:10.1088/1757-899X/277/1/012002

3.3.1. User intervace (UI)

Web interface made responsive so that can be accessed in various sizes screen without damaging appearance. here are some examples of its user interface.

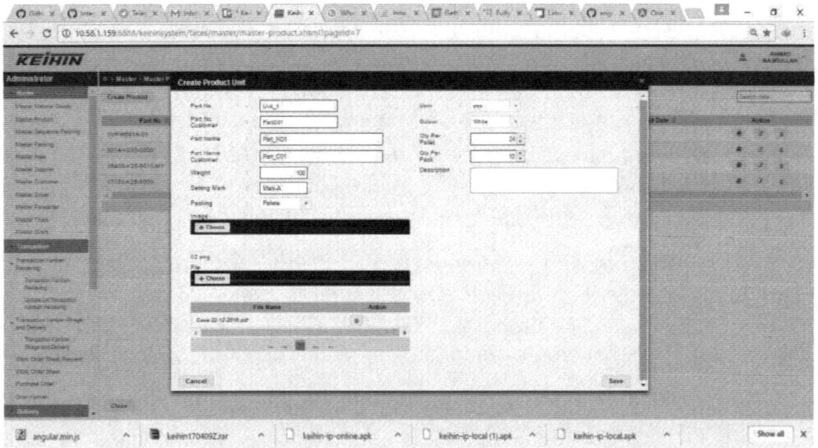

Figure 4. Master Product for Unit/Pcs

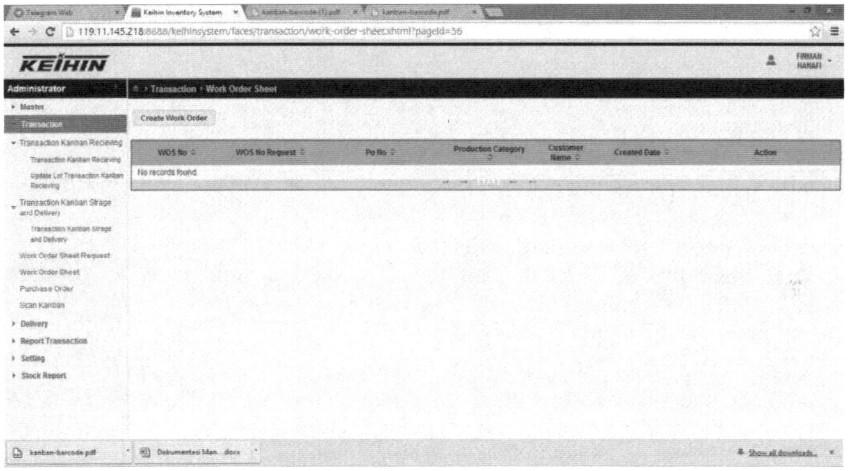

Figure 5. Work Order Sheet

4. Conclusion

The designed system of application warehouse logistics web and android based refers to the process of warehouse logistics that exists in three different factories companies, two automotive factories and one food factory (candy) and the authors conclude core of system requirement are:

1. Mapping of each location / area. Facilitated by: Master data area
2. Mapping of proses repackaging of each product. Facilitated by: Master packaging with parenting system (Parent-child packaging)
3. Systematic recording of the movement products, from where to where. Facilitated by: Android application to scan barcode of product ; scan barcode area before – scan barcode product – forwarding product – scan barcode area destination.

IOP Conf. Series: Materials Science and Engineering **277** (2017) 012002 doi:10.1088/1757-899X/277/1/012002

The application has been successfully created and implemented in one of the automotive parts industry company.

5. Reference

[1] Davis, Gordon B., BK. 1992 *Kerangka Dasar Sistem Informasi, Bagian II Struktur dan Pengembangan*, (Jakarta: PT. Pustaka Binamar Presindo).

[2] Dwi Didik Prasetyo, 2004 *Membangun Aplikasi Web* (Jakarta: PT Elex Media Komputindo kelompok Gramedia).

[3] Jiang, F., Ku, S., 2010. *How to display the data from database* by ListView on Android. *Journal of Technology*, (China: Wuhan University of Technology).

[4] Elmasri, R., Navathe, S. 200. *Fundamentals of Database Systems*, 5th edition (USA: University of Texas).

[5] Wikipedia, 2017 *Warehouse management system*, accessed on April 17, 2017, https://en.wikipedia.org/wiki/Warehouse_management_system.

[6] Infolog, 2017 *Simplifying LKogistic & Supply Chain*, accessed on April 17, 2017, http://www.infolog.com.sg/en/products/infolog-wms-warehouse-management-system

Consumer preference in using the services of bukalapak website with conjoint method

A B Batavio[1], W Tripiawan[2] and H Amani[3]

Industrial Engineering Program, Telkom University, Bandung, Indonesia
[1]arrazzaqbardo@gmail.com,[2]wawan.tripiawan@gmail.com,
[3]husniamani@telkomuniversity.ac.id

Abstract. Internet has become a necessity for society in Indonesia, is seen with the internet impact for the development of information technology in the environment of the community, one of which is E-commerce (electronic commerce) or commonly referred to as the shopping activity is done online, e-commerce is one of the utilization of internet technology is done by the community. With e-commerce, the community may make purchases or sales in accordance with his wishes. Bukalapak one of the companies in Indonesia who stand as an e-commerce company that offers a community to be able to make purchases or sales on his website for free. Bukalapak has competitors to continue to compete in the world of e-commerce in Indonesia. This research aims to know the preferences of consumers in using the services of the site Bukalapak. This research includes the types of descriptive and exploratory research, using the technique of Non-probability sampling and purposive sampling that are specific to the community of Bandung city ever doing online shopping at Bukalapak as a criterion of the respondents in this study, with the number of samples as many as 100 respondents in Bandung. The analysis of the data used is conjoint analysis. The results of this study showed that overall the respondents made the attribute method of the payment as the main preference in using the services of the site Bukalapak, and then followed by the attribute display websites, products and features.
Keywords: Conjoint analysis, E-commerce, Service, and Preferences

1. Introduction

The development of information technology in Indonesia has increased significantly from year to year, one of which is e-commerce which is the utilization of information technology. E-commerce is a buying or trading activity conducted online or online by sellers and buyers, e-commerce facilitates the community in meeting the daily needs for people who want ease of purchase or sale made on the internet with not taking as much time as the Usually done when shopping or selling conventionally.

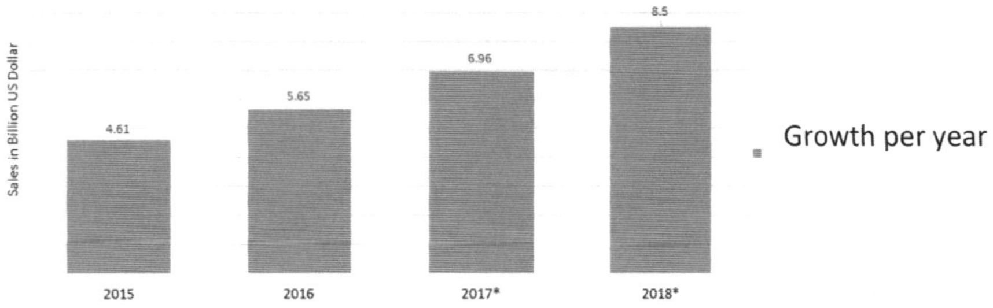

Figure 1. The Growth level of E-commerce in Indonesia (2015-2018)

As described in Figure 1 the growth for e-commerce sales is always increasing in Indonesia from 2015 to 2018, Bukalapak as one of the e-commerce industry actor, are able to get the effect of the growth, so it will help to increase the business development of the company every year.

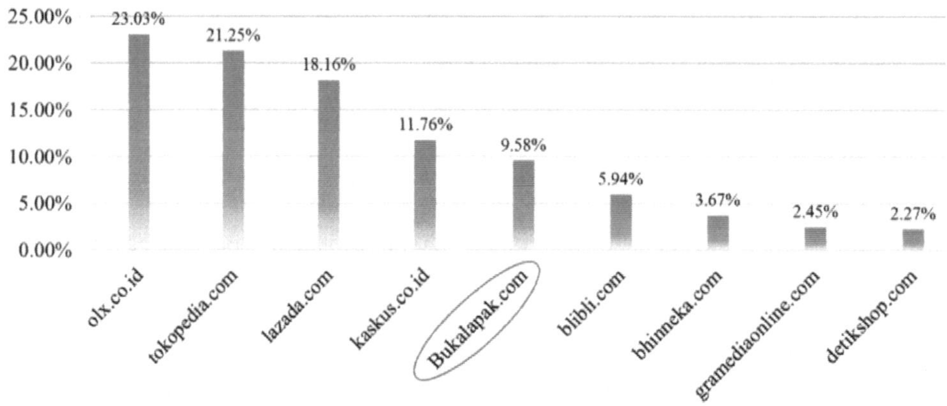

Figure 2. Positioning Map

According to data obtained from the Ministry of Communication and Informatics of the Republic of Indonesia (Kemkominfo) [1] regarding the most frequently used sites by buyers in conducting e-commerce transactions in 2015, that Bukalapak site as one of the e-commerce players in Indonesia are still under its main competitor, Tokopedia.com, with a percentage of 9.58%. This becomes the cornerstone of the problem regarding consumer preferences in choosing an e-commerce site in Indonesia that will be used by consumers in conducting online shopping activities, and the likes or interests of consumers to tend to choose a website that will be used. Therefore, companies need to know the preferences of consumers in using Bukalapak site services, so the target number of visitors per month will be achieved.

2. Methods

2.1 Costumer Behaviors
According to [2], consumer behaviour as a behaviour that displays customers in searching, buying, using, evaluating, and disposing of products and services they expect will satisfy their needs. Meanwhile, according to [3] states consumer behaviour is the study of how individuals, groups, and organizations choose, buy, use and how goods, services, ideas or experiences can satisfy their needs and desires.

2.2 Buying decision
According to [4] purchasing decisions are based on rational motives, emotional motives, or both. Emotional motives involve non-objective factors including friendliness, imitation of others and aesthetics. In consuming a product or service there are stages done by consumers where knowing about this stage can help marketers or companies to understand consumer behaviour. There are five stages for a consumer in the process of decision making are: Problem recognition, Information search, Evaluation of alternatives, Purchase decisions, Post purchase behaviour.

2.3 Ecommerce
E-commerce is a trading activity or service that is closely related to consumers, manufactures, internet service providers, and intermediary traders using electronic media. [3] define *e-commerce* as: *"E-commerce uses a Web site to transact or facilitate the sale of products and services online"*.

IOP Conf. Series: Materials Science and Engineering **277** (2017) 012003 doi:10.1088/1757-899X/277/1/012003

2.4 Product Attribute

According to [5] product attributes represent a communication over the benefits of the development of a product or service that the product or service will offer. Product attributes include product quality, product features, and product styles and design.

2.5 Preference

According to [6] preference is the process of sorting all possible consumption to be selected. According to [7] consumer preferences are closely related to consumer ratings of satisfaction or dissatisfaction towards a product or service even to a particular company. Consumer preferences can be used by a company to identify products desired by the market. According to [3] there are three patterns of preference that can be formed namely:

1. **Homogeneous Preference -** is when all consumers have the same preferences and the market does not show the natural segment. Shows a market where all customers roughly have the same preferences.
2. **Scattered Preferences -** when consumers have very varied preferences. If several brands are in the market, they tend to position themselves throughout the space and show a noticeable difference to match differences in consumer preferences.
3. **Group Preferences -** when natural market segments emerge from consumer groups with shared preferences. Indicates that markets have different groups of preferences.

2.6 Conjoint Analysis

Conjoint analysis is one multivariate analysis technique used to know consumer preference to a product that is from goods or service. There are two information that can be obtained from the conjoint analysis. First, the consumer's preference for the attributes studied. Second, the utility value of each level in each attribute. In identifying the value that respondents provide on the service-based attributes of the research object, conjoint analysis creates the function of part-worth utilities. The value of a product is the sum of the utility value of the product of each of the respondents studied. Conjoint analysis results in the form of quantitative data that can describe consumer preferences of some combination of attributes provided. The stages of conjoint analysis are spelled out into six basic steps proposed by [8]:

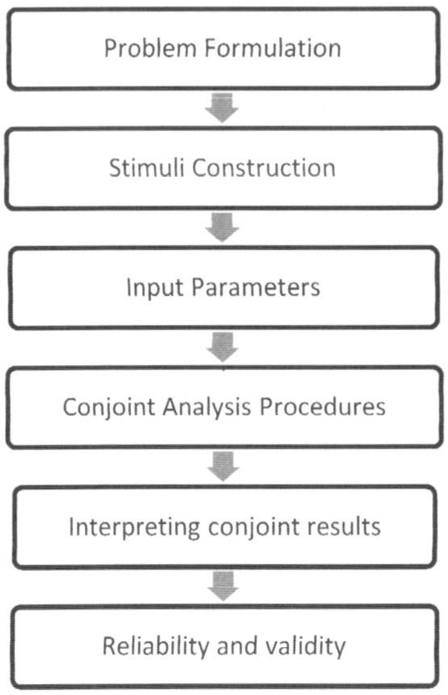

Figure 3. Stages of conjoint analysis

3. Result and Discussion

The importance of each attribute is obtained from the data processing of conjoint analysis using SPSS software. Thus, the overall importance value of each attribute is as follows.

Table 1. Value of Interest

Importance Value	
Attributes	**Value**
1. Website Interface	21.088
2. Product	17.446
3. Features	17.379
4. Payment Method	44.087

(Source: data processing)

From the table above explains that attribute payment method is an attribute that has the highest importance value that is reached 44.087. Furthermore, the attribute of the website become the second most important attribute with the importance of reaching 21.088. Product attribute with interest value 17.446. And the attribute with the lowest importance value is the feature attribute with the importance value of only 17.379.

Table 2. Total Value of Interest and Utility

Importance	Attributes	Utility	Level
21.088	Website Interface	0.020	Interesting website design
		0.013	Website navigation is easy to understand
		0.009	Innovative website appearance
		-0.042	Product ads on the website display
17.446	Product	-0.003	The product is complete and qualified
		0.012	The product matches the given description
		-0.009	The function of the product according to the needs of consumers
17.379	Features	0.051	Negotiation feature
		0.013	Returns feature
		-0.065	Pop-up chat
44.087	Payment Method	0.099	Installment without credit card
		0.305	ATM Transfer
		-0.008	BukaDompet
		0.061	Click Pay
		-0.457	Supermarket Outlets

(Source: data processing)

4. Conclusion

Based on research that has been done by the authors can be drawn conclusions about consumer preferences in the use of Bukalapak site services in the city of Bandung, sorted from the highest interest rates.

1. Payment method is the most important attribute for consumers when using Bukalapak site services with 44.087 importance value.
2. Website Interface becomes the second most important attribute for consumers when using Bukalapak website service with 21.088 importance value.
3. The product becomes the third most important attribute for consumers when using Bukalapak site service with 17.446 importance value.
4. The Features become the fourth most important attribute for consumers when using Bukalapak site services with 17,379 importance value.

5. References

[1] Kominfo 2015 *Situs Yang Digunakan Oleh Pembeli Dalam Transaksi Ecommerce* [Online] Available at: https://statistik.kominfo.go.id/site/data?idtree=430&iddoc=1462&data-data_page=2 (accessed on February 17, 2017)
[2] Schiffman L G and Leslie L K 2010 *Consumer Behavior* Tenth Edition. Pearson Education
[3] Kotler P and Kevin L K 2016 *Marketing Managemen* 15th Edition. Pearson Education, Inc.
[4] Kotler P and Gary A 2012 *Prinsip-prinsip Pemasaran.* 13th Edition. Jakarta: Erlangga
[5] Kotler P and Gary A 2009 *Manajemen Pemasaran.* Jakarta: Erlangga
[6] Frank and Robert H 2011 *Microeconomics and Behavior.* Eighth Edition, Mc.Graw Hill International Edition
[7] Fandy T and Gregorius C 2011 *Service, Quality and Satisfaction* (3rd ed). Yogyakarta: Andi
[8] Naresh K M 2010 *Marketing Research: An Applied Orientation.* Sixth Edition. Pearson Education, Inc.

Islamic value to the modification of BSC model (a case study in evaluating company performance)

Harwati[1] and Y Permana[2]

[1,2]Departement of Industrial Engineering, Universitas Islam Indonesia, Jl. Kaliurang Km 14.4 Yogyakarta
[1]harwati@uii.ac.id, [2]yogapermana174@yahoo.co.id

Abstract. Halal is the key word for the Muslim community in consuming goods or services. Providing halal products is the main focus for companies with the largest consumers from the Muslim community. Halal is not only from the ingredients but also from the process in its production. The purpose of this study is to design a model of performance measurement for a company in order to provide halal products. Balance Score Card (BSC) is used as a framework. BSC as a performance measurement system has advantages not only focus on financial aspect but also non finance. Islamic values are incorporated into four perspective of BSC. They are: Selection of sharia financial institutions in financial perspective, payment method by consumer in consumer perspective, number of product with halal certificate and menu clarify in internal business process perspective, then accuracy of payroll time in learning and growth perspective. The modified BSC model obtains 17 Key Performance Indicators (KPI). The model is applied in a culinary industry (restaurant). Structured questionnaires and scheduled interviews are used to collect the important data. The result show that these industry fail to meet 8 from total 17 KPI so performance achievement is only 70.81%, and only reach 22.58% in Islamic value aspect. This is recorded by the company as an input to improve their performance in order to compete in halal food business.
Keywords: Modified, BSC, Islamic values, Halal, Culinary

1. Introduction

Fashion, food and housing is a basic need of people. Thus, as man's physiological needs are adequately fulfilled, he will develop safety needs [4]. Therefore, business in this sector will always increase and develop at any time. Restaurants as the driving force of the food business as human daily needs are included. Food for Muslims is very important. Every Muslim must make sure what they eat is in appropriate with the religion rules or in other terms called Halal. Thus, food that are made of/or contain liquor, pork, blood of dead animal become Haram for the Muslims [9]. With a large number of restaurants, encourage all parties -like owner, government, Islamic institution- to concern on food issues, so it will be saved to be consumed by everyone. On food products, halal is not only related to the ingredient of the food, but also the business process to make that product.

The concept of halal must be completely understood by consumers or managers to achieve a good quality of life. The demand for halal food needs to be responded by the business owner. In the halal food industry began to develop - especially in Asian countries - halal standardization, traceability system and halal science center [11]. All processes to produce products especially for the needs of Muslim consumers must meet the halal criteria. The halal criteria are from the nature, origin, and method of processing the product made [5]. Every company engaged in the field of food should pay attention to all aspects of the business process to ensure halal products presented. This research is not only focus on calculation how many products are guaranteed as a halal food with the logo of halal from legal institution but also concern to measure how some principles in Islam can be applied in the whole process to serve a product for customer. In the rule of Islam, it is called by Thoyyib. That is why, in the rest of this paper the halal words is changed to Islamic value in order to avoid misunderstanding.

Restaurants as a kind of food company needs to measure their business performance to know and measure whether they have proceed with principles of Islamic value or not. To measure the performance, there is one method of planning and implementation that can be approached to, it is Balanced Scorecard method (BSC). Numerous companies and industries have adopted BSC, which meets several management needs Numerous companies and industries have adopted BSC, which meets several management needs because it can evaluate any organizations by using not only financial indicators but also considering non-financial indicators.[12],[15],[8]. (BSC) methodology focuses on major critical issues of modern business organizations: the effective measurement of corporate performance and the evaluation of the successful implementation of corporate strategy [7]. Principles of Islamic value can be added as some indicators in BSC perspective. So that we get the modified BSC that contain some Islamic value.

Many previous studies used a modified BSC to measure company performance. Modified BSC using sustainability concept was used to assess relationships between CSR and business performances to support the goals of the case hotels [16]. Modified BSC using Fuzzy Multiple Criteria Decision Making (FMCDM) approach was develop to size a banking performance evaluation [14]. BSC and AHP were also combined to measure a telecommunication company [3]. BSC was not only used to measure company performance but also can be used to select supplier [19]. BSC is common and suitable method to present performance of company because it can be applied not only organization but also in non-organization [17]. BSC also success implemented in some sectors of organization like educational, IT, SME, medical, hotels, etc, [2], [6], [1], [10], [20], [12]. The aim of this research is to build a modification model of BSC with Islamic value indicators. This modification BSC is designed to fulfil the requirement of performance measurement that need halal aspect in it. Then, the next step is applying the model in a case study to measure the performance of a food company. Performance resulted is used to suggest the improvement for the company.

2. Method

2.1 Key Performance Indicators (KPI)
KPI is formulated from some previous literature. Using four BSC perspective, KPI is developed into 17 indicators. Islamic values is inserted and combined with original indicators in each perspective of BSC. In financial perspective, Islamic value is interpreted into one indicator. It is percentage or Sharia Bank used. In customer perspective, there is percentage of order transaction as Islamic value. Islam teach its adherent who do buy and sell activities to pay the price upfront. Those rule could represent Islamic value in customer perspective. Learning and growth perspective brings three indicators as representation of Islamic values. There are 1) Percentage of defect product, 2) Percentage of Halal Certified Product, and 3) Clarity of the menu. The last perspective, internal business process perspective covers one Islamic value. It is called payroll time. Payroll time is the time when a worker is paid, this indicator is one of the factors of employee satisfaction. A good paying time according to Islam is timely in accordance with the employment agreement.

2.2 Data Collection
There are 17 modified KPI in four perspective of BSC. The next step is collecting data to measure achievement of each indicators. Data is collected from several different ways depend on its indicator. Questionnaire, interview, and document collection is developed to collect all the data needed. Technique to collect data for each indicators is shown in the following table:

Table 1. KPI and Data Collection Method

Perspective	Strategic Target	Key Performance Indicator (KPI)		Data Collection Method
		Lagging Indicator	Leading Indicator	
Financial	Increasing Sales	Sales Growth	Sales Worth	Document
	Company Ability To Earn Net Profit	Net Profit Margin	Net Income Sales	Document
	Measuring Profitability Level to Capital Investment	Return On Equity	Net profit Investment	Document
	Selecting Sharia Financial Banking	Percentage of Sharia Banking	Number of Bank	Document
Customer	Increasing Customer Satisfaction	Customer Satisfaction Index	Customer Satisfaction Survey	Survey
	Measuring Payment Order System based on Islamic Value	Percentage of Upfront Order Transaction	Number of Order Payment	Survey
Internal Business Process	Increasing Product Quality	Defect Product	Number of Defect Product	Interview
	Increasing Service Quality	Service Time	Customer Service Time Average	Survey
	Increasing Innovation	Innovation in Menu	Number of New Product each year	Interview
	Increasing Customer Trust	Halal Certified Product	Number of Halal Certified Product Total Product	Document
	Increasing Menu Clarity	Clarity of Product Menu	Number of Menu With Price List	Document
Learning and Growth	Increasing Employee Capability	Employee Training	Employee Soft Training Employee Hard Training	Interview
	Increasing Employee Motivation	Employee Retention	Number of Employee Number of Resign Employee	Interview
		Employee Satisfaction Index	Employee Satisfaction Survey	Survey
		Employee Productivity	Net Profit Number of Employee	Document
		Work Accident	Number of Accident	Interview
	Measuring Payroll Accuracy	Employee Payroll Accuracy	Payroll accuracy time	Survey

3. Result

3.1. Measuring KPI Performance

3.1.1 Financial Perspective

In financial perspective, there are four indicator to be measured. Sales growth is measured from the formula below:

$$Sales\ Growth = \frac{Sales\ (\text{Th x}) - Sales\ (\text{Th-x})}{Sales\ (\text{Th-x})}\ x\ 100\% \quad (1)$$

The data from company document and sales growth calculation is shown in the following table:

IOP Conf. Series: Materials Science and Engineering **277** (2017) 012004 doi:10.1088/1757-899X/277/1/012004

Table 2. Sales Growth Average Calculation

Year	Sales	*Sales Growth (%)*	*Average Sales Growth*
2013	Rp 5.311.000.000,-	-	
2014	Rp 7.620.000.000,-	43,47%	41.45%
2015	Rp 10.625.000.000,-	39,43%	

The second indicator is net profit margin. Measuring this indicator uses the formula below:

$$Net\ Profit\ Margin = \frac{\text{Net Profit}}{\text{Sales}}\ x\ 100\% \quad (2)$$

Data from company document and net profit margin calculation is shown in the following table:

Table. 3 Net Profit Margin Average Calculation

Year	Net Profit	Sales	Net Profit Margin	Average
2013	Rp 570.900.000,-	Rp 5.311.000.000,-	10.74%	
2014	Rp 806.400.000,-	Rp 7.620.000.000,-	10.58%	10.88%
2015	Rp 1.202.800.000,-	Rp 10.625.000.000,-	11.32%	

Return of Equity (ROE) as third indicators is measured from percentage of net profit to capital investment. All data from company and ROE calculation is shown in the following table:

Table 4. Return of Equity Calculation

Year	Laba Bersih	Modal	ROE	Average
2013	Rp 570.900.000,-	Rp 1.468.000.000,-	38.88%	
2014	Rp 806.400.000,-	Rp 2.141.000.000,-	37.66%	42.50%
2015	Rp 1.202.800.000,-	Rp 2.359.000.000,-	50.98%	

The forth indicator in financial perspective at once as a representative of Islamic value is percentage of Sharia Banking Used. This company deposits its capital in three different bank. None of these bank is categorized as sharia bank. So, performance for this indicator is 0%.

3.1.2 Customer Perspective

Customer perspective brings two KPI in it. The first is customer satisfaction. Questioners are spread to measure the score of this indicators. From 50 questioner with 28 question using RATER aspect, 49 are valid. 76.2% from customer felt satisfy with the company service. But, because the target from the company itself is 95%, it can be said that the company still under the target. The second KPI is Percentage of Upfront Order Transaction. In Islamic value, customer is suggested to pay the order in front of transaction. In this restaurant case, customer should pay the order before they eat their food and drink. The target of this indicator is 100% but in this company all order is paid in the back. So, the achievement of this indicator is only 0%.

3.1.3 Internal Process Business Perspective

In the perspective of internal business process there are five indicators covered to show company performance. That are product defect, service time, innovation menu, percentage of halal certified product and product menu clarity.

There is no defect food and drink served to the customer, so the achievement in this indicator is 100%. In KPI service time, primary data is taken by counting service time in 30 sample order. Service time average is 9.6 minutes from target 15 minutes. But from 30 data, there are 4 service order more than 15 minutes. So the achievement of this indicator is only 83%. Next KPI is innovation in menu. Every year, this company committed to create one new menu. This commitment is achieved by company. So the score for this KPI is 100%. The last KPI is clarity of menu. This is belong to Islamic value KPI because in Islamic rules sale and purchase must be done on the basis of clarity. Knowing number, type, and price of food and drink represent its clarity. From 51 kind of food and drink in menu, there is only 27 written with explanation of price. SO the performance of this KPI is 52.94%.

3.1.4 Learning and Growth Perspective

This perspective covers the most number of KPI. The first is number of employee training. There are two kind of training. The company provides not only hard skill training but also soft skill training like teamwork and spiritual training. By providing this two king of training, the company could reach 100% performance for this KPI. The second is about employee retention. To measure employee loyalty this formula is used:

$$Employee\ Retention = \frac{\text{Employee Resign}}{\text{Number of Employee}}\ x\ 100\% \quad (3)$$

All data about employee and retention calculation is shown in the following table:

Table 5. Employee Retention

Year	Number of Turnover	Number of Employee	Employee Retention	Average
2013	0	131	0,00 %	
2014	3	128	2,34 %	1.31%
2015	2	126	1,58 %	

The company target for this KPI is 0% of retention, so it can be said that the company still could not fulfill the KPI. The third is Employee satisfaction. About 15 employee were questioned by questioner and the level of satisfaction is 75.42%. The fourth KPI is work accident. For the company record there is no work accident along last three years. SO that the performance of this KPI is 100%. The fifth is employee productivity. This formula is used to calculate it:

$$Employee\ Productivity = \frac{\text{Nett Profit}}{\text{Number of Employee}} \quad (4)$$

The calculation of employee productivity is shown in the following table:

Table 6. Employee Productivity Calculation

Year	Number of Employee	Number or Resign Employee	Nett Profit	Employee Productivity
2013	131	0	Rp 570.900.000,-	Rp. 4.358.000
2014	128	3	Rp 806.400.000,-	Rp. 6.300.000
2015	126	2	Rp 1.202.800.000,-	Rp. 9.546.000

To know the level of productivity, inflation level is used as comparison. Along 2013 to 2015 the inflation rate is 3%. The average growth of employee productivity from 2013 to 2015 is 48.04%. From these comparison it can be conclude that company is success to fulfill the KPI. The last KPI is belong to Islamic value indicator. It is employee payroll accuracy. Not all of the employee feel satisfy with the payroll system. From the questioner, it is known that there are 6 from 15 employee feel unsatisfied with the payroll time. It means that performance of this KPI is only 60%.

3.2 BSC Score

From all analysis above, it can be calculated the performance score of the company as shown in the following table:

Table 7. KPI Performance Score

Perspective	KPI	Target	Performance	Score
Financial	Sales Growth	5%	41,45%	100%
	Net Profit Margin	5%	10,88%	100%
	Return On Equity	12%	42,50%	100%
Customer	Selecting Sharia Financial Banking	100%	0%	0%
	Customer Satisfaction Index	80%	76,20%	95,25%
	Percentage of Upfront Order Transaction	100%	0%	0%
Internal Business Process	Defect Product	0%	0%	100%
	Service Time	15 min	9,6 min	100%
	Innovation in Menu	1	1	100%
	Halal Certified Product	100%	0%	0%
	Clarity of Product Menu	51	27	52,94%
Learning and Growth	Employee Training	100%	100%	100%
	Employee Retention	100%	1,33%	1,33%
	Employee Satisfaction Index	80%	75,42%	94,27%
	Work Accident	0%	0%	100%
	Employee Productivity	3%%	45,20%	100%
	Employee Payroll Accuracy	100%	60%	60%
	Total Score			**70.81%**

4. Conclusion and Discussion

The focus of this research is to measure the company's performance. The results are only relevant for the case study object. Assessment elsewhere will certainly produce different results. Overall prioritization weights of criteria and sub-criteria. This study applies modified BSC that subscribed in determining the performance indicators within each perspective as financial, customer, internal, and company learning and growth as measures of organizational performance. Every perspective contains Islamic value indicator as a part of modification. Result from calculation shows that the company performance is in 70.81%. This score is quite good because still in range 60-80%. But if seen in the achievement of Islamic indicators, from the five indicators, this company is only able to achieve performance 22.58%. This score is categorized as bad category. The results obtained from the calculation have shown to a fact of a real-life problem. A big food company in majority moslem environment is not yet focus on the Islamic operation process. If the company want to fulfil their performance based on Islamic value, they have to repair from the failed indicator. Understanding the impact of each failed indicator may also help management of the company to develop their business as well. First step as a restaurant, the company should register their product to Indonesia Ulema Council (Majelis Ulama Indonesia-MUI) as the only one legal institution who have a legacy to issue halal food certificates. Next step is fixing the business process in internal operation like clarify information to consumers about products written on the menu, pay all the employee salary on time, and starting to use sharia bank as their partners.

However, there are still some limitations of this research and it can be develop in future. First, the KPI that represent Islamic value are only limited in to 6 indicators Next research can develop more indicator related with Islamic value. This can make the modification BSC model more comprehensive. Second, the weight of every indicators is assumed equal. In the real fact, it is possible if the owner or the decision maker feel that some indicators are more important than others. Multi Criteria Decision Making (MCDM) model like TOPSIS, AHP, or ANP can help to weighting the indicators [18]. Weighting process can give suggestion which indicator is prioritized to be improved.

5. References

[1] Abbas Asosheh, Soroosh Nalchigar and Mona Jamporazmey 2010 Information Technology Project Evaluation: An Integrated Data Envelopment Analysis And Balanced Scorecard Approach, *Expert Systems with Applications*, **Vol 37**, Issue 8, p:5931-5938

[2] Alireza Nazariana, Peter Atkinsonb and Pantea Foroudi 2017 Influence Of National Culture And Balanced Organizational Culture On The Hotel Industry's Performance, *International Journal of Hospitality Management* **Vol 63**, P:22–32

[3] Alexandre Veronese Bentes, Jorge Carneiro, Jorge Ferreira da Silva and Herbert Kimura 2012 Multidimensional assessment of organizational performance: Integrating BSC and AHP, *Journal of Business Research* **Vol 65,** p:1790–1799

[4] Barbaca, Domagoja Buljan and Zekan, Senka Borovac 2011 Personal Finances And Hierarchy Of Human Need, *Economy Transdisciplinarity Cognition; Bacau* **Vol 14** No 1 p: 337-348

[5] Bonne and Verbeke 2008 Religious values informing halal meat production and the control and delivery of halal credence quality. Agriculture and Human Values, **25**(1), 35-47

[6] Chang Hung-Yi Wu, Yi-Kuei Lin and Chi-Hsiang 2011 Performance Evaluation Of Extension Education Centers In Universities Based On The Balanced Scorecard, *Evaluation and Program Planning* **Vol 34**, p:37–50

[7] E. Grigoroudis, E. Orfanoudaki and C. Zopounidis 2012 Strategic Performance Measurement In A Healthcare Organisation: A Multiple Criteria Approach Based On Balanced Scorecard, *Omega*, **Vol 40**, p:104–119

[8] Eddy Cardinaels, Paula M.G. van Veen-Dirk 2010 Financial versus non-financial information: The impact of information organization and presentation in Balanced Scorecard, *Accounting, Organizations and Society*, **Volume 35**, Issue 6, Pages 565-578

[9] Ford J B, LaTour M S and Honeycutt E D 1997 An examination of the cross-cultural female response to offensive sex role portrayals in advertising, *Journal of International Marketing Review*, **Vol 14** No. 6, pp. 409-23, Electronic version.

[10] Hamid Tohidia, Aida Jafarib, and Aslan Azimi Afshar 2010 Using Balanced Scorecard In Educational Organizations, *Procedia Social and Behavioral Sciences*,**Vol:2**, p:5544–5548

[11] Hava, et.al. 2009 *New organization to sort out rotten apples in European halal food market. Sunday's Zaman.*

[12] Hao-Chen Huang, Mei-Chi Lai and Lee-Hsuan Lin 2011 Developing Strategic Measurement And Improvement For The Biopharmaceutical Firm Using the BSC Hierarchy, *Expert Systems with Applications* **Vol 38**, p:4875–4881

[13] Hsueh-Fen Chen, Ying-Hui Hou and Ray-E Chang 2012 Application Of The Balanced Scorecard To An Academic Medical Center In Taiwan: The Effect Of Warning Systems On Improvement Of Hospital Performance, *Journal of the Chinese Medical Association* **Vol 75**, p:530-535

[14] Hung-Yi Wu, Gwo-Hshiung Tzeng and Yi-Hsuan Chen 2009 A Fuzzy MCDM Approach For Evaluating Banking Performance Based On Balanced Scorecard, *Expert Systems with Applications*, **Volume 36**, Issue 6, Pages 10135-10147

[15] Jaroslava Kádárová, Michaela Durkáová and Lenka Kalafusová 2014 Innovative Approaches To The Modification Of BSC Model, *Procedia - Social and Behavioral Sciences* **Vol 143**, p:168–173

[16] Jin-Su Kanga, Chun-Fang Chiang, Kitipop Huangthanapanc and Stephen Downing 2015 Corporate Social Responsibility And Sustainability Balanced Scorecard: The Case Study Of Family-Owned Hotels, International Journal of Hospitality Management **48**, p:124–134
[17] Kaplan R and Norton D 1996 The Balanced Scorecard: translating strategy into action. (Boston: Harvard Business Press)
[18] Mohsen Varmazyar, Maryam Dehghanbaghi and Mehdi Afkhami 2016 A Novel Hybrid MCDM Model For Performance Evaluation Of Research And Technology Organizations Based On BSC Approach, *Evaluation and Program Planning* **Vol 58**, p:125–140
[19] Masoud Rahiminezhad Galankashi, Syed Ahmad Helmi and Pooria Hashemzah 2016 Supplier Selection In Automobile Industry: A Mixed Balanced Scorecard–Fuzzy AHP Approach *Alexandria Engineering Journal*, **Vol 55**, p:93–100
[20] Sangjae Lee, Sung Bum Park and Gyoo Gun Lim 2013 Using Balanced Scorecards For The Evaluation Of ''Software-As-A-Service, *Information & Management*, **Vol 50**, p:553–561

Model business intelligence system design of quality products by using data mining in R Bakery Company

R Fitriana[1], J Saragih and N Luthfiana

System & Simulation Industry Laboratory, Industrial Engineering Department, Trisakti University, Jakarta, Indonesia
[1]rinaf@trisakti.ac.id

Abstract. R Bakery company is a company that produces bread every day. Products that produced in that company have many different types of bread. Products are made in the form of sweet bread and wheat bread which have different tastes for every types of bread. During the making process, there were defects in the products which the defective product turns into reject product. Types of defects that are produced include burnt, sodden bread and shapeless bread. To find out the information about the defects that have been produced then by applying a designed model business intelligence system to create database and data warehouse. By using model business Intelligence system, it will generate useful information such as how many defect that produced by each of the bakery products. To make it easier to obtain such information, it can be done by using data mining method which data that we get is deep explored. The method of data mining is using k-means clustering method. The results of this intelligence business model system are cluster 1 with little amount of defect, cluster 2 with medium amount of defect and cluster 3 with high amount of defect. From OLAP Cube method can be seen that the defect generated during the 7 months period of 96,744 pieces.
Keywords: Model, Business Intelligence System, Data Mining, K-Means Clustering

1. Introduction

Nowadays competition becomes more intense because competitors in the same field has been increased. Many of companies flocked to improve their quality product in order to attract customers to buy their products, so did R Bakery. To face the competition with other bread company, R Bakery should be able to pay attention of the weakness factors so the company can overcome these weaknesses and be superior compared with the other competitors. The factors that R Bakery need to consider is the quality of products and how many defects of product that have been produced.

The application of model business intelligence system can make benefit for the company because business intelligence is a technique that used to transform raw data into useful data in the future. Other than that business intelligence can identify unstructured data, can improve the quality of products and create new opportunities business strategies. The purposes of this research are designing a model business intelligence system in R Bakery company and designing data mining models by using k-means clustering method.

Business Intelligence is the extraction process of operational data and collected in a datawarehouse [2]. There are many researches about business intelligence system. Business Intelligence Model in customer relationship management and quality for dairy agroindustry was made as integration of Cube, Data warehouse Model and Fuzzy system and it helps for a fast and efficient transaction in the system [7]. The topic that integrated with BI research is Supply Chain Management, Customer Relationship Management, Data Mining, Data Warehouse, Decision Support System, Performance

Scorecard, Knowledge Management, Business Process Management, Artificial Intelligence, Enterprise Resource Planning, Extract Transformation Loading, OLAP, Quality Management System, Strategic Management. [9]. The role of business intelligence systems was to enhance customer satisfaction and to develop decision support tools to improve customer satisfaction by the service provided [8]. A research consists of object-oriented analysis and design for data warehouse development in the business intelligence system has made. Modelling the core business had supported by financing and quality management model to support a modelling approach to computing intelligence and fuzzy system. [10]. State of the art of this research is the integration between business intelligence, data mining, olap cube in bread company.

The difference between the quality of data and information quality is very evident in business intelligence system. The main goal of business intelligence is to provide the quality of information for decision making [12]. Clustering is a collection of the records that have a similarity with each other of data and have dissimilarity with the record in another cluster. To do this clustering is by using k-means algorithm [5]. K-means is one of the simplest methods of grouping. The procedure is simple easy to classify the data provided through a number of clusters [4]. Distance calculation is using the Euclidean formula as follows:

$$\boldsymbol{d(x,y)} = |\boldsymbol{x}-\boldsymbol{y}| = \sqrt{\sum_{i=1}^{n}(x_i - y_i)^2} \tag{1}$$

Where: n = number of data; x_i = data-i ; y_i = centroid-i

Unified Modelling Language (UML) is a standard language to create a software design [1]. ETL (Extract, Transform, Load) process is a process that needed to make the data that ready for used in data warehouse [6]. Data warehouses are the collections of data that have the character of the subject-oriented, integrated, unchanged and have a particular time frame that supports management decision [3].

2. Research Method

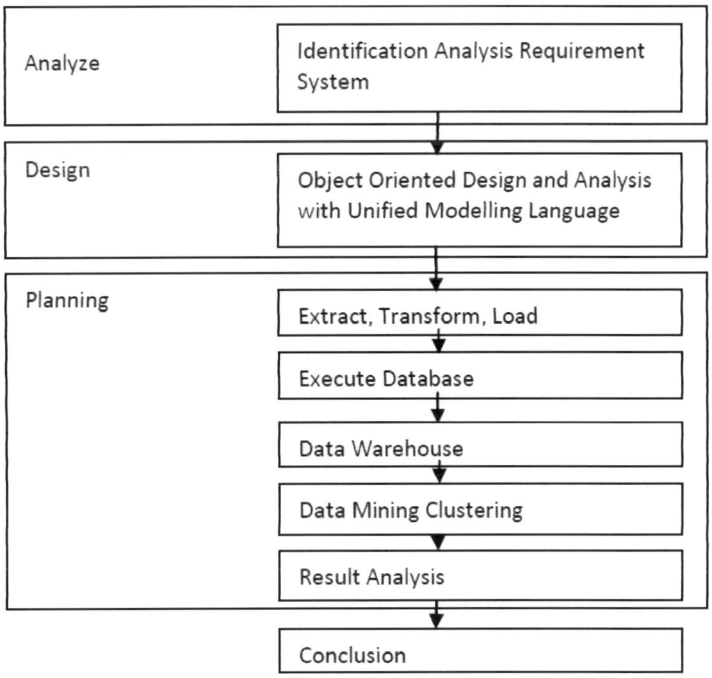

Figure 1. Methodology Flowchart

3. Result and Analysis

3.1. System Analysis

To analyse the need of system can be made by using PIECES method. PIECES method is used to analyse requirements and solutions as well [11]. Aspects that analysed are performance, information, economic, control, efficiency and customer service. This is table PIECES analysis bread company R Bakery.

Table 1. PIECES Analysis

PIECES	Old System	New System
Performance	Time needed to analyze the number of defects generated long enough because the recording is still done manually	Data processing in the form of display the number of the defects product produced faster because it is processed on based computerized
Information	The result of information is less accurate and relevant because of frequent errors and mistake during the process of determining the amount defects are generated	More accurate and relevant because did not occur because many of mistake error in the processing data
Economic	Recording is still done manually	To spend money to buy computer and training Employee to used software
Control	Control safety is less because data that produces still shaped file There is potential human error.	Easy to control, safety control Can help in making decision making With fast and right
Efficiency	Need extra cost and time.	More economical, effective and efficient
Service	Service and searching data less fast because have to open file.	Service and collect data can be faster because using computerized system

System analysis in this research is doing the interview and collecting production data of R Bakery company. From the interviews, it was found that the bakery has not had a database and information system to monitor the amount defect of bread when the production takes place so researches set out to create a database and information systems with design an intelligence business system about the quality of products in R Bakery. The data that retrieved is defective products for 7 months. Here's an overview of business intelligence system design.

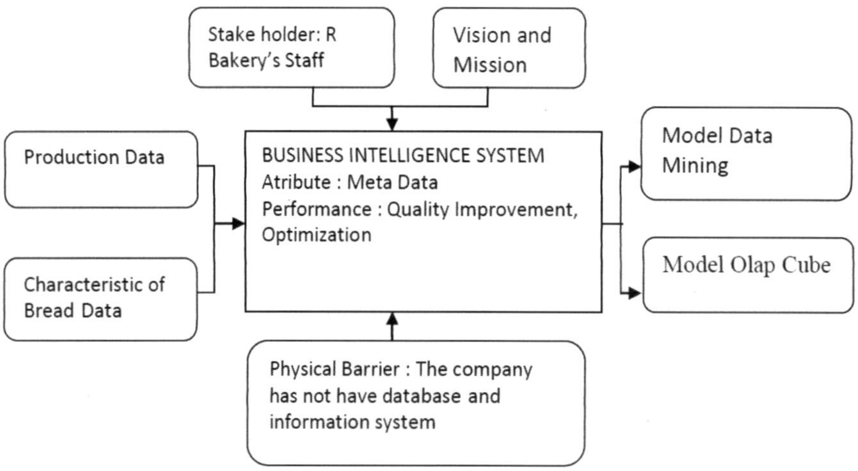

Figure 2. Business Intelligence System

IOP Conf. Series: Materials Science and Engineering **277** (2017) 012005 doi:10.1088/1757-899X/277/1/012005

3.2. UML (Unified Modelling Language)

UML is a stage design that aims to make easier to understand so when designing the system people know of how the system will be created.

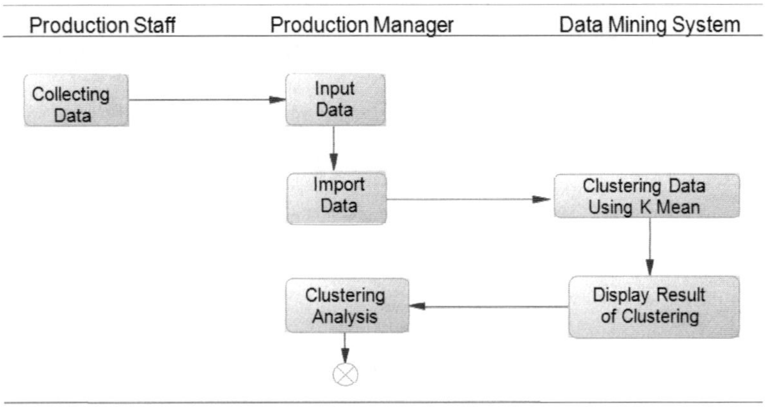

Figure 3. Clustering Activity Diagram

In the data mining activity diagram above, the employee in charge of production in gathering the required data for the establishment of intelligence business system. The required data are the amount of bread and type of defects that had been produced during 7 (months) period starting from June until December of 2016. The data that had been collected then given to the production head that in charge of making intelligence business systems. The method that used to make the system is data mining clustering by using Weka Software.

3.3. Making Data warehouse and Model Business Intelligence System

While making draft of intelligence business system, the first step that must be done is to design a data warehouse that will be created. To create the design of data warehouse is by create a snowflake schema as shown below:

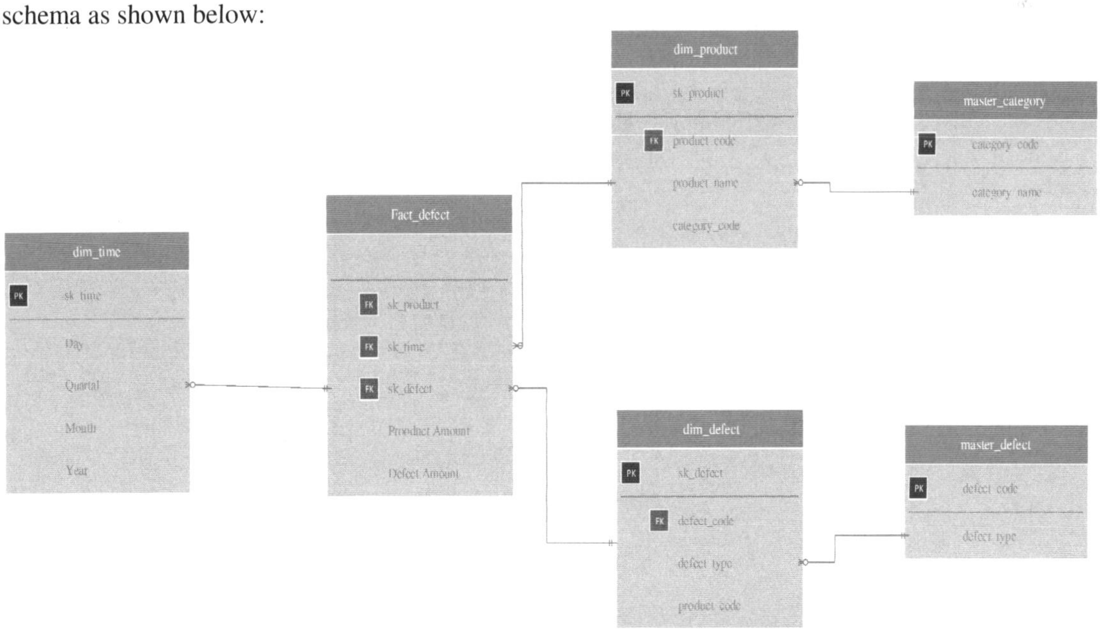

Figure 4. R Bakery Snowflake Schema

From the picture above that in making data warehouse consists of dimensions of time, product, defect, master categories and master of defects which merged later into the data warehouse production which containing a combination of attributes of each dimension. Here's the overview of product dimension and defect dimension that executed by using software called spoon.

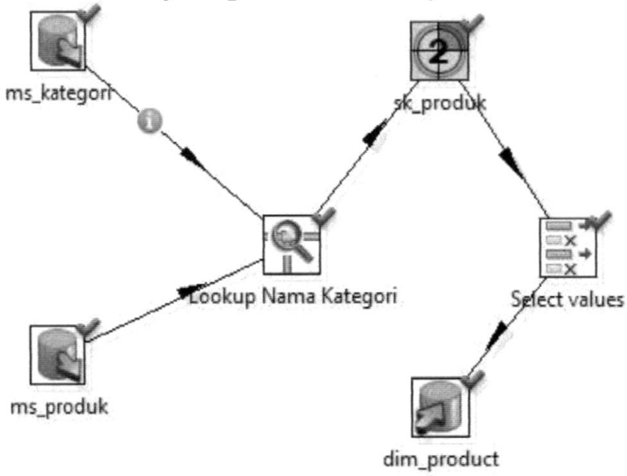

Figure 5. Spoon Product Dimension

In designing intelligence business system, the required database is product dimension, time dimension, defect dimension and fact defect. Fact defect is a merger between the product dimension, time dimension and defect dimension. As seen in the image above, the picture is the visualization of fact defect that contains data information about production data and the number of defects that had been produced during the production process.

3.4. Data Mining

Data mining aims to obtain information from various data sources to make easier in decision making. Data mining method is using k-means clustering method. The use of k-means clustering is that by classifying the product defects during the production process from June until December of 2016 consisting of three clusters include a cluster 1 with little amount of defect, cluster 2 with medium amount of defect and cluster 3 with high amount of defect.

Table 2. R Bakery's Product Defect

Product Code	Product Name	Burnt	Sodden	Shapeless	Product Code	Product Name	Burnt	Sodden	Shapeless
PD-1	Chocolate	2998	2380	2591	PD-13	Beef	933	941	947
PD-2	Coconut	1462	1249	1211	PD-14	Sausage	800	754	667
PD-3	Sarikaya	1319	1303	1244	PD-15	Milk	1192	1072	1068
PD-4	Mocha	1147	1082	1001	PD-16	5 Flavors	604	524	549
PD-5	Donuts	3095	2656	2554	PD-17	Strawberry	721	649	726
PD-6	Jelly	657	567	518	PD-18	Sausage Shredded	449	421	401
PD-7	Satay	1382	1239	1344	PD-19	KPK	566	506	512
PD-8	Choco Banana	1051	907	894	PD-20	Square Set	1542	1341	1474
PD-9	Cheese	1184	1103	1190	PD-21	Small Set	1200	1149	1061
PD-10	Choco Banana Cheese	2306	2074	2017	PD-22	Peel Bread	2819	2888	2931
PD-11	Blueberry	117	69	35	PD-23	Wheat	5012	4085	4109
PD-12	Choco Cheese	1925	1872	1861					

Example of Iteration 1:

Determine First Centre of Cluster
Assumption of first centre:
Taken data-11 as centre of cluster 1: (117, 69, 35)
Taken data-5 as centre of cluster 2: (3095, 2656, 2554)
Taken data-23 as centre of cluster 3: (5012, 4085, 4109)
Calculation Centroid of cluster
To calculate the distance between data and centroid by using Euclidean formula (1) as follows:

$$\boldsymbol{d(x,y)} = |\boldsymbol{x} - \boldsymbol{y}| = \sqrt{\sum_{i=1}^{n}(x_i - y_i)^2}$$

PD-1 = $\sqrt{(2998-117)^2 + (2380-69)^2 + (2591-35)^2}$ = 4991.549621

PD-2 = $\sqrt{(1462-117)^2 + (1249-69)^2 + (1211-35)^2}$ = 2141.121435

The number of iterations as many as three iterations. The calculation to determine the centroid of cluster was the same as before, the difference of distance of the cluster is the average value for each cluster, so the result as follows:

Table 3. Result of Clustering

Product Code	Product Name	Burnt	Sodden	Shapeless	Cluster
PD-1	Chocolate	2998	2380	2591	Cluster 2
PD-2	Coconut	1462	1249	1211	Cluster 1
PD-3	Sarikaya	1319	1303	1244	Cluster 1
PD-4	Mocha	1147	1082	1001	Cluster 1
PD-5	Donuts	3095	2656	2554	Cluster 2
PD-6	Jelly	657	567	518	Cluster 1
PD-7	Satay	1382	1239	1344	Cluster 1
PD-8	Choco Banana	1051	907	894	Cluster 1
PD-9	Cheese	1184	1103	1190	Cluster 1
PD-10	Choco Banana Cheese	2306	2074	2017	Cluster 2
PD-11	Blueberry	117	69	35	Cluster 1
PD-12	Choco Cheese	1925	1872	1861	Cluster 2
PD-13	Beef	933	941	947	Cluster 1
PD-14	Sausage	800	754	667	Cluster 1
PD-15	Milk	1192	1072	1068	Cluster 1
PD-16	5 Flavors	604	524	549	Cluster 1
PD-17	Strawberry	721	649	726	Cluster 1
PD-18	Sausage Shredded	449	421	401	Cluster 1
PD-19	KPK	566	506	512	Cluster 1
PD-20	Square Set	1542	1341	1474	Cluster 1
PD-21	Small Set	1200	1149	1061	Cluster 1
PD-22	Peel Bread	2819	2888	2931	Cluster 2
PD-23	Wheat	5012	4085	4109	Cluster 3

3.5. Processing OLAP Cube

The database used are defect dimensions, product dimensions and time dimensions. Schema depiction can be seen in the following figure.

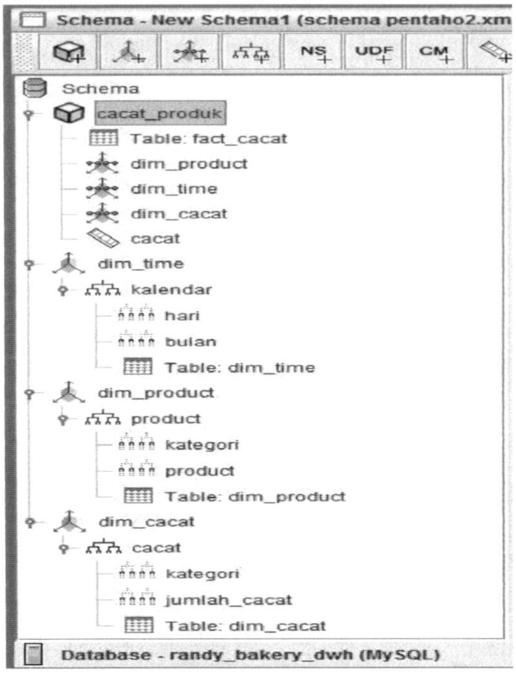

Figure 6. Merging Dimensions OLAP

From the schema is then applied to software called Pentaho Analysis. Here's the look of Pentaho Analysis based on the schema workbench that has been created.

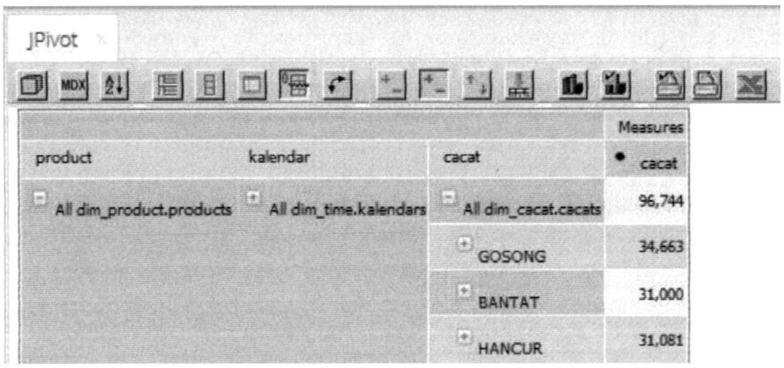

Figure 7. Pentaho Analysis

In Pentaho Analysis image above can be seen that the defect generated during the 7 months period of 96,744 pieces. Broken bread products must be given special attention because if it was left unchecked the greater the losses suffered by the company. The occurrence of the number of defects in the product can be caused by several factors:

1. Skill on human resources in the production of less good
2. Supervision on part of the production is not good.

4. Conclusion

1. The model business intelligence system is made with the Extract Transform Load process with model of data mining to improve quality product in bread factory.
2. The method of data mining is using k-means clustering method. The results of this intelligence business model system are cluster 1 with little amount of defect, cluster 2 with medium

amount of defect and cluster 3 with high amount of defect. Cluster 0 is a product with little amount of defect, cluster 1 with medium amount of defect and cluster 2 with high amount of defect. Products that are in the cluster 0 include jelly, blueberry, sausage, 5 flavors, strawberry, sausage shredded and KPK. Products that are in the cluster 1 include coconut bread, sarikaya, mocha, satay, choco banana, cheese, choco cheese, beef, milk, square set and small set. Products that are in cluster 2 include chocolate, donuts, choco banana cheese, peel bread and wheat bread.

3. From OLAP Cube Method can be seen that the defect generated during the 7 months period of 96,744 pieces.

5. References

[1] Booch, James G, et.al 2005 *The unified modeling language user guide* 2nd edition (United State: Addison Wesley Proffesional).

[2] Imelda 2011. Business intelligence. *Majalah Ilmiah UNIKOM 2011,* **Vol. 11**,No. 1.

[3] Inmon, W H 2002 *Building the data warehouse*. 3rd edition. USA: John Willey & Sons Inc.

[4] Kristanto, Christopher H, et.al 2016 Implementasi k-means clustering untuk pengelompokan analisis rasio profitabilitas dalam working capital. *JUISI,* **Vol. 02**, No.01.

[5] Larose D 2006 *Data mining methods and models*. (New Jersey: John Willey & Sons Inc).

[6] Rainardi V 2007 *Building a data warehouse with examples in sql server*. APRESS.

[7] Fitriana R, Eriyatno, Djatna T, and Kusmuljono B S 2012 Peran sistem intelijensia bisnis dalam manajemen pengelolaan pelanggan dan mutu untuk agroindustri susu skala usaha menengah. *J. Teknologi Industri Pertanian* **22** (3): 131 – 139.

[8] Fitriana R, Saragih J and Andhika M 2016 Business intelligence system models proposals to improve the quality of service at pt. gia. *Proc. 9th Int. Seminar Industrial Engineering and Management.*

[9] Fitriana R, Eriyatno and Djatna T 2011 Progress in business intelligence research: a literature review. *Int. J. of Basic & Applied Sciences IJBAS-IJENS* **11**(3).

[10] Fitriana R, and Djatna T 2012 Business intelligence design for decision support dairy agro industry medium scaled enterprise. *Int. J. of Engineering & Technology IJET-IJENS* **12**(5)

[11] Whitten, Bentley, Dittmand. 2004. *System analysis and design methods*. **6th** edition. The Mc Graw-Hill Companies.

[12] Wieder B and Luise M. 2015 *The impact of business intelligence on the quality of decision making – a mediation model*. (Australia: Elsevier B.V.)

Analysis of performance measurement at HR-GR Department using the balance scorecard method

Vienni [1] and M Bachtiar [2]

[1,2] Department of Industrial Engineering, Atma Jaya Catholic University of Indonesia, Jalan Jenderal Sudirman 51, Jakarta 12930, Indonesia
viennivien@gmail.com [1]; marsellinus.bachtiar@atmajaya.ac.id [2]

Abstract. PT. X is a company engaged in logistics service in Indonesia. Every company will certainly face a dynamic business environment. Competitors not only from domestic but also from overseas. To be successful in achieving its objectives, company should have a comprehensive measurement system as a strategy feedback that will drive the performance of company. HR-GA department is department that coordinate directly with company's management. Company through departments expect development goals in individual and also support of infrastructure will run smoothly. In 2015, company has taken steps to conduct a balanced scorecard as performance measurement. Nevertheless, a number of factors so it cannot run optimally. This study aims to analyse the current system and provided suggestions in order to give an overview to department related to its current performance. The results of data processing show that there are 8 objective strategies that have been formulated with 9 key performance indicators. Based on the results of scorecard, obtained values of 4.44 for customer perspective, 4.32 for internal business process perspective & 5.00 for learning and growth perspective. It concludes that performance based on perspectives are categorized very well
Keywords: balance scorecard, performance management.

1. Introduction

The Technology advances bring great impact in relations among Nations especially in the field of international economics. Indonesia is one of the developing countries are currently experiencing a very tight global competition. According to Gopal R, Global Vice President of transportation and logistics in the Frost and Sullivan that ASEAN logistics and transportation sector is evolving rapidly, that is because the existence of regional economic growth. Regional economic growth is due to export role is great and significant domestic demand [5].

Economic growth in Indonesia, one of which was caused by the factors of population, with the rate of population growth between 2010 and 2035 projected about 1.40% per year. According to the last study (released in 2010), Indonesia has a population of 237.6 million people. However, according to recent estimates (of the various institutions) Indonesia is estimated to have more than 255 million people by the year 2016 [2]

Population growth increased significantly affect demand for the needs of food, clothing and boards are increasingly large numbers. The demand for primary needs, secondary as well as tertiary does not only come from within the country, but along with the development of the digital age in e-commerce affects the level of the peoples consumption against importing goods especially in Indonesia. The role of the logistics service providers rather than to provide any good stuff needs shipping national and international

IOP Conf. Series: Materials Science and Engineering **277** (2017) 012006 doi:10.1088/1757-899X/277/1/012006

scale become a very important aspect in the logistics chain.

In economic development, nowadays where the business world is growing rapidly in Indonesia and an increasingly competitive rivalry. The competition made the company must have the right strategy so as to enhance sustainable competitiveness. In this regard, it is necessary to design a management system that can formulate a strategy, Strategic planning system, and the preparation of strategy programs in the form of translation of vision and mission to be achieved [3].

Companies that are ready to compete must have effective strategic management, especially in improving company performance. A strategy effectiveness can be seen from the performance of a company that continues to increase or improve. PT. X is one of the freight forwarding company that is a logistics service provider that has more than 15 years serving various business environment in Indonesia. In support of import-export activities in Indonesia, the company provides both air and sea freight services, where the process from purchasing goods, customs clearance and delivery to consignee is the responsibility of the company.

At the end of 2015, the company's management takes steps to run a performance scorecard-based performance management system that is managed online with software. But since the beginning of the system has been released to date, the company still uses the company's performance assessment conventionally. This approach still focuses on financial perspective without taking non-financial perspectives. Today's business entities require a multi-perspective performance measurement or comprehensive measurement that can reflect the needs of each stakeholder.

If company applies a comprehensive performance measurement, it is expected to assist management in measuring the extent to which the company's strategy has been achieved. Comprehensive performance assessments are also indispensable for the purpose of performing performance management so as to help integrate corporate, individual or work group objectives. Nevertheless, as for the constraints faced by the company so that the scorecard based performance appraisal system cannot be done properly. The following is a description of the constraints that companies face in implementing the balanced scorecard described in Table 1.1 below.

Table 1.1. Company Problems in Implementing BSC

No	Factor	Problems
1	System	BSC system is still fully under repair
		Socializing the BSC to each department as a system that monitors the achievement of the company strategic plan is limited to every department manager
2	Management	Although with a number of projects that are being handled by the company so that management commitment in BSC implementation can not be run optimally
3	Department	A large number of work activities within the department so that managers and employees experience obstacles in adjusting work activities with corporate strategic goals

Balanced Scorecard (BSC) is a performance management tool first appeared in 1996 by Kaplan and Norton [6][7]. BSC consists of two words namely scorecard and (2) balanced (balanced). The scorecard is the card used to record the performance score of an organization or individual score. The scorecard can also be used to plan scores to be realized in the future. Through a scorecard, the scores that the organization or individual would like to realize in the future compared to the actual performance results. The word balanced is intended to show that the performance of the organization / individual is measured on a balanced basis from two indicators, namely financial and non-financial.

The company considers that in the company's own performance that is important to build the company and walk the company's direction goes well one of which is located at the operational level of the company. The operational level of the company is located at the lower management level that is the department that supports every operational activity of the company. This research will be conducted in the HR-GA department because it has a role in supporting the implementation of corporate strategy to achieve its goals.

The HR-GA department puts its focus on the company's strategic plans that directly relate to individual development within the company as well as the management of infrastructure or infrastructure in support of business activities within the company. Of course, the performance of the HR-GA department has a significant impact on its relationship with other departments. Measurement of departmental performance is considered very important for decision making and supports internal corporate reporting activities. For management, performance measurement is an integral part of the management control system.

Based on the background, the researcher wanted to analyze how the performance of HR-GA department in support of each strategic objectives that have been declared company. This research is expected to provide an overview for the HR-GA department in particular to the performance results it has in managing every business activity. The purpose of this study include:

1. Analysing the current HR-GA department's strategy map
2. Analysing the current HR-GA department KPIs
3. Analysis of the performance of department HR-GA PT. X based on Balanced Scorecard perspective

2. Methodology

2.1. Scope of Research

Before In this study, the scope of the study is limited by measuring performance using the balanced scorecard method, since this method is a measurement system that not only assesses the financial side but overall by looking at the non-financial side. In this study limited to non-financial performance only because of the company's policy for researchers. The study was conducted in the HR-GA department and the data used were primary data from the period (January 2017-June 2017). The research location is located in the office of PT. X, Grand Slipi Tower, Jakarta.

2.2. Research Framework

The following is a flow chart to show clearly the framework and flow of research shown in Figure 2.1 below.

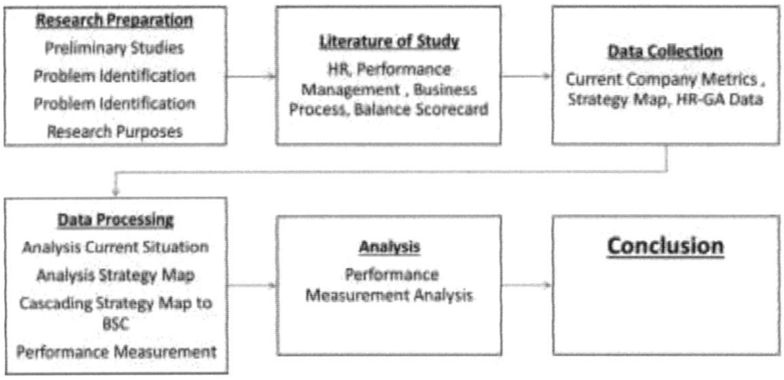

Figure 2.1. Flowchart Research Methodology

3. Result and Discussion

3.1. Analysis of Current System

The following is an explanation of a number of factors that inhibit the implementation of BSC within the company as indicated by the Fishbone diagram in Figure 3.1 below.

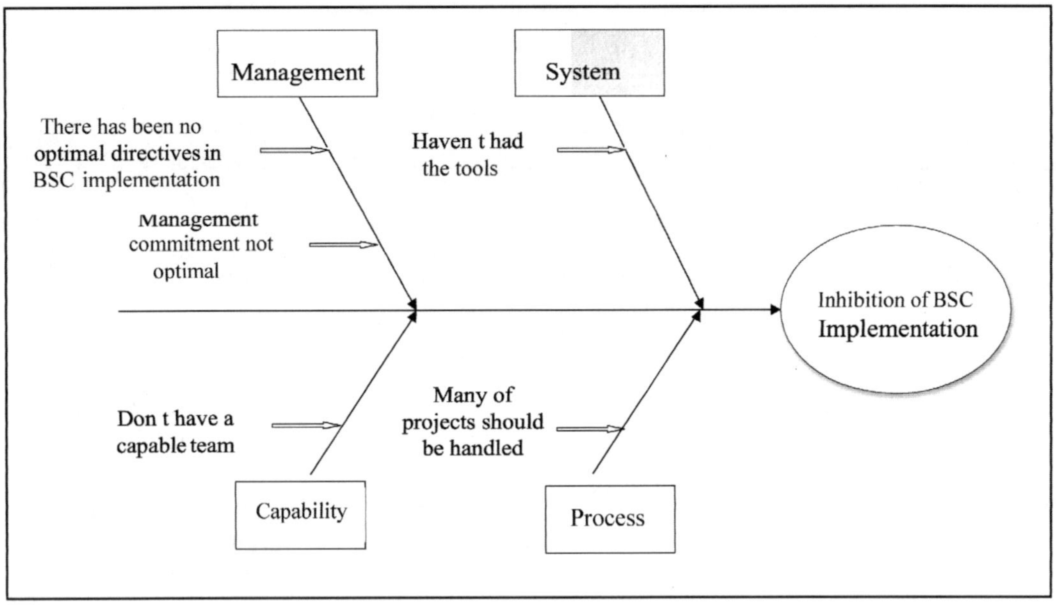

Figure 3.1. Fishbone Diagram Problem in BSC Implementation

3.2 Identify of HR-GA Department Objective

The objective is to identify The current objective of the company that is to conduct a review of the objectives previously set by the company and department together, a goal that is fixed or will not be managed and realized in the future. So it is expected HR-GA department can perform business processes within the company in accordance with the focus of the goals set. Here is a summary of the objectives shown in Table 3.1 below

Table 3.1. HR-GA Department Objective

No	*Objectives*
1	Manage operating cost efficiently
2	Develop and improve the quality of human resources in the company
3	Improve the quality of fund monitoring
4	Improving the implementation of company regulations
5	Improve the maintenance of corporate assets
6	Reduced employee turnover rate
7	Improving services between departments
8	Fulfillment of employee career ladder

3.3 Linking Strategic Objectives with Performance Measurement

Once the Strategy Map is develop , then the Strategic Objectives should cascade to performance metrics. A summary of the lag indicator and lead indicator of each perspective is shown in Table 3.2 below.

IOP Conf. Series: Materials Science and Engineering **277** (2017) 012006 doi:10.1088/1757-899X/277/1/012006

Table 3.2. Lag Indicator and Lead Indicator of Each Perspective

Perspectives	**KPI**	
Finance Perspective		
Optimized Cost	Percentage of Genex	- *turnover rate* - labour cost
Customer Perspective		
People Development	Percentage of Fulfillment	- number of training that been followed
Enliven Corp Culture, Values & Ethics	Percentage of Employee turn over	Employee satisfaction rate
	Satisfaction service level	Department service quality
Internal Business Process Perspective		
Improving recruitment quality	Percentage of Fulfill the needs of employees on time	Availability of human resources
Improving Maintenance Asset	Percentage of Maintenance of assets that are executed on schedule	- number of maintenance asset - number of complain
Improving the quality of monitoring operational funds	Percentage of Cover cash on time (H+3)	number of cash receipt
Improving the implementation of corporate regulations	Percentage of Socialization of company regulations that are executed on schedule	number of employee participated
Learning and Growth Perspective		
People Development	Percentage Training Execution based on plan	number of training needed

3.4. Scorecard Design Results

The following is the result of the scorecard design test shown in Table 3.3. In general, the determination of the number of KPIs to be set becomes the consideration of various companies. Based on best practices, KPI determination in general is between 4-10 KPIs in most industrial sectors [11]. In the financial perspective set 1 KPI % Genex Realization. From customer perspective, 3 KPIs are defined, % Employee career ladder, % Turn over employees and service satisfaction level. From the perspective of Internal Business Process, there are 4 KPIs % Socialization of PPs that are executed according to schedule,% Maintenance of assets implemented as scheduled,% Fulfillment of employee demand and% Cover on time. In the perspective of learning and growth set 1 KPI % implementation of training activities that are executed according to schedule.

Table 3.3. HR-GA Department Scorecard Result

Perspective	Strategy Objective	KPI	Polarization	Target	Achievement		KPI Weight	Result	SO Result	SO weight	Result
					Realization	Result					
Finance	*Optimized Cost*	*%Genex Realization*	Stabilize	100%	N/A	N/A	1.00	N/A	N/A	1.00	N/A
Customer	*People Development*	%Fulfillment of employee career ladder	Maximize	80%	200%	5	1.00	5.00	5.00	0.44	4.44
	Enliven Corp Culture, Values & Ethics	*%Employee turnover*	Minimize	8%	3.80%	4	0.42	1.67	4.00	0.56	
		Level of service satisfaction	Maximize	4 of 5	3.84	4	0.58	2.33			

Internal Business Process	Improving Recruitmen Quality	%Fulfill the needs of employees on time	Maximize	100%	81%	4	1.00	4.00	4.00	0.25	4.32
	Improving *Maintenance* Asset	%Maintenance of assets that are executed on schedule	Maximize	100%	95.24%	5	1.00	5.00	5.00	0.27	
	Improve the quality	%Cover the cash on time (H+3)	Maximize	100%	57.45%	3	1.00	3.00	3.00	0.21	
	Improve the implementation of corporate regulations	% Socialization of company regulations that are executed on schedule	Maximize	100%	100%	5	1.00	5.00	5.00	0.27	
Learning & Growth	*People development*	% Training Execution based on plan	Maximize	100%	92.86%	5	1.00	5.00	5.00	1.00	5

KPI Score is between 1 to 5 , based on specific performance result for each KPI and each score. The data input for this scorecard is based on calculation from current result, and then converted to BSC Score.

At this stage, researchers cannot perform data retrieval because as previously explained that the HR- GA department to date does not yet have a budget related to general costs within the department. The management of department funds by the HR-GA department is limited to the management of the casino cover. The HR-GA department in its business activities has very little control ability in managing cost. So in this study, the measurement of performance in the financial perspective is limited to the design process only. As for this, it can be an input for the HR-GA department in the coming period. Based on the objective strategy that has been established and agreed by the HR-GA department, it is expected that these performance measurements can also provide added value for the HR-GA department as well as management in controlling the costs related to the flow of cash that has been managed by employees in the company.

Based on the results of data processing on the objective strategy of employee promotion, it is known that the achievement of KPI % fulfilment of employee career ladder is 200%, exceeds the target that is 80% so the assessment scale of the realization is 5. KPI% Turn over employee is 3.8% As of June 2017, still at the lower level of the target set by the department of 8%. Based on the achievement, the scoring scale is 4.

The achievement of KPI obtained is 3.84, obtained the assessment scale 4. Based on the acquisition of course the strategic initiative for the department is to improve its service continuously. The value acquisition on the customer's perspective is derived by summing up the value and weighting between the two objective strategies in the customer's perspective. Objective people development strategy score 5.00 overall, while the objective strategy of corporate culture, values & ethics values is 4.00. Based on the data obtained that the value on the perspective of customers is 4.44, it shows that department performance based on this perspective is considered very good.

The acquisition of value on an internal business perspective is derived by summing the value and weighting of the four objective strategies in the internal business perspective. The objective strategy of improving the quality of recruitment achieves a total of 4.00, the objective strategy of improving the company's asset maintenance to the value of 5.00, the objective strategy of improving the quality of operational fund monitoring achieves 3.00 and the objective strategy of improving the implementation of the company's regulation of 5.00. Based on the data in Table 3.3 it is found that the value on the perspective of Internal Business Process is 4.32, it shows that department performance based on this perspective is considered very good.

Based on the results of data processing obtained the achievement of KPI% Implementation of training that is done that is equal to 92.86%, so the acquisition value is 5. Based on the data known in the period end of June OJT activities cannot be done due to holiday. Initiative strategies that can be done is to know the needs of employee training and evaluate each training activity. Based on the data in Table 3.3 it is found that the value in Learning and Growth perspective is 5.00, it shows that department performance based on this perspective is considered very good.

4. Conclusion

Based on the results of analysis and evaluation that have been done in the previous chapter, it can be concluded:

1. The HR-GA Department's current Strategy Map is not yet well-targeted, on the grounds that:
   - The objective Increased Profit strategy in the financial perspective, is not specific for measuring HR-GA department because this department is not profit oriented and does not have the capability to manage and generate real profits for the company.
   - From an Internal Business Process perspective, the Strategic Alliances & Networking's objective strategy is the SO that the execution is run by the company's management. In addition, GA departments do not have the authority or capability to manage the objective strategy of Improving vendor relationship because basically the relationship with the vendor is maintained by holding party. The objective strategy of Implement Risk Management is a strategy that is actually managed by a special team that manages risk management within the company so that HR-GA departments do not have control ability against this objective strategy.
   - In a learning and growth perspective, the company has previously described the objective management information system strategy. This SO is judged beyond the HR-GA department's capabilities, as IT development is a task rather than an IT department and IT departments have the capability to support this objective strategy.

2. KPI owned by the HR-GA department is currently considered to be a lack of consideration:
   - In the objective strategy Increased Profit financial perspective is considered not appropriate because the HR-GA department is not a department that generates profit (profit oriented).
   - There are similarity between KPIs in the perspective of customers with learning and growth perspectives, as the actual strategies implemented to support these objectives are equally good for internal HR-GA departments as well as for HR-GA department customers.
   - A number of established KPIs have a small control ability due to data availability and capability factors rather than those responsible for KPI.

3. Actualization of HR-GA department performance can be seen based on value earned on each balanced scorecard perspective:
   - Limitations in this study so that the financial perspective cannot be measured, this is due to the availability of data to support the actualization of the objective optimized cost strategy where both the management and the department has never made a budget.
   - Based on customer perspective, the value of 4.44 in which department performance is considered very good in this perspective.
   - Based on the perspective Internal business process obtained value of 4.32 that department performance is considered very good in perspective.
   - Based on the learning and growth perspective, the value of 5 is that department performance is considered very good in this perspective.

Suggestions proposed through this research are:

1. Consider the performance measurement that still uses traditional performance measurement to a more modern measurement using balanced scorecard, through which a number of KPI can be obtained result which will be the consideration of department in determining a decision on the upcoming strategy

2. Performance in the financial perspective can be further analysed if the department is able to create a budget cost to support the achievement of objective optimized cost strategy. Budget determination is important on the basis of consideration, as an added value for the department and management in controlling the use of operational funds as long as it is only based on the process of casino cover only where the potential losses caused by fraud either intentionally or unintentionally by the employee can be minimized
3. Strategies held in performing performance measurements should be discussed periodically both the department with the company's management because along with the dynamics of the work environment, the measurement will become more complex
4. Conduct a study / observation of employee job satisfaction within the company, as a step to evaluate employee retention and turnover rate of employees within the company
5. It takes sincerity and full support of top management in the implementation of performance measurement design to obtain optimal results for the department
6. The company's managerial evaluation can be done by measuring the performance at the level of other business units, because each business unit has a significant impact on the performance of the company

5. References

[1] Admin. 2015 *Department of Home Affairs Annual Performance Plan 2015/16*. www.home-affairs.gov.za/files/APP2015_16.pdf . Accessed on February 4th 2017.
[2] Badan Pusat Statistik 2015 *Laju Pertumbuhan Penduduk* www.bps.go.id/ linkTabelStatis/view/id/1268. Accessed on December 28 2016.
[3] Becker B E., Huselid, M A., and Ulrich 2001 *The HR Scorecard: Linking people, strategy and performance*. (Boston: Harvard Business Press)
[4] Dessler, G 2009 *Manajemen SDM buku 1* (Jakarta : Indeks)
[5] Julias F 2015 *Pasar Transportasi dan Logistik ASEAN Menjanjikan.* www.swa.co.id/swa/trends/marketing/gopal-sektor-transportasi-dan-logistik-asean-berkembang-pesat. (Accessed on December 28th 2016).
[6] Kaplan R S and Norton D P 1996 *Balanced Scorecard Menerjemahkan Strategi Menjadi Aksi* (Jakarta : Erlangga)
[7] Kaplan R S and Norton D P 2001 *The Strategy Focused Organization: How Balance Scorecard Companies Thrive in The New Business Environment.* (Massachusetts: Harvard Business School Press)
[8] Mahsun M 2006 *Pengukuran Kinerja Sektor Publik* (Yogyakarta: BPGE UGM)
[9] Mathis R L and Jackson J H 2002 *Manajemen Sumber Daya Manusia Edisi Pertama* (Jakarta: Salemba Empat)
[10] Mulyadi 2001 *Balanced Scorecard: Alat Manajemen Kontemporer untuk Pelipatganda Kinerja Keuangan Perusahaan* (Jakarta: Penerbit Salemba)
[11] Price Water House Coopers 2006 *Guide to Key Performance Indicators* www.pwc.com/gx/en/audit-services/corporate-reporting/assets/pdfs/uk_kpi_guide.pdf. Accessed on June 20th 2017.
[12] Purwanto 2008 *Manajemen Strategi* (Jakarta: PT. Raneka Citra).
[13] Robbins and Coulter 2007 *Manajemen* (Jakarta: Indeks).
[14] Sinambela L P 2012 *Kinerja Pegawai: Teori, Pengukuran dan Implikasi* (Yogyakarta: Graha Ilmu).
[15] Wibowo 2012 *Manajemen Kinerja,* Edisi 3 (Jakarta: Rajawali Pers).

Information system and website design to support theautomotive manufacture ERP system

T G Amran[1], N Azmi[2] andA A Surjawati[3]

Universitas Trisakti, Jl. Kyai Tapa No. 1, Jakarta, Indonesia
[1]tiena@trisakti.ac.id,[2]nora@trisakti.ac.id,[3]ayu@yahoo.com

Abstract. This research is to create an on-time production system design with Heijunka model so that the product diversity for all models could meet time and capacity requirements, own production flexibility, high quality, meet the customers' demands, realistic in production as well as creating a web-based local components' order information system that supports the Enterprise Resource Planning (ERP) system. The Heijunka model for equalization with heuristic and stochastic model has been implemented for productions up to 3000 units by implementing Suzuki International Manufacturing. The inefficiency in the local order information system demanded the need for a new information system design that is integrated in ERP. Kaizen needs to be done is the Supplier Network that all vendors can download and utilize those data to deliver the components to the company and for vendors' internal uses as well. The model design is presumed effective where the model is able to be utilized as a solution so that the production can run according to the schedule and presumed efficient were the model is able to show the reduction of loss time and stock.
Keywords: Enterprise Resource Planning, automotive, manufacture,information system

1. Introduction

For over two decades Indonesian automotive industry attempted to fulfillinternational standard-based market demands. Developments have been made by local vendors and Heijunka method has been studied intensively as a part of the process [2],[5].Previous researches by local automotive vendors are enough to create a base for this research that is related to the Heijunka and website based production system design. By ImplementingHeijunka iteration method the company can meet the consumer's demand accurately in terms of amount and time [15], reduce excess production [6], better production planning flow for ring gear [5],better response towards changes in demand[5],reduce the stock design [6], minimize the work load imbalance between machines [15], the machine utility value for press 300 TS machine decreases by 9 working days, production duration decreases by 0.25 hour[1]. By implementing Heijunka, the machine utility value for press 400 TS machine decreases by 10 working days and production duration decreases by 5.97 hours [1]. There are some researches that underlie this research with Heijunka's approach. Heijunka implementation can minimize the work load imbalance between machines [3],[4],[12],[13] andimplementation supports the ERP system in manufacture [14], [16].

This research emphasizes on themodified Heijunka model to calculate the production of the *skutik*(automatic scooter) model(less than 3000 units per month,produced in 3 color lots). In order to be competitive in this dynamic global market, manufacture industry need an integrated information system with the ability to deliver comprehensive information to the management, hence delivering accurate managerial decisions. Current manufacturing information system is the Enterprise Resource Planning (ERP) that combines the Manufacturing Resource Planning system (MRP II) with accounting and finance system. The ERP system is an accounting-oriented information system to identify and plan the company's resources needed to meet its customer's orders. The ERP system is also a customer oriented manufacturing management system [11],[12],[16].

The core of the ERP is MRP II or the Manufacturing Resource Planning. The MRP II process includes: production planning, rough-cut capacity planning, master production scheduling (MPS), material requirement planning (MRP), purchasing, supply management, production activity control, and performance measurements [11].

The research conducted on the production of the 4-wheeled vehicle (R4) and 2-wheeled vehicle (R2). The 2-wheeled vehicle (R2) produced are the underbone model: Smash, Shogun, and Satria, the *skutik* model: Skydrive, Skywave, Nex, and Let's, and the backbone model: Thunder. The underbone model with most demand is Satria (XB976/FU).The marketing department has revised the 2nd semester forecast for October-March for the fiscal year of April-March by approximately 30,000 units on September (end of semester 1) to 34,000-40,000 units per month. Revised marketing forecast is dominated by the Satria (FU) underbone model with production level of 60% of the total production while *skutik* (automatic scooter) model only produces 20% of the total monthly production and decreasing. The on time production system or often called the just in time (JIT) system isa part of the Suzuki Production System (SPS) that is supported by the Assembling Production Plan (RPA) making which in everyday intermingling operation known as Heijunka.

According to [7], Heijunka is a model that levels production in terms of volume and product mix (often called as leveled mixed production). Creating a product is not done based on the actual customer's demand order that might be increasing or decreasing sharply, but taking the total amount of order in a time period and averaged it so that it can be produced in the same amount and mix every day. The significant difference between production ratio of the Satria model and the *skutik* model which is over 65% with Heijunka model currently causes a model change more than 2 times a day that eventually leads to loss time and unstable quality. In order to overcome that, a more suitable Heijunka model for the large volume model and small variant model is highly needed.

2. Methodology

The tools used to model a system are:Step one: The Heijunka Model for local content program (Figure 1).Step two: Heijunka design model building (Figure 2). Step three: Information System Architecture Design(Figure 3).

The RPA made based on the Heijunka model will be used as a reference in processing RUN MRP to issue the local component's order. Currently, the output of the Run MRP is the Part Order Sheet (POS) that contains information on total order, forecast, and delivery plan for each part and each vendor, Part Tag (PT) as a component identity card, and Delivery Note (DN) as a delivery paper. POS is printed on A4 paper and there are about 220 pages for 108 vendors. The company needs about 40 boxes of paper both and total printing time for 10 working days. The next process is to put together the POS, PT, and DN based on the vendors, packing them into plastic bags or boxes based on the number of PT printed, before packing them into vendor boxes. Besides printing and preparation time, the problem faced is the need for a great amount of PT and DN printed paper form that has to be ordered and maintained, the need of a temporary storage area while waiting for the PT and DN to be picked up by the vendors, and re-print time needed were there are any vendor that loose its PT or DN. In the vendor's side, a specific time is required to pick the POS, PT, and DN up in PT. SIM, the risk of lost or damaged POS, PT, and DN, and reprint request time are the problems faced. Therefore, a web-based local component order information system is a solution to those problems.

Using Heijunka model design, *skutik* models that were not produced daily were grouped and accessible for early access in the web-based local component order information system. This design can improve the company's performance by accelerating every process thus resulted in the increase of company earning.

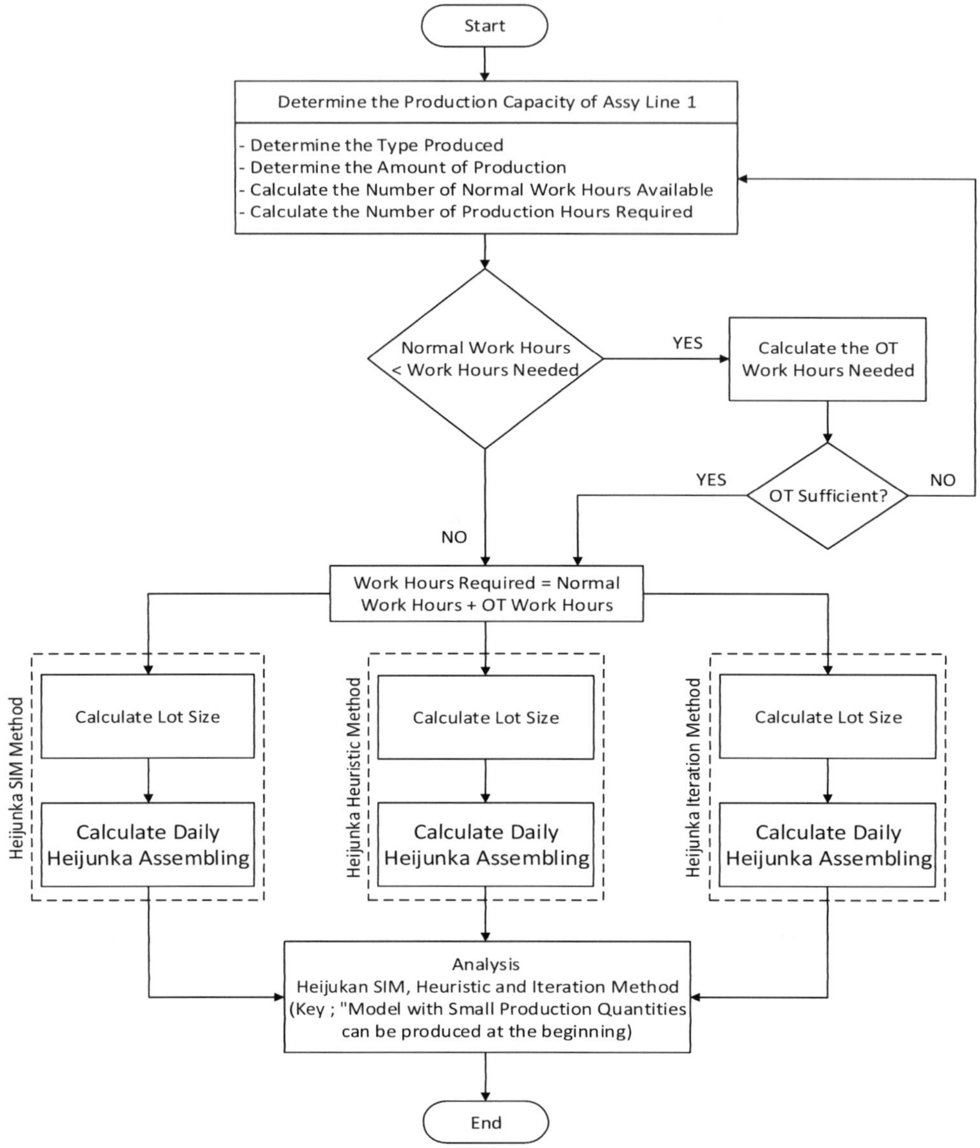

Figure 1. The Heijunka Model for local content program

2.1. Context Diagram

Context Diagram is a standalone occurrence from a data flowchart where a circle represents the whole system. This Context Diagram must be an overview that includes basic inputs, systems, and outputs. Context Diagram is the highest level of the data flowchart and only has one process that shows the system as a whole. The process is numbered as 0.

2.2. Data Flow Diagram

Basically, the steps to form a DFD are:

1. Identify all external entities involved in the system.
2. Identify all inputs and outputs involved with the external entities.
3. Create the Context Diagram.
3. Data Dictionary
4. Process Specification

IOP Conf. Series: Materials Science and Engineering **277** (2017) 012007 doi:10.1088/1757-899X/277/1/012007

2.3. Entity Relationship Diagram (ERD)

ERD is used to interpret, determine, and document the requirements for a database processing system. It shows the whole data user and the relationship between tables. In ERD, those data is described by depicting the entity symbol. In this system design, there are several interconnected entities that provide data needed by the system.

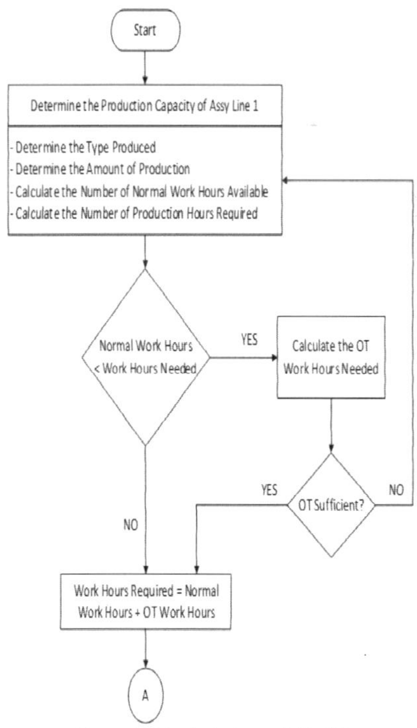

Figure 2. Heijunka design model building

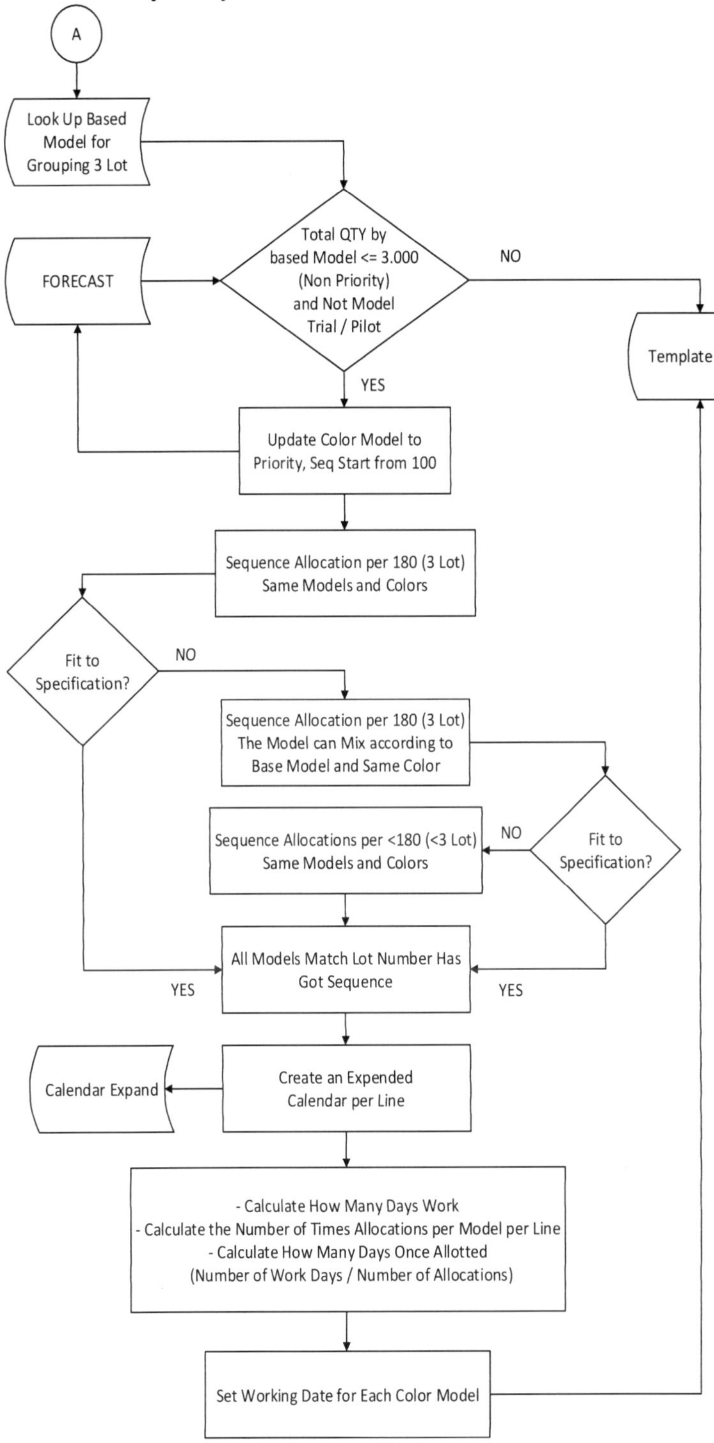
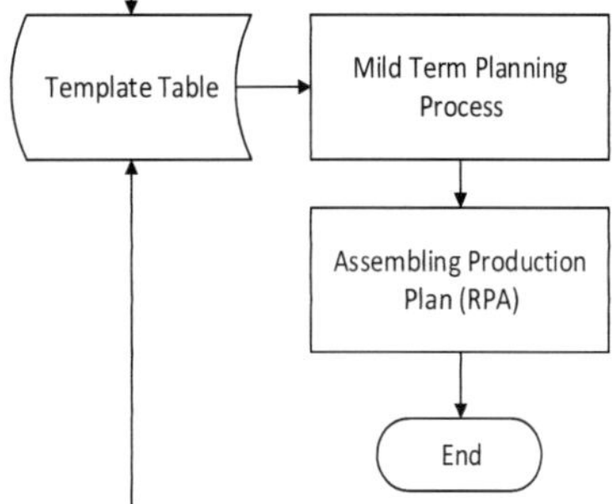

Figure 3. Information System Architecture Design

3. ResultAnd Discussion

3.1. Heijunka Results

Methods Analysis

Assembling line 1 Heijunka was chosen based on the best method which is SIM (Suzuki International Motor)compared to Heijunka stokastik and heuristic approached[1]. Production schedule of all skutik models with less or around 3000 units.

Heijunka Model Development

1. Development of Heijunka model based on Suzuki algorithm is producing the same skutik model in one color lot less than 3000 units per month.
2. Partial production schedule of all skutik models is 3 lot (180 units) and consumes the initial 8 working days.
3. Second schedule is 3 lot based on model (in this case, skutik model was planned)
4. Partial Assembling Production Plan (Rencana Produksi Asembling/RPA) of line 1 in Januaryis 3 lot. This condition affected the model change from 5 models per day to 3 models per day, leaded to quality stability.
5. Spesific model modification is clearly presented in skutir model XE312 NE P12, P14, and XE312 NB P12 that was produced according to base model XE312 with a fixed amount of 3 lot.

3.2. Information System Design

Create For Context Diagram, Context Diagram – DFD Level 0 and Level 1

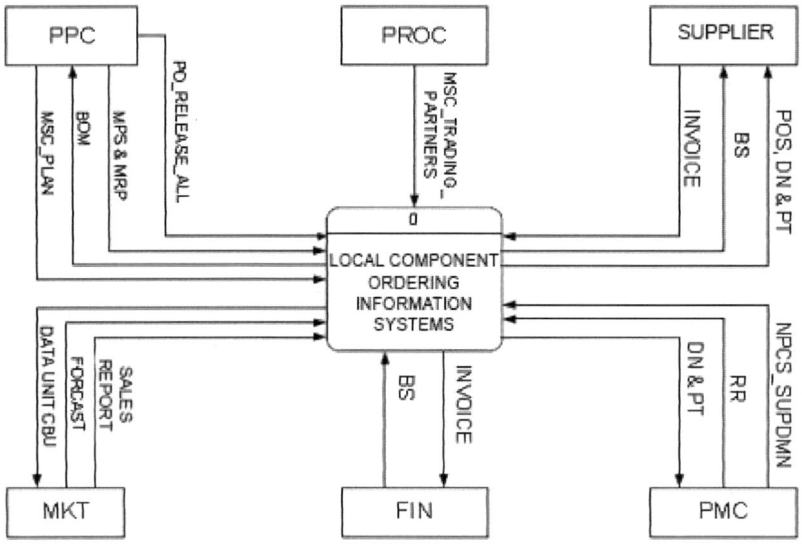

Figure 4. Local Component Order Information System Context Diagram

IOP Conf. Series: Materials Science and Engineering **277** (2017) 012007 doi:10.1088/1757-899X/277/1/012007

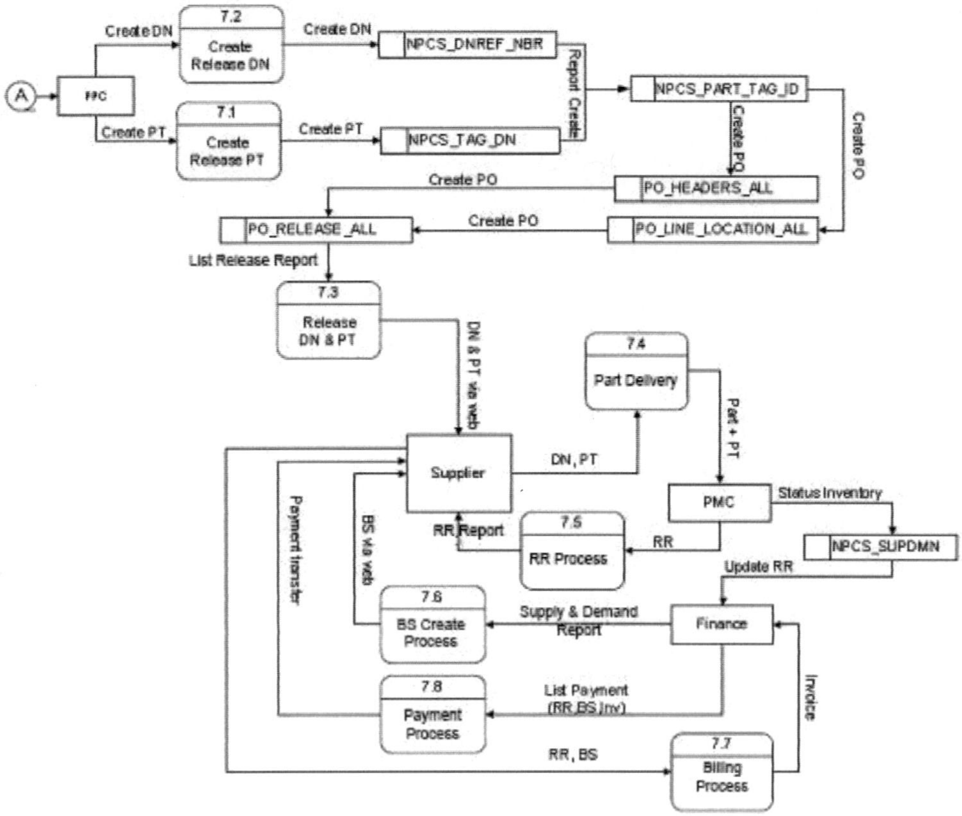

Figure 5. DFD Level 0

Create forERD PART ORDER SHEET

TABLE NAME : MSC_TRADING_PARTNERS

NO	FIELD	DATA_TYPE	LENGTH
1	PARTNER_ID	NUMBER	22
2	SR_TP_ID	NUMBER	22
3	SR_INSTANCE_ID	NUMBER	22
4	ORGANIZATION_CODE	VARCHAR2	7
5	DISABLE_DATE	DATE	7
6	STATUS	VARCHAR2	1
7	MASTER_ORGANIZATION	NUMBER	22
8	WEIGHT_UOM	VARCHAR2	3
9	MAXIMUM_WEIGHT	NUMBER	22
10	VOLUME_UOM	VARCHAR2	3
11	MAXIMUM_VOLUME	NUMBER	22
12	PARTNER_TYPE	NUMBER	22
13	PARTNER_NAME	VARCHAR2	255
14	PARTNER_NUMBER	VARCHAR2	154
15	CALENDAR_CODE	VARCHAR2	14
16	CALENDAR_EXCEPTION_SET_ID	NUMBER	22
17	OPERATING_UNIT	NUMBER	22
18	PROJECT_REFERENCE_ENABLED	NUMBER	22
19	PROJECT_CONTROL_LEVEL	NUMBER	22
20	SOURCE_ORG_ID	NUMBER	22
21	DEMAND_LATENESS_COST	NUMBER	22
22	SUPPLIER_CAP_OVERUTIL_COST	NUMBER	22
23	RESOURCE_CAP_OVERUTIL_COST	NUMBER	22
24	DEFAULT_DEMAND_CLASS	VARCHAR2	34
25	MODELED_CUSTOMER_ID	NUMBER	22
26	MODELED_CUSTOMER_SITE_ID	NUMBER	22
27	MODELED_SUPPLIER_ID	NUMBER	22
28	MODELED_SUPPLIER_SITE_ID	NUMBER	22
29	TRANSPORT_CAP_OVER_UTIL_COST	NUMBER	22
30	REFRESH_NUMBER	NUMBER	22

TABLE NAME : NPCS_SUPDMN_GR_V

NO	FIELD	DATA_TYPE	LENGTH
1	SOURCE_VENDOR_ID	NUMBER	22
2	PLAN_ID	NUMBER	22
3	TRANSACTION_ID	NUMBER	22
4	ITEM_ID	NUMBER	22
5	ITEM_NO	VARCHAR2	250
6	ITEM_DESC	VARCHAR2	250
7	FIRM_QUANTITY	NUMBER	22
8	FIRM_DATE	DATE	7
9	IMPLEMENT_QUANTITY_RATE	NUMBER	22
10	IMPLEMENT_DATE	DATE	7
11	SUPPLY_QTY	NUMBER	22
12	SUGGESTED_DOCK_DATE	DATE	7
13	DEMAND_DATE	DATE	7
14	ORG_DESCR	VARCHAR2	7
15	SUPPLIER_ID	NUMBER	22
16	SUPPLIER_SITE_ID	NUMBER	22
17	SR_ORG_ID	NUMBER	22
18	SR_INSTANCE_ID	NUMBER	22
19	ORGANIZATION_ID	NUMBER	22
20	SUPPLIER_TYPE	NUMBER	22

TABLE NAME : MSC_PLANS

NO	FIELD	DATA_TYPE	LENGTH
1	PLAN_ID	NUMBER	22
2	COMPILE_DESIGNATOR	VARCHAR2	10
3	SR_INSTANCE_ID	NUMBER	22
4	APPEND_PLANNED_ORDERS	NUMBER	22
5	ASSIGNMENT_SET_ID	NUMBER	22
6	CURR_APPEND_PLANNED_ORDERS	NUMBER	22
7	CURR_ASSIGNMENT_SET_ID	NUMBER	22
8	CURR_BILL_OF_RESOURCES	VARCHAR2	10
9	CURR_CONSIDER_PO	NUMBER	22
10	CURR_CONSIDER_RESERVATIONS	NUMBER	22
11	CURR_CONSIDER_WIP	NUMBER	22
12	CURR_CUTOFF_DATE	DATE	7
13	CURR_DEMAND_TIME_FENCE_FLAG	NUMBER	22
14	CURR_HARD_PEGGING_LEVEL	NUMBER	22
15	CURR_FULL_PEGGING	NUMBER	22
16	CURR_INCLUDED_ITEMS	NUMBER	22
17	CURR_OPERATION_SCHEDULE_TYPE	NUMBER	22
18	CURR_OVERWRITE_OPTION	NUMBER	22
19	CURR_PART_INCLUDE_TYPE	NUMBER	22
20	CURR_PLANNING_TIME_FENCE_FLAG	NUMBER	22
21	CURR_PLAN_CAPACITY_FLAG	NUMBER	22
22	CURR_PLAN_SAFETY_STOCK	NUMBER	22
23	CURR_PLAN_TYPE	NUMBER	22
24	CURR_RESERVATION_LEVEL	NUMBER	22
25	CURR_SCHEDULE_DESIGNATOR	VARCHAR2	10
26	CURR_SCHEDULE_TYPE	NUMBER	22
27	CURR_SIMULATION_SET	VARCHAR2	10
28	CURR_SNAPSHOT_LOCK	NUMBER	22
29	CURR_SPLIT_DEMANDS	NUMBER	22
30	CURR_START_DATE	DATE	7

Figure 6.ERD POS

Create : ERD Part Tag and Delivery Note
Local Component Order Information System Menu Design
Creating for Web-based Local Component Order System Design Flowchart
Below is the web-based Supplier Network menu structure design:

a. User Menu Structure
 The web-based Supplier Network menu structure for User can be seen in Figure 7.

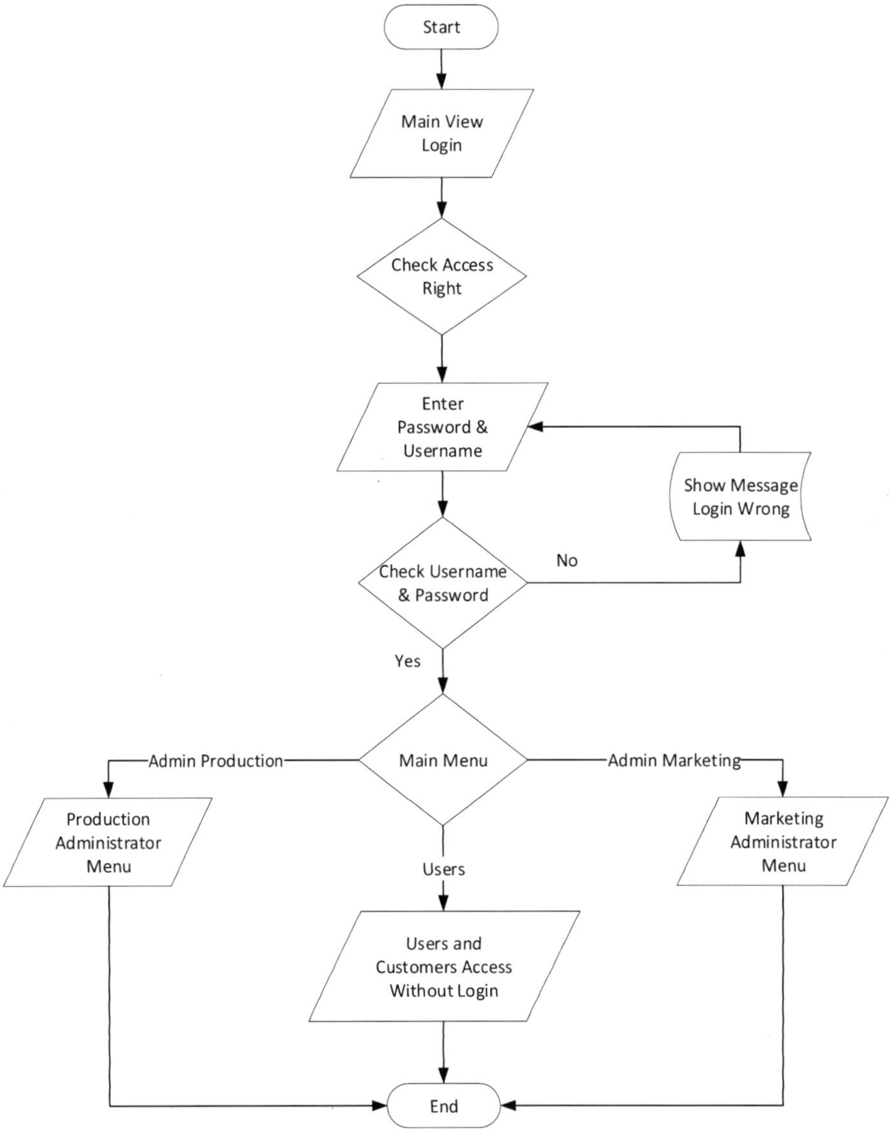

Figure 7. User Menu Structure**&** Login Flowchart

b. PPC and Finance Department Administrator Menu Structure
 The web-based Supplier Network menu structure for PPC Department Administrator.
c. Supplier Administrator Menu Structure
 The web-based Supplier Network menu structure for Supplier Administrator.

3.3. Supplier Network Procedural Design
Procedural design is a design to determine detailed algorithm that is going to be implemented into the application. The procedural design for the web-based company's Supplier Network is as followed:

1.Login Flowchart. It is a procedure that takes place when Administrator, PPC Department Administrator, and Vendor begin to access the application.
*2.Model Validation.*This research's model validation is conducted using the User Acceptance User (UAT), SIM User, and Supplier User.
3.Web-based Local Component Order Information System Model

POS – PART TAG AND DN
The Part Tag obtained is as follows

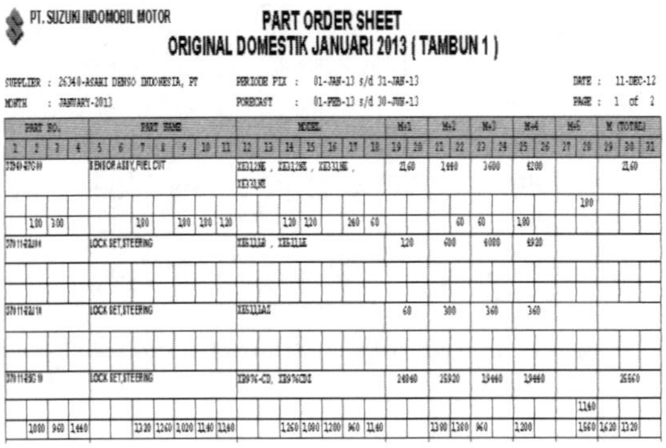

POS – PART TAG And DN
Supplier Network Website
- Login Page

Figure 8. Part Tag and Login Page

User PPC 2W Menu Display
Other Website Displays include: Administrator PPC R2 Menu Display, Upload Menu Display, Download Menu Display, Download History File 2W Menu Display, Finance Menu Display, History File Finance Menu Display

4. Model Verification and Validation
Model verification is conducted on Heijunka model by tracing the calculation algorithm based on the company's Suzuki Information Management method, i.e. by paying attention to the designed model's components and by consulting and confirming with the experts in the company which are related to the modeled system. Verification on the web-based local component order information system is conducted through the design of the supplier communication system menu, i.e. by consulting the user and supplier's needs.Model validation is conducted using the UAT where users and experts in the company, as well as the suppliers accept results.

Supplier Network System : UAT Schedule and Scenario

Data	Time	Location	PIC	Task	Remark
14-Jan-13	09.00-12.00	TB1	IT	1.Login to supplier network	
			PPC	-Change user name & password	
			Proc	2.Upload PDF file	**--> define by user what PDF will upload**
			Prod Assy	3. Download PDF file	--> by PIC Supplier
			Supplier		
		TB2	IT	1.Login to supplier network	
			PPC	-Change user name & password	
			Proc	2.Upload PDF file	**--> define by user what PDF will upload**
			Prod Assy	3. Download PDF file	--> by PIC Supplier
			Supplier		
		CKN	IT	1.Login to supplier network	
			PPC	-Change user name & password	
			Proc	2.Upload PDF file	**--> define by user what PDF will upload**
			Prod Assy	3. Download PDF file	--> by PIC Supplier
			Supplier		
		Spare part	IT	1.Login to supplier network	
			Iventory	-Change user name & password	
				2.Upload PDF file	**--> define by user what PDF will upload**
				3. Download PDF file	--> by PIC Supplier
		Fin/Acc	IT	1.Login to supplier network	
			Finance	-Change user name & password	
				2.Upload PDF file	**--> define by user what PDF will upload**
				3. Download PDF file	--> by PIC Supplier

Figure 9. Model Validation

5. Conclusion

The Architecture design suggested to be implemented in the coming year will be displayed in the Supplier Communication System (SCS) and consisted of eight main menus. They are:Spare part for Spare Part Department, Automotive for TB II R4 plant, Motorcycle for TB I R2 plant, Engine for Cakung Engine plant, Finance for Finance and Accounting Department, Administration, and the Sign out menu. The company and supplier can work more effective and efficient so that they can perform other value adding activities as well. For Heijunka model and web-based local component ordering information system model verification and validation, a test using the user and supplier UATs are conducted and the results are verified and valid for both models.

6. References

[1] Amran T G and Azmi N2014 Rancang Bangun Model Sistem Produksi Berbasis Heijunka Untuk Mendukung SistemERP Manufaktur Otomotif. *Prosiding SNTI-BKSTI*, 2-4 September 2014. Bukititinggi. ISBN 987-602-9081-4.

[2] Hüttmeir A, Trevillel S, Ackere A and Prenninge J 2009Trading Off Between Heijunkaand Just-in-Sequence.*Int. J. of Production Economics,* 118 (2), 501-507

[3] Fogarty D, Blackstone T and Hoffmann J 1991 *Production and Inventory Management*, Ohio: South-Western Publishing Co.

[4] Gaspersz V 2001 *Production Planning & Inventory Control Berdasarkan Pendekatan Sistem Terintegrasi MRP II and JIT Menuju Manufakturing* 21, Jakarta: Vincent Foundation and PT Gramedia Pustaka Utama.

[5] Groover M P2001 *Automation, Production Systems, and Computer-Integrated Manufacturing*, 2ndEd., New Jersey: Prentice Hall.

[6] Heizer J and RenderB 2004*Operation Management*, 7thEdition, New Jersey: Prentice Hall,.

[7] Liker J K 2006 *The Toyota Way: 14 Prinsip Manajemen dari Perusahaan Manufaktur Terhebat di Dunia*, Erlangga.

[8] Monden Y 1996 *Sistem Produksi Toyota Suatu Ancangan Terpadu, untuk Penerapan Just-In-Time,* Jilid 1, Jakarta: LPPM and PT Pustaka Binaman Pressindo, Seri Manajemen No 159A

[9] Monden Y 1996 *Sistem Produksi Toyota Suatu Ancangan Terpadu, untuk Penerapan Just-In-Time*, Jilid 2, Jakarta: LPPM and PT Pustaka Binaman Pressindo, Seri Manajemen No 159B.

[10] Ohno T1995 *Just-In-Time Dalam Sistem Produksi Toyota*, Jakarta: PT Pustaka Binaman Pressindo, Seri Manajemen No.168.

[11] Odang D 2008 Tutorial Membuat Website Gratis. Ilmu Komputer (http://ilmukomputer.com /Accessed on March 26, 2013: 23.46 WIB)

[12] PT Toyota Motor Company 1989 Handbook of Toyota Production System, Human Resources Development Production Control Div.

[13] Schroeder R G1993 *Operation Management, Decision Making in the Operation Function,* 4th Edition, McGraw-Hill. Inc.

[14] Shimon Y N, Wilhelm W E and Warnecke H J 1997 *Industrial Assembly*, London: Chapmant & Hall Pub.

[15] Suzaki K 1991 *Tantangan Industri Manufaktur Penerapan Perbaikan Berkesinambungan,* Saduran Ir. Kristianto Jahja, Jakarta: PQM.

[16] Gaspersz V 2001 *Desain Sistem Manufaktur Menggunakan ERP System: Suatu Pendekatan Praktis*, ISSN: 0853 – 7655,JSB No 6,**volume 01**, 77-78.

[17] Maryati W E 2001Penjadwalan Produksi campur Merata Melalui Sistem Produksi Just In Time. *Jurnal Optimum*, **2** (1), 75-85.

[18] Widagdo G A and Basri H 2005 Handout of Toyota Production System Training For PT Astra Daihatsu Motor's Vendor.*Handout PT Astra Daihatsu Motor*

Establishment proper of the balanced scorecard indicators to support decision making in a university: a case study in Institut Teknologi Indonesia

L Theresia[1], A H Lahuddin[2] and R Bangun[3]

[1]Department of Industrial Engineering, Institut Teknologi Indonesia, Serpong Indonesia, tarlind@yahoo.com
[2]Department of Industrial Engineering, Institut Teknologi Indonesia, Serpong Indonesia, abdulharislahuddin@gmail.com
[3]Faculty of Administrative Science, Universitas Indonesia, Indonesia, bangun66@yahoo.com

Abstract. Balanced Scorecard (BSC) is a powerful tool in decision making process. Nevertheless, it is not rare that the BSC does not give satisfactory results because the indicators chosen do not reflect the needs of the organization. Therefore, indicator establishment is very crucial in the utilization of BSC. This research aims to determine the indicators BSC for a university and the research is a case study in Institut Teknologi Indonesia (ITI). In this study, BSC structure and indicators, comparison made by 4 previous researchers was used as the initial guide to determine the structure and indicators of ITI. And then, questionnaires were distributed to selected respondents and a focus group discussion (FGD) was conducted in order to produce indicators of BSC based on the mental model of the ITI. It is found 15 indicators based on the mental model of ITI. Furthermore, the relationships between the indicators are seen as dynamic relationships, and by using system dynamics, some feedback loops that are considered critical to organizational success can be identified and isolated.
Keywords: Balanced Scorecard, indicators, system dynamics, decision making

1. Introduction

University is non-profit organization that operates in dynamic environment. Due to this dynamic environment, a university faces many challenges and obstacles in operating its organization. Institut Teknologi Indonesia (ITI) also faces the same problems. ITI does not have the metrics to measure the performance and relate it to the organization's strategy. To improve the quality of its management, a university needs to measure its performance. Measuring the performance has become a central issue in both academia and business [9].

To measure the performance can be used BSC. The determination of the indicator becomes important in the BSC because the proper indicators will ensure good performance of university operations. As a result, a university continuously tries to improve its weaknesses in order to ensure its competitiveness [1]. Several studies have been conducted to determine the indicator of university performance [2] and neglected the established university performance indicators [1]. Therefore, this study was conducted.

The cause and effect relationship among BSC indicators are one-way relationship [9]. So that there can be a tendency to be indifference to its implementation. While, in fact, there are different interactions between the decision and the key performance indicators. BSC does not firmly separate cause and effect in the context of time [15]. The causal relationship in the real world does not happen simultaneously, because there was a time delay. Therefore, BSC needs to be approached with system perspective because organization operates in dynamic environment. Indicators and relationships are developed based on the mental model of the organization, which illustrates the dynamic of the system in universities. There are quite rare the researches which study the dynamic of BSC, especially in a university. In this research, the dynamic of BSC in a university will be studied

A university has 3 primary functions, which are: education, research and service. A university should establish performance measurement indicators (PMI) to evaluate its performance related to resource allocated [1]. The perspectives of BSC can be determined by generating whole understanding of the organization's vision and mission, which is determined by the dominant logic of the organization. With respect to PMI, Kaplan and Norton pointed out that each BSC perspective must have at least one target item and no more than 15 items of measurement [1].

The reasons that the implementation of BSC is failed are: (i) too few measures in each perspective; (ii) to many indicators without identifying the critical view; and (iii) the failure of measures selected to depict the organization strategy [16]. In addition, the failure is also because not enough individuals involved in the determination of indicators [9]. The key performance indicators (KPIs) are distinguished into lead indicators and lag indicators. Lead indicators can identify anteriorly symptoms of organization management, and lag indicators require time to reflect the organization management performance [1]. Therefore, it is important to understand the determination of KPIs as the indicators of the performance measurement in order to be successful using them as performance measurement tool.

BSC differentiates between four perspectives regarding strategic development [15], namely: (i) financial perspective, (ii) customer perspective, (iii) internal process and learning perspective and (iv) development perspective [9]. Experiments show that managers can use information from the strategy map to assess the effectiveness of organizational strategy [11]. It has become common practice worldwide for a university to prepare strategic development plans that are accompanied by monitoring system, such as scorecard [12], which is regarded as a tool for better educational planning [6].

BSC is not only a performance measurement tool, but also the management system to promote breakthrough for competitive performance and it is most successful when used to drive the process of change. BSC as a communication tool [13], [14] helps staff members to understand their role in achieving university strategic goals and objectives. The application of the BSC in the process of strategic management is aimed: to clarify and translate vision and strategy; to communicate and link strategic objectives and measures; to plan, set targets, and align strategic initiatives; and to enhance strategic feedback and learning [7]. Therefore, BSC needs to be approached with a system perspective, which overcomes the above problems. This is because the system perspective is very suitable to support managers to learn and understand complex systems [17]. The system approach can be performed by using causal loop diagrams (CLD).

2. Theoretical Background

This research study consists of five stages which are as follows:

a. Development of ITI scorecard. Find perspective and indicators based on literature. At this stage, the perspective and indicator using BSC from previous research.
b. Find perspective and indicators based on FGD
c. Find perspective and indicators based on questionnaires.
d. Replication of the model. The next stage is the modeling of BSC dynamic model.
e. Proposed strategy. The last step was to know that the policy has a high leverage on the performance of the organization in the future.

3. Research Method

3.1 Structure of BSC Based on Literature

To develop the framework of ITI BSC, this research refers to the previous BSC research conducted by four people, namely: Mang Chuang [4], Chen, Wang and Yang [1]; Venkatesh et al., [18], Cullen, et al. [5]. This is because the research that has been done by the four researchers produced indicators that can provide benefits for the college under study to improve its performance. In addition, the study is in conformity with the needs of the research undertaken. Previous research by four researchers was the measurement of universities performance with BSC. The results of the comparison of the four researchers can be seen in Table 1.

Table 1. Comparison of the Four Researchers

Scorecard Dimension	**Framework**	**Sun Shing Chen**		**Cullen**		**Venkatesh and Kirti**		**Mang Chuang**	
		Lag	Lead	Lag	Lead	Lag	Lead	Lag	Lead
Costumer	Product	X	X	X		X	X		X
	Image	X	X	X				X	X
	Relationship	X	X	X	X	X	X	X	X
Internal Process	Innovation		X	X	X	X	X	X	X
	Services			X	X	X		X	X
	Operational	X	X	X		X		X	X
Learning and Growth	Competency of the staff members	X	X	X		X	X	X	X
	Technology					X		X	X
	Climate to act					X		X	X
Financial	Early								
	Sustainable	X	X			X	X		
	Mature			X				X	X

Comparative analyses of the four researchers are as follows:

1. Chuang; Chen; Cullen, et al.; Venkatesh et al. conduct performance measurement studies in universities referring to the BSC framework;
2. The results of the four researchers do not all thoroughly address the four frameworks of the BSC perspective, except Chuang;
3. The four researchers measure the performance of universities by using indicators that pay attention to the vision, mission and strategy in each university, so as to assess the extent to which the achievements made by the institution.

Considering the comparative analysis of the four researchers in Table 1, the structure of ITI's BSC is built with four perspectives thoroughly so that it will provide performance measurement indicators that can measure the full perspective, both internal and external, both customer satisfaction and stakeholders. Based on literature produced 4 perspectives and 57 indicators.

3.2 Find Perspective and Indicators Based on Focused Group Discussion (FGD)

The next stage was FGD consisting of 7 lecturers in ITI who understand the BSC. The result of the first stage (based on literature) is delivered in FGD to be discussed. FGD produced 4 perspectives and 65 indicators as presented in the Table 2.

IOP Conf. Series: Materials Science and Engineering **277** (2017) 012008 doi:10.1088/1757-899X/277/1/012008

Table 2. Perspective Based on the Literature

No	Research	Theme	Indicator
1	FGD	Reputation	The Number of Student
2	Chen		University reputation
3	Chen		Accreditation
4	Chuang, Venkatesh		Quality of Graduates
5	FGD		Customer satisfaction
6	Chuang		Service quality
7	Chuang		Job satisfaction
8	Chen	Organizational development	Short-term and long-term plans
9	Chuang		Organizational Capacity
10	Chen, Venkatesh	Lecturer	Competence of lecturers
11	Chen		PhD Ratio
12	FGD	Teaching resources	Further Education
13	Chuang		Training
14	Chuang		Competences of Lecturer
15	FGD		Facilities and Infrastructure
16	FGD	Finance	Operating expenses
17	Chen, Chuang, Venk		Total Income
18	Chen		Return on investment, ROI
19	Venkatesh		Number of Partnership
20	Chen, Venkatesh		The cost of infrastructure per student
21	Chen		Economic value added, EVA
22	Chen, Cullen	Research	Paper published journal per lecturer
23	Cullen		Books published per lecturer
24	Cullen		Number of studies per full-time lecturer
25	Cullen		Papers published at conference per lecturer
26	Chen, Chuang		License average per lecturer
27	Chen, Chuang		Patent average per full-time lecturer
28	Chen, Cullen		Integration of research and planning
29	Chuang, Cullen		Average research fund per lecturer
30	Chuang, Venkatesh		Alumni reputation in the market
31	FGD		Patent
32	Chen, Venkatesh	Community service	Participation in social activities
33	Venkatesh		Community service fund
34	Chen	Teaching quality	Full time student / lecturer ratio
35	Chen, Chuang	Student retention rate	Average graduation rate
36	Chen, Chuang		New student retention
37	Chen		Dropped out students
38	Chen		The number of students
39	Chen	Financial Donation	Alumni donation ratio
40	Chen, Venkatesh, Cullen		Donation from other sources

No	Research	Theme	Indicator
41	Chuang, Cullen		Fund raising increase
42	Chen	Quality of students	Requirements for new students
43	Chen, Cullen		The ratio of registering students to accepted students
44	Chen	Tutorship result	Participate student in community service activities
45	Cullen		Community service activities
46	Cullen		Career alumni monitoring
47	Chen	Continuous service	Ratio of part-time lecturers involved in promotions
48	Venkatesh, Cullen		Continuous education planning
49	Chen	Human resources	Students and lecturers ratio
50	Chen	Student structure	Full-time and part-time student ratio
51	Venkatesh	Teaching resources	Ratio between lecturer and total cost
52	Chen, Venkatesh		Financing ratio per student
53	Chen		*E-process* ratio
54	Venkatesh	Library	Average libraries cost per student
55	Chen	Curriculum	Curriculum planning
56	Cullen		Curriculum development
57	Chen		*E-curriculum* ratio
	Cullen		Curriculum improvement
58	Chen, Venkatesh Chuang	Alumni	Alumni career
59	FGD		Tuition fee
60	FGD		Grant and donation
61	FGD		Customer complaint
62	FGD		Obsolete
63	FGD		Recruitment/retirement of lecturers
64	FGD		Quality of services
65	FGD		Culture

3.3 Find Perspective and Indicators Based on Questionnaires

The next stage of the BSC is developed by collecting and processing data from questionnaires that contain draft indicators based on literature and FGD. The questionnaires were given to 30 selected lecturers in ITI. Respondents were asked to complete questionnaires with Likert Scale 1-7 (1 = strongly disagree, 7 = strongly agree). The selection of the respondents is done by purposive sampling. Six of them are lecturers who served as the leader of university and 4 others are lecturers who had served as the leader in the university. Others are lecturers who know BSC and are involved in promoting ITI performance. The next stage is to process the questionnaire data with SPSS to determine the indicators of ITI performance.

Each BSC perspective must have at least one target item and no more than 15 items of measurement [6]. From data processing, it is obtained that the structure ITI BSC consists of 4 perspectives with 15 indicators. The selection of indicators is based on ME = 0.4438 and SD = 9.89.

The indicators are as follows:

a. Costumer perspective: customer satisfaction, organizational capacity, quality of graduates
b. Financial perspective: total income, grant and donation, operating expenses
c. Learning and growth perspective: competence of lecturers, further education, research, training, job satisfaction.
d. Internal process perspective: facilities and infrastructures, the number of student, quality of services, recruitment/retirement of lecturers.

IOP Conf. Series: Materials Science and Engineering **277** (2017) 012008 doi:10.1088/1757-899X/277/1/012008

3.4 Weaknesses of BSC Framework

The above BSC framework has a disadvantage because it assumes that all indicators are independent, due to all of the indicators produced are in the same direction and every indicator is equally important. In fact, the whole indicators are not independent each other and have a circular causal relationship. The system thinking principle will cope with the problems and increase the success of BSC as a strategic management tool. The interaction among variables of the BSC in 'cause and effect relationship' is a one-way relationship [9]. So that there can be a indifference tendency to its implementation. While in fact, there are different interactions, particularly the delay between the decision and the change of key performance indicators. BSC does not firmly separate cause and effect in the context of time [15], while the causal relationship in the real world does not happen simultaneously, because there was a time delay.

Therefore, it is necessary to build a dynamic ITI BSC. The most important managerial insight of this stage is when the management realizes that the achievement of goals cannot run independently, but influencing each other. Management cannot prioritize only customer satisfaction, or internal processes or the growth of learning or cost effectiveness, but rather: how these four goals will be achieved simultaneously.

BSC dynamics aims to identify the issues in each scorecard, so it can be understood system behaviour caused by the interaction among variables. After identifying the issue, the next stage is to create feedback diagrams (Causal Loop Diagram/CLD) as well as analyse Behaviour Over Time (BOT).

3.5 Dynamic Model of ITI

The performance of the universities has a very complex inter-linkage among the variables that affect each other. The most important managerial insight from this stage is when management realizes that the achievement of goals cannot run independently, but is interrelated. At this stage dynamic BSC is build, which aims to construct CLD, prepare and analyse BOT graphs.

Through dynamic BSC, it enables leaders to identify and isolate some feedback loops that are considered critical to organizational success. This loop is called by Daniel Kim (1997) as the key success loop that needs to be monitored and managed together with key performance indicators (KPIs). Model of key success indicators representing ITI performance can be seen in Figure1.

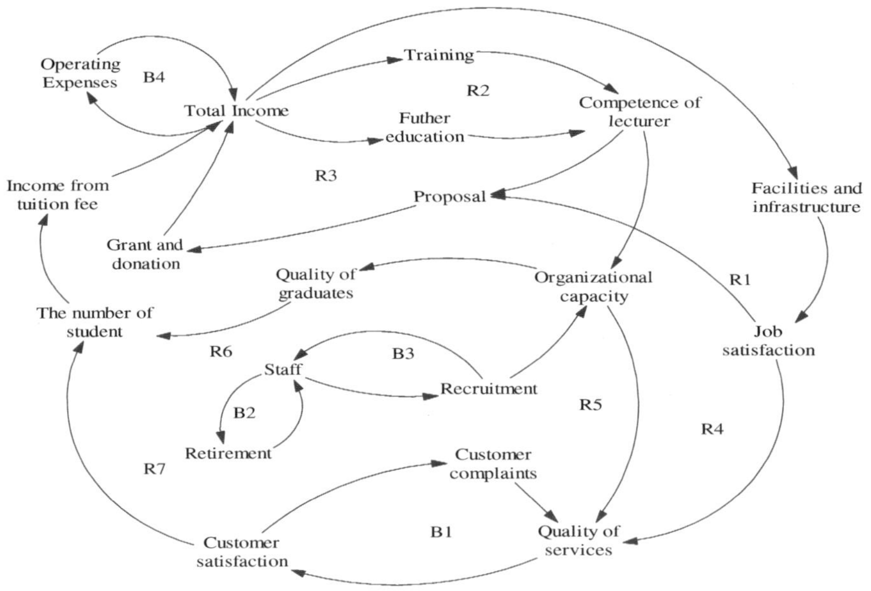

Figure 1. CLD of ITI

Figure 1 above shows that there are eleven interconnected key feedback loops that together determine the dynamic behaviour of the model, where seven reinforcing loops (R1, R2, R3, R4, R5, R6, and R7) and four balancing loops (B1, B2, B3, B4). Loop reinforcing (or positive) is a loop aimed at growth (having exponential growth as a dynamic behaviour), while loop balancing (or negative) aims for balance and static.

The reinforcing process can have a pattern with positive and negative loops. This describes that reinforcing loop not only improves growth, but also accelerates destruction. The implication is that if the target of the system works in accordance with the desired result, then the process is profitable. But sometimes in the process found the opposite conditions.

The goal of a successful strategy is to reinforce a positive feedback loop, a performance-enhancing loop, which at the same time eliminates or correctly manages other negative loops, which decrease the performance. From the results of ITI CLD performance measurements, important KPIs are obtained as leverage points that reinforce the performance, which are: total income (financial perspective), number of students (internal process perspective), job satisfaction (growth and learning perspective) and customer satisfaction (customer perspective). The determination of these variables as leverages because the variables are the principal variables in each perspective which have a positive feedback loop that improves performance. Interrelated variables and considered to be a leverage of ITI performance.

ITI is a non-profit organization, so customer perceptions with customer satisfaction variables are an important perspective for the institution. If customer satisfaction increases, then complaints on service will decrease. This will provide a good internal image, and the customers will spread their satisfaction from mouth to mouth. It acts as an advertisement that can increase the students registering to ITI.

The increasing number of admitted students will contribute to increase the income from students' tuition fee. Increase in tuition income will increase total revenue. Increased funds available will increase funding for the activities increasing income, such as training, further education so that will increase job satisfaction of lecturers/employees. Job satisfaction and lecturer competence will improve lecturers' ability, either in making proposals to earn income from grants and other sources, as well as in conducting research and community service and service to students. All of these loops are closed loops and are positive loops (active) in the system. This positive loop is the key to sustained growth and success of the institution. Therefore, management needs to identify and manage it properly.

CLD of key performance drivers form the Limit to Growth pattern. This pattern shows limited growth. The behaviour of this pattern indicates that a growth cannot be done continuously because it encounters resource constraints. This pattern is often repeated within an organization. Ultimately this pattern forms a standard structure and behaviour. The problems that form the structure of the pattern of limit to growth or limit to success has the characteristic of reinforcing in the initial process and then encounters balancing due to resource constraints to promote continuous growth.

4. Conclusion

The result shows that the BSC provides a systemic view on the strategy of higher education institution. Each university develops BSC in accordance with its vision and mission and focuses on human resources, teaching resources, research, teaching quality improvement based on the requirements of external and internal stakeholders of higher education institutions.

Although difficult and time-consuming in its development, the dynamic BSC is an efficient strategic management tool in creating the harmony among all elements of a higher education institution strategy. From the results of dynamics BSC research in ITI, it is found that the indicators becoming leverages of ITI future performance are customer satisfaction, the number of student, job satisfaction and tuition income. Therefore, management needs to identify and manage the key performance drivers correctly.

In addition, BSC can be used as a communication tool which helps staff members to understand their role in achieving the strategic goals and objectives of the organization and can provide effective organizational performance measurement. In addition, it helps not only staff members but also external stakeholders to clearly understand the strategic goals that the university wants to achieve.

5. References

[1] Chen, Wang and Yang 2009 Establishment and application of performance measure indicators for universities. *The TQM Magazine,* **Vol.21**, no.3.

[2] Chian F M and Lee C M 2001 Development and Framework of National Education Indicator System: Modern Education Indicators, *Hsue-Fu Culture*, Taipei.

[3] Chuang 2006 Balanced Scorecard: Creating a Management System for Private Technical Universities in Taiwan, *Pro-Quest Information and Learning Company.*

[4] Chuang Mang 2007 The Balanced Scorecard: Creating a management system for private technical universities in Taiwan, *Dissertations*

[5] Cullen et al. 2003 Quality in higher education: from monitoring to management. *Quality Assurance in Education*, **Vol. 11**, pp. 30-34.

[6] Kaplan Robert S and Norton David P 2001 *The strategy focused organization*, Boston: Harvard Business School Press.

[7] Kaplan S P and Miyake D N 2010 "The balanced scorecard", *The School Administrator*, **Vol. 67** No. 2, pp. 10-15.

[8] Kaplan S and Norton David P 1996 The Balance Scorecard: Translating Strategy into Action, *Harvard Business School Press*, Boston, MA.

[9] Kaplan S and Norton David P 1996b Strategic Learning and the Balanced Scorecard, Strategy & Leadership, *ABI/INFORM Global*

[10] Looy Van and Shavagatova 2016 Business Process Performance Measurement: A Structured Literature Review of Indicators, Measures and Metrics, *SpringerPlus,* 5, 1797.

[11] Lueg Rainer 2015 Strategy Map: The Essential Link between the Balanced Scorecard and Action, *Journal of Business Strategy,* **Vol. 36**, No. 2, pp 34-40.

[12] Myroslava Hladchenko 2015 Balanced Scorecard – a strategic management system of the higher education institution, *International Journal of Educational Management,* **Vol. 29** No. 2, pp. 167-176.

[13] Nickel S 2011 Strategic management in higher education institutions - approaches, processes and tools, *Leadership and Governance in Higher Education*, No. 2, pp. 1 - 28.

[14] Niven P 2003 Balanced Scorecard Step-by-Step for Government and Not-for-Profit Agencies, *John Wiley & Sons*, Hoboken, NJ.

[15] Nørreklit H 2000 The balance on the balanced scorecard a critical analysis of some of its assumptions, *Management Accounting Research*, **Vol. 11**, pp. 65-88.

[16] Rompho Nopadol 2011 The Strategic Linkages to Integrating Performance Across an Organization, Measuring Business Excellence, **Vol. 16** No. 2, pp 55-69.

[17] Sterman J D 2000 *Business dynamics: Systems thinking and modelling for a complex world.* Irwin/McGraw-Hill, Boston.

[18] Venkatesh, Umashankar and Dutta Kirti 2007 Balanced Scorecard in managing higher education institutions: an Indian Perspective. *Journal of Educational Management,* **Vol. 21**, No 1, pp. 54-67.

IOP Conf. Series: Materials Science and Engineering **277** (2017) 012009 doi:10.1088/1757-899X/277/1/012009

Dealing with dissatisfaction in mathematical modelling to integrate QFD and Kano's model

Dian Retno Sari Dewi[1] , Joana Debora[2], Martinus Edy Sianto[3]

Industrial Engineering Department, Widya Mandala Catholic University Surabaya, Indonesia

[1]dianretno@yahoo.com,[2] joana01295@gmail.com,[3] martinus.sianto@gmail.com

Abstract. The purpose of the study is to implement the integration of Quality Function Deployment (QFD) and Kano's Model into mathematical model. Voice of customer data in QFD was collected using questionnaire and the questionnaire was developed based on Kano's model. Then the operational research methodology was applied to build the objective function and constraints in the mathematical model. The relationship between voice of customer and engineering characteristics was modelled using linier regression model. Output of the mathematical model would be detail of engineering characteristics. The objective function of this model is to maximize satisfaction and minimize dissatisfaction as well. Result of this model is 62% .The major contribution of this research is to implement the existing mathematical model to integrate QFD and Kano's Model in the case study of shoe cabinet.
Keywords: QFD, Kano's model, mathematical model.

1. Introduction

Quality Function Deployment (QFD) is one of the most powerful tools in translating voice of customer to engineering characteristics. QFD also a methodology to develop a product design that enlighten the interrelationship between customer need and technical requirement/ engineering characteristic, to know exactly what is the customer wants and then assess each proposed design comprehensively [1].

Among many benefits of QFD, there were some flaws in interpreting the relation between voice of customer and engineering characteristic, the subjectivity in determining the relationship which is represented by weak, medium and strong relation in the House Of Quality [2],[3],[4]. To overcome the weakness in the QFD, a linier regression is used to modelled the relationship between voice of customer and engineering characteristics in the proposed mathematical model [5].

Conventional QFD assumes that maximizing the customer satisfaction will also effectively minimizing customer dissatisfaction, this is the theory of compensation. Meanwhile the MH-theory said that the source of satisfaction and dissatisfaction was different, so increasing the customer satisfaction was not directly minimizing customer dissatisfaction. It is very perilous to assume that the source of customer satisfaction and dissatisfaction is identical [6]. To overcome this flaw, Kano's model will contribute to the proposed mathematical model.

Kano's model concept was parallel to MH-theory. Kano's model categorised customer needs into several classifications. Customer is expected to respond questionnaire which is developed based on Kano's model. Dealing with those issues, we proposed a mathematical modelling to integrate QFD and Kano's model, objective function of this model is to maximize customer satisfaction and minimize customer dissatisfaction as well. Shoe Cabinet is used to implement the application of this mathematical model.

2. Proposed Model

The proposed model will be presented below:

$$\text{Max } Z = \sum_i w_i \times (S_i + DS_i) \quad (1)$$

The objective function is to maximize customer satisfaction and minimize customer dissatisfaction, which values lies between 0 and 1.

w_i = relative importance weight for customer need i, $0 \le w_i \le 1$
S_i = satisfaction score i, $0 \le S_i \le 1$
DS_i = dissatisfaction score i, $-1 \le DS_i < 0$
Subject to.

$$\forall i: \quad P_i = \beta_{i0} + \left(\sum_j \beta_{ij} \times x_j^c\right) + \varepsilon_i \quad (2)$$

In the conservative QFD, association strength of customer need j and the engineering characteristics are equivalent by using subjective ratings, such as 1, 3, 9. To reduce the bias of the relationship evaluations, the regression technique is demonstrated. The regression function obtained is as in equation (2). P_i represents performance level to depict the need of customer i. β_{i0} and β_{ij} is regression function that represents the relationship strength between engineering characteristics i and customer need j.

$$\forall j: \quad x_j^c = \frac{(x_j - L_j)}{(U_j - L_j)} \quad \text{for the bigger the better,or} \quad (3)$$

$$x_j^c = \left|\frac{(x_j - U_j)}{(U_j - L_j)}\right| \quad \text{for the lesser the better, } 0 \le x_j^c \le 1$$

$$\forall j: \quad L_j \le x_j \le U_j \quad (4)$$

The engineering characteristic values should be given using equation (3) to remove the effect of different scaling of different engineering characteristics. For engineering characteristic i, its values lie between upper bound U_i and lower bound L_i (4).

$$B \ge \sum_j c_j \times \left|(x_j - x_j^0)\right| \quad (5)$$

B unit organization's supply such as R&D budget are available for product development project (5). $\left|(x_j - x_j^0)\right|$ denotes the amount of development for engineering characteristics, c_j means expenses needed of making one unit improvement of engineering characteristic
We use 0.5 in case of Kano's basic parameter, 1 in case of Kano's satisfier parameter and use 2 in case of Kano's attractive parameter [7].
For each customer ($\forall i$):
If customer i is a Kano must-be/ basic type of customer :

$$S_i = 0 \quad (6)$$

$$DS_i = max\left(\left[\left(\frac{P_i}{P_i^*}\right)^{n_i} \times DS_i^*\right], -1\right) \quad (7)$$

Eq.6 and Eq. 7 clarify a basic type of customer need, score of the customer satisfaction is zero but when this basic need is not fulfilled then maximum dissatisfaction score will be -1 when it is not performed. The negative value lower than -1 for the dissatisfaction score do not have important meaning, thus it is transformed to -1. Moreover, according to Eq. 6, a basic type of customer need did not have contribution to customer satisfaction, even though they reach the maximum value.
If customer i is a Kano satisfier type of customer :

$$F_i = \propto_i \times P_i + \theta_i \,, \tag{8}$$

$$\text{if } F_i \geq 0 \text{ then } \left(DS_i = 0, S_i = min(F_i, 1)\right),$$
$$\text{if } F_i < 0 \text{ then } (DS_i = max(F_i, -1), S_i = 0)$$

Satisfier customer is the type of customer that will delighted if the need is satisfied, increasing of the product performance will also linearly increase the satisfaction, but it will affect the dissatisfaction as it is not satisfied. The satisfaction or dissatisfaction that is perceived by the customer is represent by a linear function F_i. $\propto_i$ is the slope and θ_i is the intercept of the function. The function F_i can be determined using the survey, respondents were asked to determine their emotion about at least four different levels of product performance. F_i can be in both the negative or positive value. Customer satisfaction is represented by positive value of F_i , meanwhile customer dissatisfaction is represented by negative value of F_i.

If customer i is a Kano attractive type of customer :

$$DS_i = 0 \tag{9}$$

$$S_i = min\left(\left[\left(\frac{P_i}{P_i^*}\right)^{m_i} \times S_i^*\right], 1\right), m_i > 1 \tag{10}$$

Customer do not aware for these needs, this need is categorized as an effect of surprised. It will give high satisfaction if this need is fulfilled, but it also will not affect dissatisfaction if it is not fullfilled. As a modification of Tan and Shen's satisfaction function, 2 can be chosen as an choice of n_i [7]. Score of DS_i^* and S_i^* can be obtained through surveys.

3. Numerical Example

Shoe cabinet was used for numerical examples purposes. Thirty customers as a direct user of Shoe cabinet were interviewed. The need for thirty customers as follows: have space enough (CN_1) which means that customer need a large space enough to store their shoe, sturdy (CN_2) which means that the shoe cabinet is sturdy enough to hold and keep things safely, adjustable height of rack (CN_3) which means that the customer want to adjust height, table have storage for keeping shoe accessories (CN_4) which means that customer want a space for keep their shoes accessories, easiness to clean under the cupboard (CN_5) which means that customer need a space to clean under shoe cabinet . Engineering characteristics related to each customer needs were : volume of cupboard (EC_1), thickness of material (EC_2), area of cupboard leg (EC_3), adjustable rack (EC_4), availability of drawers (EC_5), leg height (EC_6). We also used Kano's questionnaire to ask how they feel about the each customer need. It is found that CN_1 and CN_5 were classified as one dimensional category ,CN_2 was classified as must be category, CN_3 and CN_4 were classified as attractive categories.

The House of Quality is presented in Figure 1. The relative importance weight for each customer needs were obtained from the average of the customer questionnaire. The association between customer needs and engineering characteristics were scored by the expert. Three competitors were presented in figure 1, The benchmark score for the three competitors were obtained from the average of the respondent answer. Respondents were asking to give a score in the range of 0 to 5 for each customer needs. There are five customers need that should be translated to engineering characteristics. The top roof part of the HOQ is not defined, because we assumed that all engineering characteristics are independent. A linear regression model was used to portray the relationship between the customer needs and engineering characteristics.

Customer Needs	Kano	Weight	EC_1	EC_2	EC_3	EC_4	EC_5	EC_6	A	B	C
CN_1	O	0.221	9						0.46	0.66	0.87
CN_2	M	0.245		9				3	0.44	0.75	0.87
CN_3	A	0.179			9				0.47	0.72	0.83
CN_4	A	0.159				9			0	0	0.82
CN_5	O	0.194					9		0.49	0.71	0.87
		Dimension	cm^3	cm	-	-	cm	cm^2			

Figure 1. House of Quality

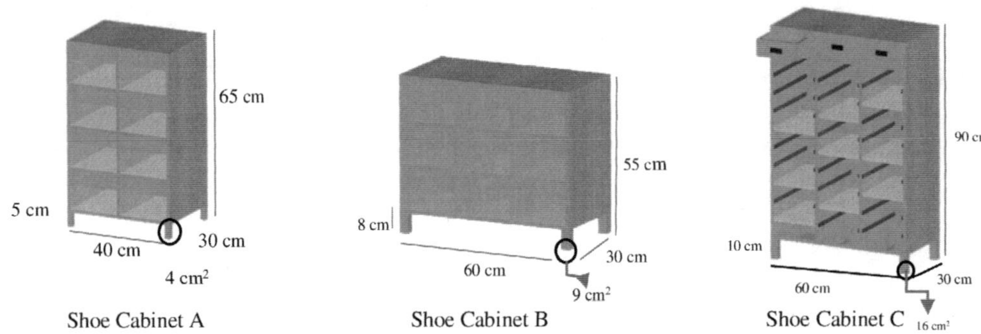

Figure 2. Product concept

Table 1. Product specifications

No.	Engineering Characteristics	Dimension	Product		
			A	B	C
1	EC_1	cm^3	78000	99000	162000
2	EC_2	cm	1.2	1.5	1.8
3	EC_3	cm^2	4	9	16
4	EC_4	-	Not adjustable	Not adjustable	Adjustable
5	EC_5	-	None	None	Available
6	EC_6	cm	5	8	10

The range of engineering characteristics are as follows: 78000 to 162000 cm^3 for EC_1, 1.2 to 1.8 cm for EC_2, 4 cm^2 to 16 cm^2 for EC_3, not adjustable and adjustable for EC_4, exist or no exist of drawer for EC_5 , 5 cm to 10 cm for EC_6. Range of engineering characteristics represent the feasible values. The relationship between the customer needs and engineering characteristics is obtained as linier regression. The dependent variables were the response of product performances while the independent variables were engineering characteristics for each product, the linier regression is as follows.

CN_1 = 0.69231 + 0.18985 EC_1
CN_2 = 0.63693 + 0.1638 EC_3+ 0.1628 EC_2
CN_5= 0.67654 + 0.19042 EC_6

All of the parameters in linear regression were significant by using $\alpha = 5\%$ and high value of R^2. The initial value of the design are: EC_1= 78000 cm^3, EC_2= 1.2 cm, EC_3= 4 cm^2, EC_4= not adjustable, EC_5 = no drawer, EC_6= 5 cm. Budget allocation for this improvement is 10% from Cost of Goods sold which is the maximum range to improve the engineering characteristics, IDR 195,000. Improvement cost for EC_1 was IDR 9, EC_2 was IDR 11,388.8, EC_3 was IDR 1,428.5, EC_4 was IDR 30,000, EC_5 was IDR 60,650, and EC_6 was IDR 250. Example of the full numerical models is as follows:

Optimal solution is initiated using Lingo 13.0 as follows: EC_1 is 86791.64 cm^3, EC_2 is 1.8 cm, EC_3 is 16 cm^2, EC_4 is 1(means adjustable rack is preferred), EC_5 is 1(means drawer is preferred) and EC_6 is 10 cm. All of the improvement reach the maximal score except EC_1. This need is categorized as one dimensional category. The score is not in the lowest score but it lies between the lowest and the highest. Since one of customer need is not fulfilled, so the overall satisfaction is not high, it is just only 62 %.

$$\text{Max} = 0.221 \text{ x } (S_1 + Ds_1) + 0.245 \text{ x } (S_2 + Ds_2) + 0.179 \text{ x } (S_3 + Ds_3) + 0.159 \text{ x } (S_4 + Ds_4) 0.194 \text{ x } (S_5 + Ds_5) \quad (11)$$

$$s.t. \quad P_{1=} 0.69231 + 0.18985 \times x_1^c \quad (12)$$

$$P_{2=} 0.63693 + 0.1628 \times x_2^c + 0.1638 \times x_6^c \quad (13)$$

$$P_{5=} 0.67654 + 0.19042 \times x_5^c \quad (14)$$

$$x_1^c = \frac{(x_1 - (78000+162000)/2)}{((162000-78000)/2)} \quad (15)$$

$$x_2^c = \frac{(x_2 - (1.2+1.8)/2)}{((1.8-1.2)/2)} \quad (16)$$

$$x_3^c = \frac{(x_3 - (5+10)/2)}{((10-5)/2)} \quad (17)$$

$$x_4^c = \frac{(x_4 - (4+16)/2)}{((16-4)/2)} \quad (18)$$

$$78000 \le x_1^c \le 162000 \quad (19)$$

$$1.2 \le x_2^c \le 1.8 \quad (20)$$

$$650 \le x_3^c \le 850 \quad (21)$$

$$5 \le x_5^c \le 10 \quad (22)$$

$$4 \le x_6^c \le 16 \quad (23)$$

$$x_3 \in (0,1) \quad (24)$$

$$x_4 \in (0,1) \quad (25)$$

$$195000 \ge 9 \times (x_1 - 78000) + 113888.8 \times (x_2 - 1.2) + 30000 \times (x_3 - 0) + 60650 \times (x_4 - 0) + 250 \times (x_5 - 5) + 1428.5 \times (x_6 - 4) \quad (26)$$

$F_1=(0.636*P_1)+0.279$ (27)

$F_1 \geq 0 \rightarrow (DS_1= 0; S_1= \min F_1,1)$ (28)

$F_1< 0 \rightarrow (DS_1= \max(F_1,-1); S_1= 0$ (29)

$S_2=0;$ (30)

$DS_2= \max (([(P_2/ 0.451)^{-2}]*-0.3968),-1)$ (31)

$X_3= 0\ (S_3 = 0; DS_3=0)$ (32)

$X_3= 1\ (S_3 = 1; DS_3=0)$ (33)

$X_4 = 0\ (S_4 = 0; DS_4=0)$ (34)

$X_4 = 1\ (S_4 = 1; DS_4=0)$ (35)

$F_5=(0.909*P_5)+0.075$ (36)

$F_5 \geq 0 \rightarrow (DS_5 = 0; S_1= \min F_5,1)$ (37)

$F_5 < 0 \rightarrow (DS_5 = \max(F_5,-1); S_5 = 0)$ (38)

4. Conclusion

The integration of linier regression to accommodate the relation between customer voice and engineering characteristics and Kano model into mathematical modelling to maximize the customer satisfaction and minimize the customer dissatisfaction has been modelled. The numerical example of shoe cabinet has already presented to perform how the mathematical modelling works. As a result, the output of this model is optimal under many constraint restrictions given in the mathematical model. The output showed that the model has valid.

Acknowledgements

This research was supported by the Ministry of Research, Technology and Higher Education of the Republic of Indonesia

5. References

[1] Cohen L 1995 *Quality Function Deployment: How to Make QFD Work for You* (New York: Addison-Wesley Publishing Company)

[2] Park T and Kim K J 1998 Determination of an Optimal Set of Design Requirements Using House of Quality *Journal of Operations Management* **16** 569

[3] Askin R G and Dawson D W 2000 Maximizing Customer Satisfaction by Optimal Specification of Engineering Characteristics *IIE Transactions* **32** 9

[4] Vanegas L V and Labib A W 2001 A Fuzzy Quality Function Deployment (FQFD) Model for Deriving Optimum Targets *International Journal of Production Research* **39(1)** 99

[5] Dewi D R S and Rahaju D E S 2016 *Proc.Int. Conf. on Industrial Engineering, Management Science and Application (South Korea)* (IEEE) p 165

[6] Dewi D R S and Rahaju D E S 2014 An Approach to Dealing With Importance Weights Of Customer Needs and Customer Dissatisfaction in Quality Function Deployment Optimization *Advanced Material Research* **931** 1636

[7] Tan K C and Shen X X 2000 Integrating Kano's Model in the Planning Matrix of Quality Function Deployment *Total Quality Management* **11(8)** 1141

IOP Conf. Series: Materials Science and Engineering **277** (2017) 012010 doi:10.1088/1757-899X/277/1/012010

Mesh control information of windmill designed by Solidwork program

T. Mulyana[1], D. Sebayang[2], A.M.D. Rafsanjani[1], J. H. D, Adani[1], and Y. S. Muhyiddin[1]

[1] School of Industrial and System Engineering, Universitas Telkom, Indonesia
tatangmulyana@telkomuniversity.ac.id
[2] Department of Mechanical Engineering, Universitas Mercu Buana, Indonesia
d_sebayang@hotmail.com

Abstract. This paper presents the mesh control information imposed on the windmill already designed. The accuracy of Simulation results is influenced by the quality of the created mesh. However, compared to the quality of the mesh is made, the simulation time running will be done software also increases. The smaller the size of the elements created when making the mesh, the better the mesh quality will be generated. When adjusting the mesh size, there is a slider that acts as the density regulator of the element. SolidWorks Simulation also has Mesh Control facility. Features that can adjust mesh density only in the desired part. The best results of mesh control obtained for both static and thermal simulation have ratio 1.5.
Keywords: mesh, control, information, windmill, solidwork

1. Introduction

The accuracy of the simulation results is influenced by the quality of mesh made. When adjusting the mesh size, there is a slider that acts as the density regulator of the element. Slide to the left to make the lowest mesh density (rough). Rough mesh is usually used for checking. Running a simulation with a short time ensures whether the input has been done in accordance with the desired or not, e.g. contact set, connector, external load, fixture, etc., so it can be anticipated if there is an error setting without having to run a long time with small element size level. Slide to the right to make the highest mesh density (smooth). If it is appropriate then mesh quality can be improved and repeat the simulation process to get results with high accuracy value. SolidWorks Simulation also has Mesh Control facility. Features that can adjust mesh density only in the desired part. Usually, mesh control is inputted before starting to "run" the simulation by accessing it from "Simulation Tree Manager> Mesh> Apply Mesh Control". But there is another simpler way to input mesh control, i.e. by accessing it through the folder "part". The trick can be done in two conditions, i.e. before meshing done or after meshing done. Using this method, mesh control is applied to the entire face of the selected part. If you want to choose a more specific face or edge can by manually selecting "entities" / desired part, then the simulation process is done [1-7].

2. Methodology

Stress concentration will be more in the sharp edges, corners, curves and in bending as well. So in order to reduce the stress concentration, fillets will be added to the edges and corners. In those regions (critical regions), the analyser has to be more careful about the mesh, If the mesh is not captured properly in those regions, it affects the solution. Hence in such regions, number of the elements are increased by changing the length of the element side (h-refinement) in order to capture the irregular regions of the geometry. Analysis done without mesh control will not give the accurate solution and will not be converged easily. So, for the same model mesh control tool is applied by splitting the face to create washer near the hole in which mesh density will be increased and entire region near the hole will be captured. There are three types of mesh refining methods, one is p-refinement, h-refinement and hp-refinement. In p-refinement, order of the element is changed (linear to parabolic or parabolic to linear) where the node numbers will be increased rather number of elements (see Figure 1).

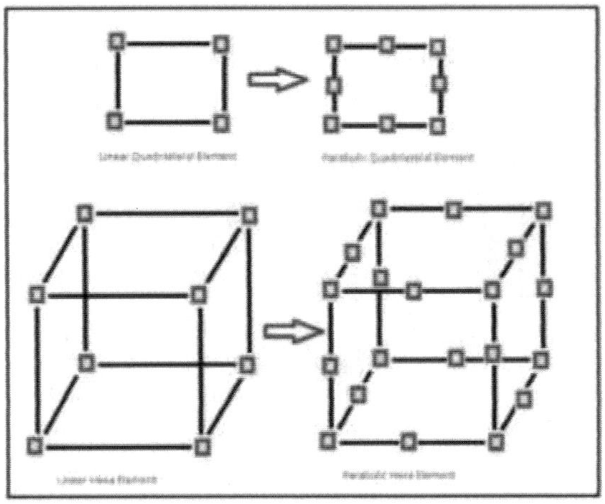

Figure 1. The p-refinement type in mesh method.

Mathematically, if the order of polynomial equation increases, then error norm also reduces unlike in single order equation. Hence it enhances the analysis in getting accurate solution. In h-refinement, sides of the elements are changed in order to increase or decrease the mesh density, node number as well as element number will be increased (see Figure 2). Mathematically, as the number of the elements increased, it captures all the region of the geometry to give more accurate solution. In hp-refinement, element size as well as order is increased which is more effective for the analysis with combination h-refinement and p-refinement.

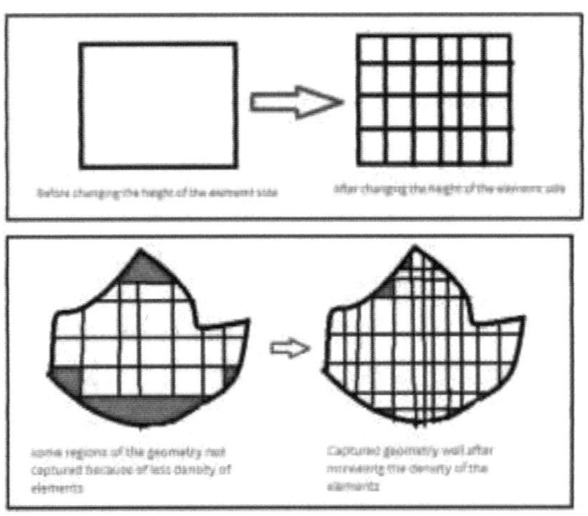

Figure 2. The h-refinement type in mesh method.

3. Result and Discussion

Mesh information for the windmill static simulation is shown in Table 1 and Table 2, and the mesh control is shown in Figure 3. While the mesh information and control for the thermal simulation is shown in Table 3, Table 4 and Figure 4.

Table 1. Mesh information for windmill static simulation.

Mesh type	:	Solid Mesh
Mesher used	:	Curvature-based mesh
Jacobian points	:	4 points
Maximum element size	:	32.6818 mm
Minimum element size	:	6.53636
Mesh quality plot	:	High
Remesh failed parts with incompatible mesh	:	Off

Table 2. Mesh information details for windmill static simulation.

Total nodes	:	146914
Total elements	:	82906
Maximum aspect ratio	:	233.29
% of elements with aspect ratio < 3	:	76.1
% of elements with aspect ratio > 10	:	2.89
% of distorted elements (Jacobian)	:	0
Time to complete mesh (hh:mm:ss)	:	00:00:11
Computer name	:	YAHYA

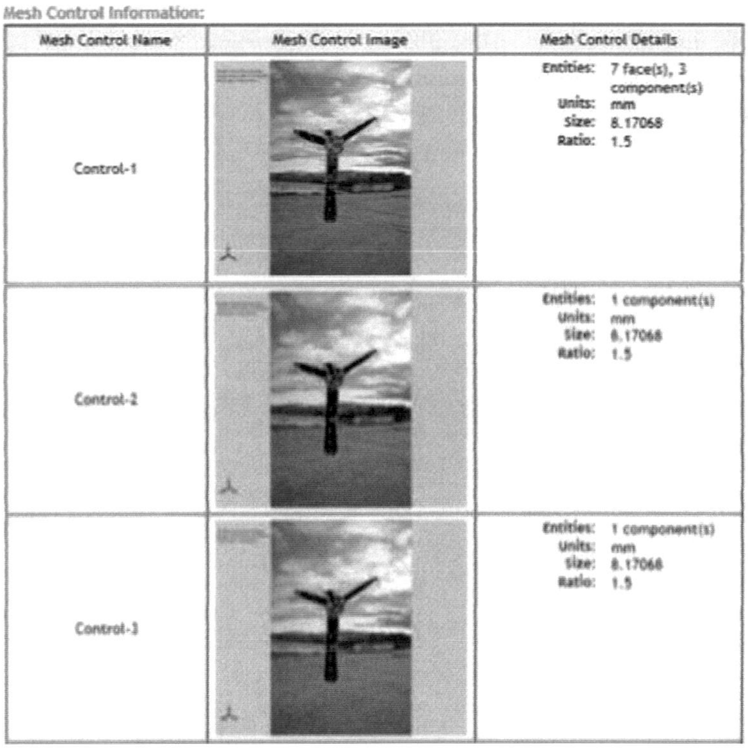

Figure 3. Mesh control information for windmill static simulation.

Figure 3. Mesh control information for windmill static simulation (cont.)

Table 3. Mesh information for windmill thermal simulation.

Mesh type	:	Solid Mesh
Mesher used	:	Curvature-based mesh
Jacobian points	:	4 points
Maximum element size	:	32.6818 mm
Minimum element size	:	6.53636
Mesh quality plot	:	High
Remesh failed parts with incompatible mesh	:	Off

Table 4. Mesh information details for windmill thermal simulation.

Total nodes	:	146914
Total elements	:	82906
Maximum aspect ratio	:	233.29
% of elements with aspect ratio < 3	:	76.1
% of elements with aspect ratio > 10	:	2.89
% of distorted elements (Jacobian)	:	0
Time to complete mesh (hh:mm:ss)	:	00:00:11
Computer name	:	YAHYA

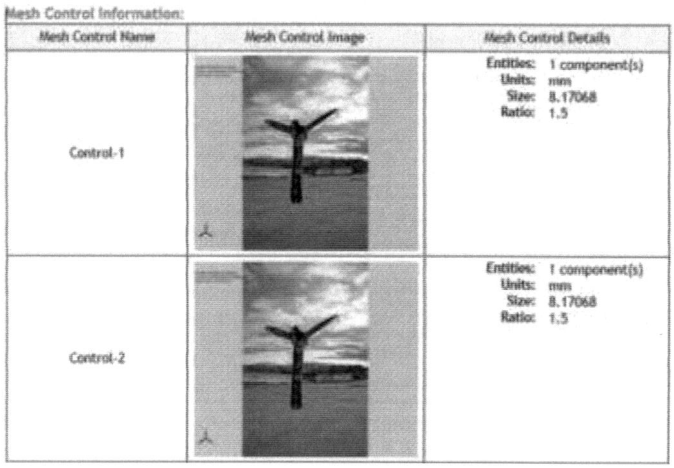

Figure 4. Mesh control information for windmill thermal simulation

IOP Conf. Series: Materials Science and Engineering **277** (2017) 012010 doi:10.1088/1757-899X/277/1/012010

Control-3		Entities: 1 component(s) Units: mm Size: 8.17068 Ratio: 1.5
Control-4		Entities: 1 component(s) Units: mm Size: 8.17068 Ratio: 1.5
Control-5		Entities: 1 component(s) Units: mm Size: 8.17913 Ratio: 1.5

Figure 4. Mesh control information for windmill thermal simulation (cont.)

From the mesh information for both simulations produce the same data, but in detail get different data. However, both mesh controls have the same ratio, which is 1.5

4. Conclusion

Based on the results obtained from the mesh information and control, it is known that the static and thermal simulations on the windmill that already designed have a ratio of 1.5. The smaller the size of the elements created when making the mesh, the better the mesh quality will be generated. When adjusting the mesh size, there is a slider that acts as the density regulator of the element. SolidWorks Simulation also has Mesh Control facility. Features that can adjust mesh density only in the desired part.

Acknowledgments

The results of the study presented in this paper are part of a study funded by a research grant from ristekdikti in post-doctoral research scheme on fiscal year 2017 study entitled design development of power plants wind-solar hybrid system.

5. References

[1] Montgomery D.C 2013 *Design and Analysis of Experiments* (New York: John Wesley and Sons)

[2] Groover M.P 2001 *Automation, Production System, and Computer-Integrated Manufacturing* (New Jersey: Pearson)

[3] Solidworks 2010 *An Introduction to Stress Analysis Applications with SolidWorks Simulation, Student Guide* (Solidworks Education - Concord, Massachusetts 01742 USA)

[4] Kunduru A.R. et al. 2015 A Brief Research, Study, Design and Analysis on Wind Turbine *International Journal of Modern Engineering Research (IJMER)* **5** p5-30

[5] Widodo et al. 2012 Design and Analysis of 5 KW Savonius Rotor Blade *Global Engineers & Technologists Review* **2** p1-8

[6] Rachmat H., Ibrahim M.R. and Hasan S 2017 Design Selection of an Innovative Tool Holder for UVAT using Finite Element Analysis Simulation *AIP Conference* 1831 p020-029

[7] Rachmat H., Mulyana T., Hasan S. and Ibrahim M.R 2016 Design Selection in UVAT using MATLAB Fuzzy Logic Toolbox *International Conference on Soft Computing and Data Mining* (*SCDM*) p538-545

Design local exhaust ventilation on sieve machine at PT.Perkebunan Nusantara VIII Ciater using design for assembly (DFA) approach with *Boothroyd and Dewhurst method*

K I Khalqihi[1], M Rahayu[2] and M Rendra[3]

Telkom University, Bandung, Indonesia
[1]khalqihi.kharin@gmail.com, [2]mira.rahayu82@gmail.com

Abstract. PT Perkebunan Nusantara VIII Ciater is a company produced black tea orthodox more or less 4 tons every day. At the production section, PT Perkebunan Nusantara VIII will use local exhaust ventilation specially at sortation area on sieve machine. To maintain the quality of the black tea orthodox, all machine must be scheduled for maintenance every once a month and takes time 2 hours in workhours, with additional local exhaust ventilation, it will increase time for maintenance process, if maintenance takes time more than 2 hours it will caused production process delayed. To support maintenance process in PT Perkebunan Nusantara VIII Ciater, designing local exhaust ventilation using design for assembly approach with Boothroyd and Dewhurst method, design for assembly approach is choosen to simplify maintenance process which required assembly process. There are 2 LEV designs for this research. Design 1 with 94 components, assembly time 647.88 seconds and assembly efficiency level 23.62%. Design 2 with 82 components, assembly time 567.84 seconds and assembly efficiency level 24.83%. Design 2 is choosen for this research based on DFA goals, minimum total part that use, optimization assembly time, and assembly efficiency level.
Keywords: local exhaust ventilation, design for assembly, maintenance process

1. Introduction

PT. Perkebunan Nusantara VIII has a tea plantation area more or less 400 hectares. Tea plantation utilized to produced black tea orthodox. Every day PT PN VIII plucked approximately 10 tons of wet tea leaves and then processed and produced dry tea powder approximately 4 tons. Every industry have a ventilation system to keep circulation of air and to vacuum the contaminants that can cause dangers for workers. PT Perkebunan Nusantara VIII will use local exhaust ventilation as artificial ventilation that will be placed in sortation area specially on sieve machine because sieve machine generate a lots contaminants like dusts that can caused dangers. Local exhaust ventilation [2] need to be maintenance routinely to keep the system well. Maintenance in PT Perkebunan Nusantara is scheduled every 1 months during work hours [4]. Maintenance process approximately 2 hours to checked all of production machines. With additional of local exhaust ventilation can caused extra time for maintenance and will caused production process delayed. So, for this research for designing local exhaust ventilation using design for assembly approach to minimize time of assembly and maintenance. Goals of design for assembly are to minimize parts, minimize time of assembly, optimize efficiency assembly level. To determine asembly time and efficiency assembly level, qualitative methods is needed. In this research use Boothroyd and Dewhurst method to determine assembly time and efficiency assembly level. Hopefully with designing local exhaust ventilation use

IOP Conf. Series: Materials Science and Engineering **277** (2017) 012011 doi:10.1088/1757-899X/277/1/012011

design for assembly approach can support maintenance process and will not caused production process delayed.

2. Methods

Structured study with do some observations by researchers in PT PN VIII that expected to determine the problems of the company and provides problem solving for company.
This research directly observe in sortation area and interviews some workers and supervisor to get know what the problem is, based on that observation researchers proposed to implemented local exhaust ventilation to solve the problems. For the design local exhaust ventilation, researchers do some observe from literature, and others media, then designing recommendation proposed design to be analyze in design for assembly approach, and choose the best design.

2.1 Design For Assembly

This method depends on the design that will be evaluated and improve continously. There are 4 primary expected in design for assembly results [3]:

1. Reduced material cost.
2. Reduce labor and automatic assembly.
3. Reduced assembly cycle time.
4. Higher product quality and reliability.

2.2 Boothroyd and Dewhurst Method

Quantative method to calculate efficiency assembly level, there are 2 steps to analyze efficiency assembly level [1].

$$E = NM. ta/TM \quad (1)$$

Where,
E = Theorical minimum item count
Ta = Actual assembly process time for the product
TM = Total assembly process time for all item.

2.3 Local Exhaust Ventilation

Local exhaust ventilation system is artifial engineering control to handle and reduce air contaminants in workplace such as dust, fog, steam and also gases [2].

3. Result and Discussion

3.1 Result

Researchers observe the area on the sieve machine, to decide the proposed design local exhaust ventilation that will be applied. Researchers proposed 2 design in this research then make different assembly methods, and parts, then analyze 2 proposed design then evaluate the design after that choose the best proposed design.

A. Design 1

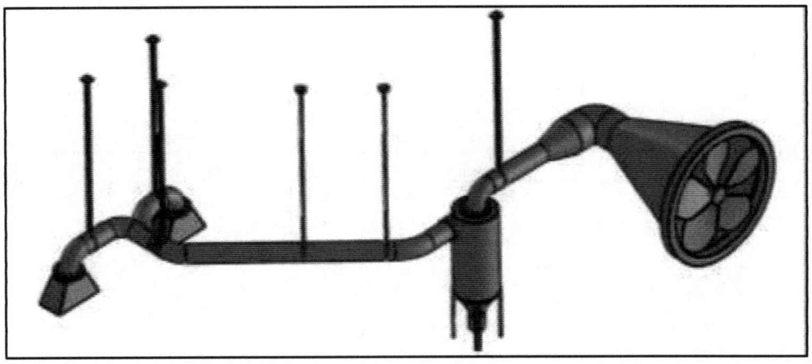

Figure 1. Proposed Design 1 Local Exhaust Ventilation

Proposed design 1, determine parts, dimension, and fastener based on benchmarking from literature, and others media.

Table 1. Proposed Design 1 Components

Design 1			
No	**Component**	**Quantity**	**Description**
1	Exhaust Fan	1	Base Part
2	Air Cleaner Pad	1	Base Part
3	Pole Support	6	Base Part
4	Exhaust Fan Pipe	1	Main Parts
5	Support Ring	6	Main Parts
6	Top outlet	1	Main Parts
7	Bottom outlet	1	Main Parts
8	Silinder vertical	1	Main Parts
9	Main Pipe	1	Main Parts
10	Hood	2	Main Parts
11	Flange	5	Fastener
12	Screw 10	12	Fastener
13	Screw 23	40	Fastener
14	Mur 23	16	Fastener

Based on the proposed design 1 there are 14 kind of parts consist of 94 components that needed for design 1. Next step is determine total operation for proposed design 1 that shown in tabel below.

Table 2. Total Assembly Operation Design 1

No	*Sub Assemblies*	Total assembly operation
1	*Exhaust fan*	4
2	Air cleaner	3
3	*Hood* and Main Pipe	8
4	*Air cleaner* dan *exhaust fan*	2
5	Pipa Pipe and *air cleaner*	1
Total		18

Boothroyd and Dewhurst Method for design 1. Total operation time for assembly 647.88 secs with efficiensy assembly level 23.62 %

Table 3. Manual Assembly Design 1

Desain 1				
A		B	G	H
ID	Part Name	Number of times the operation carried out	Operation time (TM), Bx(D+F) (sec)	Theorical minimum part (NM)
1	Assembly Exhaust Fan			
1.1	Exhaust Fan (Base Part)	1	-	-
1.2	Pipa exhaust fan	1	15.5	1
1.3	Screw 23	8	64.48	4
1.4	Tiang Support (Base Part)	1	-	-
1.5	Ring Penyangga	1	12.1	1
1.6	Screw 10	2	16.12	2
2	Assembly Air Cleaner			
2.1	Wadah Air Cleaner (Base part)	1	-	-
2.2	Pipa outlet atas	1	7.1	0
2.3	Silinder vertikal	1	5	1
2.4	Pipa outlet bawah	1	11.6	0
3	Assembly Pipa Exhaust dan Air Cleaner			
3.1	Flange	1	3.5	0
3.2	Screw 23	8	64.48	4
4	Assembly Pipa Utama dan Hood			
4.1	Pipa Utama	1	15.5	1
4.2	Flange Pipa Utama	2	7	0
4.3	Hood	2	19	2
4.4	Flange Hood	2	15	0
4.5	Screw 23	16	56.96	8
4.6	Mur 23	16	128.96	8
4.7	Tiang Support (Base Part)	5	-	-
4.8	Ring Penyangga	5	60.5	5
4.9	Screw 10	10	80.6	10
5	Assembly Pipa Utama dan Air Cleaner			
5.1	Screw 23	8	64.48	4
Total Part		**94**	**647.88**	**51**
Eficiency Formulation			3x(NM/TM)	
Eficiency			**23.62%**	

B. Design 2

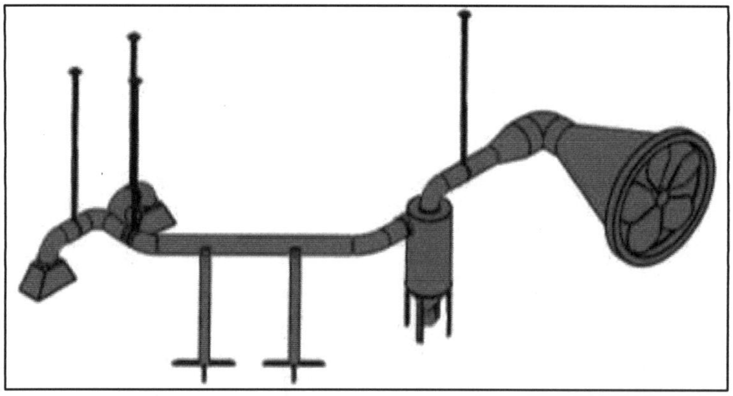

Figure 2. Proposed Design 2 Local Exhaust Ventilation

IOP Conf. Series: Materials Science and Engineering **277** (2017) 012011 doi:10.1088/1757-899X/277/1/012011

Proposed design 2, based on design 1 make an improvement from proposed design before like do simplification parts, and reduce fastener components.

Table 4. Proposed Design 2 Components

Design 2			
No	**Component**	**Quantity**	**Description**
1	Exhaust Fan	1	Base Part
2	Air Cleaner Pad	1	Base Part
3	Pole Support	4	Base Part
4	Floor Pole Support	2	Main Parts
5	Exhaust fan Pipe and top outlet	1	Main Parts
6	Support Ring	4	Main Parts
7	Bottom outlet	1	Main Parts
8	Silinder vertical	1	Main Parts
9	Main Pipe	1	Main Parts
10	Hood	2	Main Parts
11	Screw 10	8	Fastener
12	Screw 23	40	Fastener
13	Mur 23	16	Fastener

Based on the proposed design 2 there are 13 kind of parts consist of 82 components that needed for design 2. Next step is determine total operation for proposed design 2 that shown in tabel below

Table 5. Total Assembly Operation Design 2

No	Sub Assembly	Number of assembly operation
1	*Exhaust fan*	4
2	*Air Cleaner*	2
3	*Hood* and Main Pipe	8
4	*Air cleaner* and *exhaust fan*	1
5	Main Pipe and *air cleaner*	1
Total		15

Boothroyd and Dewhurst Method for design 2. Total operation time for assembly 567.84 secs with efficiency assembly level 24.83 %.

Table 6. Manual Assembly Design 2

Design 2				
A		B	G	H
ID	Part Name	Number of times the operation	Operation time (TM), Bx(D+F) (sec)	Theorical minimum part (NM)
1	Assembly Exhaust Fan			
1.1	Exhaust Fan (Base Part)	1	-	-
1.2	Pipa exhaust fan	1	15.5	1
1.3	Screw 23	8	64.48	4
1.4	Tiang Support (Base Part)	1	-	-
1.5	Ring Penyangga	1	12.1	1
1.6	Screw 10	2	16.12	2
2	Assembly Air Cleaner			
2.1	Wadah Air Cleaner (Base part)	1	-	-
2.2	Silinder vertikal	1	5	1
2.3	Pipa outlet bawah	1	11.6	0
3	Assembly Pipa Exhaust dan Air Cleaner			
3.1	Screw 23	8	64.48	4
4	Assembly Pipa Utama dan Hood			
4.1	Pipa Utama	1	15.5	1
4.2	Hood	2	19	2
4.3	Screw 23	16	56.96	8
4.4	Mur 23	16	128.96	8
4.5	Tiang Support (Base Part)	3	-	-
4.6	Ring Penyangga	3	36.3	3
4.7	Screw 10	6	48.36	6
4.8	Tiang Support Lantai	2	9	2
5	Assembly Pipa Utama dan Air Cleaner			
5.1	Screw 23	8	64.48	4
Total Part		**82**	**567.84**	**47**
Eficiency Formulation			3x(NM/TM)	
Eficiency			**24.83%**	

3.2 Discussion

Table 7. Comparison

Comparison			
No	**DFA Goals**	**Design 1**	**Design 2**
1	Total Components (piece)	94	82
2	Total Assembly Time (sec)	647.88	567.84
3	Efficiency (%)	23.6%	24.8%

The comparison proposed design between both design changes with design 1 with 94 components can be reduced untill 82 components in design 2 and assembly time 567.84 sec with efficiency level 24.8%. Its believed that design 1 is the best design to implemented in PT Perkebunan Nusantara with minimum assembly time for maintenance process. With minimize assembly time for local exhaust ventilation, its hoped that maintenance process for all machine in PT Perkebunan Nusantara VIII will not caused extra time for maintenance process, and production will continue without delay.

4. Conclusion
Based on the results of comparison proposed design, that design 2 is choosen as the best design to be implemented in sortation area for supporting maintenance process based on research problem in PT. Perkebunan Nusantara VIII. Design 2 required 82 components with assembly time 567.84 sec with efficiency 24.8%. Design 2 is fulfill DFA goals and its hoped can achieve of this research.

IOP Conf. Series: Materials Science and Engineering **277** (2017) 012011 doi:10.1088/1757-899X/277/1/012011

5. References

[1] Boothroyd and Dewhurst 1994 *Product Design for Manufacture and Assembly.* (New York: Marcel Dekker, Inc).

[2] Health and Safety Authority 2014 Local and Exhaust Ventilation Guidance. (Irlandia: Health and Safety Authority).

[3] Boothroyd, G. 1987 DFA—The key to design for manufacture, *International Journal of Advanced Manufacturing Technology,2(*3*), pp.* 3*–11, 1987* (UK: IFS Publications Ltd).

[4] PT. Perkebunan Nusantara 2008 *SOP Pembuatan Teh Hitam Ortodoks.* (Bandung: PT. Perkebunan Nusantara)

Decision Support Model for Selection Technologies in Processing of Palm Oil Industrial Liquid Waste

Aulia Ishak[1], and Amir Yazid bin Ali[2]

[1]University of Sumatera Utara, Medan, Indonesia
School Mechanical Engineering Universiti Sains Malaysia, Malaysia
[1]aulia.ishak@gmail.com, [2]meamir@eng.usm.my

Abstract. The palm oil industry continues to grow from year to year. Processing of the palm oil industry into crude palm oil (CPO) and palm kernel oil (PKO). The ratio of the amount of oil produced by both products is 30% of the raw material. This means that 70% is palm oil waste. The amount of palm oil waste will increase in line with the development of the palm oil industry. The amount of waste generated by the palm oil industry if it is not handled properly and effectively will contribute significantly to environmental damage. Industrial activities ranging from raw materials to produce products will disrupt the lives of people around the factory. There are many alternative technologies available to process other industries, but problems that often occur are difficult to implement the most appropriate technology. The purpose of this research is to develop a database of waste processing technology, looking for qualitative and quantitative criteria to select technology and develop Decision Support System (DSS) that can help make decisions. The method used to achieve the objective of this research is to develop a questionnaire to identify waste processing technology and develop the questionnaire to find appropriate database technology. Methods of data analysis performed on the system by using Analytic Hierarchy Process (AHP) and to build the model by using the MySQL Software that can be used as a tool in the evaluation and selection of palm oil mill processing technology.
Keywords: technology liquid waste processing mills, Delphi method, Sustainability, AHP (analytic hierarchy process), Decision Support System (DSS), and software MySQL

1. Introduction

The palm oil industry continues to grow from year to year. The palm oil industry processes palm oil into Crude Palm Oil (CPO) and Palm Kernel Oil (PKO). The comparison of the coconut quantity produced by both products is only 30% of the raw material. This means that 70% is industrial waste. In the production process, palm oil will contribute significantly to the processing of palm oil mills at boiling stations and destruction stations. In addition to producing products, palm oil processing also produces various types of waste, one of which is liquid waste. The amount of palm oil mill effluent will increase as the palm oil mill grows. Oil palm fruit bunch processing (FFB) for palm oil production produces several types of waste. The oil extraction, cleaning and cleaning processes in the factory produce liquid waste or palm oil mill effluent (POME). In the extraction of palm oil, there are three main processes that produce POME, among others: sterilization process, crude palm oil clearance process, which is extortion, separating and clarifying the sterilization of empty fruit bunches. The palm oil mill produces 0.7-1 m^3 of POME for every ton of freshly processed fruit bunches. The newly generated POME is generally hot (60^0-80^0C), acidic (pH 3.3-4.6), thick, brownish with solids, oils and fats, chemical oxygen demand (COD) and biological oxygen demand (BOD) tall one.

Palm mill waste is a colloidal suspension consisting of 95-96% water, 0.6-0.7% oil, a total 4-5% solid waste comprising suspended solids of 2-4% [1]. The palm oil mill effluent contains organic materials of BOD, COD, oils and fats, solid waste in varying amounts [2]. Organic materials found in wastewater cause environmental damage if waste processing technology is not available. The most common problem with choosing palm oil processing technology is that it is difficult to apply the most appropriate technology from a range of palm oil processing alternatives to a particular company. Factors such as capital costs, operating costs and maintenance and land use, are important considerations in the choice of oil waste processing technology.

Developing assessment criteria and methods that provide sustainability measurement are a prerequisite for choosing the best alternative, identifying palm oil industry's processing technology requirements, informing the integrated design of alternative show manufacturers and the effects of social environmental monitoring. The diversity of criteria and the latest gauges in this rapidly expanding field demonstrates the importance of conceptual work and methodology in this area. Development and selection of criteria requires parameters relating to reliability, suitability, practicality and size limits. Many of the weak technologies developed by the technological options are difficult because there is no database and technology valuation method. Database development can facilitate the search and identification of waste processing technologies. The results of the database become inputs in searching for and choosing waste processing technology criteria.

1.1. Waste processing technology

1.1.1. Pond System
In general, the most popular method of processing and use of liquid waste is used by more than 85% of the palm oil industry. Pool systems typically encompass sand and trap oil traps, cooling ponds, ponds, anaerobic ponds, facultative ponds and aerobic ponds. Sand and oil are manually trapped at the beginning of the processing operation unit. In the cooling pool, raw POME is cooled lower than 35 ^{0}C before transferring to the next pool [3].

1.1.2. Composting
Other common systems are conventional anaerobic and aerobic systems (digestion and ventilation openings) [4]. In the anaerobic process of processing, the organic matter contained in the wastewater is converted by bacteria into dissolved organic matter. Then, dissolved organic matter undergoes the process of acetogenesis and fermentation so that biogas consists of methane and carbon dioxide [5].

1.1.3. RANUT
The RANUT method uses a bioreactor tank where the anaerobic process takes place. The RANUT technology was developed through an increase in the organic microbial population of organic substances found in liquid waste [5].

1.2. Sustainability
Generally the criteria used in decision making have been widely discussed in research and have been widely publicized primarily for sustainability assessment. In general, the literature review of good criteria is grouped into social, economic and environmental categories [6]. Application of indicators to measure the sustainability of the manufacturing process is very difficult. Most sustainability indicators are already available which are focused on national and regional development sustainability [7]. Therefore the indicator set has been published focus on continuous product processing. In setting criteria and indicators for sustainable production at the level of process / product the performance of the energy is not taken into account. [8] proposes measures of sustainability achievement in unit processing processes such as environmental, economic, technological and social development. Most decision support system frameworks use traditional economic indicators that are not sustainability measures [7].

1.3. Decision Support System (DSS)

The main purpose of the research Decision Support System (DSS) is to provide users with tools that improve the decision-making process, so that decisions are better [9]. In a decision-making process, companies will face difficulties with the existence of various alternative options in a stage of the process to be implemented. These conditions require companies to know and understand the problems faced, alternatives and criteria for measuring or comparing each alternative to find the best alternative to choose from. DSS is the most widely used in the field of enterprise management at the functional level [10]. Turban and Aronson (2001) in [12] define the DSS as a computer-based interactive system that can assist decision makers in using data and models to solve unstructured problems. While [11] define the DSS is an interactive, flexible, and adaptable computer-based information system, specifically developed to solve non-structured management problems to improve the effectiveness of human decision-making. With the flexible and interactive characteristics of the DSS, the system helps decision makers adopt a problem-oriented approach to solve the problem of selective processing technology of waste products effectively and efficiently. This is achieved by enabling decision makers:

(a) To express linguistic preferences and

(b) To examine the relationship between the evaluation criteria, the available alternatives and the results of the selection.

Many AHP-based decision support systems have dealt with the inaccuracy and ambiguity inherent in information by applying fuzzy logic [12]. A serious practical constraint on the PHA method is the inconsistent and unclear information problem in most of the MCDM [13]. the DSS framework is designed to assist decision makers to select the most appropriate waste processing technology in a manner appropriate to the user and enable decision makers to express their needs in a linguistic manner, and to fully explore the relationship between the criteria and waste processing technologies available in the selection process. Through interactive information exchange between decision makers and DSS, decision makers can adopt a problem-oriented approach in the problem-solving process. This problem-oriented approach is essential to effectively and efficiently solve the evaluation and selection of waste processing technologies.

The rational choice of manufacturing technology and equipment is a decisive factor for the design and overall performance of the manufacturing process [14]. decision makers often face the problem of choosing alternatives. In establishing incompatible criteria requires additional technical assistance. Multi-criteria decision making is a useful tool in CMS for conflicting criteria [15]. One of the most commonly used methods is the AHP method [16]. effective in dealing with inconsistencies [17], as well as with complex and unstructured decisions [19].

2. Method

The first phase of the study was to collect questionnaires through literature reviews derived from books, references, international journals, websites related to other palm oil industries, waste processing technologies, decision making and technology selection over the last period.

Surveys were conducted to obtain the availability of palm oil mill processing technology. To obtain data on waste processing technology is done using interview techniques designed for guided interviews with members involved in the formulation of this system problem using the Delphi method.

Created a literature review to find the right criteria for waste technology. Information on these criteria is obtained from published journals and from stakeholders, as well as from expert knowledge. The results are expected to include criteria and indicators for waste technology. Created a literature review to find criteria appropriate to waste processing technology.

Furthermore, it is designed the hierarchical structure of waste processing technology consisting of objectives, criteria, sub criteria, and alternatives with AHP methods, MySQL software (database) and PHP (scripting language) with user friendly user interface.

The flow diagram in Figure 1 describes the methodology to be applied in this study.

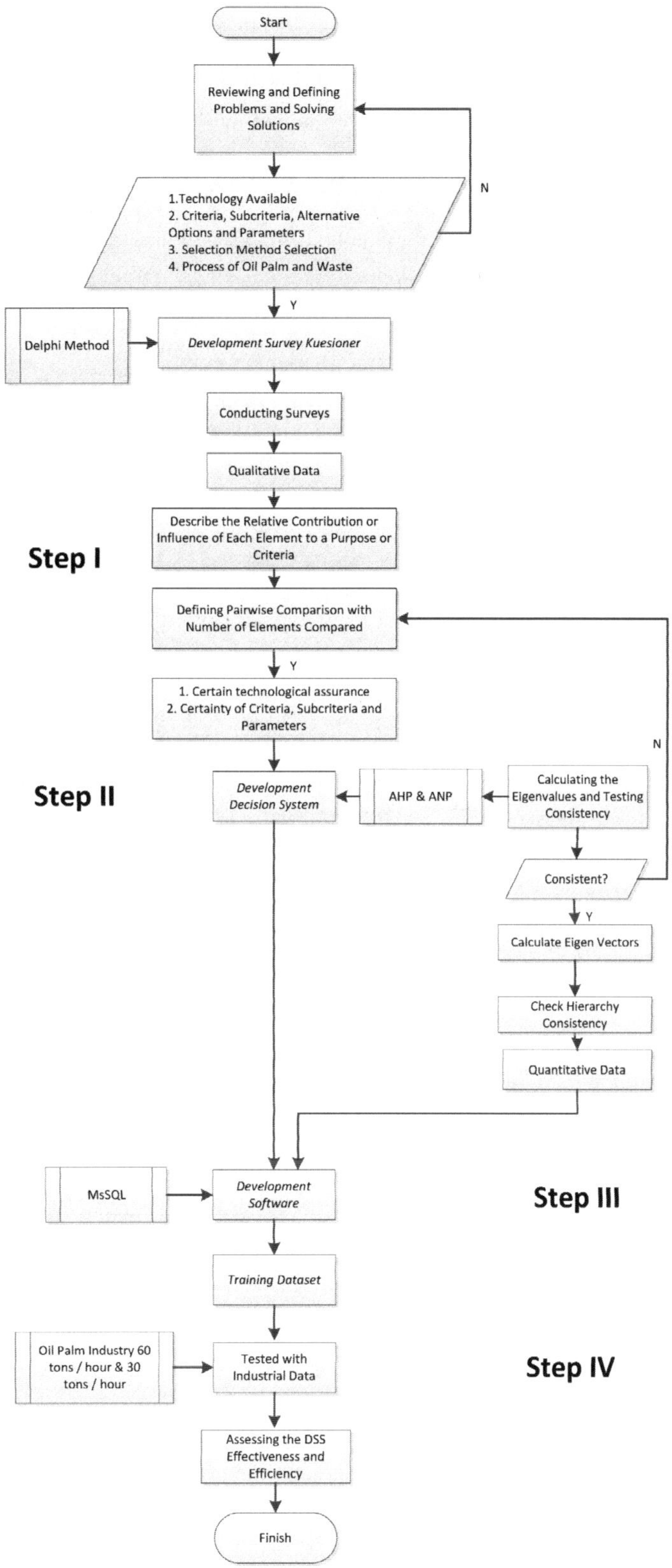

Figure 1. Flowchart of Research

3. Result and Discussion

3.1. Results of the hierarchical structure of Liquid Waste Processing Technology

The results of the questionnaire for waste processing technology aim to give users an idea of how to determine which technology will be selected for waste treatment. The result of this view is the available waste processing technology which can be used as the basis for further selection and evaluation.

In liquid waste processing technology, there are three types of liquid waste processing technology, namely:

3.1.1. Composting

Composting system is a liquid waste processing system with the utilization of liquid waste which is processed as empty stack that has been stacked and stacked in advance to produce fertilizer products.

3.1.2. Pond System

The pond system is a liquid waste processing system using several ponds that are useful for lowering quality values such as BOD, COD, and pH in liquid wastes from processing palm oil processing.

3.1.3. Ranut

Ranut is a liquid waste processing system that uses two reactors to reduce the value of waste in liquid waste and can utilize methane gas for biogas.

Figure 2. presents the hierarchy structure of the AHP hierarchy of liquid waste processing technology for the entire palm oil industry. Based on Figure 2 it can be seen that the objective is to select all palm oil mill processing technologies to be achieved.

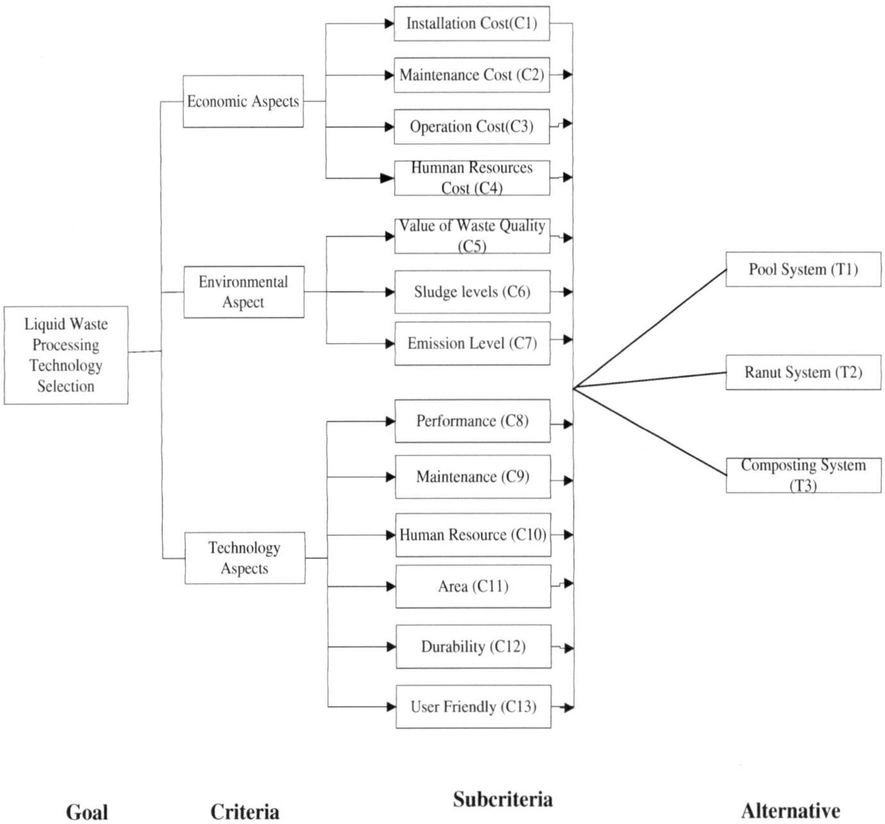

Figure 2. Hierarchy structure of liquid waste processing technology

Creation of hierarchy of management is a very important stage in the process of Analytic Hierarchy Process. At this stage each criteria and sub-criteria to be determined affect each other. Determining the relationship between the sub-criteria of influence is done through interviews with experts in the selection of waste treatment technology. The importance and dependency rating data is obtained from the survey results of squats / pairwise comparisons. The results of the overall responses from the list of questions are then formulated in the Paired Comparison Matrix. Each position in the matrix, the perspective on the left is compared with the perspective on the right value obtained from the questionnaire included in the provision. All results of the survey questionnaire recapitulation will be included in the form of Paired Comparison Matrix.

In application of Palm Oil Selling Technology, decision support systems use AHP algorithms to help determine better waste processing technology options based on criteria, sub criteria, and alternatives.

In the processing of data used in the MySQL database to store a list of data for each alternative option sub criteria. Then phpMyAdmin is used as MySQL MySQL database processing medium.

3.2. Software Design Decision Support System

The database management system is a blend of central savings, processing and data entry. This database management system has the capability to change the structure and content of data elements, ie php MyAdmin program. This database provides the convenience of adding, editing, storing and deleting in order to exchange data that can be executed as desired by the user. The database view of waste processing technology can be presented in figure 3.

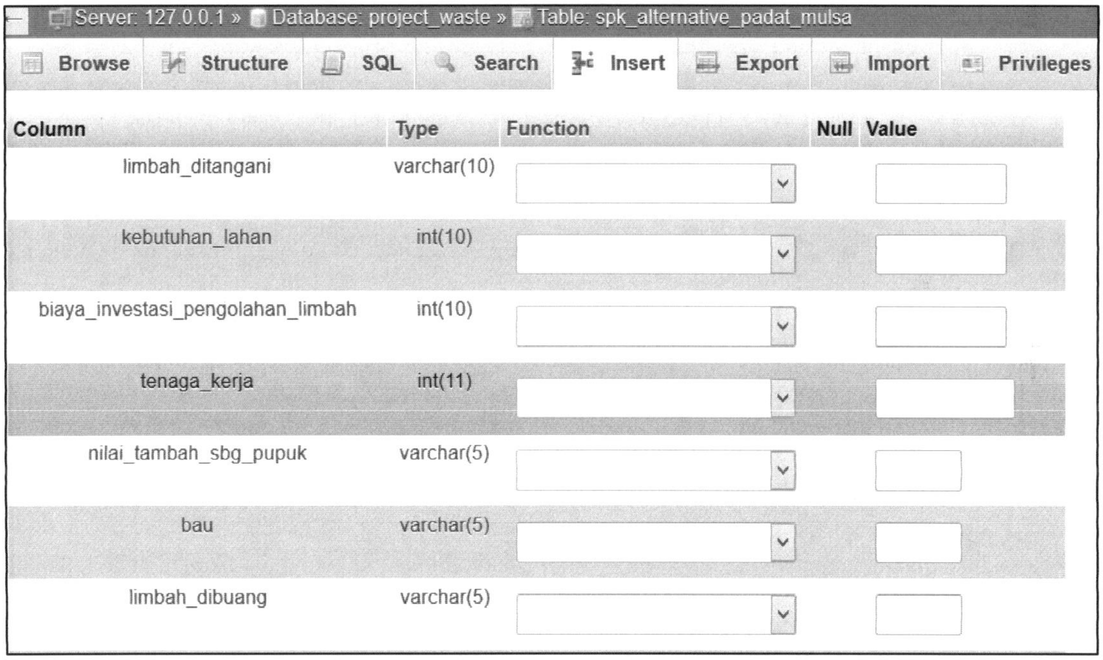

Figure 3. Waste processing technology databaseAppendices

Here are the results for each sub model of waste treatment technology selection.

- Display Process Algorithm.
  In this form, the user can see all the comparative values that have been entered in the previous two forms and see the best alternative of the input data. The process view of the algorithm can be presented in Figure 4.

IOP Conf. Series: Materials Science and Engineering **277** (2017) 012012 doi:10.1088/1757-899X/277/1/012012

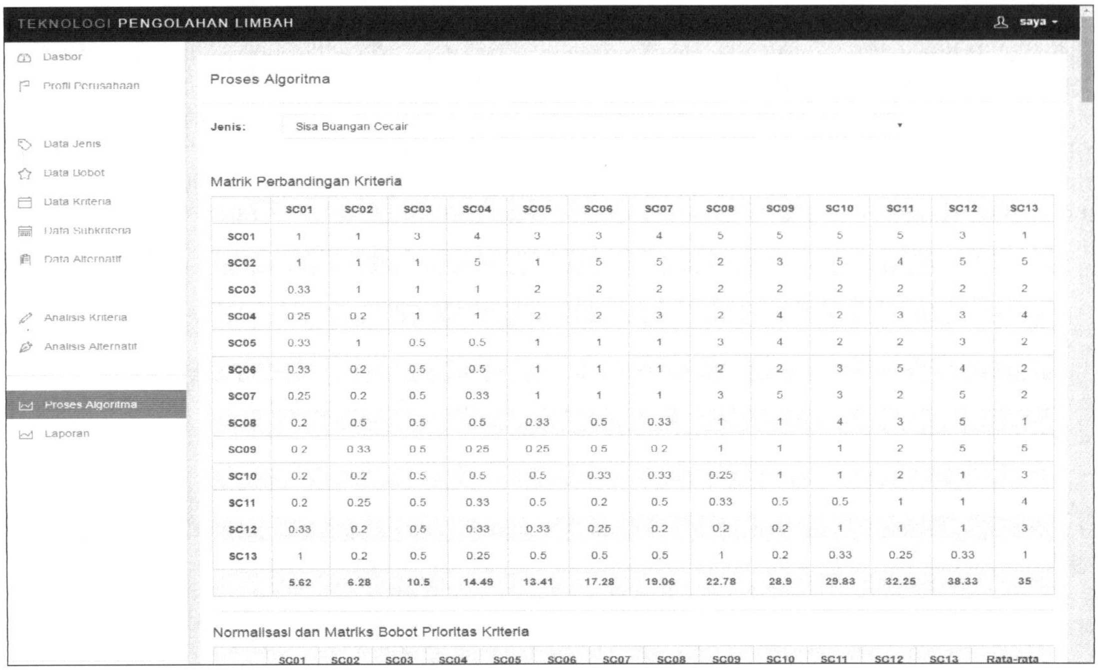

	SC01	SC02	SC03	SC04	SC05	SC06	SC07	SC08	SC09	SC10	SC11	SC12	SC13
SC01	1	1	3	4	3	3	4	5	5	5	5	3	1
SC02	1	1	1	5	1	5	5	2	3	5	4	5	5
SC03	0.33	1	1	1	2	2	2	2	2	2	2	2	2
SC04	0.25	0.2	1	1	2	2	3	2	4	2	3	3	4
SC05	0.33	1	0.5	0.5	1	1	1	3	4	2	2	3	2
SC06	0.33	0.2	0.5	0.5	1	1	1	2	2	3	5	4	2
SC07	0.25	0.2	0.5	0.33	1	1	1	3	5	3	2	5	2
SC08	0.2	0.5	0.5	0.5	0.33	0.5	0.33	1	1	4	3	5	1
SC09	0.2	0.33	0.5	0.25	0.25	0.5	0.2	1	1	1	2	5	5
SC10	0.2	0.2	0.5	0.5	0.5	0.33	0.33	0.25	1	1	2	1	3
SC11	0.2	0.25	0.5	0.33	0.5	0.2	0.5	0.33	0.5	0.5	1	1	4
SC12	0.33	0.2	0.5	0.33	0.33	0.25	0.2	0.2	0.2	1	1	1	3
SC13	1	0.2	0.5	0.25	0.5	0.5	0.5	1	0.2	0.33	0.25	0.33	1
	5.62	6.28	10.5	14.49	13.41	17.28	19.06	22.78	28.9	29.83	32.25	38.33	35

Figure 4. View Process Algorithm

- Election Result.

  Based on this decision, overall Consistency ratio value is less than 0.1, which means that all experts in completing this questionnaire are consistent. In this form the user can see the comparison of each visually assisted weighted alternative with the graph. Selection results can be given in Figure 5

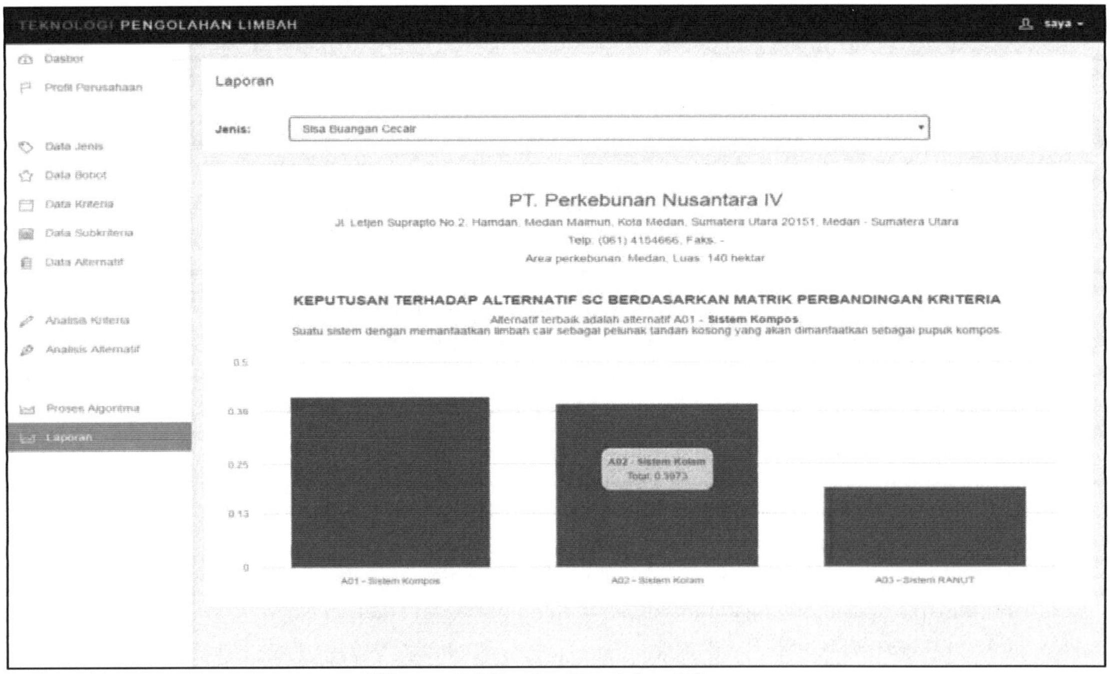

Figure 5. Election Decision View

The structure of the AHP hierarchy in selecting palm oil processing technology consists of four stages of optimum hierarchy structure of palm oil processing technology, ie objective elements, criteria, subcriteria and alternatives.

The valuation value of palm oil processing technology used in this research consists of three nominal values, namely economic point of view, environmental point of view, and technology point of view. The prospect develops into 13 sub values of cost, maintenance cost, human resources cost, operating cost, waste quality value, sludge, emission level, performance, maintenance, human resources, area, durability, and user-friendly.

4. Conclusion

This research suggests that 3 points are associated with palm oil waste processing technology:

- The database is made from data collected from the residual processing subdata of each type of palm oil industry waste. Database creation was created.
- A nominal value search for liquid waste technology is obtained, where parsecond searches use AHP (Analytical Hierarchy Process) questionnaires to help determine palm oil processing technology based on objective value, criteria, subcriteria and alternative value.
- Developing software can be created. In the design used in phpMyAdmin MySQL database software interface. Kes study has been realized.

5. References

[1] [Anonim]. 2006. *Pedoman Pengelolaan Limbah Kilang Kelapa Sawit*. Subdit Pengelolaan Alam sekitar, Ditjen PPHP, Deptan. http:// www.agribisnis.deptan.go.id. [15 January 2009].

[2] Chan, Y.J., Chong, M.F., Law, C.L., 2010. *Biological Treatment of Anaerobically Digested Palm Oil Mill Effluent (POME) using a Lab-scale Sequencing Batch Reactor (SBR).* (J. Environ. Manag). **9, 1738-1746**

[3] D. Arnott, G. Pervan, 2008, *Eight key issues for the decision support systems discipline, Decision Support Systems.* **44 (3) 657–672.**

[4] Da_gdeviren, M. 2008. *Decision making in equipment selection: an integrated approach with AHP and PROMETHEE.* (J. Intell. Manuf). **19, 397-406.**

[5] Demirel, T., Demirel, N.Ç., Kahraman, C. 2008. *Fuzzy analytic hierarchy process and its application. In: Kahraman, C. (Ed.), Fuzzy multi-Criteria Decision-making: Theory and Applications with Recent Developments.* (Springer, New York, USA), **pp. 53-84.**

[6] Durán, O., Aguilo, J., 2008. *Computer-aided machine-tool selection based on a fuzzy-AHP approach. Expert Syst.* (Appl). **34, 1787-1794.**

[7] IChemE - Institution of Chemical Engineers. 2002. *Sustainable Development Progress Metrics: Recommended for Use in the Process Industries.* (Institution of Chemical Engineers, Warwickshire, UK).

[8] Joung, C.B., Carrell, J., Sarkar, P., Feng, S.C., 2013. *Categorization of indicators for sustainable manufacturing.* (Ecol. Indic). **24, 148-157.**

[9] Kahraman, C., 2008. *Multi-criteria decision making methods and fuzzy sets. In: Kahraman, C. (Ed.), Fuzzy Multi-Criteria Decision-making: Theory and Applications with Recent Developments.* (Springer, New York, USA), **pp.1-18.**

[10] Lattimore B, Smith CT, Titus BD, Stupak I, Egnell G. 2009. *Environmental factors in woodfuel production: opportunities, risks, and criteria and indicators for sustainable practices.* (Biomass & Bioenergy). **33:1321–42.**

[11] Lohsomboon P, Palapleevalya P, Worathanakul P, Jirajjariyavech A, Liangsakul R. 2002. *Competitiveness for Thai Industri through Environmental Management Benchmarking Case Study: Palm Oil Idustry,* (Thailand Environmental Institute).

[12] Marimin. 2004. *Teknik dan Aplikasi Pengambilan Keputusan Kriteria Majemuk.* (Grasindo. Jakarta).

[13] Mohamad H, Anis M, Ridzuan R, Ropandi M, Kamaruddin H, Astimar A, et al. 2002. *MDF from oil palm biomass.* (MPOB information series), **MPOB TT No. 168.**

[14] Parveen Fatemeh Rupani, Rajeev Pratap Singh, M. Hakimi Ibrahim and Norizan Esa. 2010. *Review of Current Palm Oil Mill Effluent (POME) Treatment Methods: Vermicomposting as a Sustainable Practice*. Accessed from https://www.idosi.org/wasj/WASJ10(10)/12.pdf

[15] Rao, V. 2007. *Decision Making in the Manufacturing Environment: Using Graph Theory and Fuzzy Multiple Attribute Decision making Methods*. (Springer, London, UK).

[16] Schuchardt, F., Susilawati, E., Guritno, P.1999. *Trials About Composting of Solid Wastes From Palm Oil Mills in Indonesia - Proc. Intern. Conference ORBIT 99 on "Biological Treatment of Waste and the Environment"*, September 2-4, 1999, (Weimar, Germany), **155-164**.

[17] Schuchardt, F., Wulfert, K., Darnoko, D., Herawan, T.2007. *Effect of New Palm Oil Mill Processes on the EFB and POME Utilization*. In: Proceedings of Chemistry and Technology Conference PIPOC, **pp. 44-57** (in Kuala Lumpur)

[18] Singh RK, Murty HR, Gupta SK, Dikshit AK. 2008. *An overview of sustainability assessment methodologies*. (Ecological Indicators);**15:281–99**.

[19] Subramoniam, R., Huisingh, D., Chinnam, R.B., Subramoniam, S. 2013. *Remanufacturing decision-making framework (RDMF): research validation using the analytical hierarchical process*. (J. Clean. Prod). **40, 212-220**.

[20] S. Eom, S. Lee, E. Kim, C. Somarajan, 1998, *A survey of decision support system applications, (1988–1994)*. (The Journal of the Operational Research Society) **49 (2), 109–120**.

[21] Thanh, N.C., Muttamara, S., Lohani, B.N., 1980. *Palm Oil Wastewater Treatment Study in Malaysia and Thailand. Final Report No.114*. (International Development Research Centre, Canada)

[22] Turban, E., Aronson, J. E., Liang, T. P., and Sharda, R. 2007. *Decision support and business intelligence systems (eighth ed.)*. (New Jersey: Prentice Hall).

[23] Wang J, Jing Y, Zhang C, Zhao J. 2009. *Review on multi-criteria decision analysis aid in sustainable energy decision-making*. (Renewable & Sustainable Energy Reviews) **;13:2263–78**.

[24] Wibowo, S., Deng, H. 2012. *Intelligent decision support for effectively evaluating and selecting ships under uncertainty in marine transportation* (Expert Syst. Appl). **39, 6911–6920**.

[25] Wibowo, S., Deng, H. 2013. *Consensus-based decision support for multicriteria group decision making. Comput.* (Ind. Eng). **66, 625–633**.

[26] Wu, T.Y., Mohammad, A.W., Md Jahim, J., Anuar, N. 2010. *Pollution Control Technologies for the Treatment of Palm Oil Mill Effluent (POME) through End-of-Pipe Processes*. (J. Environ. Manag). **91, 1467-1490**.

[27] Wong, S.A. 1980. *Ponding System for Palm Oil Effluent Treatment, Palm Oil Research Institute of Malaysia PORIM Malaysia*, **pp: 18-23**

[28] Wu, T.Y. Mohammad, A.W., Md Jahim, J., Anuar, N. 2010. *Pollution Control Technologies for the Treatment of Palm Oil Mill Effluent (POME) through End-of-Pipe Processes*. (J. Environ. Manag). **91, 1467-1490**.

[29] Zakaria, Z.Z., Haron, K., Murdi, A.A. 2000. *Current Status on Land Application of POME in the Oil Palm Industri e A Survey. PORIM Occasional Paper. No. 42*. (Palm Oil Research Institute of Malaysia, Ministry of Primary Industries, Malaysia).

The Application of Data Mining Techniques to Create Promotion Strategy for Mobile Phone Shop

A U Khasanah, K S Wibowo, and H F Dewantoro

Department of Industrial Engineering
Universitas Islam Indonesia
Indonesia
annisa.uswatun@uii.ac.id, 14522280@students.uii.ac.id, 14522297@students.uii.ac.id

Abstract. The number of mobile shop is growing very fast in various regions in Indonesia including in Yogyakarta due to the increasing demand of mobile phone. This fact leads high competition among the mobile phone shops. In these conditions the mobile phone shop should have a good promotion strategy in order to survive in competition, especially for a small mobile phone shop. To create attractive promotion strategy, the companies/ shops should know their customer segmentation and the buying pattern of their target market. These kind of analysis can be done using Data mining technique. This study aims to segment customer using Agglomerative Hierarchical Clustering and know customer buying pattern using Association Rule Mining. This result conducted in a mobile shop in Sleman Yogyakarta. The clustering result shows that the biggest customer segment of the shop was male university student who come on weekend and from association rule mining, it can be concluded that tempered glass and smart phone "x" as well as action camera and waterproof monopod and power bank have strong relationship. This results that used to create promotion strategies which are presented in the end of the study.
Keyword: promotion strategy, mobile phone shop, agglomerative clustering analysis, association rule mining

1. Introduction

Mobile phone today is no longer just be a tertiary need, but it has become a primary need for almost everyone. Reported by Ericsson Mobility, in 2014 there were at least about 7 billion mobile users worldwide and this number is increasing by 7 percent annually. It is predicted that the number of smartphone users will reach 5.6 billion in the world [1]. In Indonesia, data extracted from the US Cencus Bureau by 2015 explains that mobile phone users have exceeded 281 million where as early as 2014 the number of Indonesians has reached 251 million [2]. This fact indicates that the need for communication devices especially mobile phones in Indonesia is very high and it makes Indonesia as the fourth largest mobile phone user in the world [3].

Along with the growing number of mobile phone users in Indonesia, the mobile industry is also increasingly stretched. The number of shops that sell mobile phones is growing very fast in various regions in Indonesia including in Yogyakarta. Now the store that sells mobile phones not only can be found in the city centre but almost in all corners of the city. This fact leads high competition among the mobile phone shops. The shops do not only sell mobile phone but they also sell mobile phone accessories such as power bank, action camera, monopod, case, tempered glass, earphone, and so on. Due to the high demand for mobile phones, the demand for those accessories is also high. In these conditions the mobile phone shop should have a good promotion strategy in order to survive in competition, especially for a small mobile phone shop.

Promotion is an activity that included in marketing strategy. This activity is very essential to inform the detail information about product or service and encourage the customer to buy or use it. Based on Chance [4] one important thing to be succeed in marketing including in promoting product or service is knowing the market target. This step can be conducted through customer segmentation. Segmentation is greatly embedded in a marketing strategy because different customer groups will require a different marketing mix [5]. Customer or market segmentation is the division of the market into different customer groups where specific customer groups can be selected as market targets to be achieved with a specific marketing mix [6]. After knowing the target market, a specific promoting strategy can be obtained for this specific target market.

For many shops, including mobile phone shops, customer segmentation is also essential step that must be done to create an attractive promotion strategy. Attractive promotion will encourage customer to buy their product. Usually, information that appeals to consumers is related to discounts, bonus, or product package with special prize. To create this kind of promotion, a company or shop should analysis the buying pattern of their target market, for example the company should know what kind of item that usually bough by their consumer in conjunction with another product.

The purpose of this study is to create an attractive promotion strategy for a mobile phone shop in Sleman Yogyakarta by conducting customer segmentation and analysing the product that commonly bought by the customer as the same time. This kind of analysis can be conducted using data mining techniques. Data mining is the process of automatically discovering useful information in large data repositories [7]. Data mining has several popular techniques such as clustering analysis and association rule mining. There are many methods that can be used in clustering analysis, one of them is agglomerative hierarchical clustering. In this study agglomerative hierarchical clustering, is used to conduct the customer segmentation, and association rule mining is used to discover an interesting pattern about the products that commonly bought by the customer at the same arrival time, then those results will be used as a recommendation to make an attractive promotion strategy. The rest of this paper is organized as follows, Section 2 present the literature review, Section 3 presents the research method related to this study, Section 4 shows the results and discussion. The concluding is finally made in Section 5

2. Literature Study

2.1. Agglomerative Hierarchy Clustering

Hierarchical clustering analysis is the oldest clustering method when compared to other methods, however this method is still widely used. There are two basic approach to generate this method, agglomerative approach and divisive approach, but agglomerative approach is more commonly used. Agglomerative hierarchical clustering starts with the point as individual clusters and each step, merge the closest pair of clusters until only one cluster remains [7]. This method uses a bottom up strategy and the single cluster become the hierarchy's root. In the merging step, two clusters which are close to each other are determined according to some similarity measure, then combine the two into one cluster[8]. A hierarchical clustering is usually displayed graphically using tree-like diagram called dendogram. The similarity measurement can be obtained by several methods such as min (single link), max (complete link), and group average.

There have been a lot of scholars in many different field of studies apply agglomerative hierarchical clustering. Iqbal et al [9] applied agglomerative hierarchical clustering to cluster 47 tomato genotypes into 5 clusters. Kang and Landry [10] use agglomerative hierarchical clustering to characterize, compare, and analyze eye movement sequence that occur during visual tracking of multiple moving targets. The eye movement data was collected from 25 expert air traffic controller and 40 novices through difficult scenarios on simulated radar display. The data was clustered into three different clusters, easy, moderate. While Kumar and Toshniwal [11] apply hierarchical clustering to identify the different reasons behind road accidents such as traffic characteristics, weather characteristics, road characteristics and etc. In this study trend analysis was also used to identify the increasing and decreasing accident rate in different reasons in Gujarat state India. Ip et al [12] implemented hierarchical clustering to cluster demand of electricity source for electric vehicle (BEV), the result than used to maximized the usability of battery recharging infrastructure. The plan should be designed where the charging station should serve the commuters demand and it can minimize the relevant operating cost.

2.2. *Association Rule Mining*

Association rule mining can be illustrated as follows, suppose that the set I=(i1,i2,…,ik) is the total number of items. The item set X and Y are part of the total set I. If Y happen in one transaction when X happens, as association rule can be expressed as $X \Rightarrow Y$ [13]. In market basket analysis, association rule can be can be express as "A customer who buys product X_1 and X_2 will also buy product Y with probability c%" are found [14]. For example, in a restaurant egg and coffee may be ordered together primarily between 7 am and 11 am, and maybe turkey and pumpkins are sold at the same time in the week before Thanksgiving [15]. In service industry, association rule mining can be express as a customer who use a certain service will also use another service at the same service time.

The strength of an association rule can be measured in terms of its support and confidence. Tan et al. [7] explain that support determines how often a rule is applicable to a given data set. A low support rule is uninteresting from a business perspective and sometimes it will be eliminated. While, confidence determines how frequently items in Y appear in transaction that contain X, the higher the confidence the more likely it is for Y to present in transaction that contain X. The formal definitions of these metrics are,

$$Support, s(X \Rightarrow Y) = \frac{\sigma(X \cup Y)}{N} \tag{1}$$

$$Confidence, c(X \Rightarrow Y) = \frac{\sigma(X \cup Y)}{\sigma(X)} \tag{2}$$

Some scholars have applied association rule mining. Abdulsalam et al [16] implement association rule mining which is also known as market basket analysis to reveal sales pattern in a supermarket. This study showed rules of purchasing one product would lead to the purchase of another product. This study concluded that apple and chocolate had strong correlation. This information helps the company to create marketing and advertising strategies that outshine the competitors. Verma et al [17] apply association rule mining to find out the pattern incidents in a steel plant in India. Czibula et al [18] used association rule to predict defect of a software system in order to conduct continuous improvement. Kumar, & Toshniwal [19] apply data mining techniques to analyse data related with road accident in India. Cluster analysis was used to cluster the accident data, then association rule mining was used to identify the situation when an accident occurred in each cluster. The results can be used to make some preventions for each different categories, thus it can minimize the number of accidents happened.

3. Research Method

The research was conducted using Cross Industry Standard Process for Data Mining (CRISP-DM) framework. The CRISP-DM methodology is described in terms of a hierarchical process model, consisting of sets of tasks. CRISP-DM provides a non-proprietary and freely available standard process for fitting data mining into the general problem-solving strategy of a business or research unit. This framework was firstly developed in late 1996 [20]. CRISP-DM is divided into six phases Business Understanding, Data Understanding, Data Preparation, Modeling, Evaluation, and Deployment as shown in Figure 1.

The business process describes the background and objectives of this study as discussed in introduction in Section 1. Data understanding explain the initial data. The data was customer data from "ABC" mobile shop in Sleman Yogyakarta including the gender, occupation, age, arrival date, and the product that bought by the customer. The data were collected using questionnaire, then the data was cleaned and transformed in Data Preparation Step. After data cleaning, only 50 data left. The initial data was nominal and it need to be transformed into numeric. The data for clustering analysis must be numeric since it measures distance between data. The next step is modeling step. Agglomerative Hierarchical Clustering (AHC) and Association Rule Mining are used to create the model. Clustering is used to segment the customer and association rule mining is used to know the several products that commonly bought by the customer at the same arrival time. This result that will be evaluated whether the model in fact achieves the objectives set or not. The modeling and evaluations step will further discuss in Section 4. The result from clustering and association rule will be used to create suitable promotion strategy for a certain customer segment and it will be discussed in deployment step in the last section (conclusion). Software SPSS was used to apply agglomerative clustering analysis and Rapid Miner was used to apply association rule mining.

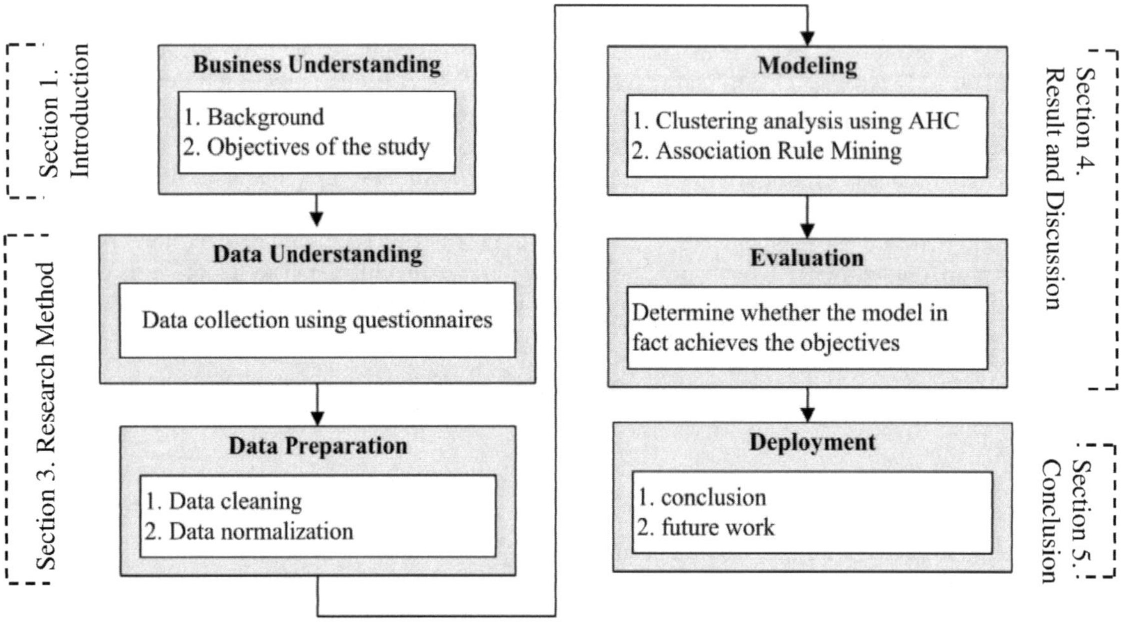

Figure 1. CRISP-DM frame work

4. Result and Discussion

4.1. Clustering Analysis

Modeling Step

In this step mobile shop customer was segmented based on several attributes including the gender, occupation, age, and arrival date. The 50 data were clustered using AHC, and some measurement distance including single linkage, complete linkage, and average linkage were used. Each measurement distance showed different output and based on the outlook of the dendogram, the output from average linkage was the most obvious to be interpreted.

Evaluation Step

The result of agglomerative hierarchical clustering is shown in dendogram (Figure 2.) The dendogram using average linkage was cut in 9,5 and it resulted 3 clusters and 1 outlier. The profile for each cluster is represented in Table 1. Cluster 1 consisted of 17 members, cluster 2 consisted of 25 members, and cluster 3 consisted of 7 members, thus cluster 2 had the biggest cluster members. There was no difference in average of age for all cluster. For cluster 1 the number of female and male were slightly difference, and it was dominated by university student who mostly come on weekdays. For cluster 2 all of the member were male, and it was dominated with university student, student, and businesswoman who come on weekend. While all members in cluster 3 were female, dominated with university student who mostly come on Sunday. From this result it was known mostly customer come on weekend and they rarely come on Monday. Then based on this result, the objective of clustering analysis in this study to know the specific customer segment was achieved.

Table 1. Cluster Profiling

Cluster	Average Age	Gender		Occupation				Arrival date
		male	female	University student	student	Employee	Business-man	
1 (17)	21.76	59%	41%	76%	0%	6%	18%	mostly on Thursday, Wednesday, Tuesday
2 (25)	21.76	100%	0%	32%	32%	4%	32%	mostly on Friday, Saturday, Sunday
3 (7)	21.57	0%	100%	43%	29%	14%	14%	mostly on Sunday

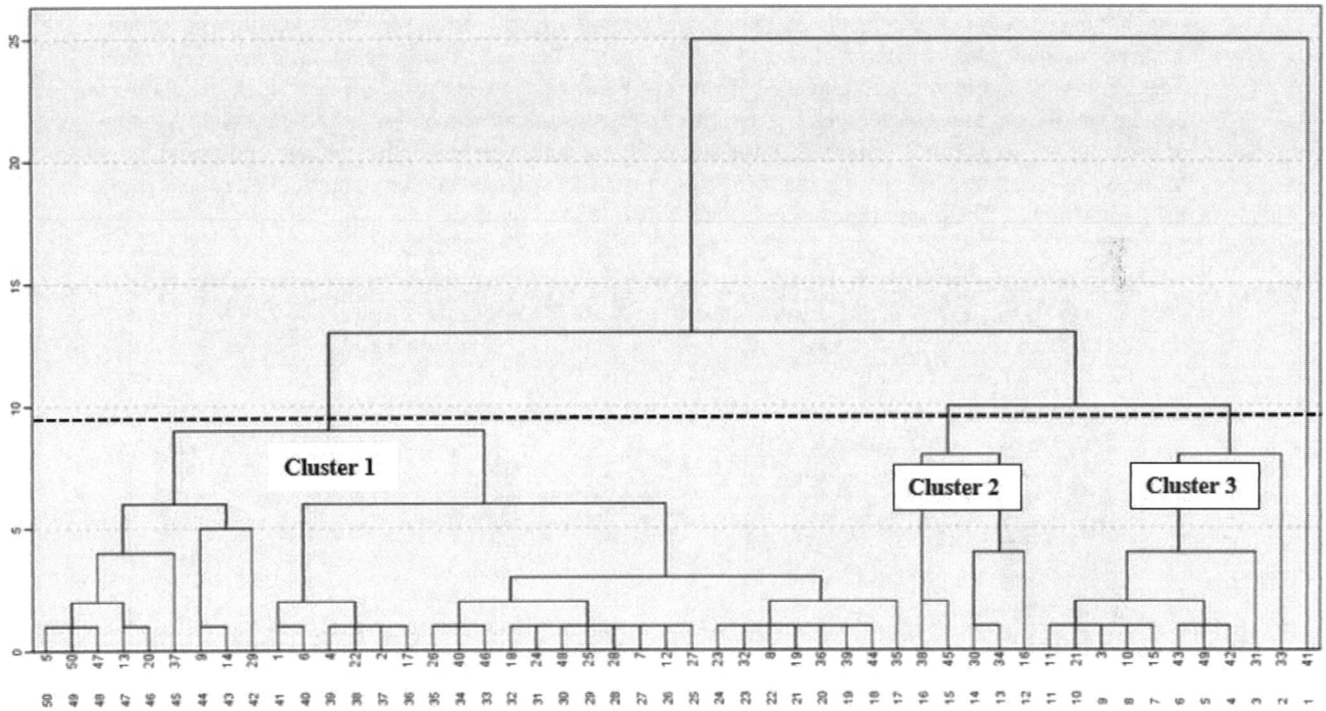

Figure 2. Dendogam using average linkage

4.2. Association Rule Mining

Modeling Step

This step was conducted to reveal the customer buying pattern on the "ABC" mobile phone shop. To get the information, the customer was asked related with what kind of products that they bought. Then the data was transformed into binary form, 0 for not buying the product and 1 for buying the product. Then, the data was put into Rapid Miner software. Minimum support was adjusted at 5% and confidence was 5%. There were lots of rules generated but only 18 rules had lift ratio bigger than 1. Lift ratio measure the correlation between item set and it can use to measure the importance of the rule. Greater lift ratio indicates stronger associations, the highest lift which means highest correlation [21,22].). From the 8 rules, there are 3 interesting rules, as follows

$$\text{tempered glass} \Rightarrow \text{smart phone "x"} \quad (s:16\%, c:53\%)$$
$$\text{action camera} \Rightarrow \text{waterproof , monopod} \quad (s:16\%, c:61{,}5\%)$$
$$\text{action camera} \Rightarrow \text{power bank} \quad (s:24\%, c:50\%)$$

Evaluation Step

The first rule means 16% of all transaction contain tempered glass and smart phone "x" and 53% of transaction have the item tempered glass also contain smart phone "x". The second rule means 16 % of the transaction contain action camera, waterproof camera case, and monopod, and 61,5% of the transaction have item action camera also contain waterproof camera case and monopod. While the third rule means 24% transaction contain action camera and power bank, and 50% of the transaction have action camera also contain power bank. Based on this result, the objectives of association rule to analyse the product that strongly related and commonly bought by the customer as the same time was achieved.

5. Conclusion

From the clustering analysis the customer segmentation can be known and from association rule mining, the important customer buying pattern can be revealed. This information can be used to make suitable promotion strategy. Not all the information is used to make the promotion strategy. From the clustering result it can be concluded that the biggest customer segment come to the shop were male university student who come on weekend, and customer rarely come Monday. While from association rule mining, it can be concluded that tempered glass and smart phone "x" as well as action camera, waterproof camera case, monopod and also action camera with power bank have strong relationship. Thus, the shop can make promotion to give free tempered glass for university student who buy smart phone "x" on Monday, or discount for water proof camera case for university student who buy action camera and monopod on Monday, or it can be also give special price for university student who buy action camera and power bank on Monday. The promotion strategies are illustrated in Figure 3.

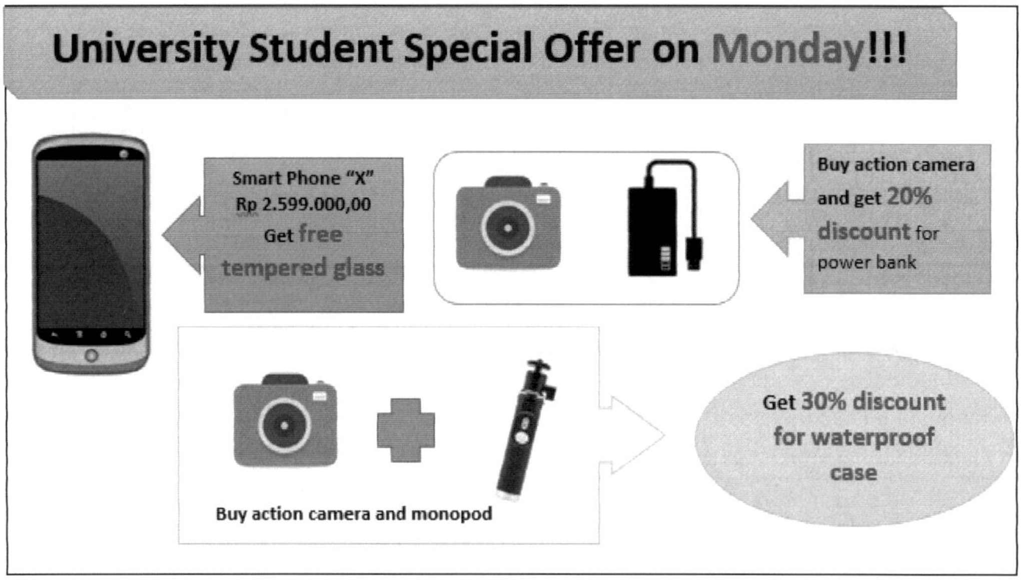

Figure 3. Promotion strategy
(Source for mobile phone picture: [23]), (source for camera picture: [24]), (Source for power bank picture: [25]), (source for monopod picture: [26])

6. References

[1] Nistanto R K 2014 2015 Pengguna "Mobile" Lampaui Jumlah Penduduk Dunia online accsess on 10 September 2017 URL// http://tekno.kompas.com/read/2014/06/04/1025003/2015. Pengguna.Mobile. Lampaui. Jumlah.Penduduk.Dunia

[2] Bursadana 2016 Jumlah Pengguna Ponsel di Indonesia online access on 10 September 2017 URL// http://media.bursadana.co.id/2016/03/18/jumlah-pengguna-ponsel-di-indonesia.html

[3] Wahyudi A 2015 Indonesia Raksasa Teknologi Digital Asia online access on 10 September 2017 URL:https://www.tempo.co/read/kolom/2015/10/02/2310/indonesia-raksasa-teknologi-

IOP Conf. Series: Materials Science and Engineering **277** (2017) 012013 doi:10.1088/1757-899X/277/1/012013

digital-asia

[4] Chance J 2017 8 Keys to a Strong Marketing Strategy online access on 10 September 2017 URL// https://www.businessknowhow.com/marketing/blocks.htm.

[5] Kahreh M S, Tive M, Babania A and Hesan M 2014 Analyzing the applications of customer lifetime value (CLV) based on benefit segmentation for the banking sector *Procedia-Social Behav. Sci.* **109** 590–4

[6] Kotler P 1980 *Marketing Management: Analysis, Planning, and Control, 4th edition* (Englewood Cliffs, N.J: Prentice-Hall)

[7] Tan P N, Steinbach M and Kumar V 2006 *Introduction to Data Mining* (MA: Pearson Education)

[8] Han J, Kamber M and Pei J 2006 *Data Mining: Concepts and Techniques, 2nd edition* (MA: Elsevier)

[9] Iqbal Q, Saleem M Y, Hameed A and Asghar M 2014 Assessment Of Genetic Divergence In Tomato Through Agglomerative Hierarchical Clustering And Principal Component Analysis *Pak. J. Bot* **46** 1865–70

[10] Kang Z and Landry S J 2015 An Eye Movement Analysis Algorithm for a Multielement Target Tracking Task: Maximum Transition-Based Agglomerative Hierarchical Clustering *IEEE Trans. Human-Machine Syst.* **45** 13–24

[11] Kumar S and Toshniwal D 2016 Analysis of hourly road accident counts using hierarchical clustering and cophenetic correlation coefficient (CPCC) *J. Big Data* **3**

[12] Ip A, Fong S and Liu E Optimization for Allocating BEV Recharging Stations in Urban Areas by Using Hierarchical Clustering 460–5

[13] Han H K, Kim H S and Sohn S Y 2009 Sequential association rules for forecasting failure patterns of aircrafts in Korean airforce *Expert Syst. Appl.* **36** 1129–33

[14] Nakhaeizadeh G, Hipp J and Güntzer U 2000 Algorithms for association rule mining - a general survey and comparison *ACM sigkdd Explor. Newsl.* **2** 58–64

[15] Lia Y, Ningb P, Wanga X and Jajodiaa S 2003 Discovering Calendar-based Temporal Association Rules G *Discovery.Csc.Ncsu.Edu* **44** 193–218

[16] Abdulsalam S O 2014 Data Mining in Market Basket Transaction : An Association Rule Mining Approach *International Journal of Applied Information Systems* **7** 15–20

[17] Verma A, Khan S D, Maiti J and Krishna O B 2014 Identifying patterns of safety related incidents in a steel plant using association rule mining of incident investigation reports *Saf. Sci.* **70** 89–98

[18] Czibula G, Marian Z and Czibula I G 2014 Software defect prediction using relational association rule mining *Inf. Sci. (Ny).* **264**

[19] Kumar S and Toshniwal D 2015 Analysing road accident data using association rule mining *Computing, Communication and Security (ICCCS)* (IEEE) pp 1–6

[20] Chapman P, Clinton J, Kerber R, Khabaza T, Reinartz T, Shearer C and Wirth R 2000 *CRISPDM 1.0 step-by-step data mining guide* (Technical report: CRISP-DM)

[21] Hahsler M, Grün B and Hornik K 2007 Introduction to arules–mining association rules and frequent item sets. *SIGKDD Explor* **2** 4

[22] El-Halees A 2009 Introduction to arules–mining association rules and frequent item sets *SIGKDD Explor* **2**

[23] Ilnanny 2012 online acsses in October 2017 URL: https://openclipart.org/detail/168116/nexus-one

[24] Sanity D 2014 online access in October 2017 URL: https://dribbble.com/shots/1741674-Camera-Icon

[25] Olike 2017 online access in October 2017 URL: http://olike.com.my/products-power-bank/

[26] Banggood 2017 online access in October 2017 URL: https://www.banggood.com/Original-Bluetooth-Remote-Control-Monopod-Selfie-Stick-for-Xiaomi-Yi-Sports-Camera-p-982396.html

IOP Conf. Series: Materials Science and Engineering **277** (2017) 012014 doi:10.1088/1757-899X/277/1/012014

Conceptual design of real time and adaptive measure of mental workload using galvanic skin respond

A Widyanti

Department of Industrial Engineering, Bandung Institute of Technology (ITB)
Ganesa 10 Bandung Indonesia
widyanti@mail.ti.itb.ac.id

Abstract. Mental workload, defined as difference between task demand and mental capacity of the worker, is crucial in providing effective mental workload that ensure productivity and safety of the worker. Thus, mental workload measures are crucially important. Besides requirements of mental workload measures that must be sensitive, diagnostic, non-intrusive, ease to implement, and acceptable; recently, there is another important requirement of the mental workload measure that is adaptive and real time. Purpose of this paper is to describe conceptual design of real time and adaptive measure of mental workload using galvanic skin response. Implication of the proposed conceptual design in discussed.
Keywords: real time, adaptive, mental workload, galvanic skin response

1. Introduction

The rapid growth of information technology and information system in daily life and work environment has changed to nature of work from physically to mentally demanding task. In the end it raised the need to take into account cognitive or mental workload, defined as the differences between cognitive demand of the task and capacity of the human [1] or mental capacity used during completing task [2], in order to ensure optimal performance of human.

There are criteria for mental workload measures. O Donnel and Eggemeier [2] stated five criteria for assessing the mental workload measures that are sensitivity (refers to the capability of measures to reflect changes in workload), diagnosticity (refers to the ability to identify changes in workload and the cause of these changes), intrusiveness (refers to the extent to which a measure interferes with normal task performance), ease of implementation (refers to easiness of implementation due to practical reason) and acceptability (refers to the degree of approval and acceptance by participants). In addition, Eggemeier et al. [3] proposed two additional criteria namely validity (refers to the ability to reflect the concept of mental workload only, not reflect other concept) and reliability (refers to the consistency of mental workload measures).

There is a body of literature describe the measurement of mental workloads that are performance measures, objective measure (based on physiological indices), and subjective measures (based on perceived load of human). Performance measures was conducted based on reaction time and error. Phsyiological indices include heart rate variability /HRV [4], Galvanic Skin Response/GSR [5], eye blink rate [6] and such advances of brain activity measures for example electroencephalogram (EEG) and functional magnetic resonance (fMRI [7]). Subjective measures includes the NASA-TLX [8,9] and the RSME [10].

It is worth noting that both objective and subjective mental workload measures mentioned above are conducted during and after the task has been completed and the more important thing is that the analysis of the result of the measures are conducted after the completion of the task. Considering that providing optimal mental workload should be conducted in real time condition, there is an urgent call of the need of real time measurement of mental workload.

In addition, there is also one new requirement for mental workload measures that is adaptive. Adaptive measures of mental workload is crucial to keep the workload in appropriate levels in order to avoid conditions of under-load or over-load, so that overall performance and safety of human can be enhanced [11]. Furthermore, in relation between adaptive and system, Djiksterhuis [12] stated that a system should adapt based on critical event driven, performance driven, and physiological driven.

In Indonesia, the need to measure mental workload is increasing recently, in particular in line with the use of technology and information technology that makes the job in industries more and more mentally demanding. Coupled with the fact that mental workload measures are influenced by factors such as culture [13], it is worth noting to develop mental workload measures that fulfill all criteria mentioned above (in particular real time and adaptive criteria) and suit with Indonesian condition.

This paper is aimed to describe conceptual design of real time and adaptive mental workload measures to be used in Indonesia. Some criteria and feasibility of the measures were described. Since subjective mental workload is culturally biased when used in Indonesia, objective measures were used as a base in developing real time and adaptive measures of mental workload in Indonesia.

2. Method

To be effectively used in Indonesia, price has been considered as an important factor. Such objective mental workload measures are high cost, for example Electroencephalograph (EEG) and Magnetic Resonance Image (MRI). In contrast, such objective measures are cost effective, for example Heart Rate Variability (HRV) monitor, Galvanic Skin Response (GSR), and eye tracker. A series of experiments was conducted before reveals that HRV, GSR, and eye blink rate has similar sensitiveness in detecting mental workload. Due to its practical in both laboratory and industrial application, the GSR was chosen as a basis of development of real time and adaptive mental workload measures.

As described in Widyanti et al. [5], GSR is a form of Electrodermal Activity (EDA) that is capable of detecting electrical changes in the skin caused by external stimuli [13]. The underlined process in GSR measure is that several layers in human skin consists of layer called epidermis which has a thick layer consisting of dead cells that result in high resistance of the skin. When the skin produce sweat, which contains about 0.3% salt NaCl, skin resistance will decrease [17]. This is because NaCl is an electrolyte and a good conductor, therefore the NaCl will leads to increased skin conductance. Since sweat glands are concentrated in the palms of the hands therefore the GSR measurements are usually performed at the palms. The GSR measure can be seen in Figure 1.

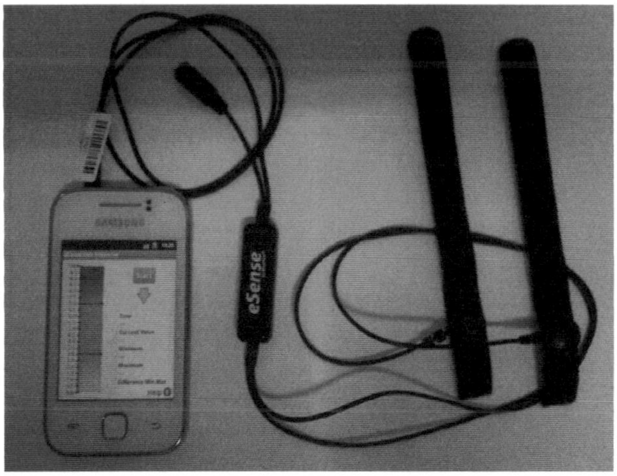

Figure 1. The GSR measure

IOP Conf. Series: Materials Science and Engineering 277 (2017) 012014 doi:10.1088/1757-899X/277/1/012014

3. Result

The conceptual design begin with analysis of the result of the GSR as can be seen in Figure 2. The development process is started by feature extraction of GSR data. Artefact removal and filtering are applied to get data that are clean from noise. After the extraction, norm of low, medium, and mental workload is proposed based on combination of previous experiment both using subjective and objective measures. The adaptive algorithm is developed based on detection of mental workload classification. Feedback mechanism is given when low or high mental workload level is achieved.

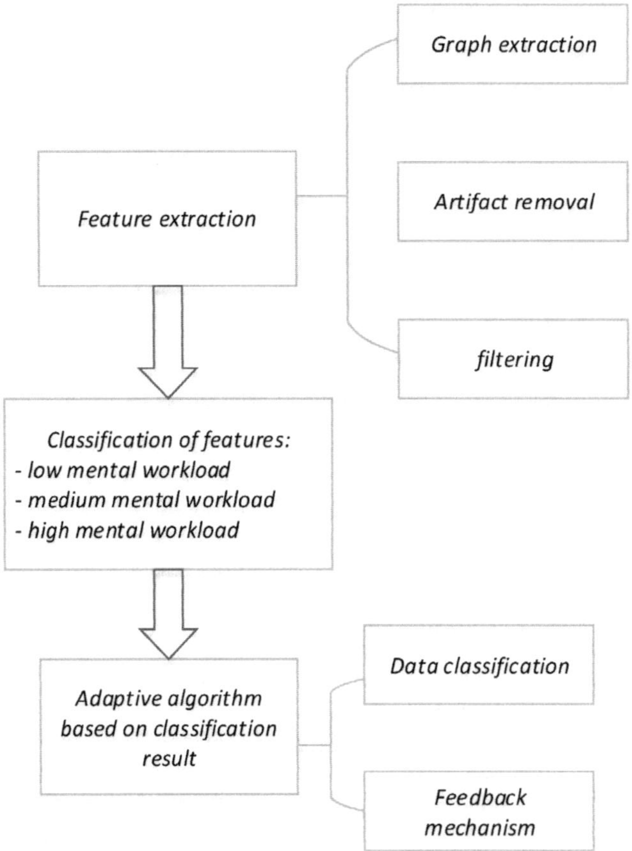

Figure 2. The conceptual design of real time and adaptive mental workload measures

Effectiveness of the proposed design will be assessed in a series of experiment. The designed adaptive measures will be activated when workload level was overload or underload.

4. Discussion

This paper is aimed to describe conceptual design of the development of real time and adaptive mental workload measures based on GSR. The design process is started with filtering data and is continued with classifying the norm of low, medium, and high mental workload. The proposed conceptual design will be implemented in the form of prototype. The effectiveness of the proposed design will be evaluated based on the success of mental workload detection and feedback mechanism.

The proposed conceptual design as well as the prototype that will be developed based on the conceptual design will be valuable in providing mental workload measures, which in the end can be used as a tool in providing optimal mental workload that enhance performance and safety of the worker.

5. Acknowledgment
Author wishes to thank to Indonesian Ministry of Research, Technology, and Higher Education for funding this research (under scheme of Hibah Kompetensi 2016-2017).

6. References
[1]. Gopher D and Donchin E 1986 Workload – An examination of the concept. In K.R. Boff, L. Kaufman, J.P. Thomas (Eds). *Handbook of perception and human performance*. New York: Wiley. Vol. 2 (pp. 41-1 – 41-49)
[2]. O' Donnel RD and Eggemeier FT 1986 Workload assessment methodology. In K.R. Boff, L. Kaufman, J.P. Thomas (Eds). *Handbook of perception and human performance*. Volume II, Cognitive process and performance. New York: Wiley. 42-1 – 42-49.
[3]. Eggemeier FT, Wilson GF, Kramer AF and Damos DL 1991 Workload assessment in multi-task environemtns. In D.L. Damos (Ed.). *Multiple-task performance* London: Taylor & Francis (pp. 207-216)
[4]. Widyanti A, de Waard D, Johnson A and Mulder B 2013 National culture moderates the influence of mental effort on subjective and cardiovascular measures *Ergonomics* 56 182-194.
[5]. Widyanti A, Hanna, Muslim K and Sutalaksana, IZ 2017 The Sensitivity of Galvanic Skin Response for Assessing Mental Workload in Indonesia *WORK* 56 111-117.
[6]. Widyanti A, Sofiani NF, Soetisna HR and Muslim K 2017 Eye blink rate as a measure of mental workload in a driving task: convergent or divergent with other measures? *International Journal of Technology* 8 283-291.
[7]. Arico P, Borghini G, Di Flumeri G, Colosimo A, Bonelli, S, Golfetti, A, Pozzi S, et al 2016 Adaptive automation triggered by EEG-based mental workload index: A passive brain-computer interface application in realistic air traffic control environment. *Frontier Human Neuroscience* 10 539
[8]. Hart SG and Staveland LE 1988 Development of NASA-TLX (Task Load Index): Results of empirical and theoretical research. In P.A. Hancock and N. Meshkati, eds. *Human Mental Workload.* Amsterdam: Elsevier, 139-184.
[9]. Hart SG 2006 NASA-Task Load Index (NASA-TLX); 20 years later. In: *proceeding of the Human Factors and Ergonomics Society 50th annual meeting*. Santa Monica, CA: Human factors and ergonomics society. 904-908.
[10]. Widyanti A, Johnson A and de Waard D 2013 Adaptation of the Rating Scale Mental Effort (RSME) for use in Indonesia *International Journal of Industrial Ergonomics* 43 70-76.
[11]. Calabrese EJ 2008 *Neuroscience and hormesis: overview and general findings* Critical Reviews in Toxicology 38 249-252.
[12]. Dijksterhuis C 2014 *Monitoring driver's mental workload for user adaptive aid*. Dissertation. University of Groningen, the Netherlands.
[13]. Johnson A and Widyanti A 2011 Cultural influence on subjective mental workload *Ergonomics* 54 509 – 518.
[14]. De Waard D 1996 *The Measurement of Drivers's Mental Workload*. Den Haag: Traffic Research Center, University of Groningen.
[15]. Boucsein W, Fowles DC, Grimnes S, Shakhar GB, Roth WT, Dawson ME, et al. 2009 Publication recommendations for electrodermal measurements. *Psychophysiology* 49 1017–1034.

IOP Conf. Series: Materials Science and Engineering **277** (2017) 012015 doi:10.1088/1757-899X/277/1/012015

Lean production design using value stream mapping and ergonomics approach for waste elimination on buffing panel upright process

M R Suryoputro[1], A D Sari[2], R Burhanudin and M Sugarindra

Department of Industrial Engineering, Universitas Islam Indonesia, Jl. Kaliurang, DI Yogyakarta, Indonesia
[1]ragil.suryoputro@uii.ac.id, [2]amarria@uii.ac.id

Abstract. This study discussed the implementation of ergonomics and value stream mapping issues to reduce the existing waste in the process of buffing upright panel in the XYZ music manufacturing company. Aimed to identify the 9 waste based on the identification in terms of production processes and ergonomic factors, namely environmental health and safety, defects, overproduction, waiting, not utilizing employee knowledge skill and ability, transportation, inventory, motion, and excess process. In addition, ergonomics factors were identified, for example posture using REBA, job safety analysis, and physical workload. This study results indicated that the process is having 21.4% of the potential dangers that could not be accepted and thus potentially lead to lost time. Continued with the physical workload, the score of % cardiovascular load value is still below 30%, which means that the physical workload is normal and allows the addition of work. Meanwhile, in the calculation of posture investigation, the REBA resulted that there was a motion waste identified on the edge buff machine and ryoto with the score of 10 and 8. In conclusion, the results shown that there were 20 overall waste produced, then thus waste were reduced based on the identification and discussion of proposed improvements.
Keywords: waste, value stream mapping, ergonomics, productivity improvement.

1. Introduction

Nowadays, in manufacturing industry each company is required to make continuous improvements in order to increase the productivity for each year. Increased productivity can be achieved through a gradual improvement with the removal of non-value added, standardization, increased flexibility of the company, and the recruitment and training of workers [1]. Activities that can be done in removing non-value added can be known through lean production approach. The approach focuses on the effectiveness of lean production processes including an increase in the value added operations, reduce waste, and meet customer needs [2]. One of the step of lean production in the reduction of waste is to assign a value stream and make it flow [3]. Value stream can be established through value stream mapping to create an image of the flow of a service, product or information [4].

This research was conducted at XYZ Company in the division of upright panel buffing process. During research, upright panel buffing process consists of several processes, namely 8 head buff, buff edge, high polish, ryoto, polisher and buff level. Based on preliminary studies that have been done, upright panel buffing process productivity decreased by 13%, so that the necessary efforts to reduce waste using value stream mapping in order to productivity can be increased according to the target company in the period 193, which was an increase in production productivity upright piano types by 10%.

Prior to this research, the use of value stream mapping at XYZ Company tends to focus on reducing waste and consider the ergonomics and musculoskeletal disorder possible risk to workers. Reduction of waste carried out aimed to increase value added activities. Whereas value added activities tend to have a higher exposure task than non-value added activities [5]. So the application of value stream mapping effort has the potential to increase the exposure on workers' task. In addition, in the other studies made clear that efforts to increase the pace of work and the workload using a lean approach can have an impact on health, stress, fatigue or injury to workers [6]. To prevent an increase in the risk of injury to workers, required the integration of ergonomics in the application of value stream mapping, especially in order to identify waste with considering the human factor. Ergonomics can be used to analyse three waste, which are waste process, excessive movement (motion), and transportation [7]. In the study Jarebrant, et al [8] showed that the use of value stream mapping can be integrated with ergonomics.

In this research, there were needs to implement and analyse the merger of value stream mapping and ergonomics as measures for the identification and reduction of existing waste in an upright panel buffing process. Tools used to identify nine waste is value stream mapping (VSM), Job Safety Analysis (JSA), 10 beats (physical work load) and Rapid Entire Body Assessment (REBA). Value stream mapping is used to describe the flow of production in a production and map the existing waste. Job safety analysis (JSA) is used to identify waste in the lost time (accident) by analysing the potential risks and dangers. While rapid entire body assessment (REBA) and physical work load analysis is used to analyse the waste in terms of motion and fatigue that occurs on the worker, so that it can be used as guidelines in the delivery of the proposal. Meanwhile, to understand the different causes of waste that appears, the research conducted the analysis and discussion with the head of the group of the division, in order to the provision of the proposal can be provided based on source and priorities. By designing lean production using value stream mapping and ergonomics are expected to make a proposal for waste reduction without increasing the risk of injury and fatigue to the operator, so that the target company to increase productivity can be achieved.

2. Methodology

Value stream mapping is an effective tool for measuring process and redesigning business processes based on the concept of lean [4]. There are four stages that could be conducted in using value stream mapping process, namely preparation, current state, future state and planning and implementation. The use of value stream mapping is required so that improvements are made can be more focused on the overall waste in the system. To support the value stream mapping, there are concept of lean and waste that should be described. Lean defined as a sustained effort to eliminate waste and increase the value added of products (goods or services), in order to provide value to the customer [9]. In applicability there are several tools that can be used to achieve a lean company. Among some of these methods are frequently used some muda (waste), value stream mapping and kaizen (continuous improvement). In addition, waste is anything that does not add value, both for the products produced as well as for consumers. There are nine of waste that need to be considered in lean, the environmental health and safety, defects, overproduction, waiting, not utilizing employee knowledge skill and ability, transportation, inventory, motion, and excess process [10]. In conclusion, four stages of value stream mapping process start from the preparation stage. This stage is the most important stage in the process of mapping the value stream so that in determining the future state to be successful. In the preparation phase of determined purpose and direction of the development is set. The next stage is the current state, describing the current state of a process. Continued with applying the manufacture of future state based on discussions in consideration with lean concepts. The final step is the planning and implementation. In general, the main objective is to achieve a predetermined future state.

In mapping the value stream mapping process flows, there are several tools that are used to determine the existence of waste. Tools are often used include lead time, cycle time, inventory, %

value added activity, and% non-value added activity. Lead time is the waiting time is calculated based on a comparison of inventory and stock overall process incoming / sent. Cycle time is the cycle time can be calculated based on the standard time using a stopwatch. %% value added activity and non-value added activity can be measured based on the results work sampling.

Move on to the ergonomics process of the research. Ergonomics is the application of scientific human science (including scientific methods to obtain such information) for design problems [11]. Safety or occupational safety is an attempt to ensure the health of workers in a company. Health and safety at work can be interpreted as a promotion, protection and improvement of health status that as the height of cover the physical, mental, and social welfare of all workers in all workplaces [12]. To reduce the chances of an accident or as a precaution against potential hazards must be identified hazards. By doing so the source of danger hazard identification can be determined so that potential accidents can be controlled. One way that can be used to identify hazards is to use a job safety analysis. Job safety analysis is an accident prevention technique that were used to identify potential risks and dangers associated with a worker and give risk control to reduce the risk and the harm [13].

For the postural analysis, rapid entire body assessment (REBA) is one of the tool recommended. The method can be used to quickly assess the posture of a worker, other than that this method is also influenced by the coupling factor, external load supported by the body as well as the activity of the workers [14]. Development of REBA conducted through six stages: observation field, select the posture to be measured, the value of work postures, calculate the scores obtained, determine the value of REBA, and confirm the value of the action level [15]. In addition, the physical workload could be conducted in several methods. One approach is to determine the severity of the workload to count the working pulse, oxygen consumption, pulmonary ventilation capacity and core body temperature [16]. To be able to measure the pulse of the work can be done using 10 beats [17].

In this study sampling was conducted using purposive sampling technique. Samples taken in this study is the work operator panel section upright buffing process that meets the criteria specified. The criteria used to select the sample is as follows:

- Have working experience in an upright panel buffing process at least 2 years
- Physically when the data retrieval process
- Operator male sex
- Range aged 20-50 years
- Working eight hours in one day

On buffing process upright panel there are 19 operators, divided into two shifts. In each shift operator is divided into four stations, namely 8 head buff, buff edge, high polish, and ryoto. The calculation of the required rate 10 and 8 operator head buff, one buff edge operator, one operator of high polish, one-person operator panel ryoto PE, 1 operator ryoto colour panel, and 1 operator side board. In the calculation takes 1 operator REBA 8 head buff, one buff edge operator, one operator of high polish, and 1 Ryoto operator.

3. Discussion

In this research, data collection job safety analysis performed directly by observation to the operator. The results of observation and then recorded into the observation sheet. The value of the risk factors (FR) will be classified becoming 2. If the value of FR <6, the potential danger is still acceptable, whereas if the value FR $\geq$ 6, the potential hazard is unacceptable. Of the four groups of machines that exist on an upright panel buffing process obtained the following results:

Table 1. Example results of operator 8 JSA head buff

Stroke	Hazard type	Impact	Likelihood	Impact	FR	Conclusion
Replacement *Fabrics Buff*	Crushed *buff cloth* when taking off the shelf	Injury	2	1	2	Unacceptable
	slipped while passing through the area beside the machine	Injury	2	2	4	Unacceptable
Top process	Maturity for tripping hydraulic cable	injuries	3	1	3	Unacceptable
	sandwiched Hand	Injury	1	2	2	Unacceptable

Based on the overall JSA results, it is known that 3 of 14 potential hazards (21.4%) included in an unacceptable condition, while 11 potential hazards (78.6%), others are still acceptable. 3 potentials which cannot be accepted that the potential mental wax to the body on edge machine buff, buff cloth potential affected hand at high machine polish, and buff cloth twisted apron were potential on ryoto machine.

Furthermore, to identify other possibilities do data processing waste lead time, work sampling, inventory, standard time, % CVL, and REBA so it can be illustrated in the following value stream mapping (Figure 1).

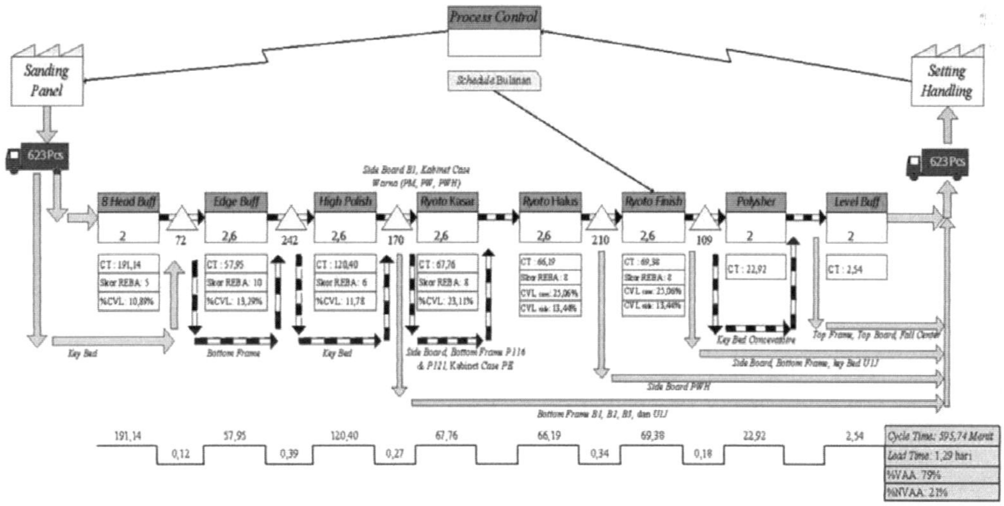

Figure 1. Value stream mapping

Value stream mapping mentioned above (Figure 1.) illustrates the process the flow of raw materials, ranging from material sent by sanding section upright panel to the cabinet sent go to handling settings. For cabinet colours: PM, PW, and PWH starting from 8 head buff, buff edge, high polish, ryoto rough, smooth ryoto, then ryoto finish. As for the cabinet PE buffing process starting from 8 head buff, buff edge, high polish, smooth ryoto, and then ryoto finish. This difference occurs because the type of paint used for the cabinet and the cabinet PE of different colours. Cabinet so that colours tend to require buffing longer for maximum results. Workmanship for colour and cabinet PE also distinguished in terms of the operator, the cabinet colour Ryoto done by two people. While 8 others do ryoto cabinet PE.

In addition to buffing process flow, of value stream mapping can also be seen that the cycle time is highest in the 8 head buff with a cycle time of 191.14 minutes. But in the machine 8 head buff has a 16 capacity, thus increasing the output of 8 head buff. Then of value stream mapping can also be seen lead times occur for buffing process upright panel is of 1.29 days. That is the target of a cabinet that went into buffing process takes 1.29 days to process up to handling settings. Lead time most processes occur at the edge buff and smooth ryoto is 0.39 days and 0.34 days. So that the necessary attention to reduce the lead time on the edge ryoto buff and smooth. The high value of the lead time is comparable with the high inventory on both processes, namely the edge 242 pcs 210 pcs on buff and smooth ryoto. In value stream mapping also shows that buffing operator panel upright conduct activities 79% value added and 21% non-value added activities. Non-value added activities consist of several activities with additional work with percentage value of 6% and the margin percentage of 11% employment. The high margin activities such work because the material is still done each grab one rack by the operators, so that operators do a lot of non-value added activities. Meanwhile, the high additional work occurred as the result of buffing there are findings of repair, so that there is additional work to repair the cabinet.

Once known waste which is attached on an upright panel buffing process, then the discussion and analysis to determine the cause of the existing waste, so that the proposed improvements that can be given a thorough and waste within their causes. There are 20 wastes identified for those 9 wastes category. The results and how the waste is identified and the recommendation described and analysed in Table 2.

Table 2. Example results of operator 8 JSA head buff

No	Process	Waste	Recommendation
1	8 Head *Buff*	Lost time potential due to slip and trip hazard	Slip and trip hazard sign provided on raw material area and machining.
2	8 Head *Buff*	Lost time potential due to strucked by compressor's cable	Compressor 5 S and neat cable.
3	*Edge Buff*	Lost time potential due to wax rebound and hands scrapped by buff cloth	Using hand tools for wax application
4	*Edge Buff*	High inventory and waiting time	Maximize the use of auto edge buff and increase the capacity into 3 pieces
5	*Edge Buff*	High REBA posture risk level 10	Reducing the usage of manual edge buff and transferring cabinets into auto edge buff. Elevating edge buff table.
6	*Edge Buff*	Process waste for performing back side on edge side board	Coordinating with spray section to replace the usage of mentory to become 1 mm
7	*High Polish*	Lost time potential due to hands scrapped by buff cloth	Auto wax addition on auto edge buff machine
8	*High Polish*	Lost time potential due to hands pinched in conveyor belt	Providing cover on the conveyor belt path
9	*High Polish*	Lost time potential due to accident of falling object from the conveyor table	Adding stopper on conveyor end
10	*High Polish*	Transport waste on waxing	Adding auto wax on high polish machine
11	*High Polish*	Defect waste due to unclean high polish	Replacing buff cloth rotation (becoming 3 buff times and 2 rinse times) and replacing wax with menzerna
12	*High Polish*	Time waste due to idle for high polish operator	Adding checking process on high polish result
13	*Ryoto*	Lost time potential due to wax bounced off to body	Using PPE (apron) and creating limit/maximum mark on using wax
14	*Ryoto*	Lost time potential due to eyes go dust splash	Providing dust filter on ryoto machine

Table 2. Example results of operator 8 JSA head buff (cont.)

No	Process	Waste	Recommendation
15	*Ryoto*	Transport waste on ryoto process	Replacing ryoto transport method from every cabinet into every rack
16	*Ryoto*	Waiting time waste on operator	Replacing ryoto transport method from every cabinet into every rack
17	*Ryoto*	Inventory waste and excess process due to excessive long and overdue process on ryoto	Direct rinse process on edge ryoto finish by replacing 6Q into *menzerna*
18	*Ryoto*	Inventory waste and excess process due to excessive long and overdue process on colour ryoto	On sanding, sander level process #800 replaced by sander level #500
19	*Ryoto*	Posture REBA score waste having high risk level 8	Concising ryoto process time by improving high polish result
20	*Polisher*	Step and process waste	Replacing polisher process with shine up and rags

In addition, the next steps in the value stream mapping is the design of future stream mapping. Future stream mapping is designed based on the discussion of the proposed improvements identified. So we get the following result (Figure 2).

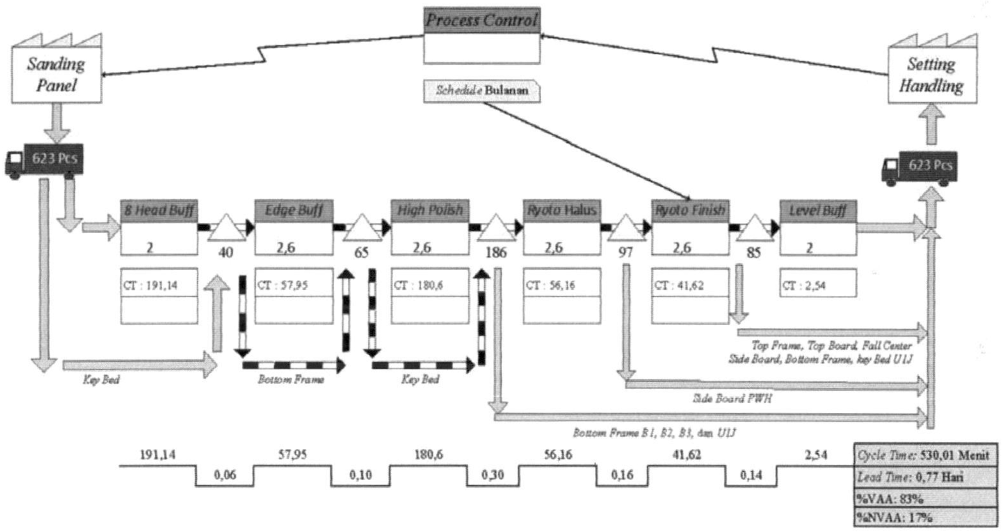

Figure 2. Future value stream mapping

Figure 2 above is a future stream mapping were designed based on the proposed improvements. Future stream mapping illustrates the reduction of rough and estimates ryoto process cycle time reduction. Reduction process and cycle time impact on operator workload so further research is needed equitable workload.

4. Conclusion

Based on the results, it can be concluded that since the calculations job safety analysis, there are three potential hazards (21.4%) which is unacceptable that are wax exposed to the edge buff operator's body, buff high polish cloth affected hand, and buff ryoto cloth entwined by apron. Based on 10 beats, resulted in the percentage of cardiovascular load is still under 30%, which can be inferred as physical

IOP Conf. Series: Materials Science and Engineering **277** (2017) 012015 doi:10.1088/1757-899X/277/1/012015

workload is still in normal condition, so that it is understandable to include additional work. In the other side, based on postural analysis using REBA, there are motion waste identified on edge buff and ryoto with the highest risk identified (10 and 8).

There are five waste contained in the process of buffing upright panels, which are potential lost time for slip, trip and falling risk, wax exposure and buff cloth contained the hands, struck in high polish, and also wax and dust exposure on ryoto. The waste reduction program identified are signage/label and neat cable trays, auto edge buff coordination, in high polish are auto wax and machine cover, shine up and cloth to replace polisher, and lastly is the wax limitation and dust exposure filtration layer additional.

5. References

[1] Dysko, D 2012 Gemba Kaizen-Utilization of human potential to achieving continuous improvement of company *The International Journal of transport & logistics*, 19-28

[2] Hines, P and Taylor, D 2000 *Going Lean* (Cardiff: *Lean Enterprise Research Centre*)

[3] Womack, J P, and Jones, D T 2010 *Lean thinking: banish waste and create wealth in your corporation* (Simon and Schuster)

[4] Locher, D A 2008 *Value Stream Mapping for Lean Development: A How-To Guide for Streamlining Time to Market* (New York: Taylor and Francis Group)

[5] Forsman, M, Neumann, P W, Palmerud, G, and Winkel, J 2014 Mechanical exposures in serial-flow car disassembly - A proactive intervention research approach

[6] Landsbergis, P A, Cahill, J, and Schnall, P 1999 The impact of lean production and related new systems of work organization on worker health *Journal of Occupational Health Psychology*, 108-130

[7] Mulyati, G T, Suharno, and Muharom, 2015 An implementation of lean-ergonomic approach to reduce ergonomic parameter waste in the manufacture of crackers *KnE Life Sciences*, 21-24

[8] Jarebrant, C, Winkel, J, Hanse, J J, Mathiassen, S E, and Ojmertz, B 2016 ErgoVSM: A tool for integrating value stream mapping and ergonomics in manufacturing *Human factors and ergonomics in manufacturing & service industries*, 191-204

[9] Gaspersz, V and Fontana, A 2011 *Organization Excellence: Systematic continues improvement and innovation* (Bogor: Vinchristo Publication)

[10] Andini, G, Prasetyawan, Y, and Supriyanto, H 2012 Pendekatan konsep lean manufaktur dalam peningkatan efisiensi pada sistem produksi kaca di PT Asahimas flat glass tbk *Jurnal Teknik Pomits*, 1-5

[11] Pheasant, S 1988 *Bodyspace: Anthropometry, Ergonomics and the Design of Work.* (London: Taylor & Francis)

[12] International Labour Organitzation 1998 *The International Labour Organization's Fundamental Conventions*

[13] Jaiswal, V, Banodha, V, and Patel, P 2014 Risk assesment in maintenance work at diesel locomotive workshop *International Journal on Emerging Technologies*

[14] Hignett, R, and McAtamney, L 2000 Rapid Entire Body Assessment (REBA) *Applied Ergonomics*, 201-205

[15] Santon, N, Hedge, A, Brookhuis, K, Salas, E, and Hendrick, H 2004 *Handbook of Human Factors and Ergonomics Methods* (New York: CRC Press)

[16] Grandjean, E 1993 *Fitting the task to the man* (London: Taylor & Francis)

[17] Kilbon, A 1992 *Measurement and Assessment of Dynamic Work*

A study of 6S workplace improvement in Ergonomic Laboratory

AD Sari[1], MR Suryoputro[2], and FI Rahmillah[3]

[1,2,3)] Department of Industrial Engineering, Faculty of Industrial Technology, Islamic University of Indonesia, Jl. Kaliurang Km 14.5 Yogyakarta, Indonesia
amarria@uii.ac.id [1],ragil.suryoputro@uii.ac.id[2], and fety.rahmillah@uii.ac.id[3]

Abstract. This article discusses 6S implementation in Ergonomic Laboratory, Department of Industrial Engineering, Islamic University of Indonesia. This research is improvement project of 5S implementation in Ergonomic laboratory. Referring to the 5S implementation of the previous year, there have been improvements from environmental conditions or a more organized workplace however there is still a lack of safety aspects. There are several safeties problems such as equipment arrangement, potential hazards of room dividers that cause injury several times, placement of fire extinguisher, no evacuation path and assembly point in case of fire, as well as expired hydrant condition and lack of awareness of stakeholders related to safety. Therefore, this study aims to apply the 6S kaizen method to the Ergonomic laboratory to facilitate the work process, reduce waste, improve work safety and improve staff performance. Based on the score 6S assessment increased audit results by 32 points, before implementation is 75 point while after implementation is 107 point. This has implications for better use for mitigate people in laboratory area, save time when looking for tools and materials, safe workplace, as well as improving the culture and spirit of '6S' on staff due to better and safetier working environment.
Keywords: 6S, laboratory, management, safety

1. Introduction

The concept of 6S is the development of the 5S introduced by Hiroyuki Hirano as a control and work environment control method designed to reduce time wastage, improve safety and optimize productivity, is an acronym of 5S + 1 (*seiton, seiton, seiso, seiketsu, shitsuke* and safety) [1] [2] [3] [4]. The 5S method (*seiri*/sort, *seiton*/set in order, *seiso*/shine, *seitsuke*/standardize, *shitkesu*/sustain) was originally developed in Japan as the basis for the creation of the principle of continuous improvement, also known as kaizen. From 5S is to create a more organized and consistent work environment with high quality, in addition 5S is also useful to reduce the ineffective time due to waste [5]. In principle, implementation of 5S concept without safety aspect is useless / waste, because safety is the main priority in work [6]. The safety concept of 6S is the implementation of safety in the work area including hazard identification, ensuring the absence of potential fire spark cable, safe working equipment free from injury, the existence of fire extinguishers that work properly, and the evacuation path is clear information in case of fire [2]. It aims to work efficiency, work quality, work safety, and improve work discipline [2] [3].

Some research related to 6S application in laboratory among others is done for laboratory management in College Laboratories [7], laboratory center in China [8], Clothing Technology Laboratory in University [9]. Similar research is the Application of 6S Management in Parker manufacturing plant Houston, Texas [2], Industry on India [3], as well as 6S practices in the silk multi-end reeling industries in Andhra Pradesh [4].

Objects in this study were developed within the University, to respond to the continuous improvement process implementation and the need to optimize existing resources. This research is a continuation of 5S implementation in Ergonomic laboratory [10]. Ergonomics Laboratory is a laboratory majoring in Industrial Engineering, with activities in it include teaching, tutorial, research and administrative activities. Tutorial activities are done more using tutorial tools such as anthropometer, lux meter, stopwatch, sound level meter and other practical fittings such as toy cars, plugs, resistors. Besides tutorial activities, business processes in the laboratory include administrative activities such as tool lending, attendance and research activities [10].

Referring to the 5S implementation of the previous year, there have been improvements from environmental conditions or a more organized workplace however there is still a lack of safety aspects. There are several safeties problems such as equipment arrangement, potential hazards of room dividers that cause injury several times, placement of fire extinguisher, no evacuation path and assembly point in case of fire, as well as expired hydrant condition and lack of awareness of stakeholders related to safety. Therefore, this study aims to apply the 6S kaizen method to the Ergonomic laboratory to facilitate the work process, reduce waste, improve work safety and improve staff performance.

2. Methods

The focus of this study is to analyse the working area of Ergonomic laboratory, Industrial Engineering Department. The first step is to identify problems by using the method of observation and checklist on the condition of every area / room available in ergonomic laboratory. Assessment 6S is applied to laboratory headroom, assistant room, lobby, classroom, equipment room, climate room and training room. The work processes that exist in the ergonomics laboratory include teaching and learning activities by lecturers, tutorial activities by assistants, research activities and laboratory equipment borrowing. The next step is to apply the 6S concept. Then in the value analysis of the assessment results 3S, 5S, and 6S so that can be given recommendations for continuous improvement (continuous improvement).

Basically, 6S method consists of 5 stages representing the meaning of five Japanese words [1] and 1 S for Safety [2] :

1. ***Seiri* equals to sort**, it means removing the waste and unused items from the working area. By sorting the unnecessary tools employee will be able to prioritize the equipment whether it is needed or not to be placed in their workplace.
2. ***Seiton* equals to set in order**, it means arranging tools in specific place and keeping it into the right position. Therefore, the items can be found easily by the workers when it is needed.
3. ***Seiso* equals to shine**, it means cleaning all working space and necessary tools. Therefore, the workers will have convenient place to work, and the equipment will always be ready to be used for other employee.
4. ***Seiketsu* equals to standardize**, it means ensuring all setup and equipment which has been arranged and cleaned in the right condition by promoting visible and modest rules.
5. ***Shitsuke* equals to sustain**, it means obeying the rules and procedures which has been issued as a respond of the first four phases.
6. **Safety,** It means make the workplace safe and free of recognizable hazards.

3. Results and Discussion

The 6S program is conducted from March to August 2017. This program is a continuation of the 5S program in the laboratory [10]. Stages of implementation of 6S in this study are as follows:

3.1. Sort

Sort is the initial stage of the 6S activity, in Japanese called *Seiri* [1]. Sort activity is identifying equipment / materials between required and unnecessary. Red tags / red labels are used to help identify the sorting, as well as given the time required for the storage of temporary goods [2] [11] ,(6S farm). The sort stages performed in this study include:

a. Identification of required / unneeded tools / materials is required [3]
b. The broken tool is separated
c. Sorting hard and soft documents of unused and expired files

Some concise tools or materials include tutorial files for the period September-February 2016, unused posters, drugs expired on first aid boxes, replacing dangerous room divider rollers. The red tagging is done on items whose status is still temporarily removed. All non-essential items located inside the laboratory are given red tags (Figure 1).

Figure 1. Sorting item and reg tagging

3.2. Set in Order

Set in order or *seiton* is the second stage in 6S, the set in order activity is physically organize the work area. Tidying these areas makes it easier for workers to find needed items [3]. Stages in *seiton* / set in order include:

a. All shelves and stored items are clearly identified
b. Good storage is organized thus it's easy to see, retrieve and restore
c. There is a clear indication of the maximum or minimum inventory amount.
d. All areas are equipped with a dividing line & all the items inside the line.
e. Document storage should be well-documented and easy to quickly find

Detailed activity on the laboratory is to color-label the tutorial class area, assistant room, exit door with yellow dotted line marks, labels for light extinguishers, and label carefully with red. Can be seen in the following picture:

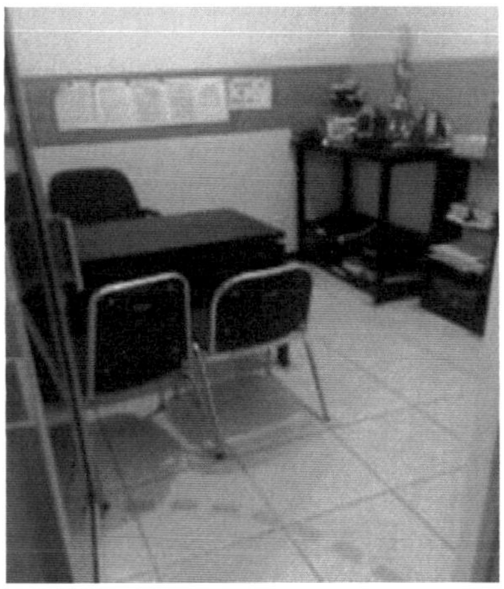

Figure 2. Sticked colored tape and labelled

3.3. Shine

IOP Conf. Series: Materials Science and Engineering **277** (2017) 012016 doi:10.1088/1757-899X/277/1/012016

Shine step includes activities like cleaning workplace, maintaining its appearance and using preventive steps to keep workplace clean [1] [3] [12]. Stages in shine activity based on 6S checklist:

a. No dust, dirt, stains, insect house / cobwebs and regularly cleaned
b. There are bins, identifiable and in accordance with its use
c. Hygiene tools enough, placement neat, protected from dirt
d. There are mechanisms for responsible hygiene

Detail Shine activity on laboratory ergonomics include, before leaving the laboratory, the staff is required to clean first in the area especially in the assistant room such as washing utensils and beverages that have been used, positioned the seat back to the place, tidied goods on the table, and throw garbage if there is garbage in the room area assistant. Meanwhile, for cleanliness coordinate with cleaning service to assist in cleaning every room in the laboratory ergonomics such as sweeping, mopping and so forth. Includes check marks on the board of janitorial activities that have been done.

3.4. Safety

The objectiveof this phase is to ensure that workers are constantly aware of the value of safety and health in the workplace and therefore adopt preventive measures against recognizable hazards. Figure 3 shows the evacuation route in ergonomic laboratory based on the location of exit door and the evacuation sign. While stages of safety activity based on 6S checklist are listed as follows:

a. Carried out hazard identification and risk assessment
b. There is a suitable fire extinguisher
c. There is a fire extinguisher sign, easily accessible (125 cm) and there is a usage way according to Minister Regulation 04 / Men / 1980
d. Checking fire extinguishers periodically
e. There is a safety sign / poster
f. There is a sign of an evacuation path
g. There is an evacuation route map
h. Available personal protective equipment that is suitable and appropriate to the standards
i. Connection cables neatly arranged, the socket is closed
j. The panel is closed
k. There is a well-functioning indicator panel
l. Available Emergency lights / Genset
m. Available first aid kit and medicines

3.5. Standardize

Standardize the ways of maintaining tidiness, orderliness, and cleanliness [1] [2]. Stages in standardize activity based on 6S checklist: The standardization of the room by making the picture on the conditions that have applied 6S as well as their poster invitation to do 6S (Figure 4). Morover, the team make slogan /poster solicitation to always implement 6S in the laboratory.

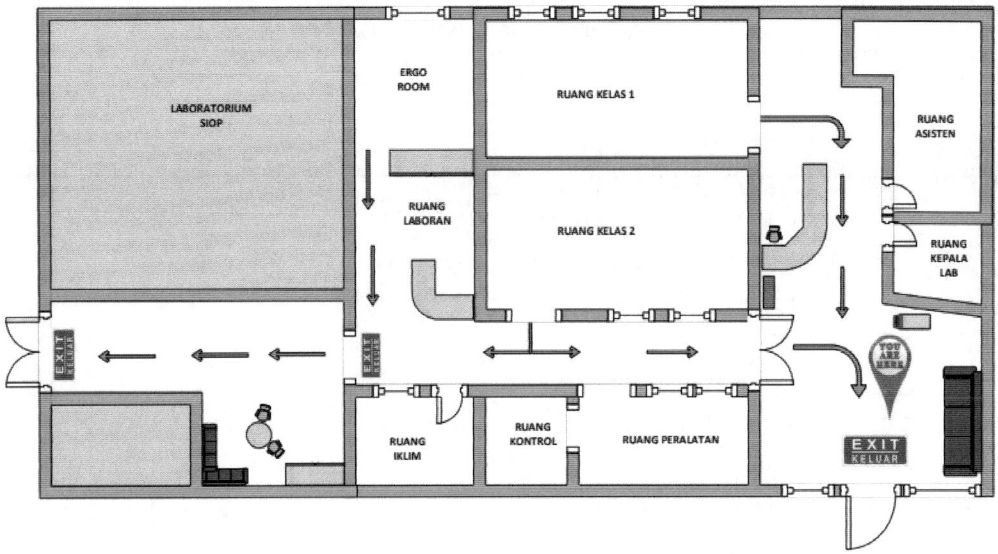

Figure 3. Evacuation route

Figure 4. 6S Poster

3.6. Sustain

Sustain means commit to 6S and practice it daily or make it a way of life [2]. The stages involve efforts for 6S learning and the involvement of every assistant, employee, and lecturer as well as make a 6S internal audit periodically, every 6 months. The 6S checklist was assessed by auditor through direct observation, interview to laboratory assistant and head of laboratory, as well as check several items such as identification sheet, checklist activity, standard sheet, audit report form, and evaluation report.

6S Checklist -Lab Area												
6S	NO	Checklist Item	Mar-17					Aug 2017				
			1	2	3	4	5	1	2	3	4	5
Sort	1	Identification of required / unneeded tools / materials is required [3]			3						4	
	2	The broken tool is separated				4					4	
	3	Sorting hard and soft documents of unused and expired files		2							4	
Set in Order	4	All shelves and stored items are clearly identified				4					4	
	5	Good storage is organized so it's easy to see, retrieve and restore				4					4	
	6	There is a clear indication of the maximum or minimum inventory amount.	1							3		
	7	All areas are equipped with a dividing line & all the items inside the line.			3						4	
	8	Document storage should be well-documented and easy to quickly find				4					4	
Shine	9	No dust, dirt, stains, insect house / cobwebs and regularly cleaned				4					4	
	10	There are bins, identifiable and in accordance with its use				4					4	
	11	Hygiene tools enough, placement neat, protected from dirt				4					4	
	12	There are mechanisms for responsible hygiene			3						4	
Safety	13	Carried out hazard identification and risk assessment	1								4	
	14	There is a suitable fire extinguisher			3					3		
	15	There is a fire extinguisher sign, easily accessible (125 cm) and there is a usage way according to Minister Regulation 04 / Men / 1980	1								4	
	16	Checking fire extinguishers periodically			3					3		
	17	There is a Safety Sign / poster			3						4	
	18	There is a sign of an evacuation path	1								4	
	19	There is an evacuation route map	1								4	
	20	Available Personal protective equipment that is suitable and appropriate to the standards			3					3		
	21	Connection cables neatly arranged, the socket is closed		2							4	
	22	The panel is closed			3					3		
	23	There is a well-functioning indicator panel			3					3		
	24	Available Emergency lights / Genset			3					3		
	25	Available first aid kit and medicines		2							4	
Standardize	26	There are efforts and mechanisms for the 4S is always implemented at all times well		2							4	
	27	There is a slogan / poster solicitation to always implement 6S	1								4	
Sustain	28	There are efforts for 6S learning and the involvement of every assistant, employee, labor, and lecturer regularly / periodically every 6 months		2						3		
	29	There are regular 6S audits	1							3		
			SCORE 75					SCORE 107				

Figure 5. 6S Assessment

Figure 5 shows a 6S checklist before and after the implementation of 6S. As a result, there is an improvement 32 points from 75 became 107. This has implications for better use for mitigate people in laboratory area, save time when looking for tools and materials, safe workplace, as well as improving the culture and spirit of '6S' on staff due to better and safetier working environment.

4. Conclusion

Based on the score 6S assessment increased audit results by 32 points, before implementation is 75 point while after implementation is 107 point. This has implications for better use for mitigate people in laboratory area, save time when looking for tools and materials, safe workplace, as well as improving the culture and spirit of '6S' on staff due to better and safetier working environment.

5. References

[1] H. Hirano 1996 *5S for Operators: 5 Pillars of the Visual Workplace*, (SteinerBooks)

[2] E. Osakue and D. Smith 2014 A 6S Experience in a Manufacturing Facility, *ASEE Anual Conference & Expostion*.

[3] S. G. Vinodkumar, R. S. Akash, N. P. Ankitkumar and D. K. Vijay 2014 Study of 6s Concept and its Effect on Industry," *International Journal of Emerging Technology and Advanced Engineering*, **vol. 4**, no. 10.

[4] Y. Sujatha and K. Rao 2016 Implementation of 6S Practices in the Silk Multi-End Reeling Industries in Andhra Pradesh, *Int. J. of Engineering Science and Computing,* **vol. 5**, no. 2, pp. 2049-2057.

[5] T. Osada 1991 *The 5S's: Five Keys to a Total Quality Environment*, (Tokyo: sian Productivity Organization).

[6] S. M. E. Elkhouly and E. A. Y. Salah 2012 "7S" As A Tool To Improve Service Quality: An Applied Study On Egypt Post," *Research Yearbook*, p. 447.

[7] G. H. Wang, B. Zong and J. Z. Wei 2007 Application of 6S Management in the College Laboratories, *Experiment Science & Technology,* **vol. 2**, 2007.

[8] D. M. Zhang, S. F. Zhang and S. Q. Zheng 2010 Application of 6S management in standard management of laboratory center, *Laboratory Science,* **vol. 3**, no. 049, 2010.

[9] H. E. Zhi-ying 2015 6S Management Mode of Clothing Technology Laboratory in University, *Journal of Fujian Commercial College,* **vol. 013**, no. 3.

[10] Sari, F. Rahmillah and B. P. Adjie 2017 Implementation of 5S Method for Ergonomic Laboratory, IOP Conference *Series: Materials Science and Engineering* **(Vol. 215)**.

[11] Krawczuk and S. Kocira 2017 Adaptation of the 6S Method in a Farm, *Journal of Research and Applications in Agricultural Engineering,* **vol. 62**, no. 1, p. 116.

[12] S. Nazarali, J. Rayat, H. Salmonson, T. Moss, P. Mathura and K. Damji 2017 The application of a "6S Lean" initiative to improve workflow for emergency eye examination rooms, *Canadian Journal of Ophthalmology*.

Machinery safety of lathe machine using SHARP-systemic human action reliability procedure: a pilot case study in academic laboratory

M R Suryoputro[1], A D Sari[2], M Sugarindra and R Arifin

Department of Industrial Engineering, Universitas Islam Indonesia, Jl. Kaliurang, DI Yogyakarta, Indonesia
[1]ragil.suryoputro@uii.ac.id, [2]amarria@uii.ac.id

Abstract. This research aimed to understand the human reliability analysis, to find the SHARP method with its functionality on case study and also emphasize the practice in Lathe machine, continued with identifying improvement that could be made to the existing safety system. SHARP comprises of 7 stages including definition, screening, breakdown, representation, impact assessment, quantification and documentation. These steps were combined and analysed using HIRA, FTA and FMEA. HIRA analysed the lathe at academic laboratory showed the level of the highest risk with a score of 9 for the activities of power transmission parts and a score of 6 for activities which shall mean the moving parts required to take action to reduce the level of risk. Hence, the highest RPN values obtained in the power transmission activities with a value of 18 in the power transmission and then the activities of moving parts is 12 and the activities of the operating point of 8. Thus, this activity has the highest risk of workplace accidents in the operation. On the academic laboratory the improvement made on the engineering control initially with a machine guarding and completed with necessary administrative controls (SOP, work permit, training and routine cleaning) and dedicated PPEs.
Keywords: SHARP, human reliability analysis, accident, control

1. Introduction

There are some discussions over the years to understand about human machine interaction [1], [2], [3], [4] and understand about accident [5], [6], [7] and [8]. Human machine interaction would involve human, machine, and environment, as a system correlated with the usage or interaction to provide its functions. Human considered as the actor who initiate and control the machine as it would receive feedbacks and provide action accordingly. The action will be based on the information processed through human information processing phase [9], [10] and [11]. In addition, there would be some characteristics that induce the human to create errors. Even though there are researches describing the accident nature [12], [13], [14], [15], human error still be a good topic to be debated over. Moreover, there are some disputes that centralized on both poles, between machine and environment [16], [17] and also human on safety [18]. Meanwhile, human behaviour is still a great consideration to implement a safety system [19], and leadership is one factor to predict this safety behaviour [20]. Therefore, there is a necessity to provide better view on human perspective as consideration on evaluating human machine interaction.

One of human perspective that can be explored is the human error or human reliability analysis. Human reliability analysis is a method for the identification of the activities conducted by human in a probabilistic assessment that considered from the safety objective [21]. Based on [22], and also [23], human reliability analysis can be conducted using SHARPR (Systematic Human Action Reliability Procedure). This method provides better comprehension on accident occurred due to human error. Accident nowadays becoming more complex and need a better comprehension in a socio-technical system point of view [24], [14].

Lathe machine is one of the machine that could be described as basic machine that consisted of complex system for turning product into a required shape. The safety analysis is required to improve the process safety and reduce the probability of accident occurred [25].

This research will be focusing on the analysis for human reliability using SHARP [22] with some detailed implementation to understand the machinery safety in Lathe machine. The purpose of this research is to help understand the human reliability analysis and to find the SHARP method with its functionality on case study of practical example and also emphasize the practice in Lathe machine. Based on the developed method, the aim is also finding out what improvement could be made to the existing safety system.

2. Methodology

There are some researches that describe the dispute about accident or incident definition. On one hand, in the field, there are employees that define the accident but not approved by the medical person or management [26]. On the other hand, there is a contribution that describing with defence could enter the dispute of definition [27]. Safety management could be described as how the safety risks are being managed and the accident that distinguished properly [28], [29]. There are serious risks that inherited from the accident were appointed as human contribution [24].

Machinery safety is the attribute that correlated with the safety aspects and issues regarding machine usage or system the machine used. There is some risk could be described for machine to understand more about machinery system, the level of hazard produced and risk prevalence/elevated should be protected with a safe precaution and correct protection [30].

Human Reliability Analysis could be described as an analysis method to identify the human action, behavior, impacts and interaction to the system (for example machine) regarding safety issues [22], [20]. Human machine interaction could be described in the safety system for the frameworks given by HARP method [23]. At first, Sharp was developed using four stages, the logical model of interaction, quantification, analysis for recovery and review form internal [23]. Subsequently, it was explained better by the origin author Hannaman and Worledge [22] that SHARP comprises of 7 stages including definition, screening, breakdown, representation, impact assessment, quantification and documentation, with the flowchart as described bellows (Figure 1.).

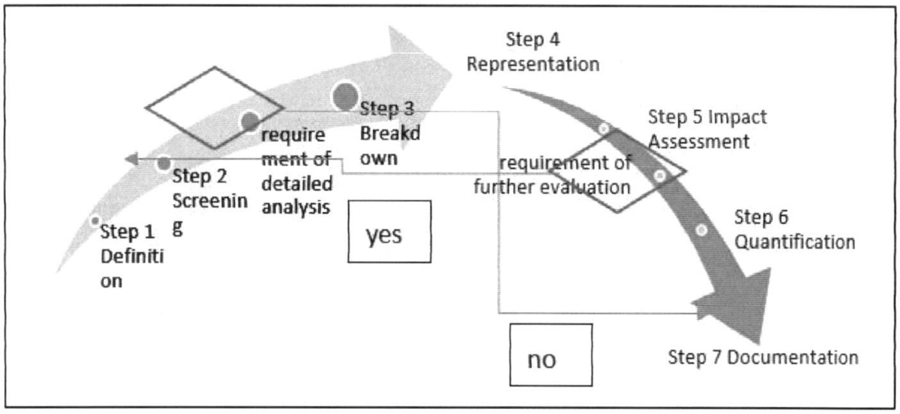

Figure 1. SHARP steps based on [23]

For overall methodology, this research stages will be explored into six continuous steps involving SHARP as the main focus of the pilot development in assessing machinery safety. The first phase of this study is the determination of concept study to be used as a reference. After the concept has been decided, the type of accident should be determined to constrict the direction of the research. To analyze the machinery hazard that have been determined, the next step is determining the method that is used as a research tool. The methods used as material of research was using seven layers of SHARP: Definition, Screening, Breakdown, Representation, Impact Assessment, Quantification and Documentation. Human reliability analysis is conducted by implementing SHARP method. The result of the implementation of SHARP method will be analyzed and assessed. Analysis of the human reliability causes will be used to evaluate and find a solution to understand the nature of the machinery safety. This analysis begins with identifying the hazard that could cause of the accidents and factors that may influence it. Analyses were performed based on seven steps of SHARP method and then the result of this method will be re-analyzed to find solutions and recommendations for the improvements of the machinery safety. The analysis result will be analyzed as a whole and then we will see which part is in need of preventive improvement. Based on the previous stage, the preventive action will be recommended to improve and standardize the practice of SHARP method (human reliability analysis).

3. Result and Discussion

The definition phase resulted from the natural use of Lathe itself. Lathes are machine tools for metal cutting processes (metal-cutting process). The basic operation of the lathe involves rotating the work piece and the cutting tool linear moves. Authentic lathe operation that is used to process the work piece with the result or the shape of circular cross section or cylindrical work piece [31]. The accident happened on the use of the lathe of them is forgotten remove lockdown lathes, fly scrap of the work piece, a broken eye chisel when the engine is operated, the eye drill could be released if it is not locked properly. Then the lack of work instructions or SOPs (Standard Operational Procedures) on a lathe, PPE (Personal Protective Equipment) is not used and the distance between one another engine narrow when used to demonstrate the practice of the practitioner.

The next phase is the screening process. In all the activities carried out by using a lathe there is human interaction with machines. Human machine interaction is a combination of one or several machines with one or several machines where interact to each other to produce outputs based on inputs obtained. In the system there is also human machine principal elements that interact with each other. Such elements include human operators, hardware, software, firmware and environment [32]. In the use of a lathe operator does not have difficulties in relationships with resulting engine easy usability and the buttons that exist on the machine are comparatively few.

Move on to the third phase, breakdown, Hazard Identification and Risk Assessment (HIRA) is one of the methods of identification of occupational accidents with risk assessment as one of the key points to implement an Occupational Safety and Health Management System (OSHMS). HIRA is being conducted to identify the potential hazards that exist in a company to assess the magnitude of the possibility of an accident or loss. Hazard identification and risk assessment and control must be done throughout the operation, including routine and non-routine activities, whether work is done by direct employees and contract workers, suppliers and contractors, as well as facilities or personal activities that go into the workplace. This step is also completed using FTA Analysis to comprehend the situation. As for the risk analysis are presented in Table 1., Table 2. and Figure 2. as below.

Table 1. Breakdown using HIRA

No	Event	Description Hazards	Potential Cause	Consumer Description	Rules	Possibility	Severity	Risk Level	P / T
1	Power Transmission Parts	Round Sleds can roll up objects such as clothing or hair that fell.	Clothes and hair that fell	To tear the operator clothes, and also can injure the operator's body part.	Y	3	3	9	1
2	Section	When at work,	Locking	Injured since	Y	3	2	6	2

No	Event	Description Hazards	Potential Cause	Consumer Description	Rules	Possibility	Severity	Risk Level	P / T
	Moving	tightening dibble less well able Causes thrown work piece.	System unfavorable	been hit by the work piece.					
3	Operating point section	scrupof the results of turning occasionally crashing / thrown out and to the workers.	The result of the erosion of workpieces slammed or thrown	Injured since been hit by the workpiece.	Y	4	1	4	3

Table 2. Breakdown using HIRA

NO	Event	Control	Possibility	Severity	Risk Level	P / T
1	Power Transmission Parts	uses a hair cover and use appropriate safety clothing sizes with its	1	1	1	1
2	Section Moving	Before conducting the work required to ensure properly secured and strong and need to be protective for the engine of the material clear	1	1	1	1
3	Operating point section	Lathe required the manufacture of the cover of material transparent in order not to affect the work	1	1	1	1

From the HIRA that has been done in analyzing the lathe at academic laboratory showed the level of the highest risk with a score of 9 for the activities of power transmission parts and a score of 6 for activities which shall mean the moving parts required to take action to reduce the level of risk. Then with the lowest score is the activities of the operating point with the score of the risk level of 4, which means the necessary monitoring to ensure existing controls are maintained. Then all the scores obtained should be controlled as much as possible to reduce the level of risk of danger to the scores level 1, which means no need for action due to the risk of injury is small.

The number of accidents that have been identified based on the findings in the field is measured by calculating the risk level of its RPN. RPN value is the product of the value of S, O, D (severity, occurrence, and detection). FMEA is a method used to examine the causes of defects or failures that occurred during the production process, evaluate priority risks that cause workplace accidents, and help take action to avoid the problems identified as the danger of accidents. Level of severity with a score of 2-3 gives a mild impact which means that work can still participate in the activity but injuries got is like a blister, heat shock, a mild sprain, lapsing or mild, bruising, mild cut and scratched. For occurrence, level score of 3 means the possibility of 1 in 15,000 treatment times which means the possibility of relatively rare, then for a score of 2 means likely to occur first time in the treatment 150.00 which means that possibility almost never happens. In addition, for detection phase, the translation for a score of 3 means the controller to detect the shape and cause high failure and a score of 2 means fatherly control device detects the shape and the cause of failure is very high. The result is seen on Table 3.

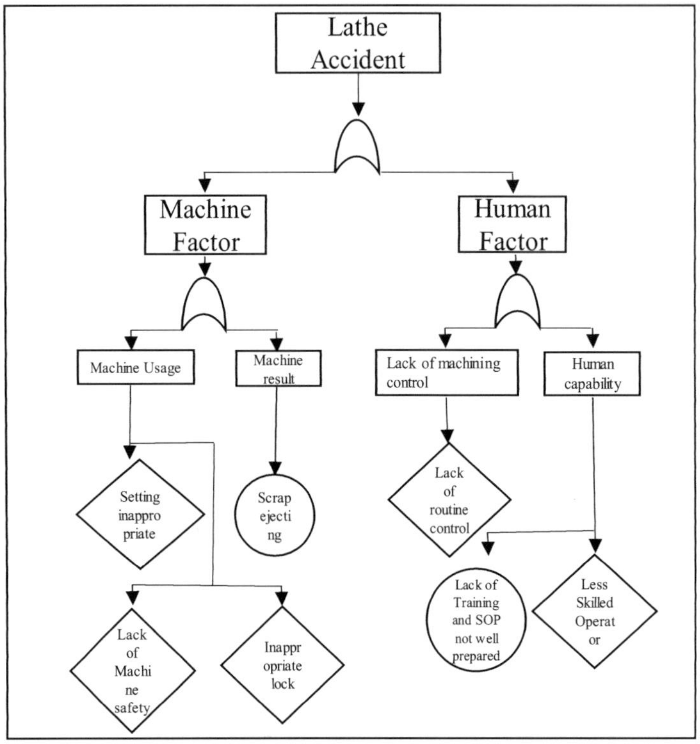

Figure. 2. FTA Analysis

Table 3. Representation using FMEA

No.	Event	Accidents	Due to	Activity	S Average	O Average	D Average	RPN
1	Power Transmission Parts	Round Sleds can roll up objects such as clothing or hair that fell.	Operators can snag clothing, and also can injure the operator's body part.	Power Transmission Section	3	3	3	18
2	Section Moving	When at work, tightening dibble less well able Causes thrown work piece.	Been injured by hit by the work piece.	Parts Moving	2	3	2	12
3	Operating Point section	scrupof the results of turning occasionally crashing / thrown out and to the workers.	Been injured by hit by the work piece.	Section Operating Point	2	2	2	8

From the results of the count value of the RPN entire workplace accidents, the highest RPN values obtained in the power transmission activities with a value of 18 in the power transmission and then the activities of moving parts is 12 and the activities of the operating point of 8. Thus, this activity is an activity with the risk of workplace accidents the highest in the operation of the lathe.

The impact assessment, as the continued phase, there is a need to define and clarify the solution to a better circumstance. From the results obtained, the necessary control of the risk of accidents will be derived from the hierarchy of risk control. In the use of a lathe with existing risks cannot be eliminated both from the material and process sequence, as elimination is the highest hierarchy. In a series of laboratory process lathe in academic laboratory using the aluminium material is a low risk of injury. Redesigning equipment used lathe as provide boundaries or machine guarding on the lathe using a coloured translucent material so that scrap or objects that are not locked properly is not thrown on the working operator. Continued with administration control, by applying strict regulations in doing work using a machine such as running SOP, permit, training to use the machine, cleaning work areas etc. In using the lathe is required to use protective equipment such as clothing, safety gloves, face shield or glasses and safety shoes.

4. Conclusion

Based on the results, it can be concluded that SHARP is reliable and applicable to analyzing the human reliability on the human machine interaction. On the academic laboratory the improvement made on the engineering control initially with a machine guarding and completed with necessary administrative controls (SOP, work permit, training and routine cleaning) and dedicated PPEs (clothing, safety gloves, face shield or glasses and safety shoes).

For further research, the method could be used combined with other accident model, systematic for example Accimaps or STAMP; JSA replacing the HIRA and also other Human Reliability Analysis method (HRA). Moreover, the combination of SHARP, HIRA, FTA and FMEA could possibly applied to understand more about human machine interaction and preventing accident to be occurred.

5. References

[1] Norman, D A 1984 Stages and levels in human-machine interaction *International journal of man-machine studies*, **21**(4), 365-375

[2] Levin, E, Pieraccini, R, and Eckert, W 2000 A stochastic model of human-machine interaction for learning dialog strategies *IEEE Transactions on speech and audio processing*, **8**(1), 11-23

[3] Hoc, J M 2000 From human–machine interaction to human–machine cooperation *Ergonomics*, **43**(7), 833-843

[4] Johannsen, G 2009 Human-machine interaction control systems, robotics and automation– Volume **XXI**: *Elements of Automation*, 132

[5] Benner, L 1985 Rating accident models and investigation methodologies *Journal of safety research*, **16**(3), 105-126

[6] Sklet, S 2004 Comparison of some selected methods for accident investigation *Journal of hazardous materials*, **111**(1), 29-37

[7] Branch, M A I, House, C, and Place, C 2004 Bridge watch keeping safety study Department for Transportation, Marine Accident Investigation Branch, (Southampton)

[8] Nahrgang, J D, Morgeson, F P, and Hofmann, D A 2011 Safety at work: a meta-analytic investigation of the link between job demands, job resources, burnout, engagement, and safety outcomes

[9] Schneider, W, and Shiffrin, R M 1977 Controlled and automatic human information processing: I. Detection, search, and attention *Psychological review*, **84**(1), 1

[10] Shiffrin, R M, and Schneider, W 1977 Controlled and automatic human information processing: II Perceptual learning, automatic attending and a general theory *Psychological review*, **84**(2), 127

[11] Sears, A, and Jacko, J A 2007 The human-computer interaction handbook: Fundamentals. Evolving Technologies and Emerging Applications
[12] Reason, J T 1997 Managing the risks of organizational accidents, Aldershot, Hants, England; Brookfield, Vt, (USA, Ashgate)
[13] Hollnagel, E 2004 Barriers and Accident Prevention, (Surrey, England: Ashgate Publishing Company)
[14] Leveson, N 2004 A new accident model for engineering safer systems *Safety Science*, **42**, 237-270
[15] Khanzode, V V, Maiti, J and Ray, P K 2012 Occupational injury and accident research: A comprehensive review (Safety Science), **50**, 1355-1367
[16] Arezes, P M, and Miguel, A S 2008 Risk perception and safety behaviour: A study in an occupational environment *Safety Science*, **46**(6), 900-907
[17] Olsen, E 2010 Exploring the possibility of a common structural model measuring associations between safety climate factors and safety behaviour in health care and the petroleum sectors *Accident Analysis and Prevention*, **42**(5), 1507-1516
[18] Vinodkumar, M N, and Bhasi, M 2010 Safety management practices and safety behaviour: Assessing the mediating role of safety knowledge and motivation *Accident Analysis and Prevention*, **42**(6), 2082-2093
[19] Rachman, S, Radomsky, A S, and Shafran, R 2008 Safety behaviour: A reconsideration. *Behaviour research and therapy*, **46**(2), 163-173
[20] Pyy, P 2000 Human reliability analysis methods for probabilistic safety assessment *VTT Publications*, **4**(2), 2.
[21] Mullen, J, Kelloway, E K, and Teed, M 2011 Inconsistent style of leadership as a predictor of safety behaviour *Work and Stress*, **25**(1), 41-54
[22] Hannaman, G W, and Worledge, D H 1988 Some developments in human reliability analysis approaches and tools *Reliability Engineering and System Safety*, **22**(1-4), 235-256
[23] Moieni, P, Spurgin, A J, and Singh, A 1994 Advances in human reliability analysis methodology Part I: Frameworks, models and data *Reliability Engineering and System Safety*, **44**(1), 27-55
[24] Reason, J T 1990 Human Error, (Cambridge University Press)
[25] Shaikh, M A 2001 Hazard perception and occupational injuries in the welders and lathe machine operators of Rawalpindi and Islamabad *Journal-Pakistan Medical Association*, **51**(2), 71-73
[26] Murphy, A W 1998 'Inappropriate'attenders at accident and emergency departments I: definition, incidence and reasons for attendance *Family Practice*, **15**(1), 23-32
[27] Niza, C, Silva, S, and Lima, M L 2008 Occupational accident experience: Association with workers' accident explanation and definition *Safety science*, **46**(6), 959-971
[28] Harms-Ringdahl, L 2004 Relationships between accident investigations, risk analysis, and safety management *Journal of Hazardous materials*, **111**(1), 13-19
[29] Howard, M E, Desai, A V, Grunstein, R R, Hukins, C, Armstrong, J G, Joffe, D, and Pierce, R J 2004 Sleepiness, sleep-disordered breathing, and accident risk factors in commercial vehicle drivers *American journal of respiratory and critical care medicine*, **170**(9), 1014-1021
[30] Mrugalska, B, and Kawecka-Endler, A 2011, July Machinery design for construction safety in practice In International Conference on Universal Access in Human-Computer Interaction (Springer Berlin Heidelberg) pp. 388-397
[31] Marsyahyo, E 2003 Mesin perkakas pemotongan logam (Malang: Banyu Media)
[32] Pulat, B M 1997 Fundamentals of industrial ergonomics (Chicago: Waveland PressInc.)

Ergonomic Analysis of the Work Condition in Aluminum Crafts Industry

H Soewardi[1,4], A Taufiq[2,5] and T Pradipta[1,6]

[1] Industrial Engineering Deparment, Faculty of Industrial Technology, Islamic University of Indonesia , Jl. Kaliurang Km. 14.5, Yogyakarta 55501, Indonesia
[2] Chemical Engineering Deparment, Faculty of Industrial Technology, Islamic University of Indonesia, Jl. Kaliurang Km. 14.5, Yogyakarta 55501, Indonesia
hartomo@uii.ac.id[4], agustaufiq@fti.uii.ac.id[5], tasyapradipta24@gmail.com[6]

Abstract. Indonesia has plenty of abundant natural wealth. This natural wealth produces natural materials that can be utilized as a craft product. The craft workers generally still work manually and the work is done repeatedly in one cycle so it is very susceptible to musculoskeletal disorders. Musculoskeletal disorders will cause decrease of worker's effectiveness and efficiency in completing the task. Thus, this fact has some adverse impact for the company. The purpose of this study is to analyze some task in manufacturing a traditional cooking tool (griddle) which encompasses mold of the griddle, refine of the griddle and the finishing touch by using the Occupational Repetitive Action Index (OCRA index) method. This method consists of high frequency movements, large forces, awkward postures, insufficient resting time and other factor such as vibration. Result of this study shows that in the molding operation, the worker has uncertain and medium risk on right hand in manufacturing small and large size of griddle with OCRA index is 2.5 and 5.3 respectively. In the refining operation the worker has light and medium risk on left hand in manufacturing small and large size with OCRA index is 3.6 and 5.8 using grinding machine and 5.8, 5.2, 6.3 and 5.0 using miser tool. In finishing operation, the worker has high risk on right hand in numbering of code griddle and has medium risk on right hand and medium on left hand in labeling with OCRA index is 9, 6.1 and 6.8 While in assembling operation the worker has no risk. Score that it indicates the risk level of the task in such job. Thus the following actions are to improve the work condition ergonomicly in all operation in alumunium craft.

Keywords: Musculoskeletal disorder, OCRA Index, Aluminum crafts, Griddle

1. Introduction

Indonesia as a developing country has many kinds of handycraft. Commonly, the crafts are manufactured from a natural material such as soil. The process of manufacture is still done traditionally which no use an advanced equipment. This is frequently called the manual handling task. Manual handling task performed repeatedly or over long periods of time can lead to fatigue and injury to muscles, tendons, ligaments, nerves, and blood vessels.This condition is known as work-related musculoskeletal disorders (WMSDs)[1] [2].

WMSDs become the most significant problem in industrialized countries especially in manual activities[3]. Based on Data from the U.S. Bureau of Labor Statistics in 2015 showed that WMSDs can cause a one third loss of working time[4] and also adversely affect the society, especially in economics terms [5].

According to the European Occupational Disease Statistics (EODS), work-related musculoskeletal disorders (WMSDs) are the most common occupational diseases (accounting for nearly 40% of the total) and an increasing trend can be observed in many member countries.It can be proved that European workers have a risk of WMSDs where its 25% experience back pain and 23% experience upper muscle pain [6] [7].

The result of the Health Department study in the profile of health problems in Indonesia 2005 indicates that 40.5% of diseases suffered by workers are related to their manual work. As many as 9482 workers in 12 cities of Indonesia experience Musculoskeletal Disorder (16%), cardiovascular disease (8%), respiratory disorders (3%), and Ears-Nose-Throat disease (1.5%) [8].

According to preliminary study in traditional cooking tools industry was found that more than 50% of workers complain about their body health when molding, refining and finishing of the griddle. However, this condition has not been identified how much risk to the worker's health. Therefore, it is very critical to know a risk level of the task that has been done so as the hazards associated with manufacturing a frying pan can be avoided.

Objective of this study is to analyze some task in manufacturing a traditional cooking tool (griddle) by using Occupational Repetitive Action Index (OCRA index).

2. Method

2.1 Occupational Repetitive Action (OCRA Index) Method

The Occupational Repetitive Actions (OCRA) is a method in ISO 11228 standard to identify the presence of upper limb work-related musculoskeletal disorders (UL-WMSD) for workers in high repeatability manual tasks. This method encompasess six variables. They are high frequency repetitiveness of actions, use of force, awkward postures, lack of recovery periods, movement of velocity and duration [2][10][11].

OCRA index is the ratio of the ratio between Actual Technical Action (ATA) and Reference Technical Action (RTA). Its procedure consists of three basic steps: (1) calculating the technical frequency of the action per minute and the total number of ATA performed in shifts and calculating the total number of RTA in the shift, (2) calculating the OCRA index, and (3) conducting a risk evaluation [11]. ATA is calculated based on several factors that are count the number of technical actions in one cycle (*nTC*), then evaluate the frequency of technical action performed per minute (*f*), and taking into account cycle time (*Tc*). It can be formulated as below :

$$f = \mathrm{n}TC \text{ x} \frac{60}{Tc} \tag{1}$$

Next, calculate the total number of technical actions performed in one shift (nATA) by using Equation (2). *t* is the total operator working time expressed in minutes.

$$\mathrm{n}ATA = f \text{ x t} \tag{2}$$

After calculating the number of ATA (n*ATA*), the next step is to calculate the amount of *RTA* (n*RTA*) by using equation (3).

$$\mathrm{n}RTA = \textstyle\sum_1^n [kf\ (Fm\ x\ Pm\ x\ Rem\ x\ Am) x\ t]\ x\ (Rcm\ x\ tm) \tag{3}$$

For calculating the number of RTA there are several factors to be considered. The first is the constant frequency (k*f*) which is the constant value for the constant frequency of the technical action, that is 30. The second is the force multiplier (F_M) using Borg-10 scale , which is intended to assess how labored the worker is doing. The third is the postural multiplier (P_M) obtained from the four upper body segments consisting of the shoulders, elbows, wrists, and type of grip. The fourth is the repetitiveness multiplier (R_{eM}), where when the task requires the performance of the same technical action for at least 50% of the cycle time, or when the cycle time is shorter than 15 seconds, $R_{eM} = 0.7$. if not, $R_{eM} = 1$. The fifth factor is the additional multiplier (A_M) obtained at least 1 additional factor that influences the work (using vibration tools, exposure temperature, light level and noise level).

Table 1. Classification of OCRA Index

Area	OCRA Index values	Risk Classification	Consequent possible actions
Green	≤ 1.5	Optimal	None
Green Yellow	1.6 – 2.2	Acceptable	None
Yellow – red	2.3 – 3.5	Uncertain or very low	Check again, if possible improve work conditions.
Red – light	3.6 – 4.5	Light	Improve work conditions, health surveillance, training
Red – medium	4.6 – 9.0	Medium	Improve work conditions, health surveillance, training
Red – high	≥ 9.1	High	Improve work conditions, health surveillance, training

The sixth is the duration of the repetitive task (t) which is the total operational time of the operator and expressed in minutes. the seventh is the recovery multiplier (R_{cM}) obtained from the total working hours at risk of overload. The last factor is the duration multiplier (t_M) obtained from the number of operators working hours.

After the calculation of the number of ATA and RTA, the next calculation of the value of OCRA index in accordance with the Press. 4 and evaluate the risk level according to table 1, which explains the risk level in accordance with the resulting OCRA index value [2].

$$\text{OCRA index} = \frac{nATA}{nRTA} \quad (4)$$

2.2 Subject and Object of Research

Focus of this study is a manufacturing process the traditional cooking tools (griddle) manually. It consists of three operations that are mold of the griddle, refine of the griddle and finishing. These operation is done by workers who have high skill for each task. Their age is about 23-60 years old and most of them are male . They works is in eight hours per day with one hour for rest and in standing position for molding and sitting position for refining and finishing.

2.3 Task

Task for manufacturing the traditional cooking tool is as follows :

1. Molding the griddles

   Molding operation consists of pouring the aluminum fluid into the mold of wok then leave it in 87.04 seconds for small size and 135.4 seconds for large size while polishing the wok with lime water for keeping the pan not sticking in the mold. And last step is remove the griddle from the mold. (see Figure 1)

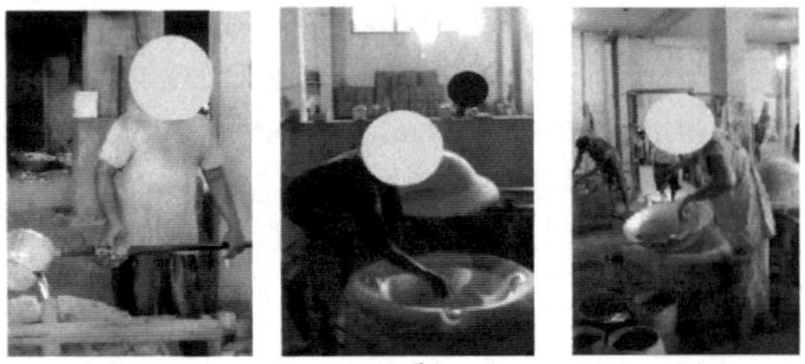

(a) (b) (c)

Figure 1. Manual Mold of The Griddle Activity : (a) Pouring the aluminum fluid into the mold of wok, (2) The griddle mold process, (3) Take a wok out of the griddle mold

2. Refine of the griddle
   Refine operation consist of refine all edges of the griddle using a grinding machine 19.61 seconds for small size and 27.2 seconds for large size. Then refine the outside and inside an unreachable griddle ear using a miser tool for 17.5 seconds for a small size and 60.9 seconds for large size. (see Figure 2)

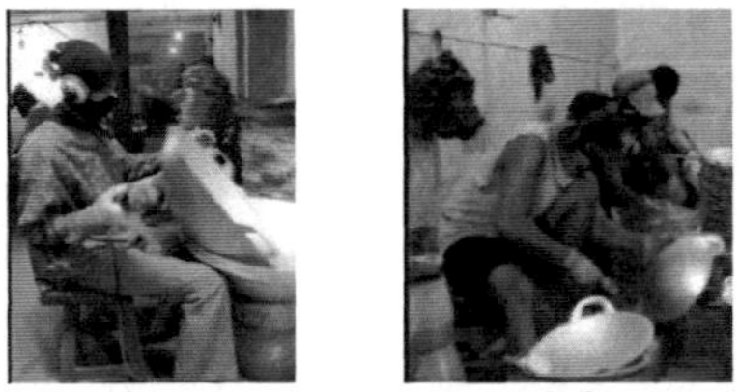

(a) (b)

Figure 2. Refine of the griddle activity : (a) Using grinding machine and (b) misery

3. Finishing the griddle
   Finishing operation consist of numbering of the griddle 3.1 seconds for small size and 7.24 seconds for large size, then label the company logo 6.3 seconds for small size and 8.58 seconds for large size , and also assemble the griddle 158.01 second for small size and 92.24 seconds for large size. (see Figure 3)

(a) (b) (c)

Figure 3. Final Finishing Activity : (a) Numbering , (b) labeling and (c) assembly of the griddle.

2.4 Analysis Method

Descriptive analysis of this study is conducted based on OCRA score to classify a risk level of each operation in a manufacture of the griddle. Futher analysis is also carry out to find a recommendation for improvement.

IOP Conf. Series: Materials Science and Engineering **277** (2017) 012018 doi:10.1088/1757-899X/277/1/012018

3. Result and Discussion

Table 2 shows scores of OCRA Index for each operation includes molding, refining and finishing. In the molding operation for manufacturing small size of griddle, score of OCRA index is 2.5 for the right hand and 1.1 for the left hand . It means this operation has uncertain and optimal risk. Uncertain risk on the right hand to make small size of griddle becauses of the force used is low as of not produce problem on arm but it is necessary to re-check the working conditions based on the analysis of movement posture on the left hand does not occur extreme movement during the time cycle of work but, on the elbow in the right hand is formed extreme flexion angle and palmar grip hand movement for 2/3 cycles, so it is necessary to re-check the tools used in molding operation to avoid muscle injury and upper body frame [13]. While for manufacturing large size of griddle, score of OCRA index is 5.3 for the right hand and 1.5 for the left hand. It means this operation has medium and acceptable risk. The medium risk occured on the right-hand is caused by use of the heavy force (at 4 level in BORG scale) and also the right hand is more dominant to take technical action on operation of mold for large size where molding tool has a weight more than 2 kg , so it can cause fatigue in the arm muscles and infammation in the joints [14]. Thus it is requaired to improve the molding method done by both hands. This effort can avoid a load on hand [16].

In operation of refining the griddle for small size using grinding machine, score OCRA index is 2.9 for the right hand and 3.6 for the left hand. It means that this operation has uncertain and light risk. The uncertain risk on right hand is moderate (with level 3 on the Borg scale) with a factor multiplier of 0.35. It because use of a grinding machine in refining requires more power than the left hand that only holding, twisting and moving a small size of griddle. However, although the right hand exerts moderate effort but no high repetition.The light risk on a left hand due to an actual technical actions (ATA) larger than Reference Technical Action (RTA) which it describes a high repetitive motion so as causing musculoskeletal disorder for long period [14] . For refining large size of griddle , score of OCRA index is 1.0 for the right hand and 5.8 for the left hand. It means this operation has optimal and medium risk. The medium risk occured on the left hand is caused by the same technical action and involves a flexion of the elbow joint exceeding 50% of the cycle time so that it will result in harmful and injury on the upper limb [14]. Thus it needs to improve work place and method based on the ergonomic principle by providing an appropriate recovery time for muscles [10] [15].

As for refining the griddle by using miser tool for small size ,score OCRA index is 5.2 for the right hand and 6.3 for the left hand. It shows this operation has medium risk. The medium risk on right and left hand is caused by frequency of repetition for every minute is large that is 20 action / minute for right hand and 34 action / min for left hand.. It means the musculoskeletal disorders will be occured due to develope the repetitive motion [14] . And also for large size of the griddle, score of OCRA index is 5.0 for the right hand and 4.7 for the left hand where it classify to medium risk . The medium risk occured on the right-hand and left hand is caused by use of the heavy force (at 3 and 4 level in BORG scale) at arms muscle so as fatigue and inflammation in joints [15]. Thus it is requaired to improve work place and method based on the ergonomic principle by providing an appropriate recovery time for muscles [10] [16].

In the finishing operation for numbering a code, score of OCRA index is 9.9 for the left hand and 0 for the right hand. It means that this operation has high and optimal . It happen because the repetitive activities are mostly done by the right hand, while the left hand is idle. For labeling of small and large size, score of OCRA index is 6.1 for the right hand and 6.8 for left hand and include risk area of medium risk. It happens because of the right hand often do pinch grip and there more opportunity there is for fatigue and inflammation to occur in the muscles and joints[14]. Thus this operation should be improved in method to avoid the excessive use of the hands. Improvements with rotating the worker in this is also done to fatigue[16].While in assembling operation the worker has no risk.

IOP Conf. Series: Materials Science and Engineering 277 (2017) 012018 doi:10.1088/1757-899X/277/1/012018

Table 2. Determining risk level of the various jobs presented in manufacturing traditional cooking tool (griddle)

Workstasion	Job type	Part of Hand	RTA	ATA	Score OCRA Index	Risk Area	Risk Classification
Mold of the griddle	small size	Right hand	1806.95	4570.3	2.5	Yellow-red	Uncertain
		left hand	3011.58	2688.4	1.1	Green	Optimal
	Large size	Right hand	555.984	2938	5.3	Red-medium	Medium
		left hand	1135.13	1728.2	1.5	Green	Optimal
Refine of the griddle	grinding of small size	Right hand	873.2	2520	2.9	Yellow-red	Uncertain
		left hand	2120.6	7710.4	3.6	Red-light	Light
	grinding of large size	Right hand	1621.6	1680	1.0	Green	Optimal
		left hand	873.18	5040	5.8	Red-medium	Medium
	Refining use a miser tool for griddle of small size	Right hand	1621.6	8400	5.2	Red-medium	Medium
		left hand	2270.3	14400	6.3	Red-medium	Medium
	Refining use a miser tool for griddle of large size	Right hand	498.96	2482.8	5.0	Red-medium	Medium
		left hand	873.2	4137.9	4.7	Red-medium	Medium
Finishing the griddle	Numbering of code at small and large size griddle	Right hand	1868.61	13632.5	9.9	Red-High	High
		left hand	0	4645.2	0	Green	Optimal
	Labeling at small and large size	Right hand	2324.08	14192.4	6.1	Red-medium	Medium

Workstasion	Job type	Part of Hand	RTA	ATA	Score OCRA Index	Risk Area	Risk Classification
	griddle	left hand	1081.49	7324.3	6.8	Red-medium	Medium
	Assembling at small and large size griddle	Right hand	2623.10	4884.25	1.9	Green-Yellow	Acceptable
		left hand	2623.10	4485.54	1.7	Green-Yellow	Acceptable

4. Conclusion

Based on the analysis, it can concluded that :

1. In operation of molding the griddle is identified that the worker has uncertain risk on right hand in manufacturing small size and has medium risk on right hand in manufacturing large size with OCRA index is 2.5 and 5.3 respectively. So, it is necessary to Re-check and improve work conditions, health surveillance as well as training for worker in molding operation.
2. In operation of refining the griddle by using grinding machine is identified that the worker has light risk on left hand in manufacturing small size and has medium risk on left hand in manufacturing large size with OCRA index is 3.6 and 5.8 respectively. While by using miser tool is identified that the worker has medium risk on left and right hand in manufacturing all size with OCRA index is 5.8, 5.2, 6.3 and 5.0 respectively. So, it is necessary to Improve work conditions, health surveillance as well as training for worker in refining operation.
3. In operation of finishing the griddle is identified that the worker has high risk on right hand in numbering of griddle and has medium risk on right hand and left hand in labeling with OCRA index is 9.9, 6.1 and 6.8 respectively. So, it is necessary to improve work conditions, health surveillance as well as training for worker in finishing operation.While in assembling operation the worker has no risk.

5. References

[1] NIOSH 2007 *Ergonomic Guidelines for Manual Material Handling*

[2] Occhipinti E and Colombini D 2007 Updating reference values and predictive models of the OCRA method in the risk assessment of work related musculoskeletal disorders of the upper limb (UL-WMSDs) *Ergonomics* **50** 1727-1739

[3] Buckle PW and Devereux JJ 2002 The Nature of Work-Related Neck and Upper Limb Musculoskeletal Disorders. *Appl. Ergon* **33** 207-217

[4] Wang J, Cui Ya, He Lihua, Xu Xiangrong, Yuan Z, Jin X and Li Zhimin 2017 Work Related Musculoskeletal Disorders and Risk Factors Among Chinese Medical Staff of Obstetrics and Gynecology. *Int. J. Environ. Res. Public Health* **14** 562

[5] Morse T F, Dillon C, Warren N, Levenstein C and Warren A 1998 The Economic and Social Consequences of Work-related Musculoskeletal Disorders: The Connecticut Upper-extremity Surveillance Project (CUSP) *Int. J. Occup. Environ. Health* **4** 209–216

[6] Boenzi F, Digiesi S, Facchini F and Mummolo G 2016 Ergonomic Improvement Through Job Rotation in Repetitive Manual Tasks in Case of Limited Specialization and Differentiated Ergonomic Requirements. *IFAC-PapersOnLine* **49** 1667-1672

[7] Eurofound 2012 Fifth European Working Conditions Survey. Publications Office of the European Union, Luxembourg

[8] Health Department of Indonesia 2005 *Profile of health problems in 2005* (Jakarta: Health Department of Indonesia)

[9] Cecchini M, Colantoni A, Massantini R and Monarca D 2010 The risk of Musculoskeletal Disorders for Workers due to Repetitive Movements during Tomato Harvesting *Journal of Agricultural Safety and Health* **16** 87-98

[10]Boenzi F, Digiesi S, Mossa G, Mummolo G, Romano VN 2013 Optimal Break Job Rotation scheduling under productivity and ergonomic risk constraints in assembly lines *Proceedings of the 22nd International Conference on Production Research (ICPR 22).*

[11]Occhipinti E 1998 OCRA, a concise index for the assessment of exposure to repetitive movements of the upper limbs. *Ergonomics* **41** 1290-1311

[12]ISO - International Organization for Standardization 2007 *ISO 11228-3:2007 - Ergonomics – Manual handling –Part 3: Handling of low loads at high frequency*

[13]Colombini D 1998 An Observational method for classfying exposure to repetituve movements of the upper limbs *Ergonomics* **41** 1261-1289

[14]Kiser D M, Murphy T J, Nielsen WJ and Rodgers S H 1986 *Ergonomic Design For People at Work* vol 2 (New york :Van Nostrand Reinhold Company)

[15]Najarkola SA M 2006 Assessment of Risk Factors of Upper Extremity Musculoskeletal Disorders (UEMSDs) by OCRA Method in Repetitive Tasks *Irian J.Pub.Health* **35** 88-74

[16]Bergamasco R, Colombini D and Girola C 1998 Guidelines for designing jobs featuring repetitive tasks *Ergonomics* 1364-1383

IOP Conf. Series: Materials Science and Engineering **277** (2017) 012019 doi:10.1088/1757-899X/277/1/012019

Development of the Strength Level on Arm for Indonesian People in Lifting Activity

H Soewardi[1], A D Prabaswari[2] and H A Muzakiroh[3]

Industrial Engineering Departement, Faculty of Industrial Technology, Islamic University of Indonesia, Jl. Kaliurang Km. 14.5, Yogyakarta 55501, Indonesia
[1]hartomo@uii.ac.id, [2]atyanti.dyah@uii.ac.id, [3]13522224@students.uii.ac.id

Abstract. Lifting is one of manual material handling activity that involves the whole part of a body. This activity is significant to contribute musculoskeletal disorder specifically on arms. It is because the arms are a major strength to lift objects. However, many people do not know the capability of their arm so that the task designed does not comply with the limitation of workers. Thus, it is required to determine a level of strength on arms. The objective of this study is to develop the strength level of arms for Indonesian people based on musculoskeletal contraction. An experimental study is conducted in the ergonomics laboratory. 24 males and 24 females was participated in this study which consists of three different ethnics. They are sixteen participants of Ethnic A, sixteen participants of Ethnic B and sixteen participants of Ethnic C. A case study of lifting consists of 4 positions of object. They are 38 cm in height, 50 cm in height, 85 – 115 cm in height for forming 90 degrees of the elbow and 100 cm in height. Back lift technique was implemented. An Electromyography is used to investigate muscle contraction on arms. Statistical analysis is done to test the hypothesis. The result of this study shows that the arm strength level for Indonesian workers has significant differences between males and females among difference Ethnic. For male, Ethnic A has 28.82% - 79.28% of MVC, Ethnic B has 17.74% - 58.67% of MVC, and Ethnic C has 22.13% - 68.67% of MVC. For female, Ethnic A has 28.28% - 84.63% of MVC, Ethnic B has 24.47% - 70.98% of MVC, and Ethnic C has 24.24% – 75.67% of MVC.
Keywords: Arm strength, Lifting activity, Musculoskeletal, MVC, Electromyography

1. Introduction

Based on data from the U.S. Bureau of Labor Statistics in 2015, approximately 31% (356,910 cases) of overexertion injury claims were caused by lifting and lowering of load. It affects injuries and illnesses by part of the body, in which 31.1% in upper extremities, 23.1% in lower extremities, 16.6% in back, 11% in multiple parts, 8.2% in head and 10% in all other parts of body. From 31.1% (358,890 cases) with injuries or illnesses to upper extremities, 143,900 cases involved the hands and 92,850 cases involved the shoulder.

Lifting is one of manual material handling activity that involves the whole part of a body. However, this activity is significant to contribute musculoskeletal disorder on arms when lifts the heavy load [1], [2]. Ref. [3] reported that heavy task had four times greaterprevalence of lumbar insufficiency, three times greater prevalence of recurrent attacks of lumbago and two times greater prevalence of pronounced sciatica than light jobs. The position of the load when lifting objects is also another factor affecting stress on arm and spine because of it configurated unnatural posture as well as a heavy moment on arm [4].The effective moment arm is related to an increase muscular contraction effort [5].

The muscle contraction occurs to improve limb stiffness and joint stability [6], [7]. To describes the muscle contraction around the arm area, it requires to quantify and analyze the muscle activity of the biceps brachii (BB) musclebased on Electromyograph (EMG) activities. It is believed that the quantity of static strength a muscle depends on the length muscle at the position of contraction[8]. Furthermore, [9], [10], [11] mentioned that a force of the static strengthwill be affected by gender, age, and posture of the subject.

EMG is a tool to identify the quantities of signal amplitude of muscle activity, especially those from deep muscles and has been used in several studies related to activity of hand [12], [13]. The EMG describes arm muscle movements within postural task [14].

The objective of this study is to investigate and develop the strength level of arms for Indonesian people based on musculoskeletal contraction by using Electromyography.

2. Method

2.1. Subject

Forty-eight university students (24 males and 24 females) was participated in this study. They are 8 males and 8 female participants respectively for Ethnic B, Ethnic A, and Ethnic C. Their age is in range between 19 to 23 years. And they had sport activities for 1 - 2 times per week and had no chronic medical history of musculoskeletal disorder on arms.

2.2. Apparatus

The main tool in this study is Electromyography (Lab Quest 2: Vernier Tech & Soft, Texas, USA) to investigate muscle contraction on arms. Raw signal from the EMG recorded by attaching electrodes on hand cleaned with an alcohol swab and normalized into a percentage data of Maximum Voluntary Contraction (MVC). A set of personal computers is used to present the raw EMG using Logger Pro Software. While, the experiments performed using an arm-leg and back dynamometer for as a trigger of muscle contraction.

2.3. Experimental Design

Experiment was conducted in Ergonomic Laboratory. Each participant stands in front of an arm-leg and back dynamometer. It requires 2 minutes for setting the equipment while 15 seconds for real experiment (5 seconds for relax, 5 seconds for reach maximum effort, 5 seconds calm down) [14]. There are 4 positions used in this study. They are 38 cm in height, 50 cm in height, 85 – 115 cm in height for forming 90 degrees of the elbow and 100 cm in height. Back lift technique was implemented. Figure below show the experiment design:

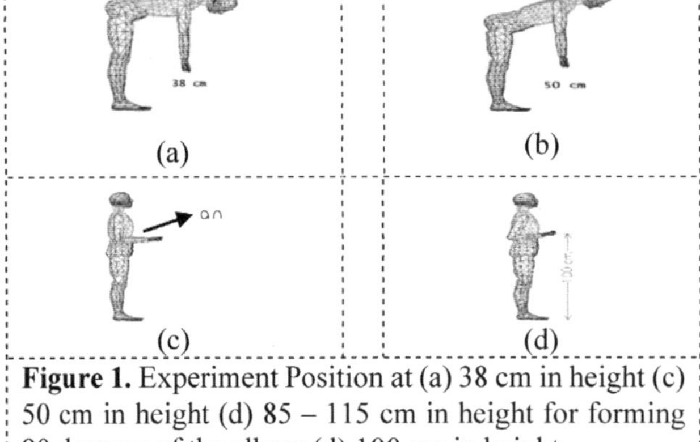

Figure 1. Experiment Position at (a) 38 cm in height (c) 50 cm in height (d) 85 – 115 cm in height for forming 90 degrees of the elbow (d) 100 cm in height

IOP Conf. Series: Materials Science and Engineering **277** (2017) 012019 doi:10.1088/1757-899X/277/1/012019

After doing the experiment, participant was instructed to pull an arm-back leg dynamometer again to normalization muscle contraction. It is needs to stabilization of the forearm in recording MVC data. Then, average data of MVC is used to analysis. Figure 2 show the effectiveness of MVC position for normalization muscle contraction.

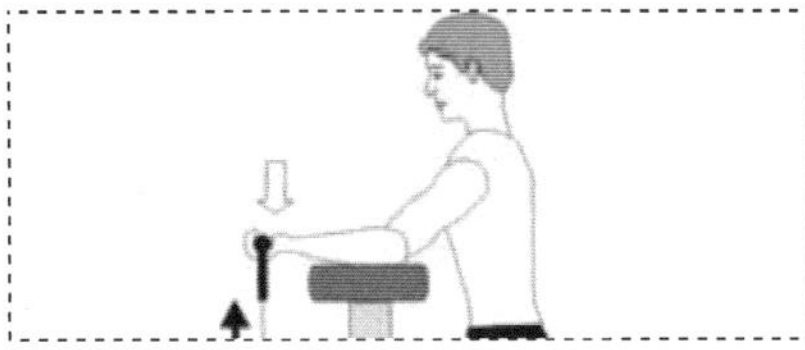

Figure 2. MVC Position of Forearm

2.4. Task

Participant was instructed to pull an arm-back leg dynamometer that it is assumed as handle of objects that has been adapted to four positions. All procedures in the process of pulling should be followed by the respondent, if not appropriate experiment procedures, respondents are asked to do the test again. Figure 3 show the task from this study.

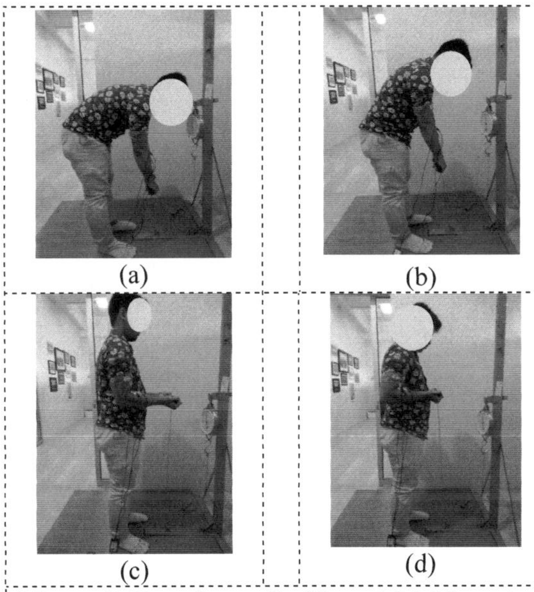

Figure 3. Experiment Position at (a) 38 cm in height (c) 50 cm in height (d) 85 – 115 cm in height for forming 90 degrees of the elbow (d) 100 cm in height

2.5. Procedure of Experiment

Identify location of the biceps muscle group, *brachioradialis* muscle. Location of *brachioradialis* muscle shown at Figure 4 which the position of electrodes is in parallel of the muscle fibres and in the middle of the muscle mass. Both electrodes are positioned as far as 2 cm [15].

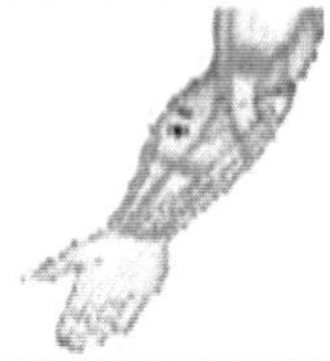

Figure 4. Electrodes Placement at *brachioradialis* muscle

2.6. Data Analysis

Data analysis was conducted based on MVC data. Statistical analysis using SPSS 24 version is to determine the significance of differences in each muscle contraction when during 4 positions of lifting activity. Non-parametric statistical analysis by using Kruskall-Wallis and Mann Whitney at 5% of significance level.

3. Result and Discussion

3.1. Analysis of EMG Signal

In Figure 5 shows RMSE signal on muscle contraction graph of *brachioradialis* muscle and amplitude score for muscle contraction (a) at 38 cm in height is 0.7228 mV for Ethnic A, 0.7134 mV for Ethnic B and 0.7241 for Ethnic C. In addition, for muscle contraction (b) at 50 cm in height is 0.7176 mV for Ethnic A, 0.7168 mV for Ethnic B and 0.7095 mV for Ethnic C. Then for a muscle contraction (c) at 90 degrees of the elbow or 85 - 115 cm in height is 0.8254 mV for Ethnic A, 0.7958 mV for Ethnic B and 0.7729 for Ethnic C. While amplitude scores for muscle contraction (d) at 100 cm in height is 0.7323 mV for Ethnic A, 0.7314 mV for Ethnic B and 0.320 mV for Ethnic C. Those indicate that there is the variation of RMSE score from each ethnic in each position. The highest score in each ethnic is at 90 degrees elbow or height 85-115 cm and Ethnic A has the highest score than the other two ethnics at that position. Then, the lowest score is at 50 cm in height and Ethnic A has the highest score. This condition happens because when pulling an arm-back leg dynamometer at 90 degrees of the elbow or 85 - 115 cm in height requires big contraction of muscle using back lift method. The graph (Figure 5) shows higher fluctuation when forearm becomes more flexes.

Figure 6 (a) shows the result of average percentage MVC (%MVC) for males between Ethnic A, Ethnic B and Ethnic C in each position. Percentage MVC scores for muscle contraction at 38 cm in height is 28.82% for males of Ethnic A, 17.74% for males of Ethnic B, and 22.13% for males of Ethnic C. In addition, percentage MVC score for muscle contraction at 50 cm in height is 34.06% for males of Ethnic A, 19.45% for males of Ethnic B, and 22.63% for males of Ethnic C. Then for muscle contraction at 90 degrees of the elbow or 85-115 cm in height is 79.28% for males of Ethnic A, 55.60% for males of Ethnic B, and 68.67% for males of Ethnic C. While muscle contraction at 100 cm in height is 78.71% for males of Ethnic A, 58.67% for males of Ethnic B, and 68.67% for males of Ethnic C. The highest score for males in each ethnic is at 90 degrees of the elbow or 85-115 cm in height. It is Ethnic A with 79.28% MVC. Then it is followed by 100 cm in height with 67.31% MVC. It means that on those positions may produce high muscle contraction and when the muscle reaches 50% MVC will cause musculoskeletal disorders and even muscle failure [15]. For 90 degrees of the elbow or 85-115 cm in height and 100 cm in height, lifting is done for 1 minute (according to [17]). So, it shows the average value of muscle contraction at 90° of elbow or 85 -115 cm in height and position at 100 cm in height must be considered because it has a value of more than 50% and unsafe for Indonesian's males of Ethnic A, B, and C in more than 1 minute in lifting activity.

Furthermore, Figure 6 (b) shows the result of average percentage MVC (% MVC) for females between Ethnic A, Ethnic B and Ethnic C in each position. Percentage of MVC scores for muscle contraction at 38 cm in height is 28.28 % for females of Ethnic A, 24.47 % for females of Ethnic B, and 24.24% for females of Ethnic C. In addition, percentage MVC scores for muscle contraction at 50 cm in height is 31.97% for females of Ethnic A, 27% for females of Ethnic B, and 25.90% for females of Ethnic C. Then for muscle contraction at 90 degrees of the elbow or 85-115 cm in height is 80.02% for females of Ethnic A, 70.98% for females of Ethnic B, and 75.67% for females of Ethnic C. While muscle contraction at 100 cm in height is 84.63% for females of Ethnic A, 59.49% for females of Ethnic B, and 66.19% for females of Ethnic C. It shows the average value of muscle contraction at 90° of elbow or 85 -115 cm in height and position at 100 cm in height must be considered because it has a value of more than 50% and unsafe for Indonesian's females of Ethnic A, B, and C in more than 1 minute in lifting activity.

Figure 6 (c) shows the result of average percentage MVC (%MVC) for both males and females between Ethnic A, Ethnic B and Ethnic C in each position. Percentage MVC scores for muscle contraction at 38 cm in height is 28.55% for Ethnic A, 21.10% for Ethnic B, and 23.19% for Ethnic C. And percentage MVC scores for muscle contraction at 50 cm in height is 33.01% for Ethnic A, 23.23% for Ethnic B, and 24.26% for Ethnic C. Then for muscle contraction at 90 degrees of the elbow or 85-115 cm in height is 79.65% for Ethnic A, 63.29% for Ethnic B, and 72.17% for Ethnic C. While muscle contraction at 100 cm in height is 81.67% for Ethnic A, 59.08% for Ethnic B, and 62.42% for Ethnic C.

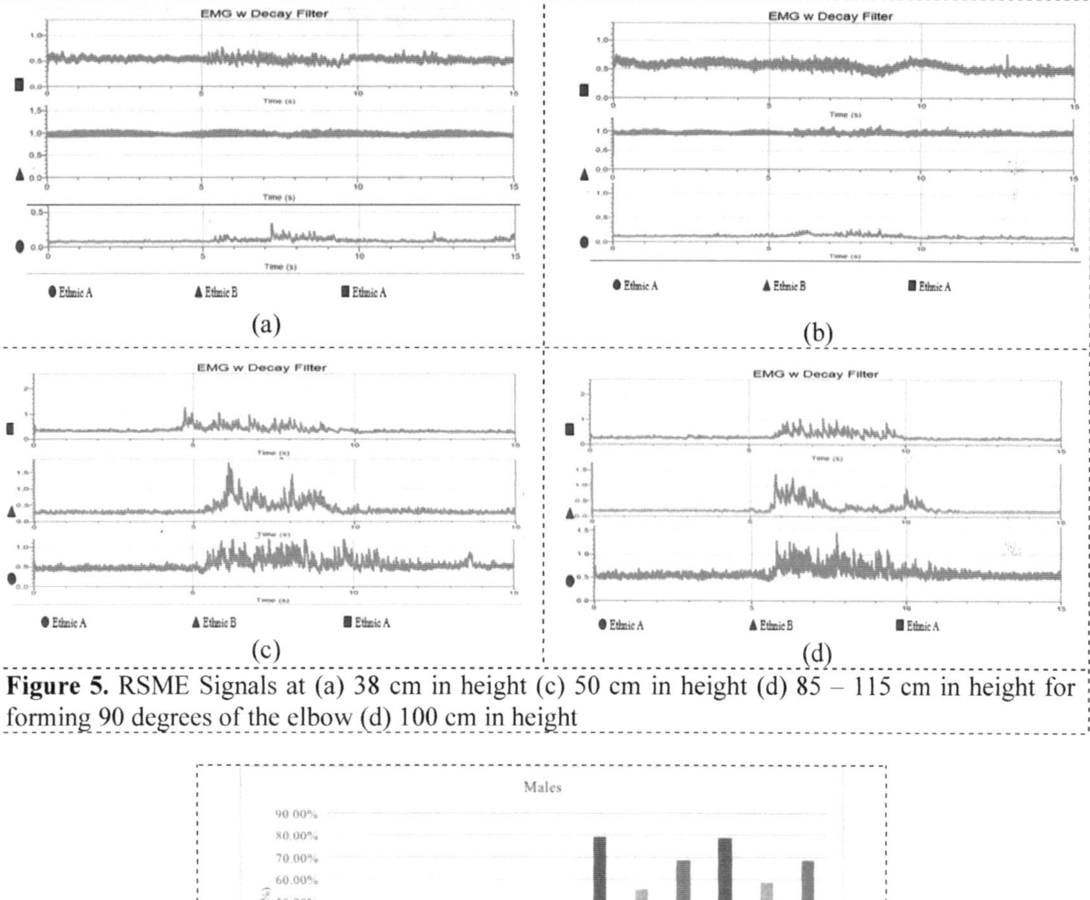

Figure 5. RSME Signals at (a) 38 cm in height (c) 50 cm in height (d) 85 – 115 cm in height for forming 90 degrees of the elbow (d) 100 cm in height

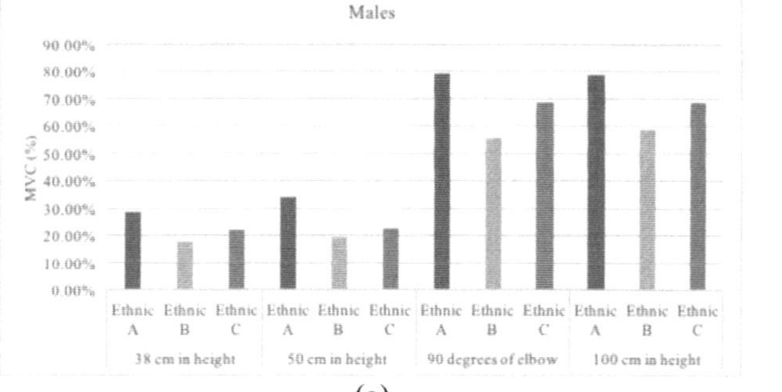

(a)

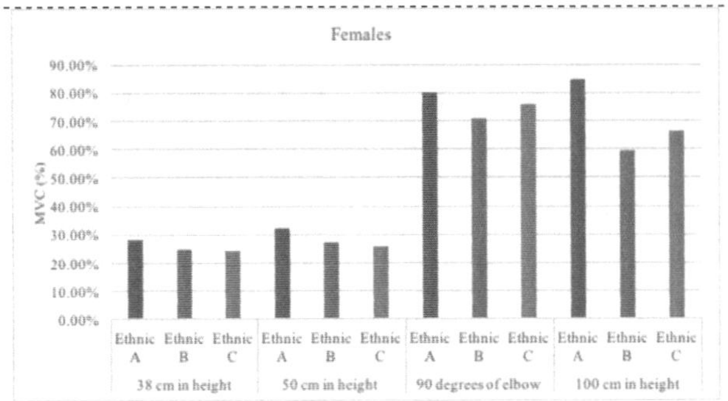

(b)

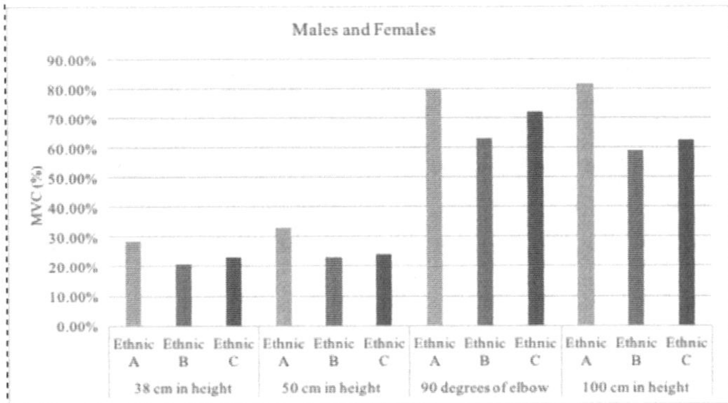

(c)

Figure 6. Percentage MVC for (a) Males (b) Females (c) Males and Females

3.2 Statistical Analysis

Non-parametric statistical analysis by using Kruskal-Wallis was performed to describe the difference significant of % MVC score between each position at 5% significant level. The results show that each position has a significant difference.

Table 1. The Result of Mann – Whitney Test for Each Position

	38 cm in height	50 cm in height	90 degrees of elbow	100 cm in height
38 cm in height		0.464	0.000	0.000
50 cm in height	0.464		0.000	0.000
90 degrees of elbow	0.000	0.000		0.190
100 cm in height	0.000	0.000	0.190	

Table 1 shows the result of Mann- Whitney test. It describes the significant difference of % MVC score among each position. There was the significant difference in percentage MVC score between 38 cm in height with 90° of the elbow or 85 -115 cm in height, 38 cm in height with 100 cm in height, 50 cm in height with 90° of elbow or 85 -115 cm in height, 50 cm in height with 100 cm in height. But, there doesn't have a significant difference between 38 cm in height with 50 cm in height and 90° of the elbow or 85 -115 cm in height with 100 cm in height.

IOP Conf. Series: Materials Science and Engineering **277** (2017) 012019 doi:10.1088/1757-899X/277/1/012019

Table 2. The Result of Kruskal-Wallis for Males

Positions	Sig.	Decission
38 cm in height	0.003	Difference
50 cm in height	0.011	Difference
90 degrees of elbow	0.041	Difference
100 cm in height	0.005	Difference

Table 2 shows result of Kruskal-Wallis test for each position between males of Ethnic A, Ethnic B and Ethnic C. It is present that there is the significant difference between males of Ethnic A, Ethnic B and Ethnic C in each position at 38 cm in height, 50 cm in height, 90° of elbow or 85 -115 cm in height and 100 cm in height. It caused by differences in height between males of Ethnic a, b and c based on anthropometric data. Males of Ethnic A has a height range of 168.58 cm in percentile 50, Ethnic B has a height range of 171.23 cm in percentile 50 and Ethnic C has a height range of 170.78 cm in percentile 50.

Table 3. The Result of Kruskal-Wallis for Females

Positions	Sig.	Decission
38 cm in height	0.047	Difference
50 cm in height	0.270	No Difference
90 degrees of elbow	0.386	No Difference
100 cm in height	0.011	Difference

Table 3 shows results of Kruskal-Wallis test for each position between females of Ethic A, Ethnic B and Ethnic C. It is present there is significant difference between females of Ethnic A, Ethnic B and Ethnic C in position at 38 cm in height and 50 cm in height. It also caused by differences in height between females of Ethnic a, b and c based on anthropometric data. Females of Ethnic A has a height range of 155.85 cm in percentile 50, Ethnic B has a height range of 156.18 cm in percentile 50 and Ethnic C has a height range of 156.2 cm in percentile 50. But, there is no significant between females of Ethnic A, Ethnic B and Ethnic C in position at 90° of elbow or 85 -115 cm in height and 100 cm in height.

4. Conclusion

Based on the analysis, it can be concluded as follows:

1. The strength level of arm based on Maximum Voluntary Capacity for Indonesian's males are difference among Ethnics considered. Ethnic A has 28.82% of MVC for 38 cm in height, 34.06% of MVC for 50 cm in height, 79.28% of MVC for 90 degrees of the elbow or 85-115 cm in height, 78.71% of MVC for 100 cm in height. Ethnic B has 17.74% of MVC for 38 cm in height, 19.45% of MVC for 50 cm in height, 55.60% of MVC for 90 degrees of the elbow or 85-115 cm in height, 58.67% of MVC for 100 cm in height. Ethnic C has 22.13% of MVC for 38 cm in height, 22.63% of MVC for 50 cm in height, 66.67% of MVC for 90 degrees of the elbow or 85-115 cm in height, 58.65% of MVC for 100 cm in height.
2. The strength level of arm based on Maximum Voluntary Capacity for Indonesian's females are difference among Ethnics considered. Ethnic A has 28.28% of MVC for 38 cm in height, 31.97% of MVC for 50 cm in height, 80.02% of MVC for 90 degrees of the elbow or 85-115 cm in height, 84.63% of MVC for 100 cm in height. Ethnic B has 24.47% of MVC for 38 cm in height, 27% of MVC for 50 cm in height, 70.98% of MVC for 90 degrees of the elbow or 85-115 cm in height, 59.49% of MVC for 100 cm in height. Ethnic C has 24.24% of MVC for 38 cm in height, 25.90% of MVC for 50 cm in height, 75.67% of MVC for 90 degrees of the elbow or 85-115 cm in height, 66.19% of MVC for 100 cm in height.

3. There are significant differences in the muscle contraction of Indonesian's males between Ethnic A, Ethnic B and Ethnic C for 38 cm in height, 50 cm in height, 90 degrees of the elbow or 85-115 cm in height, 100 cm in height. And significant differences occur in the muscle contraction of Indonesian's females between Ethnic A, Ethnic B and Ethnic C for 38 cm in height and 100 cm in height.

Acknowledgments
The Authors would like to express our appreciation for the full support to the Directorate of Research and Community Service (DPPM) Islamic University of Indonesia Yogyakarta Indonesia.

5. References

[1] Nordin M, Anderson G B and Pope M H 2006 *Musculoskeletal Disorders in the Workplace: Principles and Practice* (St. Louis, Missouri: Mosby-Year Book, Inc)
[2] Nurmianto E, Ciptomulyono U, Suparno and Kromodiharjo S 2015 Manual handling problem identification in mining industry: an Ergonomic perspective **4** 89-97.
[3] White A A and Gordon S L 1982 Synopsis: workshop on idiopathic low back pain, *Spine* **7** 141-149
[4] Chengalur S N, Rodgers S H and Bernard T E 2013 *Ergonomic Design for People at Work* (New York: J. Wiley)
[5] Kiser D M, Murphy T J, Nielsen W J and Rodgers S H 1986 *Ergonomic Design for People at Work* vol 2 (USA: Eastman Kodak Company)
[6] Latash M L 1992 Independent control of joint stiffness in the framework of the equilibrium-point hypothesis *Biol Cybern* **67** 377–84.
[7] Granata K P, Wilson S E, Massimini A K and Gabriel R 2004 Active stiffness of the ankle in response to inertial and elastic loads *Journal of Electromyography and Kinesiology* **14** 599-609.
[8] Ahamed N U, Ethnic B K, Rahman M and Islam M A 2012 Analysis of right arm biceps brachii muscle activity with varying the electrode placement on three male age groups during isometric contractions using a wireless EMG sensor *Procedia Engineering* **41** 61-67.
[9] Kroemer K H E 1970 Human Strength: Terminology, Measurement and Interpretation of Data. *Human Factors* **12** 297-313.
[10] Roebuck J A, Kroemer K H E and Thomson W G 1975 *Engineering Anthropometry Methods* (New York: J. Wiley)
[11] Harwood D L, Edwards J M, Jakobi 2011 Age- and sex-related differences for electromyography gaps during daily activity *Gait & Posture* **34** 6-12.
[12] Soewardi H, Anugraheni A R and Shabrina N 2015 Analysis of Electromyography on computer interaction device to the risk of carpal tunnel syndrome *Journal of Computer* **10** 347-352.
[13] Kim M J, Oh D W and Park H J 2013 Integrating arm movement into bridge exercise: Effect on EMG activity of selected trunk muscles *Journal of Electromyography and Kinesiology* **23** 1119-1123.
[14] Konrad P 2005 *A Practical Introduction to Kinesiological Electromyography* (USA: Noraxon Inc)
[15] Criswell E 2011 *Surface Electromyography, Second Edition* (Canada: Jones and Bartlett)
[16] Grandjean E 1988 *Fitting the Task to the Man : A Text book of Occupational Ergonomic* (New York: Taylor & Prancis)

Design of disturbances control model at automotive company

I A Marie[1], D K Sari[2], P Astuti and M Teorema

Industrial Engineering Department, Universitas Trisakti, Jakarta, 11440, Indonesia
[1]iveline.annemarie@trisakti.ac.id; [2]debbie.kemala@trisakti.ac.id

Abstract. The discussion was conducted at PT. XYZ which produces automotive components and motorcycle products. The company produced X123 type cylinder head which is a motor vehicle forming component. The disturbances in the production system has affected the company performance in achieving the target of Key Performance Indicator (KPI). Currently, the determination of the percentage of safety stock of cylinder head products is not in accordance to the control limits set by the company (60% - 80%), and tends to exceed the control limits that cause increasing the inventory wastage in the company. This study aims to identify the production system disturbances that occurs in the production process of manufacturing components of X123 type cylinder head products and design the control model of disturbance to obtain control action and determine the safety stock policy in accordance with the needs of the company. The design stage has been done based on the Disturbance Control Model which already existing and customized with the company need in controlling the production system disturbances at the company. The design of the disturbances control model consists of sub-model of the risk level of the disturbance, sub-model of action status, sub-model action control of the disturbance, and sub-model of determining the safety stock. The model can assist the automotive company in taking the decision to perform the disturbances control action in production system cylinder head while controlling the percentage of the safety stock.
Keywords: production system, disturbance control model, safety stock

1. Introduction

The sustainability of the automotive industry is influenced by the ability to react to the uncertainties and dynamics that occur. Associated with the production system, uncertainty is categorized into two groups: environmental uncertainty and uncertainty associated with production processes [1]. Uncertainty of the production system arises in the form of interruptions of the production system. The existence of a disruption of the production system leads to deviations (variance) between planning and actual conditions. Next, irregularities that occur will cause inefficiency and inhibit the achievement of corporate targets that reduce productivity in the industry.

[2] categorizes uncertainties into uncertainty of inputs and process uncertainties including: lack of /unavailability of raw materials, lack / unavailability of production operators, lack / unavailability of machine capacity, defective products and late delivery of products.

Risk and uncertainty are included in inventory analysis involving many variables. However, the most important role as risk and uncertainty is the variation in demand and lead time. Variations are absorbed in the form of safety stock or buffer stock or fluctuation stock. Safety stock is an additional stock kept as a buffer to prevent stockout due to random noise from nature or the environment [3].

To improve the Production Planning and Inventory Control (PPIC) function in the food industry, a smart and adaptive PPIC model has been developed to improve the effectiveness of PPIC functions.

The PPIC model adopts to the disturbance of the production system that occurs with the provision of safety stocks and changes periodically adjusted to the disturbance of the production system [4]. Subsequently, it is conducted a study on the use of the Disturbance Control Model consisting of Disturbance Control Action Sub Model, Advanced Control of Disturbance Policy Sub Model and Safety Stock Sub Model at the baking company [5]. The model has been applied to other food companies producing beef sausages [6]. Utilization of Disturbance Control Model that has been done in the food industry (bread-producing company and sausage-producing company) shows that the design of the disturbance control model can help the company to know the recommendation of action to be done as well as the follow-up policy.

In the process of finding a model, we use classification technique based on data mining concept which is the technique that explains and distinguishes the classes or concepts, with the aim that the model obtained can be used to predict classes or objects that have unknown class labels. The derived model is based on the analysis of the data training can be represented in various forms such as if decision tree. Decision tree algorithm works based on entropy of information contained in observation data. Training data is used by the decision tree algorithm to determine which attributes have the largest "decisive power". Determination of decisive power is arranged based on how pure or homogeneous data in each attribute to the data in the Class Attribute [7].

Considering the dynamics and uncertainty of high market in automotive industry company, it is necessary to study the utilization of Disturbance Control Model to know the policy of production inventory control with case study in automotive industry. PT. XYZ is a subsidiary of PT AI, Tbk which carries out joint venture with PT HM Corp., Ltd. The plant produces product on machining process for several type of H brand. The product studied in this research is cylinder head. Cylinder head is an important component that is part of motor engine, because it plays a major role in controlling airflow and fuel dispersal. Cylinder head is made of ingot ingredient mixture (with 40% ingot and 60% scrub of rejected component) melted at high temperature. Then the ingot is printed using jacket core mold (solid sand mold with coated sand resin material) using machine Low pressure die casting. In engine motor cylinder head functions as combustion chamber, drainage and disposal ducts, spark plug holder, cooling coat, and for valve mechanism. As one of the important components in engine motors, the cylinder head must be resistant to extreme high temperatures as it acts as a combustion chamber. In addition, the cylinder head must be free from leaks, as it is the fuel feed and discharge ducts in the engine. Figure 1. The following is a cylinder head image used on motor type A.

Figure 1. Cylinder head component for type A motorcycle

Problems of PT. XYZ is a disturbance in the production system that affects the performance in achieving KPI target company, as well as determining the percentage of safety stock that has not been in accordance with the control limits set by the company, and even tend to exceed the control limit has the potential to increase the waste of production in the company. This study aims to obtain inventory of production system disturbance that occurs and designing the control model of disturbance to obtain control action and safety stock policy in accordance with the needs of the company.

2. Research Methodology

The research activity begins with preliminary research on companies that become the object of research along with the literature study. Based on the research objectives, in accordance with data processing needs, data collection activities are conducted based on company information tracking based on interview method and open questionnaire to obtain the supporting data related to Production

Planning & Inventory Control function PPIC of to the company and find out the external and internal disturbances related to PPIC functions that occur in the company. The next stage is the design of Disturbance Control Model for PT XYZ with reference to the Disturbance Control Model that has been developed before to help companies get a solution to overcome the disturbance of production system that occurs. The design of the disturbance control model used J48 learning algorithm which is one of the variants of some decision tree algorithms that have been developed. The algorithm will be implemented in the WEKA software package (Waikato Environment for Knowledge Analysis). This stage will produce Disturbance Control Model of Production System at PT XYZ as an automotive company and will end the activity research as a whole. This research will be conducted using the following steps.

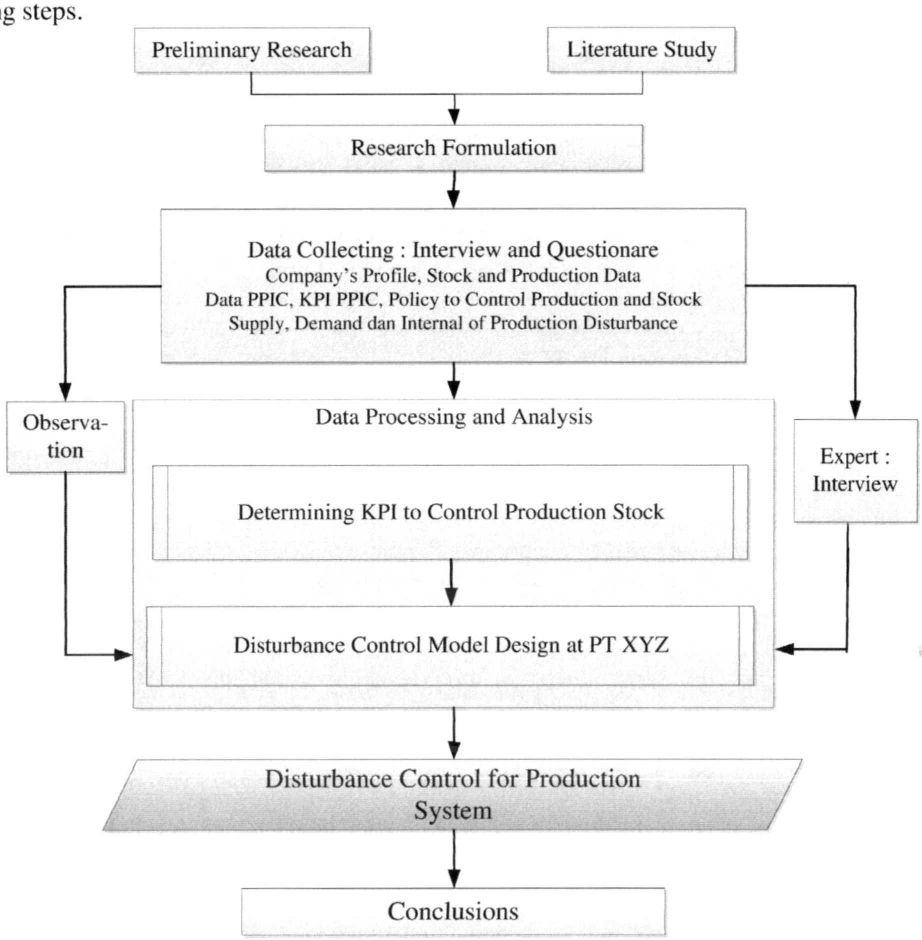

Figure 2. Flowchart of research methodology

3. Result and Discussion

3.1. Key Performance Indicator at Production Division of PT XYZ

Disturbance of production system in PT XYZ has influenced company to fulfill their target Key Performance Indicator (KPI). The design of the disturbance control model will be determined based on KPI Division of Production of PT XYZ to fit the company's needs. The following are selected KPI Division of Production based on interview with PT XYZ Production Division Head: Quality (reduced rejection & claim), Cost (reduced spoilage) and Delivery (fulfillment of Production Achievement Rate, decreasing of man hour / unit, increasing of OEE and reducing of machine downtime).

3.2. Requirement of Design of Production Disturbance Control Model at PT XYZ
The design of production disturbance control model at PT XYZ will be developed based on the previously designed Disturbance Control Model for the food industry, then adjusted to the need of controlling production system disturbance in the company to support the achievement of company's KPI Production Division. Disturbance Control Model of Production System in food industry is consisting of Disturbance Control Action Sub Model, Advanced Control of Disturbance Policy Sub Model and Safety Stock Sub Model. The Sub Model of Disturbance Control Action produces a recommendation of interference control action using protocol mechanism or rule base. The Sub Model of Advanced Disturbance Control Policy produces an advanced policy decision on fault control. The Safety Stock Sub Model produces a safety stock value using the Average Calculation method. The model is expected can control the disruption of production systems that occur in the industry [5].

3.3. Safety Stock Policy Analysis at PT XYZ
The calculation is performed to obtain a percentage of the Finished Good (FG) safety stock which is sent to the assembly department. The company policy in determining the safety stock for FG is 60% - 80% of the production plan in one work shift. In this research, safety stock policy analysis for company is only done for finished product that is cylinder head component by using data in January 2016 until June 2016. The actual percentage of safety stock during the observation period has an average exceeding the limits set by the company that is 60-80%.

Disturbance of Production System at PT XYZ
Disturbance of the production system includes internal disturbances and external disturbances. Internal disturbance of production system include production process, energy supply, labour, down time, tool change / setting, reject casting, and reject machining. The external disturbances of the production system include supply and demand disturbances. Internal disturbances caused by reject machining resulted from errors caused by machine process or errors caused by operator component errors by jig. The disturbance caused by reject casting is caused by the existence of components that do not meet the quality of the desired company and it can be known during machining process. The disturbance caused by tool change or setting is caused by tools that have exhausted life so that there is a need to change tool or the mistake in placement tool. Disturbance caused by down time caused by the disturbance that occurs in the machine caused by program error on the machine or damage to the machine. Disorders caused by labour factors can be due to lack or absence of manpower in the production line, the existence of daily or monthly production planning mistakes, or the inability of operators in machine operation. The disruption of energy supply is caused by the unavailability of energy supply that serves to support the sustainability of production process, such as power failure. Disturbance caused by the production process can be caused by the unavailability of goods available between processes on the production line, the change of the model that must be produced by the line concerned, or the existence of components of the product that stock out.

Disturbances are classified based on the classification factor of the disturbance in accordance with the results of interviews and data obtained from the company. Reject casting is the most dominant disorder that occurs in the company. The frequent disturbance due to reject casting is leakage tester, kropos, and misrun with frequency more than 150 times for 6 months. The next disturbance is the interference with the frequency of 51-150 times for 6 months, is cutting. If the company continues to experience such disturbances it will affect the achievement of the KPI, especially on reject points and lost time. The control action status is determined based on the following factors: Frequency, Severity, Disturbance Impact and KPI Achievement. Categorization of these factors in accordance with predetermined categories and has been validated to the company, can be seen in table 1.

Table 1. Category of disturbance classification factor

Frequency	1. Seldom (0-50 times / 6 month)
	2. Occasionally (51-150 times / 6 month)
	3. Often (>150 times / 6 month)
Severity	1. Not severe (trouble in casting, AE dan MC still operating)
	2. Severe (MC stopped)
	3. Extremely severe (AE stopped)
Disturbance Impact	1. AR >=67%
	2. 33%<AR<66%
	3. AR <=33%
KPI Achievement	1. Good (0-1 KPI not achieved)
	2. Moderate (2-3 KPI not achieved)
	3. Bad (4-5 KPI not achieved)

(Source : Interviewed head of production division PT. XYZ)

3.4. Design of Disturbance Control Model at PT XYZ

Disturbance control model is a model that takes into account the disturbances of the production system that occur including demand, supply and internal disturbances of the company's production system. The disturbance control model designed for handling of disturbances at PT XYZ is the result of the development of the existing Disturbance Control Model and adapted to the disturbance control requirements of PT XYZ. This model consists of sub-model of risk level of interference, sub action control model of disturbance, and sub model of determination of safety stock.

By using software WEKA 3.8.1 we obtain decision tree with class that used is level of risk of disturbances. Based on the decision tree, there are 19 rules in decision-making on the level of risk of disturbances, for example: If the severity is not severe, the risk level of the disturbance is low ; If the severity is severe, and the impact is> 6 hours, and the frequency is sometimes, and the achievement of the KPI is moderate, then the level of risk of disturbance is high, etc. The decision tree can be seen at figure 3.

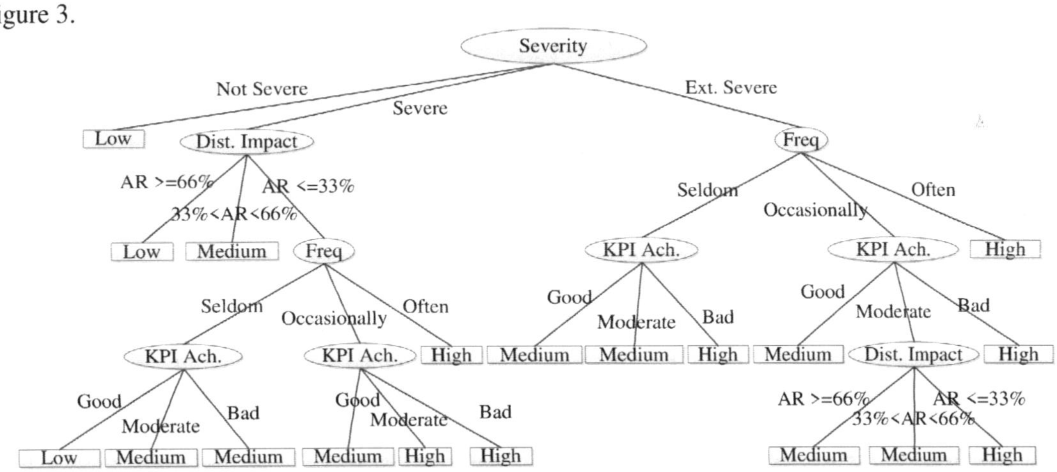

Figure 3. Disturbance risk level decision tree

The second sub model is the sub model to know the status for doing the action. This sub-model gives the result of a status for doing maintain action but no action, no immediate action, and immediate action. Decision tree obtained from the operation of WEKA software can be seen in figure 4 below. Based on the resulted decision tree, we obtained 15 rules in decision making on status for action. The following are some examples of decision rules that are formed, among others:

- If the severity is not severe, and impact <3 hours, then the status of action is maintain but no action.
- If the severity is not severe, and the impact of 3-6 hours, and the percentage of KPI achievement is, then the status of action is not immediately, etc.

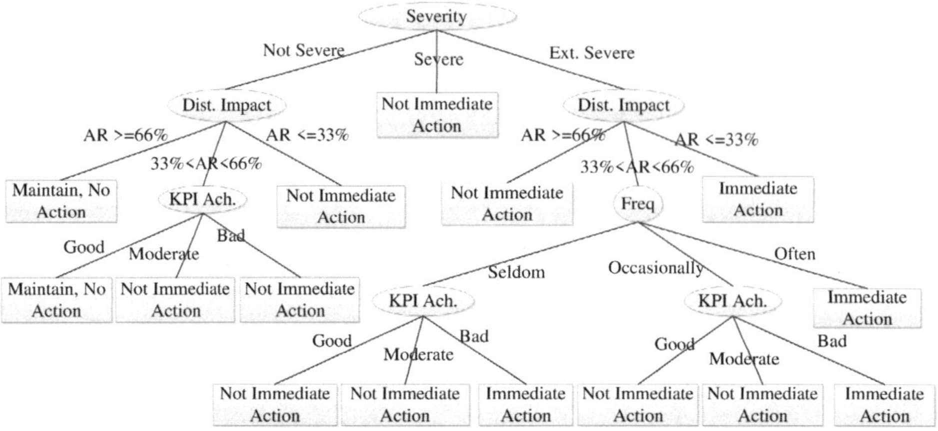

Figure 4. Status for the action decision tree

The action of disturbance control that is done based on class level in the sub-model of risk level of the disturbance and the sub-model of action status. If the level of risk of moderate disturbance and the status of action is action is not immediate, then the action is to transfer the process to another line or making PICA. If the high risk level of disturbance and the status of action is immediate action, then the action taken is the dismissal of production.

Table 2. Disturbance control action matrix

		Status for the action		
		Maintain no action	**Not immediate action**	**Immediate action**
Disturbance risk level	Low	Changing working shift	Changing level stock An addition working shift in a day	Creating PICA Meeting/ coordination
	Medium	Changing working shift Allocate MP from other line	Moving the process to other line Creating PICA	Meeting/coordination Stopping the line process
	High	Allocate MP from other line Changing level stock	Moving the process to other line Creating PICA	Stopping the line process

Table 3. Examples of disturbance control action

No	Disturbance	Keyword	Disturbance Classification				Dist. risk level	Status for the action	Recom. action
			Freq.	Severity	Dist.Impact	Achiev.KPI			
1	Empty WIP among process in a line	Empty WIP	Seldom	Severe	AR ≥ 66%	Good	Low	Maintain, no action needed	1
2	Delay caused by supplier	Supplier	Seldom	Not severe	33%<AR<66 %	Moderate	Low	Not immediate action	3,4
3	Coolant leakage	Down time machine	Seldom	Severe	AR ≥ 66%	Good	Low	Not immediate action	3,4
4	Leak test	Reject casing	Often	Not severe	AR ≥ 66%	Moderate	Low	Maintain, no action needed	1
5	Porous	Reject casting	Occasianally	Not severe	AR ≥ 66%	Moderate	Low	Maintain, no action needed	1
6	Customer complain and request a product based on the company's standard (affect the company's image)	Demand	Seldom	Severe	33%<AR<66 %	Moderate	Medium	Not immediate action	5,6

The control action in table 2 is a combination of the results of the two sub models. The action contained in the table in the form of numbers is an action that exists on the explanation of the action taken when the disturbance occurs. In table 3, there are examples of disturbances control action. In case of leaking tester trouble with low level of noise risk and maintain status, then action that need to be done is action number 1 that is change of work shift.

If continuous and repeated disturbance occurs, further action is required based on the preceding action. For example, if the disturbance of porous caused by reject casting with the increment of the number of occurrences each month, then the action previously done that changes shift work to cope with the disturbance can not be used again. The disturbance must be solved by making a change of action performed by the person responsible for carrying out the action. Table 4 shows disturbances response matrix that serves to determine what action to perform the implementing party. The preceeding action for leaking tester trouble is with the shift break changes to cope with the disorder. But if the disorder is constantly repeated, then the action is done by meeting with the suppliers and as the executor is the head of the production department.

Table 4. Disturbances response matrix

No	Disturbance	Cause	PIC	Further action
1	Delay caused by supplier	Trouble when delivering	Head of division or department	Changing level stock or meeting with supplier
2	Clamp pipe broken	Lifetime of the machine's component	Head of department	Build a team to control the machine's life time
3	Leak test	Casting process	Head of department	Meeting with supplier
4	Porous	Casting process	Head of department	Meeting with supplier
5	Misrun	Casting process	Head of department	Meeting with supplier

Sub Model of Safety Stock Determination

Determination of the safety stock value begins by calculating the percentage of the influence of the disturbance to the percentage of safety stock values that change every day based on the disturbance that occurred. The percentage value of disturbance effect is obtained from the difference of effect of each disturbance divided by the number of disturbances per day. The percentage value of safety stock this period is obtained by summing the percentage value of safety stock last period with the percentage of influence of interference. On the 1st, the percentage of safety stock last year was 58.15% which was obtained from the safety stock percentage of last period of May. Value is added with the percentage of disturbance that is 1% so the percentage of safety stock this period is 59.15%. Implementation of the control model for safety stock calculation requires that the percentage of safety stock is within the prescribed control limit of 60 - 80%.

Forecasting requirement of this period is planning of cylinder head request which for assembly section for one month is averaged. Safety stock requirement planning is the result of multiplication of percentage value of safety stock of this period with value of forecasting requirement. Actual usage is the number of units shipped to AE, and the remaining inventory is obtained by summing up the last period's inventory value forecasting and subtracting it from actual usage. For example, in the period of June,14,2016 the percentage of safety stock which is (78.72%) and forecast value of this period is 788 units, so the value of safety stock is 566 units .

The result of safety stock calculation by using sub controlling control model at table 5 shows the safety stock value of cylinder head component which is smaller than company policy. For June data, Safety stock based on company policy has a very large range, which is between 200 - 1398 units. This range is very fluctuating compared to the safety stock score based on model of 0 - 630 units. If there are many safety stocks provided by the company, it will affect the achievement of KPI cost due to stockpiling. The design of the controlling model of disturbance at PT XYZ has been based on the KPI of the company, resulting in the design of sub-model which is more effective in controlling production system disruption at PT XYZ associated with the non-achievement of KPI.

IOP Conf. Series: Materials Science and Engineering **277** (2017) 012020 doi:10.1088/1757-899X/277/1/012020

Table 5. Calculating safety stock using disturbance control model

	June,1, 2016	June,2, 2016	June,3, 2016	June,6, 2016	June,7, 2016	June,8, 2016	June,9, 2016	June,10, 2016	June,13, 2016	June,14 , 2016
Previous period stock	2109	2097	1985	2073	2061	2049	2002	1875	1913	1901
Forecasting for current period	788	788	788	788	788	788	788	788	788	788
Safety stock planning	473	481	489	495	501	507	517	528	559	566
Actual usage	800	900	700	800	800	835	915	750	800	800
Current period stock	2097	1985	2073	2061	2049	2002	1875	1913	1901	1889

4. Conclusion

The dominant disturbance in production process at PT XYZ as an automotive company comes from reject casting. The design of disturbance control model for the company consists of Sub Model of Disturbance Risk Level to know the level of risk of low, medium, or high disturbance, Sub Model Status done action to know whether the disorder requires immediate action or not or just monitoring, Sub Model of Disturbance Control Action to know the control action to be used and Sub Model of Safety Stock Calculation to know safety stock policy based on the disturbance that happened. Based on the results of analysis it is known that the range of safety stock based on disturbance control model is relatively more stable than the safety stock based on company policy.

5. References

[1] Koh, S C L, Simpson M, and Lin Y 2006 Uncertainty and Contingency Plans in ERP Controlled Manufacturing Environments. *J. of Enterprise Information Management* **19**. A6: 625-645.

[2] Marie I A, Eriyatno, Arkeman Y and Daihani D U 2011 Model Design of Adaptive Production Planning and Inventory Control in the Food Industry. *Asian Transactions on Engineering.* **01**(04).

[3] Marie I A, Eriyatno, Arkeman Y and Daihani D U 2014 Aplikasi Model Pengendalian Gangguan di PT NIC, Tbk. *Prosiding Seminar Nasional Mesin dan Industri IX.*

[4] Marie I A, Prasetyo FM and Azmi N 2015 Safety Stock Determination Based On Disturbance Control Model at PT DEE. *Int. Seminar on Industrial Engineering and Management Proceeding ISSN:1978-774.*

[5] Mula J, et.al. 2006 Models for Production Planning Under Uncertainty: *A Review. Int. J. Production Economics* **103**: 271–285

[6] Tersine R J 1994 *Principles of Inventory and Materials Management* 4th ed. (USA: Prentice Hall Inc.)

[7] Tan, Pang-Ning, et.al. 2006 *Introduction to Data Mining* 1st Edition (Boston: Pearson Education, Inc.)

Design improvement of automated gallon washing machine to minimize musculoskeletal disorders (MSDs) in CV Barokah Abadi using ergonomic function deployment (EFD) approach

Fakhriza Z[1], M Rahayu[1] and M Iqbal[1]

Telkom University, Bandung, Indonesia
[1]khalqihi.kharin@gmail.com, [2]mira.rahayu82@gmail.com

Abstract. In the production activity of Bottled Drinking Water (AMDK) in CV Barokah Abadi there is a gallon washing station. At the work station it involves three stages of activity such as washing and rinsing the outside of the gallon, spraying the inside of the gallon and rubbing the inside of the gallon which is done in a separate place. Distribution of Nordic Body Map (NBM) questionnaires showing employee complaints data at gallon washing stations where workers complained of pain in the right upper arm, right forearm and right wrist respectively 88% and workers also complained of pain in the waist and The right hand respectively by 81%. Ergonomic gallon washer is one way to minimize the risk of MSDs. The design begins with an ergonomic evaluation of the existing conditions and the concept of the initial design of the gallon washer. The evaluation is utilized for consideration of design improvements with the utilization of Ergonomic Function Deployment (EFD) in order for the product concept to conform to the ECSHE principle (Effective, Comfortable, Safe, Healthy and Efficient). The tool improvement design can minimize the risk of MSDs seen from the worker's posture while using an ergonomic washer.
Keywords: Gallon Washing, Nordic Body Map, Ergonomic Function Deployment

1. Introduction

CV Barokah Abadi is one of AMDK supplier company whose activities focus on drinking water production process of gallon packaging. Production activities CV Barokah Abadi began with mobilization gallon gross of trucks at loading bay into the washing gallon, washing gallon exterior with a detergent solution, rinsing the outside of gallons of clean water, spraying gallons inside the hose pressurized medium, scrubbing gallons inside by using a special machine, filling gallons of drinking water from the spring, closing the mouth gallon by gallon lid, sealing cap on freight gallon to gallon that has been sealed onto trucks at the loading bay for a re-marketed as can be seen in figure Figure 1. in Development of this ergonomic product, the process focused on the process of washing gallons in CV Barokah Abadi. In the existing condition of the washing section consists of four stages, which 70% of the process is still done manually, such as washing the outside of gallons of water reservoirs in an area of 2.25 m^2 by using a liquid detergent and a scouring sponge and brush to clean the outside of gallons of dirt, further gallons already transferred to the clean water tank containing clean water without detergent of the same size to rinse the outside of gallons of residual foam detergent. The next process is to clean the inside of the gallon, by spraying by high pressure water using a special hose, if there are stains that are still attached, then the inside of the gallon will be cleaned by being rubbed with a special machine and then sprayed back by using a hose.

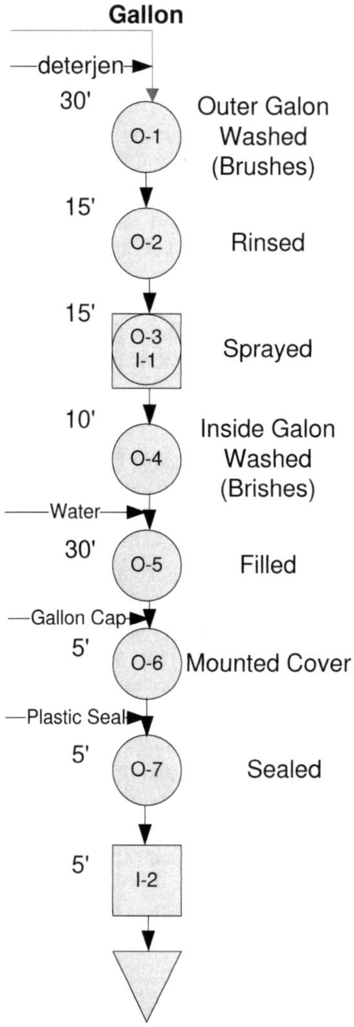

Figure 1. CV Barokah Abadi Production Flowchart

Based on the existing conditions, there are three body postures that need to be analyzed, such as body position while washing the outside of the gallon, spraying and brushing the inside of the gallon. The analysis was done by Rapid Upper Limb Assessment (RULA) method by using virtual environment based application on observation of mannequin Chinese population as can be seen in Table 1. Based on research of Anthropometry of the Singaporean and Indonesian that posture of Indonesian population have similarity with ethnic posture China with a 50% percentile rate [2].

Table 1. Rula score of existing process

Activity	RULA	Action Analysis
Washing and rinsing the exterior of Galon	7	High-risk posture, repairs should be made at this time
Spraying the inside of Galon	5	Medium-risk posture, repairs are done as soon as possible.
Rubbing the inside of the gallon	6	Medium-risk posture, repairs are done as soon as possible.

IOP Conf. Series: Materials Science and Engineering **277** (2017) 012021 doi:10.1088/1757-899X/277/1/012021

Based on the measurement of Rapid Upper Limb Assessment to the three washing activities, by looking at the score score can be concluded that the process of gallon washing in CV Barokah Abadi tends to involve awkward posture that will result Musculoskeletal Disorders (MSDs) to the workers when the work is done repeatedly. Based on preliminary studies conducted on four workers by distributing Nordic Body Map (NBM) questionnaires related to the health complaints experienced during the washing work, it was found that 100% of respondents had Musculoskeletal Disorders (MSDs) after working with six of the largest complaint ratings Can be seen in Table 2. The pain experienced by the worker is supported by the conditions and the work load is quite heavy, such as conditions where workers have to hold the gallon with one hand while the other hand spray the inside of the gallon, the condition supports muscle fatigue of the upper, lower, hand and wrist as well as The mouth of the gallon should be in a downward position so that the worker must hold the gallon continuously over and over.

Table 2. NBM questionaire result

Fatigue	Percentage
Back Pain	88%
Right-Upper Arm Pain	88%
Right-Forearm Pain	88%
Right Wrist Pain	88%
Waist Pain	81%

The pain experienced by the worker is supported by the conditions and the work load is quite heavy, such as conditions where workers have to hold the gallon with one hand while the other hand spray the inside of the gallon, the condition supports muscle fatigue of the upper, lower, hand and wrist as well as The mouth of the gallon should be in a downward position so that the worker must hold the gallon continuously over and over. The pain of Musculoskeletal Disorders on the job can certainly cause discomfort for the worker. It can trigger things like stress, decreased productivity and job satisfaction and even difficulty in outside work activities. This design focuses on the ergonomic evaluation of existing gallon washers that have been designed by previous researchers as can be seen in Figure 2. Basically, the concept of existing gallon washer tools is about how to make the washing work more static by combining three washing processes With different places from each other into one area only to shorten the material handling. In the existing gallon washing machine, the three processes are still done, in only one machine with ergonomic considerations.

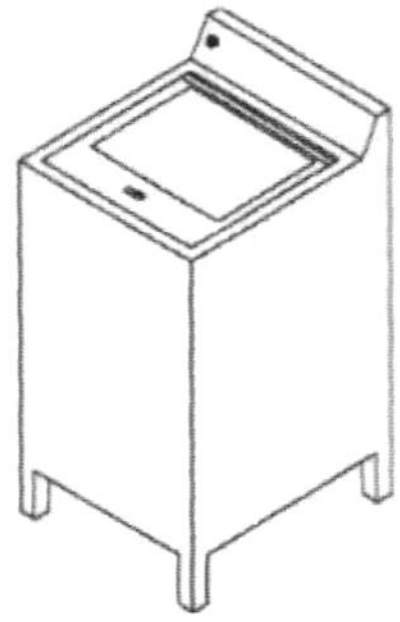

Figure 2. Initial Concept of Existing Gallon Washing Machine

To assess whether or not a design results, there are three underlying elements, namely functional, aesthetic and economic [3]. The functional basis relates to the ability of the product designed to perform certain functions in accordance with the planned including the ergonomic elements therein. The aesthetic element is related to the physical form of the product to be pleasing to the eye. The basis of the economy relates to the purchasing power of the society when the product is marketed. Based on the case studies presented, a tool is needed that can not only operate in accordance with the planned function, but also has an element of safety and comfort for workers [4]. For this purpose, an ergonomic evaluation of the existing proposed design is undertaken with the aim of ensuring that the proposed tool design simplifies leaching and reduces the risk of MSDs in workers compared to previous conditions in CV Barokah Abadi. Based on the importance of ergonomic evaluation on the product, the concept of galon washing machine development needs to be evaluated ergonomic test before the next concept is developed. In the design of the gallon washing equipment in CV Barokah Abadi an Ergonomic Function Deployment (EFD) approach was adopted. In this approach, product design concepts developed based on user desires coupled with user comfort factor became the main focus in the development of product designed specifications. Based on the background that has been presented, Ergonomic Function Deployment (EFD) method can be used in solving problems in designing gallon washing tools in CV Barokah Abadi. Using the EFD approach is an approach that refers to the principle of user ergonomics so as to produce a product design that is ENASE (Effective, Comfortable, Safe, Healthy and Efficient) in order to reduce the risk of Musculoskelatal Disorders (MSDs) in the user.

2. Methods

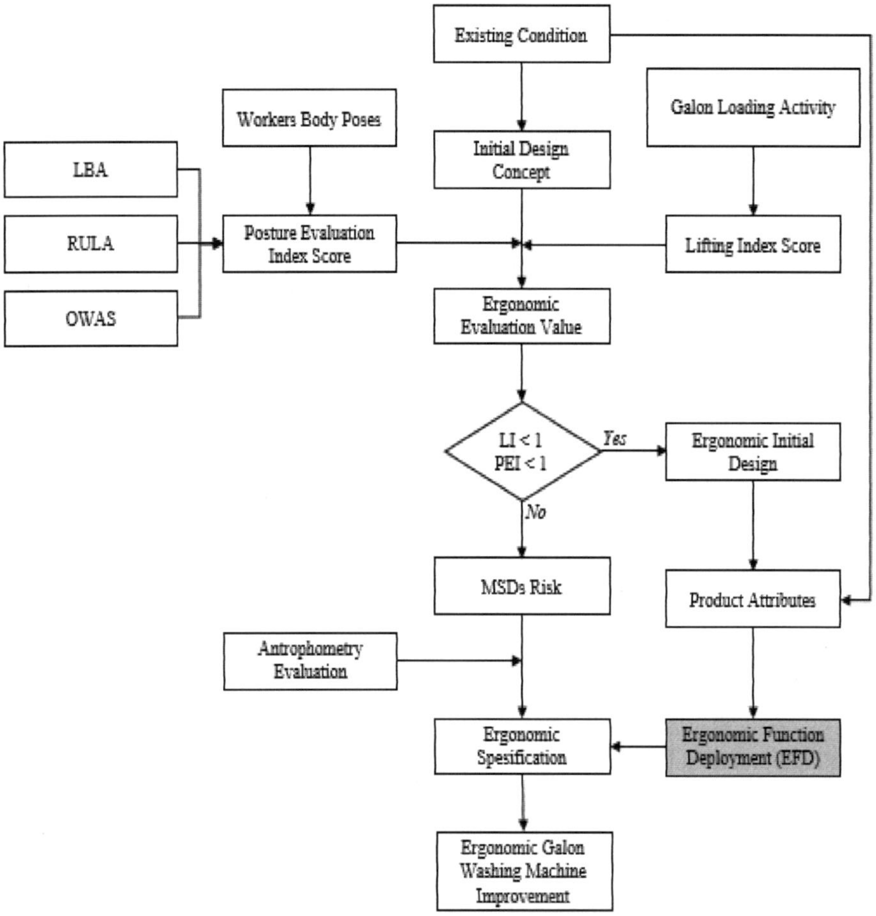

Figure 3. Conceptual model

Based on the conceptual model overview above, product development focuses on the initial concept of gallon washing tools proposed by previous researchers. This initial concept is a representation of the global voice of customer in terms of functional that has been processed by previous researchers. Existing working conditions are also examined as a reference in applying ergonomic aspects to the initial concept of gallon washer. Furthermore, further design is done by adding ergonomic aspects to the design concept by giving consideration of the comfort, safety and health of the workers. The addition of ergonomic aspects is done in the hope of reducing the risk of Musculoskeletal Disorders (MSDs) in workers. In adding ergonomic aspect, the first evaluation of initial draft concept is evaluated. The evaluation is an ergonomic evaluation that focuses on the evaluation aspect of Lifting Index (LI), Posture Evaluation Index (PEI) and anthropometric evaluation. The results of this evaluation are also used as a consideration in determining the specification of the concept of improvement of gallon washing machine design. The PEI score evaluation was performed to determine the magnitude of the risks that occur as a result of the worker's posture while performing the washing. The resulting PEI score determines the level of risk and the duration of the repair. Evaluation of manual removal is done to determine the risk of removal of gallons by workers. Anthropometric evaluation was conducted to determine the extent of the difference in the concept of the initial design dimension to the anthropometric data in question. Anthropometric evaluation will determine the dimensions of the product used to meet the aspect of ergonomics. After conducting an ergonomic evaluation of the initial draft concept, product development using the Ergonomic Function Deployment (EFD) approach is developed from Quality Function Deployment (QFD) to achieve the ergonomic proposal concept. In the EFD method, the design is based on the ergonomic aspect of the user in order that the product is designed in accordance with the principles of ENASE (Effective, Convenient, Safe, Healthy and Efficient). The applied ENASE principle is expected to help in reducing the risk of Musculoskeletal Disorders (MSDs) in gallon washing workers in CV Barokah Abadi. Proposed designs that have passed through the EFD flow will then be re-evaluated with ergonomics tools to compare aspects of ergonomics with the initial concept of the product. If the designed ergonomic aspect has been met, it is certain that the draft proposal concept is an ergonomic design concept or can be defined as an ergonomic design concept.

3. Result and Discussion

3.1. Work Process Identification of Initial Concept

To determine the evaluation of the proposed draft, it is necessary to first define the interaction that will occur between the workers and the draft concept. The definitions relate to the interaction of what happens between the worker and the parts of the tool for further observation. The following is the interaction of workers with tools can be seen in Table 3.

Table 3. Worker-machine interaction

Functional	Interaction	Parts of Worker-Related Tool
Temporarily Engine Shut Down	Workers pull the handle at the top of the door vertically.	• Lid Handle • Lid • Limit Switch
Engine Run	Workers pulled the door of the appliance until the wash chamber was closed.	• Lid • Limit Switch
Engine Stop	Workers pressed the emergency stop lever.	• Emergency Stop

3.2. Product Atributes

Research attributes were obtained by selecting and interpreting complaints and expectations of workers based on the results of observation and question and answer directly into a new concept for the gallon washer. Research attributes can also be obtained by studying the concept of similar products such as washing machines, baby taffles, block cake molding tools and gallon scrubbing machines that have been widely used by some refill drinking water companies. At this stage, the researcher can apply the concepts of those products that are relevant to the research along with the concepts obtained based on observation into research attributes that refer to the concept of ECSHE (Effective, Comfortable, Safe, Healthy and Efficient) as can be seen in Table 4 as the attribute data required in the design of an ergonomic gallon washer.

Table 4. Product atribute

Design Objective	Product Atributes
Effective	The dimensions of the washers are in accordance with the anthropometry of Indonesians.
	Washing process can be done more quickly.
	The display can give instructions to the worker.
Convenient	There is an additional handle on the door.
	The surface dimensions of the wash chamber support the worker's movement.
Safe	Emergency Stop is safe from risk.
	There is a sensor to identify the gallons in the wash chamber.
	There is a switch to turn off the device when the door is open.
	Tool safe for worker's posture
Healthy	There is a central drainage of the tool
	The use of the product does not cause injury
Efficient	Gallon can be seen by worker during washing process.
	Workers can easily do the washing
	The position of the emergency stop lever is reachable by the worker

To determine the evaluation of the proposed draft, it is necessary to first define the interaction that will occur between the workers and the draft concept. The definitions relate to the interaction of what happens between the worker and the parts of the tool for further observation.

3.3. House of Ergonomics (HoE)

In the Appendix can be seen the relationship between the matrix of attribute needs based on aspects of ergonomics with technical characteristics. Each relationship between the matrix has how big the relationship is and also there is also interaction relationship between technical characteristics that influence each other. HoE formation can be considered information to develop the product. Technical characteristics have priority which indicates higher priority, so if there is a trade-off in product development, the highest priority is not recommended to be modified in advance because it will affect product attributes and other technical characteristics.

3.4. Final Spesification

Based on the results of sorting the weight of product attributes, it was found that the 2nd alternative concept was chosen as the concept of design improvement on the gallon washing equipment in CV

IOP Conf. Series: Materials Science and Engineering **277** (2017) 012021 doi:10.1088/1757-899X/277/1/012021

Barokah Abadi. The following is the final specification of the selected ergonomic gallon washer (Table 5).

Table 5. Final spesification

No	Technical Characteristic	Unit	Target Value
1	Washing machine length	Centimeter	67,32
2	Washing machine width	Centimeter	84,56
3	Washing machine height	Centimeter	102,23
4	Washing time duration	Second	15
5	Display width	Centimeter	30,00
6	Display height	Centimeter	1,25
7	Display type	List	-
8	Handle width	Centimeter	51,63
9	Handle height from the ground	Centimeter	104,88
10	Handle diameter	Centimeter	3,00
11	Handle type	List	-
12	Wash chamber length	Centimeter	52,00
13	Wash chamber width	Centimeter	54,00
14	Maximum lid angle	Degrees	110,85
15	Emergency Stop type	List	Emergency Button
16	Emergency Stop location	List	Right Side
17	Censor type	List	Photoelectic
18	Switch type	List	Limit Switch
19	Posture Evaluation Index (PEI) score	Score	1,00
20	Drainage type	List	Convergent
21	Recommended Weight Limit (RWL) score	Kilogram	< 3,72
22	Lifting Index (LI) score	Score	< 1,00
23	Front and chamber lid material	List	Acrylic
24	Washing way	List	Handle Push
25	Emergency Stop distance from worker	Centimeter	59,76

The result of the implementation of ergonomic gear washing machine design specification which is the improvement of the design of the proposed design concept. As for the physical shape of the improvement of ergonomic gallon wash design can be seen in Figure 4.

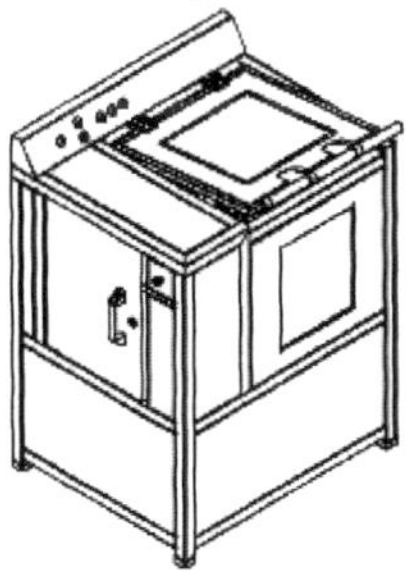

Figure 4. Ergonomic Design

4. Conclusion

Based on the results of data processing and analysis of food obtained some conclusions that refers to solve problems in this study. The conclusions are as follows:

1. Based on the purpose of research is to produce the concept of ergonomic tool improvement, the results of this study obtained the concept of ergonomic products based on product attributes that refers to the principle of ENASE (Effective, Safe, Healthy, Comfortable, Efficient)

2. This ergonomic concept of the ergonomic tea powder transporter is able to reduce the risk of MSDs by achieving some test parameters on the ergonomic evaluation performed.

5. References

[1] B. S. Perdana 2016 Desain gallon washing using ergonomic function deployment, *9th Proceeding of ISIEM*
[2] T. K. Chuan, M. Hartono and N. Kumar 2010 Anthropometry of The Singaporean and Indonesian Populations, *International Journal of Occupational Hygiene, Elsevier.*
[3] L. K. Wardani 2003 Evaluasi Ergonomi Dalam Perancangan Desain, *Dimensi Interior.*
[4] M. A. Heikal 2016 Perbaikan Rancangan Pengangkut Ceceran Pada Proses Sortasi Bubuk Teh di PT XYZ Untuk Mengurangi Risiko Musculoskeletal Disorders Menggunakan Pendekatan Ergonomic Function Deployment (EFD). *Tugas Akhir, Universitas Telkom.*

Mental workload measurement in operator control room using NASA-TLX

M. Sugarindra[1], M. R. Suryoputro[2], and A. I. Permana[3]

[1-3] Universitas Islam Indonesia, Jl. Kaliurang, Km, 14,5 DI. Yogyakarta, Indonesia
[1]sugarindra@uii.ac.id, [2] ragil.suryuputro @uii.ac.id, [3]permanaindra123@gmail.com

Abstract. The workload, encountered a combination of physical workload and mental workload, is a consequence of the activities for workers. Central control room is one department in the oil processing company, employees tasked with monitoring the processing unit for 24 hours nonstop with a combination of 3 shifts in 8 hours. NASA-TLX (NASA Task Load Index) is one of the subjective mental workload measurement using six factors, namely the Mental demand (MD), Physical demand (PD), Temporal demand (TD), Performance (OP), Effort (EF), frustration levels (FR). Measurement of a subjective mental workload most widely used because it has a high degree of validity. Based on the calculation of the mental workload, there at 5 units (DTU, NPU, HTU, DIST and OPS) at the control chamber (94; 83.33; 94.67; 81, 33 and 94.67 respectively) that categorize as very high mental workload. The high level of mental workload on the operator in the Central Control Room is a requirement to have high accuracy, alertness and can make decisions quickly
Keywords: Mental workload, NASA-TLX, Control Room

1. Introduction

Companies are always demanded to be able to achieve enhanced performance. One way is to optimize the more modern technology as well as to optimize the performance of employees. Optimizing the performance of employees will not be successful if the workload is beyond the capabilities that can be received. The workload is the difference between a worker's ability to work demands [1]. The workload can be interpreted by a number of job target to be achieved within certain time unit in normal conditions [2]. Measurement of the workload required to set a time for an employee who meets the requirements (of qualified) in carrying out certain work on the level of achievement that has been set. Human activity can be classified into physical work (muscle) and mental work (brain). Although it cannot be separated, but it still could be distinguished between work with physical domination and occupation by the dominance of mental activity [3]. If a worker's ability is higher than the demands of the job, will feel boredom. In the contrary, if the ability of worker were lower than the demands of the job, it would result in excessive fatigue [4]. Excessive fatigue will decrease performance and increase the level of work-related errors. Further increased employment will provide higher opportunities for fault accidents in the industry [5]. If the ability of workers is higher than the demands of work, will appear feeling bored. Conversely, if the ability of workers is lower than the demands of work, it will appear excessive fatigue [6]. Work fatigue will decrease performance and increase the level of work errors. Increased work errors will provide an opportunity for workplace accidents in the industry [7].

The object of this research is the company's oil processing unit that having a control chamber (Central Control Room). Central Control Room (CCR) is one of the departments in charge of monitoring the processing unit. Each worker is responsible for monitoring activity in most production processes in the company. Hours of work in this department is divided into three shifts (morning shift, afternoon shift, and night shift) with each shift consist of 8 hours. Workers in this room work with a computer for one work shift. Based on the problem formulated, this research focused on the measurement of mental workload associated with work activities and the time that workers needed to complete their tasks in accordance with the job description provided by the company.

Workload is a number of activities that require mental process or capability that must be completed within a certain period in the form of physical or psychological. NASA-TLX (NASA Task Load Index) is a subjective method for mental workload measurement. This method used to analyse the mental workload faced by workers who must perform various activities in their work which included six factors. They are Mental demand (MD), Physical demand (PD), Temporal demand (TD), Performance (OP), Effort (EF), and frustration levels (FR) [7]. Based on the calculation of the mental workload, the companies will know the condition of the worker workload.

2. Method

This study aimed to analyse the workload of operators in the main control room with descriptive and quantitative research by comparative design. Mental workload collected by method of NASA-TLX (National Aeronautics and Space Administration Task Load Index) developed by Sandra G. Hart of NASA-Ames Research Centre and Lowell E. Stave land from San Jose State University in 1981. This method used to analyse the mental workload faced by workers who must perform various activities in their work by six indicators. They are Mental demand (MD), Physical demand (PD), Temporal demand (TD), Performance (OP), Effort (EF), and frustration levels (FR). Steps measurements using NASA-TLX is as follows:

2.1. Weighting

NASA-TLX questionnaires given in pairwise comparisons. From this questionnaire counted the tally of each indicator that felt most influential. The amount of tally becomes weight for each indicator of mental load. The following table compares NASA TLX indicators.

Table 1. Comparison of Indicators

	MD	PD	TD	OP	EF	FR
MD						
PD						
TD						
OP						
EF						
FR						

2.2. Provision Rating

In this section the respondents are asked to rate the six mental workload indicators. The rating given is subjective depending on the mental workload felt by the respondent. The score is between 0 until 100. To obtain a NASA-TLX mental load score, the weights and ratings for each multiplied indicator are then summed and divided by 15 (the number of pairwise comparisons).

2.3. Calculating the value of the product

Obtained by multiplying the rating by the factor weight for each descriptor. Thus yielded 6 product values for 6 indicators (MD, PD, TD, OP, EF, and FR).

2.4. Workload calculated weight (WWL)

Sum all weighted workload of product

2.5. Calculated WWL Score
Calculate the average of weighted workload.

2.6. Score Interpretation
The score interpretation based on calculated WWL are low (0-9); medium (10-29); Rather high (30-49); High (50-79); very high (80-100).
The object of this research is the company's oil processing unit that having a control chamber (Central Control Room). Central Control Room (CCR) is one of the departments in charge of monitoring the processing unit. They work by shift for eight hours each day.

3. Results and Discussion

3.1. Data collections
3.1.1. Characteristics of respondents
Questionnaires were administered from respondents in 9 units: Distillation Treating Unit (DTU), Naphtha Processing Unit (NPU), Atmospheric Residue Hydro Demetalization Unit (AHU), Hydro treating unit (HTU), Residue Catalytic unit (RCU), Light End unit (LEU), Distribution (Dist.), Operations (OPS) and Olefins Conversion unit (OCU). Based on the questionnaire obtained from 9 respondents, the respondents can be grouped by age. From these results can be illustrate in the following table (Table 3):

Table 2. Characteristic respondent by age

Age (years old)	Sample (n)
25-30	2
31-35	5
≥ 36	2

3.1.2. Weighting indicator
NASA-TLX questionnaire is given in the form of pairwise comparisons. This questionnaire calculated from the number of tally from each indicator that felt most influential to the respondents. Total tally into weights for each indicator of mental workload. In this section, respondents asked to provide a rating of the six indicators (PD, MD, TD, OP, EF, and FR).
From this result, the data shows that OP and FR is the highest weight indicator based on the respondent in DTU and NPU unit (Table 3).

Table 3. Value Weighted Indicator

No.	Type Job	Indicator						Total
		MD	PD	TD	OP	EF	FR	
1	DTU	4	1	2	5	3	0	15
2	NPU	2	0	4	2	2	5	15
3	AHU	4	0	1	4	3	3	15
4	HTU	3	2	0	3	4	3	15
5	RCU	2	2	2	4	3	2	15
6	LEU	1	2	1	4	4	3	15
7	DIST	2	0	4	4	2	3	15
8	OPS	2	1	4	3	3	2	15
9	OCU	2	1	2	4	4	2	15

3.1.3. Provision rating

In this section, respondents were asked to give ratings against six indicators of mental workload. Assigned rating is subjective depending on the mental workload felt by respondents. Here is the data obtained for the rating value of 9 respondents (Table 4).

Table 4. Assigned Rating

No.	Type job	Indicator					
		MD	PD	TD	OP	EF	FR
1	DTU	100	80	80	100	90	10
2	NPU	80	40	90	70	70	90
3	AHU	80	30	60	80	70	70
4	HTU	100	60	80	100	100	100
5	RCU	50	50	60	60	60	50
6	LEU	70	80	70	80	70	60
7	DIST	70	50	80	100	60	80
8	OPS	90	80	100	100	100	80
9	OCU	60	40	60	80	60	40

3.1.4. Workload calculated weight (WWL) and calculate WWL score

From this result, the respondent show that in HTU and OPS unit have highest workload with score 94.67 and the lowest score is in the RCU unit with score 56 (Table 5).

Table 5. Value product, Workload Calculated Weight, WWL score and Score Interpretation

No.	Type job	Indicator						Total WWL	Score	Interpretation
		MD	PD	TD	OP	EF	FR			
1	DTU	400	80	160	500	270	0	1410	94.00	Very High
2	NPU	160	0	360	140	140	450	1250	83.33	Very High
3	AHU	320	0	60	320	210	210	1120	74.67	High
4	HTU	300	120	0	300	400	300	1420	94.67	Very High
5	RCU	100	100	120	240	180	100	840	56.00	High
6	LEU	70	160	70	320	280	180	1080	72.00	High
7	DIST	140	0	320	400	120	240	1220	81.33	Very High
8	OPS	180	80	400	300	300	160	1420	94.67	Very high
9	OCU	120	40	120	320	240	80	920	61.33	High

3.1.5. Score Interpretation

Score interpretation for this study are almost of the respondent shows the workload is very high (5 respondent) and high (4 respondent) (Table 5).

3.1.6. Statistical analysis between groups

The data of workload with 3.1.6. Statistical analysis between groups shows in Table 6. There are not significant differences between types of job with p value 0.889 ($p<0.005$). All the type job in control room have high and very high mental workload.

Table 6. Statistical analysis the workload between types of job by ANOVA

No	Type jobs	Mean ±SD
1	DTU	235±191.59
2	NPU	73.33±165.21
3	AHU	65±132.31
4	HTU	90±147.19
5	RCU	55±57.96
6	LEU	71.66±104.11
7	DIST	73.33±145.554
8	OPS	91.66±116.91
9	OCU	56.66±105.57
	P	0.889

3.2. Discussion

The height of the mental workload is found in the work involved [10]: The necessity to maintain a high level of vigilance over a certain period, the need to make decisions, the incidence of decreased concentration due to monotony and the lack of contact with other humans. The mental burden that occurs with the operator in the Central Control Room can occur because the job as an operator overseeing the course of production is required to have the thoroughness, high vigilance and demanded to be able to take decisions quickly. Based on the statistical test shows that there is no significant difference between job types. For job description of each unit is as follows:

- DTU is the primary processing that serves to separate the oil based on this boiling point. The primary job operator is to control the five main units in DTU and ensure distillation processing goes well. Mental workload in this unit is very high (WWL score 94).
- NPU aims to process and increase octane number. The primary job operator in this unit is controlling 3 main units. Mental workload in this unit is very high (WWL Score, 83.33).
- AHU processes Atmospheric Residue form Crude Distillation Unit (CDU) into a Demetallized Atmospheric Residue (DMAR) product prepared as a feedback for Residue Catalytic Cracker (RCC). Mental workload in this unit is high (WWL Score, 74.67), because this unit is secondary processing and have just a unit
- HTU is responsible for processing CDU output product, in this unit have 3 main unit. Mental workload in this unit is very high (WWL score, 94.67), because operator's duty to supervise 3 main units.
- RCU serves secondary processing to gain added value from residue. Mental workload in this unit is high (WWL score, 56), because this unit is secondary processing and not vital unit.
- Light End Unit (LEU) have 5 unit to controlling advanced processing from primary product. Processing naphtha to LPG, propylene and poligasoline. Mental workload in this unit is high (WWL score, 72).
- The process that occurs in the OCU Unit is to convert Ethylene into Propylene. This unit performs advanced processing to obtain the added value of the residue. Mental workload in this unit is high (WWL score, 61.33)
- Distribution is a unit have responsibility distribution energy needed during the production process. The operator's primary duty ensures that energy is channeled evenly to the unit in need Mental workload in this unit is very high (WWL score, 81.33)
- Operation unit is the main unit in supporting the production operation, as the supplier of electricity needs, steam, cooling water, gas N2, fuel gas system, and so on. There are 10 main units, the task of operator to ensure the supply demand have distribution to unit needed. Mental workload in this unit is very high (WWL score 94.6).

4. Conclusion

The mental workload in control room is divided into very high (5 units, DTU, NPU, HTU, DIST, and OPS) and high (4 units, AHU, RCU, LEU and OCU), but based on statistical data there is no significant difference of each unit. The high level of mental workload on the operator in the Central Control Room is a requirement to have high accuracy, alertness and can make decisions quickly

5. References

[1] Hancock P A and Meskhati N 1988 *Human Mental Workload. Amsterdam*, (North Holland: Elsevier Science Publisher B. V)

[2] Keputusan Menteri Pendayagunaan Aparatur Negara Nomor KEP/75/M.PAN/7/2004

[3] Dhania D R 2010 Pengaruh Stres Kerja, Beban kerja terhadap kepuasan kerja (Studi pada medical representative di kota kudus). (Kudus: Universitas Muria Kudus).

[4] Priscilla A D U 2008 Analisis Beban Kerja Mental Untuk Mengatahui Performansi Petugas Penjaga Pintu Perlintasan Kereta Api Di Stasiun-Stasuin Kodya Semarang. Semarang.

[5] Astuty M S, Caecillia S W, Yuniar 2014 Tingkat Beban Kerja Mental Masinis Berdasarkan NASA-TLX (Task Load Index) Di PT. KAI Daop. II Bandung. Teknik Industri Itenas, Vol.1, No.1, July 2013 *Reka Integra*, Jurnal Online Institut Teknologi Nasional (Bandung: Jurusan Teknik Industri, Institut Teknologi Nasional).

[6] Widayanti A, Jhonson A, et al. 2009. Pengukuran Kerja Mental Dalam Searching Task Dengan Metode Rating Scale Mental Effort (RSME). (Bandung: Teknik Industri ITB)

[7] Nurmianto E 2004 *Ergonomi: Konsep Dasar dan Aplikasinya* (Surabaya: Prima Printing)

[8] Hancock P A and Meshkati N 1988 *Human Mental Workload* (Elsevier).

[9] Hart S G and Staveland L E 1988 *Development of NASA-TLX (Task Load Index); Results of Emperical and theoretical research*. In P.A. Hancock and N. Meshaki (Eds.) Human Mental Workload (Amsterdam: North Holland Press)

[10] Grandjean E 1998 *Fitting the Task to the Man*, 4th ed. (London: Taylor & Francis Inc.)

Implementation of health and safety management system to reduce hazardous potential in PT.XYZ Indonesia

L Widodo[1], Adianto[1] and D I Sartika[1]

[1]Industrial Engineering Department, Tarumanagara University, Indonesia
lamtow@ft.untar.ac.id

Abstract. PT. XYZ is a large automotive manufacturing company that manufacture, assemble as well as a car exporter. The other products are spare parts, jig and dies. PT. XYZ has long been implementing the Occupational Safety and Health Management System (OSHMS) to reduce the potential hazards that cause work accidents. However, this does not mean that OSHMS that has been implemented does not need to be upgraded and improved. This is due to the potential danger caused by work is quite high. This research was conducted in Sunter 2 Plant where its production activities have a high level of potential hazard. Based on Hazard Identification risk assessment, Risk Assessment, and Risk Control (HIRARC) found 10 potential hazards in Plant Stamping Production, consisting of 4 very high risk potential hazards (E), 5 high risk potential hazards (H), and 1 moderate risk potential hazard (M). While in Plant Casting Production found 22 potential hazards findings consist of 7 very high risk potential hazards (E), 12 high risk potential hazards (H), and 3 medium risk potential hazards (M). Based on the result of Fault Tree Analysis (FTA), the main priority is the high risk potential hazards (H) and very high risk potential hazards (E). The proposed improvement are to make the visual display of the importance of always using the correct Personal Protective Equipment (PPE), establishing good working procedures, conducting OSH training for workers on a regular basis, and continuing to conduct safety campaigns.
Keywords: Potential Hazard, OSHMS, HIRARC, FTA

1. Introduction

Today, the activity of a company is increasingly widespread, thus the greater the chance of potential hazards. In every production process, there are various potential hazards that can threaten workers' safety and health. Along with the advancement of knowledge and technology, the greater the challenge that must be faced by a manufacturing company, especially to overcome the problems of Occupational Safety and Health.

Division of stamping production and casting production is a division at PT. XYZ, where the production process at the plant has different potential hazard levels. The potential danger posed in these two divisions is very large, this is because plant stamping production is very closely related to press and dies machines weighing a few tons and also the noise level reaches 109 dB. Similarly, plant casting production, where most of the activity in the working environment is closely related to the temperature conditions that reach 700°C. Based on these considerations, a good safety and health management system is needed to ensure that potential hazards in the area can be minimized.

The aim of the research done is to determine the identification of potential hazards at Stamping Division Production and Casting Production at PT. XYZ, knowing the risk assessment of the potential hazard identification results, as well as finding the root cause of why it can occur.

Based on the Manpower Minister Regulation of the Republic of Indonesia (Permenaker RI) No.5 of 1996, the Occupational Safety and Health Management System is part of the overall management system which includes the organizational structure, planning, responsibilities, implementation, procedures, processes and resources needed for the development, implementation, achievement, review and maintenance of occupational safety and health policies for the creation of safe, efficient and productive workplaces [1][2]. Hazard Identification, Risk Assessment, and Risk Control (HIRARC) are key elements in the safety and health management system that are directly related to hazard prevention and control [3][4][5]. According to the Occupational Health and Safety Assessment Series (OHSAS): 18001 (2007), HIRARC must be conducted throughout the organization to determine the organization's activities that contain potential hazards and have a serious impact on occupational safety and health [6]. HIRARC is one of the potential hazard identification methods with risk assessment as one of the key points to implement the Occupational Safety and Health Management System. The purpose of HIRARC is to identify potential hazards in a company to assess the probability of an accident or financial loss. According to AS / NZS 4360 (2004), the HIRARC method used for risk assessment can be seen in Table 1-Severity and Table 2 - Possible or Likelihood Rates [7].

Table 1. Severity Level

Level	Designation	Description
1	Insignificant	Events do not cause harm or injury
2	Minor	Inflict minor injuries, minor losses, and do not have a serious impact
3	Moderate	Severe injuries and were hospitalized, huge financial losses
4	Major	Cause severe injury, major financial loss, serious impact on production
5	Catastrophic	Resulting in dead and big losses until the cessation of all activities

Table 2. Likelihood Level

Level	Designation	Description
A	Almost Certain	May occur at any time under normal conditions
B	Likely	Occur several times within a certain period
C	Moderate	Risks can occur but not often
D	Unlikely	Sometimes it happens
E	Rare	May occur under certain circumstances

Based on the standard OHSAS 18001 (2007), Fault Tree Analysis (FTA) is a technique used to identify risks that contribute to accidents / losses. FTA is done with a top down approach, which begins with the assumption of failure or loss of a top event, then details the causes of a top event up to a basic failure [3]. Fault Tree Analysis is an effective method of finding the subject matter, as it ensures that an undesirable event or loss is caused not only to a point of failure. Fault Tree Analysis identifies the relationship between causal factors and is shown in the form of an error tree, which involves the use of simple logic gates.

2. Methods

Research begins with the identification of problems, obtained from field studies supported by literature studies. Because of the extent of the problem, the problem under study is limited to covering OSH management in one plant unit, namely Plant 2 Sunter. The next stage is data collection covering the potential hazards and implementation of OSHMS. Furthermore, a potential hazard analysis with Fault Tree Analysis (FTA) was conducted. The results concluded are taken into consideration to formulate the proposed improvement, so it is expected to reduce the risk of accident and occupational safety.

3. Results and discussion

3.1. Hazard Identification, Risk Assessment and Risk Control (HIRARC)
At this stage, hazard identification of potential hazards has been identified. Then measured risk based on severity (severity) and likelihood (level of possibility). Risk assessment provides scores of any potential hazard findings and calculates the Risk Priority Index (RPI) level for each potential hazard. Potential occupational hazards in the area of stamping production plant by using HIRARC method identified 10 potential hazards. Risk assessment on plant stamping production can be seen in Table 3. Then, the risk category is made on the risk matrix map based on the RPI value obtained from the risk assessment stage. This risk mapping indicates the position of the RPI in each potential hazard finding to find out which activities need action, based on high risk to extreme risk. Matrix analysis on plant stamping production can be seen in Table 4.

Table 3. Risk Assessment on Stamping Production Plant

No	Location	Hazards	Picture	Effects	Severity	Likelihood	RPI	Risk Category
1	Kaizen Stamping	Working at altitude does not use safety guard		Operator dropped from a height	3	C	3C	H
2	Line Z Stamping	Operator is setting the material to the die		Operator can be slashed bymaterial	2	C	2C	M
3	Line Z Stamping	Operator is fixing the die inside the stamping machine area		Operator can be hitted by die	5	B	5B	E
4	Line Z Stamping	Operator is installing a cushion pin		Operator may have a back injury	3	B	3B	H
5	Line I Stamping	Operator is operating crane		Operator can be hitted by die	5	B	5B	E
6	Line I Stamping	Operator is doing press process		Operator's hand may be pinched by die	4	B	4B	E

IOP Conf. Series: Materials Science and Engineering **277** (2017) 012023 doi:10.1088/1757-899X/277/1/012023

Table 3. Risk Assessment on Stamping Production Plant (cont.)

No	Location	Hazards	Picture	Effects	Severity	Likeli-hood	RPI	Risk Category
7	Line C Stamping	Forklift operator A puts the die, and Operator B places a mat for the die		Operator B may be strucked by die / hitted by forklift	4	B	4B	E
8	Line C Stamping	Operator is not on track to cross		Operator may get hitted by a forklift	4	D	4D	H
9	Scrap Conveyor	The noise level in the area reached 109 db		Hearing function will decrease (because noise level above 80 dB)	4	A	4A	E
10	Line H Stamping	Operator is moving the material		Operators who are on the green line may be hitted by a forklift	4	D	4D	H

Table 4. Matrix Analysis on Stamping Production Plant

	Consequence				
Likelihood	Insignificant 1	Minor 2	Moderate 3	Major 4	Catastrophic 5
A (Almost certain)	H(9)	H	E	E	E
B (Likely)	M	H	H (4)	E (6,7)	E (3,5)
C (Possible)	L	M (2)	H (1)	E	E
D (Unlikely)	L	L	M	H (8,10)	E
E (Rare)	L	L	M	H	H

In addition to plant stamping production, risk assessment of potential hazards is also done in the area of plant casting production using the HIRARC method. The results are identified as 22 potential hazards and can be seen in Table 5.

Table 5. Risk Assessment on Casting Production Plant

No	Location	Hazards	Effects	Severity	Likelihood	RPI	Risk Category
1	Line Melting Casting	Setting ladle to dolly burner using forklift	The operator may be hit by heavy objects	5	B	5B	E

IOP Conf. Series: Materials Science and Engineering **277** (2017) 012023 doi:10.1088/1757-899X/277/1/012023

Table 5. Risk Assessment on Casting Production Plant (cont.)

No	Location	Hazards	Effects	Severity	Likelihood	RPI	Risk Category
2	Line Melting Casting	Put ingots into the furnace	Operator can be exposed to heat materials	3	B	3B	H
3	Line Melting Casting	Forklift enter to pouring work area	Operator may be hitted by forklift	4	D	4D	H
4	Line Melting Casting	The liquid from furnace 6 is poured into furnace 3 through ladle transfer	Operator can be exposed to heat materials	3	B	3B	H
5	Line Melting Casting	Sprinkle the husks into the walls of the furnace (full open robuta)	Operator can be exposed to heat materials	3	C	3C	H
6	Line Melting Casting	Other operators passing forklift unloading area	Operator may be hitted by forklift	4	D	4D	H
7	Line Melting Casting	Push the dolly material into the front of the furnace across the crane path	The operator may be hit by heavy objects	5	B	5B	E
8	Line Melting Casting	Reversing the slag box	Operators may be knocked by objects for himself	2	C	2C	M
9	Line Melting Casting	Take the liquid from the furnace	Operators may be contactedwith hot material	3	B	3B	H
10	Line Melting Casting	Check the temperature of the furnace fluid	Operator can be exposed to heat materials	3	B	3B	H
11	Line Melting Casting	Sprinkle kagaraito evenly over the furnace liquid	Operator can be exposed to heat materials	3	B	3B	H
12	Line Melting Casting	Clean up the remaining charging around the furnace	Operator can be exposed to heat materials	4	B	4B	E
13	Line Melting Casting	Bring the ladle from the furnace to the pouring area	Operators can be wedged by machine	4	C	4C	E
14	Line Melting Casting	Fix the furnace and pull the hose and cross the crane path	Operators may fall because there are objects blocking	3	C	3C	H
15	Line Melting Casting	Moving the pallet forks of a fork into a wooden pallet	Operator may be hitted by forklift	4	C	4C	E
16	Line RCS Casting	When the hopper is up / down can overwrite other team members who pass	Operators can hitted by hopper	4	C	4C	E

Table 5. Risk Assessment on Casting Production Plant (cont.)

No	Location	Hazards	Effects	Severity	Likelihood	RPI	Risk Category
17	Line RCS Casting	When sacks up / down can be fall other team members who pass.	Operators can hitte by the sack	3	C	3C	H
18	Line Molding Casting	Operator standing on H-Beam with a height of ± 3meter	Operators can fall from a height> 2meter	3	D	3D	M
19	Line Melting Casting	Operator is too close to the hopper door	Operators can fall from a height> 2meter	3	C	3C	H
20	Line Melting Casting	Lifting the sand from the pit using a hoist	Operators can fall from a height> 2meter	3	C	3C	H
21	Line Core Casting	Lifting die from floor to dolly transfer	Operator may hitted by die	5	B	5B	E
22	Line Core Casting	Delivery sand to zone B with forklift, other operators working in forklift path	Operator may hitted by forklift	4	D	4D	H

Then, based on the value of RPI is made a risk matrix map. This risk mapping shows the position of the RPI in each potential hazard finding. This is to find out what activities need to be taken to reduce the risk of work accident. Matrix analysis on casting production plant can be seen in Table 6.

Table 6. Matrix Analysis on Casting Production Plant

Likelihood	Consequence Insignificant 1	Minor 2	Moderate 3	Major 4	Catastrophic 5
A (Almost certain)	H	H	E	E	E
B (Likely)	M	H	H (2,4,9,10,11)	E (12)	E (1,7,21)
C (Possible)	L	M (8)	H (5,14,17,19,20)	E (13,15,16)	E
D (Unlikely)	L	L	M (18)	H (3,6,22)	E
E (Rare)	L	L	M	H	H

3.2. Fault Tree Analysis (FTA)

The next stage is to find the cause of the incident related to the highest risk and extreme risk by using fault tree analysis (FTA). The peak events identified in plant stamping production are 9 occurrences of high risk and extreme risk potential hazards. In the casting production plant, there are 20 occurrences of high risk and extreme risk potential hazards. Here are the results of FTA analysis of some potential hazard causes in stamping production plant and casting production plant which can be seen in Figure 1 and Figure 2.

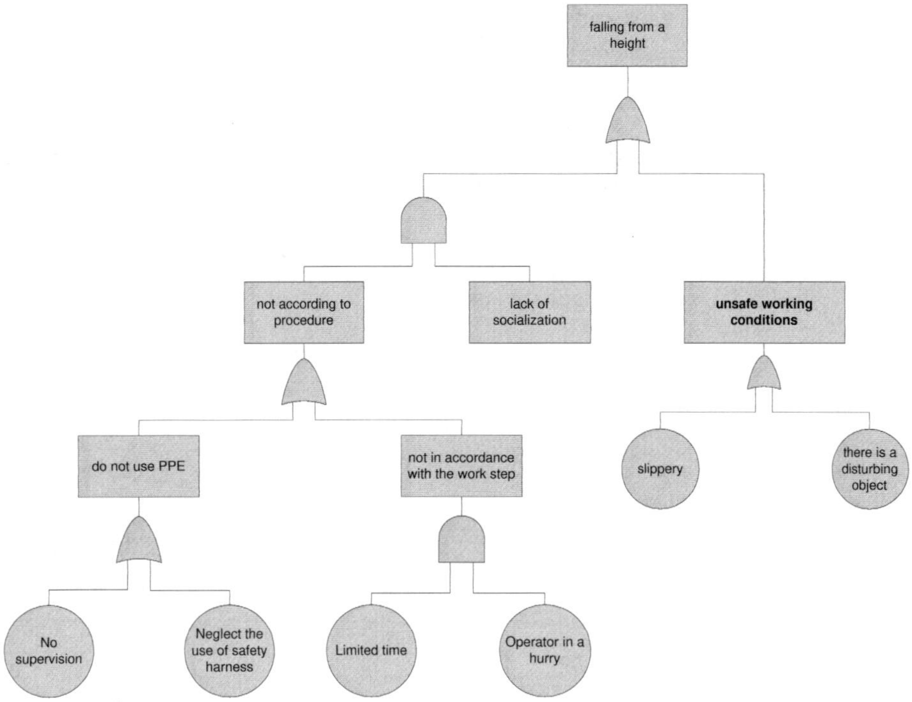

Figure 1. Causes of Potential Hazards of Falling from Height in Stamping Production Plant

3.3. Recommendation for Improvement

Based on the results of risk assessment data processing using HIRARC method, can be grouped several proposed precautions based on the assessment of the high risk and extreme risk hazards. Recommendations for improvement on stamping production plant and casting production plant can be seen in Table 7 and Table 8.

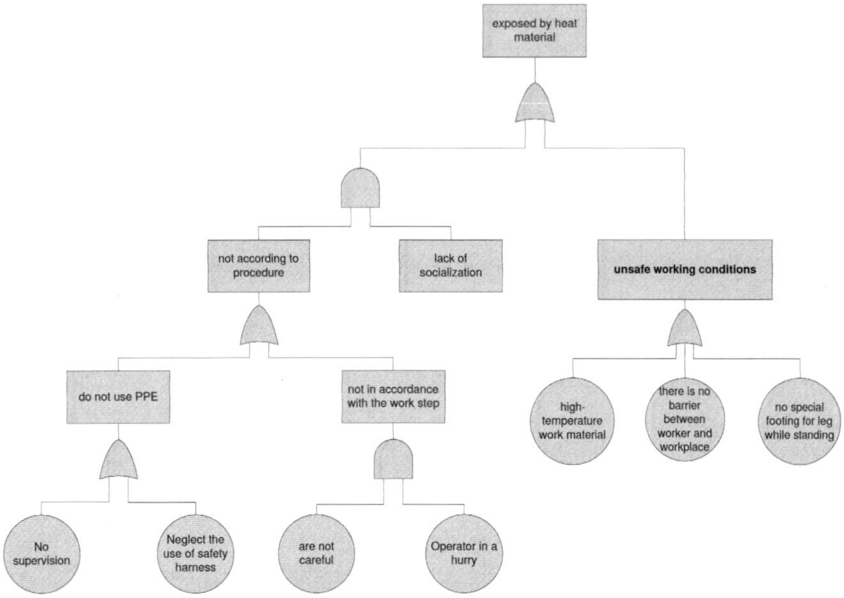

Figure 2. Causes of Potential Hazards of Hot Material Affected in Plant Casting Production

Table 7. Recommendation for Improvement on Stamping Production Plant

No.	Hazard Potential	Cause	Consequence	Action
1	Working at an elevation does not use complete PPE	1. Not using PPE 2. Low awareness and knowledge of occupational safety and health 3. No supervision from management 4. There are objects that interfere / obstruct 5. Slippery conditions 6. Work time in a hurry	Falling from a height	1. Create a visual display so that workers always wear PPE (safety harness when working at altitude) 2. Create clear and correct working procedures 3. Perform OSHMS training to workers on a regular basis
2	Operator is installing cushion pin	1. Cushion is too heavy 2. There is no cushion lifting tool, so it must be done manually 3. No supervision from management	Back injuries	Designing a system of common use cushion, eliminating the need for disassembly cushion

Table 8. Recommendation for Improvement on Casting Production Plant

No.	Hazard Potential	Cause	Consequence	Action
1	Setting ladle to dolly burner using forklift	1. No supervision from management 2. There are objects that interfere / obstruct 3. Slippery conditions 4. No boundaries	Crushed by heavy objects	1. Create clear and correct working procedures 2. Create a barrier 3. Routine supervision 4. Perform OSHMS training to workers on a regular basis
2	Put ingots into the furnace	5. Not using PPE completely 6. No supervision from management 7. No boundaries	Crushed by very hot material	5. Create clear and correct working procedures 6. Create a barrier 7. Routine supervision 8. Perform OSHMS training to workers on a regular basis 9. Using PPE completely

4. Conslussion

Based on the results of data processing and analysis that has been done, can be drawn conclusion as follows:

1. Identification of the greatest potential hazard is found in the casting production plant that is 22 potential hazards.
2. Based on Hazard Identification risk assessment, Risk Assessment, and Risk Control (HIRARC) found 10 potential hazards in Plant Stamping Production, consisting of 4 very high risk potential hazards (E), 5 high risk potential hazards (H), and 1 moderate risk potential hazard (M). While in Plant Casting Production found 22 potential hazards findings consist of 7 very high risk potential hazards (E), 12 high risk potential hazards (H), and 3 medium risk potential hazards (M).
3. The proposed improvement are to make the visual display of the importance of always using the correct Personal Protective Equipment (PPE), establishing good working procedures, conducting OSH training for workers on a regular basis, and continuing to conduct safety campaigns.

IOP Conf. Series: Materials Science and Engineering **277** (2017) 012023 doi:10.1088/1757-899X/277/1/012023

5. References

[1] Peraturan Menteri Tenaga Kerja Nomor. Per.05/MEN/1996, *Manajemen Keselamatan dan Kesehatan Kerja* (Jakarta: Kemenakertrans).

[2] PNK3 Ditjen Binwasnaker Kemenakertrans 2010 *Pedoman Praktis Penilaian Risiko Lingkungan Kerja* (Jakarta: Kemenakertrans).

[3] Ramli S 2010 *Sistem Manajemen Keselamatan dan Kesehatan Kerja OHSAS 18001* (Jakarta: PT. Dian Rakyat).

[4] Fajar N and Puspitasari D 2014 Analisis dan Usulan Perbaikan Sistem Manajemen Keselamatan dan kesehatan Kerja (SMK3) pada PT. Fumira Semarang, *Jurnal Jurusan Teknik Industri Undip*, Semarang.

[5] Suardi R 2007 *Sistem Manajemen Keselamatan dan Kesehatan Kerja*, (Jakarta: PPM)

[6] OHSAS 18001 2007 *Occupational Health and Safety Assessment Series*, OH & S Safety Management System Requirements.

[7] Australian Standard/New Zealand Standard 4360 2004, *Risk Management Guidelines* Sydney.

The importance of production standard operating procedure in a family business company

C Hongdiyanto

Faculty of Economic, Ciputra University, Surabaya, Indonesia
charly@ciputra.ac.id

Abstract. Plastic industry is a growing sector, therefore UD X which engage in this business has a great potential to grow as well. The problem faced by this family business company is that no standard operating procedure is used and it lead to problem in the quality and quantity produced. This research is aim to create a production standard operating procedure for UD X. Semi-structure interview is used to gather information from respondent to help writer create the SOP. There are four SOP's created, namely: classifying SOP, sorting SOP, milling SOP and packing SOP. Having SOP will improve the effectiveness of production because employees already know how to work in each stages of production process.
Keywords: family business, SOP, plastic industry

1. Introduction

1.1 Background of the study

Plastic is a product that we use everyday. In addition, because of its strong, lightweight, flexible, rustproof and not easily broken characteristic make plastic often used in the society. Plastic started to be used as daily appliances. In addition to being supported by the food and beverage industry, demand also comes from other sectors such as infrastructure, automotive and agriculture. Due to this, the prospect of plastic market in the future is bright and very promising to run.

The author used UD X as the object of research which is a business engaged in plastic manufacturing. UD X is a family business company located in Ambon, engaged in manufacturing related to the processing, milling and production of plastic waste. Starting from the process of sorting raw materials received in accordance with its type, then discarding parts that are not used (such as labels on plastic bottles, or separating the neck & close lid of the bottle because it is a different materials from plastic bottle). Before the milling, weighing will be done in advance to know the production capacity of each employee and as a proof of their memorandum to receive payment. The next process is the milling and washing, then it is packed into a sack and arranged.

The problem faced is the fluctuating quantity of production whereas the amount of raw materials entering the production warehouse is stable. Other than that, because the majority of the production process that occurs is done by human hands, sometimes the quality produced do not have the quality control or the right standard because the assessment of each person is different.

Quality Control is very important and required by a manufacturing company to stabilize the quality of the products they have or produce. As a manufacturing company, it is very important for UD X to keep the quality of the products sold and to stabilize the quantity of production. The need to have knowledge about the proper production system will drive the company towards business success and can later be used as a working standard within the company. Ekotama [1] mention that a business that has a good system is very easy to develop. It also states that business that has a system is always able to produce good quality and consistent so it becomes more competitive to win the competition when compared to other businesses that do not have a system. Tambunan [2] states that the system is a combination of interconnected elements on a regular basis to achieve certain objectives agreed by all parts of the system so that if one part of the system is gone, it will have an impact on the decline in performance and can not be called as one system.

Prior to this research, UD X did not have a production SOP and for the future, testing and necessary data collection will be done to make production SOP. According to Tambunan [2], Standard Operating Procedure (SOP) is basically a guide that contains standard operating procedures that exist within a company that is used to ensure that all work processes are carried out effectively, efficiently, consistently, standardized and systematic. SOP is closely related with manufacturing companies, because without a good and proper SOP, manufacturing companies will not be able to grow, compete and minimize fraud that may occur internally and externally. It shows that SOP is very important for the company.

Due to it, it is important to make SOP for UD X which will be used as a reference in the future so that employees will not have difficulties in carrying out the production process. Another impact of the making of SOP is to facilitate the company to monitor the quality of output, because by using SOP as a work guide of course the employees will work with production process steps that are always almost the same as the guide (SOP) provided so it will be easy to control the quality of output produced. No less important is to have a company SOP to save time and cost, it is said to save time because employees do not need to be confused or hesitant in doing their work because they already have a guide. And cost-effective because training every time there are new employees who joined into the company is not necessary. This research was conducted in order to know the proper production system for UD X and entitled "The Importance of Production Standard Operating Procedure in a Family Business Company"

1.2 Literature review

Based on Murtie [3], production is any activity in creating and adding the usefulness of an item that requires production factors, and production management is the overall management of that production activities. Fahmi [4] states that production management is a science that discusses comprehensively how the company's production management using the knowledge and art that can manage people to achieve a desired production result. The production function is to conduct the activity of converting and managing the production resources into a product, Haming and Nurnajamuddin [5].

According to Ekotama [1], SOP is a system designed to facilitate, tidy up and discipline work. This system is a sequence of work process from start to finish with details so that people who read it can clearly understand its meaning and purpose. SOP that originate from techniques, steps and activities that enable a job to be done well in the past, is expected to help for the moment so as to facilitate someone (both old and new people) to carry out their work if the steps or activities were written or documented as SOP, Thompson and Tracy [6]. There are several stages in the making of SOP, starting from the establishment of a SOP special team, support provided from the organization to this matter, making the plan of team task allocation, and determine the target. Finding and processing the necessary information and identifying existing alternatives to solve them. Analyze and choose the best alternative. Write SOP, review and apply SOP to be tested, rated, evaluated and implemented.

According to Kotler and Armstrong in Codita [7] quality is the ability of a product to meet consumer expectations and reach the level of customer satisfaction. According to Kotler and Armstrong in Horner and Swarbrooke [8] the product is something that can be offered / marketed to attract attention, takeover or consumption that may satisfy the wants or needs. It includes objects, people, services, places, organizations, and ideas. Product quality is a set of specifications that are professionally arranged and recognized by the society. It can be assessed from the features of the product, the sequence of a series

IOP Conf. Series: Materials Science and Engineering **277** (2017) 012024 doi:10.1088/1757-899X/277/1/012024

of production processes, and a re-check by the producers who know the correct stages. Klopic, Kuipers and Hocquette [9].

It is important to control the products produced by the company because the products with unstable quality will certainly make the consumers think twice to keep buying the products being offered. According Sunarya [10], to assess a product whether it is in accordance with the requirement / standard is to conduct quality inspection with product testing. This is where the important role of Quality Control can be seen. Quality Control is the proof that the production process can be done and operated well, without doing the inspection [11].

2. Method

The object of research is UD X located in Ambon, while the subject of research is informants who provide information about the situation and have a lot of experience in the plastic industry that are:

1. Two people from outsider who work in manufacturing companies that have a proper production system or have SOP.
2. One person from the company that is the owner (CEO) of UD X, so that he can assist in the formulation and preparation of SOP that will be done.
3. One consultant who has experience in plastic segment and production.

The author used purposive sampling technique in selecting samples. Purposive sampling is chosen because the sampling is based on certain considerations [12]. In this research, the observation method is done by visiting the manufacturing companies, studying the running production SOP [13]. Observation is chosen to know first-hand the operational activities starting from the production procedure of a product to the quality inspection to achieve quality standardized by the company [12]. Observation is done at the production site, after which internal observation is conducted to collect production data. For the interview, method used is semi structure interview. Semi structure Interview is a conversation directed to dig a predetermined topic that will aim to deepen the information obtained from the parties being interviewed. This is done to explore the issues deeper from the sample [12]. Especially for those who run the manual production SOP and understand how to maintain the consistency of quality and quantity of its products. Interview is conducted to get information about SOP that can be used as a reference for this research. Data from observation and interview is collected into one (photos, writings, or observations) from informants to be made into event records [12]. Member check is used to check the validity and reliability. Therefore, this research is mostly relied on expert opinion.

3. Result and Discussion

The production process starts from the moment the goods are derived from the L300 car and distributed to each worker according to the division of the type of plastic that has been done. After being distributed, the workers will do the process of classifying and sorting before the raw materials are deemed ready to be grinded.

Classifying means that the raw materials that come are mixed in one kind of sack and will be sorted according to its type. And then sorting is the process after the materials are done being classified, so the raw materials will be processed again, separated by type and color. After that, for some types of plastic, the cleaning the labels, stickers, or parts that still become one part but not a plastic type will be done. In addition, the sorting process also includes cutting up the plastic parts to be smaller if it cannot go directly into the head / mouth of the milling machine. After going through the process of sorting the raw material is said to be ready to be milled.

Goods are weighed first before being transported up near the mouth / head of the milling machine. After that there is one person who is tasked to spill the materials into the milling machine. Soap and caustic soda are inserted into the funnel to become one with a water spray, then the result of the grill will come out below in the first basin. There are two people on duty below, each in the first and second basin. The result of the first basin will be washed, filtered and then moved to the second basin or rinse basin with a basket in order for the result of the mill to be cleaner, then from the second basin will be washed, lifted, filtered and transferred into the sack. After the filled sack is full, it will be stitched with needle and raffia rope to not be spilled when being moved. After stitching, it will be arranged / stacked until all the same kind of mills are milled or at least the mills above is done, the purpose of this is for the water located in the sack to be drained. Then it will be weighed and moved to the drying place as well as the accumulation of ready-to-sell production materials.

For the drying process using the sun only takes one to two days depending on weather factors. The results can be said to be of high quality for the color categories as long as the production result is clean and has no dirt that sticks to it like soil, sand, glue, etc. Then for the white or transparent category, it is the same with the color categories but with the addition that the white color or transparent do not fade. To maintain the quality, the company checks on the milling process, so when it comes to the second basin of the mill and it is still not clean it will be added more soap and be milled again. Based on the interviews and observations obtained from both research subjects, their company already has SOP, but the SOP is a simple SOP from both mills factory. In addition, the production process of both companies is the same and still being done until now.

In addition to benchmarking on the production flow of two similar companies, from interviews with consultants it is found that the early stages of the production process when raw materials come is the same which is seen the goods first, classified what is necessary and what does not need to be, means that if the type of plastic is already the same it will continue directly to sorting. From the production stages that the company has and the results from interview with consultants, it provide data information that will be analyzed so that it is appropriate to be applied to UD X, and later be useful and important for the author to use in preparing / making of production SOP as a clear reference. So that later in preparing / making of SOP is not confusing for workers / others who read it.

In accordance with the interview quote that at the beginning, the raw material entry is an important / crucial stage in determining whether the goods will be good or not at the end. All respondents both internal, external and consultant also said that in the making of SOP, the classifying and sorting stage should not be missed / underestimated because it will has fatal consequences. Then there is also quote from the internal interview because it is so different than the two external companies. It is said differently because the owner of UD. X performs different actions in responding if the goods from suppliers received are already very dirty so that it will affect the quality if it is continued to the production process. Here the author finds striking differences, the striking difference between the three companies running plastic mills, the two companies said the same statements, but it is different compared to statements by internal company owners. Both external companies will still grind and sell because it is already being grinded, also because it is already used time and production costs while running. While from the internal version, instead it will be stopped as early as possible from the beginning, to maintain the quality and for the prices to remain high. UD X also does not choose to keep grinding because it feels like a loss, because it has been purchased with normal price, employees who process of course also paid normally while the output is not worth the normal price. Then it would be very detrimental to continue.

It is important to control the products produced by the company because the products with unstable quality will certainly make consumers think twice to keep buying the products offered. According to Sunarya [12], to assess a product whether it is in accordance with requirement / standard is to conduct quality inspection by product testing. This is where the important role of Quality Control can be seen. Quality Control is the proof that the production process can be done and operated well, without doing the inspection [11]. By having standardized products, it will be easier to compete in the market because buyers are not confused with non-standard products (whether of appearance, color, quality, weight). After the observation and interview it was found that the inspection / quality control is done every time after the goods were rinsed / inserted into the sack. So if the quality is bad it can still be fixed by adding soap and caustic soda. Although if it turns out after being repaired the quality of goods does not change much, at least the company can complain, get discounts, or even decide to stop cooperating with suppliers, depending on the negotiation. Also, from the consultant author found several ways / steps so that the milled product meets the standards desired by the buyer. That is by adding a filter according to the type of plastic to be milled, there are even some types of plastic that should not use a filter at all.

The importance of standardization not only determines the quality of what will come out later, but with the standardization it will ease the work and increase the effectiveness while milling. In addition, the company will have a reference in setting the quality standards of products to be achieved. The same thing constantly repeats in every interview that the process of sorting and classifying will not be eliminated from each stage / flow of production processes that occur, because the process of classifying and sorting is a crucial stage in the production process. And it should really be considered because the output of classifying and sorting will ultimately affect the result output produced by the company. Because when it is mixed in different type or color then the product will get complained and even be returned so it cannot be sold.

From all interviews and data that have been processed, the author will use it to make production SOP (Standard Operating Procedure) for UD X to make it easier to read and understand the intent and purpose of the SOP which has been made. Later there will be 4 kinds of SOP made by the author, which are: classifying SOP, sorting SOP, milling SOP and packing SOP.

3.1. Classifying SOP

1. At the moment the new goods come, it should be determined whether it is necessary to go through the classifying process or not.
2. Classifying process is done by classifying various types of plastic mixed in the sack, classified according to the type of plastic. From raw materials in one sack mixed with plastic, will be separated into several sacks that have been provided according to the type. For example: the type of PP, BLG, PET, GLASS, color mixed at this stage is not a problem.
3. PP is a type of hard plastic, easily broken and rigidly made by printing / pressing, can be in the form of chair or table. BLG is a type of plastic that is printed by blowing (BLOW) so that it has a small mouth and an expanding belly, can be in the form of jars or gallons. PET is a transparent type of plastic, can be clear or in color, made by the same way with BLOWING but the type of plastic is different, can be in the form of drinking bottles. And GLASS is a type of plastic in the form of plastic water glasses.
4. Incorrect type or mixed type is a fatal problem and should be avoided.
5. If the sack where the result of classifying process is full, it is tied with needle and raffia rope so it will not be spilled. Then it is replaced with new sack again. It is collected to be weighed every week before submitted to sorting department and also for the wage payroll for labors.

3.2. Sorting SOP

1. Sorting is the stage where the goods already being classified and will be processed to produce raw materials ready to be milled.
2. The result from classifying process or goods that do not need to be classified will go directly to the sorting process.
3. Sorting begins by separating the items that have been classified to be more specific. So that one sack will has the same type and same color of plastic.
4. Wrong type, wrong color, mixed color or mixed type is a fatal mistake and should be avoided.
5. Sorting divided in 3 departments / parts: weight sorting, glass sorting, PET sorting.
6. Sorting heavy items include PP and BLG sorting. This is done is to separate the results of classifying process in accordance with the color to be more detail. For example: black PP and red PP with its own, in this stage, aids tool in the form of a machete is used to cut if the raw material is too big to enter into the milling machine.
7. PET sorting means PET type processing. In this section, the label attached is removed using a cutter because it is considered garbage (cannot be grinded), if it is still has water left, it is emptied first. Then separate the bottle cap from the neck bottle because it belongs to a different type of plastic, then it will be separated into PET 1 and PET 2. What distinguishes PET 1 and PET 2 is PET 1 texture is more flaccid while PET 2 is more rigid, for example: PET 1 includes aqua bottle while for PET 2 is like coca cola.

8. GLASS sorting is almost like processing PET, the difference is glass label is located on the mouth that must be removed using a cutter as well. Other than that sometimes the glass comes with a straw. The straw is not thrown but also grinded because it has the same type of plastic.
9. After it is all collected, it will be collected and then weighed every week. With the purpose for the company's production record and wage payroll for labors.

3.3. Milling SOP

1. After weighing processed sorting, the goods will be lifted close to the head of the milling machine. So it is easy to pour it into the milling machine.
2. One person will be on top to pour the mill raw material and prepare and pour soap and caustic soda.
3. After soap, caustic soda and raw materials are ready, the machine will be turned on. The water pump will spray water for engine circulation. There are two people down near the basin where the grind comes out in the form of pieces of plastic flakes.
4. The two workers job is to scoop, rinse, filter out the plastic flakes that come out in the first basin, then move it to the second basin, to do the same thing. The goal is to clean the results of the mill if it is still dirty, then from the second basin, the workers in charge there will pour the results of clean mill & drained it into a sack. Occasionally hitting it to the ground so that the sack is more filled and has more space to be filled.
5. While inserting the milled product into the sack, it should be noted that no other items / goods are brought along like dirt and TUS process needs to be done or draining the water from the basin so that it does not affect the quality that will be judged by the buyer.

3.4. Packing SOP

1. After the sack is filled to full capacity then the workers whose in charge in the second basin are tasked to tie the mouth of the sack, so that if it is later be moved or weighed it will not be spilled.
2. Collected and later the milled sacks will be weighed and recorded every week to monitor weekly production results whether it is increased or decreased, and also to pay the wages of milled production workers.
3. After weighing the results of production will be arranged in a predetermined place, exposed to sunlight to dry, so that the sale price is not discounted because the product is wet (contains water).

4. Conclusion

This study aims to make SOP (standard operating procedure) for UD X. The importance of having SOP is in addition to help in explaining job descriptions of each employee, SOP also improve the effectiveness of production because employees already know how to work in each stages of production process, and also what kind of product is considered as having good quality. The SOP must be detailed so that it can be used as a reference / guide for the reader, so there is no need to always have a PIC all the time in place to supervise the new workers and old workers.

5. References

[1] Ekotama S 2013 *Cara Mudah Bikin Standard Operating Procedure* (Yogyakarta: Medpress)
[2] Tambunan R M 2013 *Pedoman Penyusunan Standard Operating Procedures (SOP)* Edisi Kedua (Bekasi: Maistas Publishing)
[3] Murtie 2012 *Belajar Manajemen dari Konsultasi Strategi* (Bekasi: Laskar Askara)
[4] Fahmi I 2012 *Manajemen Produksi dan Operasi* (Bandung: Alfabeta)
[5] Haming M dan Nurnajamuddin M 2011 *Manajemen Produksi Modern* (Jakarta: PT. Bumi Aksara)
[6] Tompshon M and Tracy B 2010 *Now, Build a Great Business!* (Unites States: AMACOM)

[7] Codita R 2011 *Contingency Factors of Marketing-Mix Standardization* (Germany: Gabler verlag)
[8] Horner S and Swarbrooke J 2012 *Leisure Marketing a Global Perspective* (UK: Routledge)
[9] Klopcic M, Kuipers A and Hocquette 2013 *Consumer Attitudes to Food Quality Products* (Netherland: Wagenigen Academic Pub)
[10] Sunarya 2012 *Standardisasi dalam Industri dan Perdagangan* (Depok: Papas Sinar Sinanti)
[11] Boyer K and Rohit V 2010 *Operations & Supply Chain Management* (United State: South-Western: Cengage Learning)
[12] Sugiyono 2013 *Metode Penelitian Manajemen* (Bandung: Alfabeta)
[13] Creswell, J 2012 *Qualitative Inquiry and Research Design: choosing among five approaches* (United States: Sage publication, inc.)

Risk management for food and beverage industry using Australia/New Zealand 4360 Standard

S Kristina[1] and B M Wijaya[2]

Industrial Engineering, Institut Teknologi Harapan Bangsa
Jl. Dipatiukur No. 80-84 Bandung, Indonesia
[1]sonna@ithb.ac.id
[2]benedictamarissa@gmail.com

Abstract. This research aims to identifying, measuring and establishing risk in food and beverage industry. The risk management is implemented by referring to Australia/New Zealand 4360 Standard which has four phases such as problem formulation, risk analysis, risk characterization, and risk management. The implementation of risk management is done by case study at Back Alley Café. Based on the risk identification result, there are 59 risks were found in Back Alley Café. The risk identification is conducted based on observation and interview with the café manager who understand the condition of Back Alley Café properly. Based on the assessment of impact and probability, the risk mapping produced four risks at extreme level, 16 risks at high level, 24 risks at medium level, and 18 risks at low level. The strategy was designed as the risk mitigation after the risk mapping for the extreme and high level. The strategy which is given as the prevention or treatment of risk in Back Alley Café is divided into three. There are reducing the risk, sharing the risk and accepting the risk. The strategy is then implemented for each relevant activities.
Keywords: Risk Management, Australia/New Zealand 4360 Standard.

1. Introduction

According to [2] risk management is defined as a process related to identification, analysis, response to uncertainty including maximizing the outcome of positive events and minimizing the impact of reverse events. The coordinated effort to direct and control the company against the risks in conducting business activities, is expected to reduce the potential loss as low as possible or if possible utilize the risk into opportunities that can increase corporate profits. In the absence of risk management, the risks that may occur will have an impact on the company itself. Risk management is a process of identifying, measuring risks, and forming strategies for managing them through available resources. The existence of risk management will assist the company in making decisions of any existing risks, so that the impact that will adversely affect the company can be minimized or even eliminated. This research discusses the understanding of the risks gained to make decisions about the next steps to be taken, where the decisions include risks that requiring risk management, which risk management activities should be undertaken, and which risks need to be prioritized in risk management.

2. Literature Review

2.1. Risks Management

According to [2] risk management is defined as a process related to identification, analysis, response to uncertainty including maximizing the outcome of positive events and minimizing the impact of reverse events. The risks involved in identifying and evaluating project risks, consider what to do with the impacts, the possibility of transferring risk to others, or how to mitigate risks. Risk management has the ultimate goal of limiting the likelihood of risk occurrence and the impact of negative project activities.

2.2. The Benefit of Risks Management

According to [2] the benefits of risk management given to the company can be divided into five main categories:

- Risk management may prevent the company from failure.
- Risk management directly supports the increase in profit.
- Risk management can deliver earnings indirectly.
- The peace of mind for managers caused by protection against pure risk, is a non-material treasure for the company.
- Risk management protects firms from pure risk, and since customer and supplier creditors prefer a protected company it indirectly helps improve the public image.

2.3. The Process of Risks Management

According to [1] the main procedures for risk management are four, among others:

2.3.1. Problem Formulation. The process of evaluating the suspicion of why an effect on a project that has already occurred, or that can occur from human activities. This stage is the initial stage of the overall project risk assessment. Some of the main things in the formulation of the problem include:

- Identify and describe the problem
- Collect and integrate available information
- Develop a conceptual model of problems
- Develop a risk analysis plan

2.3.2. Risk Analysis. [1] describes that risk is a possibility of an undesirable event that will affect an activity or object. The risk is measured in consequences (impact) and likelihood (possibly). Likelihood is a possibility within a period of time that a risk will arise. Consequence is an event of a consequence such as a loss (impact caused). Risk calculation can be formulated as multiplication of Likelihood with Consequence as in the following formula:

$$\text{Risk} = \text{Likelihood X Consequences} \quad (1)$$

As in table 1, the assessment of likelihood is divided into 5, ie, almost certain, likely, moderate, unlikely, and rare. As for the assessment of consequences, as shown in table 2, the assessment is divided into 5, ie, insignificant, minor, medium, major, and catastrophic. From these five risk analyzes, there are four levels of risk: Extreme, High, Medium, and Low, which will be described in table 3.

IOP Conf. Series: Materials Science and Engineering **277** (2017) 012025 doi:10.1088/1757-899X/277/1/012025

Table 1. The Assessment of Likelihood.

Level	Descriptor	Description
A	Almost Certain	The possibilities happen very often
B	Likely	Often occur
C	Moderate	Happened several times
D	Unlikely	Happens sometimes
E	Rare	Possibilities rarely occur

Table 2. The Assessment of Consequences.

Level	Descriptor	Description
1	Insignificant	Causes death, inflicts serious harm, and enormous financial losses
2	Minor	Needs first aid, medium financial loss.
3	Medium	Requires medical treatment, high financial loss.
4	Major	Caused widespread losses, serious injuries, and huge financial losses.
5	Catastrophic	Causes death, inflicts serious harm, and enormous financial losses.

Table 3. Risk Metrics.

Likelihood	Consequences				
	(1) Insignificant	(2) Minor	(3) Medium	(4) Major	(5) Catastrophic
(A) Almost Certain	H	H	E[a]	E	E
(B) Likely	M	H	H[b]	E	E
(C) Moderate	L	M	H	E	E
(D) Unlikely	L	L[d]	M[c]	H	E
(E) Rare	L	L	M	H	H

[a] This risk cannot be tolerated need immediate treatment
[b] This risk is undesirable and is only acceptable when risk reduction is unenforceable, special attention needs from the management.
[c] This risk is accepted with consent and requires clear responsibility from management.
[d] This risk is accepted with the consent of the management and can be addressed by routine procedures.

2.3.3. Risk Characterization. Risk characterization is the last step of a risk assessment, ie to determine the level of risk of an event. This level of risk can be determined by grouping or classifying likelihood and consequences values into a risk metrics. Having known the value of existing consequences and likelihood, can be plotted on Risk Metrics to find out how high the risk will be generated.

2.3.4. Risk Management. [1] "Risk Management" explains that the implementation of risk management is conducted to understand the risks obtained in the risk analysis phase to make decisions on the next steps to be taken, where the decisions include risks that requiring risk management, which risk management activities should undertaken, and which risks need to be prioritized in risk management. According to [1] "Risk Management" there are several types of ways to manage risk:

- Risk avoidance is to decide not to engage in risky activities at all. The decision-making process is conducted by considering the potential benefits and potential losses generated by an activity.
- Risk reduction is also called risk mitigation is a method to reduce the possibility of a risk or reduce the impact of damage generated by a risk.
- Risk transfer is to transfer the risk to other parties, generally through a contract (insurance).
- Risk deferral is the impact of a risk is not always constant. Risk deferral involves delaying the aspect of a project to the point where the probability of occurrence of the risk is small.
- Risk acceptance, although certain risks can be eliminated by reducing or transferring them, but some risks must remain accepted as an important part of the activity.

3. Methods

3.1. Research Method
This study will be carried out through some stage to analyse the risks as described above, such as identifying activities, identifying risks, setting the scale and risk criteria, assessing the impact and the probability of risk, risk evaluation and final step is to determine the acknowledgment of each risk.

3.2. Data Collection
The required data were collected through interview and questionnare. Interviews are conducted on the parties directly related to the project for validation. Questionnaires were made to obtain primary data based on existing parameters, so that the data obtained is relevant to the purpose of the study.On the determination of the evaluator will be set manager and some owner of Café Back Alley. The evaluator will be the source of the interview to obtain the company's data related to the research, the risk identification process, and to be the respondent in assessing the risks identified. The determination of this evaluator is done with the assumption that the selected evaluator knows in detail the condition of Café Back Alley.

4. Result and Analysis

4.1. Risk Identification
Risk identification is done by analyzing the business activity of each business process. From each activity there is a deeper understanding to identify the risks that exist in each activity. These identified risks were the result of interviews with evaluators at Café Back Alley and amounted to 59 risks. The 59 risk shown in Appendix 1. Here are examples of existing risks based on business activity:

Table 4. Risk Identification.

Business Activity	Possible Risks	No Risk
Conducting production process (cooking)	Not available or lack of raw materials used in the production process	R.01
	The equipment used in the production is damaged thus disrupting the production process	R.02
	Work accident in kitchen area or bar	R.03

4.2. Determination of Criteria for Impact and Probability
The determination of the criteria in the risk assessment to be performed refers to [1] made relevant to the condition of the company. The determination of this criterion is performed in conjunction with the manager of Café Back Alley who is assumed to know correctly the condition of Café. The scale used in risk assessment is one to five, which can present an identified risk. Risk assessment criteria are divided

into four criteria, namely, financial, reputation, customer satisfaction, and production processes. The criteria of risk impact assessment can be seen in table 5. Table 5 shows the risk probability criteria with qualitative and quantitative description of criteria.

4.3. Risk Evaluation
After assessing each of these risks, a risk evaluation is conducted by mapping risk into the metrics according to its impact and probability values. For this mapping do the rounded up. Metrics creation aims to determine the level of each identified risk. Determining the level of risk will be useful to determine the treatment of risk. The level of risk in the risk matrix consists of low, medium, high, and extreme. Then the risks that fall into the high and extreme levels will be given risk treatment or mitigation. Overall, risk mapping can be seen in table 7.

Table 5. Risk Assessment Criteria.

Risk Impact	Impact Scale				
	1	2	3	4	5
	Insignificant	Minor	Medium	Major	Catastrophic
Financial	Suffered a loss of <5% of start-up capital	Suffered a loss of 5% - 15% of start-up capital	Suffered a loss of 16% - 40% of start-up capital	Suffered a loss of 41% - 75% of start-up capital	Suffered a loss of >75% of start-up capital
	Costs exceed <5% of the budget	Costs exceed 5% - 20% of budget	Costs exceed 21% -40% of budget	Costs exceed 41% -60% of budget	Cost exceeds> 60% of budget
	Revenue below <10% of target	Revenue is below 10% -20% of target	Revenue below 21% - 30% of target	Revenue below 31% - 50% of target	Revenue below> 51% of target
Reputation	There was a bad publication within the company	There is bad publicity within the company and local people	Bad publications occur in fixed subscribers and suppliers	There are bad publications in market and stakeholder prey	There was a bad publication published in the media
Customer Satisfaction	Customers are disappointed with the service and product quality	Customers were disappointed with the service and product quality and showed such disappointment at the time	Customers are disappointed with the service and quality of products and do not want to go back to the café	Customers are disappointed with the service and quality of products and encourage others not to come to the cafe	Customers are disappointed with the service and product quality and write on social media
Production Process	Unavailability of materials / equipment / employees can still be replaced with existing inventory	Unavailability of materials / equipment / employees can still be replaced with materials / equipment / other employees	There was a delay in the production process <30min	There was a delay in the production process> 30min	The production process cannot be done

Table 6. Risk Probability Assessment Criteria.

Rating		Kriteria	
Angka	Deskripsi	Kriteria Kualitatif	Kriteria Kuantitatif
1	*Rare*	Possibilities rarely occur	Possible 0-10%
2	*Unlikely*	Happens sometimes	Possible 11-30%
3	*Moderate*	Happened several times	Possible 31-60%
4	*Likely*	Often occur	Possible 61-90%
5	*Almost Certain*	The possibilities happen very often	Possible 91-100%

Table 7. Risk Metrics.

Likelihood	Consequences				
	(1) Insignificant	(2) Minor	(3) Medium	(4) Major	(5) Catastrophic
(5) Almost Certain					
(4) Likely		R.59	R.35, R.42, R.46, R.57	R.47, R.48	
(3) Moderate		R.50, R.51, R.58	R.12, R.32, R.36, R.39, R.40, R.41, R.43, R.49	R.01, R.34	
(2) Unlikely	R.03, R.08, R.21, R.26, R.56	R.02, R.06, R.07, R.16, R.17, R.27, R.28, R.33, R.52, R.55, R.58	R.04, R.05, R.08, R.09 R.10, R.11, R.13, R.18, R.19, R.24, R.37, R.38, R.44, R.45, R.51, R.52	R.14, R.29	
(1) Rare	R.25, R.54		R.15, R.20, R.22, R.23, R.31		R.30

4.3.1. Risks with Low Risk Level. Small risk that easily controlled by the company, because of the impact and probability of a relatively small risk. This risk can be ignored by the company as it is deemed to be handled without an analysis of such risks. Based on the results of risk assessment, there are eighteen risks that fall into low risk.

4.3.2. Risk with Medium Risk Level. Risks that require considerable attention by the company. This risk can be accepted by the company or given an action if the company has enough data sources. Risks at this level can potentially lead to some disruption to corporate activity, although not significant enough. Based on the results of risk assessment, there are 24 risks that fall into the medium risk.

4.3.3. Risk with High Risk Level. Risks that require serious attention for the company. This risk has a significant impact on corporate activity and high probability. Under these conditions, the company needs to take measures of treatment or mitigation against risks at high levels. Based on the results of risk assessment, there are sixteen risks that fall into the high risk.

4.3.4. Risk with Extreme Risk Level. Risks with extreme risk levels are risks that require very serious concern for the company. Just like the risks at the high level, these risks have a considerable impact on corporate activity and high probability. This risk becomes the company's top priority that should be treated as soon as possible. Based on the results of risk assessment, there are four risks that fall into the extreme risk.

4.4. Determination of Risk Treatment.
After evaluating the risks and getting risks falling into the high and extreme levels, further analysis of the treatment strategy will be performed in accordance with each of these risks. Some examples of risk treatment can be seen in Table 8.

Table 8. Risk Treatment Strategy.

No Risk	Risk	Risk Level	Treatment Strategies	The Effort
R.01	Not available or lack of raw materials used in the production process	*Extreme*	*Risk Reduction*	• *Safety Stock* • *First in first out System*
R.47	The existence of changes in customer trends affecting forecasting	*Extreme*	*Risk Acceptance*	Improve forecasting by taking into account trend changes as quickly as possible
R.12	Goods received from suppliers are not in good condition	*High*	*Risk Reduction*	• *Safety Stock* • Divide reservations to multiple suppliers
			Risk Transfer	Return thing
R.46	Error in forecasting	*High*	*Risk Acceptance*	Fixed forecasting as soon as possible

5. Conclusion

In accordance with the purpose of research and the results of the analysis conducted, the following conclusions are obtained:

- Based on the business activities in the café Back Alley has identified 59 risks overall. The risks are then mapped into the metrics according to the value of impact and its probability value. Risk mapping produces four risks at the extreme level, sixteen high risk risks, 24 medium-level risks, and eighteen risks at low levels.
- The impacts resulting from the likelihood of risk are divided into four parts, namely, financial, reputation, customer satisfaction and production processes. The financial impact itself is divided into three parts that include losses from initial capital, expenses that exceed the budget, and revenue is below the target. For reputation impacts include bad publications from internal to external that will affect the company's image. The impact of customer satisfaction involves customer dissatisfaction with product and service quality. The impact of the production process involves the unavailability of goods causing delays to the production process.
- Strategies that can be provided as a form of risk prevention or treatment at Café Back Alley are divided into three that is, reducing risks, sharing risks, and taking risks. The choice of strategy is tailored to the associated risks applied with the relevant form of effort.

6. Suggestion

In this research, risk identification is done without taking into account the level of employees. Meanwhile, in reality in the café business, there are some employees who play an important role to influence the operational risks café. So in the next research can be done research by taking into account the level of employees in risk identification. Subsequent research can also be continued and complemented by measuring the success rate of risk management at Café Back Alley and making proposals of monitoring and controlling design to improve the effectiveness of risk management implementation at Café Back Alley.

Appendix A. Risk Identification.

Business Activity	Possible Risks	No. Risk
Conducting production process (cooking)	Not available or lack of raw materials used in the production process	R.01
	The equipment used in the production is damaged thus disrupting the production process	R.02
	Work accident in kitchen area or bar	R.03
Setting up and controlling product quality	The quality of the product provided does not match the customer's expectation	R.04
	There are some things that are missing in the control of product quality	R.05
Maintain personal hygiene and tidiness in production	Employees ignore personal hygiene and tidiness in production	R.06
Cleaning the cooking area (kitchen / bar)	Ignoring the cleanliness of the cooking area due to heavy production	R.07
Order raw materials to the purchasing division as inventory	Order quantity or type of goods ordered wrong (too many or few)	R.08
Ensure availability of raw materials to be purchased for production purposes	Error forecasting amount of raw material to be ordered	R.09
	Error calculating the amount of raw materials available	R.10
Controlling in and out of raw materials	Goods received from suppliers do not meet the specifications requested	R.11
	Goods received from suppliers are not in good condition	R.12
	Dishonesty of employees responsible for controlling raw materials	R.13
Serve and greet customers well	Employees make mistakes while serving and greeting customers	R.14
Recording customer orders	Error in customer order recording	R.15
Preparing cutlery and utensils	Lack of amount of cutlery and utensils	R.16
	Damage to cutlery and utensils	R.17
Lifting cutlery and utensils, and cleaning the table	The carelessness of employees when lifting cutlery and cleaning the table	R.18

Business Activity	Possible Risks	No. Risk
Take and present orders to customers	Servers misrepresentation error	R.19
Responsible for cleanliness and neatness of work area and outlet	Delay to clean and tidy outlet area	R.20
Responsible for inventory of equipment and service equipment	There is equipment and service equipment broken	R.21
	An unsuitable amount between recording and equipment quantity and service equipment	R.22
Create a daily turnover report	Error recording transactions so as to affect the report	R.23
Supervise and work on the daily cash cashier	Calculation error in doing daily cash cashier	R.24
Making financial reporting	Error calculation that impact the financial statements	R.25
Conduct a thorough analysis and create weekly report and general report	Error weekly report and general report analysis	R.26
Maintain all assets of the company through monthly stock opname	There is often a difference between recording and the number of stock of the hospital	R.27
	Error in calculation in stock opname	R.28
Pay taxes	Taxpayer calculation error	R.29
	Delay in paying taxes	R.30
Create a report on purchasing and disbursement of goods	There are unrecorded purchases or expenses that affect the report	R.31
Ensure the availability of goods / materials through audit mechanism or control stock	The ignorance of communication with the production section (kitchen and bar) about the needs of raw materials	R.32
	Error communicating with suppliers regarding ordered quantity or delivery time	R.33
Ensuring the procurement of goods operational and non-operational according to the budget, according to the time required, according to specifications and according to the desired quality	The price offered by the supplier is too high	R.34
	Fluctuations in raw material prices	R.35
	The raw material specifications do not match the original agreement	R.36
Maintaining good relationships with suppliers and stakeholders (related parties)	Communication error between supplier side with café party or stakeholder with cafe party	R.37
Trying to get the best deals from some of the proposed alternatives	Dependency occurs only on 1 supplier (Supplier monopolizes the market)	R.38
	Scarcity of raw materials	R.39

IOP Conf. Series: Materials Science and Engineering **277** (2017) 012025 doi:10.1088/1757-899X/277/1/012025

Business Activity	Possible Risks	No. Risk
Create a report on the use of goods	There is an unrecorded use of the goods that affect the report	R.40
Ensuring goods received from suppliers in good condition and ready supply	There is an in-flight delivery by the supplier	R.41
	Inaccuracy in the inspection of goods, so that goods should be returned, received and entered in the café	R.42
Determine the selling price of the product to be launched	The pricing done does not match the target market capability	R.43
Evaluate customer satisfaction from the survey results of all sales teams to ensure the achievement of target customer satisfaction is determined	Error determining criteria determination	R.44
	There are customer characteristics that are not taken into account in customer satisfaction assessments	R.45
Monitor order acquisition and summarize forecasts to ensure production capacity is optimally filled	Error in forecasting	R.46
	The existence of changes in customer trends affecting forecasting	R.47
Analyze and develop marketing strategies to increase the number of customers and areas in accordance with the specified target	The strategy is less successful in attracting the target market	R.48
Planning the utilization of existing human resources to be effective and efficient	Inefficient labour force (too many or few)	R.49
	Planning is not in accordance with the practice in the field	R.50
Ensure that all activities in all areas work well	The occurrence of things beyond the sudden planning and effect on service	R.51
Monitor any incidents that occur inside or outside of outlet-related outlets	Work accident inside the outlet	R.52
	Illegal charges made by irresponsible people	R.53
Controlling employee absenteeism	Employees quit working unexpectedly without any clear news	R.54
	Employees do not working suddenly without clear news	R.55
Payroll employees	The delay in paying employees	R.56
Conduct recruitment, selection, placement, and new employee orientation	High turn over	R.57
	Placement of employees who are not in accordance with ability	R.58
Controlling employee discipline	Employees violate the agreed rules	R.59

References

[1] Standards Association of Australia 1999 *A Basic Introduction to Managing Risk: Using the Australian and New Zealand Risk Management Standard AS/NZS 4360:1999/ Standards Australia* (Stathfield, NSW: Standards Association of Australia)

[2] Darmawi H 2012 *Manajemen Risiko* (Jakarta: Bumi Aksara)

[3] Norris C, Perry J and Simon P 2000 *Project Risk Analysis and Management* (The Assosiation for Project Management. Buckingham)

[4] Porter M E 1985 *The Competitive Advantage: Creating and Sustaining Superior Performance* (NY: Free Press)

[5] Trieschmann, Gustavson and Hoyt 2001 *Risk Management and Insurance. 8th Edition* (South-Western Collage Publishing)

IOP Conf. Series: Materials Science and Engineering **277** (2017) 012026 doi:10.1088/1757-899X/277/1/012026

Comparison between motorcyclist' violation behavior and accidents in urban and rural area in Indonesia: A comparative study

N Rahmawati[1] and A Widyanti[2]

Department of Industrial Engineering
Bandung Institute of Technology (ITB)
Ganesa 10 Bandung 40132 Indonesia
[1] novia.rahmawatiii@gmail.com
[2] widyanti@mail.ti.itb.ac.id

Abstract. Some studies stated that the main factor related to the accident was driving behavior. This study aims to explore the differences between motorcyclist‟ behaviour and repetitive violation behaviour in two different area, urban and rural area in Indonesia. Respondents were selected based on convenience sampling method in Bandung as a representative of urban area and Kulon Progo as a representative of rural area. They were asked to fill in a questionnaire about driving behaviour, consists of 10 dimensions or 51 questions with Likert scales ranging from 1 (very often) to 6 (never). The results of this study shows that the motorcyclists' behavior differ significantly between rural and urban area. Motorcyclists in the urban area (Bandung) are more committed to violations than in rural area (Kulon Progo). This result is not in line with previous studies in Australia and United States which stated that motorcyclists in rural area more frequently speeding than in urban area. Implications of the result are discussed.
Keywords: motorcyclist, behaviour, urban, rural

1. Introduction

Developments of automotive industry technology and economic conditions in Indonesia have an impact on the increased use of transportation modes, including motorcycle. The number of motorcycle in Indonesia increases about 11,08 percent per year between the periods of 2010-2014. In 2014, the number of motorcycles in Indonesia reached 92 million units [1]. The high rate of motorcycle ownership is in line with the high number of motorcycles involved in a traffic accident. In 2014, the number of traffic accidents involving motorcycles reached 108.883, or approximately 71.6% of total traffic accidents in Indonesia [2].

Some studies in other countries revealed that the main factors related to the accident were driving behavior of motorcyclist [3] [4] [5] [6]. Motorcyclists are often regarded as "speed rebels" and "risk takers" [3]. The behavior of motorcyclists who speeding on the road [3] [4], obey pedestrian space, and lack of obedience to traffic rules is considered as the main cause of accidents [5]. The facts that motorcyclist‟ violation behavior is a factor that affects the accident has been the focus of research in several countries, both in developed and developing countries such as Turkey [7], the UK [4] and Hong Kong [3].

Considering that motorcyclist behavior is not universal depends on the context and places, several studies focuses in motorcyclist behavior differences in rural and urban area. A study stated that there are significant differences between driver behavior and accident characteristics between urban and rural area in Utah, United States [8]. Speeding, alcohol involvement, safety belt wearing and various road characteristics become a factor prevalent in the rural area. In line with previous study, study in the rural Queensland showed that road users tend to be involved in fatal accidents due to speeding, drunk and violation of road rules [9]. Otherwise, a study in the rural New South Wales concluded that the majority of young people in the rural area are not high-risk takers or sensation seekers [10]. Respondents in the rural area tend to driving over the speed limit or speeding. This condition is because the average distance that must be taken by young people in the rural area farther than in the urban area.The fact that Indonesia suffers from high rate of motorcyclist accident underlined a comprehensive approach in effort to reduce motorcyclist accident. However, in author knowledge, only limited research has been conducted in Indonesia related to motorcyclist behavior, that are influencing factors on motorcycle accident in urban area [5]; and reasons underlying behavior of motorcyclists disregarding traffic regulations in urban area of Indonesia [11]. Considering that behavior approach works in changing behavior in other field of study in Indonesia [12] (e.g., Widyanti et al., 2014), this approach is worth noting to implement in driving behavior.

This study aims to observe and to compare motorcyclist behavior and accident between urban and rural area as a first step in a series of efforts to reduce motorcyclist accident in Indonesia. With a hypothesis that there are differences in motorcyclist behavior and accident between rural and urban area in Indonesia, different approach should be taken.

2. Methods

The research was conducted in one of urban area in Indonesia, i.e., Bandung and one of rural area, i.e., Kulon Progo, Yogyakarta, Indonesia. Bandung is one of the metropolitan cities in Indonesia and the capital of West Java province. While, Kulon Progo is one of the district in Yogyakarta, Indonesia. The number of respondents in Bandung are 32, whereas Kulon Progo are 33. Respondents were selected based on convenience sampling method.

The data collection was conducted using questionnaire. The questionnaire consists of three parts, (1) demographic data of respondents such as gender, age, employment status, marital status, frequency of the motorcycle usage per trip and travel distance per trip, (2) the motorcyclist behavior of respondents. Variables used in the questionnaire were developed based on the concept of Theory of Planned Behavior [13] and other factors from literature studies [3] [4]. The scale used for the driving behavior question is a Likert scale with a range of 1-6 ranging from "very often" to "never".

The result of the questionnaire is independent data between respondents in the urban and rural area. The scale used is ordinal scale and the data are not following normal distribution. Therefore, the data were analyzed by using Mann Whitney U Test to evaluate significant differences among 10 dimensions of driving behavior (repetitive violation behavior, ordinary violation, aggressive violation, subjective norms, perceived behavior, believe and attitudes, speed violation, traffic errors, safety equipment and stunts). All statistical analyses were performed using the IBM SPSS program version 20.0 for Windows.

3. Result and discussion

Demographic characteristics of the respondents in each area can be seen in Table 1. Behavior of motorcyclist of Bandung and Kulon Progo respondents are describing in Table 2. In addition, Mann Whitney U Test is applied to observe mean differences between the groups. H0 is that motorcyclist's violation behavior in Bandung and Kulon Progo is not different, while H1 is that motorcyclist's violation behavior in Bandung and Kulon Progo is different. H0 is accepted if the alpha cronbach more than 0.05, otherwise H0 is rejected.

Table 1. Demographic of sample of respondents

Characteristics		Proportion (%)	
		Bandung	Kulon Progo
Gender	Male	40.6	45.5
	Female	59.4	54.5
Age	Aged between 17 and 29	93.8	75.8
	Aged between 30 and 39	3.1	3.0
	Aged between 40 and 49	3.1	12.1
	Aged 50 or more	0	9.1
Employment	Student	81.3	27.3
	Full-time employed	12.5	57.6
	Entrepeuner	0	6.1
	Unemployed	3	0
	Other	3.1	9.1
Marital status	Single	90.6	69.7
	Married	9.4	30.3
Frequency of the motorcycle usage per day	Once	15.6	3.0
	Two times	31.3	30.3
	Three times	12.5	24.2
	Four times	3.1	6.1
	More than four times	37.5	36.4
Travel distance per trip	Less than 1 km	3.1	0
	1-2 km	6.3	9.1
	2-3 km	9.4	0
	3-4 km	6.3	6.1
	More than 4 km	75.0	84.8
Driving license ownership	Have a driving license category A	6.3	0
	Have a driving license category B	0	3.0
	Have a driving license category C	37.5	66.7
	Not own any driving license	21.9	0
	Have more than one driving license	34.4	30.3

Based on the result obtained in Table 2, there are 13 statements having significant difference between motorcyclists violation behavior in Bandung and Kulon Progo, that are repetitive violation behavior (2, 8, 11, 12, 13, 14, 15, 17), ordinary violation behavior (2, 7), subjective norms (2) and perceived behavioral control (1,2).

Table 2. Different behavior of motorcyclist of Bandung and Kulon Progo
(a)

Dimensions		Variables	Mean		Asymp. Sig. (2-tailed)
			Bandung	Kulon Progo	
Repetitive violation	1	Ignore traffic lights at junctions	1,625	1,697	.752
	2	Do not respect the stop line at junctions	2,062	1,606	.031*
	3	Do not turn on head-lights during the day	1,344	1,394	.365
	4	Do not turn on turning-lamp (indicator light) when turning	1,969	1,546	.162
	5	Speeding	1,321	2,606	.979
	6	Overtaking other vehicles on the wrong side	1,153	2,656	.305
	7	Deliberately do not allow the other vehicles to overtake	1,247	1,103	.067
	8	Cutting ones' way of move or take short cuts	1,304	.708	.024*
	9	Do not bring the required paperwork and driving license whilst driving	1,680	.809	.246
	10	Carrying overweight goods or too many passengers	1,198	.882	.117
	11	Driving slow on fast lanes	.896	.658	.035*
	12	Pushing the motorcycle through a (very) narrow gap	1,319	1,237	.001*
	13	Listening to music whilst driving	1,515	.933	.002*
	14	Eating, smoking or drinking whilst driving	1,306	.415	.046*
	15	Calling or texting whilst driving	1,318	.822	.003*
	16	Chatting with other riders and/or passengers whilst driving	1,436	1,479	.275
	17	Not wearing safety equipment such as helmets	1,216	.833	.023*
	18	Not wearing additional safety equipment such as jackets and gloves	1,496	1,034	.107
	19	Riding on pedestrian overpasses	1,014	.364	.526
Ordinary violation	1	Driven through an amber light when it was about to turn red	2,812	2,606	.876
	2	Made an illegal U-turn	2,187	1,424	.003*
	3	Imitated the postures and movements used in a motorcycle race while driving on the public road, such as touching the ground with a knee when turning	1,218	1,272	.759
	4	Driven as fast on a wet road surface as on a dry one	1,968	1,606	.203
	5	Driven too close to the car in front	2.25	2,393	.601
	6	Sounded your horn to show your annoyance to another road user	2,468	2,333	.536
	7	Driven the wrong way down a one-way street	1,875	1,363	.005*

(b)

Dimensions		Variables	Mean		Asymp. Sig. (2-tailed)
			Bandung	Kulon Progo	
Aggressive violation	1	Exceeded the speed limit in an urban road	3,187	2,878	.250
	2	Crossed double white lines to overtake even though there were vehicles approaching in the opposite lane a short distance away	2.0	1,757	.202
	3	Crossed double white lines to overtake on a sharp bend	1,562	1,575	.953
	4	Overtaken the driver in front when he/she was driving at the speed limit	1,843	2.06	.414
	5	Overtaken even when close to turning	1,718	1,606	.697
	6	Driven on the pavement when caught in a traffic jam	2.0	1,757	.491
Subjective norms	1	Using high beam on congested places	1,906	1,848	.798
	2	Switching engine on in small alleys	2,875	2.03	.023*
Perceived behavioral control	1	The traffic condition	2,312	1,727	.025*
	2	The number and width of road lanes	2,406	1,697	.005*
	3	The road geometric (e.g. slopes and turns)	2,093	1,636	.056
	4	The road side friction and junction side conditions (e.g. "spilled-over" hawkers on the street)	1,937	1,575	.040
	5	The Weather Condition	2,218	1,909	.304
Believe & attitudes		Driven above the speed limit in order not to be late for an appointment	3.25	3,303	.979
Speed violation	1	Exceed the speed limit on a residential road	2,781	2,454	.498
	2	Disregard the speed limit late at night or in the early hours of the morning	2,968	2,333	.076
	3	Race away from traffic lights with the intention of beating the driver/rider next to you	1,281	1,151	.594
	4	Get involved in unofficial 'races' with other riders or drivers	1,218	1.06	.560
	5	Ride so fast into a corner that you scare yourself	1,375	1,272	.762

(c)

Dimensions		Variables	Mean		Asymp. Sig. (2-tailed)
			Bandung	Kulon Progo	
Traffic errors	1	Ride so fast into a corner that you feel like you might lose control	1,468	1,484	.944
	2	Fail to notice that pedestrians are crossing when turning into a side street from a main road	2,031	2,242	.852
	3	Attempt to overtake someone that you had not noticed to be signaling a right turn	1,812	1,667	.328
Safety equipment		Wear body armor	3,593	3.0	.252
Stunts		Intentionally do a wheel spin	1,187	1.06	.560

The results of this study show that there are several behavior items that differ significantly between motorcyclists behavior in Bandung and Kulon Progo, that are repetitive violation behavior item (2, 8, 11, 12, 13, 14, 15, 17), ordinary violation behavior item (2, 7), subjective norms (2) and perceived behavioral control (1,2). Based on the results of the questionnaire, motorcyclists in Bandung more committed these violations than in Kulon Progo. In urban area, the range "often" is more dominant than in rural area.

As stated in previous study, the majority of motorcyclists in the rural area are not high risk takers or sensation seekers [10]. However, in this study, motorcyclists in the rural area do not have significant differences in factor of speeding with motorcyclists in urban area. This result does not in line with previous studies in Australia and United States which stated that motorcyclists in rural area more frequently speeding than in urban area. Presumably, this result was influenced by some of the respondents in Bandung were young motorcyclist. Young motorcyclists are most likely to have a high level of violation behavior such as speeding [3] [4] [11] [14]. Older motorcyclists tend to be acting cautious when riding a motorcycle especially motorcyclists over 50 years old.

Result of this present study gives valuable sight of the differences between urban and rural area. This results imply that different approach must be taken if policy or regulation are determined in Indonesia related to efforts to reduce number of motorcyclist accident in Indonesia. One of the differences in approach that can be emphasized is an education about safety riding for motorcyclists in urban and rural area. Differences in the characteristics of the road and accident can be the factor to be considered.

The present study has some limitations. First, limited numbers of respondents were involved in this study. Much more respondents are needed to draw more rigorous conclusion. Second, respondents in this study are dominated by young motorcyclists (i.e., 17-29 years old). More diverse respondents from the age categories will enrich the result as well. More diverse respondents from other demography categories in Table 1 is needed, for example, since merit and single status of respondents is hypothesized to influence motorcyclist behavior and accident, more representative respondents in both merit and single status is needed.

4. Conclusion

The study found that H0 is rejected and this means that there are significant differences in motorcyclist behavior and accident between rural and urban area in Indonesia. Motorcyclists in the urban area (Bandung) are more committed to violations than in rural area (Kulon Progo).

Government should promote safety riding for young motorcyclists and make regulations to prohibit motorcyclists who do not have a driver's license as an effort to reduce accidents. Because most offenders are young motorcyclists.

5. References

[1] Statistics Indonesia 2015 *Land Transportation Statistics 2014* (Jakarta) p 37-39
[2] Directorate General of Land Transportation 2015 *Land Transportation in Figures 2014* (Jakarta) p 495
[3] Cheng A S K, Liu K P Y and Tulliani N 2015 Relationship between driving-violation behaviours and risk perception in motorcycle accidents *Hongkong Journal of Occupational Therapy* 25 32-38
[4] Elliott M A, Baughan C J and Sexton B F 2007 Errors and violations in relation to motorcyclists‟ crash risk *Accident Analysis and Prevention* 39 491-499
[5] Indriastuti A K and Sulistio H 2010 Influencing factors on motorcycle accident in urban area of Malang, Indonesia *International Journal of Academic Research* 2 252–255
[6] Ozkan T and Lajunen T 2005 Multidimensional Traffic Locus of Control Scale (T-LOC): factor structure and relationship to risky driving *Personality and Individual Diferences* 38 533-545
[7] Ozkan T, Lajunen T, Dogruyol B, Yıldırım Z and Coyma A 2012 Motorcycle accidents, rider behaviour, and psychological models *Accident Analysis and Prevention* 49 124-132
[8] Donaldson A E, Cook L J, Hutchings C B and Dean J M 2006 Crossing county lines: the impact of crash location and driver‟s residence on motor vehicle crash fatality *Accident Analysis and Prevention* 38 723– 727
[9] Siskind V, Steinhardt D, Sheehan M, O, Connor T and Hanks H, 2011 Risk factors for fatal crashes in rural Australia *Accident Analysis and Prevention* 43 1082–1088
[10] Knight P J, Iverson D and Harris M F 2012 Early driving experience and influence on risk perception in young rural people *Accident Analysis and Prevention* 45 775-781
[11] Susilo Y O, Joewono T B and Vandebona U 2015 Reasons underlying behaviour of motorcyclists disregarding traffic regulations in urban area of Indonesia *Accident Analysis and Prevention* 75 272-284.
[12] Widyanti A, Sunaryo I and Kumalasari A D 2014 Reducing the dependency on rice as staple food in Indonesia – A behavior intervention approach *Journal of International Society of South East Asian Agricultural Science* 20 93-103
[13] Ajzen I 1991 The *Theory of planned behavior, Organizational Behavior & Human Decision Processes* 50 179-211
[14] Lucidi F, Giannini A M, Sgalla, R, Mallia L, Devoto A, and Reichmann S 2010 Young novice driver subtypes: relationship to driving violations, errors and lapses *Accident Analysis and Prevention* 42 1689-1696

Competition preparation guideline in undergraduate program of information system school of Industrial Engineering Telkom University based on knowledge conversion

F R Darmawan[1], R P Soesanto[2], A Kurniawati[3], and M T Kurniawan[4]

School of Industrial Engineering, Telkom University, Bandung, Indonesia
febyboy90@gmail.com[1],raysoesanto@telkomuniversity.ac.id[2],
amelia.kurniawati@telkomuniversity.ac.id[3], teguhkurniawan@telkomuniversity.ac.id[4]

Abstract. The role of higher education in the development of science and technology is not only from the contribution of the high-quality alumni but also from the research and relevant competition with the needs of development in such a country. In a competition, the student can improve their soft skill and academic skill such as analytical and critical thinking, communication skills and mental. The number of relevant competition by students is also included in accreditation clause, therefore student involvement in competition is seen as important for the undergraduate program in University. The most problem in university is the high turnover from the student. Bachelor program in Indonesia usually takes 4 years to complete, and the high turnover causes the student come and go as they are a graduate from the institution without preserving the knowledge and experience from the competition to other students. This research aims to develop a guidance for competition preparation in the university by using knowledge conversion. The object of this research is an information system undergraduate program in the school of industrial engineering Telkom University. The best practice selection is done by using factor rating method. Delphi method is used to identify the criteria, and AHP method is used to calculate the weight of each criterion. From the factor rating result it is known that from 3 respondent, best practice from respondent A (7.321) is used for preparing the programming competition in an undergraduate program of information system in the school of industrial engineering Telkom University. FGD is done to disseminate the selected best practice into the process stakeholder which is head of the student affair of the school of industrial engineering, students, and laboratory assistants. Future research can be done to create more comprehensive criteria for selecting the best practice.
Keywords: Analytical Hierarchy Process, Competition Guideline, Delphi Method, Knowledge Conversion

1. Introduction

The role of higher education in the development of science and technology is not only from the contribution of the high-quality alumni but also from the research and relevant competition with the needs of development in such a country. The quality of a university can be seen from the quality of alumni that can compete and survive in the development of science and technology. The graduate from a university must have competencies to support from and act as a competitive advantage from such person. The company usually want the fresh graduate that not only good at the academic but also good at soft skill such as communication, thinking, loyalty, competition and etc.

Table 1. Graduate Competency Criteria

Higher Education Graduate	Academic Skill
	Competent
	Implement the Study
	Analytic Thinking Critical Thinking Good Communication Professional Ethics Competition

University can provide the student to improve their soft skill by giving them a chance to compete in a competition. In a competition, the student can improve their soft skill and academic skill such as analytical and critical thinking, communication skills and mental. It is often said that competition is strong motivators that help students learn the real world design [1]. Institution must have a good accreditation to increase the relevance of programmes [2]. According to Ministry of Research, Technology and Higher Education of Indonesia (RISTEKDIKTI) number of relevant competition by students are included in accreditation clause, therefore students involvement in competition is seen as important for undergraduate program in University. The most problem in university is the high turnover from the student. Bachelor program in Indonesia usually takes 4 years to complete, and the high turnover causes the student come and go as they are a graduate from the institution without preserving the knowledge and experience from the competition to other students. Knowledge consists of experience, value, contextual information, and expert insight that develop a framework to evaluate and combine the experience with new information [3]. According to [4], there is only 20% of knowledge that is preserved in the organization, the rest of 80% lies within people. If the knowledge is not preserved in program study level, there is slight chance that the continuous improvement process for the competition is not well managed and can affect to the study program accreditation. Preparation phase for entering the competition is considered as an important thing. University involvement in such process can help the students to gain the desired result in competitions [1].

Telkom University is one of leading private university in Indonesia. Consists of 7 faculties that range from art, science to management the university provide the students with various competencies and activities. One of the faculties is School of Industrial Engineering (SIE). SIE consist of three study program which is an undergraduate program of industrial engineering, undergraduate program of the information system, and magister program of industrial engineering. SIE needs to manage the competition to maintain the accreditation of each program. For the undergraduate program, only information system study program that is given "B" for the accreditation. The number of participation in the competition is relatively high and the competition won by the students is also high, but there is no assurance nor guarantee that the process can keep the participation is high. This research aims to develop a guidance for competition preparation in the university by using knowledge conversion by [5]. The object of this research is an information system undergraduate program in the school of industrial engineering Telkom University. Previous research about the knowledge conversion practice is the research from [6] that focused on multiple organizations. Research from [7] focused on knowledge conversion application in Railway company. Research from [8] focused on knowledge conversion in military organizations. Another research of knowledge conversion application is the research from [9] that focused on knowledge conversion in research institution.

2. Methods

The purpose of this research is to develop a guidance for competition preparation in the university by using knowledge conversion. [5] describe the four phase of converting knowledge which is socialization, externalization, combination, and internalization. Socialization consists of sharing process and creation of tacit knowledge through interaction and direct observation. Externalization consists of articulation of tacit knowledge into explicit knowledge through dialogue and reflexion. Combination consists of the conversion process of explicit knowledge into new explicit knowledge through systemization and information. Internalization consists of the learning process and knowledge acquisition by organization member towards explicit knowledge that spread into the organization through self-experience so it can be tacit knowledge of organization member. Figure 1 shows the SECI Phase.

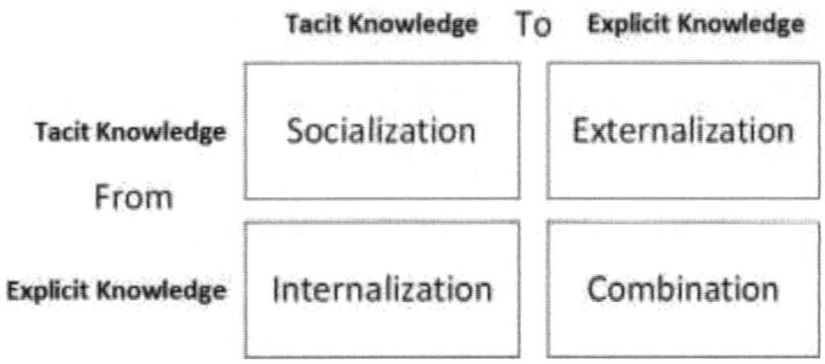

Figure 1. SECI Phase

2.1. Socialization
In this phase, the knowledge of competition preparation that is still tacit knowledge will be transferred to the interviewer for the externalization phase. The transfer process from tacit to explicit is done by interview, the main topic of the interview is the identification of business process of competition preparation and identification of the tacit and explicit knowledge from the preparation. The socialization phase is finished when the data about the process business of competition preparation and their tacit and explicit knowledge is identified.

2.2. Externalization
In this phase, the business process of competition preparation is documented. The documenting process starts from the preliminary activities, a grouping of students, practice session, supporting activities preparation (book, software, and administration) and tutoring phase.

2.3. Combination
In this phase, the process business will be converted into explicit knowledge so there will be new knowledge in form of best practice of competition preparation. The best practice will be presented and confirmed by the stakeholder which is head of the student affair in the school of industrial engineering. Factor rating method is used in the selection process. First of all the data will be reconfirmed to the stakeholder. After the data is re-confirmed the next step is to create criteria for selection process using Delphi. After the criteria are identified the next step is weighing process of each criteria using AHP (Analytical Hierarchy Process). After the instrument is finished, the next step is an assessment of each business process to choose the best of the best practice.

2.4. Internalization
In this phase, selected best practice from the previous phase is socialized to the academic community of the school of industrial engineering Telkom University so the best practice can be used as a guidance for preparing a competition.

3. Result and Discussions
In this section, the result of knowledge conversion process is presented.

3.1. Socialization

Socialization is a sharing process and the creation of tacit knowledge through interaction and live experience. Socialization is a capturing process of competition preparation from the former competition participants. The interview is done with 3 people which are a former participant of programming competition. The competition participant has a knowledge about what to be done and prepared when participating in a competition. The experience from the former participant is still in a tacit form that will help another participant in next competition. Figure 2 shows the socialization phase scheme.

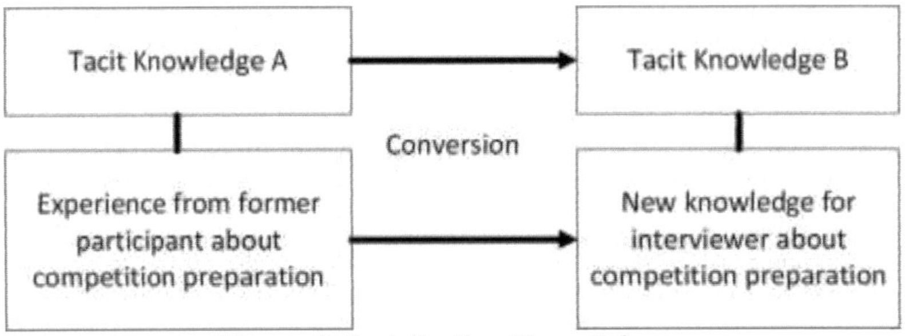

Figure 2. Socialization Phase Scheme

3.2. Externalization

In this phase, the business process of competition preparation is documented. From the result of the interview it is known that there is distinct business process which is business process for “PIMNAS PMKT”, “PIMNAS PKMM” and programming. The three business process describes the activities that must be done, the process owner, and the flow from the beginning of the process. Figure 3 shows the externalization phase scheme, example of the business process is presented in Appendix A. In externalization phase, the tacit and explicit knowledge can be identified. The identification is gained from the interview process.

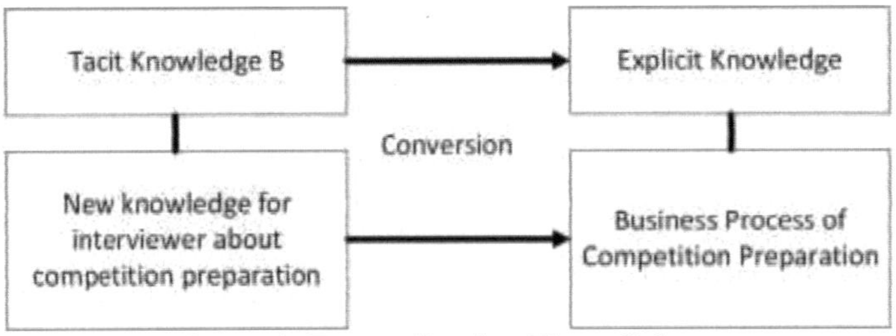

Figure 3. Externalization Phase Scheme

3.3. Combination

In this phase, the process business will be converted into explicit knowledge so there will be new knowledge in form of best practice of competition preparation. Figure 4 shows the combination phase scheme.

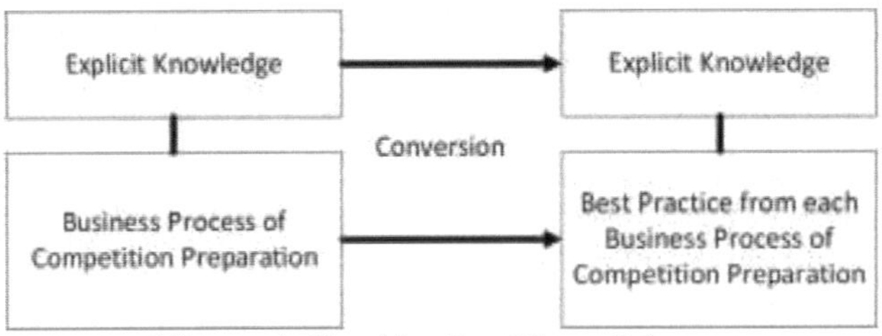

Figure 4. Combination Phase Scheme

3.3.1. Criteria Selection. In combination phase Delphi method is used to identify criteria for best practice selection, the method ends when every respondent have the same answer and consensus. The questionnaire is used to gain the criteria. The respondent is asked which criteria is considered important in selecting the best practice of a process. Table 2 shows the default Delphi criteria list.

Table 2. Default Delphi Criteria List

No.	Criteria
1	Number of Activity in a Business Process
2	Time needed in one activity
3	Number of person in activity
4	Availability of tacit knowledge
5	Availability of explicit knowledge

The questionnaire is given to 3 respondent. The criteria are obtained after the respondent consensus in three cycles of Delphi. Table 3 shows the final criteria list. A number of activity in the business process is a set of actions that must be done by process performer to prepare for the programming competition. A process performer needs explicit knowledge in form of documentation for the competition which is guidance of the competition, availability of relevant reference for competition, supporting article, and availability of software/hardware for the competition. Availability of tacit knowledge is needed to help the competition process.

Table 3. Final Criteria List

No.	Criteria
1	Number of Activity in a Business Process
2	Availability of tacit knowledge
3	Availability of explicit knowledge

3.3.2. Criteria weighting. AHP is used to identify the weight of each criterion, the AHP questionnaire is given to the respondent. The respondent is the former participant of programming competition because they are considered as an expert in the competition and understand the business process of programming competition. Table 4 shows the weighting result for a programming competition.

Table 4. Weighting of Programming Competition Preparation Business Process

No.	Criteria	Weight
1	Number of Activity in a Business Process	0.374
2	Availability of tacit knowledge	0.339
3	Availability of explicit knowledge	0.286

The consistency test is done to know the consistency of the result of the questionnaire. The result is shown in Table 5. From Table 5 it is known that the value of CR is below 0.1, it means that the AHP questionnaire is consistent [10].

Table 5. Consistency Test Result

Criteria	Value
CI	0.018
CR	0.032

3.3.3. Best Practice Selection. After the weight is identified, the next step is the selection process of business process forms each respondent. The comparison between criteria is shown in Table 6.

Table 6. Criteria Comparison Result

Respondent	Number of Activity	Explicit Knowledge	Tacit Knowledge
A	12	Well documented and can be used	Available but incidental
B	13	Well documented and can't be used	Available and managed
C	8	Not well documented and can't be used	Available but incidental

Factor rating method is used to select the best of best practice. The result from table 6 then transforms into a quantitative value. The quantitative result is shown in Table 7.

Table 7. Quantitative Criteria Comparison Result

Respondent	Number of Activity	Explicit Knowledge	Tacit Knowledge
A	12	10	5
B	13	5	10
C	8	0	10

The result of quantitative comparison then normalized and multiplied by each criteria weights. From factor rating result, it is known that the total weight of respondent A is the largest between three respondent, therefore the best practice that will be used is the best practice from respondent A. Table 8 shows the factor rating result.

3.4. Internalization

Internalization phase is the last step in knowledge conversion. In this phase, the explicit knowledge which is the best practice is disseminate into tacit knowledge. Figure 5 shows the internalization phase scheme.

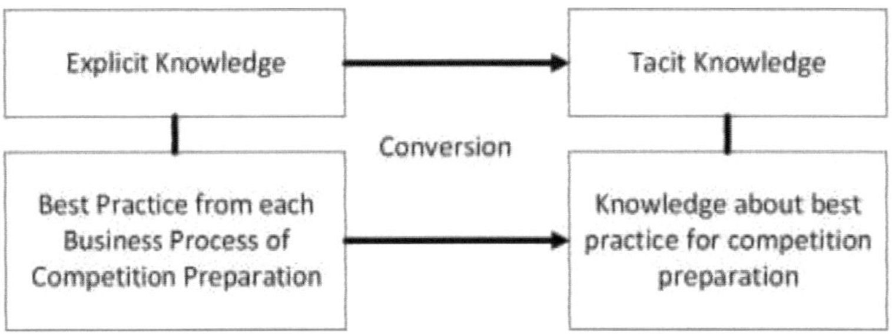

Figure 5. Internalization Phase Scheme

In this phase, Focus Group Discussion (FGD) is done to discuss the best practice result to the process stakeholder, which is the head of students affair of the school of industrial engineering, students, and laboratory assistants. The purpose of FGD is to convert explicit knowledge in form of best practice to tacit knowledge of each process stakeholder. From the FGD result, it is known that the best practice can be applied in an undergraduate program of the information system.

Table 8. Factor Rating Result

Respondent	Weight	Number of Activity 0,374	Explicit Knowledge 0,339	Tacit Knowledge 0,286	Total
A	Normalized	6,667	10	5	7.321
	Weight	2.495	3.396240806	1.4302	
B	Normalized	6.153	5	10	6.862
	Weight	2.303	1.698	2.860	
C	Normalized	10	0	10	6.603
	Weight	3.743	0	2.860	

4. Conclusions

This research intends to develop a guidance for competition preparation in the university by using knowledge conversion. The documentation of knowledge from tacit knowledge is documented. The best practice selection is done by using factor rating method. Delphi method is used to identify the criteria, and AHP method is used to calculate the weight of each criterion. From the factor rating result it is known that from 3 respondent, best practice from respondent A (7.321) is used for preparing the programming competition in an undergraduate program of information system in the school of industrial engineering Telkom University. FGD is done to disseminate the selected best practice into the process stakeholder which is head of students affair of the school of industrial engineering, students, and laboratory assistants. Future research can be done to create more comprehensive criteria for selecting the best practice.

5. Reference

[1] Wankat P C 2005 Undergraduate Student Competitions *J. of Eng. Edu.* 343-47

[2] Berkeens E 2010, Global models for the national research university: adoption and adaptation in Indonesia and Malaysia *Globalisation, Societies and Education,* **8(3)** 369-91

[3] Davenport T H and Prusak L 1998 *Working Knowledge: How Organizations Manage What They Know* (Boston: Harvard Business School Press)

[4] Faust B 2007 Implementation of Tacit Knowledge Preservation and Transfer Methods in *Proc. International Conference on Knowledge Management in Nuclear Facilities*, (Vienna)

[5] Nonaka I and Takeuchi H 1995 *The Knowledge-Creating Company* (New York: Oxford University Press)

[6] Finley D and Sathe V 2013, Nonaka's SECI Framework: Case Study Evidence and an Extension *Kindai Mgt. Rev.* **1** 59-68,

[7] Nisa A S, Kurniawati A and Pratami D 2013, Knowledge Conversion Pada Proses Perencanaan Proyek Di PT. Len Railway System Untuk Standardisasi Proses Dengan Metode SECI *J@TI* **8(1)** 27-36,

[8] Lis A 2014, Knoweldge Creation and Conversion in Military Organizations: How the SECI Model is Applied Within Armed Forces *J. of Entrepreneurship Mgt and Innov. (JEMI)* **10(1)** 57-78

[9] Nugraha N W, Kurniawati A and Hediyanto U Y K S 2015, The Design Of Best Practice On The Media Transfer Activities And Preservation Based On Knowledge Conversion With SECI Method in *Proc. of Industrial Engineering and Service Science* (Yogyakarta,)

[10] Hsu P F, Chiang H Y and Wang C M 2011 Optimal selection of international exhibition agency by using the delphi method and AHP *J. of Info. and Opt. Sci.* 1353-69

IOP Conf. Series: Materials Science and Engineering **277** (2017) 012028 doi:10.1088/1757-899X/277/1/012028

The analysis of composite properties reinforced with particles from palm oil industry waste produced by casting methods

Tugiman[1], F Ariani, F Taher, M S Hasibuan and Suprianto[2]

Department of Mechanical Engineering, Faculty of Engineering, University of Sumatera Utara, Medan 20155, Indonesia
[1]tugiman.karmani@gmail.com, [2]suprianto@usu.ac.id

Abstract. Palm oil processing industries are very attractive because they offer plenty products with high economic value. The CPO factory processes not only produces crude palm oil but also generates fly ash (FA) particles waste in its final process. The purpose of this investigation to analyze and increase the benefits of particles as reinforcement materials for fabricating aluminum matrix composites (AMC's) by different casting route. Stirring, centrifugal and squeeze casting method was conducted in this study. Further, the chemical composition of FA particles, densities and mechanical properties have been analyzed. The characteristics of composite material were investigated using an Optical microscope, scanning electron microscope (SEM), hardness (Brinell), impact strength (Charpy). The pin on disc method was used to measure the wear rate. The results show that SiO_2, Fe_2O_3, and Al_2O_3 are the main compounds of fly ash particles. These particles enhanced the hardness and reduce wear resistance of aluminum matrix composites. The squeeze method gives better results than stir and centrifugal casting.
Keywords: composite; fly ash particles; casting

1. Introduction

The palm oil processing yield shell products are used as energy sources to generate electrical. The results of a combustion process in the chamber not only create the electricity energy but also fly ash particles waste from palm oil. The main composition particles consist of SiO_2 reach 55.20% [1] and the additional oxide compound. The elements have an attractive characteristic with high melting point temperatures such as oxide materials. Fly ash from different source has been applied as reinforcement particles of composite material [2],[3],[4],[5]. The strength of composites influenced by particles shape, type, such SiC [6] and a number of FA elements [5]. The addition of fly ash improves the hardness, otherwise decrease the density of composites [7],[8] and reduce wear rate [9]. The process production technique composite reinforced by fly ash very diverse such as common stir casting [10],[11] squeeze casting [2]. From previous research known that fly ash used as reinforcement material of composites have been successful. Fly ash particles from palm oil processing are a potential candidate for the composites material cause it primary compositions are is different compared with fly ash from coal. The purpose of this research to investigate fly ash characteristics and fabrication of composites reinforced with fly ash from palm oil by several casting technique.

2. Methods

2.1. Preparation of fly ash particles

Particles fly ash are used in this research taken from palm oil industry located in North Sumatera. The grinding process was conducted to achieve small size about 0.074μm and more homogenous particles. This element heated at 800°C to eliminate the moisture and an element that can still burn. The gravimetry and spectrometry methods were conducted to analyze the composition of fly ash particles. the group compound of fly ash particles was investigated by FTIR (Fourier Transform Infra Red) method. The characteristic of density reinforcement material was examined by briquettes method with 85, 90, 95, 100 and 105 pressure ratio.

2.2. Composite production process by casting route

Aluminum matrix composites (AMC's) production reinforced by fly ash particles use stir casting, centrifugal and squeeze casting technique was conducted. 2.5wt.% fly ash elements added to aluminum matrix when the semi-solid phase followed by stirring slowly for 5 minutes. The Magnesium element about 5wt.% added to the composite melt to increase of wettability. Further, pouring temperature at 760°C. Centrifugal casting has been done with 366 rpm of speed and molding temperature at 450°C. Additionally, squeeze casting with 4 MPa pressure and mold temperature at 300°C was conducted.

2.3. Aluminum matrix composites (AMC's) product testing

Composite products have been tested to know the characteristics of the composite material. The examination consists of an optical microscope (OM), scanning electron microscope (SEM) to see the microstructure composite, Brinell hardness test using 500 kg of load. The impact test by Charpy method and pin on disc test to investigate the wear resistance with varying in sliding speed 0.3534, 0.4712, 0.589, 0.7068 and 0.8246 m/s. the wear test for composites that manufactured by squeeze casting technique with 2.5wt.% fly ash compared with raw material (A356).

3. Result And Discussion

3.1. The result of composition test

The testing of particles was conducted to knows the main elements of fly ash from palm oil. The result as shown in Table 1 below.

Table 1. The composition of fly ash from palm oil particles

compound	wt.%	methods
SiO_2	36.73	Gravimetry
Al_2O_3	6.83	Spectrometry
Fe_2O_3	0.69	AAS
CaO	0.53	AAS
MgO	3.11	AAS
K_2O	6.54	AAS
SO_3	0.92	Gravimetry
Others	balance	-

The main composition of particles are oxide compounds such SiO_2, Al_2O_3, Fe_2O_3, CaO, MgO, K_2O and SO_3, the oxide formation in the particles due to this product is the residual of shell combustion in the boiler combustion chamber, in this process the oxygen element tends to interact with the elements contained in the fuel. Silicon dioxide (SiO_2) is the highest about 36.73%, This is appropriate with previous research silicon dioxide, ferric oxide and aluminum oxide are the main contents of fly ash [1]. This oxide compound between aluminum matrix will act as reinforcement material in the composite materials. The addition of oxide particles improve the hardness and wear resistance of composite materials.

IOP Conf. Series: Materials Science and Engineering **277** (2017) 012028 doi:10.1088/1757-899X/277/1/012028

3.2. FTIR test result

The group compound of fly ash particles has been analyzed by FTIR method. The result (figure 1) shows the material consist type of compound such as alkane at a frequency of 3001.24 (1/cm) with strong absorption, alkane compound with high intensity at the frequency of 2877.79 (1/cm). The frequency area 802.39 (1/cm) is an aromatics ring with low absorption.

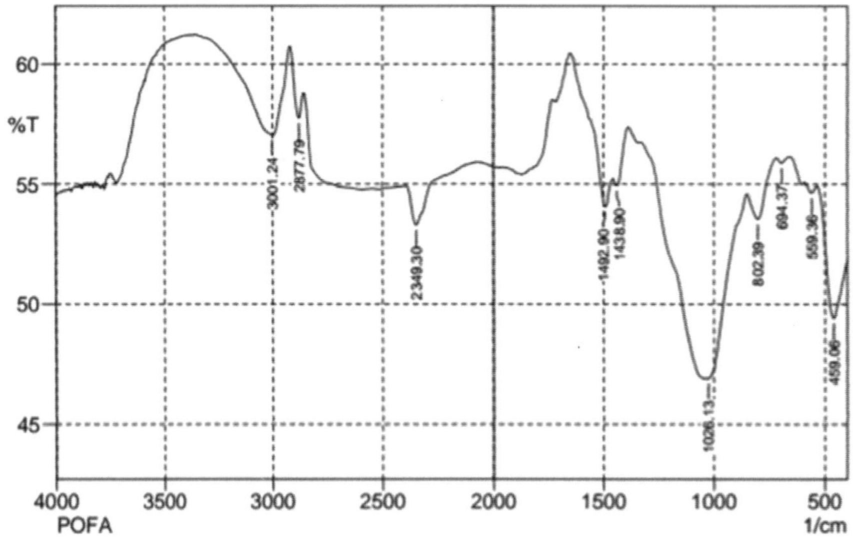

Figure 1. FTIR test result of fly ash particles from palm oil

3.3. Density of fly ash particles

The density of fly ash particles was examined by briquettes technique, the results as shown in figure 2 below.

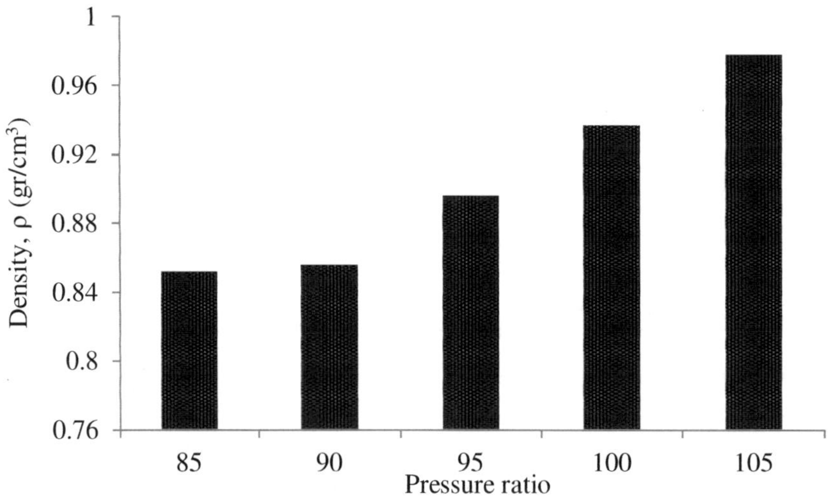

Figure 2. Effect of pressure ratio on density of fly ash particles

Figure 2 shows the increase with increasing the pressure ratio, the maximum density of particles obtained 0.978 gr/cm^3 at 105 pressure ratio. This value lower than the density of aluminum matrix

IOP Conf. Series: Materials Science and Engineering **277** (2017) 012028 doi:10.1088/1757-899X/277/1/012028

about 2.78 gr/cm^3. The density of reinforcement material and matrix are very different, this causes both of them are difficult to coalesce homogeneously by conventional casting technique. Otherwise, the advantage fly ash particles as reinforcement material are reducing the density of the composite material. The density composite is the important factor in design technique of material to reduce energy consumption.

3.4. Microstructure test results

The microstructures examine was conducted to knows the distribution of fly ash particles between aluminum matrix. Optical microscope results as shown in figure 3 below.

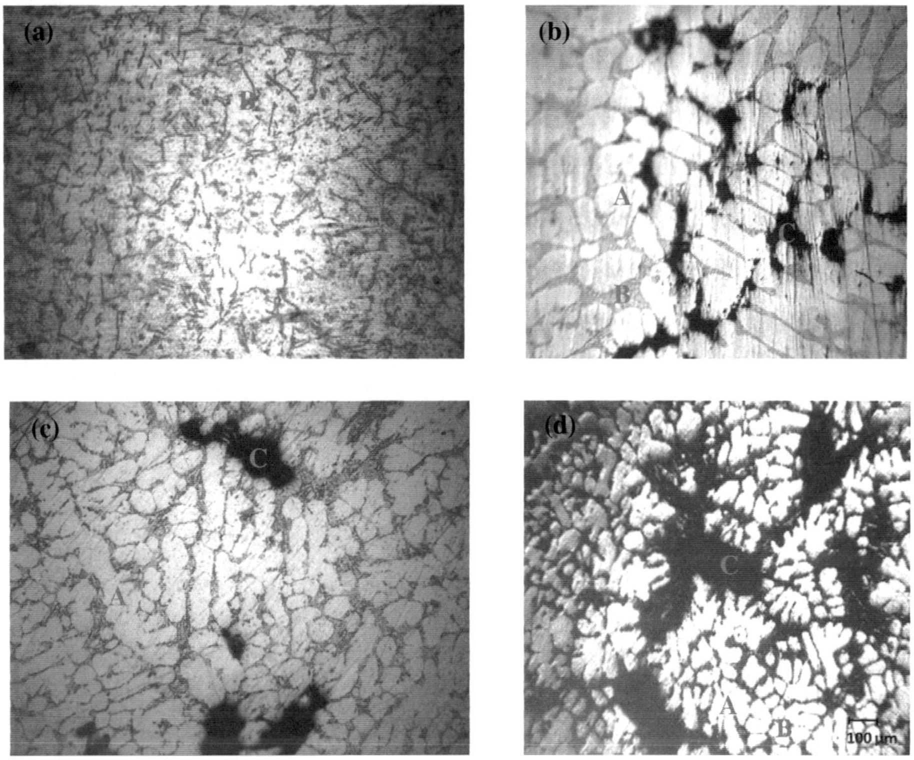

Figure 3. microstructure of composite material with 2.5wt.% fly ash (a) raw material, (b) stir cast, (c) squeeze cast and (d) centrifugal cast method

Figure 3a shows the microstructures of an aluminum cast, this microstructure consists of α-aluminum with bright color (A) and silicon particle (B) gray color between aluminum matrix. Silicone elements play an important role to improve yield strength, microhardness and reduce of ductility [12]. The addition fly ash particles by using stir casting technique (figure 3b) exhibit successful inserted fly ash particles into the aluminum matrix. The particles tend to form the cluster between α-Al matrix, and solid solution phase difficult to formed with aluminum due the different of characteristics. Microstructures produced by squeeze (figure 3c) and centrifugal casting (3d) route have successfully inserted fly ash particles (C) into solution with the dark color. The squeeze casting technique obtained smaller α-aluminum grain size, in this procedure more rapid core forming process occurs due to the pressure on mold promote heat transfer quickly. The pressure applied to this process alter the size of the silicon structure [4], reduce the number of porosity [13]. Scanning electron microscope (SEM) and Energy-dispersive X-ray spectroscopy (EDS) were conducted to know the microstructures and composition of the composite material. The test result is shown in figure 4, the dark point (A) shows the main component consists of aluminum, silicon, and oxygen. Al element act as composite matrix with the main alloy element is silicone, the presence of Si element tends to improve the hardness composite. Chinese script (B) structure Formed on this alloy and the bright color (C) is α-Aluminum.

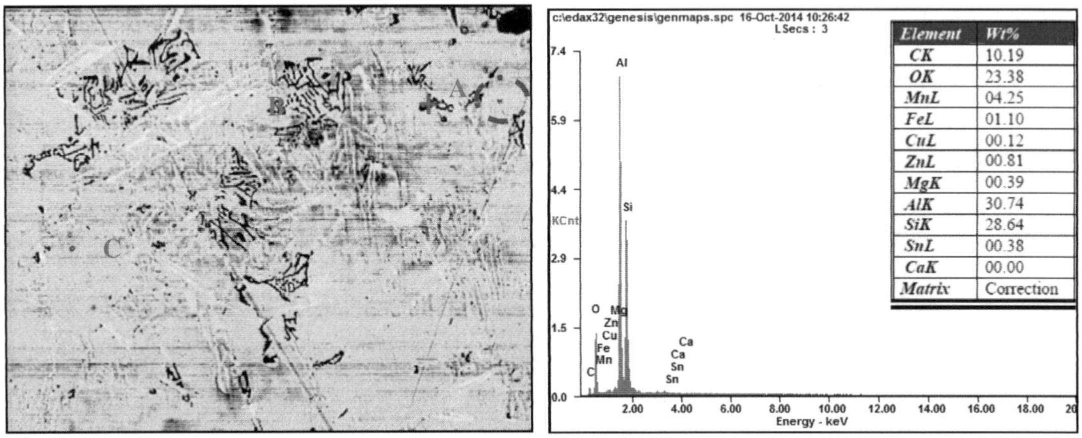

Figure 4. SEM and EDS test result of MMC reinforced by 2.5wt% fly ash

3.5. Hardness test result

The hardness test was obtained by Brinell method on raw material A356 alloy and aluminum matrix composites (AMC's) reinforced by 2.5wt% fly ash. The result as shown in figure 5 below

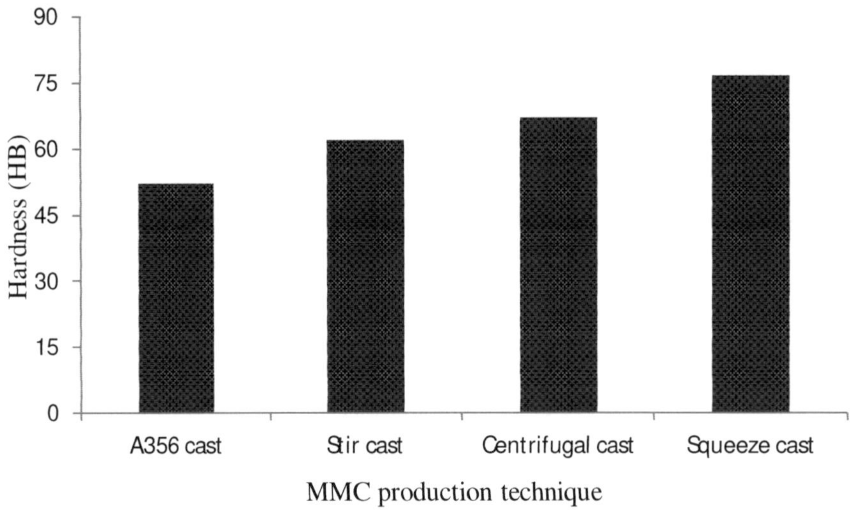

Figure 5. The hardness of MMC's reinforced by fly ash with different route

The addition of 2.5wt% particles on composite production has been successful to increase the hardness by various casting route. The increase of fly ash composition has been successful improve the hardness of composite [14]. Silica compound contained in the fly ash (FA) acts an important role to enhance the hardness of material [15]. The highest hardness about 76.85 BHN obtained by using squeeze method with 4 MPa pressure. This method increases the hardness up to 32% than raw material (A356) due to oxide particles dispersed between Al, these element acts as a barrier to dislocation movement.

3.6. Impact test result

The impact test was conducted by the Charpy method to know the effect of casting procedure on strength of composites material with load suddenly at room temperature. The results as shown in figure 6 below.

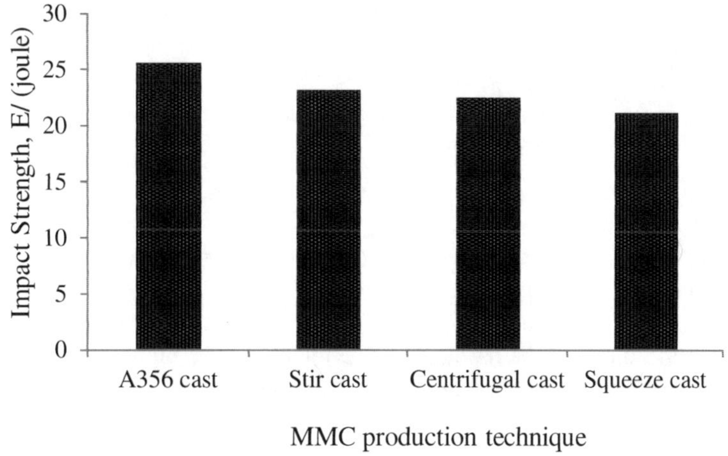

Figure 6. Impact strength for various casting technique with 2.5wt% fly ash and raw material

The addition of particles tends to reduce impact strength of the composite material for all casting procedure. it means, the toughness of materials decreased. This corresponds to diminish in composite toughness with enhanced in reinforcing particles [16]. The highest toughness obtained on A356 material produced by casting with value about 25.57 joules. Additionally, aluminum matrix composite has the lowest strength about 21.13 joule by squeeze casting technique. The phenomenon of decreasing impact strength in this case due to the addition of particles wouldn't be formed solid solution state with α-Al and the porosity in others side were formed. Furthermore, this phenomenon affects the strength of materials such as metal matrix composite (AMC) material [17].

3.7. Wear test result

Wear resistance of composite materials obtained by pin on disc test method in dry sliding condition for various sliding speed.

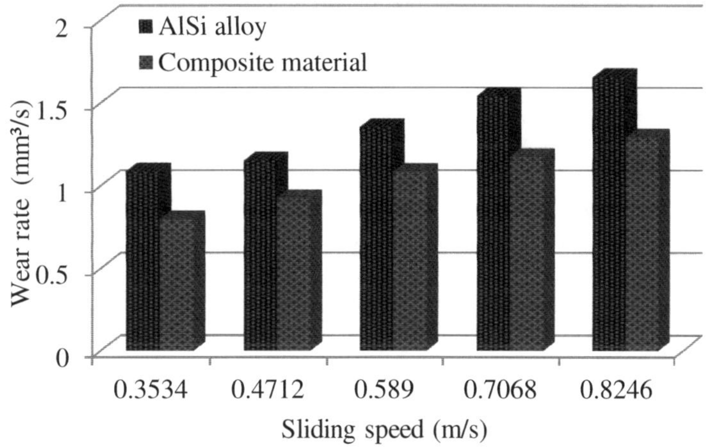

Figure 7. Effect of sliding speed on wear rate of composite and raw material

The wear test result as shown in figure 7 above shows the increase sliding speed tends to increase wear rate both of raw and composite material. This phenomenon correlated with wear resistance [18]. The increase sliding speed generates the temperature and enhance ductility [19] between pin and surface material. The increase temperatures promote wear more easily formed. From the figure 7 obtained that composite material reinforced with fly ash from palm oil has lower wear rate than the raw material in various sliding speed.

4. Conclusion

The metal matrix composite products by various casting route were conducted. From the data of this research, it can be concluded in some points

- The main composition of fly ash particles contains SiO_2, Fe_2O_3, Al_2O_3, and alkane as the main group compound.
- The density of materials increases when the pressure ratio increase.
- The addition of fly ash particles have been effectively dispersed into aluminum matrix.
- The production of aluminum matrix composites (AMC's) by casting route has been successful to increase the hardness of the material. The highest hardness obtained by squeeze cast method. This technique also reduces wear rate material compared with raw material.
- The impact strength of AMC's slight to decrease in various casting method compared with the A356 cast material.

5. Referrences

[1] Abdullah K and Hussin M 2010 Fire Resistance Properties of Palm Oil Fuel Ash Cement Based Aerated Concrete *Conrete Research* **vol.1 (3)** pp 107-114

[2] Bienias J, Walczak M, Surowska B and Sobczak J 2003 microstructure and corrosion behaviour of aluminum fly ash composites *Journal of Optoelectronics and Advanced Materials* **Vol. 5** No. 2 pp 493 - 502

[3] Lokesh G N, Ramachandra M and Mahendra K V 2014 Evaluation of Al-4.5%Cu alloy reinforced fly ash and SiC by stir and squeeze casting with rolled composites *International Journal on Mechanical Engineering and Robotics (IJMER)* **vol.2** pp 6-11

[4] Prasad K N P and Ramachandra M 2013 Effect of Squeeze pressure on the hardness and wear resistance of Aluminium flyash composite manufactured by stir-squeeze casting *International Journal of Engineering Inventions* **vol 3** pp 01-08

[5] Suragimath P K and Purohit G K 2013 A Study on mechanical properties of aluminium alloy (LM6) reinforced with SiC and fly Ash *IOSR Journal of Mechanical and Civil Engineering (IOSR-JMCE)* **vol.8** pp 13-18

[6] Meena K L, Manna A, Banwait S S and Jaswanti 2013 An analysis of mechanical properties of the developed Al/SiC-MMC's *American journal of mechanical engineering* **Vol. 1** pp14-19

[7] Lokesh G N, Ramachandra M, Mahendra K V and Sreenith T 2013 Effect of Hardness, Tensile and Wear Behavior of Al-4.5wt%Cu Alloy/Flyash/SiC Metal Matrix Composites *International Journal of Modern Engineering Research (IJMER)* **vol.3** pp 381-385

[8] Mahendra K V and Radhakrishna K 2007 Fabrication of Al–4.5% Cu alloy with fly ash metal matrix composites and its characterization *Materials Science-Poland* **Vol. 25** pp 57-68

[9] Suresh N, Venkateswaran S and Seetharamu S 2010 Influence of cenospheres of fly ash on the mechanical properties and wear of permanent moulded eutectic Al–Si alloys *Materials Science-Poland*, **Vol. 28** pp 55-65

[10] Hashim J 2001 The production of cast metal matrix composite by a modified stir casting method *jurnal Teknologi* **35(A)** pp 9-20

[11] Singla D and Mediratta S R 2013 Evaluation of mechanical properties of al 7075-fly ash composite material *International Journal of Innovative Research in Science, Engineering and Technology* **Vol. 2** pp 951-959

[12] Gupta M and Ling S 1999 Microstructure and mechanical properties of hypo/hyper-eutectic Al–Si alloys synthesized using a near-net shape forming technique *Journal of Alloys and Compounds* **Vol. 287** pp 284-294

[13] Prasad K N P and Ramachandra M 2013 Evaluation of factors affecting sliding wear behaviour of Al-fly ash metal matrix composites by using design of experiments *International Journal of Modern Engineering Research (IJMER)* **Vol.3** pp 2591-2599

[14] Kesavulu A, AnandRaju F and Deva Kumar, M L S 2014 Properties of aluminium fly ash metal matrix composite *International Journal of Innovative Research in Science,Engineering and Technology* **Vol.3** pp 17160-17164

[15] Subarmono, Jamasri, Wildan M W and Kusnanto 2008 Pemanfaatan limbah abu terbang sebagai penguat aluminium matrix composite *Jurnal Teknik Mesin* **Vol. 10** pp 109–114

[16] Alaneme K K and Bodunrin M O 2013 Mechanical behaviour of alumina reinforced aa 6063 metal matrix composites developed by two Step – stir casting process *Acta Technica Corviniensis* **Vol.3** pp 105-110

[17] Aqida S N, Ghazali M I and Hashim J 2004 Effects of porosity on mechanical properties of metal matrix composite: an overview *Jurnal Teknologi* **Vol.40(A)** pp 17–32

[18] Kandeva M, Vassileva L, Rangelov R and Simeonova S 2011 Wear-resistance of aluminum matrix microcomposite materials *Tribology in industry*, **Vol. 33** pp 57-62

[19] Kumar V, Dev Gupta R and Batra N K 2014 Comparison of Mechanical Properties and effect of sliding velocity on wear properties of Al 6061, Mg 4%, Fly ash and Al 6061, Mg 4%, Graphite 4%, Fly ash Hybrid Metal matrix composite *Procedia Materials Science* **Vol.6** pp 1365 – 1375

IOP Conf. Series: Materials Science and Engineering **277** (2017) 012029 doi:10.1088/1757-899X/277/1/012029

Promotion bureau warehouse system design. Case study in University of AA

N Parwati[1] and M Qibtiyah[2]

[1,2]Industrial Engineering Department, Fakulty of Science and Technology, University of Al Azhar Indonesia, Masjid Agung Al Azhar, Jalan Sisingamangaraja Road, Kebayoran Baru, South Jakarta 12110, Indonesia
[1]niken_parwati@yahoo.com

Abstract. The warehouse becomes one of the important parts in an industry. By having a good warehousing system, an industry can improve the effectiveness of its performance, so that profits for the company can continue to increase. Meanwhile, if it has a poorly organized warehouse system, it is feared there will be a decrease in the level of effectiveness of the industry itself. In this research, the object was warehousing system in promotion bureau of University AA. To improve the effectiveness of warehousing system, warehouse layout design is done by specifying categories of goods based on the flow of goods in and out of warehouse with ABC analysis method. In addition, the design of information systems to assist in controlling the system to support all the demand for every burreau and department in the university.
Keywords: warehouse, ABC analysis, layout, information systems

1. Introduction

In a service company such as university,it is required to have a warehouse to store all goods that supporting the business activities. Warehouse is a place to store various goods ranging from raw materials, production goods, production tools, work-in-process goods, to finished good [1]. The Promotion Bureau in University of AA, has an objective to have targeted number of intake college student, through sales and marketing activities such as exhibitions, high school visits, and events (open house, teacher seminars, and others). In supporting this objective, the bureau warehouse has to guarante that the merchandises are always available when needed. In this case, the warehousing system at the Promotion Bureau had not been arranged sistematicaly and properly. Especially in inflow outflow system that need an arrangement of warehouse layout and IT support system. In this research, the layout arrangement was using ABC Analysis method. IT system was also designed for the warehouse so that the good inflow and outflow could be record and done properly.

IOP Conf. Series: Materials Science and Engineering **277** (2017) 012029 doi:10.1088/1757-899X/277/1/012029

2. Methods

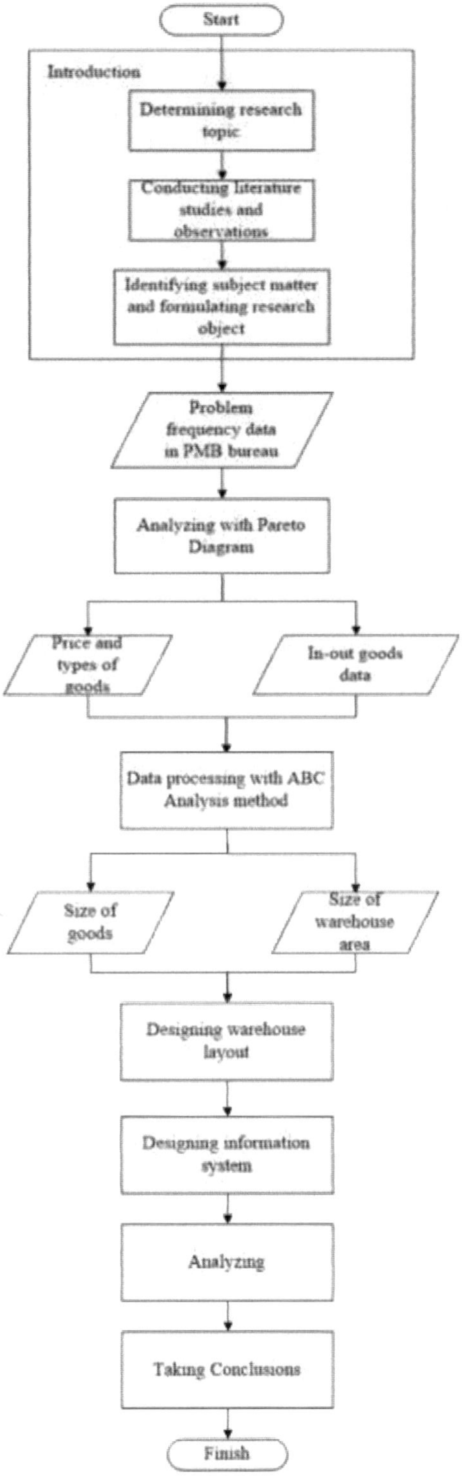

Figure 1. Research Methodology

3. Result and Discussion

3.1.Data Collection
Data collection was done by conducting interviews and observations at the research sites. In addition, the data collected also comes from the recapitulation of data on the receipt and demand of goods, delivery order, the price of goods, as well as data warehouse and size of goods in the warehouse.

3.2 Pareto Diagram Analysis
Pareto diagram analysis is conducted to find out the causes of problems that often occur in the Promotion Bureau associated with inventory or inventory. In this pareto diagram analysis, the researchers conducted interviews with Mr. Nurfadillah, S. Kom as the Promotional Officer who have access in control of inventory and warehousing. Interviews were conducted to determine the types of problems that often occur as well as the frequency of occurrence of the problem.

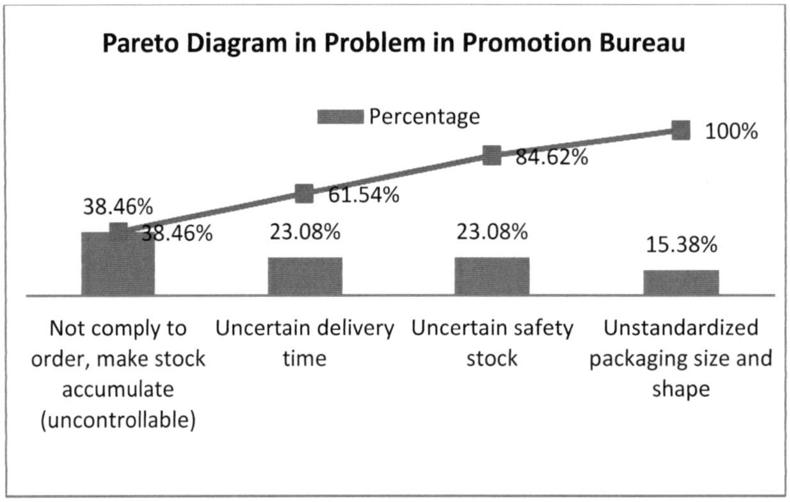

Figure 2. Pareto Diagram Analysis Results

3.3 Data Processing

3.3.1 ABC Classification
ABC Classification is a method of group creation or classification based on the rank of value from the highest value to the lowest value, and is divided into 3 major groups called groups A, B, and C [6]. In ABC method classification, was done by making a list of inflow and outflow during one period from January 2016 - January 2017. After that, the list of goods was ranked based on demand quantity.
Because of different supplier and different time of order, each type of goods, the price is not fixed, so we use average price. Then we calculate the total price in every category.
From the processing by using ABC classification method, it can be concluded that there are 7 types of goods that fall into category A with the total product price of Rp 285,916,875, -, 9 types of goods that fall into category B with the total product price of Rp 55,081. 025, -, and 30 types of goods that fall into category C with total product price of Rp 19,378,000, -. From these results it was known that the total of goods included in the categories A, B and C there are 46 items with the total price of the product as much as Rp 360.375.900.

Table 1. Classification of Goods on Promotion Bureau Warehouse with ABC Method

ABC Classification	Number of Goods	Total Price	Total Price Percentage	Cumulative Percentage
Category A	7	Rp285,916,875	79.34%	79.34%
Category B	9	Rp55,081,025	15.28%	94.62%
Category C	30	Rp19,378,000	5.38%	100.00%
TOTAL	46	Rp360,375,900		

After grouping of goods in the warehouse by the ABC classification method, the researcher did the design of the warehouse layout based on the result of ABC classification. Goods in Category A are placed near the exit / entrance of the warehouse to facilitate the operator in the process of expenditure and the entry of goods, it is because Category A has a high intensity in the outflow / incoming goods in the Promotion Bureau warehouse. Goods in Category B are placed not so far from the entrance / entrance of the warehouse, it is because the goods in this category have the intensity of being in the outflow / incoming goods on the warehouse of Public Relations Bureau and PMB. Goods in category C are placed away from the warehouse entrance / entrance, this is because the goods in this category have a low intensity in the outflow / incoming goods in the Promotion Bureau warehouse

This re-arrangement expected to assist the staff, especially the operators, in the process of taking the needs or in filling stock.

The area in the warehouse was divided into 3 areas, namely area A, which belongs to the category of Golden Zone, area B which belongs to the category of Silver Zone and area C which belong to Bronze Zone category.

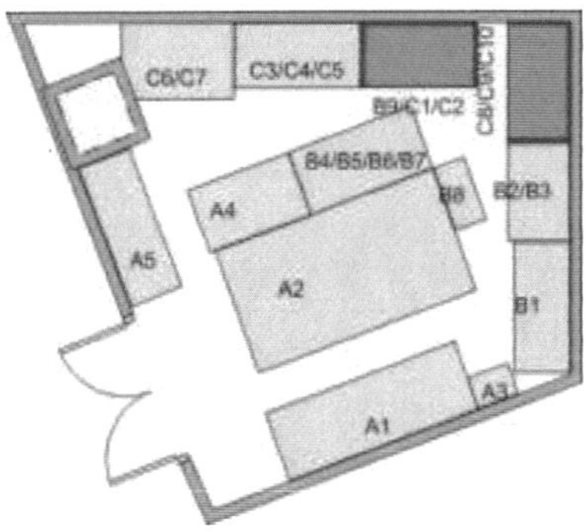

Figure 3. Proposed Layout of Promotion Bureau Warehouse

After designing the layout, the collection of category A items from the warehouse was simulated, to calculate the efficiency and effectiveness level of the layout changes in the warehouse. The simulation results show that the improved layout design can decrease the collection time of goods up to 50% from the previous oneThe calculation results can be seen in Table 2.

Table 2. Simulation of Goods Taking Time

Taking Goods Process	Before Layout Designing	After Layout Designing	Efficiency Percentage
Goody Bag	15	6	40%
Notes	20	5	25%
University Brocure	20	10	50%
Pen	15	7	47%
Blocknote	15	7	47%

IOP Conf. Series: Materials Science and Engineering **277** (2017) 012029 doi:10.1088/1757-899X/277/1/012029

3.3.2 Information System Design

Information systems can be defined as a systems made by humans in general composed of a set of integrated computer-based components as well as manual components established to collect, store and organize data and provide output information for its users [3]. In this study, besides proposing the warehouse layout design, the researcher also made the proposal of information system design. The objective was that employee can locate goods efficiently, also could control all goods in the warehouse effectively and efficiently.

3.3.2.1. Identifying Actors

In the design of information systems, first identification of actors involved in the information system. In the warehouse information system of Promotion Bureau, it is designed to have 2 actors directly involved in the information system. The actor is Requester and admin. In addition, there are also two actors who are not directly involved in the information system, namely the leadership and the procurement department.

3.3.2.2. Mapping Chart of System Information Design

In the design of information systems, Mapping Chart serves to describe the flow of documents, data and information that will run on the system. The design of Mapping Chart warehouse information system was made based on the flow of documents, data and information that goes in the process running in the information system.

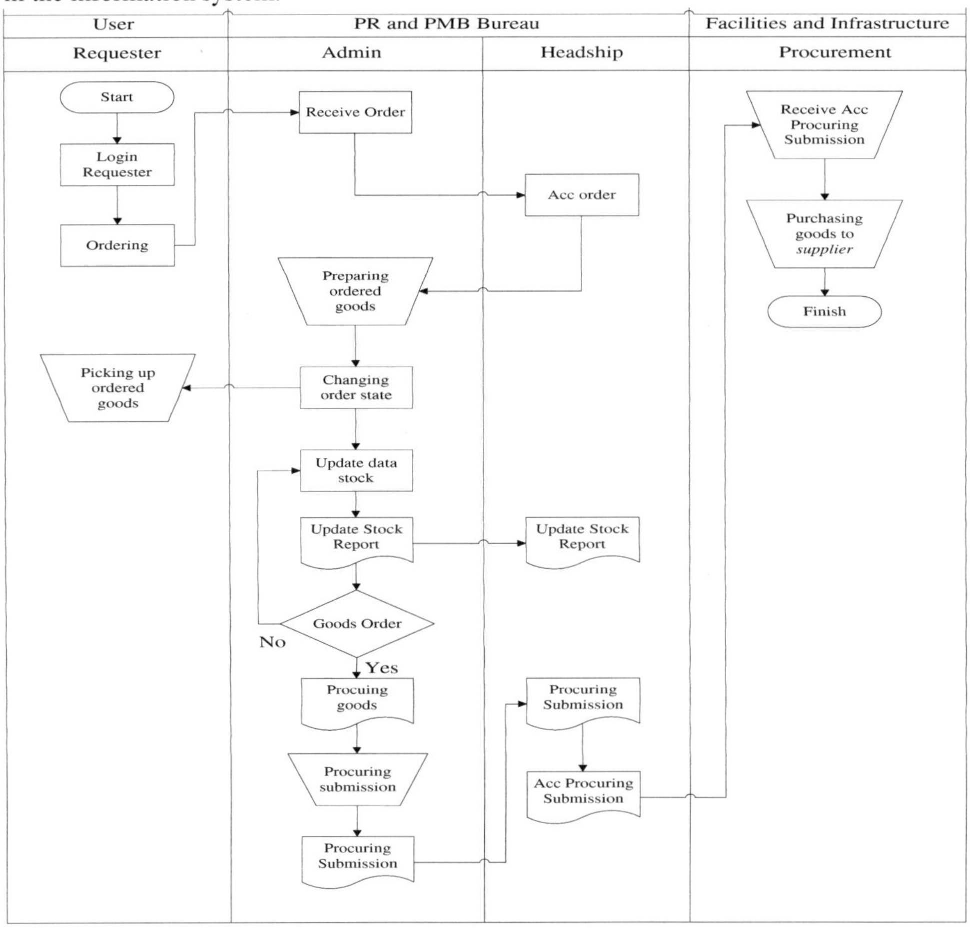

Figure 4. Promotion Bureau Warehouse Mapping Chart Information System Design

3.3.2.3. Data Flow Diagram

Data Flow Diagram (DFD) is a tool used to describe the flow of existing data in the system and a process performed by a system. To be able to make DFD required symbols to describe the flow of data [4]. In this study, DFD is described to be 3 levels:level 0, level 1, and level 2. At DFD level 0, the data flow in the information system is described thoroughly. In the picture above can be seen that DFD has 2 entities that are directly related to the system that is Requester and admin and there are 2 entities that become part support system (external entity) that is part of procurement and leadership.

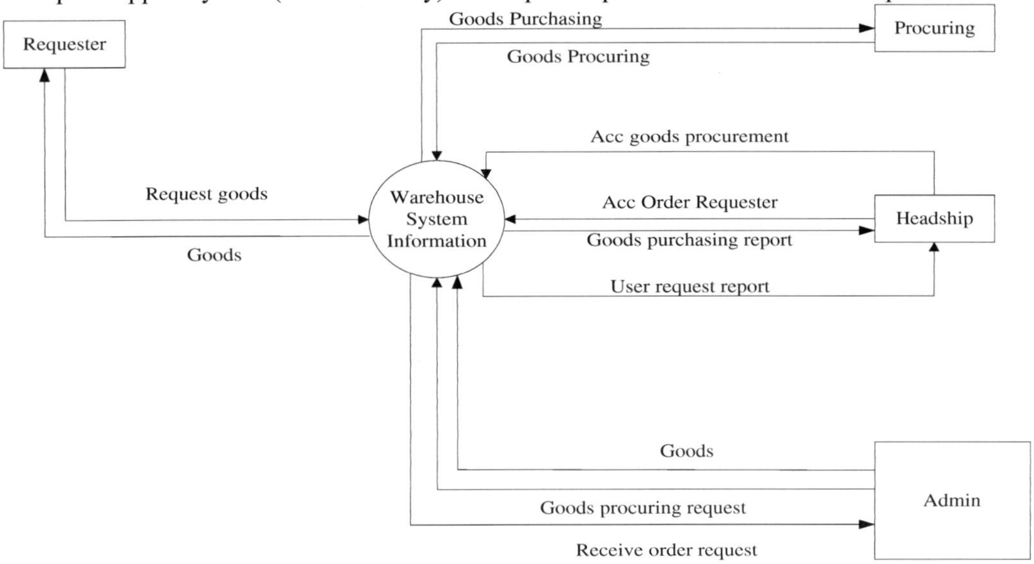

Figure 5. DFD Level 0 Design of Information Systems for Promotion Bureau Warehouse

In the DFD level 1, described the flow of existing data in the warehouse information system. It can be seen in that the DFD level 1 has 4 process data flow with 2 entities that directly relate to information systems and 2 external entity as a support system information. In addition, there are 3 databases that are used, namely data access, order data, and goods data.

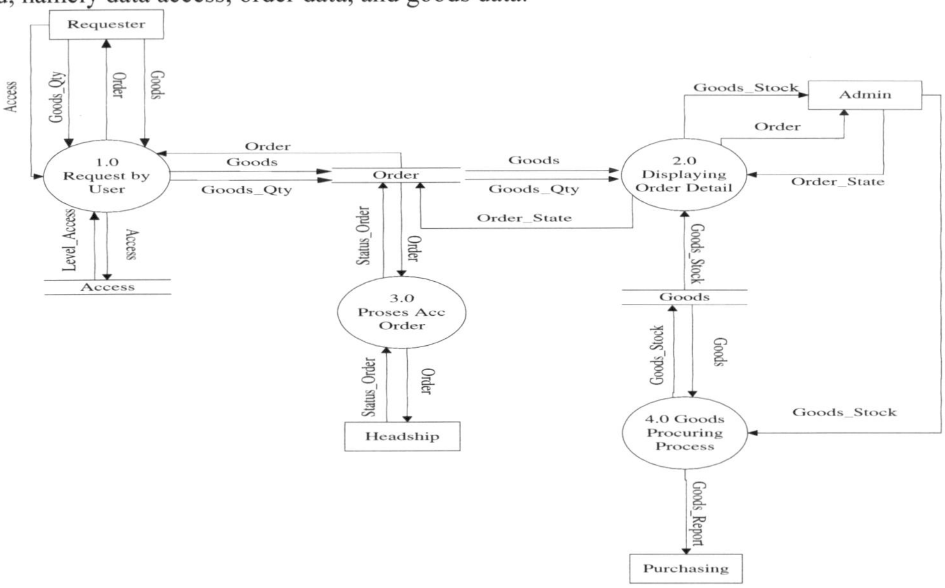

Figure 6. DFD Level 1 Information Systems Design

DFD level 2 describes the process that exists in level 1 DFD more specifically. In this case, DFD level 2 describes the Request by User process in DFD level 1. In the figure can be seen DFD level 2 has 1 entity that is Requester with 3 interconnected processes, and the database used is data access and order data.

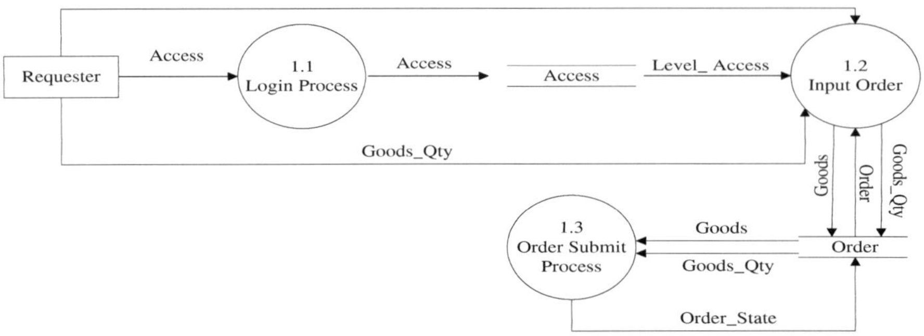

Figure 7. DFD Level 2 *Request by User*

3.3.2.4. Entity Relationship Diagram (ERD)

Entity Relationship Diagram (ERD) is a set of ways or tools to describe data or objects created based on and derived from the real world called entities (entities) and relationships between entities using several notations [2]. In information systems, Entity Relationship Diagram (ERD) is made to see the relationship between entities contained in an information system. Entities in information systems designed by researchers that divided into 3, namely entity goods, Order, Requester, and Access.

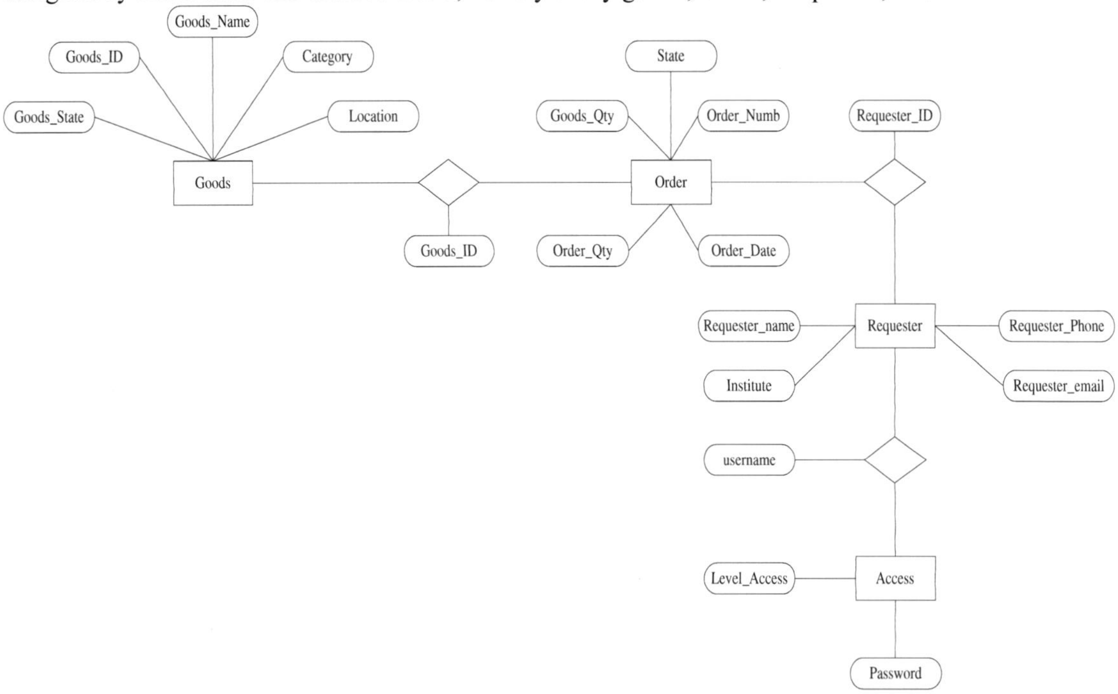

Figure 8. ERD Warehouse Information System of Public Relations and PMB

3.3.2.5. User Interface

[5] argued that the interface consists of 2 kinds of interface system and user interface. The interface system is a display for other systems, while the user interface is the view for the user. From the design that had been built, the user interface were created. For access as a requester, users can directly order.

In the order menu there is information that must be filled by requester to order goods, such as name, need, and due date or the required date of the goods. In the order menu, requester can also choose their own items needed and see the amount of goods provided by the Admin. In the order menu there is also a sign submit to submit order if the requester has finished doing the selection of goods for the order. Display menu order on the information system can be seen in Figure 9.

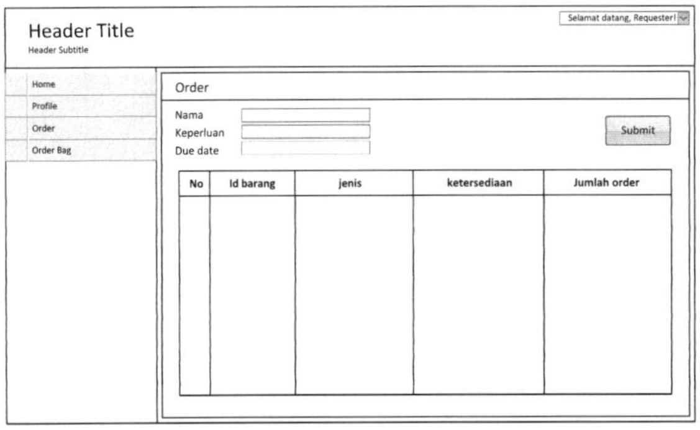

Figure 9. User Interface Layout Page *Order For Requester*

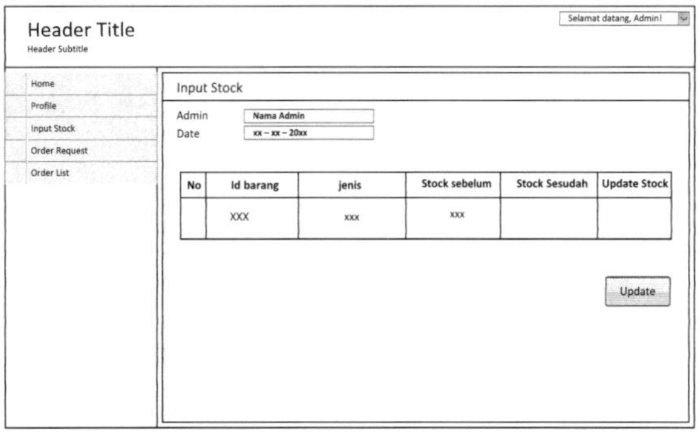

Figure 10. User Interface Layout Page *Input Stock For Admin*

Figure 10 shows the input page for the admin. When going to input, admin must first write the name and date of input stock. After that the admin can do input stock on the system. After the input is complete, the admin then selects the update menu to include stock updates on the system.

4. Conclusion

Based on the analysis and discussion on the warehousing system of Promotion Bureau, the conclusion was below:

1. The more effective system for Promotion Bureau warehouse can be built by changing the layout and develop information system to facilitate the staff in controlling the warehousing system.
2. From the results of ABC Analysis can be concluded that:
   a. From 46 types of goods, there are 7 items included in category A, 9 items belonging to category B, and 30 types belonging to category C.
   b. Proposed layout were made Warehouse layout changes can decrease the retrieval time up to 50% of the load time before the layout changes.
3. To ease the control of goods in the warehouse, the information systems was design by making ERD and DFD for information system.

5. References

[1] Apple J M translated by Nurhayati Mardiono 1990. *Tata Letak Pabrik Dan Pemindahan Bahan* (Bandung: ITB Press)
[2] Edi D and Bethsani S 2009 Analisis Data dengan Menggunakan ERD dan Model Konseptual Data Warehouse *Jurnal Informatika*, Bandung **Vol.5**, No.1, 71 – 85
[3] Gelinas UJ and Dull R B 2010 *Accounting Information Systems* (Ohio: South-Western Cengage Learning)
[4] Whitten J L, Bentley L D, Dittman K C, and Translated Team 2004. *Metode Desain dan Analisis Sistem* (Yogyakarta: Andi)
[5] Mathiassen L 2000 *Object Oriented Analysis and Design* (Denmark: Makro Publishing)
[6] Vollmann T E, Berry W L, Clay Whybark D, and Jacobs F R 2004 *Manufacturing Planning and Control Systems for Supply Chain Management* (New York: McGraw Hill)

Training needs analysis for MSMEs: how to improve training effectiveness

Y Rohayati[1,2] and S Wulandari[1,3]

[1]Industrial Engineering, Telkom University
Jl. Telekomunikasi, Terusan Buah Batu, Bandung 40257, Indonesia
[2]yatirohayati@telkomuniversity.ac.id, [3]sariwulandariit@telkomuniversity.ac.id

Abstract. The study aims to analyze training needs for MSMEs in the area of Kabupaten Bandung by selecting the case of MSMEs joined in Association for Agricultural Product Process, focusing on marketing as the main topic of the training. The needs analysis was required to improve training participation and effectiveness. Both aspects are important to notice since making MSMEs participate in training is not an easy task. Similarly, the needs analysis was carried out to anticipate participants' thoughts that the training does not give any benefits for them or is ineffective because it does not meet their needs although it was actually to help MSMEs improve their marketing knowledge expected to lead to their success. This research involved 100 MSMEs with business ages starting from less than five years to more than 15 years. Those involved MSMEs were dominated by MSMEs targeting local marketing areas. The data were collected by survey and judgmental sampling technique. By conducting a descriptive analysis, it can be concluded that the needs of SMEs on marketing training materials should focus on improving marketing skills such as product development, sales, and use of marketing media as well as discussing legal aspects such as the need for certification and product brand. The results of the study also concluded that there is a need for training that is supplemented by making visits to more successful SMEs as well as practices with on the job training methods.
Keywords: MSMEs, marketing, training need, Kabupaten Bandung

1. Introduction

To strengthen and advance MSMEs business, MSMEs actors who have similar business get together to form an association. In Kabupaten Bandung, one of the associations for MSMEs is known as Agricultural Product Process which is also the main subject of the research. This organization consists of micro and small business actors and is initiated by society running businesses in the sector of agricultural process. In 2014, this association was successful to gather 160 MSMEs both individual and small group of MSMEs business actors, made up of 400 people as a whole. This association has also been acknowledged by the local government of Kabupaten Bandung through Agriculture Department. The target of the association itself is to carry out business within the group, especially in the sectors of fostering and marketing. This association is open for others to cooperate, particularly with those supporting the aim of the association formation, intended for fostering and marketing.

MSMEs started their businesses with limited finance and knowledge whereas in a dynamic business area, an association was demanded to survive and grow to face the increasing competition from time to time [1]. MSMEs business actors often experience ups and downs in building and running their businesses because of the lack of investments including finance, accesses for support to market and government, accesses for cooperation, and other factors contributing to business success. To anticipate challenges and limitations in the running of the business, Association for Agricultural Product Process facilitated its members with training and market information. Based on the initial analysis by performing an interview with some members and staff of Association for Agricultural Product Process, it shows that the biggest problems were that their businesses did not get any development and advances due to limited knowledge, particularly in marketing sector [2]. Staffs of Association for Agricultural Product Process planned to create a fostering program regarding marketing both generally and specifically, depending on the needs of its members. The program was initiated by the exsistence of the fostering programs given by the local government, or other parties which provided material that was deemed unnecessary or not a major need so that it resulted ineffectiveness. In spite of this, so far the association has never conducted problems identification such as what aspects relate to marketing expected to solve first that become the most need by its members. Therefore, this research was conducted to analyze needs of training for MSMEs actors in Association for Agricultural Product Process. In this research, the need analysis of training focused on the need for marketing training to fill in the gap of previous studies.

Many researchers have performed the training need analysis. However, the majority of the research focused on training requirements to develop ability and competence of human resources, particularly in those MSMEs with workforces management as the main problem. Some research related to the topics above were implemented by [1], [3], [4], and [5]. In their research, [3] used human resource management approach to identify a type of training suited for MSMEs. Furthermore, [4] utilized MSMEs' perspective and Human Resource (HR) to develop sustainability organizational model for MSMEs.

2. Research Methods

2.1. Problem Formulation

Based on the purpose of this research, hence an analysis of the requirement of training according to the problem faced by UMKM was performed. To be able to capture the factors that influence the effectiveness of the training was done by identifying the problems that they encountered in marketing. The identification was conducted by interviewing the trainees who have participated in the previous marketing training. By the identification results, then the data extraction of needs was performed by distributing questionnaires to 100 UMKM. The data obtained was then analyzed with descriptive data analysis approach. MSME conditions considered in determining the training according to the needs are:

a. Problem area faced by MSMEs
b. Priority or level of needs to solve problems faced by MSMEs
c. MSMEs' experience – training topic which they have learned within the last three years as well as their opinions regarding the needs level of training that they have participated previously.
d. MSMEs' experience – regarding the benefits of the training topics that have been previously joined

In addition to paying attention to the training materials in accordance with the needs, the information about the important factors related to the implementation of training which can improve the UMKM participation to the training was also extracted, namely:

a. Time adjustment,
b. Training duration time,
c. Training location,
d. Certificate provision,
e. Early on information regarding the training,
f. Fee (whether it is paid or free),
g. Motivation level of the training participants,
h. Times allowed in the implementation of training,
i. Suitable training method.

2.2. Data Collection and Processing

To find out the needs of the training, the data were collected by using questionnaire and by using judgmental sampling technique in which the sample criteria used were 100 MSMEs in Kabupaten Bandung, running businesses in the sector of agricultural product. After the data were collected, the data were analyzed quantitatively by using descriptive statistic analysis, describing whatever the data reveals. The results of the analysis are outlined below.

3. Result and Discussion

3.1. Analysis of Descriptive- Quantitative
In following the marketing training, based on the data, it is shown that the ineffectiveness of the training was caused by one of the materials that were not by the needs. SME needs of marketing training are expected to help provide answers to the problems they have.

3.1.1. Problem area of MSMEs and Needs Level of Problem Solving
Problem area faced by MSMEs with the need level of problem-solving can be seen in Figure 1. Description for each coding of problem area in Figure 1 is as follows.

- AP.1 – the change in the use of Marketing Media
- AP.2 – Product innovation (from product variation to new product)
- AP.3 – Service improvement
- AP.4 – Change of the market needs
- AP.5 –The decrease of Market demands
- AP.6 – The increase of production cost
- AP.7 – ineffective marketing activities
- AP.8 – Difficulties in Marketing enlargement
- AP.9 – Product quality improvement
- AP.10 – the decrease of customer satisfaction
- AP.11 – the difficulties of government standard fulfillment
- AP.12 – the increase of competition
- AP.13 – the difficulties to get qualified raw materials.

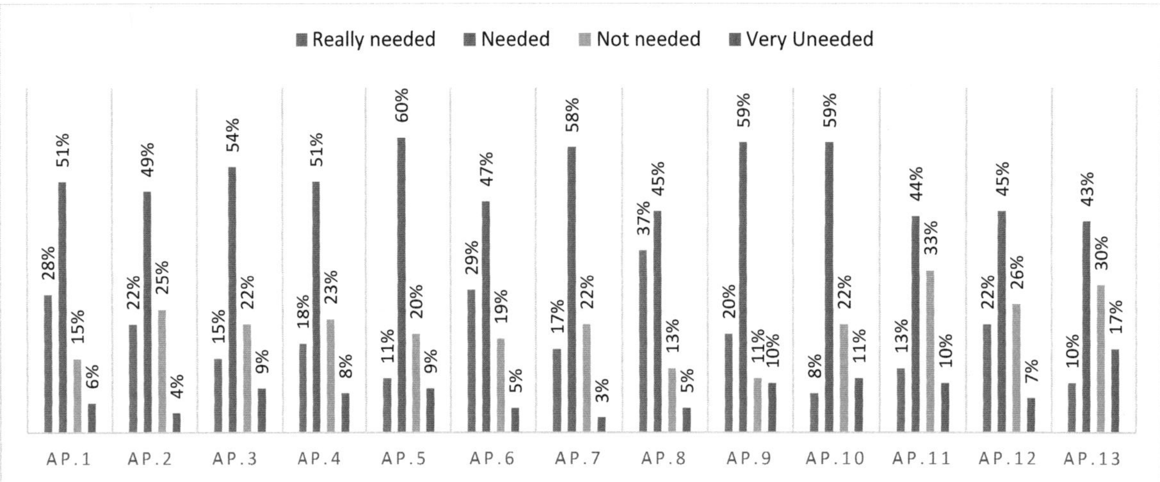

Figure 1. Problem area of MSMEs and needs level of problem-solving

Data in Figure 1 demonstrates that there are five areas of problems faced by MSMEs with high needs level of problem-solving. Those problems were in the categories of 'needed' and 'really needed.' The problems include AP 8 (Difficulties to enlarge market) with the percentage of 82%, AP 1 (change in the use of marketing media) and AP 9 (Product quality improvement) with the percentage each of 79%, AP 6 (the increase in production cost) at 24% and AP 7 (ineffective marketing activities) at 75%.

Besides, the data also reveals the problems that are in the least needed and in the least to solve. The needs –that are not needed involved three problem areas with the highest percentage respectively of 33 %, 30 %, and 26 % in the areas of AP. 11 (difficulties to fulfill government standard), AP.13 (difficulties to get qualified raw materials) and AP.12 (the increase of competition). Further, the needs level-category 'not needed at ll' went to AP.13 (difficulties to get qualified raw materials) with the percentage of 17%.

Thus it can be concluded that all this time, the development of MSME was constrained by not having knowledge about how to develop the market. This condition was supported by their ignorance in using appropriate media for marketing so that marketing activity has been considered ineffective.

Furthermore, UMKM business development was constrained by ignorance of the importance of making efficient product quality improvements, so it has resulted in substantial cost increases.

3.1.2. Training Areas that Have Been Participated and The level Benefits gained by MSMEs

The results that show the experience of MSMEs in the participation of training in the last three years can be seen in Figure 2 and Figure 3. In this case, respondents were asked to provide information regarding the experience of training by using 18 topics of training as outlined in Table 1 below.

Table 1. Topics of MSMEs training

Training topics 1-9		Training topics 10-18	
TP.1	Market trends	TP.10	Product patents
TP.2	General marketing	TP.11	Specifies STP
TP.3	Government policy	TP.12	Methods of Brand Making
TP.4	Marketing & sales using the internet	TP.13	Selling Strategies
TP.5	Packaging	TP.14	Methods of customer management
TP.6	Requirements for IPRT Certificates	TP.15	Management of quality & standard
TP.7	Requirements for **halal* Certificates	TP.16	Preparation of business plan
TP.8	Product development	TP.17	Identification of product life cycle
TP.9	Product quality improvement	TP.18	Product pricing

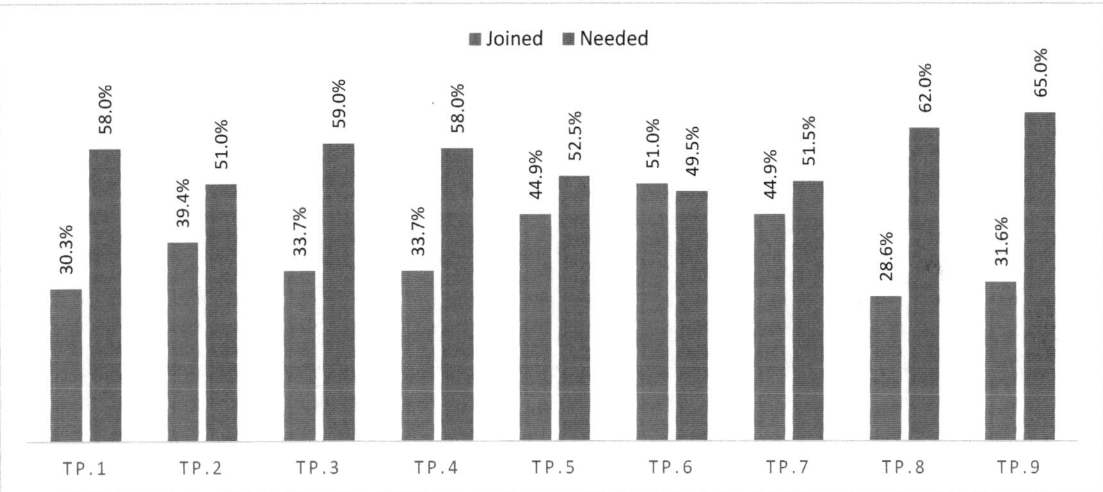

Figure 2. Training that has been joined and that is needed by MSMEs-Training topics 1-9 (in the last three years)

Figure 2 illustrates that there were four training topics that had been participated by respondents in the last three years, and all of them were the highest percentage. Those included TP.6 (requirements for PIRT certificates) at 51%; TP.7 (requirements for **halal* Certificates), and TP.5 (package) each of 44.9%. Furthermore, TP.2 (general marketing) at 39.4% was the fourth highest percentage regarding training that has been joined by respondents. Those four topics had the 'low needs value' whereas those that had 'high value' in the level of needs for training topics lied on TP.8 (product development) and TP.9 (product quality improvement).

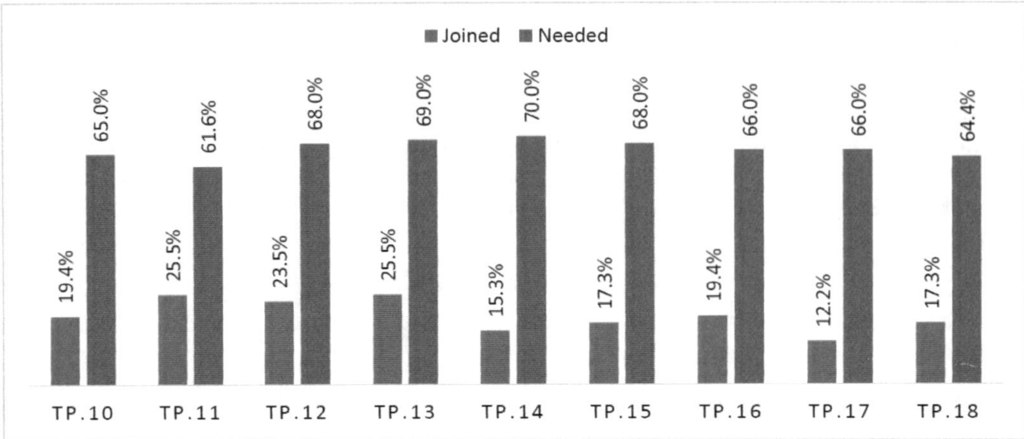

Figure 3. Training that has been joined and that is needed by MSMEs'Training topics 10-18 (in the last three years)

Furthermore, the data in Figure 3 shows MSMEs' experience for training topic 10 to 18. The data describes that MSMEs' experiences about topic 10 to 18 were categorized to be low, which was under 30% in average. However, there were three highest percentage values of needs level about the training topic including TP.14 (methods of customer management) with the needs level of 70%, TP.13 (Selling Strategies) with the needs level of 69%, and TP. 12 (Methods of Brand Making) and TP. 15 (management of quality and standard) respectively at 68%. Furthermore, the data also indicates other topics for training. Those are outlined below.

1. Training on technology
2. Consultation on business and investment
3. Training on simple administration
4. Seminar on Agriculture Safety and entrepreneurship spirit, production of Household Industrial Agriculture
5. Training on Production tools
6. Training on Mushroom Cultivation
7. Training on Production Management
8. Accounting for Enterprise Management, Franchise
9. Export

Apart from giving information on the training topic experiences, respondents were also asked to provide information on the benefit level gained from every training topic, as can be seen in Figure 4 and Figure 5. Data in Figure 4 reveals that the respondents averagely thought that the training which they had participated in were 'useful' and 'very useful' in a category. In this case, the highest percentage for 'useful' value was at 59.6 % in TP2 while TP1 and TP3 were at 57.4% and 53.2% respectively. Furthermore, the highest percentage of 'very useful' value was TP.7 of 59.6%, TP.6 of 59.3%, and TP.5 of 53.8%.

Data in Figure 5 indicates that the respondents averagely thought that the training which they had participated in were 'useful' and 'very useful' in category. In this case, the highest 'useful' value was in TP14 of 56.7%, TP.15 of 53.1%, and TP.16 of 50% while the highest percentage of 'very useful' value was in TP13 of 65%, TP.11 of 50%, and TP.18 of 48.5%. On the other hand, the highest values of topics for training in the category of 'not useful' are as follows: TP.12 as much as 13.2% and TP.10 as much as 10.5%. Both were above the percentage of 10 %.

Thus, the benefit factor of the training materials is a factor influencing the effectiveness of the training. This benefit factor is a reinforcement of their needs. The material is said to be useful if it can provide knowledge to overcome the problems experienced.

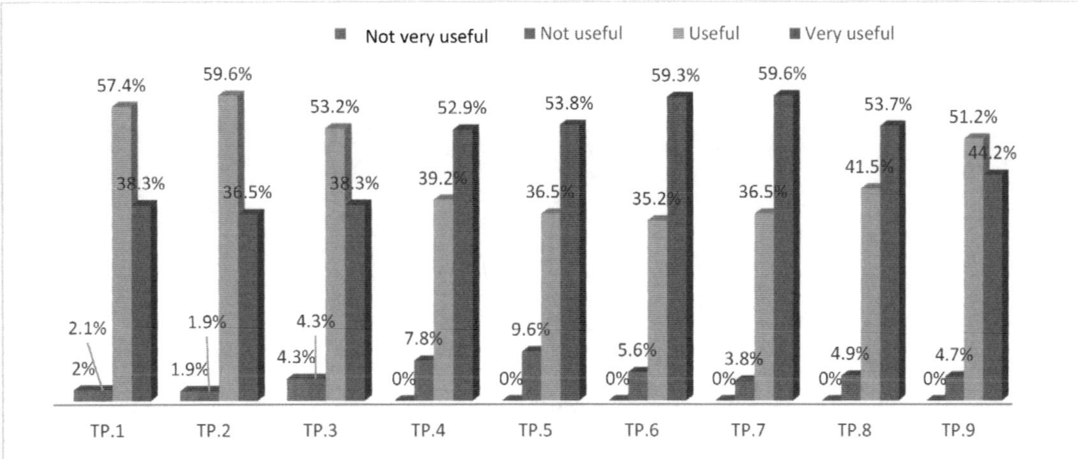

Figure 4. Percentage of Training Benefits Level (Training Topics 1-9)

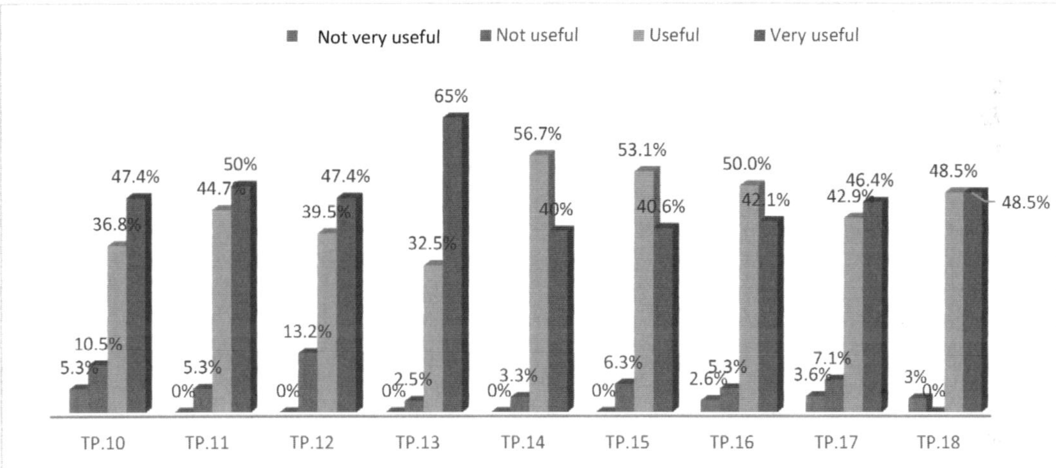

Figure 5. Percentage of Training Benefits Level (Training Topics 10-18)

The important factors in the the training implementation can be seen in Figure 6. The data in the figure shows that there were three factors categorized as 'very important,' namely; certificate provision, early on information regarding the training, and fee. These three factors gained the highest value of 'very important' category as much as 51%, 53%, and 43% respectively. Furthermore, the other factors categorized as 'very important' and high in value were: the training source person (48%), and location (31%).

Two factors that were categorized as 'important.' Those were the participant's motivation with the percentage of 77 % and training time with the percentage of 65%. On the other hand, a factor categorized as 'unimportant' which had the highest percentage of the other 'unimportant' factor was an adjustment with the percentage of 17%. Thus, the other thing to be considered in supporting the effectiveness of the training is their level of importance to the training that is held formally, as indicated by the certifier. In addition, the provision of information about the training needs to be socialized beforehand to provide an initial idea of whether or not to participate in the training, and the effectiveness of the training is also strongly influenced by the participants' motivation in the training.

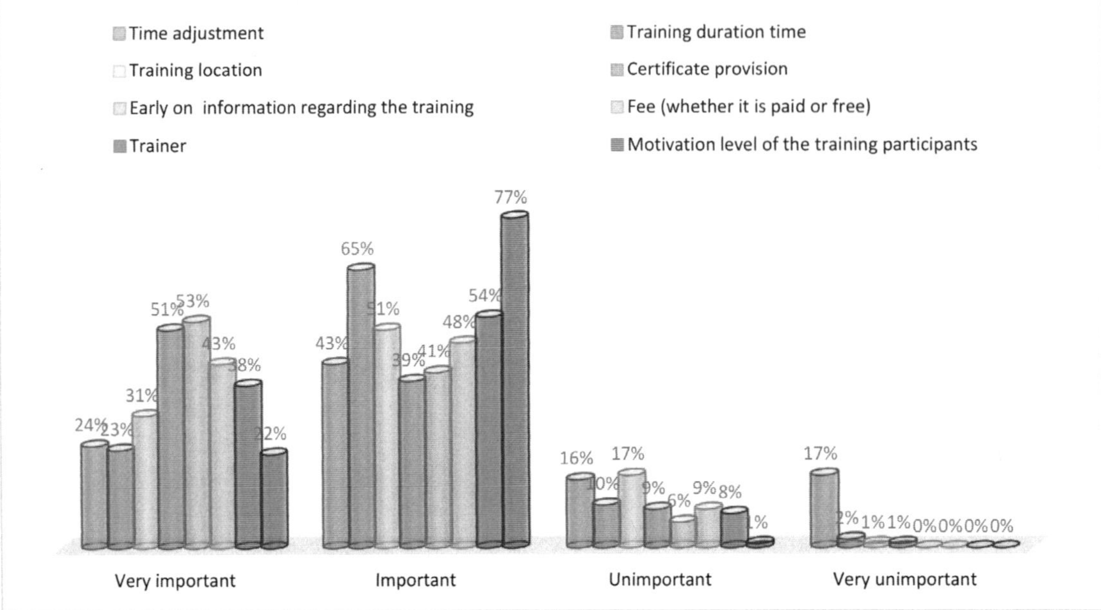

Figure 6. Percentage of Importance level of Training Implementation Factor

3.1.3. Training Methods for MSMEs

The method mostly wanted by MSMEs based on figure 7 was study tour- a visit to successful MSMEs. It is indicated by the percentage of those who 'really agree' as much as 37% and 'agree' as much as 54% and the total of 91 %. Another method mostly chosen by respondents was the practice on the job training with the percentage of those who 'really agree' of 32% and 'agree' as much as 58%, 90 % in total. The other method wanted by the participants was the simulation with the percentage of those who 'really agree' of 31% and 'agree' of 54%, becoming 85 % in total. Further, another alternative method with the smallest percentage of all alternative methods was a combination of teaching and interactive approaches wiht the total precentage of 78%.

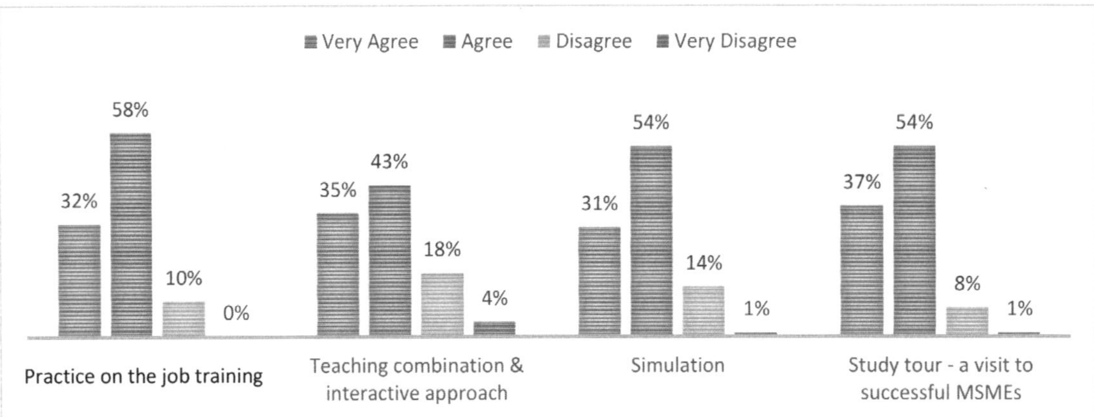

Figure 7. Percentage of Teaching Method for MSMEs Training

4. Conclusion

This research was conducted to fill the research gap in identifying the training needs based on the problems faced by MSMEs whereas the previous research has focused more on identifying training needs based on the level of competence of human resources of MSMEs, with the ultimate goal of developing human resource capability as the study which has been done by [3], [4], [1] and [5]. The facts indicate that the problem of MSMEs about the training needs is influenced by experience and the level of importance in problem solving. Therefore, this study is different from others because it does not consider the competence of human resources.

Thus, the focus of this research is to determine the effectiveness of training that can be seen from two sides, that are from the materials and the training methods which are adjusted to the problems faced by MSMEs. Training material provision as needed and acceptable training methods are the factors that can be considered to carry out the training for MSMEs. Based on the results of research, it is shown that the problems of MSME which required priority in problem solving were difficulties for market expansion, change of way of using marketing media, improvement of product quality and production cost, and ineffective marketing activity. This indicates that MSMEs are still having difficulty in acquiring customers and doing basic marketing activities such as selling and developing products (product quality and cost).

Based on the necessary materials of training besides marketing training, MSMEs actors also need a type of training that can support legal aspects such as the use and the set on brand and product certificates while from the training implementation aspect, it is found that motivation is the most important factors in the participation of training. Another factor to consider when performing a training is the initial announcement given about the training and the trainer. The most expected training methods to obtain SMEs are to make a visit to SMEs that have succeeded in their business and to get a training with on the job training and simulation.

5. References

[1] Lakshminarayanan S, Pai Y P, and Ramaprasad B S 2016 Competency need assessment: a gap analytic approach *Industrial and Commercial Training* **Vol 48** Issue: 8 pp 423-430

[2] Rohayati Y, and Wulandari S 2016 Development of Marketing Training Model for SMEs *Proceeding Sixth International Conference On Industrial Engineering and Operation Management* pp 2501-2505

[3] Thassanabanjong K, Miller P, and Marchant T 2009 Training in Thai SMEs *Journal of Small Business and Enterprise Development* **Vol 16** Issue: 4 pp 678-693

[4] Darcy C, Hill H, McCabe TJ, and McGovern P 2014 A consideration of organizational sustainability in the SME context: A resource-based view and composite model *European Journal of Training and Development* **Vol 38** Issue: 5 pp 398-414

[5] Agostini L, Nosella A, and Filippini R 2017 Does intellectual capital allow improving innovation performance? Quantitative analysis in the SME context *Journal of Intellectual Capital* **Vol 18** Issue: 2 pp 400-418

IOP Conf. Series: Materials Science and Engineering **277** (2017) 012031 doi:10.1088/1757-899X/277/1/012031

Design of customer knowledge management system for Aglaonema Nursery in South Tangerang, Indonesia

D Sugiarto[1], I Mardianto[2], TS Dewayana[3] and M Khadafi[1]

[1]Faculty of Industrial Technology, Information System Program, Trisakti University, Jakarta 11440, Indonesia
[2]Faculty of Industrial Technology, Informatics Program, Trisakti University, Jakarta 11440, Indonesia
[3]Faculty of Industrial Technology, Master of Industrial Engineering Program, Trisakti University, Jakarta 11440, Indonesia
dedy@trisakti.ac.id

Abstract. The purpose of this paper is to describe the design of customer knowledge management system to support customer relationship management activities for an aglaonema nursery in South Tangerang, Indonesia. System. The steps were knowledge identification (knowledge about customer, knowledge from customer, knowledge for customer), knowledge capture, codification, analysis of system requirement and create use case and activity diagram. The result showed that some key knowledge were about supporting customer in plant care (know how) and types of aglaonema including with the prices (know what). That knowledge for customer then codified and shared in knowledge portal website integrated with social media. Knowledge about customer were about customers and their behaviour in purchasing aglaonema. Knowledge from customer were about feedback, favorite and customer experience. Codified knowledge were placed and shared using content management system based on wordpress.
Keywords: aglaonema, customer knowledge management, content management system

1. Introduction

Based on South Tangerang statistics [1], aglaonema is an ornamental plant with the fourth largest area after orchids, adenium and philodendron with area 10570 m^2. Aglaonema plant species together with the type of anthurium is a type of plant that is currently experiencing pressure in terms of marketing and selling prices. The pressure influenced by the growing number of plants are imported from Thailand which is known as a developed country in terms of horticultural crops. This rising pressure require aglaonema nursery to redesign their business model. One possible solution is to focus on the processes of the customers through customer relationhip management (CRM) supported by knowledge management (KM). Several studies about customer knowledge management have been conducted. Research by [2],[3] illustrate the use of information technology to provide knowledge support to CRM sub processes (campaign management, lead management, offer management, contract management, complaint management and service management) and KM takes the role of a service provider for CRM, managing the four knowledge aspects content, competence, collaboration and composition to satisfy customer requests within stated budget restrictions.

There are several types of customer knowledge management. They are knowledge about customer, knowledge from customer and knowledge to customer [3]. Knowledge about customer are type of customer-oriented knowledge comprises information about the customer regarding his purchase and payment behavior, his motivation, buying habits and demands. Knowledge from customer mostly arrives at the company in a direct way, the customer informs the organization about his experiences with products, services, processes or his expectations. Knowledge to customer when the customer shares his with another company this company is knowledge gaps and to further develop the customer's 'non-knowledge' [3]. CRM can provide interpersonal messaging services through the website, analysis of customer purchasing data so that it can be determined who the potential customer is, consultation through the FAQs forum, complaint service and suggestions and delivery of promotional materials to customer email [4]. Management systems can be defined as systems that support the creation, collection, storage, maintenance and deployment of firms and corporate knowledge [5].Customer knowledge management system was created for Diah Nursery at South Tangerang. The decreasing sales occurred on Diah Nursery in 2015-2016. The purpose of this research was to describe the process of designing customer knowledge management system to support some of customer relationship management activities for an aglaonema nursery in South Tangerang, Indonesia

2. Methodology

The steps started with knowledge identification (knowledge about customer, knowledge from customer, knowledge for customer. and knowledge capture using in-dept interview to the owner and staff at Diah Nusery, focus group discussion with the aglaonema nursery community and the customer. Knowledge for customer codified using decision tree. Steps are then followed by the analysis and design of system requirement, creating use case diagram and activity diagram and creating content management system using wordpress. Knowledge about customers obtained from sales data entered using the facility of google form connected with the portal. Knowledge from customers also obtained by using the facilities of google form plugin to wordpress

3. Result and discussion

The result showed that some key knowledge for customers were about supporting customer in plant care and types of aglaonema including with the prices. Knowledge of plant care related to the fertilization process, the treatment process of damaged plants, the preparation process of planting media, the watering process and the process of preparing the plant location. Knowledge of plant selection is packed through a plant decision tree that begins with questions about the desired price range, leaf color and leaf shape.

Table 1. Requirement of knowledge management system and technology support needed

No	Customer Knowledge Management System Requirement	Web tool
1	A media that can be used to write articles about aglaonema production process and plant care, new product information and gallery (knowledge for customer)	Page management
2	Media that can be used for knowledge codification and application (knowledge for customer)	Decision tree and decision support module
3	Media that can be used for discussion among Diah Nursery staff and customer or nursery community (knowledge from customer)	Discussion forum
4	Media that can be use for customer feedback and survey (knowledge from customer)	Google form plugin to CMS
5	Media for customer analytics (knowledge about customer)	Google form plugin to CMS

The result of the user requirement for the development of the aglaonema customer knowledge management portal can be seen in table 1. The features created enable the process of acquisition, dissemination and storage of knowledge. The acquisition, knowledge sharing and application can be mediated through google form module, forum, page management and decision support module.

IOP Conf. Series: Materials Science and Engineering **277** (2017) 012031 doi:10.1088/1757-899X/277/1/012031

In the usecase diagram (figure 1), we know that there are 3 main roles that are guest, customer and admin. Each user has different role in the application. First, guest is the role that has no login information. Hence, guest can only access 5 part of this program that are login authentication, sign up for member, read public article, give a comment in article section and open the gallery. Second, customer is the role that has access to private page. Customer can access 4 part of the program that are read the private and public article, give a comment in article section, open gallery and use the decision tool. Third, admin is the role for maintaining the application. Admin has access to all part of the program. Spesifically, admin can add new user, make a decision tool, input sales data and update the gallery

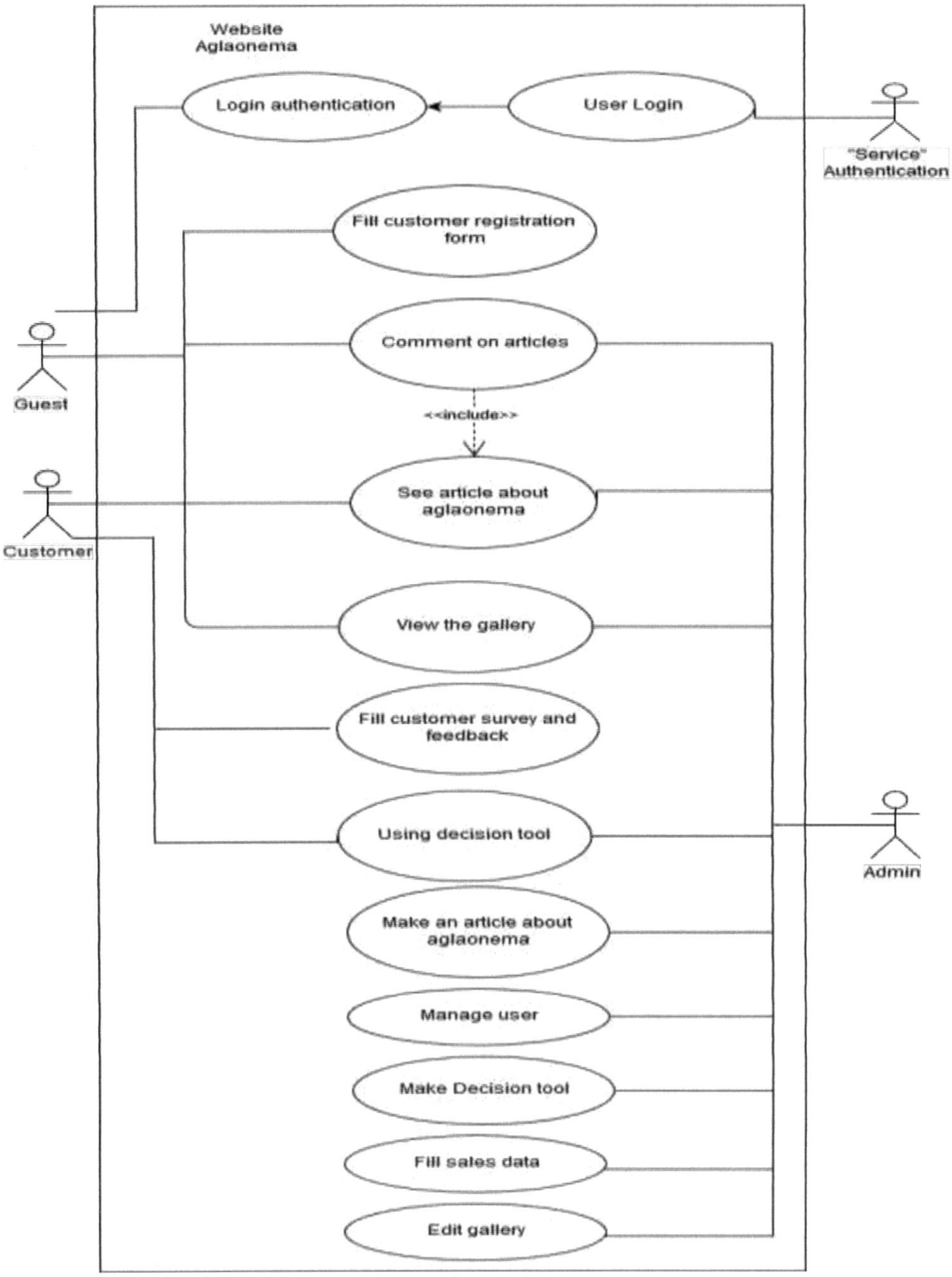

Figure 1. Use Case Diagram

Several activity diagram were created to describe use case diagram in detail. Create Article activity diagram (figure 2) aims to sharing knowledge about aglonema production and plant care. This activity is created by admin to be read by the customer. Decision Tool activity diagram (figure 3) aims to assist customers at making decisions in buying process and aglonema plant care.

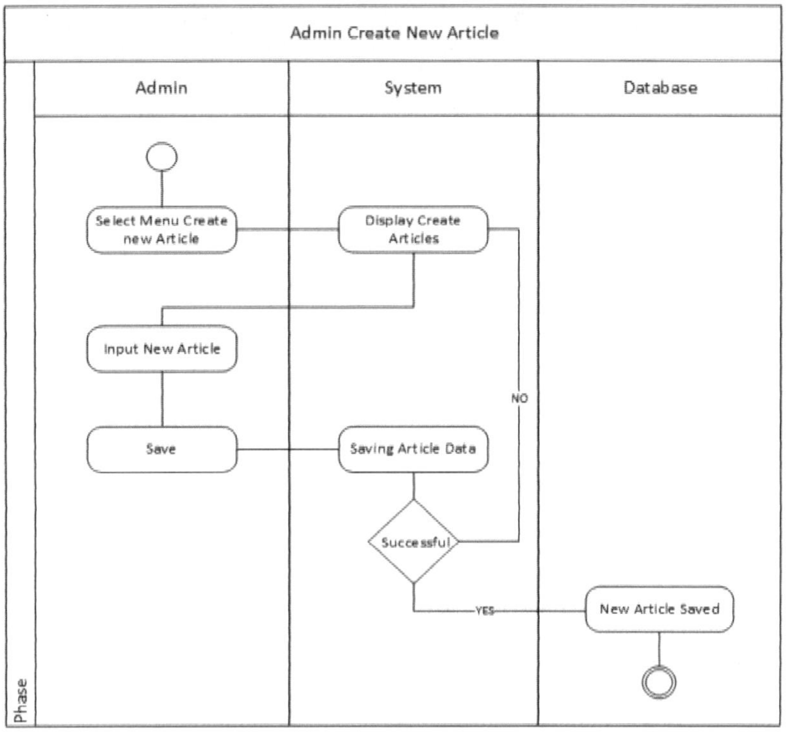

Figure 2. Create article activity diagram

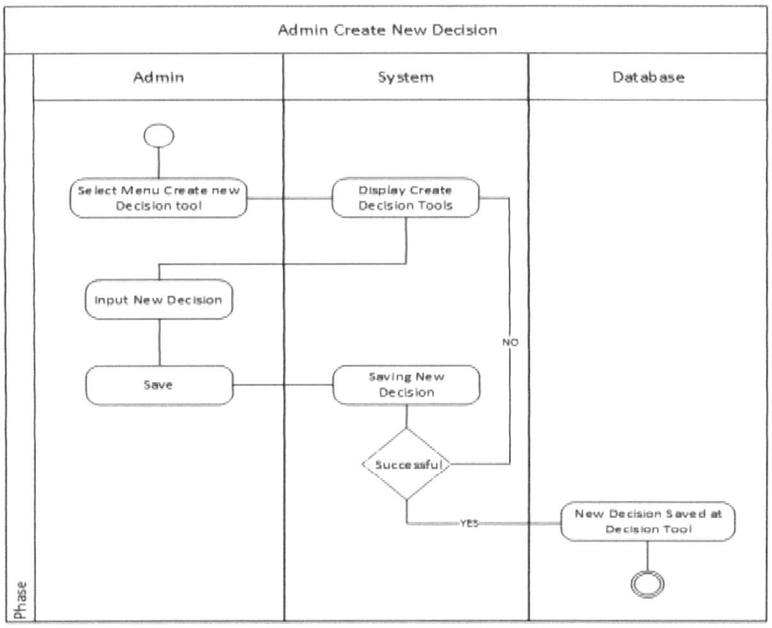

Figure 3. Decision tool activity diagram

Customer knowledge management portal (figure 4) was created with the ability to connect to youtube and give link to facebook, It is significant considering nowdays facebook as a tool to market their product.

Figure 4. Customer knowledge management portal

Decision tool (figure 5) can guide customer in purchasing process. Decision tree begins with questions about the desired price range, leaf color and leaf shape. The end of decision tree is the name of the type of plant that corresponds to the customer decision option and its price.

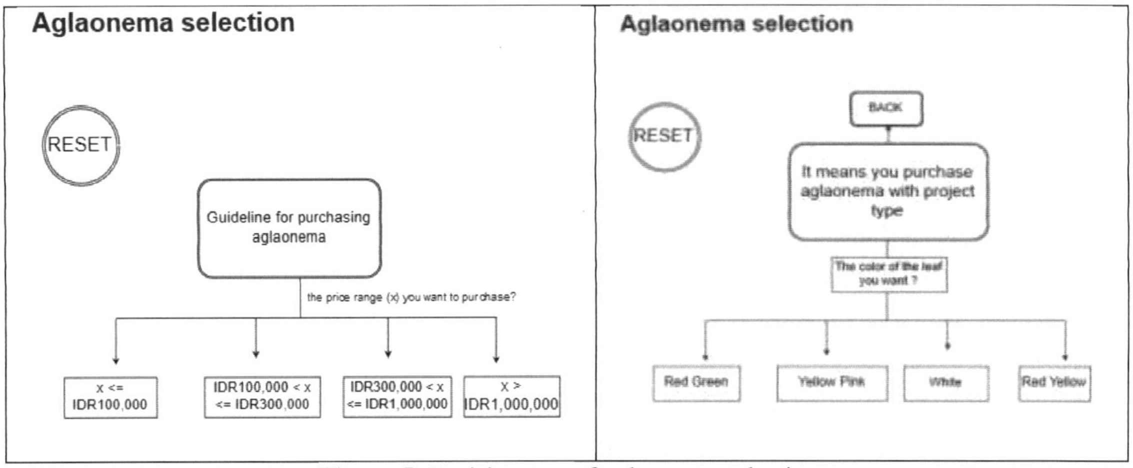

Figure 5. Decision tree of aglaonema selection

IOP Conf. Series: Materials Science and Engineering **277** (2017) 012031 doi:10.1088/1757-899X/277/1/012031

4. Conclusion

Development of customer knowledge management system which is integrated of customer relationship management system and knowledge management system was built for Diah Nursery at South Tangerang. Three types of knowledge (about customer, from customer and for customer) were identified and installed to portal based on content management system. The acquisition, knowledge sharing and application can be mediated through google form module, forum, page management and decision support module.

Acknowledgement

This research was supported by the Directorate General of Higher Education, Ministry of Education and Culture of the Republic of Indonesia

5. References

[1] Badan Pusat Statistika 2015 *Tangerang Selatan dalam Angka Tahun 2015.*

[2] Bueren A, Schierholz R, Kolbe L and Brenner W 2004 Customer knowledge management - improving performance of customer relationship management with knowledge management *Proc. Int. Conf. on System Sciences.*

[3] Gebert H, Geib M, Kolbe L and Riempp G 2001 Towards Customer Knowledge Management: Integrating Customer Relationship Management and Knowledge Management Concepts *Proc. Int. Conf. on Electronic Business.*

[4] Wilde S 2011 *Customer knowledge management – Improving customer relationship through knowledge application.* (Springer-Verlag Berlin Heidelberg)

[5] Tiono, W. J. H. 2015. Sistem Informasi Manajemen Hubungan Pelanggan Apotek X Berbasis Web. *Calyptra: Jurnal Ilmiah Mahasiswa Universitas Surabaya*, **Vol. 4**. No 1

[6] Laudon KC, Laudon and Laudon JP 2012 *Management Information Systems – Managing The Digital Firm.* 12th Edition. (Boston: Prentice Hall).

An overview of knowledge management (KM) issues for implementation in consultant firms in Malaysian construction industry

Azlan Othman [1], Syuhaida Ismail [2] and Khairulzan Yahya [3]

[1,3] Faculty of Civil Engineering, Universiti Teknologi Malaysia, Johor Bharu, Malaysia
[2] UTM Razak School of Engineering and Advanced Technology, Universiti Teknologi Malaysia, Kuala Lumpur, Malaysia
[1]irazlan@gmail.com, [2]syuhaida.kl@utm.my, [3] khairulzan@utm.my

Abstract. In the past few years, there has been a growing interest in treating knowledge as a significant organisational resource. Thus, effective development and implementation of KM requires a foundation in several rich literatures. As a preparation for the competitive industrial nation, KM is an important countenance that should be the point of convergence for the industry players. This paper wishes to draw the attention on the current situation of KM practice, focusing on consultant firms in Malaysian construction industry. Questionnaires were distributed to about 200 respondents working in the industry, with the objective of appraising the KM implementation amongst consultant firms working in construction industry in Malaysia. This paper also gives the overview on KM definition, process, understanding and challenges in construction industry, besides the critical success factor of KM implementation. The literature is restricted on the recent KM study of 17 years research from 2000 to 2017. Finally, this paper proposes the conceptual ideas of relationship between KM process, KM understanding and KM challenges with critical success factor of KM implementation.
Keywords: Knowledge management (KM); consultant firm; Malaysian construction industry; KM implementation.

1. Introduction

Construction Industry is considered to be one of the important sources of economic development, growth and activities. Malaysian economy has to perform efficiently and to play effective role in making true the dream of developed nation status. According to Prime Minister of Malaysia, Dato' Sri Najib bin Tun Abdul Razak the importance of the construction industry to Malaysia's economy cannot be overstated, where construction in Malaysia has traditionally been a substantial driver of growth, and looking to the future, thus it is expected that this trend to will continue to expand [1].

A successful construction can only be achieved with good civil engineering design and consultations, which require decent engineering knowledge and experience. However, there is a gap in managing the civil engineering knowledge and expertise amongst the consultant firms in the construction industry. In general, a consulting firm is a business of one or more experts (consultants) that provides professional advice to an individual or an organization for a fee. The primary purpose of a consulting firm is to provide access to industry-specific specialists/consultants and subject-matter experts. There are different types of consultant firms serving different sectors. This paper mainly focuses on construction industry, which falls under civil engineering fields.

In the recent time, many challenges faced by the engineering consultant firm because construction project have turn into more complicated scenario. Hence, knowledge is seen as one of the most important resources contributing towards decision making in any organization and to maximize competitive advantage of organizations to survive in construction industry. In general, knowledge management (KM) is the process of creating, modifying, sharing and applying knowledge to create value from organization's knowledge assets [2]. KM systems and related initiatives have become a popular focus in many firms, yet many KM systems initiatives fail to achieve their goals. Therefore, there is a need of continuous efforts to improve the use of KM integration in construction industry and overcome projects diversity, complexity and non-standard production methods.

2. Process of knowledge management (KM)

In order to better understand how to share and manage knowledge, an operational framework should be initially developed since knowledge management (KM) is about people and the process that used to share information and build knowledge [3]. The effective implementation of KM requires sequent processes that ensure and encourage the knowledge flow [4]. Table 1 shows the literature summary of KM process based on questionnaire create in this paper.

Table 1. Processes of knowledge management (KM) practice.

No	Knowledge Management Process Elements					
1	**Identification**	[5]	[6]	[7]	[8]	[9]
B.1	Knowledge assets (information, skills) is identifiable		X			
B.2	Identification of the different types of knowledge (tacit, explicit) available in the organization	X				
B.3	Core values and competencies identified in the organization				X	
B.4	Opportunities for innovation identified to create competitive advantage.				X	X
2	**Capture**					
B.5	Knowledge and information is captured in minutes of meetings			X		
B.6	Knowledge and information is captured in seminars			X		
B.7	Knowledge and information is captured in presentation sessions			X		
3	**Dissemination**					
B.8	Knowledge dissemination is encouraged through workshop					
B.9	Knowledge dissemination is encouraged through meetings					
B.10	Documents are disseminated to authorized personnel through appropriate channel	X				
4	**Utilization**					
B.11	Best practices from previous project have been utilized in the new projects	X				
B.12	Lesson learnt from previous project will be utilized in the future projects	X				
B.13	New knowledge utilized by team members in the company/organization	X				
5	**Sharing**					
B.14	Knowledge sharing (best practices/lesson learnt) is encouraged through meetings			X		

IOP Conf. Series: Materials Science and Engineering **277** (2017) 012032 doi:10.1088/1757-899X/277/1/012032

Table 1. Processes of knowledge management (KM) practice (cont.)

B.15	Knowledge sharing (best practices/lesson learnt) is encouraged through presentations			X	
B.16	Knowledge sharing (best practices/lesson learnt) is encouraged through publications				X
B.17	Knowledge is shared among peers	X			
6	**Storage**				
B.18	Documents and reports properly stored in a softcopy database system		X		
B.19	Documents and reports stored in hardcopy in the library		X		X
B.20	Physical storage uses established records management or archival practices, with adequate shelving durable boxes, folders, labeling etc.				X
B.21	The repository for electronic resources has adequate capacity for long-term storage				
B.22	The repository for electronic resources is backed up routinely, based on established and enforced procedures and protocols.				
B.23	The location of stored resources, physical and electronic is reasonably convenient and accessible				X
B.24	Documents securely stored and protected		X		
7	**Strategizing**				
B.25	KM process has been well strategized by the company/organization				
B.26	Strategic knowledge well protected				
B.27	Training of workers required to meet the strategic objectives in the organization			X	
B.28	Employees well understood and aware of KM strategies in the organization				X
B.29	Relevant knowledge assets created in the organization				

3. Understanding of knowledge management (KM) implementation

The understanding of knowledge management (KM) will produce learning and facilitate knowledge sharing culture and environment, provide vision and effective leadership to overcome learning barriers. In fact it will help an organization to be transformed into a learning organization, where it is open to learn new techniques and continuously changes itself based on learned knowledge. Table 2 shows the literature summary of KM understanding based on questionnaire create in this paper.

Table 2. Understanding of knowledge management (KM) implementation.

No	Measure of Understanding the Importance of Knowledge Management (KM)	[10]	[11]	[12]
1	**Attitude**			
C.1	Information and knowledge are confidential and therefore need to be kept safely in personal storage drive			X
C.2	KM processes are very time consuming	X	X	
2	**Ignorance**			
C.3	KM has no benefits to the company/organization			
C.4	I do not know what KM is			

IOP Conf. Series: Materials Science and Engineering **277** (2017) 012032 doi:10.1088/1757-899X/277/1/012032

Table 2. Understanding of knowledge management (KM) implementation (cont.)

C.5	I am not aware if KM exists in my company/organization			
C.6	Company/organization does not need KM-based KPI			
C.7	Staff/Individuals do not need KM-based KPI			
3	**Motivation**			
C.8	KM improves employee's performance	X	X	
C.9	KM encourages innovations			X
C.10	KM enhances better decision making		X	
4	**Benefit**			
C.11	KM creates competitive advantage			X
C.12	KM generates revenue growth			X
5	**Need**			
C.13	Training on KM is necessary			X
C.14	Coaching on KM is necessary			X

4. Challenges of knowledge management (KM) implementation

In implementing knowledge management (KM), barriers and challenges are inevitable. According to [13], barriers to knowledge building mostly stem out from the existence of poor organisational culture. Furthermore, barriers to knowledge building are classified into three categories, namely individual (people), organisational, and technological barriers. In order for KM to be fruitful for the organization, these three barriers should be integrated in a way they complement each other. Table 3 shows the literature summary of KM challenges based on questionnaire create in this paper.

Table 3. Challenges in knowledge management (KM) implementation.

No	Challenges of Knowledge Management (KM)	[14]	[15]	[16]	[17]	[18]	[19]	[13]	[20]	[21]	[22]
1	**Organizational Barriers**										
D.2	No sharing culture										X
D.3	Difficult to implement									X	
D.8	Lack of social network				X	X					
D.9	Differences in culture				X	X				X	
D.14	Lack of interaction				X	X		X		X	
D.15	Fear of not receiving recognition				X	X	X	X			
2	**Individual Barriers**										
D.5	Difficult to capture knowledge during informal discussion		X								X
D.6	KM is not well understood				X	X		X			
D.7	Lack of communication skills				X	X		X		X	X
D.10	Lack of time				X	X				X	
D.11	Lack of trust				X	X		X			
D.12	Lack of motivation				X	X	X	X	X		

Table 3. Challenges in knowledge management (KM) implementation (cont.)

D.13	Lack of awareness of the benefit of knowledge sharing				X	X				X	X
D.17	Fear of causing internal conflicts						X				
3	**Technological Barriers**										
D.1	Technology limitation	X		X			X	X			X
D.4	Document contents are difficult to understand		X				X		X	X	
D.16	Fear of knowledge insecurity						X		X		
D.18	Keeping data accurate and relevant						X				
D.19	Difficult to determine where KM belongs to (HR/IT/etc)						X	X			

5. Critical success factors (CSFs) of knowledge management (KM) implementation

The main use of critical success factors (CSFs) concept is as the focus for implementing organisational transformation by supporting beneficial change. The benefits of identification of CSFs allow a simple understanding of the amount of information, help the management to focus on major concern, easy to monitor and can be used in connection with strategic planning purposes. CSFs not only represent critical strategic thinking, but it should be used in conjunction with a planning process as CSFs are very important in keeping people focus. Table 4 shows the literature summary of CSFs in knowledge management (KM) based on questionnaire create in this paper.

Table 4. Summarised critical success factors (CSFs) associated with knowledge management (KM)

No	Critical Success Factor in Knowledge Management Implementation	[23]	[24]	[25]	[26]	[27]	[8]	[28]	[29]
E.1	Continuous organisation support		X		X	X	X	X	
E.2	Execution of plan					X			
E.3	Knowledge and sharing culture		X		X	X	X		
E.4	HR development							X	
E.5	Financial resources	X							
E.6	Technology		X	X	X	X	X	X	X
E.7	Infrastructure		X	X	X				
E.8	Clear strategy on knowledge management process	X	X	X		X	X		
E.9	Leadership demonstration by senior staff/management	X	X	X				X	
E.10	Knowledge management processes should be precise								X

Table 4. Summarised critical success factors (CSFs) associated with knowledge management (KM) (cont.)

E.11	Strong link to a business imperative								
E.12	Continuous learning							X	
E.13	HR management			X					
E.14	Quality control and assurance			X					
E.15	Marketing			X			X		
E.16	Environmental influences namely competition, markets, time pressure, governmental and economic climates	X					X		
E.17	Training of staff should be made compulsory				X				
E.18	Measurement of knowledge management in term of KPI		X		X				
E.19	Benchmarking of a successful knowledge management system nationwide				X				X
E.20	Elimination of constraints that defer knowledge management implementation				X				X
E.21	Knowledge management processes should be easy to follow								

6. Framework development

For a successful project, the implementation of framework with certain guidelines is necessary to achieve the objective of the project [30]. The development of the framework of knowledge management (KM) will need to emphasis on all the critical activities involved in the KM process. A framework with the list of all critical processes involved and also takes into account other factors, such as level of understanding KM, challenges of the implementation and CSFs in implementing KM in construction industry, will be developed. Once the conceptual framework is developed, it will be verified by the expert focus group. Figure 1 shows the initial description on relationship between all objectives of this paper. The CSFs are set to be dependent variables while a set of independent variables were constructed to be the process, understanding and challenges. The further analysis on the relationship between all variables is discussed in further study.

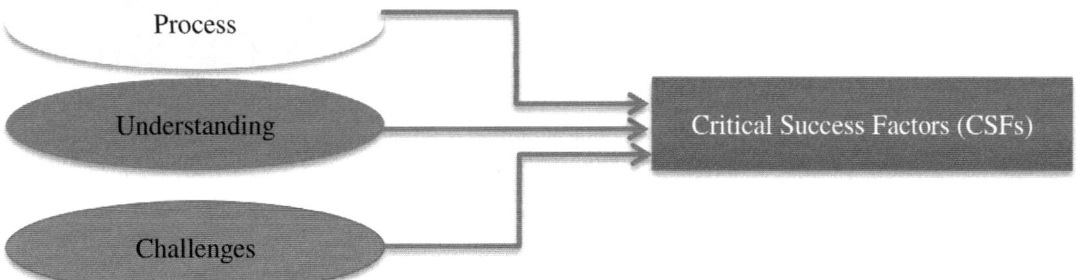

Figure 1. Relationship between critical success factors (CSFs) and process, understanding and challenge of knowledge management (KM).

7. Conclusion
This paper provides an overview of knowledge management (KM) issues from years 2000 to 2017. The findings show that KM implementation in the Malaysian consultant firms are still lacking and less giving consideration. This paper has successfully achieved its aim in appraising KM implementation amongst consultant firms working in construction industry in Malaysia. This paper also gives some ideas of developing a conceptual framework of KM for consultant firm in the Malaysian construction industry by taking into account the quantitative elements of KM processes, KM understanding, KM challenges and KM critical success factors (CSFs). It is hoped that the discussion would lead to the successful implementation of KM practice in the construction industry.

8. Acknowledgements
The authors would like to express their since gratitude to the Ministry of Education, Universiti Teknologi Malaysia (UTM) and the Research Management Centre (RMC) of UTM for providing the financial support for this paper to be published. This paper is financed by UTM Razak School Grant for research funding under Cost Centre No. R.K130000.7740.4J290.

9. References
[1] CIDB 2015 Construction industry transformation programme pp 2016- 20
[2] Davenport T H and Prusak L 1998 Working knowledge: how organizations manage what they know (Boston, MA: Harvard Business School Press)
[3] Hanley S S 1999 A culture built on sharing. Informationweek
[4] Liebowitz J N, Ayyavoo H, Nguyen D Carran and Simien J 2007 Cross-generational knowledge flows in edge organizations *Industrial Management and Data System* **107(8)** pp 1123-53
[5] Yang Xu, Jing Xi and Alain Bernard 2013 Knowledge management in e-commerce mass customisation. *IFIP Int. Conf. on Product Lifecycle Management (PLM2013)* pp 259-67
[6] Tseng S M 2014 The impact of knowledge management capabilities and supplier relationship management on corporate performance *Int. J. of Production Economics* **154** pp 39-47
[7] Fischer M and Marsh T 2014 Recognizing intellectual capital as an asset *J. of Business and Economics Research (Online)* **12(2)** pp 177
[8] Obeidat B Y, Obeidat B Y, Al-Suradi M M, Al-Suradi M M, Masa'deh R E, Masa'deh R E and Tarhini A 2016 The impact of knowledge management on innovation: an empirical study on Jordanian consultancy firms *Management Research Review* **39(10)** 1214-38
[9] Masa'deh R E, Masa'deh R E, Shannak R, Shannak R, Maqableh M, Maqableh M and Tarhini A 2017 The impact of knowledge management on job performance in higher education: the case of the University of Jordan *J. of Enterprise Information Management* **30(2)** pp 244-62
[10] Lin J M C and Wang P Y 2008 A knowledge management system for Chinese language arts teachers *British J. of Educational Technology* **39(5)** pp 935-43
[11] Todorović M L, Petrović D Č, Mihić M M, Obradović V L and Bushuyev S D 2015 Project success analysis framework: a knowledge-based approach in project management *Int. J. of Project Management* **33(4)** pp 772-83
[12] Abd Rahman A, Imm Ng S, Sambasivan M and Wong F 2013 Training and organizational effectiveness: moderating role of knowledge management process. *Eur J of Training and Development* **37(5)** pp 472-88
[13] Chong C W and Besharati J 2014 Challenges of knowledge sharing in the petrochemical industry *Knowledge Management and E-Learning: An Int. J. (KM&EL)* **6(2)** pp 171-87
[14] McAdam R and Reid R 2001 SME and large organisation perceptions of knowledge management: comparisons and contrasts *J. of Knowledge Management* **5(3)** pp 231-41
[15] Collison C and Parcell G 2001 Learning to fly: Practical lessons from one of the world's leading knowledge companies (Oxford: Capston Publishing limited)
[16] Riege A 2005 Three-dozen knowledge sharing barriers managers must consider *Journal of Knowledge Management* **9(3)** pp 18-35

[17] Jain K K, Sandhu M M and Sidhu G 2007 Knowledge sharing among academic staff: a case study among business schools in Klang Valley Malaysia *J. for the Advancement of Arts and Science* **23**

[18] Ling C W, Sandhu M S and Jain K K 2009 Knowledge sharing in an American multinational company based in Malaysia *J. of Workplace Learning* **21(2)** pp 42-125

[19] Dzone 2013 What is a knowledge management system? (NC, USA: Dzone Software)

[20] Niazi M, Mahmood S, Alshayeb M and Hroub A 2015 Empirical investigation of the challenges of the existing tools used in global software development projects **9(5)** pp 135-43

[21] Morteza-Shokri G, and Chileshe N 2014 Knowledge management: barriers to capturing lessons learned from Australian construction contractors perspective *Construction Innovation* **14(1)** pp 108-34

[22] Lotti Oliva F 2014 Knowledge management barriers, practices and maturity model *J. of Knowledge Management* **18(6)** pp 1053-74

[23] Holsapple C W and Joshi K D 2000 An investigation of factors that influence the management of knowledge in organizations *J. of Strategic Information Systems* **9(2)** pp 235-61

[24] Hasanali F 2002 Critical success factors of knowledge management (USA: KM Advantage)

[25] Chourides P, Longbottom D and Murphy W 2003 Excellence in knowledge management: an empirical study to identify critical factors and performance measures. *Measuring Business Excellence* **7(2)** pp 29-45

[26] Hammer A C 2015 A study of business projects in innovation brokering in vri. Master in Management and Technology (Trondheim: Trondheim Business School, Sor-Trondelag University College)

[27] Luo S H and Lee G G 2015 Applying failure mode and effects analysis for successful knowledge management *Total Quality Management and Business Excellence* **26(1-2)** pp 62-75

[28] Inkinen H and Inkinen H 2016 Review of empirical research on knowledge management practices and firm performance *J. of knowledge management* **20(2)** pp 230-257

[29] Kim T H, Lee J N, Chun J U and Benbasat I 2014 Understanding the effect of knowledge management strategies on knowledge management performance: a contingency perspective. *J. of Information and Management* **51(4)** pp 398-416

[30] Baccarini D 1999 The logical framework method for defining project success *Proj. Manage. J.* **30(4)** pp 26–32.

An overview of turbomachinery project in Malaysian oil and gas industry

Harris Abd. Rahman Sabri [1], Abd. Rahman Abdul Rahim [2], Wong Kuan Yew [3] and Syuhaida Ismail [4]

[1,2,4] UTM Razak School of Engineering and Advanced Technology, Universiti Teknologi Malaysia, Kuala Lumpur
[1] harris.ars@gmail.com, [2] rahmanar@utm.my, [3] m-wongky@utm.my, [4] syuhaida.kl@utm.my

Abstract. Being the most demanding, challenging and exciting engineering and technological advances has provided escalated interests amongst the engineers at large to venture into the oil and gas (O&G) industry. Although claimed as the most expensive industry in the world via the utilisation of critical equipments, the O&G industry is still recording notorius failures in its project management especially due to turbomachinery issues, the heart equipment of any O&G project. Therefore, it is important for this paper to review turbomachinery project as one of the long lead items during project executions that is also proven to be the most costly and expensive equipment. This paper therefore discusses the gaps in turbomachinery studies via literature review in highlighting its application in O&G projects. It is found that the main components of turbomachinery are driver and driven equipment, which are applied for mechanical equipment, Electric Power Generation and heat generation for Combined Cycled Configuration. Important variables for turbomachinery selection include: (1) process requirement; (2) site location; (3) driver selection; (4) equipment sparing philosophy; (5) efficiency and reliability; (6) operability and maintainability; and (7) cost. It is hoped that this paper would lead to the successful project management of turbomachinery in the O&G industry.
Keywords: Turbomachinery; Malaysia; oil and gas industry; overview.

1. Introduction

Oil and gas (O&G) industry contributes to Malaysian economic as one of the most important sectors. The country's O&G industry has developed from mere production of crude for export involving the front-end engineering design (FEED) of oil production facilities to the design and construction of chemical plants [1]. Therefore, there are opportunities of the expansion of local engineering positions in the market.

Due to its unique requirement, the O&G industry in Malaysia has developed standards and practices almost at par with the high standards of requirements of foreign countries, for instance the United States of America (USA) and the United Kingdom (UK). Nevertheless, looking at the Malaysian O&G practices, which are split into upstream and downstream sectors, the upstream is observed as the most critical sector [2] to be improved via these standards given that the upstream sector is the first ever sector traditionally developed in the industry [3]. However, for both sectors, turbomachinery is observed as the heart of this sector project management as the success and failure of the sectors operation is eventually determined by the turbomachinery delivery.

This is due to the fact that although major turbomachinery service provider or Original Equipment Manufacturer (OEM) in Malaysia are recognised internationally and based worldwide, major issues related to turbomachinery project management are still being recorded. These have eventually led to the additional cost incurred and loss of business opportunity to the client side [4] as well as failed the O&G facilities project in Malaysia as a whole.

Hence, based on turbomachinery that is proven to be the most costly/expensive equipment [5], there is a significant need to review the management of turbomachinery project in Malaysia, whilst at the same time are capable of driving the O&G project towards success. These reviews, integrating with the project management of various stakeholders of multi-disciplinary (including contractors) throughout the project life cycle, ranging from procurement, engineering, construction and commissioning stages [6] will be able to properly manage the turbomachinery projects in the O&G industry.

2. Gaps in turbomachinery project research

With a larger demand for gas and new technologies, which has made the oil and gas (O&G) industry becoming financially attractive, O&G project industry has became the focus of many studies around the globe, for instance [7] who investigate the relational study of supply chain agility, competitiveness and business performance in the oil and gas industry, [8] who identifies the economics, execution and management of complex offshore projects and [9] who determines the list of criteria for success in O&G industry. However, literature review focusing on the turbomachinery project execution, the most critical equipment in any O&G facilities project [10] is missing.

Even though as mentioned beforehand that turbomachinery is the heart of any O&G facilities project, no academic study has been carried out pertaining to its application [11]. Most of the studies available around the globe focus on the overall O&G project execution, namely platform design and construction with no interest to turbomachinery.

Therefore, it is seen by this study that Malaysia, and in particular PETRONAS, would need to develop a support base that can effectively and efficiently manage the turbomachinery project in O&G industry. This is because a lot of project challenges are faced by PETRONAS while managing the turbomachinery during the conceptual phase (Original Equipment Manufacturer (OEM) selection), front-end engineering design (FEED) and detailed design (technical specification, procurement, vendor data engineering review and incorporation, interface with disciplines, namely process, electrical, instrument, piping, structural, safety, equipment inspection and critical testing), construction (installation, pre-commissioning and commissioning) and finally the operation and maintenance of the turbomachinery [12].

3. Turbomachinery application in oil and gas projects

The non-existence of the turbomachinery leads to the impossibility of the facilities to function and eventually fail the operation. Apart from that, the turbomachinery, which consists of compressors, generators and turbine units, are identified as critical packages by [13] due to its complexity. As highlighted by [14], the design, procurement, manufacture, test and documentation of a turbomachinery system itself becomes a unique project within the larger overall project.

The main component of turbomachinery can be divided into two, namely the driver and driven equipment. For turbomachinery, the selected driver, or also known as the prime mover, will be a turbine (steam or gas turbine), as other drivers, such as electric motor and internal combustion engine, are not classified as turbomachinery. The most commonly used for O&G facilities project is gas turbine. According to [15], gas turbine is divided into five groups: (1) frame type heavy duty gas turbine; (2) aeroderivative gas turbine; (3) industrial type gas turbine; (4) small gas turbine; and (5) micro turbine.

For the driven equipment, it depends on the application of the turbomachinery itself. It can vary from centrifugal compressor, electric generator and pump. A typical schematic of turbomachinery package is shown in Figure 1 Depending on the operating speed of the driver and driven equipment, a gearbox might be required in between to ensure smooth operation.

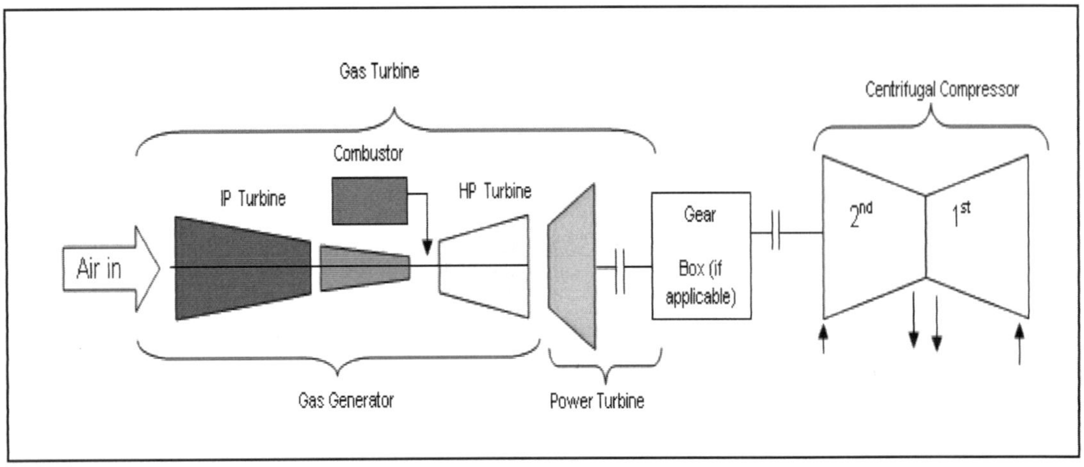

Figure 1. A typical schematic of turbomachinery package [11].

3.1. Typical turbomachinery application

It is noteworthy to underpin that the turbomachinery involves in each of the aforementioned phases. First of all is mechanical drive. The turbomachinery configuration is selected to drive the main mechanical equipment, such as centrifugal compressors and pumps, to be used for oil and gas (O&G) facilities process requirement. This varies from gas compression system, gas re-injection system, gas lifting system, refrigeration system, liquid pumping system and various process requirements. There are also increasing demands of turbomachinery application for Carbon Capture and Storage (CCS), Enhanced Oil Recovery (EOR), Coal Gasification and offshore high pressure Carbon Dioxide (CO2) Compression [16].

Next is the application for Electric Power Generation. In this application, the turbomachinery main function is to generate power for internal O&G facilities consumption or even for external by exporting the electricity via national power grid. The most common equipment identified for the power generation is the turbogenerator or gas turbine generator. A multi-train configuration consists of aero-derivative gas turbines is commonly considered compared to industrial type gas turbines due to the advantage of weight reduction, space requirement and maintenance logistic reasons [17].

Finally is the Combined Cycled Configuration, which means that the exhaust heat is being recycled by adding additional equipment on the exhaust system. As explained by [15], the combined cycled configuration generates up to 70 per cents of heat efficiency compared to the simple cycle which can only provide between 30 per cents to 40 per cents of heat efficiency. The exhaust can be utilised to generate steam or heat other heat transfer medium by adding a heat recovery steam generator and waste heat recovery unit at the exhaust system of the driver equipment.

3.2. Turbomachinery selection

There are various variables need to be taken into consideration while selecting turbomachinery in any oil and gas (O&G) platform project execution. As described by [11], these variables consist of: (1) process requirement; (2) site location; (3) driver selection; (4) equipment sparing philosophy; (5) efficiency and reliability; (6) operability and maintainability; and (7) cost. These variables are substantially discussed in the following subsections.

3.2.1 Process requirement. Process requirement is basically the main technical requirement that determines whether the turbomachinery meets the technical criteria or not. Under process requirement, the first requirement is to identify the type of services/application. Each project needs to determine the main function of the turbomachinery, whether to be used for mechanical drive and/or electrical power generation. Each service will have a different production profile of the process requirement. Depending on the application, some services require a constant production profile and some of the applications require a dynamic/broad production profiles, which will change according to time.

Next requirement is about the oil and gas (O&G) field production profile. As highlighted by [18], the trend for a production oil and gas facilities generally foresees a peak production within the initial production years, where the gas production profile will decrease subsequently. Changes of production characteristic, such as gas properties and lowering of production inlet pressure in the process system in order to maintain the production at an acceptable level and/or introduction of new field development within the area, will result in changes in production profile with various operating cases. Therefore, it is critical to optimise the driven equipment design to cover at least majority of the operating cases, if not all cases.

Despite traditional approach of utilising reciprocating compression system to handle high carbon dioxide (CO2) application, according to [19], centrifugal compressors have been favoured due to capacity of most CO2 recovery system exceeds the reciprocating compressor operating range, maintenance intensive of reciprocating compressors, CO2 high density which may lead to reciprocating compressor high velocities problems at the valves and require high strength foundation, which lead to high capital and operating expenditure. Centrifugal compressor, which usually can match the high speed driver and combine under turbomachinery, provides superior efficiency, offer higher capacity/operating range and maintenance wise it has an acceptable extended interval between overhauls.

There is also an option to pre-engineer for future compressor bundle/impeller design to cover the balance of future operating cases. According to [20], due to depleted gas reservoir after operating for few years, the facility production will decline well below the design flow rate of the centrifugal compressor. Therefore, the most effective approach would be to increase the production by reducing the well backpressure and raising the centrifugal compressor polytropic head capacity. This can be undertaken by performing the retrofit of the compressor aero bundle. By performing this, usually the gas turbine driver and auxiliaries system will remain the same. This enables to lower down the overall life cycle cost.

At this stage, it is very important not to oversize the turbomachinery equipment as it will lead to the unnecessary increase of energy consumption, operation and maintenance cost. According to [21], the application of excessive safety margin in the equipment sizing to accommodate unforeseen system changes is equivalent to an expensive insurance policy, which leads to oversized equipment.

Next criteria is to check the Criticality of Equipment/Service. If the equipment is out of service and resulted in reduction or loss of oil and gas facilities production, the service is deemed critical. Therefore, it will impact the selection by introducing equipment sparing during the selection. The same goes for power generation application, where it might result in loss of production if the turbomachinery is unavailable due to the scheduled or unscheduled maintenance. Hence, an adequate power margin and equipment sparing is necessary to mitigate the issue.

The final criterion identified under process requirement is the material selection. The material selected must be suitable for each application selected for the turbomachinery equipment. The compressed medium composition is vital in determining the material selected for the mechanical drive section of the turbomachinery. Factor that leads to the selection of super alloy materials is due-creation of corrosive environment, such as high carbon dioxide content, chloride content and traces of mercury.

For the driver equipment, continuous research and development by gas turbine manufacturer leads to the evolution of new material in order to achieve a higher combustion firing temperature, thus increasing the overall power output of the turbomachinery. As mentioned by [22], one of the main active areas in research and development of the gas turbine is the material itself; covering the superalloys, coatings and ceramic.

3.2.2 Site location. The site location of the oil and gas (O&G) facilities plays a significant role during conceptual engineering in order to decide the type of turbomachinery to be selected for each particular project. The main factor would be whether the facility is based at onshore or offshore as well as site rating. For offshore application, it is very unlikely that the frame type heavy duty gas turbine can be applied due to the space concerns and weight limitation.

Due to site specific geographical location for each O&G fields, different site ambient temperature and site elevation will result in changes in the driver power output, especially gas turbine. Therefore, site rating is very important to ensure that correct prime mover is selected.

A prime mover is rated at the International Organisation for Standardisation (ISO) standard condition, which is equivalent to ambient temperature 15 degree Celcius, 1 Bar and 60 per cents relative humidity. As each O&G facilities are located at a different site condition, the prime mover needs to be leveraged according to the site rating to reflect the actual condition. With reference to Petroliam Nasional Berhad (PETRONAS) Technical Standard PTS 20.180L, the gas turbine will be de-rated at 0.8 per cents for every 1 degree Celcius increment above the ISO condition, 4 per cents for fouling and wear; and finally another four percents due to the intake and exhaust losses [22].

3.2.3 Driver selection. In general, there are multiple choices for driver selection, ranging from gas turbine, steam turbine, electric motor and even internal combustion engine. This study focuses more on gas turbine selected as the prime mover, where the reason behind this selection is discussed in the later part. According to [23], the main advantages of selecting gas turbine compared to steam turbine are as the following: (1) gas turbine application provides a smaller footprint or space required compared to a steam turbine application at the same power requirement, thus lesser weight; (2) gas turbine application provides a better purchase cost per unit of power; and (3) gas turbine overall project installation is generally involved faster and quicker delivery.

3.2.4 Equipment sparing philosophy. In reference with Petroliam Nasional Berhad (PETRONAS) Technical Standard PTS 20.180L, all critical equipments are required to have a N+1 philosophy, with N is the total number of trains required to have a total of 100 per cents production [22]. This sparing philosophy needs to be adjusted according to the availability of prime mover in the market. For example, take a case of 2 x 100 percent configuration, where one unit operates at 100 percent capacity, whilst another one unit is a spare. If the process requirement resulted with one unit requires a total power of 45 MW, whereas the current prime mover model is not able to provide the power required; it is not feasible to select the 2 x 100 percent philosophy. The project needs to explore for other configuration such as 3 x 50 percent, 4 x 33 percent or more. This exercise needs to be balanced between the overall project cost as well as the available prime mover model in the market.

3.2.5 Efficiency and reliability. An efficient turbomachinery leads to lower fuel consumption and in a long run, resulted in lower total cost of ownership. In order to increase the overall efficiency, an option that can be taken is to go for dual cycle configuration. Another point to ponder is the capability for fuel flexibility. This is vital especially during first or black start up scenario, where the main fuel might not be available yet. Instead of utilisation of hydrocarbon gas as the main fuel, some gas turbines can also accept dual fuel capability by having a backup fuel such as diesel or heavy oil.

High carbon dioxide (CO_2) gas is also a challenge [24][25], which needs to be considered in the selection of an appropriate turbomachinery. Since the technology becomes more mature, the turbomachinery original equipment manufacturer (OEM) is able to adapt their machines to accept these high CO_2 gases as the main fuel source without jeopardising the power and ability to start up the machine. However, to ensure high reliability during start up, the option of having a dual fuel capability need to be considered and balanced with the overall project cost.

On top of high CO2 fuel gas, listed down several challenges for upcoming offshore CO2 compression system such as space restriction, high reliability, very high compression pressure which relates to high density and high molecule weight of CO2 and turbomachinery rotordynamic behaviour [16].

3.2.6 Operability and maintainability. For mechanical drive, [27] has stated that the centrifugal compressor can be controlled via few methods, including variable speed control of the driver, adjustable inlet guide vane, suction or discharge valve throttling as well as flow recycling. The main purpose is to ensure the operational flexibility in case of production upsets or even unscheduled shutdown.

The turbomachinery equipment design must incorporate the maintainability of this turbomachinery equipment by having a modular configuration, having a quick replacement of parts and allowing a one-to-one replacement in order to reduce the equipment downtime or production loss. For upstream oil and gas (O&G) facilities, the direct impact of production loss would be the loss of revenue. Apart from that, there might be additional penalties imposed by the buyer which lead to damage of reputation to the company. Additional utilities system required to support the operability of the turbomachinery needs to be considered during the design stage, as the requirement will have a significant impact on other equipment and also overall cost.

3.2.7 Cost. Apart from the technical aspect, the most important and influential decision during the final selection of the turbomachinery for each project is the cost. Material and equipment cost accounts for a notable portion from the overall cost profile of an oil and gas (O&G) company [28]. As highlighted by [29][21], the overall or Life-Cycle Cost (LCC) shall be considered by including the capital expenditure (CAPEX) and operating cost (OPEX). Traditionally, some equipment procurements are still based on capital expenditure (CAPEX) instead of LCC, which is according to [30], this was due to the precisely known price of the CAPEX, whereas the LCC is based on theoretical cost. Another main reason is due to the deeply rooted practice by certain company, which is solely based on CAPEX. Therefore, the cultural resistance is identified as an obstacle in order to adopt the LCC [30], which obviously provides a wider perspective to the end user and provides a broader view of overall cost expenditure.

According to [27], CAPEX for a project covers the initial cost and installation cost of the newly purchased equipment. The initial cost shall not only cover the cost of the turbomachinery equipment, but also the auxiliaries system required to ensure the successful operation of the equipment. This includes, for example the structural platform to house the turbomachinery, separators, interconnecting piping, control system, coolers and other associated equipment.

[28] also suggested that LCC to be calculated based on different contributing factors, such as initial equipment purchase, installation, operation, maintenance, inventory and training. On top of that, the LCC calculation must also take the Net Present Value (NPV) at a chosen interest, for example PETROBRAS has established 15 per cents per year for their feasibility study [28]. Thus, the NPV interest might vary according to each company.

According to [31], the major compressor and gas turbines manufacturers are also experiencing the same issue of cost incremental. It was recorded that between 2006 and 2007 alone, the machinery cost jumped between 8 percent to 20 percent and 15 percent to 25 percent, respectively.

4. Conclusion

This paper provides an overview of turbomachinery project in the oil and gas (O&G) industry in Malaysia. The findings show that turbomachinery is the critical equipment of any O&G facilities. Even so, there is a huge gap in terms of literature review related to the management of turbomachinery in O&G projects that eventually lead to many unsuccesful turbomachinery project execution. The main components of turbomachinery are driver, which is commonly used for gas turbine, and driven equipment, which can vary from centrifugal compressor, electric generator and pump. Generally, turbomachinery is applied for main mechanical equipment, Electric Power Generation and heat generation for Combined Cycled Configuration. In addition, turbomachinery selection is very critical in any O&G project, thus seven variables need to be focused on during its selection: (1) process requirement; (2) site location; (3) driver selection; (4) equipment sparing philosophy; (5) efficiency and reliability; (6) operability and maintainability; and (7) cost. It is hoped that the discussion would lead to the successful project management of turbomachinery in the O&G industry.

Acknowledgements

The authors would like to express their since gratitude to the Ministry of Education, Universiti Teknologi Malaysia (UTM) and the Research Management Centre (RMC) of UTM for providing the financial support for this paper to be published. This paper is financed by UTM Grant for UTM Razak School under Cost Centre No. R.K130000.7740.4J290.

5. References

[1] Mehden F R V D and Al Troner 2007 PETRONAS: A national oil company with an international vision (USA: James A. Baker III Institute for Public Policy of Rice University)

[2] Martén I, Whittaker P and de Bourio A M 2015 *Government take in upstream oil and gas: framing a more balanced dialogue* (USA: The Boston Consulting Group)

[3] Petroleum Services Association of Canada (PSAC) 2017 Industry overview (Toronto, Canada: PSAC)

[4] Parker R W, Collier D H, Rudisel D A, Rutan C R and Pyle A S 1996 Planning, preparation, and implementation of turbomachinery turnarounds Proc. of the Twenty-Fifth Turbomachinery Symposium (Texas, USA: TEES)

[5] Heighes B 2013 *Life extension of turbo machinery packages* (USA: Dresser-Rand)

[6] Robb D 2017 *The evolution of field service International Turbomachinery* (Norwalk, CT: UBM PLC)

[7] Yahaya Y, Gunasekaran A, Musa A D, Mohammed El-Berishy, Nagham M and Cang S 2012 A relational study of supply chain agility, competitiveness and business performance in the oil and gas industry *Int. Jour. of Production Economics* **147** pp 531-43

[8] Doherty P M 2011 The economics, execution and management of complex offshore projects, Offshore Technology Conference, OTC 21878, Houston Texas, USA 2-5 May 2011

[9] Martin 1998. An ERP Strategy Fortune **2** pp 95-97

[10] Schneider Electric 2016 *Turbomachinery solutions for compressor control* (Rueil-Malmaison, France: Schneider Electric)

[11] Harris A R S, Abd Rahman Abd Rahim and Haris A Majid 2013 Concept of turbomachinery selection for upstream oil and gas facilities project. *2nd Int. Conf. on Mechanical, Automotive and Aerospace Engineering*

[12] Nasir H, Saiful N and Yuzaini Y 2004 Challenges to improve turbomachinery availability. SPE Asia Pacific Oil and Gas Conf. Exh. (Perth, Australia: Society of Petroleum Engineer)

[13] Oeystein K, Borgersrud and Dagfinn Ellingsen 1998 Systematic approach to working environment in design and construction of major offshore projects (Caracas Venezuela: SPE 46756)

[14] Hoppock G W 2002 Evolution toward effective project management in turbomachinery project *Project Management Institute Annual Seminars and Symposium* (San Antonio, Texas: PMI)

[15] Boyce M P 2002 *Gas turbine engineering handbook second edition* (USA: Gulf Professional Publishing)

[16] Rolf Hable 2011 Advanced compression solution for CCS, EOR and offshore CO2 Offshore Technology Conf. (Rio De Janeiro, Brazil: OTC 22659)

[17] Alvarado C S and Wagner J V 2002 Next generation FPSO: combining production and gas utilisation Offshore Technology Conf.

[18] Marchini R, Calvarano M P, Cocchi G and Botta M 2011 A new approach for gas fields development within uncertain business context *Int. Petroleum Techonology Conf.*

[19] Bovon P and Habel R 2007 *CO2 compression challenges ASME Turbo Exp.* (Montreal, Canada: CO2 Compression Panel)

[20] Constantini S and Grifoni L 2011 Maximize offshore reservoirs productivity with the latest design techniques and advanced centrifugal compression technologies *Offshore Technology Conf.*

[21] Mohamed Soliman 2012 *Pitfalls of oversizing plant rotating equipment* (Abu Dhabi, UAE: Society of Petroleum Engineers)

[22] Najjar S and Yousef H 1995 Some trends and progress in gas turbine technology and research *Energy Convers. Management*

[23] Petroliam Nasional Berhad (PETRONAS) 1991 Petronas technical standard PTS 20.180L

(Kuala Lumpur: PETRONAS)

[24] Soares C M 2007 *Gas turbine: a handbook of air, land and sea application* (United States: Butterworth-Heinemann)

[25] Abdul Rahim Samsudin and Mohd Sabri Mohamed 2001 Peninsular malaysia gas utilisation project – challenges in gas supply planning *Society of Petroleum Engineers*

[26] Sriyanta Hadi, Ambok Malek Meri, Ahmad Zulhairi Mahadi, Ezatul Shima Abd Ghafar, Siti Nurul Aimi Mohamed Shariff Din and Soh I Fen 2013 Operational challenges in managing high CO2 content gas production in peninsular Malaysia operations. *Int. Petroleum Technology Conf.*

[27] Kurz R and Lubomirsky M 2011 Concepts in gas compressor station configuration *Int. Petroleum Technology Conf.*

[28] Ribeiro E M, Farias M A and Dreyer S R B 1995 *Equipment cost optimization Offshore Technology Conf.* (Houston Texas, USA: OTC 7761)

[29] Project Management Institute (PMI) 2013 *Project Management Book of Knowledge* **5** (USA: PMI)

[30] Juran J M 1980 *Quality planning and analysis* (USA: McGraw-Hill)

[31] Tom Phalen and Jim Scotti 2008 *Update on LNG facility construction Offshore Technology Conf.* (Houston, Texas: OTC 19306)

Implementation of enterprise resource planning using Odoo module sales and CRM. Case study: PT Ecosains Hayati

A Terminanto[1], R Hidayat[2], and A N Hidayanto[3]

[1,3]Faculty of Computer Science, Universitas Indonesia, Depok, Indonesia. [2] Industrial Engineering Department, Faculty of Engineering, Pancasila University, Jakarta, Indonesia.
[1]agung.terminanto@gmail.com, [2]rianhidayat53@gmail.com , [3]nizar@cs.ui.ac.id

Abstract. Marketing is the most important part of PT Ecosains Hayati as a distributor company. Sales looking for prospective buyers and provide product price quotations. Quotations are made by accessing various data in a separate document. It makes the work process less efficient. Implementation of ERP system could improve the efficiency of sales work. It used RAD method that faster than other methods. The selected ERP system is Odoo, which contains various business application programs. Gap and efficiency analysis were performed to compare business processes before and after using Odoo. User Acceptance Test (UAT) is conducted to determine user acceptance of the applications and features available in Odoo module. After implementation of Odoo, there was an increase in the efficiency of the quotations business process by 63% in numberf of activity and by 50% in number of the actors involved. Odoo customization is done on 26 of the 41 module menus used. Based on UAT results, the implementation of Odoo meets the usability aspect with the overall average value 3.7. This indicates that users have a good level of understanding in the use of Odoo, and the features on Odoo can meet the needs of users.
Keyword: ERP, Odoo, RAD, UAT

1. Introduction

1.1. Background

PT Ecosains Hayati is one of distributor laboratory equipment in Indonesia. The sales/marketing division is the most important part of PT Ecosains Hayati. Every salesman is responsible for purchasing from prospective buyers. Before obtaining a purchase, a salesman must make a quotation by accessing various data such as product price, goods availability data in the warehouse, data sequence number, and product information on the supplier's official website. The number of documents that must be accessed makes the sales work process less efficient. The unintegrated data of the customer makes the sales have their respective customer databases unknown to each other so that there is often overlap of customers who make the same customers can be contacted by different sales.

An information system is a system that records and organizes data to produce useful information to support an organization whether employees, customers, suppliers, or other co-workers [1] Enterprise Resource Planning (ERP) is one concept that can be applied to improve the efficiency of sales work of PT Ecosains Hayati. ERP is a concept to plan and manage enterprise resources, in the form of integrated programming packages and multimodules designed to serve and support various functions within the company, so that work becomes more efficient and can provide better service to consumers [2].

IOP Conf. Series: Materials Science and Engineering **277** (2017) 012034 doi:10.1088/1757-899X/277/1/012034

Implementation of ERP system at PT Ecosains Hayati using RAD (Rapid Application Development) method, since it faster than other method. The selected ERP system is Odoo. Odoo is a fully integrated and open source ERP application that contains a variety of business application programs including Sales, Customer Relationship Management (CRM), Project Management, Warehouse Management, Manufacturing, Finance and Accounting, Human Resource, and so on. The use of Odoo can improve the purchasing, inventory, and sale information systems of the Wholesale and Retail Emi Stores [3].

1.2. Project Purposes

Purposes to be achieved in this project are:

- Improve sales business process workflow of PT Ecosains Hayati.
- Implement ERP Sales Management and CRM systems to enable the sales process to be integrated with other sections.
- Adjustment of the Sales Management module features of selected ERP Software.

2. Methods

RAD method is the development of an information system with a relatively short time. Development of a normal information system takes at least 180 days, but with RAD method a system can be completed in just 60-90 days [4]. Figure 1 depicts three phases of RAD [5].The steps Implementation of Odoo at PT Ecosains Hayati using (RAD) method, presented in Figure 2.

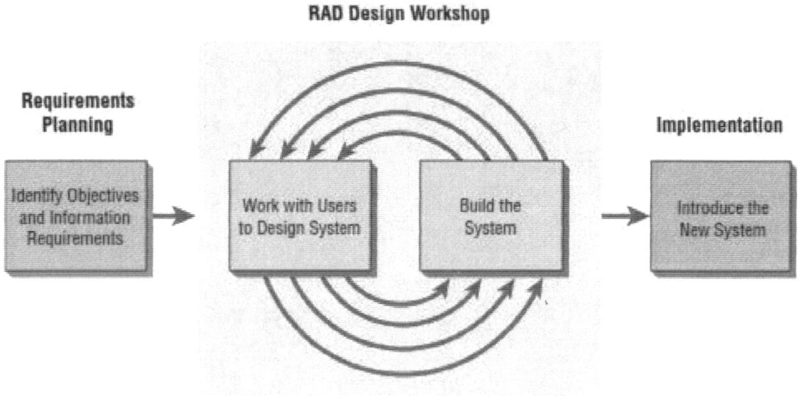

Figure 1. Three phases of RAD

3. Result And Discussion

3.1. Business Process Sales / Marketing Division Using Odoo

Odoo is used to improve the business processes of the actual sales / marketing division to run more efficiently and effectively. Improvements made include:

- Use an ERP system so that data from each division is stored in one system and integrated well. With the data stored in a system is expected to coordinate between divisions can run more efficiently.
- Using the ERP system to create a quotation form. With the ERP system, sales do not need to open various types of documents to complete the data on the offer form (Figure 3). After opening the system, sales can directly enter the data needed to make a quotation from the database provided by the system. Sales also do not need to save the quotation form in the form of Microsoft Excel because the quotation form created automatically stored in the system.

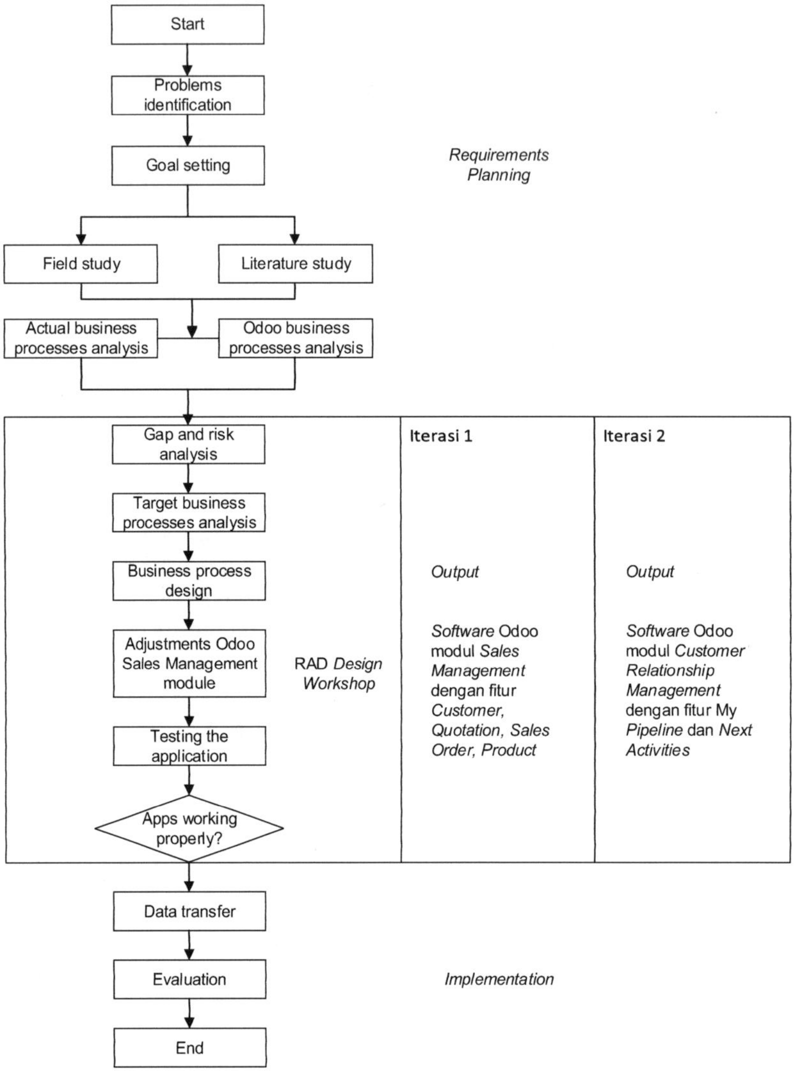

Figure 2. RAD Method

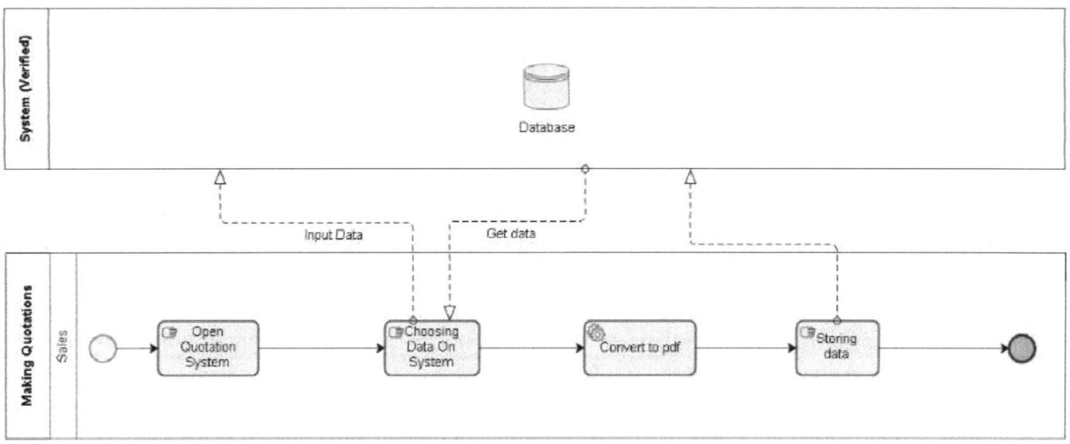

Figure 3. Quotation Making Process using Odoo

3.2. Gap Analysis

The gap analysis is done to find the gap that exists between the current business processes with the existing business processes on the Odoo, as well as adding the needs of the users. Table 1 shows the gap analysis obtained.

Table 1. Gap Analysis

No.	Business Process	Needs	Fullfilment			Existing	Odoo
			N	P	F		
1	The Division Of Sales/ Marketing	System Integration of data that can be monitored in real time by sales / marketing manager		√		Each sales have their own data that can not be monitored by sales / marketing manager	Sales Module can integrate data between sales and can be monitored in real time by sales / marketing manager
2	Make a Quotation	The quotation system with the data already entered into the system database and already through the verification of sales / marketing manager, so in making the quotation no longer open a lot of documents		√		Quotations are made in a Microsoft Excel document with a prepared quotation template. Data entered by copy and paste into available templates.	In the Sales Management module there are quotations feature to make a quotation with data already entered and verified by the previous sales / marketing manager, so in the process of making the data quotation stay selected without having to open many supporting documents again.
3	Make a Report	Reporting system for every sales activity that can be monitored by sales / marketing manager and director of operations in real time, and there is a data analysis that supports for the assessment of sales performance		√		Each salesman writes a report on Google Sheets about week-long activities and plans for the following week	In the CRM module there is My Pipeline feature that will record the sales activity automatically when following the Sales Flow that has been determined. There are 2 Report features, namely Activities Analysis and Pipeline Analysis that can be used to view sales performance, and on the Dashboard CRM sales module can see its own performance

3.3. Analysis of Effectiveness and Efficiency of Bidding Procurement Business Process

The business process of making quotation done now (actual) has been effective but can't be said to be efficient. Sales must open multiple documents to get the data and enter them one by one into the quotation form by copy-paste. The number of documents to be opened makes the quotation process take a long time. The flowchart making of the offering with the Odoo is presented in Table 2. The comparison between the actual quotation-making business process and with the Odoo is presented in Table 3.

$$\text{Increased efficiency} = \frac{\text{Actual amount of activity - Amount of activity of the ERP system}}{\text{Actual amount of activity}} \text{ x } 100\% \quad (1)$$

$$\text{Increased activity efficiency} = \frac{11 - 4}{11} \text{ x } 100\% = 63\%$$

$$\text{Increased efficiency of the actor} = \frac{2 - 1}{2} \text{ x } 100\% = 50\%$$

Table 2. Flowchart Business Process Making Quotation (Odoo)

Job : The Making Of The Quotation (Proposal)
Follow the : ☐ Product ■ Method ☐ Material ☐ Form
Chart Begins : Open The Quotation Tamplate
Chart Ends : Store data
Charted By : Rian Hidayat

No : 1
Page : 1
Of : -

Summary			Manager (M) / Sales (S)
○ Operations	8	1	
⇨ Transportations	-	10	
☐ Inspections	2	-	-
D Delays	-	-	-
▽ Storages	1	-	-
Totals	11	11	

	Details of ~~Present~~ Proposed Method	Operation	Transport	Inspection	Delay	Storage	Pelaku S	Pelaku M	-	-	Notes
1	Open Quotation Making System	○	⇨	☐	D	▽	X	-	-	-	Ms Excel Document
2	Choose The Data On A Quotation Making System	○	⇨	☐	D	▽	X	-	-	-	Supporting data already in the input in a database and verified
3	Changing the extensions document Quotation already approved into pdf	○	⇨	☐	D	▽	X	-	-	-	Changed by the system
4	Save The Document	○	⇨	☐	D	▽	X	-	-	-	In a database system

Table 3. Comparison of the Quotation Business Process

	The Business Process of Making the Quotation				Increased efficiency
	Business Process Modelling		*Flowchart*		
	Actual	Odoo	Actual	Actual	
Number of Activities	11	4	11	4	63%
Number of Actor	2	1	2	1	50%

Based on the comparison of actual business processes and with Odoo it appears that:

- There was a 63% increase in work efficiency by reducing activity from 11 activities to 4 activities.
- There was an increase in the efficiency of the number of actors by 50% by reducing the actor from 2 actors to 1 actor.

3.4. Customize Odoo

The customization of Odoo is done on 26 of the 41 module menus (63%). Customization is done mainly on the quotation view. Figure 4 shows the quotation view after customization. View changes that occur after customization include:

- The quotation number is in accordance with the applicable provisions in PT Ecosains Hayati.
- Appearance of Pricelist, Delivery Method, and the number of discount.
- Enable IDR, EUR, USD.
- Set the Decimal Accuracy to round numbers on offer.
- Add terms and conditions in accordance with the terms.

In addition to the quotation view, customization is also done on the My Pipeline feature by adding a Lead column to detail the sales activity. Step Lead is added by selecting Settings> Activate the developer mode> Sales Module> Configuration> Stages> Create. Settings adjusted to the needs of the company. My Pipeline display after customization is shown in Figures 5.

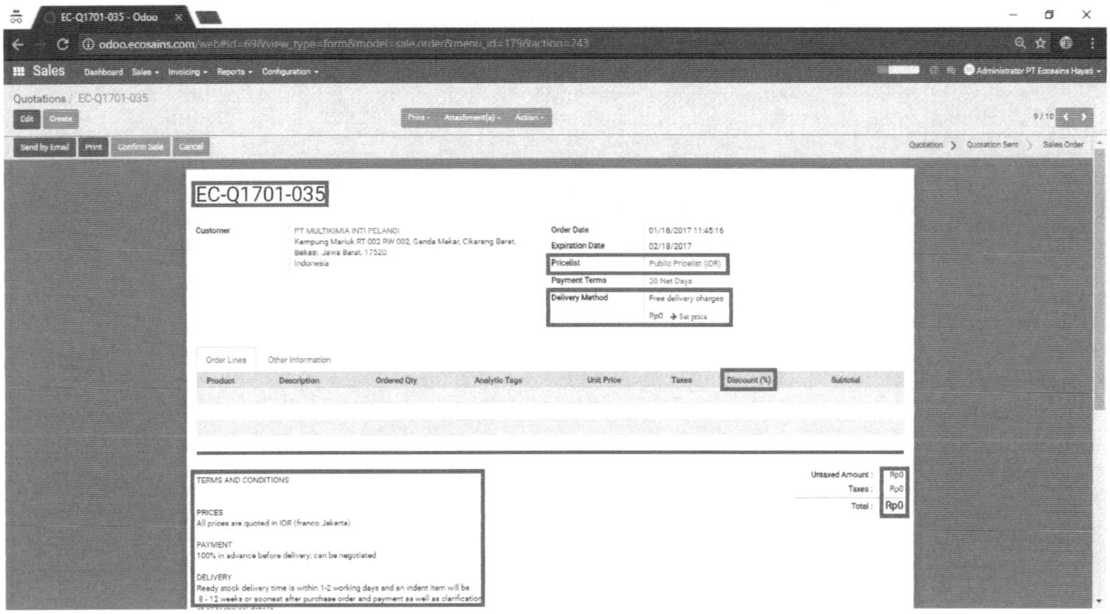

Figure 4. Quotation Form after Customization

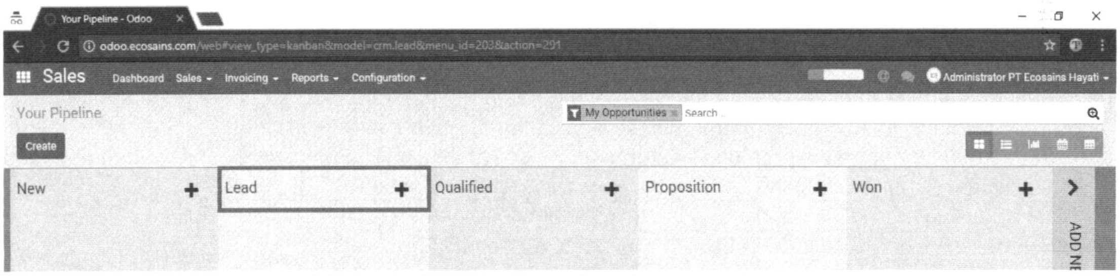

Figure 5. My Pipeline after Customization

3.5. User Acceptance Test (UAT)

User Acceptance Test (UAT) is done to find out if the user can use the applications and features that are available well, and whether the application can meet the needs of users. UAT respondents are 5 employees of PT Ecosains Hayati, the operational director, sales / marketing manager, and 3 sales people. Test results obtained after performing User Acceptance Test through the filling questionnaire are as follows:

- Users can understand and operate the Odoo properly.
- The menu display and symbols used by Odoo are well understood by the user.
- After customization, the features that Odoo has in accordance with the needs of the sales and able to accelerate the process of making the offer.

The result of a usability value recap shows the overall attribute has an average value of 3.7 so that it can be said that the user understands the Odoo system that has been implemented, and the features available on the Odoo can meet the needs of the user.

Table 4. Recapitulation of Questionnaire Usability Values

No	Questions	Grade	Percentage
Systems Aspect			
1	Is the look of the Odoo app easy to understand?	4.2	84%
2	Is the Odoo app easy to operate?	3.8	76%
User Aspect			
3	Is the in-app menu easy to recognize?	3.8	76%
4	Are image symbols easy to understand?	3.6	72%
Interaction Aspect			
5	Is the lead features appropriate for the application of customer qualification before being converted to Opportunity and forwarded to sales?	3.2	64%
6	Does customer data collection use the Odoo system according to company needs?	4.0	80%
7	Is the customization of sales flow on the existing Odoo system in the My Pipeline feature matching with the company's needs?	4.0	80%
8	Is the use of My Pipeline sales activity can be handled properly?	3.8	76%
9	Is the sales reporting system using My Pipeline easy to use?	3.8	76%
10	Is the Quotation feature integrated with My Pipeline useful?	3.8	76%
11	Are Mark Won, Mark Lost, and Log Activity features useful for sales reporting?	4.2	84%
12	Is the Next Activity feature and Expected Closing integrated with My Pipeline useful?	3.4	68%
13	Are the Customer and Sales Person features integrated with My Pipeline useful?	3.8	76%
14	Is the Management Quotations feature on the Odoo system make it easy for sales?	4.0	80%
15	Is the automatic quotation numbering system in accordance with the company's needs?	4.2	84%
16	Is the look of the Odoo Quotation form easy to understand?	3.2	64%
17	Are automated dating systems, customer data integration, price lists, Payment terms, Delivery methods, products, Term and Condition, and automatic calculations for discounts, VAT and Total make it easy to make Quotation?	3.4	68%
18	Is it faster making a quotation using Odoo than the previous system?	3.6	72%
19	Is the integrated email delivery feature with the Odoo quotations system work properly?	3.0	60%
20	Is the conversion feature (Print) on the Odoo quotation system fit the company's needs?	3.4	68%
21	Is the conversion result fit the company's needs?	3.6	72%
22	Is the Confirm Sale feature change Quotation into Sales Order useful?	3.6	72%
23	Is the Sales Order feature as a Purchase Order (PO) management important?	3.8	76%
24	Is the Product feature on the Sales module, the product data can be recorded well?	3.8	76%
25	Is Odoo system help Sales activities?	4.0	80%
Average		**3.7**	**74%**

4. Conclusion

After implementing Odoo on the Sales / Marketing division of PT Ecosains Hayati, it can be concluded that:

- Data storage between divisions can be integrated with the application of Odoo.
- Implementation of Odoo improves the efficiency of the quotation business process in the sales division by 63% in terms of activity, and by 50% in terms of actors involved in the business process.
- Customized Odoo on 26 of the 41 module menus used (63%), so the module meets the needs of the sales division of PT Ecosains Hayati.
- Based on the results of User Acceptance Test (UAT), it appears that the features of Odoo in accordance with the needs of the sales and able to accelerate the process of making the quotation. The overall attribute has an average value of 3.7 so it can be said that the

IOP Conf. Series: Materials Science and Engineering **277** (2017) 012034 doi:10.1088/1757-899X/277/1/012034

implementation of ERP Odoo in PT Ecosains Hayati can be well understood by the users and has a usability aspect value.

5. References

[1] Whitten JL and Bentley LD 2007 *Systems Analysis & Design Methods* (New York: McGraw-Hill Irwin)
[2] Wijaya F, Santo, Darudiato S 2009 *ERP dan Solusi Bisnis* (Yogyakarta: Graha Ilmu)
[3] Akbar R, Juliastrioza, Arici YR 2015 Penerapan Enterprise Resource Planning (ERP) untuk Sistem Informasi Pembelian, Persediaan, dan Penjualan Barang pada Toko Emi Grosir dan Eceran. *Teknosi*. 01. 7-17
[4] Mishra A and Dubey DA 2013 Comparative Study of Different Software Development Life Cycle Models in Different Scenarios, *IJARCSMS*, 1, 64-69
[5] Kendall and Kendall 2010 *Systems Analysis and Design* (New Jersey) Pearson Education

Product design for cable waste peeler machine

B Cahyadi[1,3] **and Y Kurniawan**[2,4]

[1] Department of Industrial Engineering, Pancasila University, Srengseng Sawah, Jagakarsa, Jakarta 12640, Indonesia
[2] Mechanical Engineering Department, Pancasila University, Srengseng Sawah, Jagakarsa, Jakarta 12640, Indonesia
[3] bambang.cahyadi@univpancasila.ac.id
[4] yani.kurniawan@univpancasila.ac.id

Abstract. The design of the cable waste peeler is a product design that aims to reduce the impact of health losses. In general, this is due to the activity of taking the metal in a cable that is done by burning the cable. The design process begins with the identification of customer needs. The customer referred to in this study is the informal sector. Based on interviews that have conducted on several respondents obtained four primary needs that must be illustrated in product design. Then, to explain the four needs are available three types of design variants. The results showed weighting measurements and calculations on load values of approximately 4.15 for the second variant. The results of these calculations are higher than the other two types of design variants. Thus, the design of the variant is chosen for the next design process.
Keywords: wire waste peel design, customer need, design variants

1. Introduction
Cable plays a very important role in the electronic and electrical of product applications today. The use of cables has increased significantly in line with the increasing use of electronics and other electrical products such as computers, the internet, cable television and increased power services worldwide. The cable has the function to connect an electrical current from one component with other electrical and electronics components. Furthermore, the cable could consist of dimensions of size and material type. Generally, the cable consists of a core made of copper or fiber, insulators, and cable cover. The high use of cables in various electronic products has an impact on increasing electronics waste. This cable is one of the recyclable electronics wastes and still has material that has economic value. In developing countries, the waste recycling process still uses very simple methods such as burned or manually shelled to obtain the metal in a cable. Thus, this has the potential to have a negative impact on the environment and human health. For example, dioxins produced from the burning process of cables [2], a potential trigger for cancer [1]. Therefore, in order to solve the problem, it is necessary to design an environmentally friendly waste cable stripper machine design and can meet the needs of the stakeholders of electrical wastewater management.

2. Methods
The Voice of Customer method is one of the product development techniques that used to describe the process of capturing customer needs arranged in a hierarchical structure and then set the priority requirement [3]. The output and benefits of the VOC are able to recognize a detailed understanding of customer needs, as crucial inputs for setting the right design specifications for the product as well as a useful stepping stone for product innovation. There are three methods commonly used to obtain VOC,

for instance, interviews, focus groups discussion and product observations when operated [4]. However, the method used in this study was an interview to conduct the discussions with prospective customers such as the activists of the cable waste processing in the informal sector. Once the information is obtained, the next activity that can be done is to organize needs into hierarchies and then establish the relative importance of each need [4].

In general, the product development process is divided into six phases, such as phase 0 planning, is the process of determining what products and how to be made/developed. Product planning activities include the discovery of product ideas and policies of product. Phase 1 concept development, in this phase, has been done activities to identify the needs of the customer (activities of the voice of customer). As a result obtained alternatives product concepts and conduct evaluation or selection of one or more concepts to be developed and conducted further experiments. Phase 2 system level design, a phase that includes the definition of product architecture and product description into subsystems and components. Phase 3 system design, including complete the specifications of product models, materials, and tolerances of all components of the product and identification of all standard components purchased from suppliers. Phase 4 testing and repair, involves the construction and evaluation of various versions of initial product production. Phase 5 product launch, the product is made using the actual production system. [4].

The design process begins with the identification of customer needs by interviewing in two different places, namely Bekasi and Bojonggede in West Java. Once the needs are identified, the next activity is to create a hierarchy of consumer needs. Then, determine the specification matrix tool that will be designed and then terminated and give weight of interest assessment. In order to know the advantages and disadvantages of tools to be designed then benchmarking of similar products must be done. The final step is to assess several design variants including the components to be used. Variety of designs that have the highest scoring value will be chosen to be developed into a complete design form.

3. Result and discussion

The research begins with the data collection activities of customer needs from the cable waste activist, especially informal sectors. Furthermore, the data is interpreted into the form of specifications and product design. There are some considerations of a shortage of existing cable stripper machine products such as less of secure due to the stripping process is not equipped with a safety cover. Furthermore, the existing machine is less of comfortable due to the equipment rotation speed cannot be changed. In addition, the machine is also not portable due to of its large and heavy dimensions. Therefore, the mission statement of the cable waste stripper to be developed comprises product descriptions, business objectives, primary and secondary markets and stakeholders related to the utilization of such tools.

Table 1. Cable wire peeler mission

Mission Statement	: Cable wire stripper machine
Product description	: Stripper Cable Machine
Business Target	: Middle and low Customer
Primary Market	: Customer of cable waste collectors
Related Stakeholder	: Any user

In the identification stage, especially the customer needs several stages are required to obtain data for the development of the tool, [4]:

1. Collecting raw data from customers or users, in this research, as a tool of data collection is used the interview process to the cable waste collectors, interviews are conducted in two places such as Bekasi and Bojonggede.
2. Interpretation of raw data into the needs of customers or users, the results of the interview process then translated into the table of a statement from the customer. Moreover,

interpreted to be the needs desired by consumers to be implemented into a tool that will be developed.

3. Organizing the needs into a hierarchy, this activity sequences the need for visible levels of needs, as summarized in Table 2. below:

Table 2. List of hierarchy of needs

Level of needs	Needs
	The cable waste stripper is convenient to use
Hight	Rotation speed cannot be changed
Moderate	Machine is not portable due to of its large and heavy dimensions
Hight	Cable stripper machine components are easy to unload
Moderate	An ergonomic form design of cable stripper
	Waste Cable stripper is safe to use
Hight	The process of stripping cable equipped with a cover
Low	The stripping cable is rolled automatically
Hight	The stripping process can be done by using the motor and manually
Moderate	The stripping process is not noisy
Low	Vibration of machine is not large
	Durable stripping machine
Hight	Knife blades are not easily dulled
Moderate	Stripping blades are able to withstand axial loads
Hight	Knife blades are able to withstand dynamic loads
Moderate	Rust stainless blades
	Cable wire Striper is an affordable price
Moderate	Using DC Motor
Moderate	Cable wire stripper can be made easily
Moderate	Cable wire stripper is made with simple technology

There are several processes or stages in the manufacture of product specification targets to be developed [4]:

a. Prepare image metrics using the metric needs that have been previously obtained
b. Gathering information about competitors
c. Set an ideal and marginal target value that can be achieved for each matrix.

A good matrix is one that directly reflects the value of the product that satisfies the customer's needs [4]. The relationship between these needs and matrices is at the main of the process of defining specifications which the assumption is to translate customer needs into a set of precise and measurable specification values. Thus this will be able to produce satisfaction with customer needs. The cable stripper equipment design specification based on user needs is shown in the table below.

Table 3. The cable welder equipment design specification based on user needs

User needs \ Matrices		1	2	3	4	5	6	7	8	9	10
		Dimension of machine	Machine model	Total weight of machine	Sharpness knife to split cable	Strength to withstand dynamic loads	Corrosion resistant material	Device lifespan	Production cost	Precision manufacturing technology	Energy saving
1	Rotation speed can be change										•
2	The dimensions of the cable waste stripper can be moved	•	•	•					•		
3	The components are easy to uninstall	•	•	•					•	•	
4	Stripper machine is ergonomic	•	•	•					•		
5	The process of stripping the cable using a safety cover		•								
6	The stripping cable is rolled by automatically		•								
7	The stripping process can be done by using the motor and manually		•						•		•
8	The stripping process is not noisy				•	•					
9	Vibration of machine is not large					•					
10	Knife blades are not easily dulled				•		•	•	•		
11	Stripping blades are able to withstand axial loads					•					
12	Rust stainless blades				•		•	•	•		
13	Using DC Motor								•		•
14	Cable wire stripper is made with simple technology								•	•	

After the cable stripper design specification matrix had been completed, then the next activity was ranking the weight of metrics. The process of assessment was performed by the researcher.

Table 4. Cable welding equipment design specifications

Number		Metrics	Interest
Matrices	Needs		
1	2,3,4	Dimension of machine	4
2	2,3,4,5,6,7	The model of machine	4
3	2,3,4	Total weight of machine	4
4	8,10,12	Sharpness knife to split cable	4
5	8,9,11	Strength to withstand dynamic loads	4
6	10,12	Corrosion resistant material	4
7	10,12	Device lifespan	4
8	2,3,4,7,10,12,13,14	Production cost	4
9	3,14	Precision manufacturing technology	4
10	7,13	Energy saving	4

The next stage is to collect data specifications of competitors or products that already exist. This becomes the basis of the target product specification that will be developed.

Table 5. Specification data from competitors of cable wire peeler

Number		Matrices	Interest	Competitor
Matrices	Need			China products
1	2,3,4	Dimension of machine	4	120 x 40 x 120 mm
2	2,3,4,5,6,7	Model of machine	4	There is > 1 knife that can be used in any different cable dimensions
3	2,3,4	Total weight of machine	4	> 50 Kg
4	8,10,12	Sharpness knife to split cable	4	Less sharp
5	8,9,11	Strength to withstand dynamic loads	4	Yes
6	10,12	Corrosion resistant material	4	Yes
7	10,12	Device lifespan	4	> 3 years
8	2,3,4,7,10,12, 13,14	Production cost	4	> 50 million
9	3,14	Precision manufacturing technology	4	Yes
10	7,13	Energy saving	4	Yes

Designing of product concept is an activity of developing ideas into a concept of the design solution. This stage is part of the design concept and design detail [4]. The product concept is done by first categorizing design specifications that have been determined in the previous stage. This stage is a step where all the needs and desires of customers can be documented, this is the result of the determination of the customer [3]. In this study used a simple information system that can be compiled by using a list of needs. Based on this information will be obtained the right data in order to determine the specifications of new or developed products. The product specifications should be different with competing products and should be feasible in technically and economically. Highlights the clarity of technical specifications can be obtained after knowing the presence of competitors and similar products [3]. In addition, this steps also to find out the problems and weaknesses of the product. Therefore, it is expected that this design can modify existing products and closer to the needs of users. In addition, user satisfaction is expected to be fulfilled and provide solutions to all complaints and problems. Based on the specified specifications, it can be made the concept of the solution. From that point, alternative parts design and part of the cable stripper to be selected can appear.

Table 6. Variant concept of waste cable peeler component

Component	Solution 1	Solution 2	Solution 3
Motor	Motor DC & Manual  1-1	Motor AC & Manual  1-2	Motor fuel & Manual  1-3
Power transfer	Pully  2-1	Gear  2-2	Gears  2-3
Knife	Jagged 3-1	Without Jagged 3-2	
Frame	Steel Rectangle  4-1	Steel right angle  4-2	Steel  4-3
	Variant 1 Variant 2	Variant 3	

Based on several concepts of existing solutions, then to find the best solution concept must develop a judgment for each concept that exists. The criteria used to assess the concept of the shape, size, and the ability to split the cable. Selection criteria obtained from the need metrics are assigned weighting assessments based on interview results or survey results obtained from cable stripper consumer [4]. The stage of selection or determination of product concept consists of several stages, for example:

1. Preparing a selection matrix, some results of product concept ideas that have been obtained based on the identification of needs, target specifications and concepts that have been created using tables.
2. Assess the concept, the assessment stage of the results of existing concepts conducted by the team or experts.
3. Ranking the concept, after an assessment of the existing concepts, then the next stage is to rank or rank of the concepts from the highest to the lowest.
4. Combining and refining concepts, concepts with the highest ranking value (one or more concepts) are re-analyzed to ascertain whether the concept is appropriate, if not yet fixed until the concept can be realized
5. Choosing the concept, from the stage of improvement of the concept with the highest ranking has been done, the next stage is to choose or determine a concept that is really convincing and has fulfilled aspects or specifications that have been predetermined.

IOP Conf. Series: Materials Science and Engineering **277** (2017) 012035 doi:10.1088/1757-899X/277/1/012035

6. Reflecting on results and processes, the final step of the concept selection stage is to reflect it with agreement and consolidation of the chosen concept.

By using the selection criteria of the final product specifications that have been obtained previously, then a concept screening metric is made of the components of the cable stripper that has been generated. The method quantifies or weighs the criteria of the target specification to be achieved from each of the design alternatives. The weighting is evaluated using the assumption that the alternative can meet the final product specifications and in this study, the assessment is performed by experts in the field of wire stripping. The assessment criteria used in this study are shown in Table 7 below.

Table 7. Scale weighting assessment

Relative performance	Score
Very bad compared to references	1
Poor than reference	2
Equal as references	3
Better than references	4
Very much better than reference	5

Based on the weighting and calculation of the load values of the design variant for the waste of cable stripper, it was concluded that the second variant of the 2nd waste of cable stripper has the highest value approximately 4.15. Thus, this will be the design chosen and will be developed for the waste of cable stripper.

Table 8. An appraisal matrix of the concept of a wastewater peeler component

No	Criteria Selection	Load	Reference		Model Varian waste of cable stripper					
					Variant 1		**Variant 2**		**Variant 3**	
			Weight	**Load Value**	**Weight**	**Load Value**	**Weight**	**Load Value**	**Weight**	**Load Value**
1	Easy to operate	15%	4	0.6	4	0.6	5	0.75	3	0.45
2	Easy to move	15%	3	0.45	3	0.45	4	0.60	3	0.45
3	The storage space is small	10%	4	0.4	4	0.4	4	0.40	4	0.4
4	Result chunks	20%	3	0.6	4	0.8	4	0.80	4	0.8
5	Ease of manufacturing	20%	3	0.6	3	0.6	4	0.80	3	0.6
6	Ease of assembly	10%	3	0.3	3	0.3	4	0.40	3	0.3
7	Cost production	10%	3	0.3	4	0.4	4	0.40	4	0.4
		Total value	3.25		3.55		4.15		3.40	
		Level	4		2		1		3	
		Continue?	No		No		Yes		No	

From the chosen design variant is then drawn in such a way that the components can simultaneously create a cable stripper machine in the motion of each component does not collide with each other. Thus, the resulting product can perform its function properly as expected.

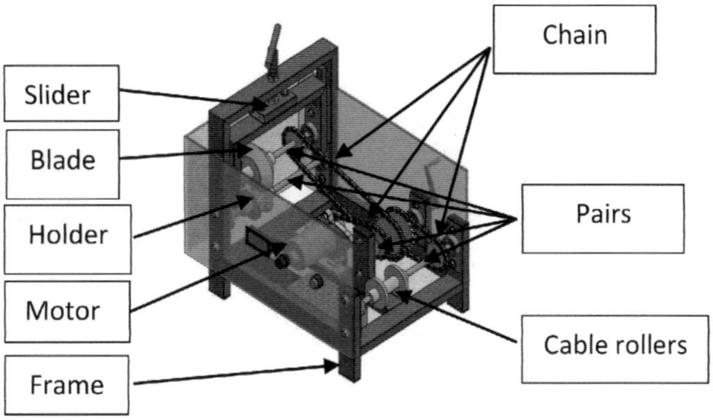

Figure 1. Cable peeler machine design

4. Conclusion

In the process of designing the wastewater peeler, this cable consists of six different stages of activity performed. However, these activities are interconnected. The process begins with the identification of customer needs and ends with a tool design. The design of the developed tool is that has a weighted value and calculation of the load value of the highest variant. The highest value weighs around 4.15 is the second design concept variant. The success of a design is the suitability of the design made with what is required by the customer.

5. References

[1] Anonim 2002 Environmental, Health and Safety Issues in the Coated Wire and Cable Industry *Technical Report No. 51* (Massachusetts: University of Massachusetts Lowell)

[2] Gaidajis G, Angelakoglou K and Aktsoglou D 2010 E-waste: Environmental Problems and Current Management *J.Eng.Sci.and Tech.Rev.* **3(1)** 193-99

[3] Katz G 1993 The Voice of the Costumer *The PDMA Toolbook 2 for New Product Development* (New Jersey: John Wiley & Sons) eds Belliveau P, Griffin A and Somermeyer S

[4] Ulrich K T and Eppinger S D 2000 *Product Design and Development* (Irwin: McGraw-Hill)

IOP Conf. Series: Materials Science and Engineering **277** (2017) 012036 doi:10.1088/1757-899X/277/1/012036

Configurations and implementation of payroll system using open source erp: a case study of Koperasi PT Sri

A Terminanto[1,3], H A Swantoro[2,4] and A N Hidayanto[1,5]

[1]Faculty of Computer Science, Universitas Indonesia, Depok, Indonesia.
[2] Department of Industrial Engineering, Pancasila University, Jakarta, Indonesia.
[3]agung.terminanto@gmail.com
[4]haris.as@univpancasila.ac.id
[5]nizar@cs.ui.ac.id

Abstract. Enterprise Resource Planning (ERP) is an integrated information system to manage business processes of companies of various business scales. Because of the high cost of ERP investment, ERP implementation is usually done in large-scale enterprises, Due to the complexity of implementation problems, the success rate of ERP implementation is still low. Open Source System ERP becomes an alternative choice of ERP application to SME companies in terms of cost and customization. This study aims to identify characteristics and configure the implementation of OSS ERP Payroll module in KKPS (Employee Cooperative PT SRI) using OSS ERP Odoo and using ASAP method. This study is classified into case study research and action research. Implementation of OSS ERP Payroll module is done because the HR section of KKPS has not been integrated with other parts. The results of this study are the characteristics and configuration of OSS ERP payroll module in KKPS.
Keywords: ERP, ODOO Payroll, ASAP, Open Source System, Human Resource Management

1. Introduction

The everlasting business environment necessitates that most of business activities are supported by integrated Information Systems (IS). ERP is a software which completely integrates IT and core and basic business functions [1]. The object of this research is Koperasi Karyawan PT Suryaraya Rubberindo Industries (KKPS) which a retail business entity established to develop member welfare. KPPS has a store and saving-loan business unit. Various kinds of store activities such as ordering goods, receipt, storage, sales, delivery, payment invoice supplier and receipt of customer invoice. KKPS has several problems in running business processes, especially in Human Resources (HR) that is: the absence of an integrated payroll system in supporting the running of business processes. Information and data between each department is not accurate. Payroll calculation and management is still done manually without any management or database system adopted and the output cannot real time. Based on the existing problems in KKPS, Enterprise Resource Planning (ERP) HR system as one of the integrated application solutions, becomes the framework to change and help the process of building the foundation of information systems within the company [2]. The method used is the ASAP (Accelerated SAP) method, because according to the research the method significantly accelerates the deployment process of Enterprise Resources Planning and can directly end up in accordance with the project plan and in accordance with the needs of existing resources [3]. This study has scope up to Go Live and Support stage.

IOP Conf. Series: Materials Science and Engineering **277** (2017) 012036 doi:10.1088/1757-899X/277/1/012036

Purposes to be achieved in this research are the improvement of HR business process workflow of PT SRI Cooperative, the implementation of ERP Payroll systems to enable the process to be integrated with other sections, and the adjustment of the Salary Rule and Structure in Payroll module in features of selected ERP Software.

2. Methods

2.1. Determination of Problems and Purposes

The company's current business system is less able to support the information in company's activities to run smoothly. It require a new system in accordance with the current condition of the company, especially in the HR division. The solution will be implemented is to use an ERP system so that business processes can be integrated and adjustment of Salary Rule and Structure into the Payroll module could be done in near future. The identification of business processes could be done by using Business Process Management (BPM) by analyzing existing procedures and activities inside company to evaluate the relation between divisions so that business processes can be improve to be more efficient than ever. The identification of business processes is also done to facilitate the implementation of business processes against Software ODOO. Information on the flow of data is obtained through interview to all stakeholder and observation with related divisions.

2.2. Identification of Research Framework

This study is classified into case study research as it requires a special focus on a particular area, which is a KKPS retail business which has two business units with different characteristics. In addition to case study research, this study is also included in action research (Action Research). Data is collected from users to assess or improve ongoing processes. This action research is very relevant because in this research resulted identification, configuration and implementation of OSS ERP from various parts of business function.

2.3. Integrating ODOO system and ASAP Method

This research is done using ASAP (Accelerate SAP) method, because in Odoo does not have a special method then adopt method of SAP which significantly accelerate the process of Enterprise Resources Planning deployment and can directly end in accordance with project plan and according to requirement of existing resources. This study is limited to Go Live and Support stage, because at this stage is the final stage of implementation and maintenance while the research is limited to the design of its Enterprise Resources Planning.

2.4. Odoo Design and ASAP Implementation

In order to succeed this Project to be implement, there is several step to be adapted:

2.4.1. Project Preparation. This stage is adapted to identify the background of the problems that occur and raises the problem that needs to be done research and determined also the stages undertaken in the work of this project. Based on the problems found, the goal set is to build an HR system using Odoo on KKPS so that the HR system in the company is not centralized and each part is integrated with the HR department.

2.4.2. Business Blueprint. This stage is adapted to identify the company's business processes by understanding the current business processes and business processes target and then conducted an analysis of existing gaps and generate system design to be developed.

2.4.3. Realization. This stage is done by creating project work in accordance with the established goals and stages of workmanship

2.4.4. Final Preparation. At less precise founded then it need to be repaired. Next step is make the transfer of existing data from the old system to the new system. After doing so, it need to provide training for users who will use the system.

2.4.5. Go Live and Support. This stage is the last stage in the ASAP method. Once the system is implemented, supervision of the system is already in place

3. Result and Discussion

3.1. Project Preparation
Project Preparation stage will be executed before applying ERP system into KPPS by preparing related things to do, that are project goals, project scope, project team, and project timeline.

3.2. Analysis of the Current State
In Payroll Process, KPPS also still uses a system of feed that excel in employee paycheck. Plot by looking at data from excel file containing employee name data, employee contract and calculated in excel file to make paycheck. The problem is most of it didn't integrated and it was applied by calculating salary of employee one by one, starting from arrival allowance, wage, income tax, insurance cost, and other components, all of this done by calculating manually. This manual method allows for large space of errors when calculating and payroll, it also will take a lot of time because before printing paycheck, the slip calculations should be checked back to avoid any errors. The payroll system uses the excel system with the following process:

a. *Collecting Overtime and Debt Form.* Overtime data from each employee written on overtime form, at the end of each month will be collected in order to be calculated with the payment to be paid. In addition, existing forms related to employee debts are also collected to be taken together.
b. *Calculates Employee Debt.* Last month's debt coupled with this month's debt which could possibly arise with a cash-out form in the form of Cash Receipt will result in new debt. This new debt is then reduced by the amount of various obligations to be paid, eventually resulting in outstanding debt.
c. *Make a Payroll Recapitulation.* The calculation of overtime and deductions from each employee combined with basic salary will ultimately result in a net salary to be received in the month. The results of such calculations are written on a piece of paper, and repeated again so that all the net salaries for each employee can be reported on a piece of paper.
d. *Creating Pay Slip For Each Employee.* After a salary recapitulation can be generated, the next step is to make a paycheck for each employee. This pay slip is made in 2 (two) copies, in one of this sheet will be delivered to the employees along with the money they receive, and the next one will be deposited as financial records.
e. *Transferring money* to employees via bank
f. After the transfer Every Employee already. *Employees are then summoned one by one to receive their salary*, and are asked to sign the two paychecks already provided.

3.3. Payroll Function in implementing Odoo
In order to implement ASAP to PT SRI Cooperation, it needs to details the function the payroll cycles corresponding to the Odoo workflow. It can be viewed in figure bellow:

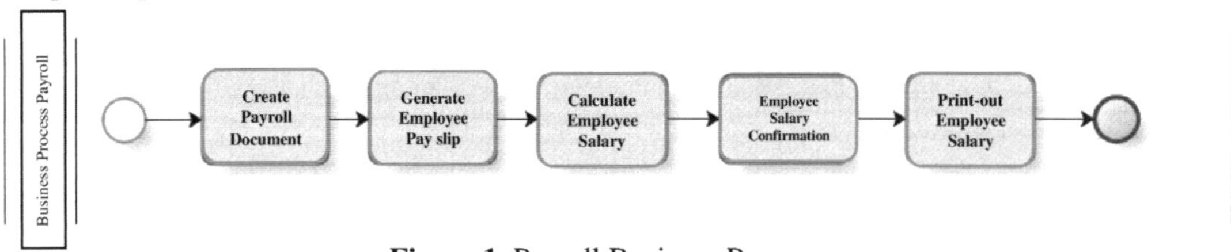

Figure 1. Payroll Business Process

3.4. Technical Configuration in Odoo Payroll

After the general picture of the Payroll business process has been described, the next step is how to configure the Payroll module in Odoo. In order to create Employee salary first thing to do is setting salary rule and salary structure. (Human Resource>Payroll>Salary Sturcture>Create)

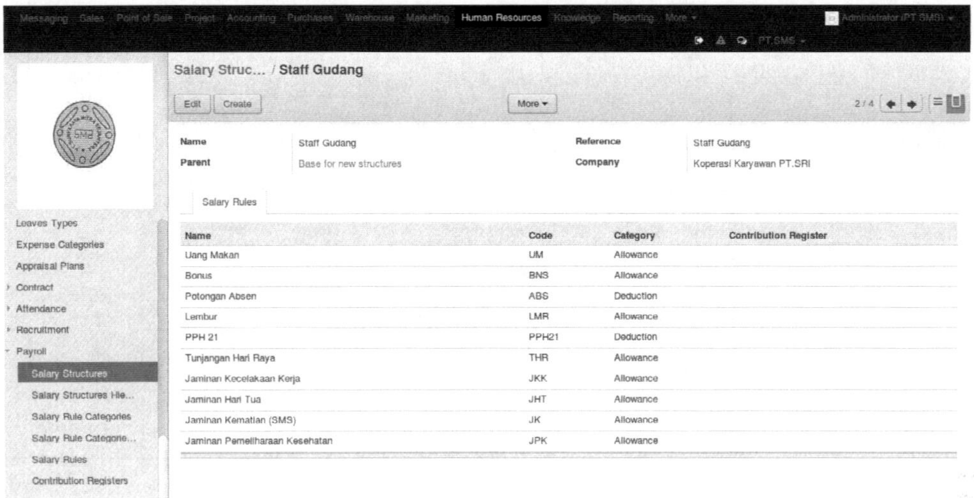

Figure 2. Salary Structure

An example from figure above here are layouts the existing salary structure in KPPS, i.e. Warehouse Staff salary structure. There are several components of the Warehouse Staff salary structure that were used by company, as well as the explanation;

Table 1. Details of Payroll example in Odoo

Name	Information
Basic Salary	Salary according to work contract
Bonus	An additional bonus of 10% from basic salary
Meal Allowance	Additional meal allowance, fixed value for example 200,000 per month
Overtime	Additional overtime pay, the value is 10,000 per hour
Absent Cuts	Absentee deduction, the value of basic salary divided by 173 then it times to the number of abscess hours, so the number of hours of absentee inputted at the time of his pay slip issued
PPH 21	Indonesian Income Tax from Governance

3.5. Creating Salary Paycheck

Once employee data, salary structure, and contracts are established, then we can start making pay slips of each employee. Go to menu: Human Resources> Payroll> Employee Payslip. It will seen layout list of employee salary slip data that already exist. Click the Create button to create new pay slip data.

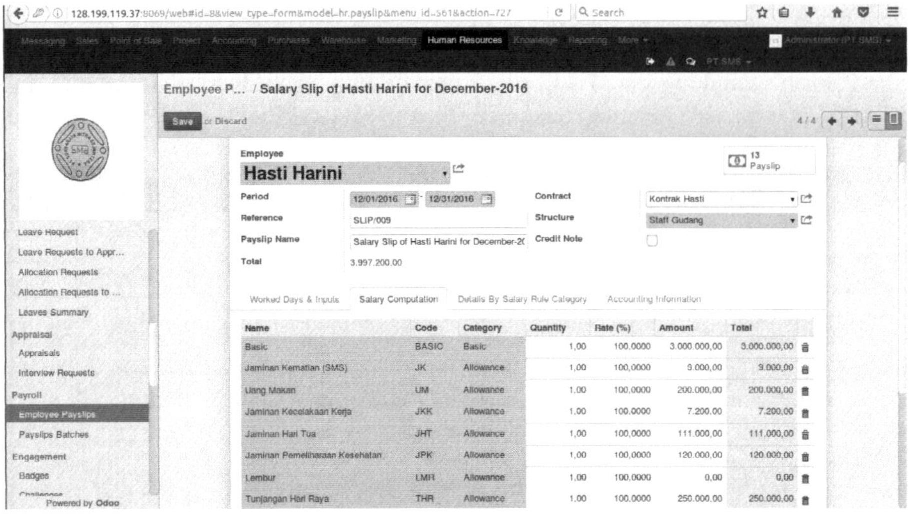

Figure 3. Employee Payslip in Odoo

An example from figure above here are layouts the existing salary payslip of one employee whose name is Hasti Hartini. There are several components of the detail in salary payslip with specified term that were used by company, as well as the explanation;

- Employee: select the employee name to make the paycheck.
- Period: select the start and end date of the payroll, or leave as is, for this period of the month e.g. 12/01/2016 - 12/31/2016.
- Contract: automatically filled with selected employee contract data, e.g. Hasti.
- Reference: will be automatically filled with the payroll document number.
- Structure: automatically selected employee salary structure selected in accordance with his contract, eg Warehouse Staff.
- Payslip Name: automatically filled with pay slip information, e.g. Salary Slip of Hasti for December-2016. Editable if necessary.
- Credit Note: check if this slip is a deduction from the previous slip.
- Click the Compute Sheet button. Salary calculations will be made by Odoo according to the pay structure and the results can be seen on the Salary Computation tab.

4. Conclusion

This project report will shows that Implementation the human resource module using the Odoo is expected to assist in the processing of employee data which will could make cooperatives to provide information easily, cheaply and accurately related to the Human Resources section like every employee profile is recorded with accurate data. It also employee salary management could be done easily without fear of error data in the calculation, the existence of Payroll module with Odoo on KPPS will facilitate the employee in searching about information of their salary details, Salary Structure and Salary role in Payroll module can reduce the miscalculation significantly and reduce the human error inputs, and Last, the existing system in Odoo can be integrated with the attendant in company local database such as update the status of employee attendance. This project report will conclude that the ERP Payroll System should be implemented by KPPS in order to solve several problems in running business processes, especially in Human Resources (HR).

5. References

[1] Rajnoha R, Kádárová J, Sujová A, and Kádár G 2014 Business Information Systems: Research Study and Methodological Proposals for ERP Implementation Process Improvement *Proc.Soc. Behav.Sci.* **109** 165-70

[2] Grubisic I 2014 ERP in clouds or still below *J.of Systems and Information Technology* **16(1)** 62–76

[3] Gruman G 2007 *Is Open Source the answer to ERP?* CIO **20(9)** 23–28

[4] Gurusamy K and Campbell J 2012 Enablers of Open Source Software adoption: A case study of APS organisations *Australasian J.of Information Systems* **17(2)** 67-73

IOP Conf. Series: Materials Science and Engineering **277** (2017) 012037 doi:10.1088/1757-899X/277/1/012037

Application of scl - pbl method to increase quality learning of industrial statistics course in department of industrial engineering pancasila university

M Darmawan[1] **and N Y Hidayah** [2]

[1]Department of Industrial Engineering, Pancasila University, Jagakarsa, Jakarta 12640, Indonesia
[2]muchtar_darmawan@univpancasila.ac.id
[3]nurhidayah@univpancasila.ac.id

Abstract. Currently, there has been a change of new paradigm in the learning model in college, ie from Teacher Centered Learning (TCL) model to Student Centered Learing (SCL). It is generally assumed that the SCL model is better than the TCL model. The Courses of 2nd Industrial Statistics in the Department Industrial Engineering Pancasila University is the subject that belongs to the Basic Engineering group. So far, the applied learning model refers more to the TCL model, and field facts show that the learning outcomes are less satisfactory. Of the three consecutive semesters, ie even semester 2013/2014, 2014/2015, and 2015/2016 obtained grade average is equal to 56.0; 61.1, and 60.5. In the even semester of 2016/2017, Classroom Action Research (CAR) is conducted for this course through the implementation of SCL model with Problem Based Learning (PBL) methods. The hypothesis proposed is that the SCL-PBL model will be able to improve the final grade of the course. The results shows that the average grade of the course can be increased to 73.27. This value was then tested using the ANOVA and the test results concluded that the average grade was significantly different from the average grade value in the previous three semesters.
Key words: SCL-PBL, Final Grade, 2nd Industrial Statistics Course, ANOVA

1. Introduction

True learning is based on discovery guided by mentoring rather than the transmission of knowledge. So based on this opinion the actual learning is learning based on self-discovery which is assisted by mentor mentoring rather than by knowledge transfer. Therefore, to get this kind of learning, it is necessary to shift the paradigm that is learning by teacher centered model into student centered. With the Student Centered Learning (SCL) model, the learner will be placed as the focus of the learning process, responsible for what is learned and can take the initiative because of the nature of active learning. In SCL model, teachers more function as a facilitator for the learners. The course of the 2nd Industrial Statistics in the Department Industrial Engineering Pancasila University with 3 credits weight is part of Basic Engineering subject group. Therefore this course is an important, because it will be the basis for other engineering courses. So far, the applied learning model refers more to the Teacher Centered Learning (TCL) model, and the fact of the field shows that the learning outcomes of the 2nd Industrial Statistics courses are less satisfactory. Data on learning outcomes (final score) in the last three semesters, ie even semester 2013/2014, 2014,2015 and 2015/2016, to be each giving an average score of 56.0; 61.1 and 60.5. Based on the background that has been described then the main issues to be raised in this study are the average grade of Industrial Statistics Course is low in Department Industrial Engineering and how to apply Student Centered Learning model in this course.

The objective of this research is to apply Student Centered Learning model in 2nd Industrial Statistics course in Department Industrial Engineering so that the average grade can be increased. The learning method is by using Problem Based Learning (PBL) method.

2. Methods

2.1. Analisis of variance (Anova)

Anova is a procedure to test whether the mean of several populations is equal, with the number of populations tested more than two. In the hypothesis statement can be written as follows [5]:

$$\text{Ho}: \mu_1 = \mu_2 = \mu_3 = \ldots = \mu_k \quad (1)$$

with k is the number of groups (population) to be tested. While the alternative hypothesis is at least one population average is different from the other population. So:

$$H_a: \mu_i \neq \mu_k \text{ for an i and k} \quad (2)$$

The procedure to test the hypothesis H_o versus H_a is basically an analysis of the variation of the values of measurement results. There are three types of variation values :

- Variation of values within each group, $x_{11} - \overline{x_1}$
- Variations between the group average and the total mean, $\overline{x_1} - \overline{x}$
- Total Variation, $x_{11} - \overline{x}$

Total Variation = variation in group + variation between groups (see Figure 1)

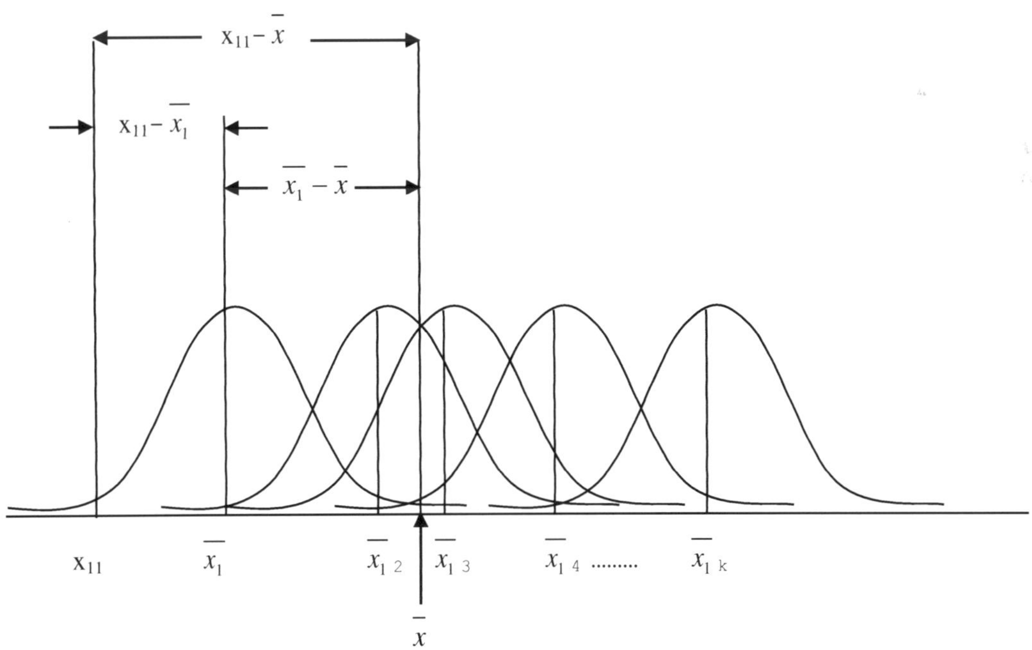

Figure 1. Variation in the anova

From Figure 1 can be derived the following formula.

$$\sum_{k=1}^{k}\sum_{i=1}^{n_k}\left(x_{ik}-\overline{x}\right)^2=\sum_{k=1}^{k}\sum_{i=1}^{n_k}\left(x_{ik}-\overline{x_k}\right)^2+\sum_{k=1}^{k}n_k\left(\overline{x_k}-\overline{x}\right)^2 \quad (3)$$

The above formula explains that :
The sum of squares of the whole deviation (SS_T) = The sum of squares of all deviations in the group (SS_W) + The sum of squares of the entire deviation between the groups (SS_B), where :
SS_T = Sum of Square Total
SS_W = Sum of Square With-in
SS_B = Sum Square Between

Variation = The sum of squares of deviation divided by degrees of freedom. The degree of freedom of variation between groups is = k - 1, whereas the degree of freedom of variation within the group is = N - k, where N = n1 + n2 + n3 + ... n_k.
So the amount of variation between groups is :

$$MS_B = \frac{SS_B}{k-1} \quad (4)$$

MS_B = Between Group Mean Square

While the amount of variation within the group is :

$$MS_W = \frac{SS_W}{N-k} \quad (5)$$

MS_W = Within Group Mean Square.

Further, the differences between the two variations can be tested by using F test which is a comparison between two variations. So the statistical test for Anova is :

$$F = \frac{MS_B}{MS_W} \quad (6)$$

Based on the above description, a summary of variance analysis is presented in Table 1.

Table 1. Anova formula

Source of Varation	Sum Square (SS)	Degree of Freedom	Variation Value	F Test
Between Group	$\sum_{k=1}^{k} n_k\left(\overline{x_k}-\overline{x}\right)^2$	k - 1	MS = $\frac{SS_B}{k-1}$	$\frac{MS_B}{MS_W}$
Within Group	$\sum_{k=1}^{k}\sum_{i=1}^{n_k}\left(x_{ik}-\overline{x_k}\right)^2$	N - k	MS = $\frac{SS_W}{N-k}$	
Total	$\sum_{k=1}^{k}\sum_{i=1}^{n_k}\left(x_{ik}-\overline{x}\right)^2$	N - 1		

2.2. Research methodology

The research methodology will be carried out using the following steps [4] :

Step 1. Determination of research hypothesis

H_o : Application of SCL-PBL method in Dept. Industrial Engineering can not raise the average grade of 2nd Industrial Statistics

H_a : The application of SCL-PBL method in Dept. Industrial Engineering can increase the average grade of 2nd Industrial Statistics

Step 2. Designing learning with SCL-PBL method

a. Preparing Semester Learning Plan (SLP)
   The prepared SLP is the same as the SLP that has been approved by Dept. Industrial Engineering namely SLP 2nd Industrial Statistics which applied to the Curriculum Year 2016.
b. Designing of Implementation Learning Plan (ILP)
   ILP is a further elaboration of SLP, which is adapted to the learning method used, in this case is the SCL-PBL method. Therefore, in the SCL-PBL method, the students will be more active and become the focus of learning by solving cases of real problems, while the lecturers are more facilitators. Therefore, lecture activities of the lecturers will be replaced more by group discussion and student presentation
c. Designing an assessment form
   In the SCL-PBL method, students/student groups are also given the opportunity to assess the other groups of students. Therefore, the existing assessment form so far will be added so as to be as follows [1] :
   - Group discussion assessment form (assessment by student)
   - Group presentation assessment form (assessment by student and lecturer)
   - Group assignment report assessment form
   - Mid semester exam assessment form
   - Final semester exam assessment form
   - Final value recapitulation assessment form (from faculty)
d. Designing a case study
   The SCL-PBL method requires real case studies (triggers) to be solved by the students. Therefore, it is necessary to prepare the case study. The case study is planned to be taken from final project at Department of Industrial Engineering related to statistical problems.
e. Designing assessment weights
   The weight of scores used during this course in 2nd Industrial Courses is as follows:
   - Attendance : 5%
   - Assignment : 25%
   - Mid Semester Exam (MSE) : 30%
   - Final Semester Exam (FSE) : 40%

   In the SCL-PBL method, the volume of student assignment is increasing, therefore the weight of the percentage of assignment scores needs to be increased, so that the design of the assessment weight is changed as follows :
   - Attendance : 5%
   - Assignment : 30%
   - MSE : 30%
   - FSE : 35%

Step 3. Data collection

The data to be collected are as follows:

a. Scores on the learning result of 2nd Industrial Statistics three semesters before.
b. Scores of 2nd Industrial Statistics Even Semester 2016/2017 according to SCL-PBL method includes the following:
   - Attendance score

- Students score in group discussions
- Student score in assignment presentation
- Student score in the assignment report
- Score of Middle Exam Semester and score of Final Exam Semester

Step 4. Data processing and analysis
Score data that has been collected is processed and then analyzed. The analytical method used is the Analysis of Variance (Anova) method. Anova is a hypothesis test of differences in mean values of some populations (more than two populations) by comparing their variance values, ie the variance values in groups with inter-group variance. The assumptions used in the Anova method are [2]:

- The random samples taken are independent of each other
- The population distribution from which the sample is taken is normal
- Variants of population distribution are considered equal

The next steps of hypothesis testing using Anova are as follows [2]:

a. Set the hypothesis test
   $H_o : \mu_1 = \mu_2 = \mu_3 = ... = \mu_k$
   $H_a : \mu_i \neq \mu_k$ for an i and k
b. Calculate statistical test value
   Anova test is comparing the variance value, the test statistic used is F test (eq.6).
c. Set the H_o acceptance / rejection area
   With the level of confidence given (C, in%) and the degree of freedom according to the number of data owned is known then the acceptance/rejection zone H_o can be determined (Table F, $F_{\alpha;v1;v2}$).
   Note: α = error rate = 1 - C
   v_1, v_2 = degree of freedom
   In this study, to ensure that the results are convincing there will be three different levels of confidence : 90%, 95%, and 99%.
d. Take the conclusion
   H_o will be accepted if the F test statistic is inside the H_o acceptance area (F-test value < F-table value). Conversely H_o will be rejected if the value of statistical test F is within the rejection region H_o. (F-value test > F-table value). If H_o is accepted then H_a automatically rejected, otherwise if H_o is rejected then H_a is accepted.

Using Anova method do hypothesis test that the average grade of 2nd Industrial Statistics Even Semester 2013/2014, 2014/2015, and 2015/2016 (three semesters) are the same. The allegation of equality of the average value is based on the fact that in the three semesters the learning of 2nd Industrial Statistics uses the same method, that is a Teacher Centered Learning method. So our hope here is H_o will be accepted. As noted earlier, the assumption that needs to be taken is that the variations in data from these three groups are considered equal.
After the final value of the 2nd Industrial Statistic in the Even Semester of 2016/2017 is obtained, then do the next Anova test to prove that the average grade of 2nd Industrial Statistics courses in Even Semester of 2013/2014, 2014/2015, 2015/2016, and 2016/2017 (four semesters) is to be not the same. The allegation of average value inequality is based on the fact that the method of study of the subject of 2nd Industrial Statistics is now used unequal method, ie in Even Semester 2013/2014, 2014/2015, 2015/2016 used Teacher Centered Learning method, while in Even Semester 2016/2017 used the Student Centred Learning method. So here we hope will reject H_o which means accept Ha. Assumptions that need to be taken in this case are also the same ie the data variations of these four groups are considered the same.
Step 5. Conclusion and suggestion
Based on the results of the analysis that has been done, it will be given a conclusion and suggestions in accordance with the results.

3. Result and discussion

In accordance with the research steps outlined above, after the research hypothesis has been established, the design of the appraisal form and case study is prepared then the implementation of this study will begin with data collection and then proceed with the processing and analysis of the data.

3.1. Data collection

The learning process of 2nd Industrial Statistics course with SCL-PBL method in Even Semester 2016/2017 implemented in accordance with SLP and ILP as follows :

a. Twelve meetings designed to discuss 4 (four) case studies allocated into 4 (four) cycles where each cycle consists of three meetings, ie
   - The first meeting was filled with lecture activities from lecturers that explaining the theories to be used in solving case studies.
   - The second meeting was filled with group discussions by students discussing case studies provided by lecturers and how to solve the case study. At this meeting the student group also prepared a presentation of what was discussed to be presented at the next meeting.
   - The third meeting is filled with the presentation of each group and then the results of this presentation are reported in writing to the lecturer. Figure 3. below is a presentation of tasks performed by students.

b. Two meetings are designed to explain other theories that are not directly related to the case studies that have been given to the students.

c. One meeting for Mid Semester Exam and one meeting for Final Semester Exam. So the total number of meetings is 16 times.

From the results of the lecture consisting of theoretical explanations by the lecturers, group discussions, task presentations by students and then the implementation of Mid Semester Exam and Final Semester Exam then the evaluation score data is collected consisting of data score of discussion, presentation score, Mid Semester Exam score, and score of Final Semester Exam. The entire score data is combined to produce the final score.

3.2. Processing and analysis of data

In accordance with the methodological steps outlined before, the data processing and analysis are carried out as described below.

a. Analysis of Variance (Anova) for the score of 2nd Industrial Statistics three semesters previously.
   Hypothesis test procedure is as follows:
   - Statement of H_o and H_a

     H_o: $\mu_1 = \mu_2 = \mu_3$

     H_a: $\mu_i \neq \mu_j$ for i = 1, 2, 3 and j = 1, 2, 3

     Where :

     μ_1 = The average value of the Class population of 2nd Industrial Statistics Even Semester 2013/2014.

     μ_2 = The average value of the Class population of 2nd Industrial Statistics Even Semester 2014/2015.

     μ_3 = The average value of the Class population of 2nd Industrial Statistics Even Semester 2015/2016.
   - Value of F-Test

     Using the data score and the Anova formula in Table 1, F-test values can be obtained as follows:

Table 2. Anova summary table

Variation Source	SS	DF	MS	F-Test	F-Table		
					0.01;2;98	0.05;2;98	0.10;2;98
Between Group	480.95	2.00	240.47	1.06	4.79	3.07	2.35
Within Group	22225.17	98.00	226.79				

- Determination of Acceptance / Rejection Area of H_o (Df_1 = 3-1=2 and Df_2 = 101-3 = 98).
  With 99% confidence level (or α = 1%) then obtained F-table = 4.79.
  With a 95% confidence level (or α = 5%) then obtained F-table = 3.07.
  With a confidence level of 90% (or α = 10%) then obtained F-table = 2.35.
- Conclusion :
  Since the F-test value < F-table for all confidence levels, then the F-test value is at the acceptance area of Ho. So Ho is accepted which means the grade point average for Even Semester 2013/2014, 2014/2015, and 2015/2016 is not significantly different, so it is considered the same.

b. Anova for 2nd Industrial Statistics courses score three previous semesters plus semester 2016/2017. As previously, hypothesis test procedure is as follows:

- Statement of H_o and H_a
  H_o: $\mu_1 = \mu_2 = \mu_3 = \mu_4$
  H_a: $\mu_i \neq \mu_j$ for i = 1, 2, 3, 4 and j = 1, 2, 3, 4
  Where :
  μ_1 = The average value of the class population of 2nd Industrial Statistics Even Semester 2013/2014
  μ_2 = The average value of the class population of 2nd Industrial Statistics Even Semester 2014/2015.
  μ_3 = The average value of the class population of 2nd Industrial Statistics Even Semester 2015/2016.
  μ_4 = The average value of the class population of 2nd Industrial Statistics Even Semester 2016/2017.
- Value of F-test
  Using the data score and the Anova formula in Table 1 then F-test values can be obtained in Table 3.
- Determination of Acceptance / Rejection Area of H_o (Df_1 = 4-1=3 and Df_2 = 145-4 = 141).
  With 99% confidence level (or α = 1%) then obtained F-table = 3.78.
  With a 95% confidence level (or α = 5%) then obtained F-table = 2.60.
  With a confidence level of 90% (or α = 10%) then obtained F-table = 2.08.
- Conclusion: Since the F-test value > F-table for all confidence levels, then the F-test value are inside the H_o's rejection area. So H_o is rejected which means that the average value for Even Semester of 2013/2014, 2014/2015, 2015/2016, and 2016/2017 are not significantly the same, so are considered different. This difference occurs due to the entry of the average score of Even Semester 2016/2017 ie 73.27. Rejection H_o here means that the SCL-PBL method can improve the previous learning method, ie TCL, so as to increase the average grade. The average value of Even Semester 2016/2017 is the highest, which is 73.27 compared to the previous three semesters, ie. 56.1, 61.1, and 60.5.

Table 3. Anova summary table

Variation Source	SS	DF	MS	F-Test	F-Table		
					0.01;2;141	0.05;2;141	0.10;2;141
Between Group	6325.28	3.00	2108.43	9.87	3.78	2.60	2.08
Within Group	30133.48	141.00	213.71				

4. Conclusion

Based on the explanations and explanations that have been stated previously it can be given the following conclusions:

- Hypothesis testing that the learning method of Student Center Learning - Problem Based Learning (SCL-PBL) can improve the learning result of Industrial Statistics course in Department Industrial Engineering of Pancasila University is true.
- The truth of hypothesis test above is limited to the assumptions taken in this study such as the value data assumed to be normally distributed, samples that have been taken are independent of each other and the variation of value data is considered equal.

5. References

[1] Rahmawati H and Farida Y 2006 *Upaya Peningkatan Kualitas Belajar Mahasiswa Melalui Model Pembelajaran Koperatif Kepala Bernomor Pada Pembelajaran Topik Reaksi Kimia Mahasiswa Semester Satu di Fakultas Farmasi Universitas Pancasila* Laporan Penelitian Tindakan Kelas (Jakarta: Universitas Pancasila)

[2] Hinkle, Dennis E, Wiersma, William, Jurs and Stephen G 2003 *Applied Statistics for Behavioral Sciences Fifth Edition* (Houghton Mifflin Company)

[3] Mitra A 2008 *Fundamentals of Quality Control and Improvement Third Edition* (John Wiley & Sons)

[4] Priyatmojo and Achmadi 2010 *Student Centered Learning (SCL) dan Student Teacher Aesthethic Role-Sharing (STAR)* Buku Panduan Pelaksanaan Pusat Pengembangan Pendidikan (Yogyakarta: Universitas Gajah Mada)

[5] Walpole R E 2011 *Probability and Statistics for Engineers and Scientists* (Prentice Hall International)

IOP Conf. Series: Materials Science and Engineering **277** (2017) 012038 doi:10.1088/1757-899X/277/1/012038

Optimization benefits analysis in production process of fabrication components

R Prasetyani[1,2], A Y Rafsanjani[1] and D Rimantho[1]

[1]Industrial Engineering Department, Pancasila University, Jakarta
[2]rini.prasetyani@univpancasila.ac.id

Abstract. The determination of an optimal number of product combinations is important. The main problem at part and service department in PT. United Tractors Pandu Engineering (shortened to PT.UTPE) Is the optimization of the combination of fabrication component products (known as Liner Plate) which influence to the profit that will be obtained by the company. Liner Plate is a fabrication component that serves as a protector of core structure for heavy duty attachment, such as HD Vessel, HD Bucket, HD Shovel, and HD Blade. The graph of liner plate sales from January to December 2016 has fluctuated and there is no direct conclusion about the optimization of production of such fabrication components. The optimal product combination can be achieved by calculating and plotting the amount of production output and input appropriately. The method that used in this study is linear programming methods with primal, dual, and sensitivity analysis using QM software for Windows to obtain optimal fabrication components. In the optimal combination of components, PT. UTPE provide the profit increase of Rp. 105,285,000.00 for a total of Rp. 3,046,525,000.00 per month and the production of a total combination of 71 units per unit variance per month. **Keywords:** Optimization, Fabrication Component, QM for Windows

1. Introduction

Mining is one of an important pillar of development in Indonesia. This sector has long been the main sector contributing to the cash inflow of the country. However, starting in 2011 to date, the sector is showing a downward trend. The price of mineral and coal mining commodities is declining and has not shown any signs of increasing. Indonesia as one of the countries rich in mining commodities also feels the impact. The impact of the declining mining trend also has an impact on the manufacturing industry and heavy equipment distributor, where the number of heavy-duty equipment sales decreased dramatically. PT. United Tractors Tbk, as the core company of PT. United Tractors Pandu Engineering throughout 2016 must make several revisions to its sales target. The original sales target was 4,494 units of heavy mining equipment, then revised to 3,800 units. The Company has decreased approximately 10% by year on year based on 2016 PT United Tractors Annual Report. Faced with these conditions, PT. United Tractors Pandu Engineering (PT UTPE), which focuses on manufacturing heavy equipment supporting units such as dump vessel, backhoe bucket, grader blade, semi-trailer, water tank, etc. to perform efficiency measures on all business lines.

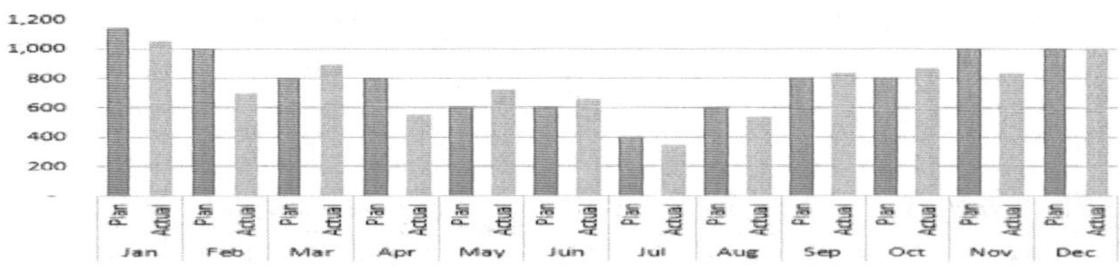

Figure 1. Plan and actual sales (total) fabrication component at PT. UTPE (million rupiahs)

The graph above shows the comparison of plans and actual sales of the fabrication components at PT. UTPE in million rupiahs during January to December 2016. Furthermore, the graph shows the value of fluctuating sales figures. Because of its fluctuating value, it cannot be concluded directly about the optimization of production of fabrication components in PT. UTPE just by looking at the graphs only. Based on these conditions, this paper will conduct further tests to determine whether the production of fabrication components at PT. UTPE is currently providing optimal benefits or not. The optimum benefits are influenced by the condition of factors affecting the production of the fabrication component, namely: production cost, raw materials, production hours, production equipment, and market demand. This study aims to analyze maximum profitable and the amount of optimum product combination.

2. Literature review

2.1. Optimization

Every company will try to achieve the optimal state by maximizing profits or by minimizing costs incurred in the production process. The company expects the best results with limited resources, in solving problems with optimization techniques to produce the best solution [1].

2.2. Linear Programming

Linear programming is a method of calculation for the best planning between possible actions that can be done. The best determination is that there are many alternatives in planning to achieve specific objectives on limited resources. The Linear program consists of two kinds of functions, namely the purpose function and function constraints. The purpose function is a function that describes the goal or goal in the sources to obtain maximum profit or minimum cost. While the constraint function is a mathematical representation of the available constraints that will be allocated optimally to various activities. In general, the linear programming model can be expressed as follows [3]:

- Function Objectives, maximizing or minimizing:

$$Z = C_1X_1 + C_2X_2 + \dots + C_nX_n \quad (1)$$

- Eligible Constraints

$$a_{11}X_1 + a_{12}X_2 + \dots + a_{1n}X_n \quad (=, \leq, \geq)\ b_1 \quad (2)$$

$$a_{21}X_1 + a_{22}X_2 + \dots + a_{2n}X_n \quad (=, \leq, \geq)\ b_2 \quad (3)$$

$$a_{m1}X_1 + a_{m2}X_2 + \dots + a_{mn}X_n \quad (=, \leq, \geq)\ b_m \quad (4)$$

$$x_1, x_2, \dots, x_n \geq 0 \quad (5)$$

The limiting function can be either (=) or inequality ($\leq$ or $\geq$). The limiting function is also called a constraint. Constants (both as coefficients and right values) in the constraint function and on the destination are said to be model parameters. The symbols $x_1, x_2, \dots, x_n$ (x_i) show the decision variables. The number of decision variables (x_i) therefore depends on the number of activities or activities undertaken to achieve the goal. The symbols $C_1, C_2, \dots, C_n$ are the contribution of each decision variable to the goal, also called the coefficient of the objective function in the mathematical model. The symbols $a_{11}, \dots, a_{1n}, \dots, a_{mn}$ are the per unit use of the decision variable of the limiting resource, also called the functional constraint coefficient of the mathematical model. The symbols b_1, b_2, ..., b_m denote the number of each resource available. The number of constraint functions will depend on the limited number of resources. The last inequality (x_1, x_2, ..., $x_n \geq 0$) shows non-negative limits [3].

3. Result and discussion

3.1. Model Formulation

Formulation of the model in this research is to obtain the purpose function and constraint function. The purpose function of optimization of fabricated component products at PT. UTPE is maximizing profits or finding the difference between the selling price and the production cost. The costs include fixed costs and non-fixed costs. To find the value of decision variables in the form of profit on the purpose function (1) which has been described in Chapter III, the data used is the selling price and production cost of each product. The selling price, production cost, and net income of each fabrication component can be seen in the Table 1 below.

Table 1. Selling price, production cost, and net income of 2016

Product Type	Price (Rp/Unit)	Production Cost (Rp/Unit)	Profit (Rp/unit)
Liner plate dozer	84,460,000	67,031,740	17,428,260
Liner plate vessel	144,900,000	120,709,700	24,190,300
Liner bucket PC 4000	344,900,000	265,307,700	79,592,300
Liner bucket PC 2000	299,500,000	233,984,375	65,515,625
Liner bucket PC 1800	199,880,000	156,156,250	43,723,750

The coefficient on the purpose function (1) is the net income (profit) per unit of each type of fabrication product obtain from the sale. After the input parameters for each product is known then the purpose function to maximize the profit can be formulated as follows:

$$\text{Maximize } Z = 17{,}428{,}260X_1+24{,}190{,}300X_2+79{,}592{,}300X_3+65{,}515{,}625X_4+43{,}723{,}750X_5 \quad (6)$$

3.2. Formulation of constraining function

Several obstacles that often occur in the production process, especially in manufacturing fabrication components at PT. United Tractors Pandu Engineering consists of: production cost constraints, raw material constraints, production hours constraints, time constraints on the use of production equipment, and market demand constraints.

3.2.1. Cost production constraint. The availability of financial capital is a way of allocating the average capital available to commercialize each type of fabrication product. The availability of working capital for fabrication component business in PT.UTPE monthly average of 2016 is Rp. 11,470,231,000. The constraint function for production costs based on Table 1 can be formulated below:

$$67{,}031{,}740X_1+120{,}709{,}700X_2+265{,}307{,}700X_3+233{,}984{,}375X_4+156{,}156{,}250X_5 \leq 11{,}470{,}231{,}000 \quad (7)$$

3.2.2. Raw material constraint. In the formulation of the function of raw material constraint coefficient function of raw material, the constraint is the amount of raw material per unit of product. In addition, the amount of raw material availability per month is shown on the right value of the raw material constraint function. Raw materials used in producing fabrication components at PT. UTPE is a sheet iron with a grade of Hardox grade 400. The amount of iron sheeting is a factor to be considered. The constraints on the sheet iron of each sheet per unit are obtained from data provided by the production department of PT. UTPE is 300 sheets/month. Based on the data of the amount of iron usage, the formulation of the iron sheet raw material constraint function in manufacturing the fabrication component is as follows:

$$3X_1 + 2X_2 + 8X_3 + 5X_4 + 4X_5 \leq 300 \quad (8)$$

3.2.3. Production hour constraint. Production workforce is a person who contributes to assisting the production process in turning raw materials into finished materials. In the production process, the fabrication component has a total of 10 persons of production labor, the number of working hours is 6 hours/day, and the number of working days is 22 days in one month. From the data obtained, it is known that the availability of labor production time per month is 1320 hours/month. Based on the 2016 monthly average labour data, the formulation of the constraint function of production working hours in the manufacture of fabrication components are as follows:

$$12X_1 + 10X_2 + 24X_3 + 18X_4 + 13X_5 \leq 1320 \quad (9)$$

3.2.4. Production equipment usage constraints. The constraints of production equipment are calculated based on the amount of time available for an apparatus to produce a product. This is influenced by the number of working hours available. The units used to know the constraints of production equipment are hours/units. PT. UTPE in running its business using production equipment namely Plasma Cutting Machine, Big blue Air pack Welding Machine, and Crane. The production equipment is used to perform the production process on each type of fabrication component.

Plasma cutting machine is used to assist the manufacturing process of fabrication components especially in producing pieces of iron sheets (liner plate) according to the design made. The availability of plasma cutting machine working hours at PT.UTPE is 420 hours/month. This result is obtained from multiplication of available plasma cutting machine 3 units with production time 7 hours/day, and a number of working days 5 days/week and 4 weeks in a month. Based on the 2016 monthly average data of cutting time, the formulation of time constraint function of plasma cutting machine in making of fabrication component is as follows:

$$4X_1 + 4X_2 + 8X_3 + 6X_4 + 5X_5 \leq 420 \quad (10)$$

Welding machine big blue air pack is used to assist the welding process between liner plates after cutting. Availability of welding machine big blue air pack at PT. UTPE 540 hours/month. This result is obtained from the multiplication of the Number of Welding Machine Big blue Air pack available 3 units with production time 9 hours/day, and the number of working days 5 days/week for 4 weeks in 1 month. Based on monthly welding time, formulation of time constraint function of welding machine big blue air pack in the manufacture of fabrication component is as follows:

$$6X_1 + 5X_2 + 10X_3 + 8X_4 + 8X_5 \leq 540 \quad (11)$$

Crane is used to assist the process of moving or changing the position of the liner plate from post 1 to the next post. Crane clock availability at PT. UTPE is 280 hours/month. This result is obtained from the multiplication of the Number of cranes available is 2 units with production time 7 hours/day, and the number of working days 5 days/week for 4 weeks in 1 month. Based on the monthly crane usage time data, the formulation of crane availability constraint function in manufacturing fabrication component is as follows:

$$X_1 + X_2 + 5X_3 + 3X_4 + 3X_5 \leq 280 \quad (12)$$

3.2.5. Market demand constraints. Market demand constraints are used to determine the minimum constraints and the maximum limitations of production that must be generated. Demand is a number of goods consumers want to buy. The amount consumers want to buy is not always the same as the number actually purchased by consumers. The amount consumers want to buy is called a potential demand, while the amount actually purchased by consumers is called actual demand. Therefore, the potential demand is a demand that has not been followed by purchasing power, while the actual demand is demand followed by purchasing power. Potential demand is greater than real demand but can be equally large. Market demand constraints are used to determine the minimum and maximum limitations that PT.UTPE must produce to meet the market demand.

IOP Conf. Series: Materials Science and Engineering **277** (2017) 012038 doi:10.1088/1757-899X/277/1/012038

The constraints of market demand referred to in this constraint function are actual sales based on company data. The data is obtained from the total units that have been sold and operate the field that can be referred to as total demand. From the total demand available, sales potential can be determined by referring to the market share. The size of market share, influenced by several factors such as price, product quality, after-sales service, etc. The formulation of monthly market demand constraint function is formulated as follows:

$$X_1 \leq 10$$
$$X_2 \leq 10$$
$$X_3 \leq 18$$
$$X_4 \leq 10$$
$$X_5 \leq 10 \quad (13)$$

3.3. Primal Analysis

In the formulation of the optimization model, the expected results are the optimal results that can be achieved by the company in accordance with the objectives and constraints that become the limit in the production. Based on data processing with Microsoft Office Excel program and POM QM has been known optimal analysis results. The optimal results show optimal solutions consisting of optimal product combinations, optimal resource status, and sensitivity analysis.

From the results of the targeted product fabrication analysis set by PT. UTPE earned a profit of Rp 2,941,240,000 per month with the production of a total combination of 58 unit/month unit variants, consisting of 10 units of X1 components, X2 of 10 units, X3 of 18 units, X4 of 10 units and X5 of 10 units.

Table 2. The Use of resources at PT. United Tractors Pandu Engineering

Product Type	Availability	Optimal usage	Slack/surplus
Production Cost	11,470,240,000	10,554,360,000	915,881,000
Sheet steel	300	284	16
Man hour (hour)	1,320	962	358
Time usage for plasma cutting machine (hour)	420	334	86
Time usage for welding machine (hour)	540	450	90
Time usage for crane (hour)	280	190	90

3.3.1. Raw materials usage analysis. The raw materials that become a constrain in this study was an iron sheet of Hardox 400. In Table 2 it can be seen that the use of sheet metal has slack/surplus value of 16 sheets/month. This indicates that the sheet iron used is still excessive during the production process which means that the use of sheet iron is not yet optimal.

3.3.2. Labor time usage analysis. Based on the optimum results in Table 2, the use of production labor time has an overtime time of 358 hours which means the use of labor time of this production is not yet optimal. Under optimal conditions, the production workforce used is 962 hours from the availability of 1,320 hours.

3.3.3. Production equipment usage analysis. Production equipment that becomes an obstacle in this research is time constraint of using plasma cutting machine, welding machine big blue air pack, and crane. There is an excess time of use of production equipment which means not optimal when the use of production equipment. In the use of plasma cutting machine has an excess time of 86 hours/month with optimal use of 334 hours/month from the availability of 420 hours/month. Use on welding machine big blue air pack 450 hours/month from the availability of 540 hours/month and has an excess time of 90 hours/month. Similarly, the use of cranes during the production process has a usage time of 190 hours/month from the availability of 280 hours/month and has an overtime of 90 hours/month.

3.4 The Dual Analysis

The dual analysis is done to determine the assessment of resources by looking at the value of slack/surplus and its dual value. The value of slack/surplus is also related to the magnitude of the effect of adding or decreasing the number of resources available and the results can be used for management decision-making. Analysis of the status of these resources can be seen in the Table 3 below.

Table 3. Analysis of resource status at PT. UTPE

Constraint	Dual Value	Slack/surplus	Original Val	Lower Bound	Upper Bound
Production Cost[a]	0	915.88	11,470.240	10,554.36	Infinity
Raw material	0	16	300	284	Infinity
Man hour	0	358	1,320	962	Infinity
Time usage for machine PC	0	86	420	334	Infinity
Time usage for machine WM	0	90	540	450	Infinity
Time usage for machine CN	0	90	280	190	Infinity
Total Demand X1	17,428,260	0	10	0	15
Total Demand X2	24,190,300	0	18	0	18
Total Demand X3	79,592,300	0	10	0	20
Total Demand X4	65,515,620	0	10	O	13
Total Demand X5	43,723,750	0	10	0	14

[a] = in million

Table 3 above provides information related the available resources are still excessive. Excess resources have dual value = 0 and slack/surplus value> 0, this indicates that there is no deficiency in the use of raw material business capital, labor time, and time of equipment during the production process of fabrication component at PT. UTPE. Thus, there is no limit on the availability of raw materials.

3.5. Sensitivity Analysis

Sensitivity analysis is performed to find out how far the answer of an optimal solution can be applied if there is a change in the model. This analysis shows the lapse of changes in the amount of resource availability that does not cause the change in the value of the dual constraints concerned. The hose also indicates the importance of a resource, where the smaller the hose the more important the resources are of the condition. The sensitivity interval is indicated by the minimum and maximum value of the allowable inventory. Table of sensitivity analysis for business capital resources, raw materials in the form of an iron sheet, labor time, and machine usage time is as the Table 4 below.

Table 4. Analysis of resource sensitivity at PT. UTPE

Constraint	Dual Value	Slack/surplus	Original Val	Lower Bound	Upper Bound
Production Cost[a]	0	915.88	11,470.240	10,554.36	Infinity
Raw material	0	16	300	284	Infinity
Man hour	0	358	1,320	962	Infinity
Time usage for machine PC	0	86	420	334	Infinity
Time usage for machine WM	0	90	540	450	Infinity
Time usage for machine CN	0	90	280	190	Infinity
Total Demand X1	17,428,260	0	10	0	15
Total Demand X2	24,190,300	0	18	0	18
Total Demand X3	79,592,300	0	10	0	20
Total Demand X4	65,515,620	0	10	O	13
Total Demand X5	43,723,750	0	10	0	14

[a] = in million

From Table 4 above it can be seen that the status of excess resources. Thus, that if resources such as business capital, raw materials, labor time, machine usage time are reduced as much as allowed in accordance with the minimum limit in Table 4 then it will still get the optimal benefit. This is because the available resources have advantages over the intended use of the fabrication component manufacturing process. In addition, the availability of venture capital, raw materials, labor time of production, machine usage time has an infinite increase (infinity), it is because the optimal conditions are achieved resources that are not used up so that the addition of resources will not affect the value of the dual and does not change its optimum production. By knowing at the minimum value and maximum limits of each resource constraint function, the management of PT. United Tractors Pandu Engineering may obtain information in making adjustments to the procurement of resources.

3.6. Verify Combination Targets

Based on the results of optimization and analysis that has been done, it can be concluded that the target combination of production of fabrication components that is determined by the company has not been optimal. Under these conditions, the study re-tested the combination of products, by increasing the number of market demand and without changing the function of objectives and the availability of venture capital, raw materials, labor time, and time of use of existing machinery. The optimum iteration table with customized market demand constraints is shown below:

Table 5. Optimal iteration with an increase in the number of market demand

Constraint	X1	X2	X3	X4	X5	RHS	Dual
Maximize[a]	17.42	24.19	7.96	65.51	43.72		
Production cost[a]	67.03	120.71	265.31	233.99	156.15	11.47	0
Raw material	3	2	8	5	4	300	0
Man hour	12	10	24	18	13	1,320	0
Time usage for machine PC	4	4	8	6	5	420	0
Time usage for machine WM	6	5	10	8	0	540	0
Time usage for machine CN	1	1	5	3	5	280	0
Total Demand X1	1	0	0	0	0	13	17,426,260
Total Demand X2	0	1	0	0	0	22	24,190,300
Total Demand X3	0	0	1	0	0	12	79,592,300
Total Demand X4	0	0	0	1	0	13	65,515,620
Total Demand X5	0	0	0	0	1	11	43,723,750
Solution	13	22	12	13	11	3,046,520,000	

[a] = in million

Table 6. Analysis of resource sensitivity at PT. UTPE with increasing market demand:

Variable	Value	Reduced	Original	Lower	Upper
X1	13	0	17,428,260	0	Infinity
X2	22	0	24,190,300	0	Infinity
X3	12	0	79,592,300	0	Infinity
X4	13	0	65,515,620	0	Infinity
X5	11	0	43,723,750	0	Infinity
Constraint	Dual	Slack	Original	Lower	Upper
Production cost[a]	0	0	11,470.24	11,470.23	Infinity
Raw Material	0	12	300	288	Infinity
Man hour	0	279	1,320	1,041	Infinity
Time usage for machine PC	0	51	420	369	Infinity
Time usage for machine WM	0	40	540	500	Infinity
Time usage for machine CN	0	91	280	189	Infinity
Total Demand X1	17,428,260	0	13	0	13
Total Demand X2	24,190,300	0	22	0	22
Total Demand X3	79,592,300	0	12	0	12
Total Demand X4	65,515,620	0	13	O	13
Total Demand X5	43,723,750	0	11	0	11

[a] = in million

From the Table 5 above, after an increase in the number of market demand then PT. UTPE earns a profit of Rp 3,046,525,000 per month (profit increase of Rp 105,285,000) with the production of a total combination of 71 units per unit variance, consisting of X1 = 13 units, X2 = 22 unit, X3 = 12 units, X4

= 13 units and X5 = 11 units. From the combination result if continued with the primal-dual analysis and sensitivity analysis, then the condition as illustrated in the Table 5 above.

In Table 6 it can be seen that the use of production cost is optimal, it is because of the value of slack/surplus = 0 besides the value for the upper limit of the number of requests X_1, X_2 ..., X_5 is equal to the original value, which is optimal. But there are still excess resources, although there is a decrease in the amount of raw material with the initial value of 16 drops to 12, the labor time with the initial value of 358 drops to 279, the machine usage time A with initial value 86 drops to 51, with an initial value of 90 down to 40, and an increase in value at the time of use of machine C with an initial value of 90 rising to 91. With the end result, the combination of fabrication components that have increased the number of market demand can be concluded close to optimal.

4. Conclusion

From the study conducted can be summed up that the results shows that the target combination of fabrication component products set by the management of PT. United Tractors Pandu Engineering currently could earns a maximum profit of Rp 2,941,240,000 per month, with 10 unit monthly production of Liner Plate Dozer, Liner Plate Vessel, Liner Plate Bucket PC 2000, and Liner Plate Bucket PC 1800.

Based on the analysis conducted, the result depict that the sheet iron used is still over 12 sheets/month, production labor time has an overtime time of 279 hours/month. The usage of plasma cutting machine has 51 hours/month, the wicking machine big blue air pack has 40 hours/month, and the use of crane has 91 hours /month which all of them stood under capacity.

The result of production optimization model shows that the company will gain maximum profit at Rp 3,046,525,000, - per month which has to produce monthly 13 units of Liner Plate Dozer, 22 units of Liner Plate Vessel, 12 units of Liner Plate Bucket PC 4000, 13 units of Liner Plate Bucket PC 2000, and 11 units of Liner Plate Bucket PC 1800.

5. References

[1] Herjanto E 2008 *Manajemen Operasi Edisi Ketiga* (Jakarta :Grasindo)
[2] Mulyono S 2007 *Riset Operasi* (Jakarta: Fakultas Ekonomi Universitas Indonesia)
[3] Nasendi B D E and Anwar *Program Linear dan Variasinya* (Jakarta: Gramedia)
[4] Wahyuni T and Nuharini D 2008 *Matematika Konsep dan Aplikasinya* (Surakarta: Departemen Pendidikan Nasional)
[5] Handoko T H 1999 *Dasar-dasar Manajemen Produksi dan Operasi* (Yogyakarta: BPFE)

Risk assessment of underpass infrastructure project based on IS0 31000 and ISO 21500 using fishbone diagram and RFMEA (project risk failure mode and effects analysis) method

Bambang Purwanggono[1] **and Anastasia Margarette**[2]

[1,2]Industrial Engineering Department, Faculty of Engineering, Diponegoro University, Semarang, Indonesia
b.purwanggono@undip.ac.id[1] and anastasia.m2104@student.undip.ac.id[2]

Abstract. Completion time of highway construction is very meaningful for smooth transportation, moreover expected number of ownership motor vehicle will increase each year. Therefore, this study was conducted with to analyze the constraints that contained in an infrastructure development project. This research was conducted on Jatingaleh Underpass Project, Semarang. This research was carried out while the project is running, on the implementation, this project is experiencing delays. This research is done to find out what are the constraints that occur in execution of a road infrastructure project, in particular that causes delays. The method that used to find the root cause is fishbone diagram to obtain a possible means of mitigation. Coupled with the RFMEA method used to determine the critical risks that must be addressed immediately on road infrastructure project. The result of data tabulation in this study indicates that the most possible mitigation tool to make a Standard Operating Procedure (SOP) recommendations to disrupt utilities that interfere project implementation. Process of risk assessment has been carried out systematically based on ISO 31000:2009 on risk management and for determination of delayed variables, the requirements of process groups according to ISO 21500:2013 on project management were used.
Keywords: risk, fishbone diagram, RFMEA, ISO 31000, ISO 21500

1. Introduction

The lifestyle of society that continues to develop in the following times results the government to develop in order to adjust the needs of the community. For the example is construction of road infrastructure. The development needs to be done, because it is estimated that the number of motor vehicle ownership will increase every year. Which is will causes congestion if the available road isn't adequate.

One of the road construction example that expected to cause traffic congestion is on Jatingaleh, Semarang, Central Java. According to kompas.com, Jatingaleh road which is only 1.5 km long takes about 20 minutes to pass through. Finally, Semarang city's government issued a policy to undertake Jatingaleh underpass development project. Properly on June 18th 2015, the parties involved in this project signed the contracts.

The initial agreement for the project was completed on October 13th 2016, but the project was unable to comply the targets of the contract. The extension of PHO (Provisional Hand Over) is until December 15th 2016. However, still cannot be resolved and finally renewed until July 31th 2017. Delay in completion of the project shows the fact that the implementation has not prepared yet in facing the risks may occur. Therefore it is necessary to perform a risk management that embraces the process of risk management of ISO 31000 with the addition of ISO 21500 (Project Management). Where in conducting risk analysis using the method Project Risk Failure Mode and Effects Analysis (RFMEA) and in forming appropriate recommendation tool made fishbone diagram which will discuss the factors of delay. Therefore, this study is expected to manage the risks that occur in road infrastructure projects in order to ensure that what has been planned in the implementation of a project can run smoothly so as not to cause further delays.

Based on preliminary studies that have been done, it was found that the improvement of infrastructure is indeed needed (such as Jatingaleh underpass development project). In fact there is a delay in the settlement. Factors that cause delays can be seen by observing man power, machine, environment, finance, method, and material. So we need to identify and assess risk that can overcome the delays in infrastructure projects in general.

The purpose of this research is to know the factors causing the delay of infrastructure project, to know the state of project risk management and to provide general prevention advice to be used on the upcoming infrastructure projects.

2. Methods

Research methodology is used as a reference in conducting research. This methodology contains the research flow, data collection techniques, research variables, and data analysis.

2.1 Research Flow

This study was initiated by conducting preliminary studies on the object of research. After that, make the formulation of the problem in accordance with the field studies and existing literature. Followed by the determination of research objectives. After the goal set, research continued with the data collection and processing which adopted risk management process according to ISO 31000. Where the first step are communicating and consulting with stakeholder, then making the context setting with fishbone to find out the root of existing problem, and making the risk assessment by spreading the questionnaire which will be processed according to RFMEA method. The RFMEA process is done as follows.

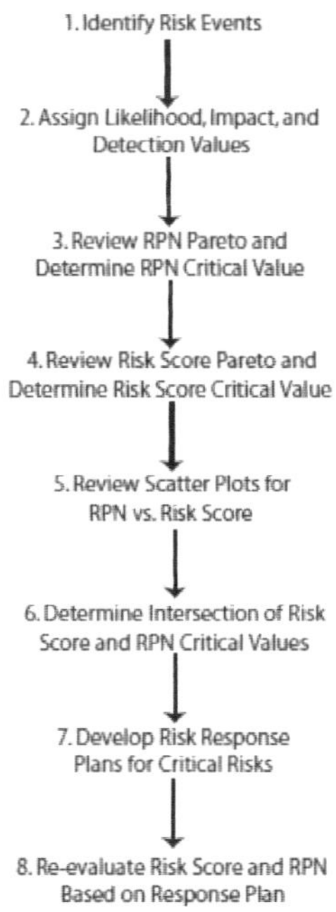

Figure 1. This figure explains the steps of RFMEA.

However, in this study the eighth step is not implemented because there are limitations in this research. And the last is treating the risks. After the data processing has been done then proceed with the data analysis to determine the response of the mitigation that allows to be applied. But, before the recommendations are made it is necessary to verify and validate the selected mitigation facilities so they can be applied to upcoming infrastructure projects.

2.2 Research Variables

This study required variables associated with project activities in accordance with ISO 21500 to determine the risk event categories that occur, such as (ISO, 2013):

Table 1. Definition of some variables that used in this research.

Variable	Definition
Initiating	Undertaken Activities in defining the project
Planning	Activities undertaken when detailing and planning project goals
Implementing	Activities integrate all required resources by implementing what is already planned
Controlling	Activities undertaken to periodically measure and monitor the progress of the project and also to identify whether there any abuse of the implementation of the planned plan

3. Result and Discussion

Based on the data collection that has been done, the following data processing along with the recommendations to be used were obtained.

3.1 Identification of Root Causes

Based on the results of the reference factors causing the delay, this research obtained the description of fishbone as follows.

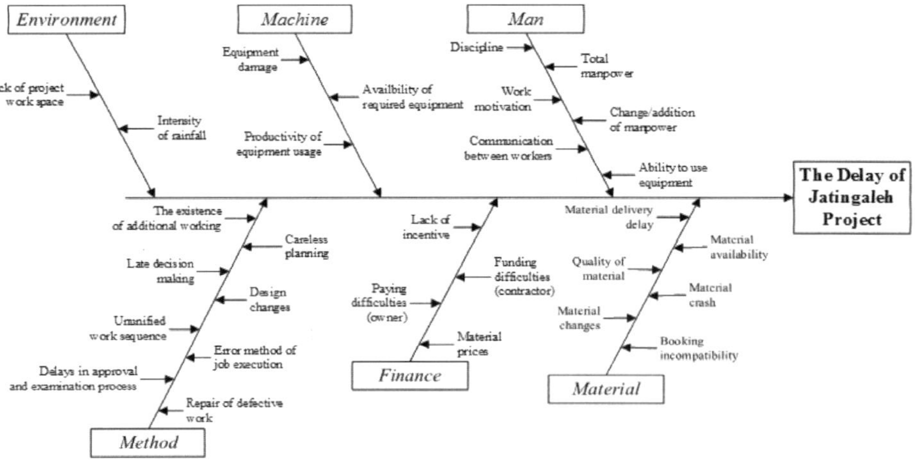

Figure 2. Fishbone Diagram shown the factors can caused the delay of Jatingaleh Project.

3.2 Identification of Risk with RFMEA

By adopting the RFMEA process [1], the risk score was obtained based on the multiplication of likelihood with impact. Here are the pareto results for the risk score on the Jatingaleh project.

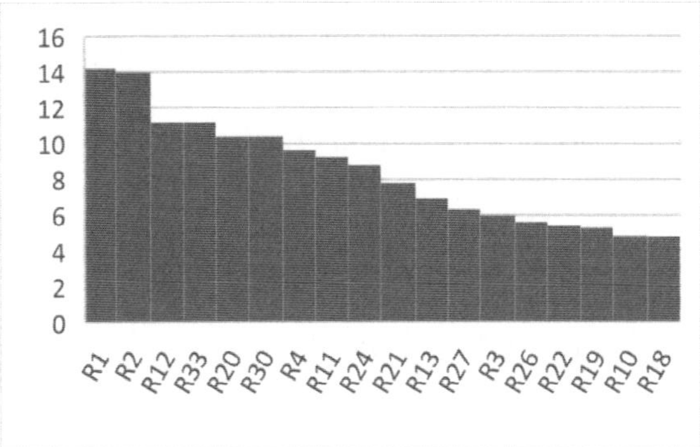

Figure 3. Shown the pareto of Risk Score

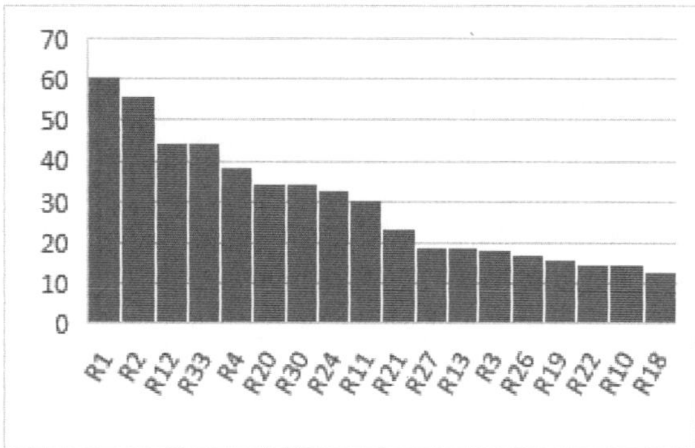

Figure 4. Shown the pareto of RPN

In Figure 3, it was found that R1 is the risk with the highest risk score value. In accordance with the pareto risk score (RS) results, critical value was selected to identify the most important risk on the project. In a study conducted [5], it was stated that there was no systematic procedure in determining criteria for critical RS values as noted [1]. Comparing the maximum value recorded can be done to determine the critical value of Risk Score. Based on existing pareto results, the maximum risk value is almost 0.4 times than early studies [1], where the critical value is 20. So the critical value that selected for this study is 8.

Next step is add a third factor to each risk in the risk list, i.e. detection. Followed with, multiplication between RS with detection value for each Risk ID to obtain Risk Priority Number (RPN) value. As with RS, a pareto chart is created using existing RPN values as follows.

Similar to RS, critical value for RPN in Figure 4 is also needed to identify the most important risks in this project. The determination of critical RPN value is also not explained in detail in the previous studies [5]. Based on the existing RPN pareto, the selected RPN critical value is 31.25. The value is selected based on the maximum RPN value in this study which is considered to be about 0.25 times the maximum value of RPN in early studies [1].

The RFMEA technique introduces a way of prioritizing risk by using two attributes, the RS and RPN attributes. By creating a scatter diagram where the X axis plots the RS and Y axes plots the RPN value, the major risk on the project can be visually identified and it provides a stronger priority mechanism. This scatter diagram shows all the risks that have RS and RPN values greater than the predetermined critical value (the risk that is in the top right quadrant on the scatter diagram). So values RS and RPN are described as follows.

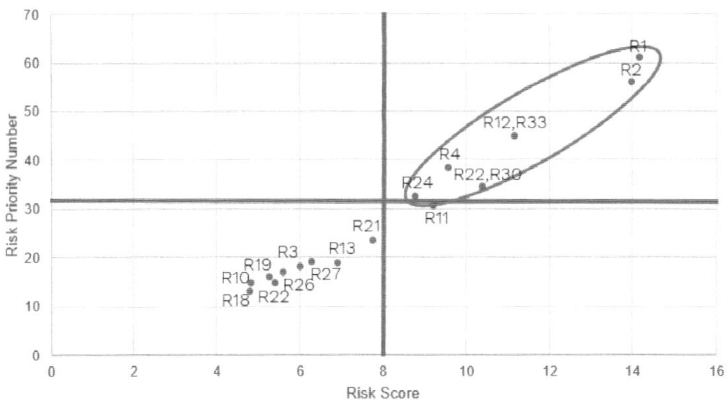

Figure 5. This figure shown the distribution of risk event based on results of matrix RPN vs Risk Score

After obtaining the intersection point between the RS and RPN critical values, it can be found that R1, R2, R4, R12, R20, R24, R30, and R33 are in the red circle on the scatter diagram (Figure 5) and categorized as project risk with highest priority countermeasures. And the final step in the RFMEA process for this research is developing a response plan to address the identified key project risk that will be discussed further in next subsequent chapter.

3.3 Analysis of Data Processing Results
The result of data processing that has been done is adjusted to risk management process at ISO 31000 as follows.

- Communicate and consult
  Based on the results of communications and consultations that have been done, the element arrangement that causes the project risk is obtained. Using the previous research, a list of project risk elements that usually occurs in road infrastructure research is acquired. The list is screened to obtain the relevant project risk in Jatingaleh project. This research used 4 process group according to ISO 21500 as category of project risk group. And 37 elements of project risk that may occur in the Jatingaleh project.
- Establish Context using Fishbone Diagram
  Context establishing needed to be done to identify and uncover the objectives which this research wants achieve, to assess the nature and complexity of the risks. Based on the results of the discussions that have been done, the results conclude that the method factor is the root of the main problems of project delays. Inadequate planning, uncooperative work sequences and errors in the method of job execution which are the causes of method factors slow down the completion of the Jatingaleh project.
- Project Risk Assessment (RFMEA)
  Appropriate with ISO 31000 process, risk assessment process is carried out by method of Project Risk FMEA (RFMEA). The way of processing data using RFMEA method itself is by assessing the risk score and risk priority number on each risk event. The next step is creating pareto for each category of assessment, in order to get the critical value for each category of assessment. Determining the critical value for the risk score and the RPN is followed by placing risk events in each quadrant according to the value of risk scores and existing RPN.
- Treat Risks
  Based on the results of data processing with fishbone diagram, it is found that the means of countermeasures that allow to be recommended is making a SOP. Where the SOP recommendation's title is addressed to resolve the critical risks that exist in the critical area of risk score vs RPN matrix of RFMEA method. From the results of data processing with RFMEA there are 8 critical risks for this Jatingaleh project. In order of importance level to be addressed immediately are R1, R2, R12, R33, R20, R30, R4, R24. However, since this study also has restrictions, R1 and R2 can be eliminated because the risk handling is beyond the boundary control of the researcher. So obtained R12, R33, R20, R30, R4 and R24 that allows to be used as SOP title. By considering the value of risk scores and existing RPN, R12 and R33 share the same value and based on the discussion that has been done, R33 is agreed to be prioritized compared to R12. Because R33 concerns the association with groups outside the contractor, consultant and project owner, it will spend longer consolidation time compared to R12 which have links within the internal group of contractor, consultant and project owner.
  Therefore, possible critical risks are treated by establishing SOP related to the settlement of risks concerning the existence of utilities owned by PLN, PDAM, Telkom which interfere with this project. This recommendation is also made in general, that later this SOP can be used on other road infrastructure projects to prevent any further delays.
- Monitoring and Review
  This stage is done after the risk treatment has been determined and implemented. if the chosen treatment after being implemented is not appropriate or useful to overcome the critical risk, it

needs to be reviewed. So, the step of establish context until risk treatment need to be checked again. However, due to the limitations in this study, the researcher did not carry out this stage and fully submitted the results of risk behavior recommendations to the project owners, consultant and contractor for monitoring and review.

3.4 Designing Standard Operating Procedure

Then the result of SOP design as follows.

Table 2. A head table for SOP Recommendation

Utility Clearance XYZ Project	SOP Number Revision Number Effective Date Approved By SOP Names — **Utility Clearance**
SOP DESCRIPTION	**QUALIFICATION AND LIST OF IMPLEMENTATIONS**
SOP Utility Clearance is a guide used by infrastructure project implementers to handle utility problems that interfere with project completion. **The Purpose** of this SOP is to guide the problem solving to be as effectively and efficiently as possible, with an asset record and also directed and completed information	IMPLEMENTERS • *Owner* • Contractor • Consultant • Vendor (related) EXECUTOR QUALIFICATION • Good technical and interpersonal skills • The ability to cooperate and coordinate with other parties • A good communication techniques • A broad understanding and knowledge about the development of road infrastructure and the placement/removal of road utilities
LINKAGES	
To assist in the rapid handling of related utilities that disrupt the implementation of road infrastructure development projects	
REFERENCES	
• SNI 03-2850-1992, Tata Cara Penempatan Utilitas di Jalan	
• Pd. T-13-2004-B, Pedoman Penempatan Utilitas pada Daerah milik Jalan	**EQUIPMENTS / REQUIREMENTS**
• Peraturan Menteri PU no. 20/PRT/M/2010 Pasal 43, Pemindahan dan Pembongkaran	• Permit for removal of utilities form • Disturbing utilities form
WARNING	**RECORDING**
• If this SOP is not implemented, then the handling related to the disruption of utility placement on the road infrastructure project will be delayed. • Delayed handling will impact the disruption of operational implementation of the project	• Note the requirement of standard quality according to step • Note detail the required of standard quality

Table 3. The description of SOP Recommendation

(a)

	DESCRIPTION STEP OF SOP *Utility Clearance*	EXECUTOR: Owner	Consultant	Service Provider	Vendor	STANDARD QUALITY: Requirement	Output	Times
1	Comparing the soft drawing to real condition on project					*Soft drawing* project	Field Engineering	± 2-4 weeks
2	Are there any utilities that interfere with soft drawing execution? **Yes**. Coordinate for displacement of related utilities. **No**. Can proceed to next job-description					Field Engineering	List of utilities that interfere with soft drawing	-
3	Conduct a coordination meeting about the name of utilities that interfere the job-description					List of utilities that interfere with soft drawing	Inventory constraints on the field	± 1-2 days
4	Giving invitation regarding coordination meetings for displacement of the utilities to vendors (per-sector / individual) which interfere the job-description					Removal License	Meeting invitations with related vendors	± 1-5 days

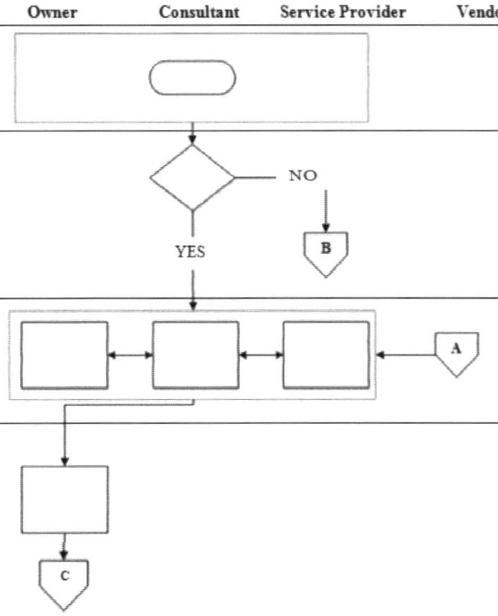

(b)

	DESCRIPTION STEP OF SOP *Utility Clearance*	EXECUTOR: Owner	Consultant	Service Provider	Vendor	STANDARD QUALITY: Requirement	Output	Times
5	Confirmation of utility placements that interfere the job-description (clarification and coordination) Is it tolerable? **Yes**. Can proceed to next job-description **No**. Related utilities process of displacement should be done	C		B ← YES	NO	1. Confirmation the arrival of the coordination meeting 2. Availability of cost and labor by vendor	Explanation of the boundaries for Road Owned Area	± 1 day
6	Scheduling of utilities displacement by each vendor, tailored to the project schedule					*Project schedule*	Timetable for implementation of utility displacement	± 1 day
7	Implementation of utilities displacement					*Soft drawing*	Utility displacement	Adjust the implementation of the job-description related
8	Coordinate the results of execution for utility displacement.					The process of displacement utility has been completed	The utility does not interfere with the next job-description	± 1 week
9	Are there still any disturbance the soft drawing? **Yes**. Return to second procedure **No**. Then proceed to the next job-description			A ← YES; B	NO	-	-	-

3.5 SOP Examination

After obtaining SOP recommendation, the next step is examination against SOP submitted. The goal is to ensure that the designed SOPs can be used and implemented. So it is necessary to do verification and validation test by giving one scenario in each test, as follows.

Table 4. Shown the SOP Examination in this research

(a)

	GOAL	METHOD	AIM
Scenario I	Verify related documents, to prove the truth of the information contained in the SOP-utility clearance	Interview	• Owner • Consultant

(b)

	GOAL	METHOD	AIM
Scenario II	Validate related documents to see if the SOPs can run under existing conditions and to find out the shortcomings of the proposed SOP to be improved then applied.	SOP simulation and survey	Implementing SOP, i.e.: • Contractor • Owner • Consultant

4. Conclusion
This study discusses how to overcome the existing problems in road infrastructure projects, especially those causing delays in project completion. The selected countermeasures of mitigation results a SOP recommendation. The provisions of the SOP title obtained from the critical risks that need to be addressed immediately in this Jatingaleh project, namely the utilities that interfere the project work. The selected recommendation, namely general SOP utility clearance that later can be used for other infrastructure projects.

From the result of this research, it is expected further research can conduct similar research by solving other critical risks, using likelihood assessment based on direct data results, and discuss the results of implementation of SOP utility clearance.

5. References
[1] Thomas A Carbone and Donald D Tippet 2004. Project Risk Management Using the Project Risk FMEA. *Engineering Management Journal* **16** 29
[2] ISO 2009 *ISO 13000:2009—Risk Management: Principles and Guidelines* (Genève: International Organization for Standardization)
[3] ISO 2009 *Risk Management - Principles and Guidelines* (Genève: International Organzation for Standardization)
[4] ISO 2013 *ISO 21500: Guidance on Project Management – A Pocket Guide* (Genève: International Organzation for Standardization)
[5] R Luppino, Hosseini, M Reza, and R Rameezdeen 2014. Risk Management in Research and Development (R&D) Projects: The Case of South Australia. *Asian Academy of Management Journal* **19** 67–85
[6] N G D Mahadipta 2010 *Analisis Risiko pada Proyek Pembangunan Sentral Parkir di Pasar Badung* Denpasar: Universitas Udayana)
[7] Pd. T-13-2004-B *Pedoman Penempatan Utilitas pada Daerah milik Jalan.*
[8] Peraturan Menteri PU no. 20/PRT/M/2010 Pasal 43 *Pemindahan dan Pembongkaran.*
[9] Permen PAN Nomor: PER/21/M.PAN/11/2008 *Pedoman Penyusunan Standar Operasional Prosedur Administrasi Pemerintahan.*
[10] B Proboyo 1999 *Keterlambatan Waktu Pelaksanaan Proyek: Klasifikasi dan Peringkat dari Penyebab-Penyebabnya* (Jakarta: Dimensi Teknik Sipil)
[11] SNI 03-2850-1992 *Tata Cara Penempatan Utilitas di Jalan*

Linking human capital and enterprise sustainability in Indonesian medium-sized food manufacturing enterprises: the role of informal knowledge sharing practice

O Sunardi[1]

[1] Department of Industrial Engineering, Krida Wacana Christian University, Jakarta 11470, Indonesia
oki.sunardi@ukrida.ac.id

Abstract. Medium-sized food manufacturing enterprises in Indonesia are significant in a number of contexts, in terms of their part to the national production (GDP) and their establishment to the employment. In term of their role to national production, manufacturing sector contributes the highest GDP by 85%. In this sector, food manufacturing subsector contributes the highest GDP. Nevertheless, they faced the same common problems: quality of human capital and sustainability issues. Previous government supplementary programs have been established to expand the human capital capability amongst medium enterprises. Adequate amount of fund has been apportioned to develop human capital, though, the medium enterprises sustainability is still in question. This study proposes and examines the human capital role from informal knowledge sharing perspective. By conducting qualitative approach through interviews to four informants in Indonesian medium-sized food manufacturing enterprises, a set of hypotheses is derived from this study for future quantitative study. This study indicates that human capital traits (diverse education background, employee skills, and employee experience) could leverage the practice of informal knowledge sharing. Constructs such as mutual trust and reciprocal intention could play as mediating variables, and cultural interpretation perspective could act as moderating factor to informal knowledge sharing effectiveness. In final, informal knowledge sharing is indicated to play as moderating variable for human capital policy and practice to support enterprise sustainability.
Keywords: human capital, informal knowledge sharing, sustainability, medium enterprise.

1. Introduction

Medium-sized enterprises in Indonesia are significant in a number of contexts, in terms of their part to the national production (GDP) and their establishment to the employment. As twice the economic crisis hit Indonesia in 1997 and 2008, medium enterprises have indicated their ability to endure better than small and large enterprises [1][2]. In term of their role to national production, manufacturing sector contributes the highest GDP by 85%. In this sector, food manufacturing subsector contributes the highest GDP. This underlines the important of manufacturing sector to the development of Medium Enterprises, and the stability of Indonesian economic as a whole.

Previous studies as regards to medium enterprises have revealed that the quality of human capital still become the key problem to most Indonesian MEs [3][4][5]. Adequate amount of fund has been apportioned to many programs to develop human capital within medium enterprises. The programs included of training and development in production, quality management, general management, and entrepreneurship. Nevertheless, sustainability still become a difficult target.

The development of human capital amongst medium enterprises still become the main objective of the Indonesian Ministry of Cooperative, Small, and Medium Enterprises (Kemenekop). The strategic plan stresses that the human capital development programs will be focusing on education and training, in terms of basic production, technical support, entrepreneurship, and managerial development. In agreement with Kemenekop development policy, the Ministry of National Development Planning (BAPPENAS) recommends to increase the negotiation capability of medium enterprises by accelerating technical proficiency and entrepreneurial ability of human capital. The key interest of the plans is to assist medium enterprises to develop and sustain. Three success parameter were grouped from 70 previous findings on the relationship between business success and human capital: profitability, business size, and growth [6]. However, previous findings neglected to see 'sustainability' as a criterion of success, while sustainability should be considered as fundamental criterion to any business achievement [7][8][9], especially for Small and Medium Enterprise.

Most of Indonesian enterprises, specifically medium scale enterprises, tend to fully implement what so called "the western human resource management models" in order to manage their human capital [10][11]. The main motivation is to be able to competitive in global competition. Previous study found that the improvement of intense economic actions in the Asia-Pacific zone emphasized the imperfection of todays' Western Human Resource Management (HRM) approaches [12]. The approval of HRM initiatives is rather to management fashion than research, both by practitioner and academics. In fact, todays' HRM models are greatly constructed based on large enterprises in UK and USA [6][12][13].

In fact, people management model and practices cannot be generalized between one nation and another, because each nation has its own environment characteristic, which is different one to another [11][14]. It is observable that the need for comprehensive human capital model in medium enterprises is perceived, since most human capital studies relied on resource-based view of the firms, which examine human using every element of production factor, product life cycle theory, strategic management, financial management and accounting, economics perspective, technology management, and organizational ecology [6][15][16][17][18].

Therefore, this study proposed a more comprehensive model of managing human capital for medium-sized food manufacturing enterprises in Indonesia from knowledge-based perspective. By examining previous studies, this study characterizes essential human capital programs and policies within medium-sized food manufacturing enterprises. These programs and policies are then synthesized to a model that simplifies how human capital can be managed in relation to enterprise sustainability through informal knowledge sharing practices [19].

The structure of this paper is constructed as follows. First, conceptual model regarding the relationship between selected human capital traits and informal knowledge sharing is examined to form the research framework. Second, research framework is then reorganized by addressing the issues concerning with human capital policies and initiatives, with regards to enterprise sustainability. Third, a set of hypotheses is derived by using qualitative approach. The final part of this paper incorporates conclusion and implications for further studies.

2. Human Capital Traits and Informal Knowledge Sharing

Previous work [19] conceptualizes another perspective in knowledge sharing initiative by involving certain human capital traits to informal knowledge sharing activity. The study suggests that particular human capital traits, for example: employee experience, education background, and employee skills will likely to influence the effectiveness of informal knowledge sharing. Furthermore, the work provides four major propositions regarding the relationship between these human capital traits and informal knowledge sharing. Thus, this study tries to explore more of the propositions in order to build a research framework. The exploration is piloted by conducting interview to three informants from one leading medium-sized food manufacturing enterprise. The informants consist of individuals from all organizational level: top manager, middle manager, and staff. The use of these different level of employment aims to obtain views from different perspectives.

First proposition discusses that "*when individuals in the organization reciprocally focus on building each other's new skills, the exploration of employee skills amongst individuals affect positively to the informal knowledge sharing process*". The first proposition suggests that knowledge sharing should be seen from the perspective of individual behavioural intention. 'Skill' is defined as "the ability to master the concepts of a discipline or domain, and to apply this knowledge appropriately in new situations" [20]. 'Reciprocity' indicates that individual who involves in informal knowledge sharing will attempt the best to discover the skills other individual has, and attempts to take benefits to each other's necessities. The clarifications to the proposition are as follow:

> *"There is no free lunch... If you have reliable skills, people will come to you. In my experience, others will exchange theirs for mine" (Informant 1, Production Staff).*

> *"I would say, knowledge sharing related to skills exploration is mostly done outside formal occasions. Lunch times, sport sessions, shift breaks, and locker rooms become the most preferable time and place. Interestingly, I never see more than two people discuss their skills together in one group. It seems that this knowledge sharing activities become transactional and reciprocal, though conducted in informal ways. To me, it doesn't matter, as long as they can perform better in their works....and they do" (Informant 2, Factory Manager).*

The above discussion also confirmed that if an organization provides supporting environment to any knowledge sharing activities, it will encourage interaction and collaboration among individuals, which is very important to enable knowledge flow, and supports individuals with the ability to self-organize their networks and share knowledge [21]. The discussion also confirmed the previous finding [22], which argues that informal conversation might be the most effective mechanisms of transferring individual's tacit knowledge. The above discussion concludes the following set of hypotheses:

Hypotheses 1: *Employee skills have positive direct effect to reciprocal intention*
Hypotheses 2: *Reciprocal intentions have positive direct effect to informal knowledge sharing process*

The second proposition argues that "*when mutual trust exists among individuals in the organization, the use of employee experience improves the quality of knowledge shared in the informal knowledge sharing process*". The argument reflects that individuals is taking a risk while sharing their experience to others. The sharer could be in position of being taken advantage. Nevertheless, in order to gain trust, each individual should take the risk, so that he or she could also gain benefit of others' experience.

> *"Often I think, my accumulated experience becomes my key source of knowledge. Any Knowledge from formal education doesn't significantly contribute. This is also a limitation for me. Since I didn't attend college, I must compete with others just using my skill and experience. Skill and experience become my ultimate weapons. However, many times I must share my experience to other younger workers. Many times I let my co-workers to learn specific skill by themselves. I only share small portion of knowledge to them. Nevertheless, I have two or three friends in this company to whom I'll be willing to share the best of my experience. Only two or three.....because I trust them. And because I believe they trust me also....and we share our knowledge usually not at the work station, because I don't want other people take advantages of it. We usually share our experience in the weekends.....at lunch time after we play soccer". (Informant 1, Production Staff)*

The opinion portrays that risk emerges by trusting other party in any interactions. In reciprocity perspective, individual would deliver certain benefits to others if he/she judges that the opposite is dependable, and able to offer what he/she need [19]. Further, knowledge quality depends on the level of trust within interacting individuals. The following hypotheses are derived:

Hypotheses 3: *Employee experience has positive direct effect to the enhancement of mutual trust.*
Hypotheses 4: *Mutual trust has positive direct effect to informal knowledge sharing.*

Third proposition believes that "*reciprocity norm and mutual trust have a reciprocal positive effect one to another*", while "*when level of trust increase within individuals, reciprocity level will also increase, and both promote the proper level of informal knowledge sharing process*". The arguments believe that mutual trust can be leveraged if individuals have the eagerness to provide benefits to others. These arguments also implied that trustworthy individuals will leverage the trust of among partners. The next discussions reflect the supporting facts:

> *"In my experience, if we can boost the trust level to mutually trust stage, personal relationship will be built. In many cases, not every employee has the ability to build a good interaction environment. For example.....myself.....I'm not a good at creating a lively interaction with my co-worker...but I always try to show them that I'm a useful person, an individual they can rely on something. That way, I can gain their trust. That will be the starting point to ask back from them. I do it slowly. If they respond, I make sure that if they give what I need, they will get the best of what I know....Don't ever expect this condition will be built through formal meetings. I do it 100% outside the office hour.....and it takes one to two years to be in this stage. Many times, it costs me some money". (Informant 1, Production Staff)*

> *"At first, I don't believe that trust should be mutually built before any 'knowledge exchanges' are made. In my experience, sometimes one individual share his knowledge because he wants to show others that he is a good person...and competent. This individual doesn't care whether he will receive benefit or not. I do that also. However, when we do it in formal interaction, the responds tend to be positive. But when we do it in informal occasions, the other parties tend to negatively responded. They said that I do that because I have hidden agenda. I've tried this many times....and the outcome is obvious, they don't like me". (Informant 2, Factory Manager)*

The discussions support the next hypotheses:

Hypotheses 5: *Mutual trust fosters reciprocal intention of individuals.*

The next proposition suggests that "*when organizations consist of employees from diverse educational background, the use of cultural stories, examples, and metaphors improve the knowledge formulation in the informal knowledge sharing process*". The proposition reflects that in formal knowledge sharing, people often challenged with complication to interact with others, particularly with individuals whose education background are different. The difference of education background can be in form of cultural background during education, education level, and or geographical area. The proposition implies that diversity of education background would affect informal knowledge sharing. The following dialogue discovered the following information:

> *"This is interesting. And it happens most of the time. Our employees come from many different areas in Indonesia: Sulawesi, Java, Sumatera, Madura and some from Eastern part. Well, I must be frank with you....we have diverse level of ability to absorb new things, because the education level and type in each area is different...and you cannot force them to understand at the same pace one to another. In fact, you cannot force them to learn new thing in formal training. You need to do it step by step.....and informal occasions are the best time. Yes.....in informal occasions. And you must use stories and examples so that they can understand better and faster. It is difficult....but it is effective. I've done it many times. And the most important thing is...diverse education background means diverse skills. And we need that..." (Informant 2, Factory Manager)*

The dialogue shows that diverse education background (education level, education type) plays important influence to the reliability of knowledge. It affects the study and resources capacity of individuals. The analysis is constructed the next hypothesis:

Hypotheses 6: *The diversity of education background has positive direct influence to informal knowledge sharing.*

Hypotheses 7: *The cultural interpretation perspective moderates the positive direct influence of diverse education background to informal knowledge sharing.*
Hypotheses 8: *The more diverse education background, the richer variation of skills is acquired by individuals.*

3. Linking Human Capital Policies & Initiatives and Organizational Sustainability: the role of Informal Knowledge Sharing

In Indonesian perspective, specifically in small and medium scale enterprises, the adoption of "western human resource management models" tend to be widely acceptable However, previous studies also revealed that human capital is still the main problem to enterprise sustainability issue [10][11]. This condition justify the previous argument [14] which reflect that current human resource management models view a human as an object (as one of the factors of production), not an actor. In other words, there is a need to develop a more comprehensive model that is more suitable to Indonesian condition.

Previous studies revealed that many enterprises still not aware of the importance of governing human capital traits and organizational knowledge in relation to enterprise sustainability [16][23][24], though, in the knowledge economy era, knowledge becomes the key enablers to many organizations. Previous studies regarding Indonesian small and medium enterprises revealed that "poor human capital management initiatives" become the most critical issue. Moreover, the Indonesian Ministry of National Development Planning (BAPPENAS), stated that the government schemes to leverage the negotiating capability of Indonesian Medium Enterprises (MEs), purposely, by assisting the entrepreneurship skill and technical competence of the MEs [25].And again, the government focus is still in the area of skill training, instead of developing a more appropriate human capital management system.

Indonesian medium-sized manufacturing enterprise still adopt the generic human resource management model in governing their human capital. Interestingly, previous study [25] revealed that only certain human capital related policies and initiatives were truly necessary, such as: corporate social responsibility policy, compensation policy, personnel administration initiative, and training and development initiative. The study also suggest that "where human capital experience is acquired into organisational knowledge, the use of knowledge management framework will help the process of knowledge acquisition, and therefore will directly impact the enterprise sustainability". This is stimulating argument because most medium-sized enterprises were not sure of the benefit of employing knowledge management initiatives.

> *"Yes, we have many policies and programs in our company. But, I don't think we really apply all of those. We usually sharing our skills or experience or idea in relations to training. For some managers, sharing knowledge is also in relations to development program. I don't see any relations between knowledge sharing and CSR or personnel administration. With compensation policy.....well, maybe." (Informant 3, Senior Production Staff)*

> *"I see no connection between human capital programs and knowledge sharing, except with training. But, I think compensation policy could matched, since it is a delicate area. In practice, top management once discussed the bonus policy with me during lunch. I gave them suggestion based on my experience, and happily, they acknowledged my thinking. Yes, it could work." (Informant 2, Factory Manager)*

> *"I believe if compensation policy is also discussed through informal occasions, the acceptance level will be increased, and the spirit of people will be better. It's good for the company. I think training program should be also accompanied by many informal and relaxing discussions to be effective. I believe the innovativeness of people can be better." (Informant 4, QC Manager)*

The dialogues reflects two preliminary evidences. First, the informants suggest that only particular practice and policy related to human capital are relevant to be exposed informal knowledge sharing: compensation policy and training and development program. Second, innovativeness of individuals is a crucial outcomes of informal knowledge sharing, in relation to any training program. This judgement is related to previous findings [26], which claims that innovation is correlated with knowledge sharing. Though, in manufacturing perspective, enterprise ability to adapt is unavoidable [27]. In manufacturing sector, which involves continuous technological changes, must retain the capacity to deal with fast evolving environment, in line with the ability to reconstruct and upgrade enterprise capacity to grow. The dialogues support the following hypotheses:

Hypotheses 9: *The relationships between enterprise sustainability and human resource policy and initiative is strengthen by employee perceptions of informal knowledge sharing.*

Figure 1 represents the research framework, based on the propositions, discussions, and theories.

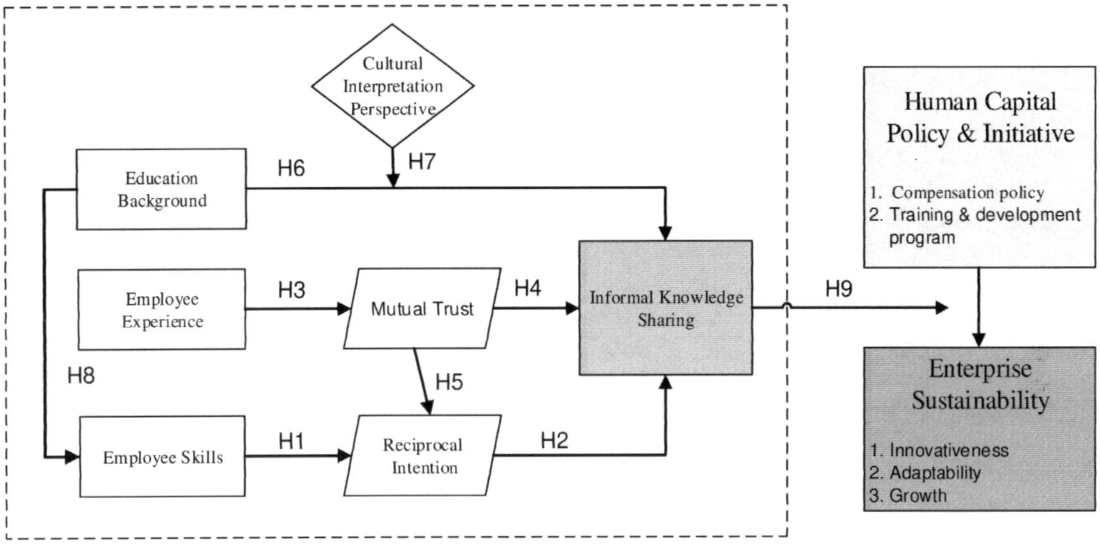

Figure 1. Research framework

The interview sessions occasionally provide the researchers with some fascinating outcomes of informal knowledge sharing practice. For example, when Informant 2 met with a local mechanic in the rest room, Informant 2 mentioned about the incapability of Taiwanese mechanic to solve troubleshooting of CNC machine. The local mechanic explained (while urinating) that the repair order was not correct. Further, the suggestion from the local mechanic was then informed to the Taiwanese mechanic by Informant 2. Curiously, the Taiwanese followed the suggestion and in only 10 minutes, the troubleshooting was fixed.

Another example of how informal knowledge sharing helps the organization to sustain occurs in "candy factory", the organization of Informant 4. Formally, the candy factory conduct monthly formal meeting to discuss about new candy product. Optimally, six new designs were introduced. When the new factory manager joined the organization, 40 new designs are introduced to market, without any formal meetings. The key is that each production staff is encourage to experiment and discuss the experiment during lunch. In fact, lunch time is more effective to exploit employees' ideas compare to formal meetings when most of top management are present.

4. Conclusions

This study proposes a set of hypotheses for future research. Important finding that differentiate this study from previous one [19] is that 'mutual trust' should be considered as enabler to the formation of 'reciprocal intention' amongst individuals in organization. The direct positive effect of 'employee skills' to informal knowledge sharing is believed to be intervened by 'reciprocal intention' of people in the organization, by first developing mutual trust within them. Another interesting finding is that in order to be effective, 'employee skills' is also influenced by 'education background' of the individuals within the informal knowledge sharing network.

Moreover, this study also reveals that among people in Indonesian medium-sized food manufacturing enterprise, 'compensation policy' and 'training & development program' are believed, practically, the most influential policy and initiative to them. By delivering informal knowledge sharing practice, the effect of compensation policy and training related programs to innovativeness of the enterprise is believed to be more advantageous. Future study will examine the interrelation between human capital policy and initiative, enterprise sustainability, and informal knowledge sharing, using quantitative approach. Future study will focus on how these hypotheses can describe the nature of informal knowledge sharing practice amongst medium-sized sustainable enterprises. Next study will employ quantitative survey to nearly 32 confirmed sustainable Indonesian medium-sized food manufacturing enterprises.

5. References

[1] Tambunan T 2009 Export-oriented small and medium industry clusters in Indonesia *J. Enterprising Communities: People and Places in the Global Economy* **3** 25
[2] Tambunan T 2011 Development of small and medium enterprises in a developing country: the Indonesian case *J. Enterprising Communities: People and Places in the Global Economy* **5** 68
[3] Djamhari C 2006 Factors affecting the development of SMEs to dynamic clusters (in Bahasa Indonesia) *INFOKOP* **21** 83
[4] Firdanianty 2009 Pictures of human capital in SMEs: the key is attitude and mentality (in Bahasa Indonesia) *Human Capital* **63** 7
[5] Tambunan T 2012 Pasar bebas ASEAN: peluang, tantangan dan ancaman bagi UMKM Indonesia *INFOKOP* **21** 13
[6] Unger J M, Rauch A and Frese M 2011 Human capital and entrepreneurial success: a meta-analytical review *J. Bus. Venturing* **26** 341
[7] Drucker P F 1954 *The Practice of Management* (New York: Harper and Row Publishers, Inc.)
[8] Hannan M T and Freeman J 1977 The population ecology of organizations *American Journal of Sociology* **82** 929
[9] Suarez F and Utterback J M 1995 Dominant design and the survival of firms *Strategic Management Journal* **16** 415
[10] Kuncoro M 2009 *Ekonomika Industri Indonesia: Menuju Negara Industri Baru 2030?* (Yogyakarta: Andi Publishing)
[11] Hartanto F M 2012 *The Development of Human Capital in National, Regional and Global Perspective* (Bandung: Graha Ilmu)
[12] Price A 2011 *Fundamentals of Human Resource Management* (Hampshire: Cengage Learning)
[13] Bratton J and Gold J 2012 *Human Resource Management: Theory and Practice* (London: Palgrave Macmillan)
[14] Gonzalez S M and Tacorante D V 2004 A new approach to the best practices debate: are best practices applied to all employees in the same way? *Int. J. Human Resource Manage.* **15** 56
[15] Snell S A, Youndt M A and Wright P M 1996 Establishing a framework for research in strategic human resource management: merging resource theory and organizational learning' in Ferris, G. (Ed.) *Research in personnel and human resource management* **14** 61
[16] Nafukho F M, Hairston N and Brooks C 2004 Human capital theory: implications for human resource development *Human Resource Dev. Int.*. **7** 545
[17] Baron A and Armstrong M 2007 *Human Capital Management: Achieving Added Value through People* (London: Kogan Page)
[18] Tjiptono F 2012 *The Development of Human Capital in National, Regional and Global Perspective*, in Bahasa Indonesia, (Bandung: Graha Ilmu)
[19] Sunardi O, Tjakraatmadja J H and Bangun Y R 2015 Human capital traits and informal knowledge sharing: the role of reciprocity norm, mutual trust, and cultural interpretation perspective *Int. J. Knowledge Management Studies* **6** 123

[20] Sanchez R, Aime H and Howard T 1996 *Dynamics of Competence-Based Competition: Theory and practice in the New Strategic Management* (Oxford: Elsevier)
[21] O'Dell C and Grayson C 1998 If only we knew what we know: identification and transfer of internal best practices *California Management Review* **40** 154
[22] Gupta A K and Govindarajan V 2000 Knowledge flows within multinational corporations *Strategic Management Journal* **21** 473
[23] Nafziger E W and Terrell D 1996 Entrepreneurial human capital and the long-run survival of firms in India *World Development* **24** 689
[24] Amayah A T 2013 Determinants of knowledge sharing in a public sector organization *J. Knowledge Management* **17** 454
[25] Sunardi O and Tjakraatmadja J H 2013 Enablers to knowledge management implementation in Indonesian medium-sized manufacturing enterprises: a preliminary study *Paper presented at the 2013 Int. Decision Science Institute and Asia Pacific Decision Science Institute Conf.* (Bali)
[26] Pawlowsky P and Schmid S 2012 Interrelations between strategic orientation, knowledge management, innovation and performance. Empirical findings from a national survey in Germany *Int. J. Knowledge Management Studies* **5** 185
[27] Teece D J 1997 Explicating dynamic capabilities: the nature and micro-foundations of (sustainable) enterprise performance *Strategic Management Journal* **28** 1319

Reliability analysis of component of affination centrifugal 1 machine by using reliability engineering

N Sembiring[1,2], E Ginting[1,3] and T Daniello[1,4]

[1]Department of Industrial Engineering, Faculty of Engineering, Universitas Sumatera Utara
Jalan Almamater Kampus USU, Medan, Indonesia 20155
[2]nurhayati4@usu.ac.id
[3]ir.elisabethginting@gmail.com
[4]timbul.daniello@yahoo.com

Abstract. Problems that appear in a company that produces refined sugar, the production floor has not reached the level of critical machine availability because it often suffered damage (breakdown). This results in a sudden loss of production time and production opportunities. This problem can be solved by Reliability Engineering method where the statistical approach to historical damage data is performed to see the pattern of the distribution. The method can provide a value of reliability, rate of damage, and availability level, of an machine during the maintenance time interval schedule. The result of distribution test to time inter-damage data (MTTF) flexible hose component is lognormal distribution while component of teflon cone lifthing is weibull distribution. While from distribution test to mean time of improvement (MTTR) flexible hose component is exponential distribution while component of teflon cone lifthing is weibull distribution. The actual results of the flexible hose component on the replacement schedule per 720 hours obtained reliability of 0.2451 and availability 0.9960. While on the critical components of teflon cone lifthing actual on the replacement schedule per 1944 hours obtained reliability of 0.4083 and availability 0.9927.
Keywords: Reliable, Reliability, Maintenance

1. Introduction

In 2015 Indonesia's national sugar demand reached 5.7 million tons and is projected to increase 5% per year. Increased demand for sugar in line with population growth and number of food, beverage and pharmaceutical industries (Ministry of Agriculture, Directorate General of Plantation). High consumption levels are not matched by the availability of available sugars. Some of the contributing factors are the availability of limited raw materials due to minimal land. Besides the limited processing plants in Indonesia and the company's unpreparedness of factory operation. Operational unpreparedness is caused by shortages or delays in raw material supply, downtime due to damage to production machinery, and stalled energy supply. As a manufacturing company, consumers' complain should be avoided. Improvement in quality and quantity of product have to be prioritizied [1].

This research is conducted in a refined sugar manufacturing company. The company has only been operating for 3 years, but downtime or loss of production time due to machine failure is still common. Table 1 shows the breakdown frequency ≥ 4 times and downtime of production machine in 2016.

Table 1. Breakdown Frequency and Downtime of Production Machine in 2016

Machine	Breakdown Frequency	Total *Downtime* (minute)	Category
Stirrer For Affination Mingler	5	480	A
Affination Centrifugal 1	15	1020	B
Affination Centrifugal 2	14	540	B
Affination Centrifugal 3	15	560	B
Pressure Filters For Second Filtration 1	5	435	B
Pressure Filters For Second Filtration 2	7	525	B
Brown Liquor 2 Pump 1	7	1635	C
Spent Regenerant Pump 1	4	250	C
Spent Regenerant Pump 2	5	325	C
Refined Massecuite Receiver 1	6	300	A
Refine Massecuite Centrifugal 1	9	660	B
Refine Massecuite Centrifugal 2	16	1110	B
Refine Massecuite Centrifugal 3	13	740	B
Refine Massecuite Centrifugal 4	4	170	B
A-Massecuite Centrifugal	5	175	B
C-Massecuite Centrifugal 2	4	200	B
Vibrating Grade B	7	1600	C
Co2 Gas Compressor With Gas Sparator	4	370	B

Table 1 shows the centrifugal affination machine of one of the machines with the highest damage frequency. In addition, the machine is also included in category B where if the damaged machine can reduce the amount of production. Based on these considerations, research is needed on the reliability of the centrifugal affination machine. Loss of production time (downtime) can have an impact on reducing the amount of production which it should be targeted, until it caused financial loss and lost profit opportunities. Decreasing production time happen because of the production process stopped while maintenance team repairing or replacing the damage machine and component [2].

The company has been implementing maintenance system in preventive maintenance and corrective maintenance to eliminate downtime. However, machine maintenance with the replacement of machine components is still corrective because there is no regular replacement scheduling. Replacement of components by correctively will show in a loss of production time is longer than preventive because it has not been predicted before so there is no preparation step repair for damage to the machine. Some studies shows that high frequency oh downtime happen due to applied traditional maintenance [3],[4],[5].

Preventive maintenance of the company is still based on experience and predictions without any theoretical calculations. This leads to a decrease in machine reliability due to unpredictable and uncontrollable the failure rates. Therefore it is necessary to schedule regular component replacement scheduling and identification of component damage sources to improve machine reliability [6].

Improved reliability (reliability) machine / component can be done by lowering the rate of failure of the machine / component through the approach of Reliabilty Engineering. Reliabiliy Engineering method is a statistical approach in analyzing the reliability of a machine / equipment. The lower failure rate achieved, the higher reliability obtained. This can have an impact on reducing maintenance costs due to increased reliability. In addition it can reduce the loss of production time (downtime) and production targets achieved [7].

Research on the reliability of tuber and bottomer machine at PT. X uses the machine age distribution data to know the pattern of machine damage. Test results obtained by tuber machine with optimum preventive time every 27 days and minimum cost of 40.92 million rupiah while the bottomer machine has optimum preventive time every 72 days and minimum cost of 55.61 million rupiah [8].

The research analyzes the damage of Screw Press machine based on the calculation of reliability function, failure rate and Mean Time Between Failure (MTBF). The results obtained an optimum time interval for the components of Screw Press machine. Decrease in cost before and after scheduling ranged from 6.96 to 45.57% [9]. Based on previous research references, reliability analysis of production machine at this company is performed to obtain the optimum component replacement time interval and decrease maintenance cost based on machine failure analysis.

2. Methodology

The type of research used is causal research, is research that aims to describe systematically, factually and accurately about the facts and properties of a particular object or population to investigate the causal relationships that occur and possible factors (cause) causing the consequences. The results of this type of research are expected to minimize downtime and maintenance costs.

The object of research observed is an Affination Centrifugal 1 machine, covering the extent of damage and maintenance activities performed on the machine. Data collection methods used in this study is through direct observation and interviews with the company. Data processing methods used in this study are [10],[11],[12] :

1. Identify the machine and its critical components and sources of damage.
2. Testing distribution patterns, parameter calculations, and MTTF.
3. Determination of the level of reliability and rate of damage critical component.

This research stage can be described briefly through block diagram. Block diagram of this research can be seen in Figure 1.

3. Results and Discussion

3.1. Identification of Critical Types and Components as well as the Failure Sources

The selection of critical components that will be the focus of research is done using the Pareto 80/20 rule. Figure 2 shows the Pareto Diagram of the damage of the Affination Centrifugal 1 machine component. Pareto diagrams are used to identify the most important issues. The diagram above shows 80% loss of production time caused by 20% broken components [11]. Thus there are 2 types of critical components that become the focus of research problems that is flexible hose and teflon cone lifthing. The result of the identification of the cause of the damage of the flexible hose component on the affination centrifugal 1 machine can be seen on the Cause and Effect Diagram (Fish Bone Diagram) [11]. The Cause and Effect Diagram can be seen in figure 3.

3.2. Distribution Testing, Parameter Calculation and MTTF

The distribution test is performed on the time interval data between the damage and the average time of repair of critical machine components. The distribution types used are normal, lognormal, weibull, and exponential distributions [12]. The best distribution is the one with the greatest index of fit. Table 2 shows the recapitulation of selected distributions of time data between damages of critical machine components.

Table 2. Recapitulation of the Best Distribution Time between Critical Components Damage

Component	Distributions	Index of Fit	Selected Distribution
Flexible Hose	Normal	0.8972	Lognormal
	Lognormal	0.9831	
	Eksponensial	0.9772	
	Weibull	0.9688	
Teflon Cone Lifthing	*Normal*	0.9766	*Weibull*
	Lognormal	0.9443	
	Eksponensial	0.9243	
	Weibull	0.9803	

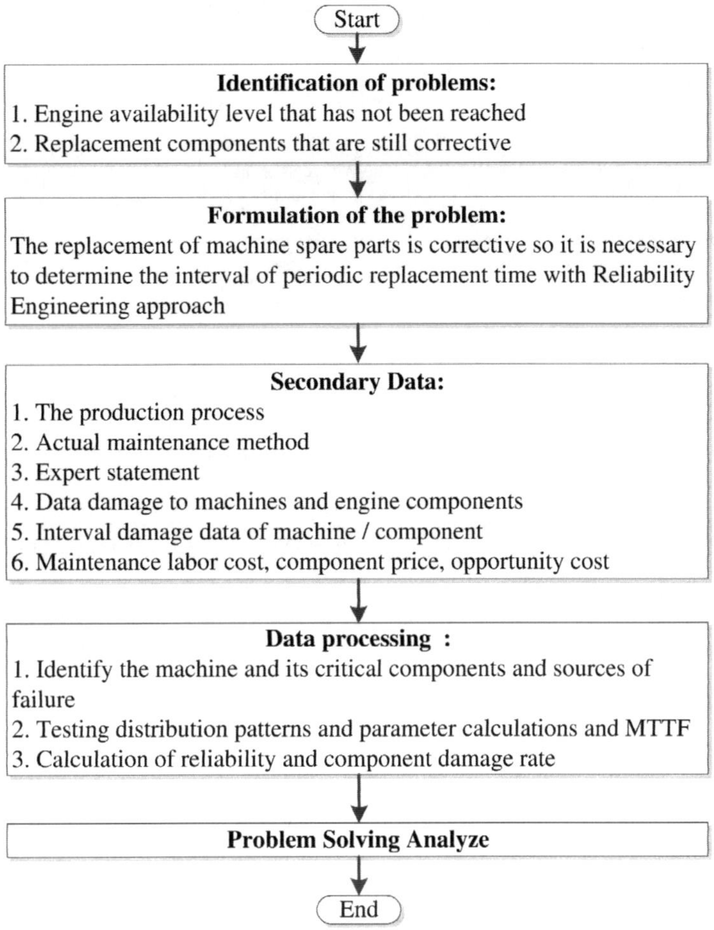

Figure 1. Block Diagram of Research

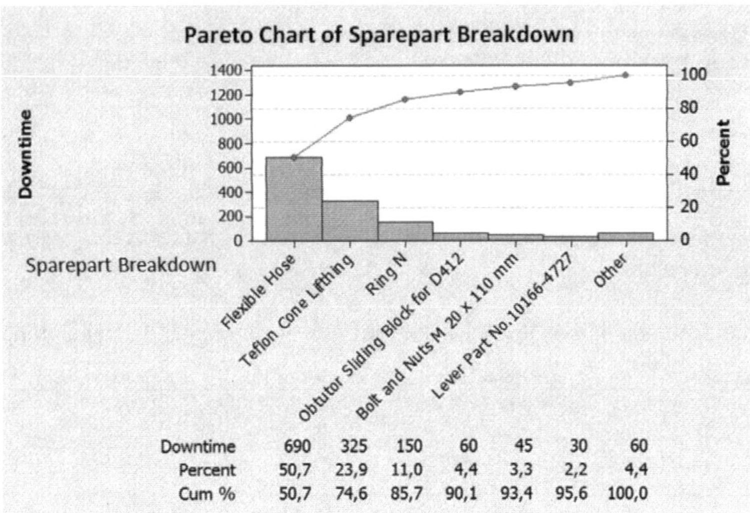

Figure 2. Critical Components of Pareto Diagram

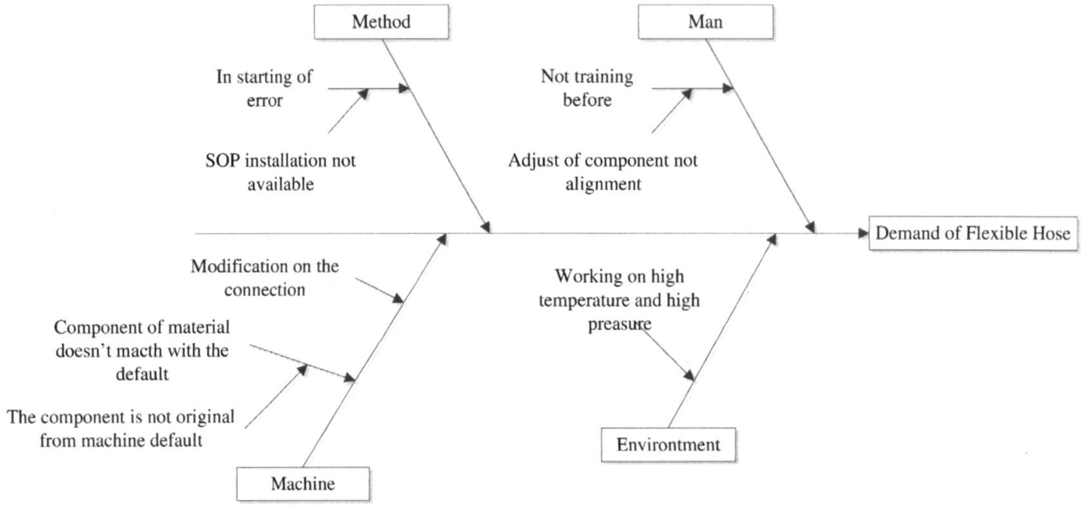

Figure 3. Cause Diagram Due to the Cause of Flexible Hose Component Damage

The identification of the cause of damage to the flexible hose component in the affination centrifugal 1 machine can be seen on the Cause and Effect Diagram (Fish Bone Diagram) in Figure 4.

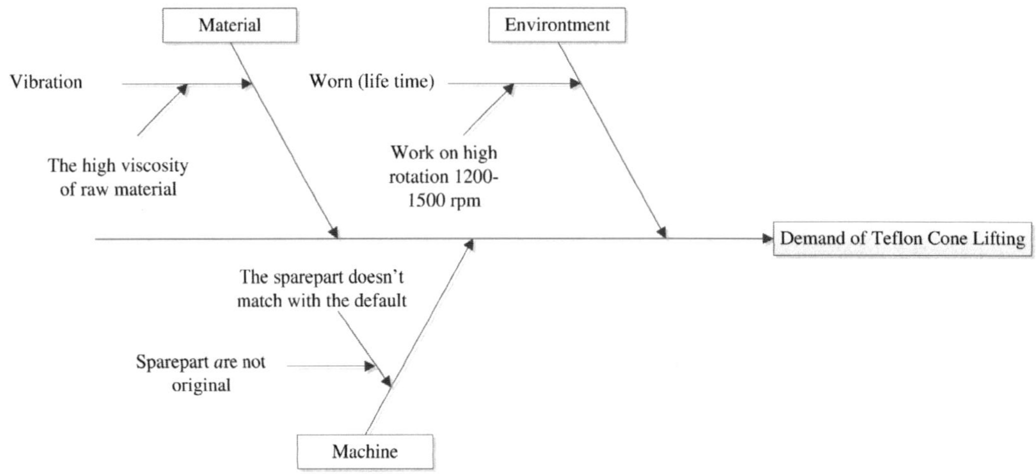

Figure 4. The Cause Due to Damage Diagram of Teflon Cone Lifthing Component

Table 3 shows the best recapitulation of the data distribution of the average time of repair of critical machine components.

Table 3. Recapitulation of the Best Distribution Time Average of Critical Component Repair

Component	Distributions	Index of Fit	Selected Distribution
Flexible Hose	*Normal*	0.7378	Eksponensial
	Lognormal	0.7378	
	Eksponensial	0.8435	
	Weibull	0.6446	
Teflon Cone Lifthing	*Normal*	0.5746	*Weibull*
	Lognormal	0.5746	
	Eksponensial	0.3644	
	Weibull	0.6746	

Distribution parameter calculations are performed on the best distribution of time between damage and average time of repair. Table 4 shows the parameters and MTTF distribution results of the selected time between critical component breakdowns.

Table 4. Results Calculation of Parameters and Time Interruption (MTTF) Critical Components

Time Interruption (MTTF)			
Component	**Distribution**	**Parameters**	**MTTF (hour)**
Flexible Hose	Lognormal	$\hat{s}$ = 1.4027 $\hat{t}_{med}$= 271.59	726.31
Teflon Cone Lifthing	Weibull	β = 1.3042 θ =2115.3665	1953.9641

Table 5 shows the best parameters and MTTF distribution results for the average time of critical component repair.

Table 5. Parameter Calculation Results and Average Time Improvement (MTTR) Critical Components

Average Repair Time (MTTR)			
Component	**Distribution**	**Parameters**	**MTTR (hour)**
Flexible Hose	Eksponensial	λ = 0.6476	1.5441
Teflon Cone Lifthing	Weibull	β = 13.0461 θ =10.1008	9.7381

3.3. Calculation of Reliability Value and Rate of Machine Damage

Calculation of reliability and damage rate using the distribution parameters of MTTF The calculation of the reliability value of each component is as follows [13],[14],[15],[16] :

1. Flexible Hose Components
   Time interval data between damage of Lognormal distributed components
   Parameter : MTTF= 726.31 hour
   : $\hat{s}$ = 1.4027
   : $\hat{t}_{med}$ = 271.59

   Then calculation of reliability value of machine component is:
   a. Probability Density Function f (t).
   $$f(t) = \frac{1}{\sqrt{2\pi}st}\exp\left[-\frac{1}{2s^2}\left(ln\frac{t}{\hat{t}med}\right)^2\right]$$
   $$= \frac{1}{\sqrt{2\pi}1.4027(726.31)}\exp\left[-\frac{1}{2(1.4027)^2}\left(ln\frac{726.31}{271.59}\right)^2\right]$$
   $$= 0.0003$$
   b. Cumulative Distribution Function
   $$F(t) = \Phi\left(\frac{1}{s}ln\frac{t}{\hat{t}med}\right) = \Phi\left(\frac{1}{1.4027}ln\frac{726.31}{271.59}\right)$$
   $$= \Phi(0.7012)$$
   $$= 0.7580$$
   c. Reliability Value Function
   $$R(t) = 1-\Phi\left(\frac{1}{s}ln\frac{t}{\hat{t}med}\right) = 1-0.7580$$
   $$= 0.2419$$
   d. Damage Rate Function h (t)
   $$\lambda(t) = \frac{f(t)}{R(t)} = \frac{0.0003}{0.4721}$$
   $$= 0.0006$$

Based on calculations obtained that after 726.31 hours (30 days) the use of flexible hose components reliability value is 0.2419.

2. Teflon Cone Lifthing Components

Time interval data between damage of Weibull distributed components

Parameter : MTTF= 1953.9641 hour (81 days)

: β = 1.3042

: θ = 2115.3665

Then calculation of reliability value of machine component is:

a. Probability Density Function f (t).

$$f(t) = \frac{\beta}{\theta}\left(\frac{t}{\theta}\right)^{\beta-1} e^{-(\frac{t}{\theta})^{\beta}} = \frac{1.3042}{2115.3665}\left(\frac{1953.9641}{2115.3665}\right)^{1.3042-1} e^{-(\frac{1953.9641}{2115.3665})^{1.3042}}$$

$$= \frac{1.3042}{2115.3665}\left(\frac{1953.9641}{2115.3665}\right)^{1.3042-1} e^{-\left(\frac{1953.9641}{2115.3665}\right)^{1.3042}}$$

$= 0.00024$

b. Cumulative Distribution Function

$$F(t) = 1- e^{-\left(\frac{t}{\theta}\right)^{\beta}} = 1- e^{-\left(\frac{1953.9641}{2115.3665}\right)^{1.3042}}$$

$= 1- 0.4058 = 0.5941$

c. Reliability Value Function

$$R(t) = e^{-(\frac{t}{\theta})^{\beta}} = e^{-(\frac{1953.9641}{2115.3665})^{1.3042}}$$

$= 0.4058$

d. Damage Rate Function h (t)

$$\lambda(t) = \frac{f(t)}{R(t)} = \frac{\beta}{\theta}\left(\frac{t}{\theta}\right)^{\beta-1}$$

$$= \frac{1.3042}{2115.3665}\left(\frac{1953.9641}{2115.3665}\right)^{1.3042-1}$$

$= 0.0006$

Based on calculations obtained that after 1953.9641 hours (81 days) the use of components of teflon cone lifthing reliability value is 0.4058.

4. Conclusion

Based on the research that has been done, obtained the following conclusions:

1. The critical machine that became the research priority is the Affination Centrifugal 1 machine is one of the machines that have the greatest damage frequency during the period of January 2014-January 2017 with flexible hose flexible component and teflon cone lifthing.
2. Testing the distribution of intermediate time data between damage (MTTF) flexible hose component is obtained by lognormal distribution while component of teflon cone lifthing is weibull distribution. The test of the distribution of the mean time of improvement (MTTR) obtained flexible hose component is exponential distribution and the teflon cone lifthing componentis weibull distribution. MTTF calculation obtained maintenance schedule with component replacement per 726.31 hours usage for flexible hose and per 1953.9641 hours usage for teflon cone lifthing.
3. Calculation of the reliability of critical components by using the best distribution parameters obtained by 0.2419 (24.19%) with a 0.0006 damage rate for the flexible hose component. While for component of teflon cone lifthing obtained reliability equal to 0.4058 (40.58%) with damage rate 0.0006.

Acknowledgment

The authors wish to express sincere gratitude to the manufacturing company that gives permission to make this research.

IOP Conf. Series: Materials Science and Engineering **277** (2017) 012041 doi:10.1088/1757-899X/277/1/012041

5. References

[1] J P Rishi, Dr. Ramachandara C G and Dr. T. R Srinivas 2016 Keys to Succeed in Implementing Total Preventive Maintenance (TPM) and Lean Strategis *IJMTST* **2** pp.25-30

[2] Yang Li, Xiaobing Ma, Rui Peng, Qingqing Zhai and Yu Zhao 2017 A Preventive Maintenance Policy Based On Dependent Two-Stage Deterioration And External Shocks *Reliability Engineering And System Safety* **160** pp.201-211

[3] Kumar Saureng, Raj Bhushan and Shubham Swaroop 2017 Study of Total Productive Maintenance & It's Implementation Approach in Steel Manufacturing Industry: A Case Study of Equipment Wise Breakdown Analysis *IRJET* **4** pp.608-613

[4] Wang Zhenyou, Cuntao Xiao, Xianwei Lin, and Yuan-yuan Lu 2017 Single Machine Total Absolute Differences Penalties Minimization Scheduling with a Deteriorating and Resource-Dependent Maintenance Activity *The Computer Journal* pp.1-11

[5] Maleki Hamed and Yang Yiangjie 2017 An Uncertain Programming Model For Preventive Maintenance Scheduling *Grey Systems: Theory and Application* **7** No.1 pp.111-122

[6] Aven, Terje 2017 Improving the Foundation and Practice of Reliability Engineering *Journal of Risk and Reliability* **23 I** (3) 295-305

[7] Compare, Michele, Michele Bellora and Enrico Zio 2017 Aggregation of Importance Measures For Decision Making In Reliability Engineering *Journal of Risk and Reliability* pp.1-13

[8] Satria Hikmawan, *et.al.* 2016 Optimasi Preventive Maintenance pada Mesin Tuber dan Bottomer dengan Metode Analisis Reliabilitas di PT.X *Journal FMIPA Institut Teknologi Sepuluh November (ITS)*

[9] Yuhelson, *et.al.* 2010 Analisis Reliability dan Availability Mesin Pabrik Kelapa Sawit PT. Perkebunan Nusantara 3 *Journal Universitas Sumatera Utara* No. **6**

[10] Besterfield, H. Dale 1995 *Quality Control* (Southern Illinois: College of Engineering Southern Illinois University)

[11] Charles E. Ebeling 1997 *An Introduction to Reliability and Maintanability Engineering* (Singapore: Thetatayo McGraw-Hill Book Co.)

[12] Kapur, KC. and Lamberson, L. R. 1977 *Reliability in Engineering Design* (New York: John Wiley & Sons)

[13] Lina Alhmoud and Bingsen Wan 2017 A Review Of The State-Of-The-Art In Wind-Energy Reliabilty Analysis *Renewable and Sustainable Energy Reviews*

[14] Miguel Angel Navas, Carlos Sancho and Jose Carpio 2016 Reliability Analysis in Railway Repairable System *Int. Journal of Quality & Reliability Management* **34** Issue: 8 pp.1373-1398

[15] Rajkumar B Patil, Basavraj S. Kothvale, Laxman Yadu and Shridhar G. Joshi 2016 Reliability Analysis of CNC Turning Center Based on The Assessment of Trends in Maintenance Data *Int. Journal of Quality & Reliability Management* **34** Issue:9 pp. 1616-1638

Challenges in detecting drowsiness based on driver's behavior

V Triyanti [1 2] and H Iridiastadi [1]

[1] Faculty of Industrial Technology, Institut Teknologi Bandung, Jl. Ganesha No. 10, Bandung 40132, Indonesia
[2] Department of Industrial Engineering, Atma Jaya Catholic University of Indonesia, Jalan Jenderal Sudirman 51, Jakarta 12930, Indonesia
vivi.triyanti@student.ti.itb.ac.id [1]; vivi.triyanti@atmajaya.ac.id [2]

Abstract. Drowsiness while driving has been a critical issue within the context of transportation safety. A number of approaches have been developed to reduce the risks of drowsy drivers. The mechanisms in detecting fatigue and sleepiness while driving has been categorized into three broad approaches, including vehicle-based, physiological-based, and behavior-based approaches. This paper will discuss recent studies in recognizing drowsy drivers based on their behaviors, particularly changes in eyes and facial characteristics. This paper will also address challenges in capturing aspects of natural expressions, driver responses, behavior, and task environment associated with sleepiness. Additionally, a number of technical aspects should be seriously considered, including correctly capturing face and eye characteristics from unwanted movements, unsuitable task environments, technological limitations, and individual differences.
Keywords: facial expression, driver's drowsiness, driver's behavior.

1. Introduction

The number of traffic accidents has increased from year to year in Indonesia [5]. Factors affecting road traffic accidents vary, ranging from driver factors, vehicle factors, and environmental factors. In Indonesia, accidents on the highway more than 75 percent occurs due to human factors [27]. This finding is also supported by the data of [6]. Loss of control and error during driving may be caused by fatigue. Most researchers have agreed that driving is activities that can cause both physical and mental fatigue for drivers. Fatigue due to transportation can be caused by many factors, including factors related condition in the transportation [18], health factors [10], psychosocial factors and even psychological factors [4].

The first approach uses the measurement of physiological activities of the human body such as brainwave (using electroencephalogram - EEG) [2], [19], heart rate (using electrocardiogram – ECG) [3] and electric signals from muscle cells (using electromyogram – EMG) [22]. The measured signal is usually reliable enough to detect drowsiness because its correlation with driver alertness is relatively strong. However, this technique has some limitations due to the necessity to apply tools to the user's body which is somewhat intrusive.

Driver's fatigue-related behavior status can also be detected on the basis of vehicle behaviors, such as steering movements, accelerations and braking, a speed of vehicles, and lane crossings [3],[13],[15],17]. Unfortunately, this approach can only be used for very limited conditions, because it could be a sudden movement that occurs due to road conditions, not because of the sleepy condition of the driver [3],[17].

A third approach to estimate driver drowsiness is by employing a set of cameras and image processing technique to detect changes in driver behavior, such as movement of eyes, yawning [1],[16], head movement [7], and facial expression [11],[12],[13],[14],[23],[25]. The appearance and behavior of a sleepy person are usually different compared to fit drivers. This technique is increasingly popular, because of its non-intrusive approach in monitoring the alertness of a driver. A set of camera and image processing technology (computer vision) can be used to capture and process video or pictures. Based on driver's facial appearance, a technology is used to extract visual characteristics to define driver's alertness level.

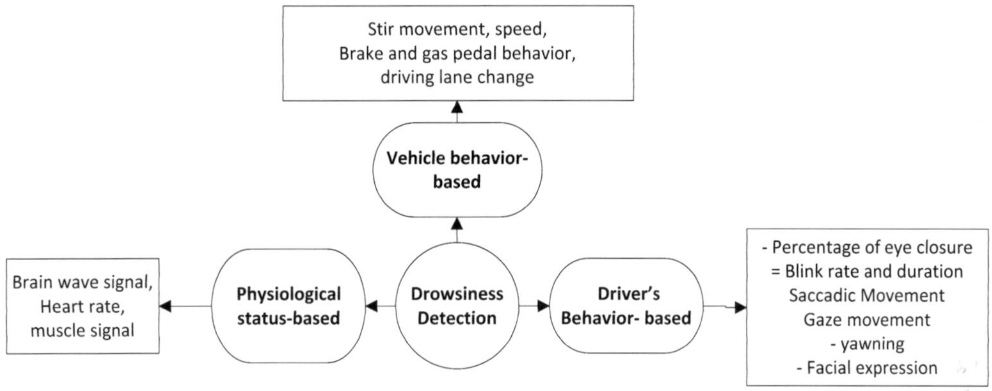

Figure 1. Approach to Detect Drowsiness in Driver

This paper mainly discusses the third approach – behavior drowsiness detection system – since the approach has become popular for its non-intrusive characteristics. There are a lot of studies that have been done to detect and extract characteristics of face and eye for detecting drowsiness. However, only a limited number of studies have focused on model development to build robust formula, calculation, and threshold to define drowsiness status based on face and eyes characteristics.

Defining drowsiness status is challenging because there are many characteristics that can be alternatives in defining drowsiness status. We have to carefully select which characteristics that mainly represent driver's drowsiness status effectively and efficiently. While effective means the characteristics have a small possibility to provide a false alarm of drowsiness, efficiently means those characteristics should be relatively easy to capture and extract. Additionally, a number of factors can affect the driver's behavior, both from outside and inside the car. These factors should also be considered in order to build a valid model.

2. Behavior-based drowsiness detection system

2.1. Detecting driver behavior

Before we can measure the drowsiness level based on driver's behavior, several substantial issues need to be addressed. The first thing is the behavior itself. Generally, we can divide driver's behavior into two big groups, i.e. behavior in driving and the natural behavior when the driver is in fatigue or drowsiness condition. The natural behavior of drowsiness points toward to the natural reaction of the body when it is in fatigue or drowsiness condition.

We may divide this behavior into two categories: eye behavior and face behavior. When a man is in a sleepy condition, the eye tends to have specific behavior; for instance, the faster blink rate, longer eye closure time, slow gaze movement, etc. In behavior-based detection system, researchers mostly have used the criteria related to eyes' natural behavior. Not only eyes, overall face also can be used to detect sleepiness, i.e. expression and mouth. Several studies have reported that when we are in sleepy/drowsy condition, there were changes in face's appearance. For instance: slow eyelid movement [21],[26] decreased eyes openness level [9],[14], sudden nod [7], frequent yawning [1],[16], limited gaze movement area [20],[21],[24], "dropped" facial expression [11],[12],[23],[28], and feeble body posture [15],[23]. These and other changes in patterns of behavior of a driver are manifested when the person is feeling drowsy.

There are points on an individual's face that can be observed to detect our facial expression. For instance, Ekman [8] has developed 44 Action Unit that indicated single or group of face muscles movement that related to certain expression. This classification was widely used in a lot of research related to facial expression detection and interpretation. Including in the Action Unit is yawning. Yawning can be very common criteria that can be easily detected when someone in a drowsy condition. However, we have to be careful to use this phenomenon because it can be mistakenly detected or misinterpreted. Researcher usually detects yawning as a condition when the mouth opens [12],[23]. However, the condition that requires opened-mouth can be when the driver is talking or singing. Therefore, we have to be careful in interpreting which open-mouth condition represents the real yawning.

Other body parts also can be used to detect fatigue. Although the criteria are rarely used (compared to eyes and face criteria), they are sometimes used as supporting criteria. The examples of this category are head movement and body posture. When a driver is in a drowsy condition, the head may nod frequently. Often, it is followed by eye closure movements. Usually the nodding time is relatively short because the driver tried to maintain the head position to normal position, so he can monitor traffic condition.

Driving behavior is behaviors of a driver when face certain driving condition, for instance, how the driver deals with traffic condition, whether he is a risk taker or safety driver, and whether he is a smooth-driving or reckless-driving type. Risk taker driver usually is easily observed by the faster driving speed, frequent frequency of changing driving lane, and sudden brake pedal pushing [13],[15]. On the other hand, risk-averse/safety driver usually has behavior that opposite of risk taker driver. Driving behavior is unique for each driver. However, a researcher should consider the driver's driving behavior because it may affect the approach to measure and the interpretation of another type of behavior – which is a natural behavior of fatigue condition.

2.2. Detecting performance in driving

Behavior monitoring is useless unless we also measure driver's performance during driving. There are some performances that can be monitored during driving, including response ability (reaction time), vigilance, balance, vision ability, etc. [3],[24]. By correlating driver's behavior and performance, we can identify precisely fatigue and drowsiness behaviors that have a strong relation to driver's performance. In a laboratory environment, performance driving parameter usually is measured by putting some stimulus in driving scenarios. The stimulus can be changes in weather, road condition, traffic condition, etc. Driver's responses when reacted to those stimuli are observed and measured. Sometimes the driver is asked to do certain activity when facing certain stimulus (example: press a button). The activity of pressing a certain button indicates if he still in aware condition. On the other hand, he also can make a mistake by falsely press the button when it is not necessary.

Beside direct driving performance, sometimes driver is also asked to do certain task after driving in a certain period of time. Objectively measured, the method that mainly used was Psychomotor Vigilance Task (PVT), which can measure reaction time of person based on certain stimulus [10]. For subjective measurement, respondent can self-assess their level of sleepiness using certain questionnaire. There are certain questionnaires that widely used, for instance, KSS (Karolinska Sleepiness Scale), FAS (Fatigue Assessment Scale) or Observer Rating of Drowsiness (ORD) [15]. Each questionnaire has its own scales and meaning

2.3. Considering driving environment

When we discuss behavior, it is important to consider the environment related to the behavior. When driving, a driver may face a certain condition that is challenging, or on the other hand, boring. That condition may affect certain natural and habitual behavior of the driver.

The environment can be classified into two main categories, which are inside-car condition and outside-car condition. Inside car condition means the condition inside the car that may affect driver's behavior. Temperature plays important role in determining person's behavior. In driving, temperature is determined by Air-Condition. Cool and fresh air may comfort the driver so that he tends to drive calmly. However, comfort condition can make the driver get sleepy easier than in uncomfortable condition. Uncomfortable condition (for instance when the temperature is relative hot) generates more aggressive behavior. Another inside-car condition that should be considered is sound. Sound can be produced by music, people voice (from radio, talking, etc.). Calm, slow, and soft sound (or even silent condition) tends to make drivers feel comfort and sleepy. When it is combined with comfort temperature, it may increase the driver sleepy effect.

From the outside-car, things such as road and traffic condition can affect driver behavior. Bumpy road, traffic jam, or challenging track may be something undesirable for the driver. But these conditions can awaken drivers and make driver more alert. On the other hand, smooth and easy track, with light traffic, tend to make driver comfortable, raise the car speed, and fall sleepy easily.

Driving in comfortable condition with the long monotone condition potentially worsens the effects. Sit in the same position, in the same condition of road and traffic, in the same condition of inside-car, for long period of time (can be hours), sometimes in night hours, making the possibility of getting drowsy while driving is getting bigger.

Long driving is a challenge for the driver. On one side we want to make driver comfort while doing car driving, but on the other hand, the comfort may have bad impacts due to its drowsy tendency. In this condition, car manufacturer sometimes puts an alarm in the car that can give a signal (sound, light, or vibration) when driving condition indicators show unsafe condition [15]. Safe and not safe condition was judged by a set of equipment that was placed in front of the driver to assess face behavior status. Further technology also was developed by car manufacturer so the equipment not only triggers a warning signal but also make some adjustment in the vehicle (such as decrease the speed and adjust temperature).

The monotony of a certain task can reduce the concentration of person and may cause distractions. Monotony can be caused by doing a repetitive task for a long time. Prolonged driving on highways with flowing traffic also causes monotony. In this case, the driver is not really fatigued. But the monotony makes his concentration will gradually be decreased and the driver will not have a careful control of the vehicle.

2.4. Behavior to handle fatigue and drowsiness

When the body was in fatigue and sleepy condition, along with comfortable monotonous condition, the driver usually manages some actions to avoid the worsening effect of drowsiness or to make the body back to fresh condition. The actions include music listening, singing, talking, head shaking, and body stretching. Driver sometimes makes a little dancing while singing and listen to music to boost more energy and spirit. Other actions that aimed to cope with sleepy symptoms, for example, exercise the eyelid, rub the eyes area using hand or handkerchief, close the mouth using the hand when yawning, etc. Although from human sight, these behaviors clearly are seen as the effort to refresh the body, in terms of fatigue and drowsiness detection, these behaviors complicated the monitoring process. Since there are usually certain algorithm to continuously detect eye and face status, all the behaviors disturbed the monitoring process, especially in detecting face and eyes area, and in extracting some characteristics from face and eyes

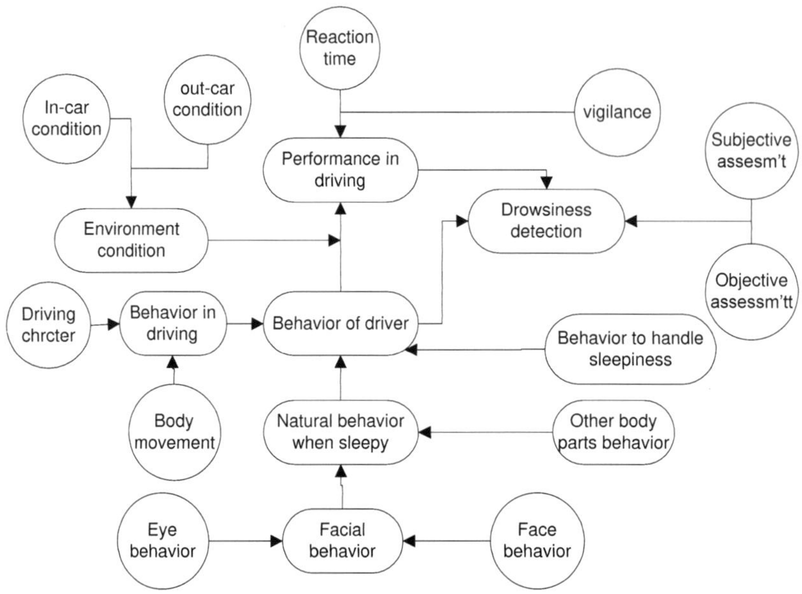

Figure 2. Factors Related to Driver's Behavior

3. Approach in developing behavior-based real time monitoring system

We have discussed factors that related to driver behavior in detecting drowsiness while driving. Next, the big question is how we use it to detect drowsiness. Like it was displayed in figure 3, to use them in detecting drowsiness, surely the first thing to do is building the method that can help us determining driver's drowsiness status. Based on the method, we can build another method and algorithm to capture those behaviors from a real driving condition and use them for determining driver's drowsiness status (using previously developed method). The third challenge in building this system is to construct it in the real comprehensive equipment.

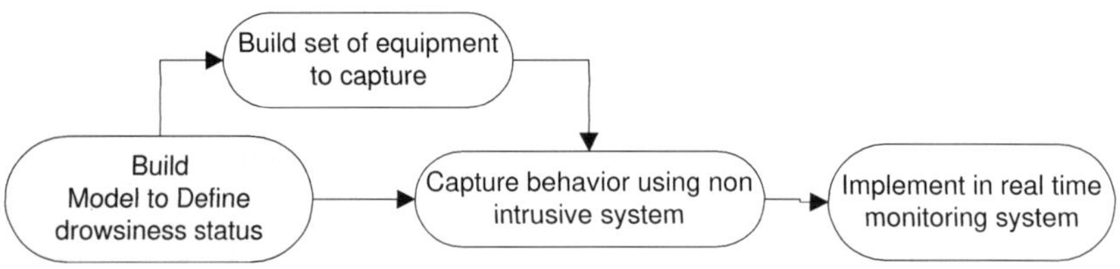

Figure 3. Main Steps in Developing Behavior-Based Real-Time Monitoring System

In this paper, we only discussed the first step, which is modelling drowsiness detection. Another step is presented further in another paper. For modeling drowsiness, the safest approach is by doing laboratory experiment, in which we measure drowsiness value on a certain eye and facial behavior conditions. Objective and quantitative measurement should be applied for determining face and eye status. For example, for the blink of the eye, how many frequencies blinks or duration of each blink that is considered as drowsy. The model is formed by considering the correlation of each indicator used, as well as the subjective level of sleepiness of the participants.

Observations and experiments are usually performed in the laboratory, using a driving simulator. Example of controlled variables: room condition (temperature, light, humidity, etc.), driving behavior (average speed), and time period of driving. We can use some variables as independent (manipulated) variables, for instance: Time of driving (example: night and day), duration of driving, outside environment (ex: bumpy and smooth road), and difficulties in driving (traffic jam versus light traffic). As dependent variables, we use selected characteristics of face and eyes to be measured. Eye movements can be captured using a number of tools such as camera, eye tracker, and EOG. Several scenarios of challenging events for the normal driver (not sleepy) and a sleepy driver can be designed. Using the scenario, the driver is asked to give a certain reaction when certain events occur. To compare the results of the drowsiness model and the actual level of drowsiness that the driver feels, then the subjective sleepiness measurement model such as Karolinska Sleepiness Scale (KSS) can be conducted before and after the test. Using psychomotor vigilance task (PVT) instrument, psychomotor performance can also be performed to measure the driver's reaction level before and after the experiment.

A number of statistical analyses should be used to determine the association among all drowsiness indicators to form a driver drowsiness model. The model of sleeping threshold value determination can be based on standard deviation, development of regression model between 2 or more characteristic, or development of regression model between characteristics and performance. Based on the developed model, it also can be analyzed how well the drowsiness level can predict driving performance (ex: reaction time). Of course, the model needs to be validated to see whether the model has been able to accurately detect drowsiness or not. If there is still a mistake, then the calculation or experiment should be done again with necessary adjustments. Calculations and experiments will be carried out until a valid drowsiness detection model is obtained.

4. Discussion

In the process of modeling, the researcher has to handle many challenging conditions that sometimes disturb and distract the process of gathering and analyzing data. There are a lot of conditions that make detecting process become difficult. For devices that have to capture the face and eye appearance, any condition that can cover face are undesirable. For instance, when the driver does not face towards the front, when lighting conditions are lacking, when driver wears accessories such as hats and glasses; the camera will not be able to capture the characteristics of the eyes and face.

Beside camera, there are types of equipment attached to the face and other body regions to detect muscle contraction and eye activities (blinking, iris movement, etc.). This device usually cannot distinguish between natural/spontaneous expression or activity due to fatigue and intended (purposive) expression and movement. For instance, it is difficult to distinguish between yawning and talking activity, eyelid closure due to sleepiness or intended expression while talking or singing.

When the model is completely built and be used for the real monitoring system, the device used for monitoring is only a set of cameras. But when we are in the process of model building (to define appropriate formula and threshold to define alertness level), we have to use an intrusive device in the driver. The device usually is like a cap with wires and pads that should be patched on the face and head area. This device can be used to measure the eyes characteristics precisely instead of only counting on visual monitoring. This device also can disrupt image capturing process.

Last but not least challenges of these processes are individual differences. Everybody has his own characteristics, including an eye and face. For instance, there is a person with round eyes versus slanted eyes, long versus short eye lace, and also expressive versus nonexpressive person. These differences make model formulation could be more complicated

5. Conclusion

Building driver's drowsiness status based on driver's behavior cannot be done by only considering the natural behavior of eye or face when the driver is in the drowsy condition. A number of other factors should also be considered, including the environment and driving behavior itself. Many factors can influence the drowsy level since there are also differences among drivers in responding to sleepiness.

6. References

[1] Alioua N, Amine A and Rziza M 2014 Driver's fatigue detection based on yawning extraction, *Int. J. Vehicular Technology*, Volume **2014** 7 pages

[2] Basim N M A, Sathyabalan P and Suresh P 2015 Analysis of EEG signals and facial expressions to detect drowsiness and fatigue using Gabor Filters and SVM linear classifier *Int. J. Computer Applications* **115** (11)

[3] Bhatt P P and Trivedi J A 2017 Various methods for driver drowsiness detection: an overview *Int. J. Computer Science and Engineering (IJCSE)* **9** (03 Mar 2017) 70-74

[4] Biggs, Charles H, Dingsdag DP and Stenson N 2009 Fatigue factors affecting metropolitan bus drivers: a qualitative investigation. *Work: A J. Prevention, Assessment, And Rehabilitation*, **32** (1) 5

[5] Biro Pusat Statistik, 2016, https://www.bps.go.id/linkTableDinamis/view/id/1134

[6] Djaja S, Widyastuti R, Tobing K and Lasut D, Irianto J 2016 Gambaran kecelakaan lalu lintas di Indonesia tahun 2010-2014, *J. Ekologi Kesehatan*, 15 (1): 30 – 42

[7] Diddi V K and Jamge S B 2014 Head pose and Eye state Monitoring (HEM) for driver drowsiness detection: Overview, *IJISET – Int. J. Innovative Science, Engineering & Technology*, 1 (9): November 2014

[8] Ekman P and Rosenberg E L 2005 *What the face reveals: Basic and Applied Studies of Spontaneous Expression Using the Facial Action Coding System (FACS) Second Edition* (Oxford: Oxford University Press)

[9] Fernández A, Usamentiaga R, Carús J L and Casado R 2016 Driver distraction using visual-based sensors and algorithms *Sensors 2016,* **16** (1805)

[10] Gertler J.,Popkin S, Nelson D and O'neil K 2002 Transit Cooperative Research Program (TCRP) *Report 81: Toolbox For Transit Operator Fatigue* (Washington, D.C: National Academy Press)

[11] Hachisuka S, Ishida K, Enya T and Kamijo M 2011 Facial expression measurement for detecting driver drowsiness *Int. Conf. on Engineering Psychology and Cognitive Ergonomics, EPCE 2011: Engineering Psychology and Cognitive Ergonomics* pp 135-144

[12] Ilbeygi M and Hosseini H 2012 A novel fuzzy facial expression recognition system based on facial feature extraction from color face images *Engineering Applications of Artificial Intelligence* **25** (1):130-146.

[13] Jabon ME, Bailenson JN, Pontikakis E, Takayama L and Nass C 2011 Facial-expression analysis for predicting unsafe driving Behaviour Face and Gesture Recognition O*ctober–December 2011 Pervasive Computing*

[14] Jayswal A S and Modi R V 2017 Face and eye detection techniques for driver drowsiness detection *Int. Research. J. Engineering and Technology (IRJET)* **04** (04) : Apr 2017

[15] Karchani M, Mazloumi A, Saraji G N ,Gharagozlou F, Nahvi A, Haghighi K S, Abadi B M and Foroshani AR 2015 Presenting a model for dynamic facial expression changes in detecting drivers' drowsiness *Electronic Physician* **7**(2):1073-1077

[16] Kumar N and Barwar N. 2014 Detection of eye blinking and yawning for monitoring driver's drowsiness in real time, *Int. J. Application or Innovation in Engineering & Management*, **3** (11): November 2014

[17] Li Z, Li S E , Li R, Cheng B, and Shi J 2017 Online detection of driver fatigue using steering wheel angles for real driving conditions *Sensors 2017*, **17** 495; doi:10.3390/s17030495

[18] Lim S M and Chia S E 2015 The prevalence of fatigue and associated health and safety risk factors among taxi drivers in Singapore *Singapore Med J.* **56** (2): 92–97

[19] Mardi Z, Ashtiani S N M, Mikaili M. 2011 EEG-based drowsiness detection for safe driving using chaotic features and statistical tests. *J Med Signals Sens* **1** (2): 130–137

[20] Maruta J, Heaton KJ, Maule AL, Ghajar J, 2014 Predictive Visual Tracking: specificity in mild traumatic brain injury and sleep deprivation *Military Medicine: Brief Reports*, **179** (6):619

[21] Pascual MF and Zapirain BG 2015 Assessing visual attention using eye tracking sensors in intelligent cognitive therapies based on serious games *Sensors 2015*, **15**:11092-11117

[22] Sahayadhas A, Sundaraj K, Murugappan M 2014 Electromyogram signal based hypovigilance detection. *Biomedical Research: An International Journal of Medical Sciences* ISSN: 0970-938X

[23] Shamian M, Khan I, Abdullah H 2014 Efficient drowsiness detection by facial features monitor *Research J. Applied Sciences, Engineering and Technology* **7** (11): 2376-2380

[24] Sigari MH, Pourshahabi Soryani Fathy M 2014 A review on driver face monitoring systems for fatigue and distraction detection *Int. J.of Advanced Science and Technology* **64** (2014) 73-100

[25] Sigari MH, Fathy M, Soryani M 2013 A driver face monitoring system for fatigue and distraction detection *Hindawi Publishing Corporation Int. J. Vehicular Technology* **2013**, ArticleID 263983, 11 pages

[26] Singh MS., Bhavana BG., Pooja S S and Ashish M M 2015 Eye tracking based driver drowsiness monitoring and warning system *Int. J. Technical Research and Applications* **3** (3) 190-194

[27] Wismadi A., J. Soemardjito And H. Sutomo 2013, 'Transport situation in Jakarta', in Kutani I. (Ed.), Study on energy efficiency improvement in the transport sector through transport improvement and smart community development in the urban area. *Research Project Report* 2012-29, pp.29-58.

[28] Ying LT. Kanade T, Cohn JF. 2001 Recognizing Action Units for facial expression analysis, *IEEE Trans Pattern Anal Mach Intell.,* **23**(2) 97–115

Risk assessment of manual material handling activities (case study: PT BRS Standard Industry)

Deviani[1], V. Triyanti[2]

[1,2] Department of Industrial Engineering, Atma Jaya Catholic University of Indonesia, Jalan Jenderal Sudirman 51, Jakarta 12930, Indonesia
devianiwijaya@gmail.com[1], vivi.triyanti@atmajaya.ac.id[2]

Abstract. The process of moving material manually has the potential for injury to workers. The risk of injury will increase if we do not pay attention to the working conditions. The purpose of this study is to assess and analyze the injury risk level in manual handling material activity, as well as to improve the condition. The observed manual material handling activities is pole lifting and goods loading. These activities were analyzed using Job Strain Index method, Rapid Entire Body Assessment, and Chaffin's 2D Planar Static Model. The results show that most workers who perform almost all activities have a high level of risk level with the score of JSI and REBA exceeds 9 points. For some activities, the estimated compression forces in the lumbar area also exceed the standard limits of 3400 N. Concerning this condition, several suggestions for improvement were made, improving the composition of packing, improving body posture, and making guideline posters.
Keywords: Manual Material Handling (MMH), 2D Planar Static Model, Compression Force, Rapid Entire Body Assessment (REBA), Job Strain Index (JSI)

1. Introduction

Manual Material Handling (MMH) is one of the most physically demanding works due to its repetitive movements, awkward postures, contacted stresses, or forceful exertion [15]. Increased rate of musculoskeletal injuries on worker's back was reported especially injuries that caused by for work that required manual lifting/lowering, carrying, or pushing/pulling of heavy materials by laborers [8], [10], [16]. Back injuries - particular injuries to the lower back - occurred with alarming frequency [2]. Approximately, back injury is one of the most common (22% of all accidents that occurred) and most required a fee for treatment. One cause of this injury is incurred by overload force of the spine (>60%) and 60% of the overload is caused by lifting, 20% pushing or pulling goods and 20% as a result of carrying goods [15]. Occupational injuries can occur suddenly, unanticipated, and unwanted events during work leading to harm or damage to at least one part of the body [4]. Risk of injury is increased when job strength requirements exceed worker capabilities and overextension injuries are costly [6]. The injuries incurred by workers can be detrimental to both workers and industry. Workers who injured will reduce the performance of the work and the productivity of the industry. The application of ergonomic principles would help to increase performance and productivity, but mostly help a human operator to be comfortable and secure [1].

There is a considerable amount of literature on manual materials handling. Consider, for example, reference [1] is to highlight on ergonomic intervention of material handling work system in an Indian manufacturing plant, reference [15] showed that repetitive work activities, particularly involving high force, extreme joint postures, and exposure to vibration that resulted in musculoskeletal disorders, especially at the distal upper. In this literature describes how to reduce the risks that occur in the automotive industry using the strain index, reference [3] showed that ergonomic problem not only in manual materials handling activity but also occurs in other activities such as surgery. Maintaining the same posture for a long time during surgery is one of the reasons the issue of ergonomics. Reference [5] showed that the magnitude of the compression force is influenced by posture during loads lifting. In this paper investigated three types of loads lifting. Reference [4] showed an analysis of the influences and effects of different postures performed within tasks of waste collection. While reference [8] showed analyze manual material handling working posture of the operators using 3D Static Strength Prediction Program (3DSSPP) software and to identify major areas causing long last injury of operators.

There are several studies related to manual material handling especially lifting. Some of the commonly used methods are Job Strain Index, REBA, and 2D Planar Models Methods [1], [3], [4], [5]. For this paper, those methods were applied to Indonesian workers, especially at PT BRS Standard Industry. Although the company performed frequent manual material handling activities in its daily activities, there was no investigation of lifting risk that had been done.

The study was conducted to analyze the manual material handling work system of the operators and to identify major areas causing injury of operators that related to the activities. The main purpose is to minimize the risk of operators' physical and physiological disorders caused by poor MMH system.

2. Methods

This research was conducted to 8 workers who perform manual handling material activity, especially the activity of lifting the pole and loading the goods. Nordic body map questionnaire was used to identify symptoms, complaints, disorder, and injuries experienced by workers. Measuring tools that used to perform data collection were the camera and the meter. Data collected included:

1. Anthropometry: height and weight, and body segment length (upper arm, forearm, and torso).
2. Joint Angles, including data related to angles of between 2 body segments or between body segment and horizontal/vertical line during work, such as legs back neck arms and wrists.
3. Material Specification, including size, type, and weight of materials to be transported.
4. Hand Loads show load received by hand.
5. The intensity of labor usage, duration, working speed, and length of work per day.
6. Worker photos and videos while performing manual material handling activities.

The next step is to perform data processing by Job Strain Index, REBA, and Chaffin's 2D Planar Static Model. Some variables needed to calculate the score on Job Strain Index Method. After JSI method, processing data was continued using REBA method.

Joint angle was determined using the measured angle from pictures that were taken during activities. Angle was measured with the help of Image Analysis in Ergofellow software. The angle values then were inputted to Upper Extremity Analysis [1] by ErgoIntellegence to perform REBA analysis.

In REBA method, assessment of operator's work posture was done to know the possibility of risks due to work postures performed by the operator. If the REBA score showed "undesired" score, then some necessary improvements were suggested.

The software of 3D SSPP is used to find out the maximum force suffered by the worker in doing such a job [4], [8]. The assessment was based on the results of Sagittal Plane Low back Analysis Compression Force on L5 / S1. If the total compression is still within the specified limit or less than 3400 N, then it can be concluded that the maximum force that has to be accepted is still acceptable for the worker's body.

For every score that beyond the specified limit, work improvement was suggested, especially to improve the work posture. Moreover, further analysis was done to estimate the requirement of supporting tool. If it was necessary, a new tool design was suggested. Suggested work posture then is re-assessed using the same posture analysis method.

3. Result and Discussion

Using JSI, REBA, Planar 2D Static Model, the recapitulation of the results presented in table 1:

Table 1. Recapitulation of result

Task	Job Strain Index Score (lower better)	REBA Score (lower better)	Compression Force (N) (lower better)
Research Subject A			
Pole Lifting	7.3	6	2060.5
Research Subject B			
As Loading - worker A	9.8	10	3729.4
As Loading - worker B	9.8	10	2548.5
As Loading - worker C	4.9	10	2457.1
Leaf Slat Loading - worker A	7.3	12	3931.4
Leaf Slat Loading - worker B	7.3	7	2602
Pole Loading- worker A	7.3	12	2660.3
Pole Loading- worker B	14.6	11	3019.5

When maximum compression force for lifting is 3400 N, for JSI and REBA, low-risk work is achieved when the score below 3. On the other hand, an activity is determined to have high risk of injury when the score is more than 7. Based on the results of table 1, almost all activities have risk levels according to JSI and REBA. The activity that has relatively high level of risk of all three methods is the activity of slat leaf removal by worker A.

Figure 1 shows worker position when he wants to take leaf slat. It can be seen that the body is bent and slightly spun. The body angle for each segment of the body is quite large, resulting in a high score. The value of the compression force obtained exceeds 3400 N, indicating that this activity is dangerous.

Figure 1. Working posture of Leaf slat loading – worker A

Based on these results, the proposed improvements are:

1. Proposed improvement by *Engineering Controls*
   In the proposed improvement, a suggestion was made on the leaf slats packing arrangement so that the pack was easier to lift by the workers. In the current situation, 2 workers are working together to lift 1 pack of leaf slats (1 pack consists of 10 slats). However, the worker has to widely stretch out his arms to lift the edge of the pack. This overextended posture causes high risk for arms. In suggested system, Using the proposed system, 5 slats were stacked together. In one lift, a pair of workers can lift 2 or 3 packs, depend on the weights of each slat. This arrangement is safer for workers since they did not have to overextend their arms.

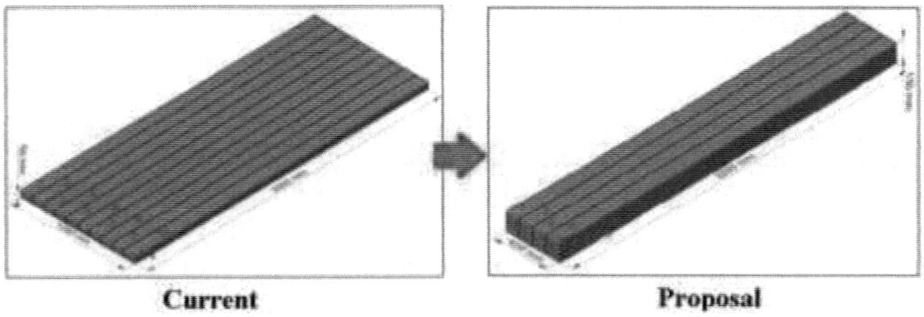

Figure 2. Proposed Leaf slat packing

2. Proposed Improvement by *Management Controls*
   Each worker should be trained on safety working, especially in manual material handling activities. The proposed lifting position was presented in figure 3. In this position, legs should be opened wide and bent with 1 leg was slightly toward the other. Do not forget to keep maintain back in straight position. When the object was started to lift, put the objects as close to the body. If it is possible, put the object between his legs. Body move towards slightly with still keeping the back straight position and begin to lift. It is important to keep all lifting activities as smooth as possible and avoid jerk lifting.

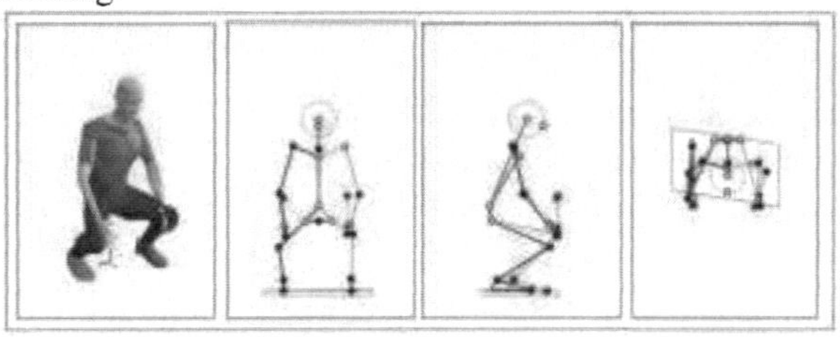

Figure 3. Position of Proposed posture

3. Proposed Improvement by *Work Practice Controls*
   It also suggested making lifting guidelines posters. The posters should be put in the area where the lifting activity occurs, in the location that can be easily seen by workers.

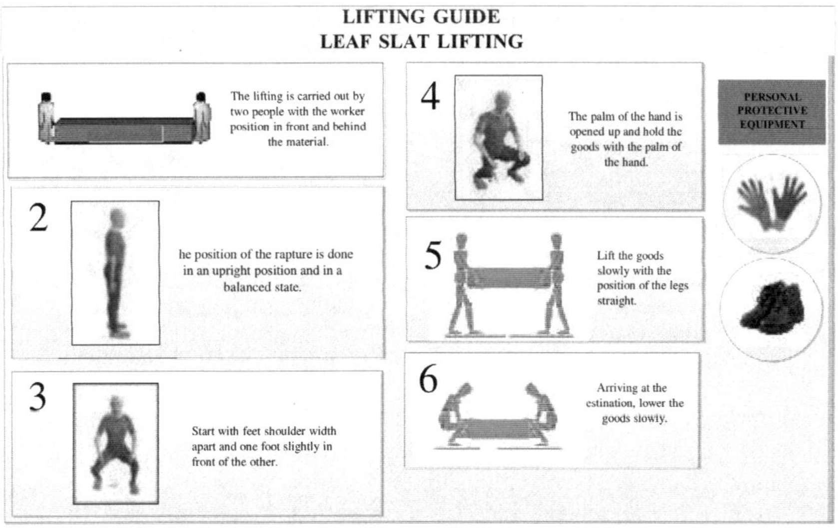

Figure 4. Example of lifting guide poster

Proposed postures and improvements later were analyzed using the same methods. From table 2 it can be seen that the proposed situation reduce both times of lifting and risk level of injury.

Table 2. Performance comparison of current and proposed condition

	Pole Lifting		As Loading		Leaf Slat Loading		Pole Loading	
Criteria	**Current**	**Proposal**	**Current**	**Proposal**	**Current**	**Proposal**	**Current**	**Proposal**
Time of lifting (s)	21.75	16.86	17.8	12.4	13	9	13.4	9.93
Score JSI	7.3	1.5	9.8	3.375	7.3	4.5	14.6	4.5
Score REBA	6	4	10	5	12	5	12	5
Compression Force(N)	2060.5	1378	3729.4	2594	3931.4	2975	3019.5	2975

The proposed improvements were suggested for all activities. One improvement that implemented was tried is work posture changing like suggestion number 2. When the trials were conducted, workers have confusion when applying the new lifting method, often the position only persisted for some replication only and returns to its original position. However, after some repetition finally, the workers understand. Table 2 shows the comparison of value between the current situation and proposed situation score. Proposed situation score was estimated based on trials after the workers can apply new lifting and loading posture relative persistently. The results of the table show that the proposed method was able to reduce the risk of injury and shorten activity time.

Figure 5, 6, 7 show the comparison between current and proposed system. In all current methods, the main flawless is the back position. Workers tend to their back instead of their leg. Therefore, maintaining straight back position has been the main improvement. Second thing, in the current system, workers tend to perform the activity individually. Similar to lifting process in figure 6, worker lifted the slat alone by putting the slat on shoulder one by one. In the proposed system, the activities should be done by two persons, lifting 2 slats at a time. Therefore the productivity is still maintained.

Figure 5. Comparison of the current and proposed lifting position

After the trials were conducted, an interview was conducted with the workers who had performed the proposed condition, where the results found that the worker felt awkward at the beginning of the new position but then felt more comfortable with the position. Workers give no complaint at the waist and shoulders. From the trial processes, it can be concluded that basically, the proposed method improved the work qualities. However, further training is needed to make sure workers aware about the safe lifting procedure and how to apply it in their daily working activities.

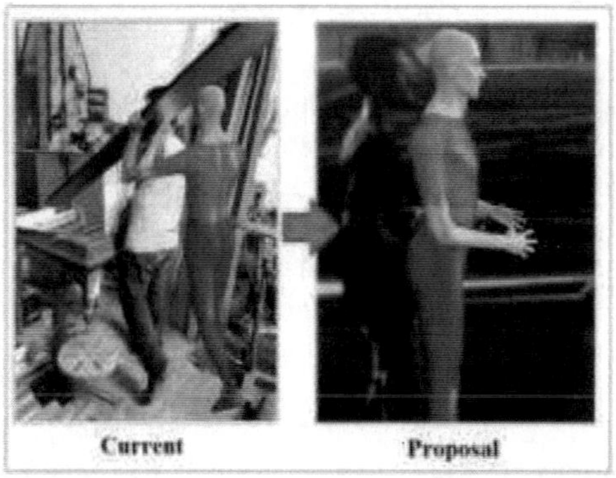

Figure 6. Comparison of the current and proposed lifting position to pole lifting

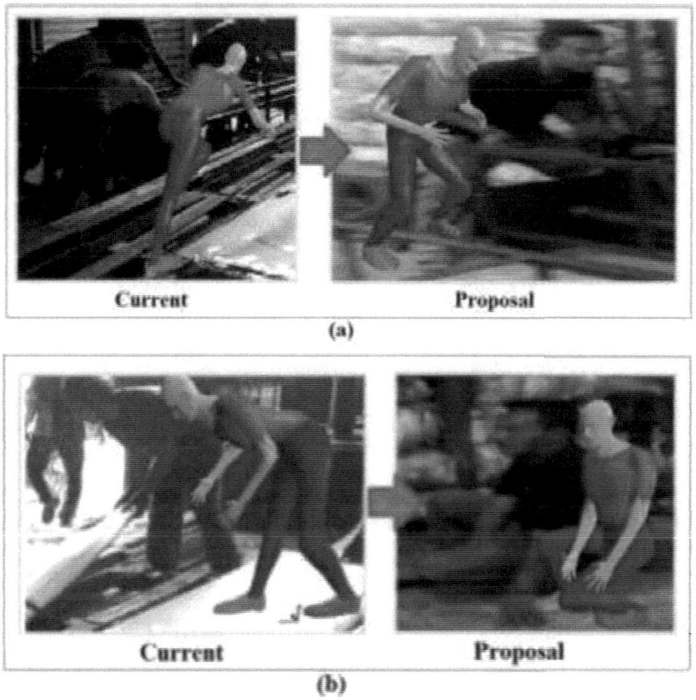

Figure 7. Comparison of loading position (a) As (b) Leaf Slat and Pole

4. Conclusion

Using JSI, REBA, and 2D Planar Models Methods, Manual material handling activity identify that almost all activities have high-risk levels, with the highest was found in the loading of slat leaf goods activity. Body position greatly affects the occurrence of high risk, especially the back position. There are three improvements applied to reduce the risk of injury to the activity, including improving the composition of packing, improving body posture, and making guideline posters.

IOP Conf. Series: Materials Science and Engineering **277** (2017) 012043 doi:10.1088/1757-899X/277/1/012043

5. References

[1] Ansari N A and Sheikh M J 2014 Evaluation of work Posture by RULA and REBA: A Case Study *IOSR Journal of Mechanical and Civil Engineering* **11**(4) 18-23

[2] Ayoub M M 1989 *Manual Materials Handling: Design And Injury Control Through Ergonomics* (Florida: CRC Press)

[3] Bartnicka J, Ziętkiewicz A and Kowalski G 2013 Analysis of selected ergonomic problems in the use of surgical laparoscopic tools *Zeszyty Naukowe/Akademia Morska w Szczecinie* **34** (106) 19-26

[4] Cakit E 2015 Assessment of The Physical Demands of Waste Collection Tasks. *Global Nest Journal* ***17*** (2) 426-438

[5] Devassy V and Rengaraj R M 2015 Biomechanical Analysis of Spinal Compression during Load Lifting in Multi Postures *International Journal for Research in Technological Studies* **2** (10) 19-22

[6] Hoffman S G, Chaffin D B and Wagner D W 2006 Strength and Balance Guided Posture Selection during a Battery Maintenance Task *SAE Technical Paper* (No. 2006-01-0698) (Warrendale: SAE International)

[7] Levy B S, Wegman D H and Halperin W E 2000 Recognizing occupational disease and injury. *Occupational Health: Recognizing and Preventing Work-Related Disease and Injury* **4** 99-122

[8] Matebu A 2014 Design of Manual Material Handling System Through Computer Aided Ergonomics : A Case Study at BDTSC Textile Firm *International Journal for Quality Research* **8** (4)

[9] Mital A and Das B 1987 Human strengths and occupational safety *Clinical biomechanics* **2**(2) 97-106

[10] Mital A 1997 *Guide to manual materials handling* (Florida: CRC Press)

[11] Moore J S and Garg A 1995 The Strain Index: A proposed method to analyze jobs for risk of distal upper extremity disorders *Journal of the American Industrial Hygiene Association* (56) 457-458

[12] Nurmianto E 1996 *Ergonomi konsep dasar dan aplikasinya* (Jakarta : Guna Widya)

[13] Nurmianto E 2004 *Ergonomi, Konsep Dasar dan Aplikasinya: Tinjauan Anatomi, Fisiologi, Antropometri, Psikologi dan Komputasi untuk Perancangan Kerja dan Produk* (Guna Widya : Surabaya)

[14] Pulat B M 1996 *Fundamentals of Industrial Ergonomics* (Illinois : Wafeland Press)

[15] Rajesh R, Babu R V and Ramachandran S 2013 Ergonomic Redesign of a Material Handling Worksystem in a Manufacturing Plant *International Journal of Innovative Research in Science Engineering and Technology 2*

[16] Santoso D 2006 Kapasitas Angkat Beban untuk Pekerja Indonesia *Jurnal Teknik Industri* **8** (2) 148-155

IOP Conf. Series: Materials Science and Engineering **277** (2017) 012044 doi:10.1088/1757-899X/277/1/012044

Coloring your information: How designers use *Theory of Color* in creative ways to present infographic

C R Lucius[1] **and A Fuad**[2]

[1] Department of Visual Communication Design, Universitas Esa Unggul, Jakarta, 11150, DKI Jakarta, Indonesia
[2] Department of Visual Communication Design, Universitas Esa Unggul, Jakarta, 11150, DKI Jakarta, Indonesia
[1] christophera.lucius@esaunggul.ac.id
[2] ahmad.fuad@esaunggul.ac.id

Abstract. Various methods of data presentation is now visualized through engaging infographics and perform the presentation techniques a new kind of storytelling. Geometric elements for infographics perform interesting data, which is developed with color harmony. There are categories of colors based on color circle from the theory of color design: primary color, secondary color and tertiary color. This color circle allows a designer to visualize the balance and harmony of colors when they are side by side. These composition of colors can be formed as a harmonious dyad, triad, or tetrads. A harmonious dyad is formed from two diametrically opposed colors on the color circle, which known as contrast complementary and works best in color harmonious if one of the colors is dominant. A harmonious triad is represented by three colors from the color circle which positions with an equilateral triangle. An triangle of yellow-red-blue shows the most powerful of harmonious triad and call as the fundamental triad. A harmonious tetrad is developed from two pairs of complementary colors, which can be formed by rectangle or square on the color circle. It help to figure out how objects are connected on presenting data. To create an efficiency infographic, presenting data has to prepare with some strategic. The color circle has the power to perform the infographic when it is made for a fascinating design.
Keywords: Infographic, Color Harmony, Color Circle, Theory of Color

1. Introduction

Presenting data is now changing with the new media how the way informations content are produced and influence the viewer´s understanding of the informations. Various methods of data presentation is a new kind of storytelling and make the presentation techniques more shorter and visualized through engaging infographics. [4] state the infographics should be viewed as a communication medium used by the organization to connect with different target audiences. Presentation techniques are done by visualizing data and reporting stories from new perspectives. According to [6] data visualization is a very broad concept, referring mainly to graphical methods to clearly and effectively distribute and communicate information. Geometric elements for infographics develop the attraction of data and can be enriched with emotional color. Using color is one of the easiest ways to take data visualization to a level of competence. There is a need for color selection in presenting data when dealing with complex topics. Color selection for data presentation is very important and is reflected in infographics.

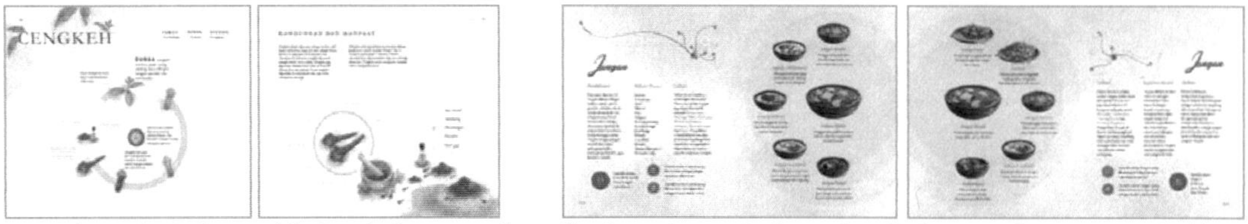

Figure 1. Infographic book from student final projects: Indonesian Spices and Javanese Culinary.

According to [5] graphics can be designed to have at least three viewing depths: what is seen from a distance, what is seen up close and in detail, what is seen underlying the graphic. In the field of graphic design, there are categories of colors based on color circle from the theory of color design: *primary color, secondary color and tertiary color.* [1] developed the 12-hue color circle based on primary colors: *yellow - red - blue*. These colors are placed in a equilateral triangle, yellow at the top, red at the lower right, and blue at the lower left. A circle is circumscribed about this equilateral triangle, in which a regular hexagon is inscribed. Three mixed colors which composed of two primary colors are placed between contiguous sides of the hexagon. It mentions as secondary colors: *orange - violet - green*. The six tertiary colors are located between the primary and secondary colors they are made from: *yellow-orange - orange-red - red-violet - violet-blue - blue-green - green-yellow*. This color circle allows a designer to visualize the balance and harmony of colors when they are side by side.

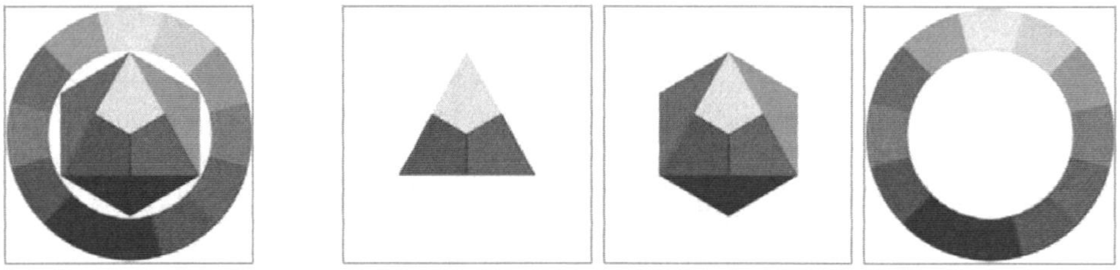

Figure 2. Theory of color by [1]:
primary colors, secondary colors, tertiary colors, and color circle.

2. Methods

According to [1] color harmony means the craft of developing themes from systematic color relationships capable of serving as a basis for composition. These composition of colors can be formed as two (dyads), three (triads), or four (tetrads) colors.

2.1. Dyads (two-color combination)

Two diametrically opposed colors on the color circle form a harmonious dyad and well-known as contrast complementary. These complementary pairs show powerful combination. An number of harmonious dyad can be created on the color circle.

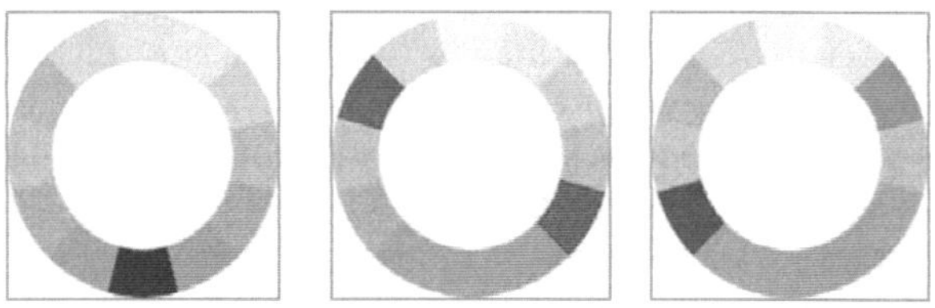

Figure 3. Harmonious dyads:
(*yellow* + *violet*) or (*red* + *green*) or (*blue* + *orange*).

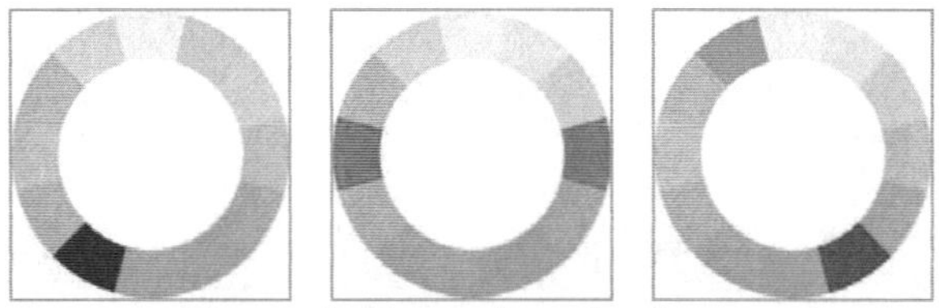

Figure 4. Harmonious dyads:
(*yellow-orange* + *violet-blue*) or (*orange-red* + *blue-green*) or (*red-violet* + *green-yellow*).

2.2. Triads (three-color combination)

Three colors from the color circle which positions with an equilateral triangle represent a harmonious triad. An triangle of yellow-red-blue shows the most powerful of harmonious triad and call as the fundamental triad. The other triads are made up of these color combination from the secondary colors. Another particular triads can be formed by one color in the complementary dyad which replaced by its two neighbour colors.

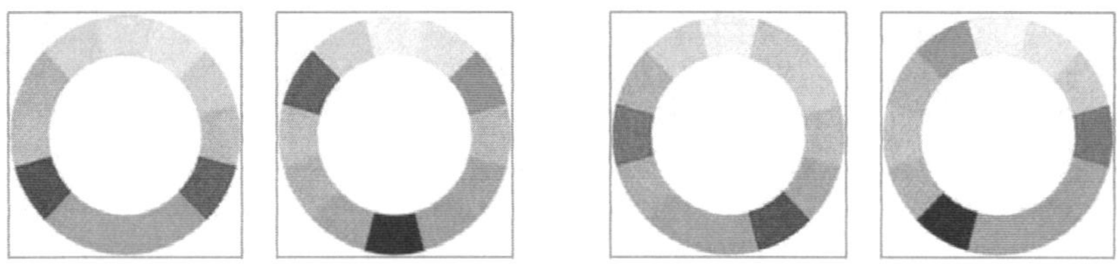

Figure 5. Harmonious triads:
(*yellow* + *red* + *blue*) or (*orange* + *violet* + *green*)
or (*yellow-orange* + *red-violet* + *blue-green*) or (*orange-red* + *violet-blue* + *green-yellow*).

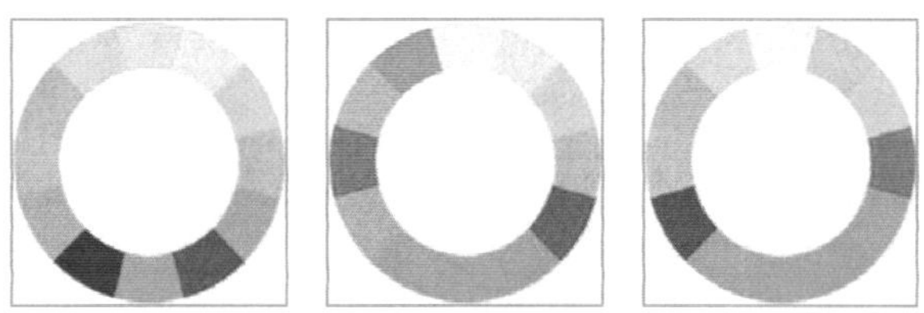

Figure 6. Harmonious triads:
(*yellow* + *red-violet* + *violet-blue*) or (*red* + *blue-green* + *green-yellow*)
or (*blue* + *yellow-orange* + *orange-red*).

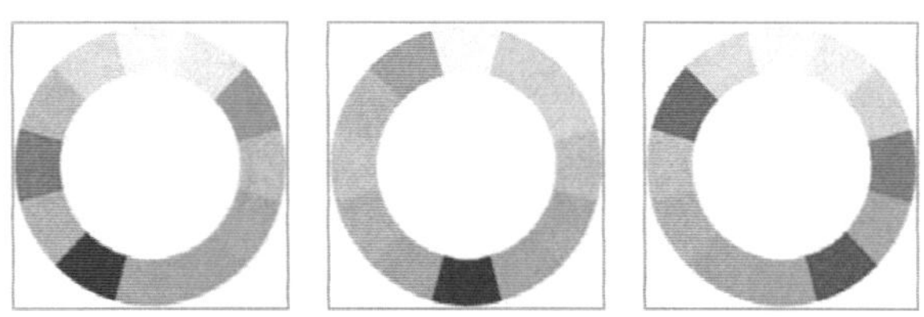

Figure 7. Harmonious triads:
(*orange* + *violet-blue* + *blue-green*) or (*violet* + *green-yellow* + *yellow-orange*)
or (*green* + *orange-red* + *red-violet*).

2.3. Tetrads (four-color combination)

Two pairs of complementary colors show a harmonious tetrad. This color combination can be formed by rectangle or square on the color circle. Tetrads based on square are made up of two complementary pairs which connecting diameters are perpendicular to each other. Tetrads based on rectangle are obtained with a rectangle containing two pairs of complementaries.

IOP Conf. Series: Materials Science and Engineering **277** (2017) 012044 doi:10.1088/1757-899X/277/1/012044

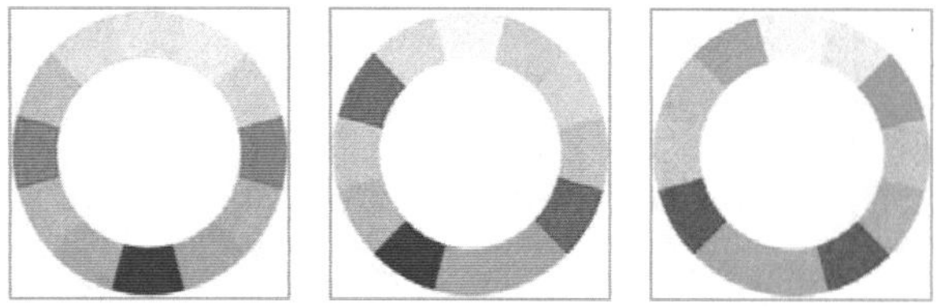

Figure 8. Harmonious tetrads based on square:
(*yellow* + *orange-red* + *violet* + *blue-green*) or (*red* + *violet-blue* + *green* + *yellow-orange*) or (*blue* + *green-yellow* + *orange* + *red-violet*).

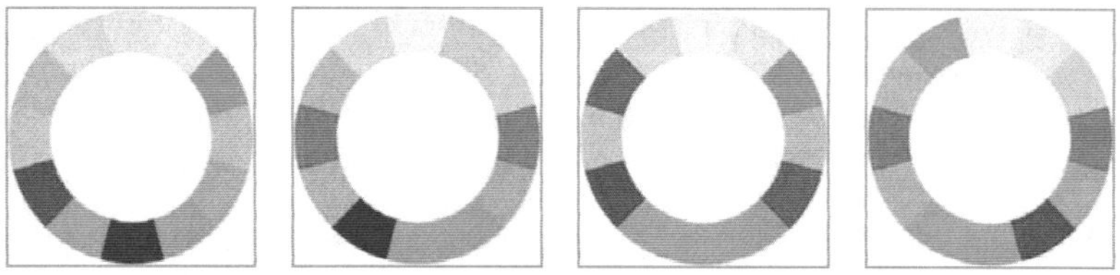

Figure 9. Harmonious tetrads based on rectangle:
(*yellow* + *orange* + *violet* + *blue*) or (*yellow-orange* + *orange-red* + *violet-blue* + *blue-green*) or (*orange* + *red* + *blue* + *green*) or (*orange-red* + *red-violet* + *blue-green* + *green-yellow*).

3. Result and Discussion

3.1. Harmonious Dyads for Infographic Design

Complementary pairs may be the most frequently used color combination on infographic design. The idea of choosing contrast complementary is behind making graphical element on infographic effective. Using a harmonious dyad gives these infographic an efficiency of interpretation found in no other data composition. This type of combination works best in color harmonious if one of the colors is dominant and the other serves to create contrast. Because both colors are integrated, they need to blend together definitely. Graphical element with a harmonious dyad is the place to help viewers understand the infographic data better. The perfect harmonious dyad can attract attention the viewer and complement the data, all in one consistent information design. The success of complementary pairs compared to the black white visual gives viewer a lead on how graphical elements can communicate complex information easily.

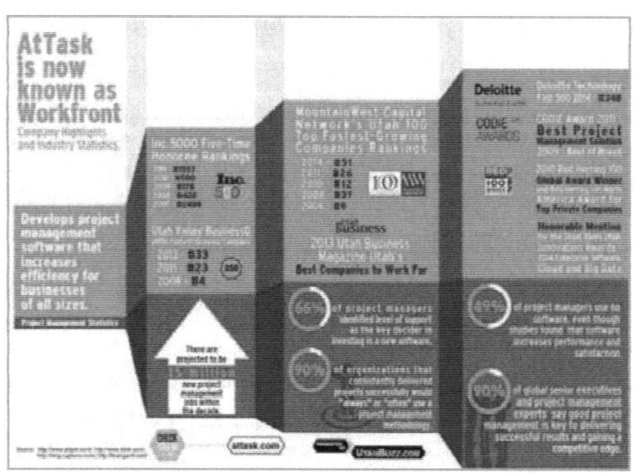

Figure 10. Infographic data using two-color combination: [7]

3.2. Harmonious Triads for Infographic Design

The goal of a triad harmonious is to make complex data easier to understand. The infographic shown is integrated through its data content and separated several different information. Setting objectives does not have to be formal to be effective. The information needs to present some value to the viewer consuming the content. It will be better at gathering and transforming information into infographics. The important thing to consider is the integrity of the data cause the data represent a complex concept within a specific information. This approach involves the viewer to think about presenting the data and showing the information with the infographic. The three-color combination should take those data in the viewer who find the informations in the infographic more clearly. The rule triads makes reasonable graphical information. It is important that the rule follow from the idea of maximizing the combination of colors. In these infographics, the concept of color theory are taken and present three popular and easy-to-use color-combinations to experiment with.

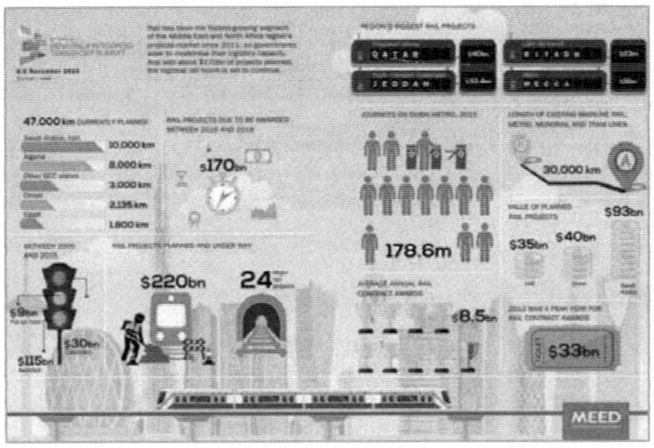

Figure 11. Infographic data using three-color combination: [8]

3.3. *Harmonious Tetrads for Infographic Design*

Color selection for presenting complex data is very important, because the colors chosen can help viewers understand the presentation more quickly in order to show information. If the viewer don´t see how objects are connected, it is difficult to explain a good processes that will impact understanding the information. A harmonious tetrad help to figure out which infographics are working and which ones are not. Thus design techniques for adding graphical clearness against of data complexity is developed with element of color. The content theme is reflected on infographics by the use of four colors as the combination color. This color combination guide to create objects and content the infographic communication to become more relevant and effective on presenting data. Different colors show different pieces of data. The four-color series perform the different sampled infographics shown.

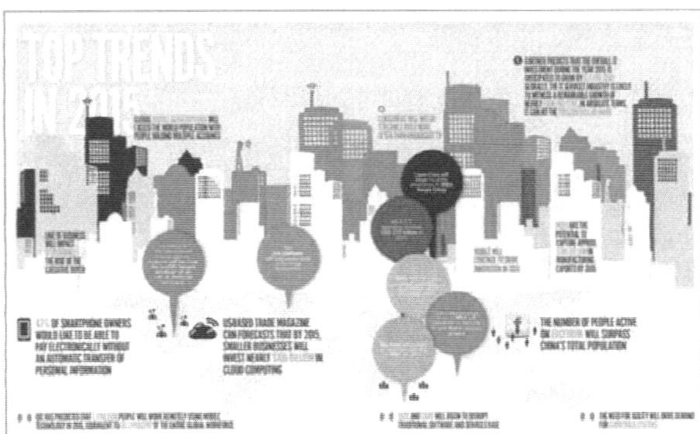

Figure 12. Infographic data using four-color combination: [9]

4. Conclusion

Data selection can influence the viewer´s understanding of the information. Different ways of presenting data obtain different informing information. [4] state thinking visually helps viewer synthesize the information more easily. Using infographics to visualize how data are related support understanding of the connection that are significant to information. According to [5] the complexity of multifunctioning elements can sometimes turn data graphics into visual puzzles, crypto-graphical mysteries for the viewer to decode. To develop the efficiency of infographic, presenting data should be prepared in some strategic. Using color to deliver data on infographic help to impresse the point of information. [6] state when making infographics, rational use of geometric elements can increase the appeal of data, enriching the data with emotional color. The color circle has the power to perform the information and support the data on infographic when it is made for a fascinating design.

5. References

[1] Itten J 1970 *The Elements of Color* (Ravensburg: Otto Maier Verlag) pp 29-74
[2] Lucius C R and Astuti F 2016 *The Final Year Project Indonesian Spices* Department of Visual Communication Design (Jakarta: Universitas Esa Unggul)
[3] Lucius C R and Lim C 2017 *The Final Year Project Milang Warisan* Department of Visual Communication Design (Jakarta: Universitas Esa Unggul)
[4] Smiciklas M 2012 *The Power of Infographics Using Pictures to Communicate and Connect with Your Audiences* (Indianapolis: Que Publishing)
[5] Tufte E R 2001 The Visual Display of Quantitative Information (Connecticut: Graphics Press)
[6] Yikun L and Zhao D 2015 Visual Storytelling Infographic Design in News (Victoria: The Image Publishing) pp 7-14
[7] www-03.ibm.com
[8] www.meed.com
[9] www.ssmeadvisor.com

IOP Conf. Series: Materials Science and Engineering **277** (2017) 012045 doi:10.1088/1757-899X/277/1/012045

The performance and emissions of diesel engines with biodiesel of sunan pecan seed and diesel oil blends

F. Ariani[1], T. B. Sitorus[1] and E. Ginting[2]

[1]Mechanical Engineering, Universitas Sumatera Utara, Medan, 20155, Indonesia
[2]Industrial Engineering, Universitas Sumatera Utara, Medan, 20155, Indonesia
arianiida@yahoo.com

Abstract. An observation was performed to evaluate the performance of direct injection stationary diesel engine which used a blends of biodiesel of Sunan pecan seed. The experiments were done with diesel oil, B5, B10, B15 and B20 in the engine speed variety. Results showed that the values of torque, power and thermal efficiency tend to decrease when the engine is using B5, B10, B15 and B20 , compared to diesel oil. It also shown that the specific fuel consumption is increased when using B5, B10, B15 and B20. From the results of experiments and calculations, the maximum power of 3.08 kW, minimum specific fuel consumption of 189.93 g/kWh and maximum thermal efficiency of 45.53% when engine using diesel oil. However, exhaust gases were measured include opacity, carbon monoxide and hydrocarbon when the engine using biodiesel B5, B10, B15 and B20 decreased.
Keywords: performance, emissions, sunan pecan seed

1. Introduction

Nearly 26-27% of energy consumption met by fossil fuels in the transport sector should be replaced in 2050 [1], [2]. The increasing energy demand and environmental deterioration are two major factors which drive the investigation on alternative fuels to replaced the conventional fuels for internal combustion engines. One of these alternative fuels is biofuels in the form of alcohol or biodiesel which are recommended as an alternative fuel for internal combustion engines [3], [4], [5]. Biodiesel is an alternative expectant fuel because it is renewable, biodegradable, non-toxic and environmentally friendly [6], [7]. Biodiesel can be used in its pure form or also as a blend form. It can also be used as a diesel fuel additive to improve its properties. Biodiesel has been widely used in diesel engines mainly because the raw materials can be obtained from a wide variety of oils such as plants, animals, and waste oil [8], [9], [10]. As noted, the fuel type is also affecting the performance of the diesel engine. One of the biodiesel raw material is sunan pecan seed . The populations of Sunan pecan seed are widely available in Indonesia, particularly in the area of Garut, Majalengka, West Java, Central Java, East Java, East Tenggara Nusa, West Tenggara Nusa, Riau, Jambi, Bangka, and East Kalimantan [11]. Currently, biodiesel of Sunan pecan seed was rarely used as an alternative fuel in diesel engines. Exceedingly, Sunan pecan seed biodiesel was potentially be used because they are not consumed by humans, therefore do not interfere with food needs. This experimental study aims to determine the performance of single-cylinder engines diesel which use Sunan pecan seed-diesel oil.
The parameters studied for engine performance were included of power, specific fuel consumption, thermal efficiency, effective mean pressure and exhaust gases produced.

IOP Conf. Series: Materials Science and Engineering **277** (2017) 012045 doi:10.1088/1757-899X/277/1/012045

2. Methods

Please follow these instructions as carefully as possible so all articles within a conference have the same style to the title page. This paragraph follows a section title so it should not be indented.

2.1. Parameter of Internal Combustion Engine

The performance of the internal combustion engine is specifically indicated by the value of the parameters of the engine. Some of these parameters can be described as follows [12]:

2.1.1. Power. The power is influenced by engine speed and torque that produced by the engine. In most cases, in the internal combustion engine, it is known two type of power that consist of shaft power and power indicator. As practice, it uses only the shaft power. The shaft power or effective power is the power that produced by an engine on the output shaft or it commonly known as brake horse power which is calculated by the equation:

$$\dot{W} = \frac{2\pi \,.\, N \,.\, \tau}{60000} \quad \text{kW} \tag{1}$$

where N is engine speed (rpm) and τ is engine torque (Nm).

2.1.2. Specific Fuel Consumption. The specific fuel consumption is the amount of fuel that should consumed per unit of power which produced per hour of operation. Indirectly, the specific fuel consumption is an indication of the engine efficiency to generating power from fuel combustion. The value of specific fuel consumption can be defined as follows:

$$Sfc = \frac{\dot{m}_f}{\dot{W}} \,.\, 3600000 \quad \text{gr/kWh} \tag{2}$$

where $\dot{m}_f$ is mass flow rate of fuel (kg/s).

2.1.3. Thermal Efficiency. The thermal efficiency of an internal combustion engine constituted the ratio between the output energy and the chemical energy of fuel, that can be stated as:

$$\eta_t = \frac{\dot{W}}{\dot{m}_f \,.\, Q_{HV} \,.\, \eta_c} \tag{3}$$

where Q_{HV} is calorivic value (kJ/kg) and η_c is combustion efficiency 0,97.

2.1.4. Mean Effective Pressure. Mean effective pressure is a good parameter to compare engines for design or output because it is independent of engine size and/or speed. The value of mean effective pressure can be defined by using equation:

$$mep = \left(\frac{4\,\pi\,\tau}{V_d}\right) x\ 1000 \quad \text{kPa} \tag{4}$$

where V_d is displacement volume (cm^3).

2.2. Material

The experiments were using a blends of diesel oil, the biodiesel of Sunan pecan seed with the composition of, B5 (diesel that contained 5 vol.% of biodiesel), B10 (diesel with 10 %vol.), B15 (15 vol.% of biodiesel) and B20 (contained 20% vol). The blended fuel is limited to B20 because higher biodiesel are not in common use.

2.3. Main Equipments

The main equipments used in this study consisted of:
1. Single cylinder stationary diesel engine
2. Calorimeter bomb to determine the calorific value of the fuel
3. Automotive emission analyzer to determine the composition of exhaust emissions

Figure 1. Single cylinder stationary diesel engine

Tabel 1. Engine specifications [13]

Type	ROBIN - FUJI DY23D
Valve position	Overhead
Valve rocker clearance	0,10 mm
Cylinder volume	230 cm^3
Bore	70 mm
Stroke	60 mm
Compression ratio	21
Number of cylinder	1
Maximum power	4.2 kW / 3750 rpm
Maximum torque	11.2 Nm / 3500 rpm
Injection time	23^0 BTDC

2.4. Scheme of Experimental

Figure 2 shows the scheme of experimental. The ROBIN FUJI DY23D engine that used in this research is the type of a four stroke one cylinder direct injection stationary diesel engine which specifications are summarized in Table 1. All experimental data were taken after the engine was properly warmed up 15 minutes after starting. An automotive emission analyzer HG-510 with accuracy of ±5% were used to measure the exhaust gases which placed on the exhaust gas manifold. For each fuel, different emission data were taken over three repetitions. The testing was under the constant engine loads of 4 kg at the engine speed variety from 1000 rpm, 1300 rpm, 1600 rpm, 1900 rpm, 2200 rpm, 2500 rpm and 2800 rpm. Loads were supplied by an eddy current dynamometer. Maximum torque 11.2 Nm at speed of 3500 rpm and maximum power 4.2 kW at 3750 rpm. The Accuracy of torque measurement is about ±0.17 Nm, and the acccuracy of rotation controlling ±10 rpm. The engine was started by using diesel fuel, where as it was warmed up, should switched to the biodiesel, the diesel blends (B5, B10, B15 and B20). After switching fuel from one type to another, engine was running about 10 minutes to get stable condition with the new fuel before the measurements were taken. When the load decreased, the engine rpm tends to increase, therefore the amount of fuel supply reduced by adjusting the fuel control screw to maintained the engine speed. Engine torque and fuel consumption were measured to calculate the power, specific fuel consumption and thermal efficiency of the engine. A 8 ml burette and a timer are used to measure the fuel consumption. The accuracy of the burette about ±0.1 ml and the timer about ±0.01 s.

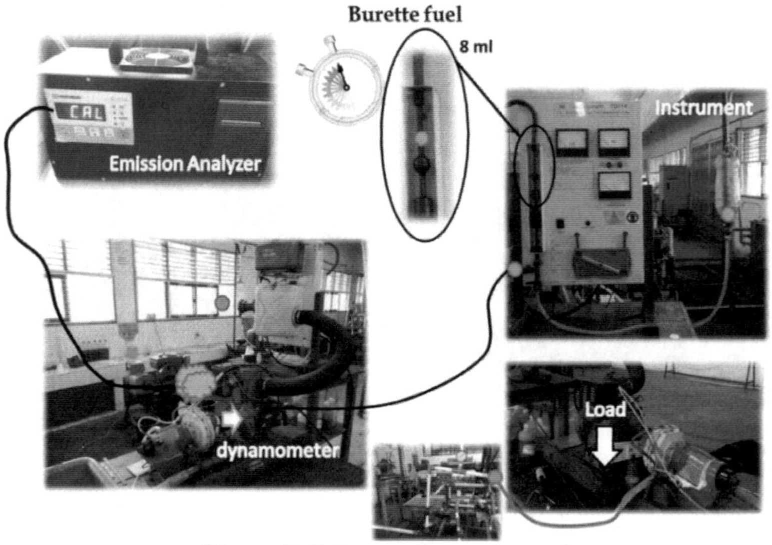

Figure 2. Scheme of experimental

3. Results and Discussions

3.1. The Calorivic Value of Fuels

The calorivic value of fuel was examined by using a bomb calorimeter apparatus with the accuracy of ±5%. To get the calorific value of diesel oil, B5, B10, B15 and B20, were the experiments carried out which each fuel as many as five sample. Then the average results of the five samples were to obtain the calorific value of the fuel. The Observation results obtained that the highest calorific value is in diesel oil that reached about 43142 kJ/kg and the lowest is B20 about 36819 kJ/kg. The calorific value of fuel shows the energy that generated during the combustion process of fuel per unit mass which is influenced by the composition of the fuel constituent. The higher the concentration of biodiesel in the blends, will decreased the calorific value of the fuel. Table 2 shows the calorivic value of fuels which obtained by using a calorimeter bomb.

Table 2. The calorivic value of fuels

Fuel	Calorivic Value (kJ/kg)
Diesel oil	43142
B5	41231
B10	40201
B15	39025
B20	36819

3.2. The Performance of Diesel Engine

3.2.1. Power. The experimental data in figure 3 shows that the maximum power occurs in 3.08 kW when the engine is using diesel oil at speed of 2800 rpm. The minimum engine power obtained 0.74 kW when engine using B20 at the speed of 1000 rpm.

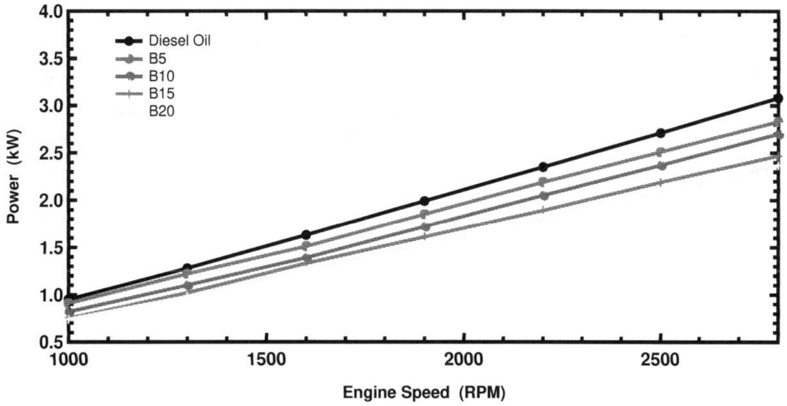

Figure 3. The engine power during experiments

The most influential of the power that generated by the engine shaft is torque. When the engine torque more greater, the engine power generated is also getting bigger and vice versa. This is due to the changing of engine power which is equally linier with engine torque. Another factor that allows the reduction in engine power is when the engine using B5, B10, B15 and B20 due to less precise injection timing settings, this is appropriated to the cetane number of the biodiesel. The cetane number of Sunan pecan seed is higher than diesel oil so that the required ignition delay is shorter than using diesel oil. Therefore, it is necessary to advance the injection timing in order to obtain the optimum power. As noted, during the experiments, no modification to the diesel engine used.

3.2.2. Specific Fuel Consumption. The experimental results in figure 4 shows that when the diesel engine using B20 at the speed of 1000 rpm , the maximum specific fuel consumption is obtained 439.26 g/kWh. The minimum value of specific fuel consumption when the engine using diesel oil at speed of 2800 rpm is obtained to 189.93 g/kWh. The average of specific fuel consumption during experiments is 284.38 g/kWh.

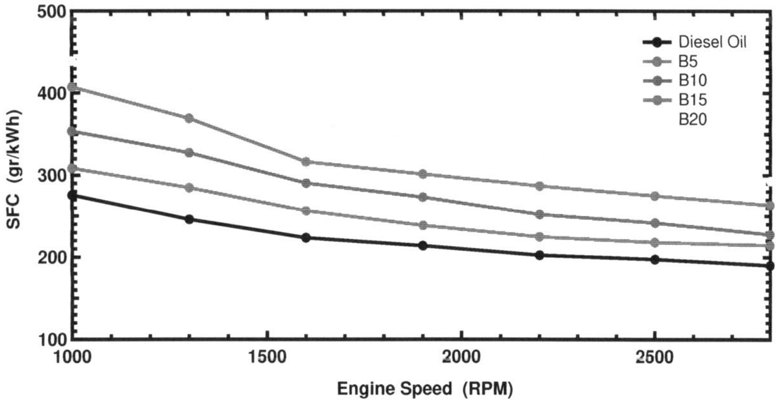

Figure 4. Specific fuel consumption during experiments

In general, the specific fuel consumption when the diesel engine using B5, B10, B15 and B20 increased. The calorific value of B5, B10, B15 and B20 were lower compared to the diesel oil because the presence of oxygen in biodiesel which result the air fuel ratio becomes a lean mixture. These conditions make the fuel becomes more needed than when using diesel oil. Higher specific fuel consumption Sunan pecan seed- can be attributed to the lower the heating values of these blends compared to diesel oil.

3.2.3. Thermal Efficiency. Figure 5 shows the value of maximum thermal efficiency 45.53% was reached by using diesel oil at the speed of 2800 rpm.

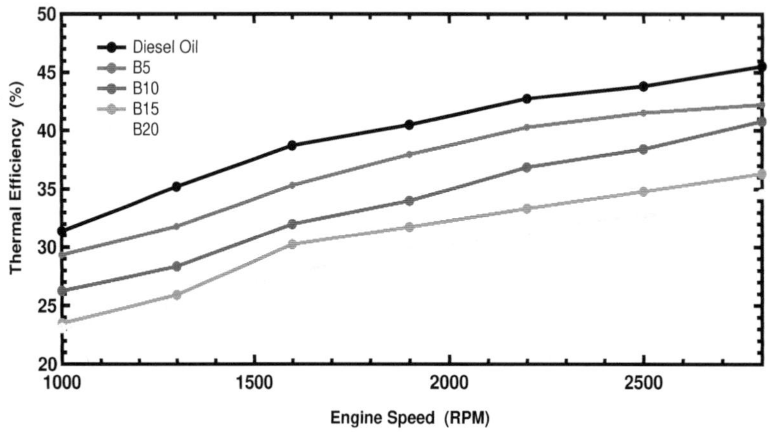

Figure 5. The thermal efficiency of engine during experiments

The value of minimum thermal efficiency is 23.07% by using B20 at speed of 1000 rpm. The average value of thermal efficiency resulting from the experiments conducted 34.08%. The decreasing of the thermal efficiency when the diesel engine is using blends fuel is about 9% - 35%. The average decreasing of the thermal efficiency occurs when the engine using blends fuel is 26.21%.
As noted, that the thermal efficiency of the internal combustion engine are influenced by several parameters such as engine power, specific fuel consumption and the calorific value of the fuel that be used. The three parameters simultaneously affect the achievement of the engine thermal efficiency.

3.2.4. Mean Effective Pressure. The experimental data shown the maximum mean effective pressure (mep) is 571.97 kPa which occur when the speed constantly 2800 rpm using diesel oil as shown in Figure 6. The minimum mean effective pressure is obtained 382.11 kPa by using B20 on the speed of 1000 rpm. As noted that the maximum mean effective pressure of good engine designs is well established, and is essentially constant over a wide range of engine sizes. The actual mean effective pressure is a particular engine develops that can be compared with this norm and the effectiveness with which the engine's displaced volume can be assessed. For design calculations, the engine displacement required to provide a given torque or power at a specified speed that can be estimated by assuming the appropriate values for mean effective pressure for that particular application [14].

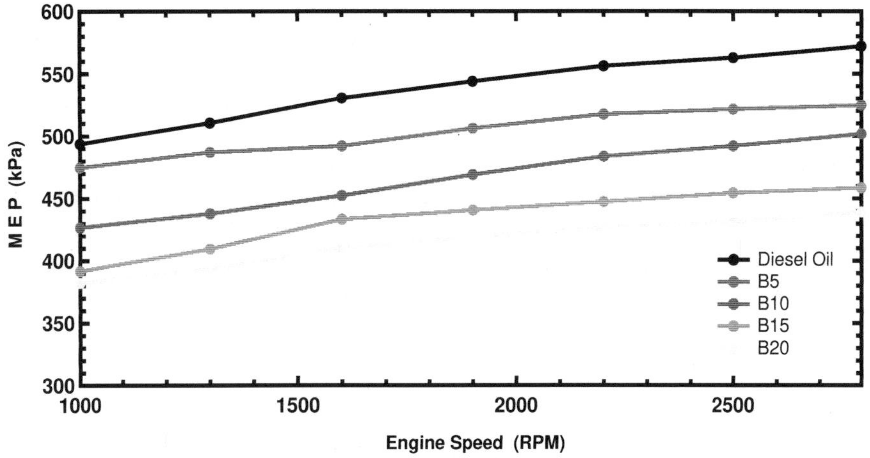

Figure 6. The mean effective pressure of engine during experiments

3.2.5. Emissions Measurement. In this study, the emissions studied were opacity, hydrocarbons (HC) and carbon monoxide (CO) by using automotive gas analyzer. The data measurement of exhaust gases can be seen in Table 3.

Table 3. The measurement result of exhaust gases

Load (kg)	Fuel	Opacity (%)	CO (%)	HC (ppm)
	Diesel oil	18.018	0.058	29.728
	B5	15.799	0.048	26.855
4	B10	15.078	0.048	21.878
	B15	14.448	0.038	19.264
	B20	13.818	0.038	16.909

Based on the results of the experiments when the engine is using B5, B10, B15 and B20, it were obtained the reduction in exhaust gases of carbon monoxide in the ranges of 13% - 27% for variations of speed. The exhaust gases of CO is caused by lack of oxygen produced, therefore the combustion process is not perfect because a lot of the carbon atoms are not getting enough oxygen to form carbon monoxide gas. The otherwise, this condition would make carbon dioxide emissions have increased. It has argued that biodiesel has an excess of oxygen atoms so that biodiesel is an oxygenate fuel that have the ability to bind a molecule of carbon monoxide to be the carbon dioxide. The reduction in exhaust gases of hydrocarbon ranges about 10% - 44% for all engine speed variation. The presence of hydrocarbon emissions is caused by unperfect of the combustion process. The fuels of B5, B10, B15 and B20 which have the OH bond in the molecular arrangement makes the combustion process of fuel in the combustion chamber is better so that the exhaust gases produced is becoming more environmentally friendly than diesel oil.

4. Conclusions

The performance and emission of a single cylinder direct injection CI engine fuelled with Sunan pecan seed and its blends have been observed and analyzed comparing to the base line of diesel oil. Experimental result analysis has shown that the specific fuel consumption increases as the increase of biodiesel in the blends fuel. On the other hand, CO and HC emissions can be reduced efficiently using Sunan pecan sees-diesel oil at speed.variation. HC emissions can be decreased significantly simultant with the increase of biodiesel in blends fuel under all engine speed. The experimental data show that the maximum power is obtained to 3,08 kW, specific fuel consumption 189.93 g/kWh and maximum thermal efficiency 45.53% when using diesel oil. The specific fuel consumption increases with the increasing of the percentage of Sunan pecan seed in the blends due to the lower calorific value of Sunan pecan seed. The amount of calorific value of a fuel is influenced by the composition of the constituent elements in it. The higher the concentration of biodiesel in the blends, the smaller calorific value of the fuel. The experimental results show that B20 has the smallest value of calorific value.

Acknowledgments
The authors acknowledge University of Sumatera Utara through TALENTA project 2017 for financial support to this research.

5. References

[1] Imdadul H. 2016 *Higher alcohol-biodiesel-diesel blends: An approach for improving the performance, emission, and combustion of a light-duty diesel engine*. Energy Conversion and Management. Vol. 111; pp. 174-185.
[2] Imdadul H, Masjuki H, Kalam MA, Zulkifli N, Rashed M and Rashedul H 2015 *A comprehensive review on the assessment of fuel additives effects on combustion behavior in CI engine fuelled with diesel biodiesel blends*. RSC Adv.
[3] Murari M. R., Wilson W and Justin Bujold 2013 *Biodiesel production and comparison of emissions of a DI diesel engine fueled by biodiesel-diesel and canola oil-diesel blends at high idling operations*. Applied Energy vol.106 pp. 198-208.
[4] Agarwal A.K 2007 *Biofuels (alcohols and biodiesel) applications as fuels for internal combustion engines*. Prog Energy Combust Sci. vol. 33 pp. 233-71.
[5] Demirbas A. 2007 *Progress and recent trends in biofuels*. Prog Energy Combust Sci.vol 33 pp. 1-18.
[6] Rashedul K.H, Masjuki H., Kalam A.M, Y.H. Teoh, H.G. How, I.M and Rizwanul Fattah 2015 *Effect of antioxidant on the oxidation stability and combustion-performance-emission characteristics of a diesel engine fueled with diesel-biodiesel blend*. Energy Conversion and Management. Vol. 106 pp. 849-858.
[7] Sajjad H, Masjuki H, Varman M, Kalam M, Arbab M and Imtenan S. 2014 *Comparative study of gas-to-liquid fuel, B5 diesel and their blends with respect to fuel properties, engine performance and exhaust emissions*. RSC Adv vol. 44 pp 452-459.
[8] Chih-Cheng Chou, Ying-Wei Lin, Chia-Jui Chiang and Yong-Yuan Ku 2014 *Experimental Analysis of a Turbo-Charged Common-Rail Diesel Engine Fueled with Biodiesel*. Energy Procedia. Vol 61 pp. 1167 - 1170.
[9] United States Environmental Protection Agency 2002 *A Comprehensive Analysis of Biodiesel Impacts on Exhaust Emissions*. EPA420-P-02-001.
[10] Mohd Hafizil Mat Yasina,Rizalman Mamata and Ahmad Fitri Yusop 2015 *Comparative study on biodiesel-methanol-diesel low proportion blends operating with a diesel engine*. Energy Procedia. Vol 75 pp. 10 - 16.
[11] Maman Herman, Muhammad Syakir, Dibyo Pranowo, Saefudin and Sumanto 2013 *Kemiri Sunan Tanaman Penghasil Minyak Nabati dan Konservasi Lahan*, ISBN : 978-602-1250-35-2, IAARD Press.
[12] Willard W. Pulkrabek 2004 *Engineering Fundamentals of The Internal Combustion Engine*. University of Wisconsin, Prentice Hall New Jersey.
[13] Manual Book 2000, *TD 110-115 Test Bed Insrumentation for Small Engines*. TQ Education and Trainning Ltd-Product Divition.
[14] John B. Heywood 1988 *Internal Combustion Engines Fundamentals*. Masschussets Institute of Technology.

Efficient distribution of toy products using ant colony optimization algorithm

S Hidayat[1,2] and C A Nurpraja[1]

[1]Industrial Enginering, Faculty of Science and Technology, University Al Azhar Indonesia. Komplek Masjid Agung Al Azhar, Jalan Sisingamangaraja, Kebayoran Baru, Jakarta Selatan 12110
[2]syarif_hidayat@uai.ac.id

Abstract. CV Atham Toys (CVAT) produces wooden toys and furniture, comprises 13 small and medium industries. CVAT always attempt to deliver customer orders on time but delivery costs are high. This is because of inadequate infrastructure such that delivery routes are long, car maintenance costs are high, while fuel subsidy by the government is still temporary. This study seeks to minimize the cost of product distribution based on the shortest route using one of five Ant Colony Optimization (ACO) algorithms to solve the Vehicle Routing Problem (VRP). This study concludes that the best of the five is the Ant Colony System (ACS) algorithm. The best route in 1st week gave a total distance of 124.11 km at a cost of Rp 66,703.75. The 2nd week route gave a total distance of 132.27 km at a cost of Rp 71,095.13. The 3rd week best route gave a total distance of 122.70 km with a cost of Rp 65,951.25. While the 4th week gave a total distance of 132.27 km at a cost of Rp 74,083.63. Prior to this study there was no effort to calculate these figures.
Keywords: Ant Colony Optimization, Vehicle Routing Problem, wooden toys, delivery route.

1. Introduction

CV. Atham Toys (CVAT) is a small and medium industry (IKM) company engaged in the production and distribution of wooden toys for children. The company was founded by Mr. Thamrin in 2009. The company operates in Ciledug, Tangerang area. In this study, the researchers compared two of the five types of Ant Colony Optimization (ACO) algorithms namely the Ant System (AS) and Ant Colony System (ACS) to determine the best algorithm in order to minimize the distribution costs in CVAT. Currently CVAT plans its delivery routes by determining the nearest travels intuitively, but the selected route is not optimal and the cost is high. With this study CVAT can plan the minimum cost delivery path. Figure 1 shows the spread of the distributors (numbered 1-13) for Jakarta and Tangerang areas, and the central warehouse (point 0) of CVAT. CVAT has more distributors in other provinces like West Java and Central Java. The main motive for this study is to find the best route to deliver the goods such that CVAT can minimize costs and use the saving for other more productive activities.
The following are assumptions and contraints used for this study:

1. The schedule is done only for distributors in Tangerang regional area,
2. Delivery activities are done within one business day,
3. One liter of gasoline can be used to travel as far as 12 km,
4. The types of toys for delivery are always available in the warehouse.

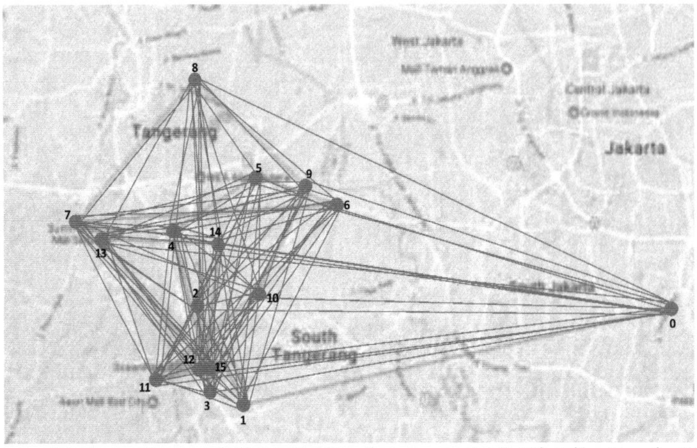

Figure 1. Geographic Locations of CVAT Warehouse and 13 Distributors in Tangerang area

2. Methods

The method used for this study is based on the mathematical formula of ACO.

2.1. Ant Colony Optimization (ACO)

ACO is inspired by the behavior of ants in determining the shortest path to finding and collecting food. The remainder of the pheromone left by the ants on each route is a communication medium for ants to exchange information [1]. ACO offers five types of algorithms: Ant System (AS), Elitist Ant System (EAS), Rank Based Ant System (ASRank), Min-Max Ant System (MMAS), and Ant Colony System (ACS). This study is conducted in the following steps:

2.1.1. Tour Construction

In this rule the ant will explore nodes that have never been visited. This is the probability of the m ants at point i going to point j or other.

$$P_{ij} = \begin{cases} \dfrac{[\tau_{ij}]^{\alpha}[\eta_{ij}]^{\beta}}{\sum_{h \in F_k(i)} [\tau_{ih}]^{\alpha}[\eta_{ih}]^{\beta}} & if\ h\ \in\ F_k(i) \\ \\ 0 \quad if\ h\ F_k(i)\ otherwise \\ \quad u\ \in F_k(r) \end{cases} \tag{1}$$

At this stage an ant is positioned at node i to select city j with an implementation rule called pseurandom proportional rule, as follows:

$$s = \begin{cases} \arg\max_{u\ \in F_k(r)} \{[\tau(i,h)].[\eta(i,h)]^{\beta}\}, \text{if } q \le q_0 \text{ (eksploitation)} \\ \\ S, \text{ otherwise (eksploration)} \end{cases} \tag{2}$$

where:

τ = *pheromone* value
η = Inverse distance d, worth of 1 / dij
F_k = A collection of towns to be visited by ants in the city i
α = Pheromone intensity control constant ($\alpha > 0$)
β = Visibility control constants ($\beta > 0$)
ρ = Pheromone evaporation coefficient ($0 \le \rho \le 1$)
q = Random numbers are uniformly distributed, ranging from 0 to 1
q_0 = Cumulative probability ($0 \le q_0 \le 1$)
j = Random variables are selected based on the Pij's opportunity distribution

When an ant in town i must choose a city j to pass by giving an example of random value $0 \leq q \leq 1$. If $q \leq q_0$, then the best side j will be selected (exploitation).
With the pseudorandom proportional rule, the ant will follow a route with larger pheromone. Global renewal takes place after all the ants have completed their journey [2],[3]. The pheromone level is updated according to following formula:

$$\tau_{ij}^{new} = \rho.\tau_{ij} + \sum_{k=1}^{n} \Delta\tau_{ij}^{k} \quad (3)$$

Where $\Delta\tau_{ij}$ is a change of pheromone value between nodes by ants based on the following equation:

$$\Delta\tau(i,j) = \begin{cases} Q/_{L_k}, & if\ (i,j) \in best\ global\ travel \\ 0, & otherwise \end{cases} \quad (4)$$

2.1.2. Mathematical Formula

The objective formula for obtaining the shortest distance can be written as follows:

$$FT = Z_{min} = \sum_{k=1}^{k} \sum_{i=0}^{V_i=1} \sum_{j=1}^{V_j=15} d_{ij}X_{ij}^{k} \quad (5)$$

The purpose of making this model to minimize the distance that must be taken by the vehicle when distributing.

Constraints:

$$\sum_{k=0}^{k} \sum_{j=1}^{V_j=15} X_{ij}^{k} = 1 \quad k = 2,3,4 \quad (6)$$

Constraint (6) To ensure that distributors are visited only 1 time.

$$\sum_{i=0}^{V_i=1} \left(q_i \sum_{j=1}^{V_j=15} X_{ij}^{k}\right) \leq W, \quad W = 24\ koli \quad (7)$$

Constraint (7) to ensure that the load of goods transported is less than equal to the maximum capacity of the vehicle.

$$\sum_{j=1}^{V_j=15} X_{0j}^{k} = 1 \quad (8)$$

Constraint (8) is a vehicle only departs once from the warehouse or depot.

$$\sum_{i=0}^{n+1}, X_i^k, (n+1), \quad (9)$$

Constraint (9) is the vehicle used also must end in the depot.

$$\sum_{i=0}^{V_i=1} X_{il}^{k} - \sum_{j=1}^{V_j=15} X_{lj}^{k} = 0 \quad (10)$$

Constraints (10) of the same vehicle will arrive and depart from the customers served.

$$q_j = 0 \rightarrow X_{ij}^{k} = 0 \quad (11)$$

Constraint (11) if the demand for goods in the city i does not exist or is zero then the city is not selected in the travel route.
Where:
Z = Minimum distance gained
k = The number of fleets of vehicles
i = 1, 2,.. 15
j = 1,2,...15

V_i = Vertex city *i*
V_j = Vertex city *j*
q_i = Demand for goods in *i*
X_{ij}^k = Has a value of 1 if a vehicle k runs from distributor i to distributor j, and has value 0 otherwise.
n = Number of cities / distributors
W = Maximum capacity of vehicle
d_{ij} = Distance from city i to city j

2.2. Flowchart of the method
Figure 2 display the flowchart of study.

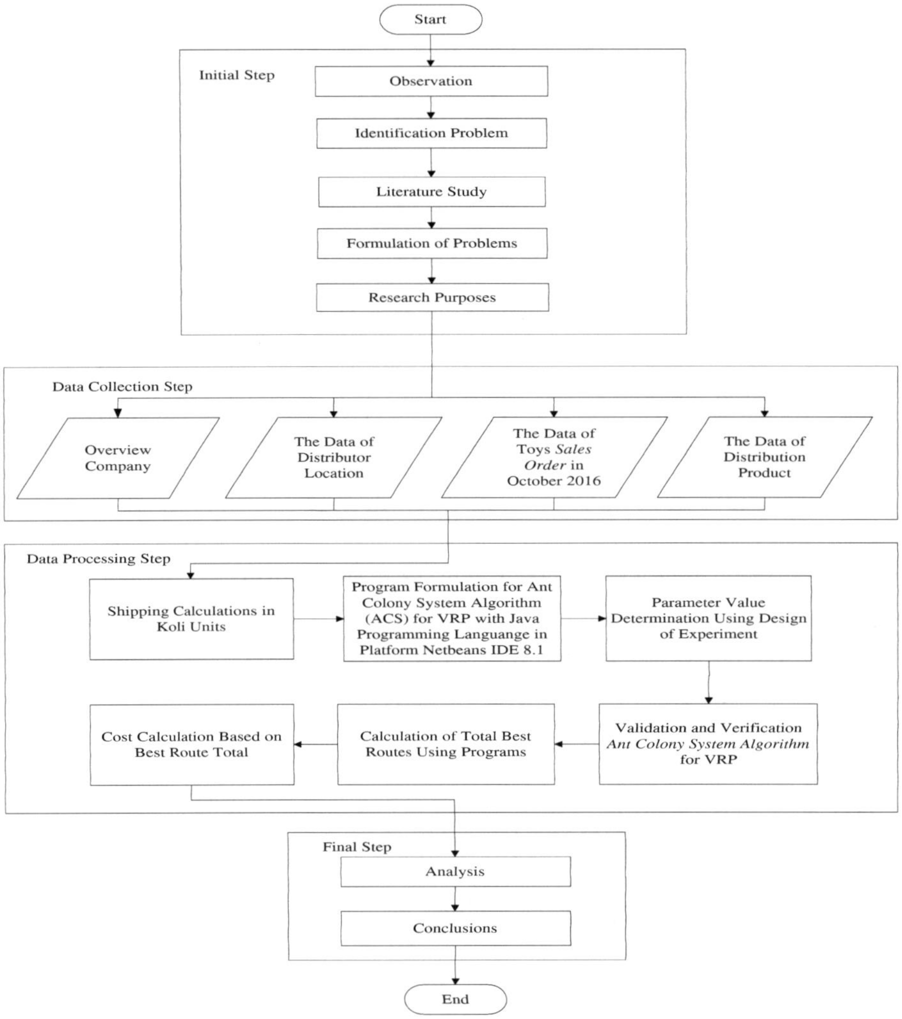

Figure 2. Flowchart of Study

The ACS program is created with the Java programming language using the Netbeans IDE 8.1 Platform. The following is a description of the ACS program.

Identification and determination of parameter values α, β, ρ, and m, the researchers used the Taguchi Design of Experiment (DOE) method [4].

3. Execution of experiment and discussion of results

3.1. Basic data from CVAT

Distribution operational activities of CVAT starts at 08.00 am until 17.00 pm. The fuel cost is Rp 6,500/liter. Vehicle capacity for 1 route of 24 carton consisting of 15 units of toys. Distribution cost in 1 day delivered by CVAT is Rp 150,000. Vehicles currently owned by CVAT is 1 unit of Grand Max Blind Van 1300cc car. CVAT may have to hire a second car.

The study was only conducted on 15 distributors in Tangerang area. Table 1 shows the coordinates of their locations. Table 2 shows the distributors orders in October 2016. Toys must be transported in units of cartons1 carton can take 15 units of toys. Number of less than 1 is rounded up. On the calculation of sending unit sales order CVAT October 2016 is divided into 4 weeks: Week1 (October 3-9), Week2 (October 10-14), Week3 (October 17-21), and Week4 (October 24-28). The table shows that the largest order is in week4 with a total order of 38 cartons or equal to 491 toy units.

Table 1. Coordinate Distributor CVAT in Tangerang Region

No	Distributor's Name	Coordinat	
		X (Easting)	Y(Northing)
0	Gudang CV. Atham	705.79963	9306.62807
1	Popplay BSD	686.029439	9302.64486
2	Popplay VMM	683.82541	9306.91772
3	Popplay GS	684.49476	9303.2274
4	Popplay AS	682.719023	9309.94893
5	Popplay Cipondoh	686.624012	9312.22912
6	Popplay Ciledug	690.451186	9311.06506
7	Popplay Kelapa Dua	678.20562	9310.3712
8	Popplay Taman Royal	683.811962	9316.33106
9	DS Toys	689.029877	9311.76542
10	Rosaria	686.70899	9307.35121
11	Jesicca	681.85277	9303.70999
12	Silvia	684.244129	9304.19808
13	Susan Prelix	679.498932	9309.67848
14	Toys Palace	684.921529	9309.43017
15	Trixie	684.250123	9304.20392

Table 2. CVAT Distributor Orders in Tangerang Region in October 2016

No	Distributor's name	Order size							
		Week1		Week2		Week3		Week4	
		Toys	Carton	Toys	Carton	Toys	Carton	Toys	Carton
1	Popplay BSD	17	2	20	2	16	2	20	2
2	Popplay VMM	28	2	35	3	20	2	21	2
3	Popplay GS	17	2	27	2	18	2	25	2
4	Popplay AS	30	2	20	2	20	2	30	2
5	Popplay Cipondoh	20	2	18	2	15	1	25	2
6	Popplay Ciledug	16	2	15	1	18	2	35	3
7	Popplay Kelapa Dua	35	3	21	2	11	1	28	2
8	Popplay Taman Royal	22	2	25	2	18	2	16	2
9	DS Toys	0	0	0	0	0	0	28	2
10	Rosaria	121	9	43	3	49	4	109	8
11	Jessica	36	3	78	6	70	5	53	4
12	Silvia	0	0	0	0	17	2	0	0
13	Susan Prelix	0	0	9	9	0	0	0	0
14	Toys Palace	0	0	0	0	0	0	101	7
15	Trixie	0	0	0	0	117	8	0	0
	TOTAL	342	29	311	34	389	33	491	38

Both tables 1 and 2 are used as inputs for the developed ACS-VRP program. Table 1 provided the distance matrix, while table 2 provided the demands for each distributor.

3.2. Data processing and calculations.

3.2.1. Program Formulation for Ant Colony System Algorithm (ACS) for VRP

Amongst Ant Colony Optimization (ACO) algorithm family, ACS algorithm is the more optimal algorithm compared to Ant System algorithm (AS). ACS algorithm can produce the best tour in faster time compared to AS. In addition, ACS was proven to be able to solve bigger problems (in terms of numbers of cities) compared to AS. The number of cities that ACS can test is ≥ 30 cities, while for the AS the number of cities that can be tested is <30 cities [5].

Based on calculations conducted in this study, the results of the calculation algorithm AS result of the smallest total distance is equal to 63.90 km. While using the total ACS algorithm the smallest distance obtained is 60.47 km. So it can be analyzed that on the same problem calculation by using ACS algorithm yield more optimal result compared to using AS. The formulation of an ACS program for solving VRP problems using Java programming languages can be analyzed that, the use of the program can facilitate the completion of the calculation. This is because the greater the number of distributors calculated then the time required for completion will be longer [2].

3.2.2. Parameter Value Determination Using Design of Experiment (DoE)

In this study 4 parameters are defined. α is the parameter for ant pheromone intensity trace controller, β is visibility control parameter, ρ is pheromone evaporation coefficient, and m is the number of ants. The combination of these factors can influence the determination of the best route and the total route followed by the ants. To facilitate testing, the researchers defined three levels of each factor, i.e. low, medium, and high. In order to get accurate results, the researcher conducted 9 experiments (Pn) and each combination was tested 3 times. Using the Minitab 16 software, Table 3 shows the combination of the four parameters.

Table 3. Combination of Parameters with Taguchi

Experiment	Level			α	β	ρ	m
(Pn)	1	2	3				
1	1	1	1	1	2	0.1	2
2	1	2	2	1	4	0.45	3
3	1	3	3	1	6	0.9	4
4	2	1	2	2	2	0.45	4
5	2	2	3	2	4	0.9	2
6	2	3	1	2	6	0.1	3
7	3	1	3	3	2	0.9	3
8	3	2	1	3	4	0.1	4
9	3	3	2	3	6	0.45	2

Figure 3 presents the output results of the Taguchi method in Table 3. The Minitab 16 program results initially as the number of known factors with the Factors description and the number of combinations of parameters represented by the description of Runs. Symbols P1, P2, etc are the number of experiments conducted by researchers. So the experiments performed have a signal to noise value (S/N) and means shown in Figure 3. Parameters with the smallest means of each factor are selected.

From the experimental results of determining the parameters for the VRP settlement, the parameters with the smallest means are obtained, i.e. the parameter levels are: α at level 1, β at level 3, ρ at level 1, and m at level 1.

To verify the correctness of the mathematical formulation, the researchers run the program using different combinations of parameters. This is shown in Table 4. With parameter values (1) of $\alpha = 1$, $\beta = 6$, $\rho = 0,1$; and $m = 2$, the total distance solution produced is 120.84 km. While the parameter values (2) of $\alpha = 3$, $\beta = 4$, $\rho = 0.9$; and $m = 4$, the total distance solution produced is 130.83 km. This is consistent with an experiment with the Taguchi method, wherein the higher value of the selected parameter using the Taguchi method produces the least total distance, i.e. alternative (1). α is the parameter for ant pheromone intensity trace controller = 1, β is visibility control parameter = 6, ρ is pheromone evaporation coefficient = 0.1, and m is the number of cars = 2.

Table 4. Parameter Combination Results

No	α	β	ρ	m	Total Distance (km)
1	1	6	0.1	2	120.837
2	2	2	0.45	3	127.799
3	3	4	0.9	4	130.825

```
———— 5/30/2017 9:36:12 PM ————

Welcome to Minitab, press F1 for help.

Results for: Worksheet 2

Taguchi Design

Taguchi Orthogonal Array Design
L9(3**4)

Factors:  4
Runs:     9

Columns of L9(3**4) Array

1 2 3 4

Taguchi Analysis: P1, P2, P3 versus Alpha, Beta, Rho, Semut

Response Table for Signal to Noise Ratios
Nominal is best (10*Log10(Ybar**2/s**2))

Level  Alpha   Beta    Rho  Semut
1      22.14  22.82  19.85  25.08
2      26.50  25.98  20.78  21.84
3      21.85  21.69  29.86  23.57
Delta   4.65   4.30  10.01   3.24
Rank       2      3      1      4

Response Table for Means

Level  Alpha   Beta    Rho  Semut
1      131.0  137.7  131.9  133.1
2      138.0  134.8  137.1  135.5
3      132.6  129.1  132.7  133.1
Delta    7.0    8.6    5.2    2.4
Rank       2      1      3      4
```

Figure 3. Output of Taguchi Method in Determining Parameter Experiment with Minitab 16

3.2.3. *Calculation of Total Best Routes and Cost Using Developed Program*

Table 5. Results Calculation of Total Best Routes Using Programs

Week	1st route		2nd route		Total distance (km)	Total Cost (Rp)	Efficiency (%)
	Route	Cost (Rp)	Route	Cost (Rp)			
1	0-9-6-5-14-4-13-7-12-15-1-3-11-2-8-0	46,171	0-10-0	20,533	124.11	66,704	56
2	0-1-3-12-15-2-14-4-13-7-9-5-6-0	36,878	0-8-10-11-0	34,217	132.27	71,095	53
3	0-6-9-5-14-4-13-7-8-2-15-12-3-1-0	39,458	0-10-11-0	26,493	122.7	65,951	56
4	0-5-9-6-10-14-4-2-1-0	34,830	0-13-7-8-11-15-0	39,254	137.83	74,084	51

From the interviews with resource persons, the current daily distribution cost incurred by the CVAT is Rp 150,000. Table 5 shows the result of route calculation in October 2016. The shortest route is indicated for week 3, with a total distance of 122.70 km, and efficiency of 56%.

4. Conclusions

Of the five types of ACO algorithms to solve CVAT problem, the researchers concluded that the Ant System (AS) algorithm is better than the other algorithms. Due to the selection process the next point

IOP Conf. Series: Materials Science and Engineering **277** (2017) 012046 doi:10.1088/1757-899X/277/1/012046

becomes easier based on pheromone updates on the edge which is the best tour. The number of ants used in the US and ACS should be more than one. ACS algorithm can solve more problems that is ≥ 30 cities.

The parameters designed in the ACO algorithm are the value of the pheromone instability control constant (α), the visibility control constant (β), the pheromone evaporation coefficient (ρ), and the number of ants (m). The found optimal parameters are $\alpha = 1$, $\beta = 3$, $\rho = 1$, and $m = 1$.

The optimal route path and minimum cost for CVAT for October 2016 are as follows. For the 1st week total distance of 124.11 km at a cost of Rp 66,703.70, and efficiency= 56%. At week 2nd the total distance is 132.27 km at a cost Rp Rp 71,095.13, and efficiency=53%. In the 3rd week the total distance is 122.70 km at a cost Rp 65,951.25, and efficiency=56%. At week 4th the total distance of 137.83 km with a cost of Rp 74,083.63, and 51% efficiency.

The optimum number of vehicles owned by CVAT to support its distribution activities is 2 four-wheeled vehicles with a capacity of 24 cartons.

Acknowledgement
The writer of this study acknowledges and is grateful that the fund for this study is subsidized by the Institution of Research and Public Service (Lembaga Penelitian dan Pengabdian Masyarakat = LP2M) of University Al Azhar Indonesia (UAI).

5. References

[1] Dorigo M and Stützle T 2010 *Ant Colony Optimization* (London: The MIT Press)
[2] Martua P Bangun, Komarudin M, Armand O and Hidayatno A 2011 Perancangan Algoritma Ant Colony Optimization (ACO) Untuk Penyelesaian Vehicle Routing Problem (VRP). *Prosiding Seminar nasional dan Workshop Pemodelan dan Perancangan Sistem* ISBN: 978 602 19492 0 7, p 251-260
[3] Rosyid M, Harun A, Suprajitno, H, Miswanto 2013 Algoritma Ant Colony Optimization (ACO) dengan Mutasi dan Local Search Untuk Menyelesaikan Vehicle Routing Problem. *Jurnal Matermatika* ISSN: 977 2337 310008, Volume 1 p 40-49
[4] Telaumbanua A, Siregar K, Sinaga T Sarma 2013 *Analisis Pengendalian Kualitas Dengan Pendekatan Metode Taguchi Pada PT. Asahan Crumb Rubber. E-Jurnal Teknik Industri FT USU* **3**, No 5 p 1-7
[5] Refianti R and Arnesia P Dewi 2009 Aplikasi Ant Colony System (ACS) Pada Travelling Salesman Problem *SNASTI 2009* ISBN: 978-979-8968-303 p 10-19

Implementation of fuzzy logic to determining selling price of products in a local corporate chain store

S P D Kristiana

Faculty of Industrial Engineering, Atma Jaya Catholic University of Indonesia, Jakarta 12930, Indonesia
stefani.kristiana90@gmail.com

Abstract. Corporate chain store is one type of retail industries companies that are developing growing rapidly in Indonesia. The competition between retail companies is very tight, so retailer companies should evaluate its performance continuously in order to survive. The selling price of products is one of the essential attributes and gets attention of many consumers where it's used to evaluate the performance of the industry. This research aimed to determine optimal selling price of product with considering cost factors, namely purchase price of the product from supplier, holding costs, and transportation costs. Fuzzy logic approach is used in data processing with MATLAB software. Fuzzy logic is selected to solve the problem because this method can consider complexities factors. The result is a model of determination of the optimal selling price by considering three cost factors as inputs in the model. Calculating MAPE and model prediction ability for some products are used as validation and verification where the average value is 0.0525 for MAPE and 94.75% for prediction ability. The conclusion is this model can predict the selling price of up to 94.75%, so it can be used as tools for the corporate chain store in particular to determine the optimal selling price for its products.
Keywords: Corporate chain store, Fuzzy Logic, Selling price

1. Introduction

Retail industry is one type of industries that are developing growing rapidly in Indonesia. Competition among retail industries is very tight and the competition does not only occur among modern retail industries but also modern with the traditional retail. One example of retail industry is corporate chain store. Corporate chain store is a network of two or more stores owned and controlled jointly who make centralized purchasing and trading and sells similar product lines [11]. Corporate chain stores as modern retail industry have to compete with many other retail industries not only local and national, but also international. Corporate chain store X is one type of local chain stores in Yogyakarta area, where it is always trying to develop itself in order to survive in competition with other retail industries. It should evaluate its performance continuously in order to compete with the competitors. One aspect that is used to evaluate the performance of retail is consumer preferences aspects which include human resources, merchandise, and price. Human resource associated with the services provided, merchandise related to the number of products available, diversity of products, as well as the diversity of brands sold, whereas the price associated with a low price [3]. In the previous study [6] only discusses about merchandise aspect, namely how to determine the optimal product allocation from the central warehouse to each branch so the products required by the customer can be provided. Price is a very essential attribute and gets the attention of many consumers. Consumer put "low price" in the first rank among the top ten attributes of service for all commodities in the traditional and modern markets

IOP Conf. Series: Materials Science and Engineering **277** (2017) 012047 doi:10.1088/1757-899X/277/1/012047

[3]. Every retail try to sell their products at low prices but still provide optimum benefits. Many factors affect the retailers to determine their selling price so the pricing of the product is a quite difficult decision [1]. Factors affect the determination of the selling price is divided into two factors, namely cost factors and non-cost factors [5]. This research will specifically discusses about the determination of selling price by considering factors related to cost, namely the purchase price of the product from supplier, holding costs, as well as transportation costs. Result of this study is expected to provide tool in determining the optimal selling price.

2. Methods

Study case of this research is a local chain store "X" in the region of Yogyakarta. Model is built using fuzzy inference system especially fuzzy mamdani with the help of MATLAB R2008a software. Fuzzy logic approach is selected to solve the problem because this method can consider complexities factors. It has the potential of combining costs into selling price decision making. The step of this research starts with work activity identification and data collecting at PT.X. Input for the model is holding cost, transportation cost, and purchase price from supplier. Calculating holding cost is done by considering the costs involved in the process of storage product, such as cost of electricity, building cost, salary of employees, property taxes, damage cost, order cost, etc. Whereas transportation cost was calculated by considering shipment distance, maintenance cost, and fuel price. Next step is determining membership function and fuzzy rules to illustrate the relationship between the purchase price, holding cost, transportation cost, and selling price. The model is built by make changes to the membership function curve to obtain a smooth graph, where the selected membership function was gauss2mf and gbellmf because based on previous research both provide more optimal result than other forms. There are 27 rules are used, namely:

1. If purchase price “L”, holding cost “L” and transportation cost “L” then selling price “L”
2. If purchase price “L”, holding cost “L” and transportation cost “M” then selling price “L”
3. If purchase price “L”, holding cost “L” and transportation cost “H” then selling price “M”
4. If purchase price “L”, holding cost “M” and transportation cost “L” then selling price “L”
5. If purchase price “L”, holding cost “M” and transportation cost “M” then selling price “M”
6. If purchase price “L”, holding cost “M” and transportation cost “H” then selling price “M”
7. If purchase price “L”, holding cost “H” and transportation cost “L” then selling price “M”
8. If purchase price “L”, holding cost “H” and transportation cost “M” then selling price “M”
9. If purchase price “L”, holding cost “H” and transportation cost “H” then selling price “M”
10. If purchase price “M”, holding cost “L” and transportation cost “L” then selling price “L”
11. If purchase price “M”, holding cost “L” and transportation cost “M” then selling price “M”
12. If purchase price “M”, holding cost “L” and transportation cost “H” then selling price “M”
13. If purchase price “M”, holding cost “M” and transportation cost “L” then selling price “M”
14. If purchase price “M”, holding cost “M” and transportation cost “M” then selling price “M”
15. If purchase price “M”, holding cost “M” and transportation cost “H” then selling price “H”
16. If purchase price “M”, holding cost “H” and transportation cost “L” then selling price “M”
17. If purchase price “M” holding cost “H” and transportation cost “M” then selling price “H”
18. If purchase price “M”, holding cost “H” and transportation cost “H” then selling price “H”
19. If purchase price “H”, holding cost “L” and transportation cost “L” then selling price “M”
20. If purchase price “H”, holding cost “L” and transportation cost “M” then selling price “M”
21. If purchase price “H”, holding cost “L” and transportation cost “H” then selling price “H”
22. If purchase price “H”, holding cost “M” and transportation cost “L” then selling price “M”
23. If purchase price “H”, holding cost “M” and transportation cost “M” then selling price “H”
24. If purchase price “H”, holding cost “M” and transportation cost “H” then selling price “H”
25. If purchase price “H”, holding cost “H” and transportation cost “L” then selling price “H”
26. If purchase price “H”, holding cost “H” and transportation cost “M” then selling price “H”
27. If purchase price “H”, holding cost “H” and transportation cost “H” then selling price “H”

** L= Low, M= Medium, H= High*

Selling price is output from this data processing. Validation model is done towards four products namely Maya Sardines Chili 155 g, Fortune Refill 1lt, Happytos Real Corn Chips/Red 170 g and Cap Lang MKP 30 ml.

3. Result and Discussion

Costs that were used as inputs to obtain the selling price were purchase price, holding cost, and transportation cost. Table 1 shows the result of costs calculation that were used to build the model.

Table 1. Data Input

Product	Purchase Price/pcs (Rp)	Holding Cost /pcs (Rp)	Transportation Cost /pcs (Rp)
Blue Poci Tea 40 g	1517	9.074	85.020
Tea Box Sosro 200 ml	1850	1.759	212.549
Red Happytos	6700	14.731	318.824
Prenagen Mommy Cocholate 200 g	32225	6.221	850.197
Honey Chil Kid 200 g	30475	7.165	850.197
Cap Lang MKP 60 ml	10396	9.330	87.951
Indocafe Coffeemix 20 g Sachet	883	0.612	9.109
Vanilla Poci Tea Bag 10*5*25 S	3450	2.893	221.791

That costs be normalized to the total sum of all costs, so the numbers were put into fuzzy input was the cost that has been normalized. Costs should be normalized to determine the applicable range for a characteristic variable in this design. If the range is too small or too large may have impact on the overall system performance. It is a temptation to have wide membership functions on the right or left to capture the extreme input values.

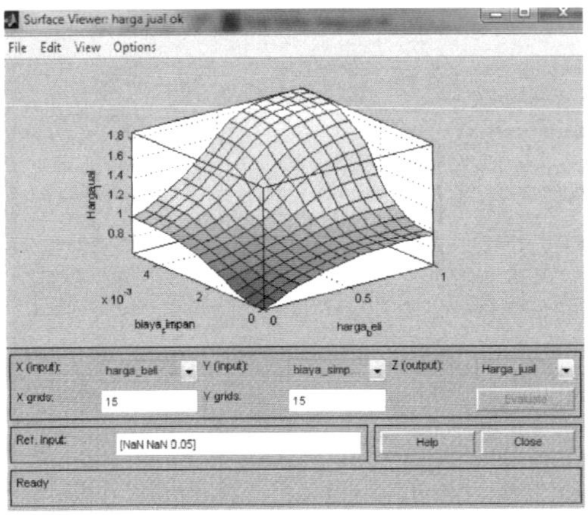

Figure 1. Relationship between purchase price, holding cost, and selling price

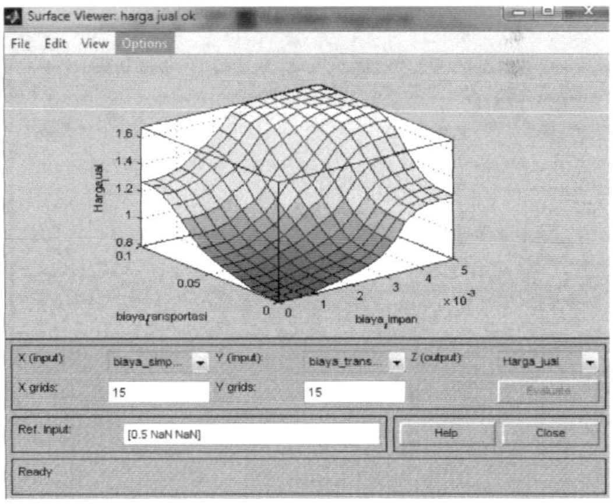

Figure 2. Relationship between transportation price, holding cost, and selling price

IOP Conf. Series: Materials Science and Engineering **277** (2017) 012047 doi:10.1088/1757-899X/277/1/012047

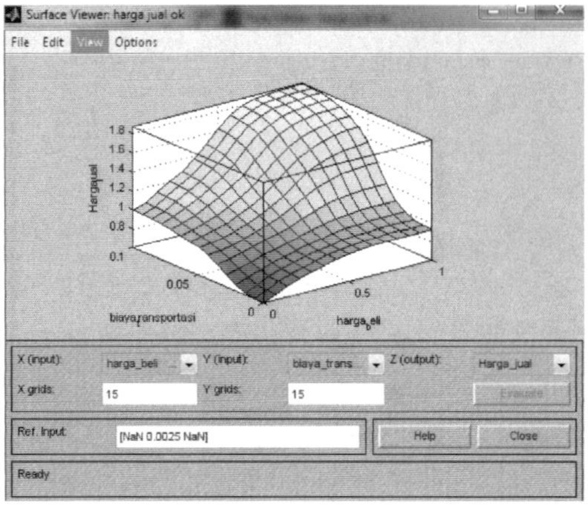

Figure 3. Relationship between transportation price, purchase price, and selling price

Changes to curves are intended to produce a better model seen from the graph, MAPE and predictive ability. The curve shows that if purchase price is high so the selling price is also high. Likewise also to holding cost and transportation cost, if the holding cost and transportation cost are high, value of the selling price is also high. Based on the rules that have been made and the limits of the universe of discourse namely value of variables are used, then obtained the optimal graph model that describes the relationship between the purchase price, holding cost, transportation cost, and selling price. They are shown in Figure 1, 2, and 3.

The model should be validated and verified. Validation is done by calculating the MAPE whereas verification is done by calculating predictive ability for some products that not include in process making the model. MAPE is calculated based on formula (1).

$$\text{MAPE} = \frac{\text{ABS}(\text{Y} - \text{Y}')}{\text{Y}} \tag{1}$$

Description:

MAPE = *Mean Absolute Percentage Error*

Y = selling price

Y' = selling price of fuzzy logic result

Predictive ability is obtained by formula (2).

$$\text{Predictive ability} = 1 - \text{MAPE} \tag{2}$$

Result of calculating MAPE and prediction ability of 4 products is shown in Figure 4 and Figure 5. Based on these calculations, the average value of predictive ability is 94.75%. This value is high enough, so this model can be used to determine selling price of product.

There are some differences in fuzzy logic prediction ability for every product. It is because the different pattern of sales the product, but the average ability of the model prediction have been built is high enough because fuzzy logic can combine the various possibilities in its Fuzzy Inference System (FIS), such as combination of curves, combination of rules, and change the curve shifts. However, this fuzzy model still has deficiency that because of the limits of universe of discourse is used, so that if the input value is entered outside from the universe of discourse, the result is not necessarily valid. In addition, the process to determine the rules and FIS in producing the best model takes a long time. So if the variable is used more and more, the time required to obtain the optimal model also will be longer.

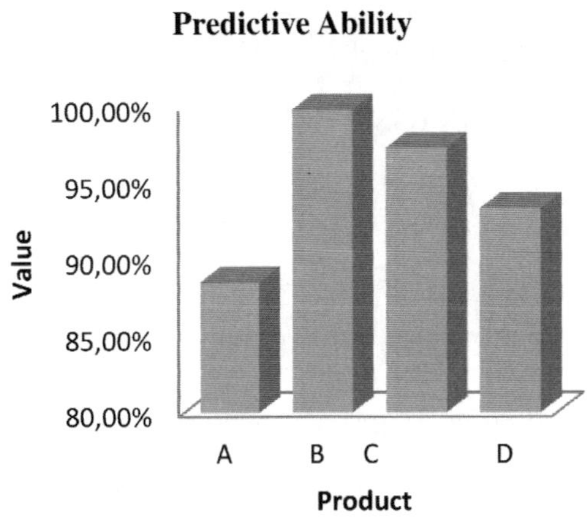

Figure 4. Value of predictive ability

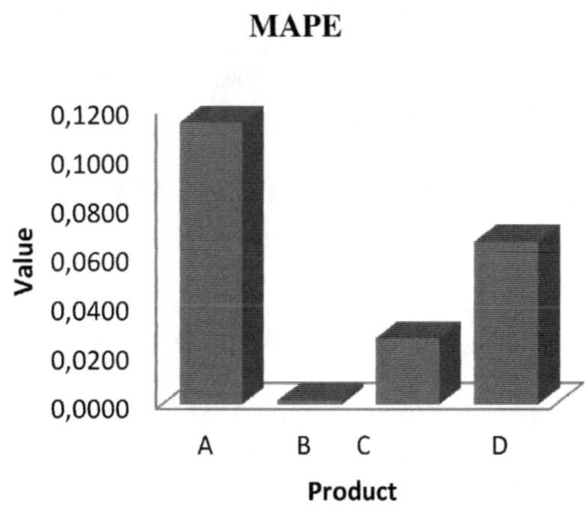

Figure 5. Value of MAPE

Description:
A = Happytos Real Corn Chips 170 g
B = Fortune Refill 1 l
C = Cap Lang MKP 30 ml
D = Maya Sardines Chili 155 g

4. Conclusions

Model of determining the selling price of product has been produced by using fuzzy logic method. Fuzzy model has three inputs namely purchase price, holding cost, transportation cost, and one output namely selling price. Each input and output has three fuzzy set. They are Low (L), Medium (M), and High (H). There are twenty seven rules that are used, they are combination of input and output fuzzy membership function. Model validation was done by calculating the MAPE and the ability of the model prediction, where the average value is 0.0525 and 94.75%. Therefore, fuzzy model that has been built can predict the selling price of up to 94.75%, so it can be used as tools for the corporate chain store in particular to determine the optimal selling price for its products.

5. References

[1] Arismunandar, I. R 2009 Programming and Simulation Price Determination of Light Food Industry Products Using Fuzzy Logic Approach by Paying Attention to Consumer Perceptions *Thesis* Gadjah Mada University, Yogyakarta, Indonesia
[2] Hapsari, Y. T and Sudiarso, A 2014 Determination of Pricing Product by Considering Production Cost and Intangible Factors with Fuzzy Logic Approach *Master's Thesis* Gadjah Mada University, Yogyakarta, Indonesia.
[3] Indef 2007 Economic Impact Assessment Hypermarket Presence on Retail/Traditional Market: Executive Summary. *Cooperation with Puslitbang Domestic Trade Ministry of Trade*
[4] Komariah, Ainur 2013 Development of Product Pricing Model by Considering Cost and Intangible Factors *Thesis* Gadjah Mada University, Yogyakarta, Indonesia

[5] Kotler, Philip 1996 *Marketing Manajement: Analysis, Planning, Implementation, and Control 6th Edition*. Jakarta: Erlangga

[6] Kristiana, S.P.D and Sudiarso, A 2014 Optimization Allocation of Product Distribution and Order Quantity with Fuzzy Logic Approach at Central Warehouse Local Chain Store *Master's Thesis* Gadjah Mada University, Yogyakarta, Indonesia

[7] Kurniawan, V. R. B 2012 Analysis of Pricing Product by Considering Intangible Factors *Thesis* Gadjah Mada University, Yogyakarta, Indonesia.

[8] Lestariningsih, Kusriniarti 2013 Determination Price of Artwork Product *Thesis* Gadjah Mada University, Yogyakarta, Indonesia

[9] Mulyono, Muhammad 2014 Implementation of Fuzzy Logic to Determination Price of Used Toyota Avanza 1,3 G M/T *Thesis* Computer Science Faculty, Dian Nuswantoro University, Indonesia

[10] Pradipta, Pitra A 2013 Determination of Selling Price of Used Smartphone by Applying Mamdani Fuzzy Logic Method *Thesis* Computer Science Faculty, Dian Nuswantoro University, Indonesia

[11] Saladin, Djaslim 2003 *Marketing Management: Analysis, Planning, Implementation, and Control 2nd Edition* Bandung: CV. Linda Karya

[12] Supriyono 2001 *Management Accounting 3: Management Control Process 1st Edition* Yogyakarta: STIE YKPN

[13] Widiyantoro, Andreas 2013 Application of Mamdani Fuzzy Logic to Determine Selling Price of Batik *Thesis* Computer Science Faculty, Dian Nuswantoro University, Indonesia

IOP Conf. Series: Materials Science and Engineering **277** (2017) 012048 doi:10.1088/1757-899X/277/1/012048

Optimization for routing vehicles of seafood product transportation

I A Soenandi[1],**Y Juan**[2], and **M Budi**[3]

[1,2,3]Department of Industrial Engineering Krida Wacana Christian University
Tanjung Duren Raya No. 4, West Jakarta 11470, Indonesia
[1] iwan.as@ukrida.ac.id

Abstract. Recently, increasing usage of marine products is creating new challenges for businesses of marine products in terms of transportation that used to carry the marine products like seafood to the main warehouse. This can be a problem if the carrier fleet is limited, and there are time constraints in terms of the freshness of the marine product. There are many ways to solve this problem, including the optimization of routing vehicles. In this study, this strategy is to implement in the marine product business in Indonesia with such an expected arrangement of the company to optimize routing problem in transportation with time and capacity windows. Until now, the company has not used the scientific method to manage the routing of their vehicle from warehouse to the location of marine products source. This study will solve a stochastic Vehicle Routing Problems (VRP) with time and capacity windows by using the comparison of six methods and looking the best results for the optimization, in this situation the company could choose the best method, in accordance with the existing condition. In this research, we compared the optimization with another method such as branch and bound, dynamic programming and Ant Colony Optimization (ACO). Finally, we get the best result after running ACO algorithm with existing travel time data. With ACO algorithm was able to reduce vehicle travel time by 3189.65 minutes, which is about 23% less than existing and based on consideration of the constraints of time within 2 days (including rest time for the driver) using 28 tons capacity of truck and the companies need two units of vehicles for transportation.
Keywords: Optimization, Indonesia Marine product, Stochastic Vehicle Routing Problem with Time Windows and Capacity Windows, Ant Colony Optimization.

1. Introduction

Based on a news article written by Deny in 2015[1], President Joko Widodo has goals to restore Indonesia's maritime heyday. It is becoming a reference for the various parties to participate in achieving goals. To achieve these goals has a lot of ideas that was blasted out. The idea of the toll of the sea, such as building improvements to regulation, legal affirmation, to attract investors to participate build a sector maritime Indonesia begins to look. This is evident with the start of the work of the project with a fantastic nominal [2].

Minister of marine and the fisheries from Indonesia, Susi Pudjiastuti has several times issued its policy regarding fishing and fish were able to reduce the theft of up to 50 percent it will increase the productivity of the fishermen and this situation needed the transportation to send the products to customer such as fresh market. In fact, the catch increased indirect impact on improving the welfare of fishermen. This can occur if the selling price of the catch from the fishermen to the dealer/company is still low, while the selling price of the catch to consumers. Increased costs occurring during the process of distribution of the catches. As expressed by the Chairman of the Association of Logistics and Forwarders Indonesia (ALFI), the increase in food prices caused by the high cost of distribution. Increasing in price that is too high can also affect the purchasing power of consumers will be the products of the sea. An increase in the harvest of the sea also spawned new obstacles for the company engaged in the export of seafood products. The company may experience constraints in terms of transporting the catch. This can happen if the constraints almost all locations experienced a harvest [3], but the transport resources owned by a limited company.

The easiest solution that can be offered is the purchase of a vehicle to transport as well as the recruitment of drivers. However, this requires more costs [4]. More costs are for the purchase of the vehicle and the driver's salary (even though the vehicle is not used if the sea after harvest). This is the problem that we want to solve in this research. During this time the company has not used the scientific method in determining the route of the transport vehicle. The company still uses the conventional way in determining the route of the vehicle transport, namely by looking at the distance on the map. As the vehicle routing is a multi-objective decision that not simple to solve [5]. Thus, companies require finding the optimum route results with methods that can be used to complete the VRP based research from Golden in 2008 [6]. In this research, to solve this VRP problem [7] for the seafood products, we want to compare the optimization with several algorithms as previous research with time uncertainty [8] and stochastic demands [9]. In this research, we compared the optimization of this VRP problem in seafood products on another metaheuristic methods such as Ant Colony Optimization [10], compared to methods modeling in WinQSB software as Branch and Bound, Dynamic Programming, Two-Way Exchange Improvement, Cheapest Insertion, Nearest Neighbor[11] to find which one could get the best result for optimization.

2. Research methodology

The stages of research briefings can be seen in figure 1.

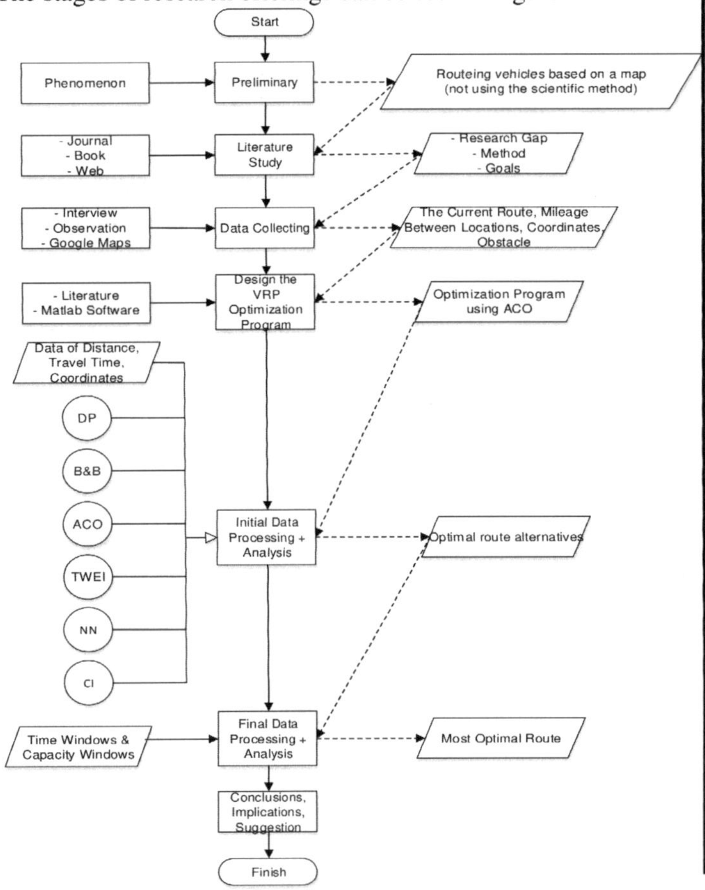

Figure 1. Research methodology

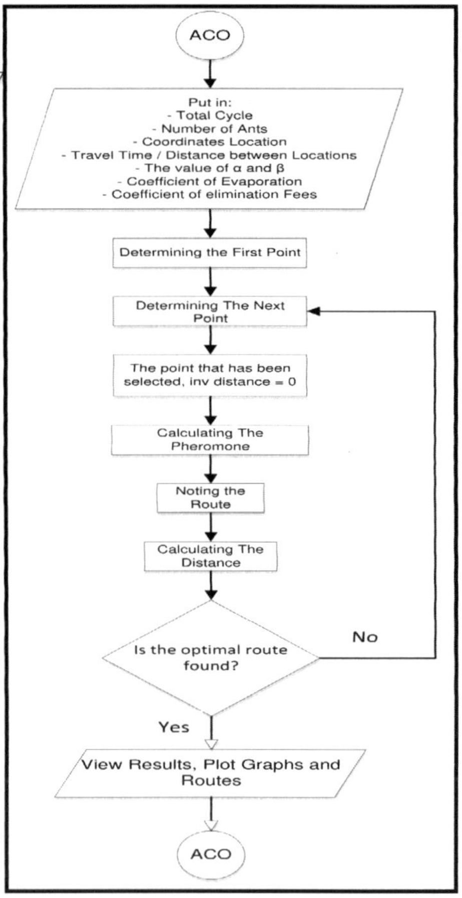

Figure 2. ACO Algorithm used in MATLAB

IOP Conf. Series: Materials Science and Engineering **277** (2017) 012048 doi:10.1088/1757-899X/277/1/012048

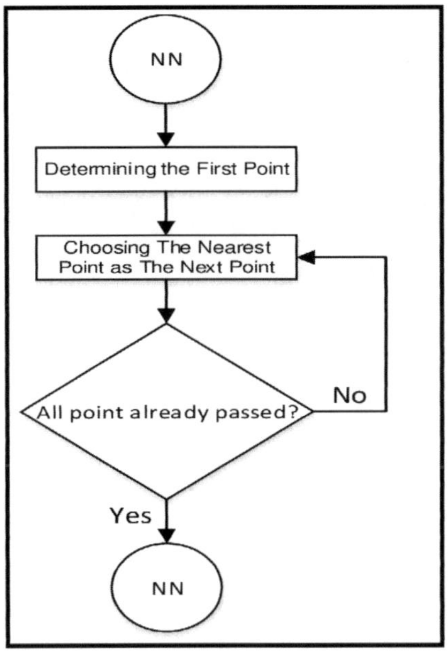

Figure 3. Nearest Neighbor Algorithm used in Software WinQSB

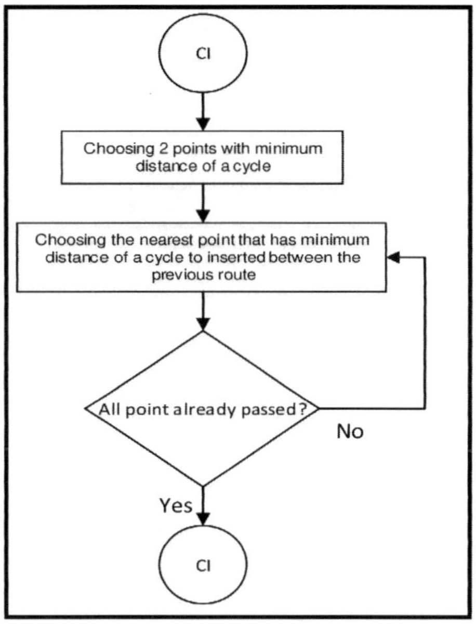

Figure 4. Cheapest Insertion Algorithm used in WinQSB

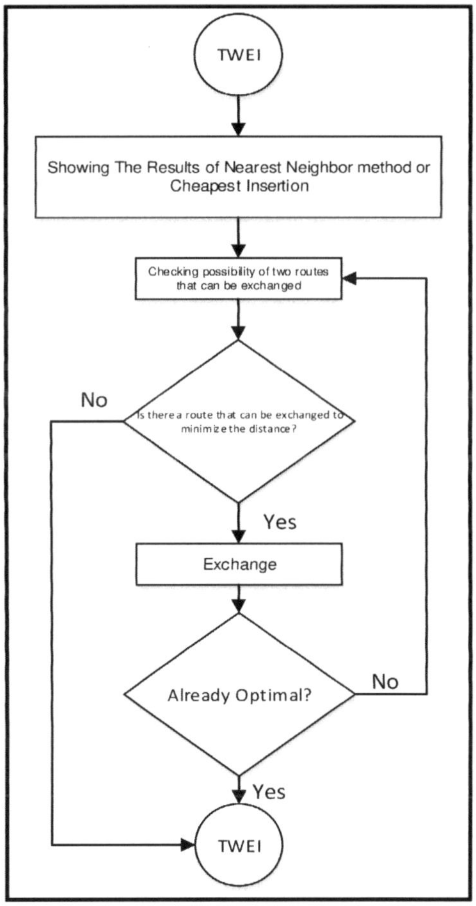

Figure 5. Two-Way Exchange improvement used in WinQSB

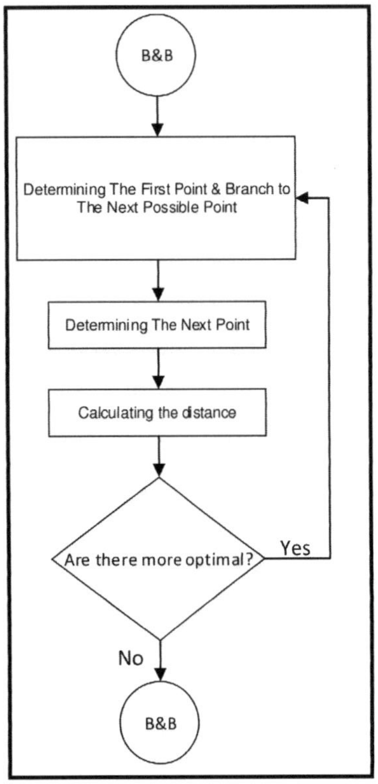

Figure 6. Branch and bound algorithm used in WinQSB

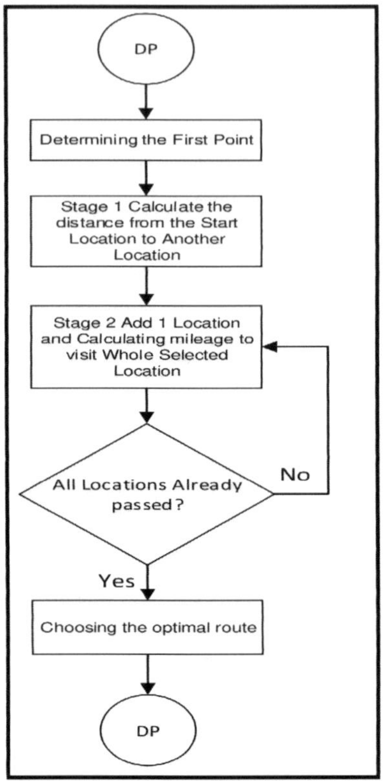

Figure 7. Dynamic programming algorithm

A flow diagram can be seen on the research methodology starting from the discovery of the phenomenon, to getting the results and conclusions of the research. For the step of the algorithm we described in figure 2 as the flow diagram of ACO algorithms according to our reference[12] that we run the algorithm using MATLAB data processing with licensed version 2013b from Mathworks. In figure 3 until figure 6 we showed the four algorithm flowchart shows the methods contained in the software i.e. WinQSB as Nearest Neighbor, Cheapest Insertion, Two-Way Exchange Improvement, and Branch and Bound as a comparison of ACO algorithm. Also for comparing the result of optimization in figure 7, we showed the algorithm of the Dynamic Programming using analytic method.

3. Data collection

The object of the research is a national company exporter of marine products such as fish, prawn, and seaweed was located at West Nusa Tenggara. We have collected data from January 2015 till March 2015 such as the location of source marine products as seaports or fishing village, the travel time between location and capacity of the vehicle are obtained from the interview with the company. The location of the selected capture is not the entirety of the location of the fishing area in Indonesia but represented by area that was rated to do research on the determination of transport vehicle lines. The area is selected with consideration of the following:

1) The distance between locations is not the same. The distance between the location of more or less the same, even the same votes is not too important to do the research because the route of the vehicle can smother them or pick up the catch is sorted from most locations do end up to the most approaching the main warehouse (location 1) as described in figure 8.

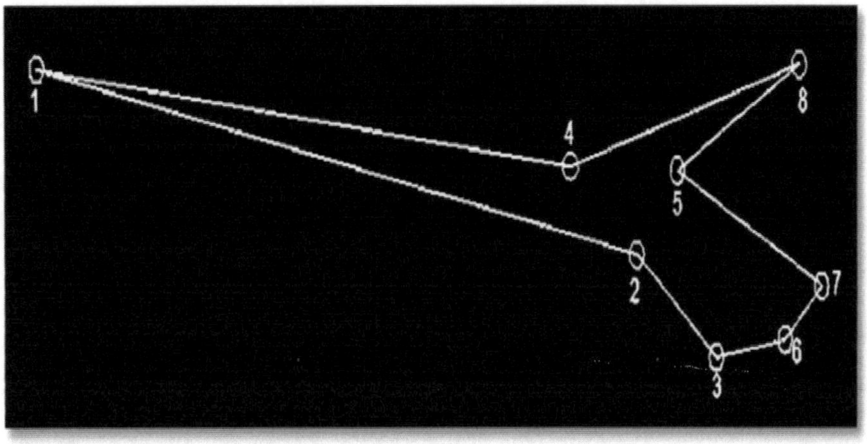

Figure 8. The current route

2) The region is one of the regions with a total harvest per day, more and the number of the location of the fishing area more than other regions. Location data obtained, in the plot into the map coordinates. The coordinates of a location can be seen in table 1. The first location was the main warehouse, whereas the other location is the location of the source of marine products.

Table 1. The coordinates of each location

Coordinates / Location	X	Y	Capacity (tons/day)
1 (Warehouse)	1,7	6,9	-
2	23,4	2,7	5
3	26,8	0,4	10
4	20,6	4,7	6
5	25,1	4,6	4
6	29,7	0,8	5
7	31,3	2	4
8	30,3	7	6

The distance between the locations as we get from Google Maps and shown in table 2 in units of kilometers.

Table 2. The distance between locations (kilometers)

Distance	1	2	3	4	5	6	7	8
1	0	1551.8	1576.4	1538.9	1564.64	1601	1612	1708
2	1551.8	0	40.33	34.35	15.6	62.36	72.32	165.54
3	1576.4	40.33	0	57.43	24.03	22.03	31.99	125.21
4	1538.9	34.35	57.43	0	20.94	45.74	54	48.4
5	1564.64	15.6	24.03	20.94	0	33.42	33.01	32.46
6	1601	62.36	22.03	45.74	33.42	0	9.96	103.18
7	1612	72.32	31.99	54	33.01	9.96	0	93.22
8	1708	165.54	125.21	48.4	32.46	103.18	93.22	0

Travel time between locations which are obtained from the results of the interview is shown in table 3. in units of minutes. In fact, it takes is not always the same in each time travel, but the travel history recorded by the company showed that the difference between each trip not too large. Thus to simplify this, we used average travel time to complete the matrix.

Table 3. Travel time between locations (minutes)

Travel time	1	2	3	4	5	6	7	8
1	0	1446	1448	1445	1559	1450	1451	1464
2	1446	0	90	180	90	120	180	240
3	1448	90	0	45	35	20	30	75
4	1445	180	45	0	20	40	20	19.65
5	1559	90	35	20	0	30	40	10
6	1450	120	20	40	30	0	25	300
7	1451	180	30	20	40	25	0	240
8	1464	240	75	19.65	10	300	240	0

4. Results and discussion

From this research, we obtained the three approach of positioning the search space has several advantages and disadvantages. The advantage of the optimization based on the Cartesian coordinate is can be the route that does not tangle. The shortcoming of this optimization is to ignore the condition of the road should turn/rotate, the difficulty of the terrain, and traffic congestion. This condition resulted in the route gained can require relatively high travel time. Approach distance with coordinates should not be done in the process of optimization of the VRP as shown in table 4 with the capacity of vehicle set for 28 tons. That approach distance was assessed quite reasonable because will consider aspects of the terrain of the road but has a weakness that is ignoring the aspect of time.

Table 4. Comparison of the optimization result without constraint

The optimization results	Approach		
	Coordinate	Distance	Time
Distance(km)	3309.64	3276.89	3323.62
Travel time (minutes)	29357	29182	3189.65

Approach distance was judged well done if the condition of the distance with the travel time is directly proportional. The short distance that would not necessarily require less travel time compared to long distances. Therefore, the best approach is to consider the travel time that figured the situation of traffic, rest time for the driver and loading time. As additional reason, the approach it takes is rated very fit with the case because of a few things. First, the route gained will absolutely optimal. This can be seen by comparing the time it takes between the three approaches. In addition, the product, in this case, products that get from the sea was required special handling such as cooling and will quickly damage if not quickly deposited in special warehouses. One of the obstacles faced is time constraints. With the same unit, will facilitate the optimization is done.

Table 5. Result Comparison of Six Method that Used in This Study

Methods	Branch and bound	Dynamic programming (Analytic)	Ant Colony (Metaheuristic)	Two-way exchange improvement (Heuristic)	Cheapest insertion (Heuristic)	Nearest neighbor (Heuristic)
Travel time (minutes)	3189.65	3189.65	3189.65	3189.65	3380.65	3293.65

The optimize route obtained a results with data in the form of distance in units of length are the same as the route that carried the company at this time. Through research with data usage and ACO algorithms in the form of a time that has been done at this stage, the travel time, travel can be pressed up to about 23% of the time it takes the current (routes obtained from a distance on a map). In this research we run the algorithm using various alpha value and beta value for testing the optimization result. In the ACO algorithm, the alpha value is the rank value for the ant's sight. The value of the ant's sight is describing the ability of ants to see the trace or pheromone and the beta value is the value of rank for tracers effect, this phenomena was used to find the optimization. By using this algorithm the optimum travel time can be seen in table 5. It has indicated a reduced time of 4147 minutes into 3189.65 minutes. This result has more efficient in time and supported the factor of capacity and time windows as objective in this transportation system of marine products.

The value of the trace's effect describes the ability of ant leaves a trace or pheromone. In accordance with the logic of the algorithm, if the value of alpha and beta to be promoted, should affect the most optimal results quickly found. It also fits with the findings of the authors during the process of research. Comparison of method by using the WinQSB software optimization with calculation and Dynamic Programming is used on the condition without any constraints of time and capacity. Optimization results by using the methods of the ACO to be compared with the results of the four methods that are there are WinQSB software and methods on Dynamic Programming. As said by our literature [13], an analytic method is able to get the most optimal though takes a long computation but based on comparisons of methods of testing, which is done, do not cover the possibility of other methods (heuristic and metaheuristic) can find the optimum results and speed in obtaining the results of each method, the analytic method takes more time needed.

Another case with a heuristic method result is not always a global optimum [14]. The theory is evident from the length of time to run the program. The pace looks not so different when the program starts, but if we study each software used will we find differences that affect the speed of the process. The difference is the ability in the core utilization on the processor of the computer. In this research by using WinQSB utilized the entire core of processor, which is different with MATLAB that are only utilized one core to compute the algorithm. In practice, we compared WinQSB and MATLAB is installed on a computer with a number of four-processor core.

Application of stochastic on this issue, not very influential on the value of optimization, it is apparent from the discovery global optimal route at the output of the program MATLAB to ACO. This can occur because the algorithm is designed to obtain routes intact (the beginning and end of the route at one point). If the identification is done more in taking into account the time constraints and capacity, then schedule the assignment of optimal transport vehicles can be obtained.

Table 6. Optimal Route Generated by Six Methods

Methods	Branch and bound	Dynamic programming (Analytic)	Ant Colony (Metaheuristic)	Two-way exchange improvement (Heuristic)	Cheapest insertion (Heuristic)	Nearest neighbor (Heuristic)
Travel time (minutes)	3189.65	3189.65	3189.65	3189.65	3380.65	3293.65
Optimal route	1-2-3-6-7-4-8-5-1 1-2-3-7-6-4-8-5-1 1-5-8-4-7-6-3-2-1 1-5-8-4-6-7-3-2-1	1-2-3-6-7-4-8-5-1 1-2-3-7-6-4-8-5-1 1-5-8-4-7-6-3-2-1 1-5-8-4-6-7-3-2-1	1-2-3-6-7-4-8-5-1 1-2-3-7-6-4-8-5-1 1-5-8-4-7-6-3-2-1 1-5-8-4-6-7-3-2-1	1-2-3-7-6-4-8-5-1	1-5-7-6-2-3-8-4-1	1-5-4-6-3-7-2-8-1

Companies are advised to use analytical methods or methods of metaheuristic optimization results because its global optimum value found as the result shown in table 6. Based on our computational result, we do not recommend the company to use heuristic methods. In addition to the results of the analytical method and optimization of the metaheuristic better than the heuristic methods.

5. Conclusion

In this research, we have considered the seafood product delivery routing problem in Indonesia as transportation with time and capacity windows and we confirmed that some algorithms are proved to solve these problems effectively and at reasonably running times. In accordance with the research objectives, the results of the research can be listed as follows: The company has 4 alternative routes with a total minimum travel time i.e. 3189.65 minutes and with the constraints of time within 2 days and 28 tons capacity for each vehicle we get solution that needed at least two trucks as route 1-2-3-6-7-1 and route 1-4-8-5-1. And for the optimum alternative, the first alternative vehicle utility level is very good because every vehicle transporting corresponds to minimum time and maximum capacity.

From the results calculated by using the method of the ACO, the presence of stochastic, not too influential in the global optimum value, this is shown by the discovery of the remains of a global optimum. Stochastic has had an impact on the priority of the location in terms of transport. Changes in priorities will certainly have an impact on the issue of optimization results are not optimal based on travel time. Next, by doing experimental computation from six methods. Therefore, the proposed method such as ACO can be considered as an effective method for the seafood product delivery routing problem.

We hope this presented research can offer another researcher different insights into the issue of VRP, and promote the exploration of new algorithm to solve the VRP problem in more efficient computational time needed.

6. References

[1] Deny S 2015 *Ambisi Jokowi Kembalikan Kejayaan Maritim Nusantara* from http://bisnis.liputan6.com/read/2293616/ambisi-jokowi-kembalikan-kejayaan-maritim-nusantara. (Accessed on March 20th, 2015)
[2] John B 2017 Indonesian Maritime Strategy in the 21st Century http://asiafoundation.org/video/indonesian-maritime-strategy-21st-century/. (Accessed on January 15th, 2017)
[3] Kang M H, Hyung R C, Hyun S K and Byung J P 2012 Development of a maritime transportation planning support system for car carriers based on genetic algorithm. *Appl Intell* **36** p 585
[4] Alt B and Ulrich W 2011 *A stochastic multiple areas approach for public transport network design*. Springer – Verlag
[5] Ibaraki T, Imahori S and Dubo M 2005 Effective Local Search Algorithms for Routing and Scheduling Problems with General Time-Window Constraints. *Transportation Science* **39**(2), 206-232
[6] Golden B, Raghavan S, Wasil A and Edward 2008 *The Vehicle Routing Problem: Latest Advances and New Challenges*. Springer
[7] Kumar S N and Panneerselvam R 2012 A Survey on the Vehicle Routing Problem and Its Variants. *Intelligent Information Management* **4**, 66
[8] Zhang J, William H K and Bi Y C 2013 A Stochastic Vehicle Routing Problem with Travel Time Uncertainty: Trade-Off Between Cost and Customer Service. *Netw Spat Econ* **13**, 471
[9] Juan A A, Jorba J, Caceres J and Marquès J M 2011 Using parallel & distributed computing for real-time solving of vehicle routing problems with stochastic demands. *Ann Oper Res* **207,** 43.
[10] Benslimane M T, Mohammed T and Youssef B 2013 Ant colony algorithm for the multi-depot vehicle routing problem in large quantities by a heterogeneous fleet of vehicles. *INFOR* **51,** 31
[11] Reger B 2014 *Using WinQSB modeling software for logistics course: 1st part theory*, http://portal.zmne.hu/download/bjkmk/bsz/bszemle2010/3/03.pdf. (Accessed on March 22nd, 2015)
[12] Dorigo M and Blum C 2005 Ant colony optimization theory: a survey. *Theoret Comput Sci* **344,** 243
[13] Marti R and Reinelt G 2011 *The Linear Ordering Problem, Exact and Heuristics Methods in Combinatorial Optimization*. Springer-Verlag Berlin Heidelberg
[14] Bouthillier A and Crainic T G 2005 A cooperative parallel meta-heuristic for the vehicle routing problem with time windows. *Computers and Operations Research* **32**(7), 1685-1708

IOP Conf. Series: Materials Science and Engineering **277** (2017) 012049 doi:10.1088/1757-899X/277/1/012049

An urban transportation performance model with spatial structure transit

A. Rachmat
Faculty of Engineering, Atma Jaya Catholic University of Indonesia, Indonesia
Leand.agatha@gmail.com

Abstract. This research has developed a mathematical model to measure an urban transportation system performance which includes spatial structure. The result shows this method of measuring urban transportation system is also able to be used to project the next step transportation system by changing its variable. To be able to represent each city and its support areas a further research are to be made, including variable related to service.
Keyword: urban transportation, mathematical model, system performance.

1. Introduction

In developing country, urban transportation is crucial part of the solution to the nation's economic, energy, and environmental challenges. Challenges in urban transportation such as traffic congestion, longer commuting time, public transport inadequacy and difficulties for non-motorized transport often are to be tackled to provide better transportation mode option. These challenges are linked to each other which resulted with added complexity within urban transportation system. Urban transportation is highly complex because of the modes involved, the multitude of origins and destinations, and the amount and variety of traffic. Traditionally, the focus of urban transportation has been on passengers as cities were viewed as locations of utmost human interactions with intricate traffic patterns linked to commuting, commercial transactions and leisure/cultural activities. Conceptually, the urban transport system is intricately linked with urban form and spatial structure. In most developing cities, a similar structure is found, which it shows most cities are supported by satellite city or suburban on the outer layer to support millions of city residents.

Urban transportation's challenges such as traffic congestion, longer commuting time, public transport inadequacy and difficulties for non-motorized transport often are to be tackled to provide transportation mode option with higher quality. These challenges are linked to each other which resulted with added complexity within urban transportation system. In order to tackle those challenges and improve the performance of the system a research to build a performance model is needed. In developing country such as Indonesia, fare consistency and on-time performance are not being considered seriously and is suggested to provide more than one general level of transportation services[6]. Meanwhile in urban transportation, the attribute on-time performance and safety at stops are considered influential[5].

In previous research on Transit-oriented development (TOD) and multimodal transportation, it is suggested to use TOD as an effective approach, as key nodes provide accessibility of integrated multi-modal public transportation system. These key nodes or centres are to reduce reliance on automobile (P Mess, 2015). Most of previous research focus on statistical model to understand the perceived image of public transportation performance which resulted with the understanding This research develops a mathematical model of urban transportation which includes integrated multi-modal transportation.

2. Methods

This research develops a mathematical model of urban transportation which includes integrated multimodal transportation, spatial structure and transport interchange. The model is developed into a generic model that represents the actual concept of a city with supporting suburb area, travelling time based performance, service variables. This objective of the model is to maximize performance based on total trips and service rate of each transportation mode. Objective function of this model is to maximize performance:

$$\sum_i \sum_j \sum_l P_{ijl} = \sum_i \sum_j \sum_l \{S_{ijl} \times R_{ijl} \times X_{ijl}\} \quad (1)$$

The model is developed to measure the current development of urban transportation by using transport interchange concept to integrate urban multimodal transportation. Equation (2) is set to limit each mode of transportation by its capacity.

$$\sum_i \sum_j \sum_l R_{ijl} \leq \sum_i \sum_j \sum_l C_{ijl} \quad (2)$$

To incorporate each big city supporting area which are usually used as housing complex for those who work in central are city, a spatial structure is used. Nodes are used to represent the spatial area as shown in Figure 1. Then each node is assigned a number as demand of each area of daily trips which shown in model by variable T. Equation (3) represents transportation demand from each origin to destination, and the failure to facilitate trips demand is shown.

$$\sum_i \sum_j R_{ij} + \sum_i \sum_j F_{ij} = \sum_i \sum_j T_{ij} \quad (3)$$

2.1. Spatial Structure

Each city has its own spatial structure characteristics, in this model a grid based spatial structure is developed as shown in figure 1. Each node represents an origin or a destination of trips. This grid or node contains its attributes of service variable. To apply spatial structure in the model a data of city structure is needed, to represent city center, business district, and housing complex or city support (satellite city, suburb). Within this spatial grid, necessary attributes of trips such as demand, service variables are divined.

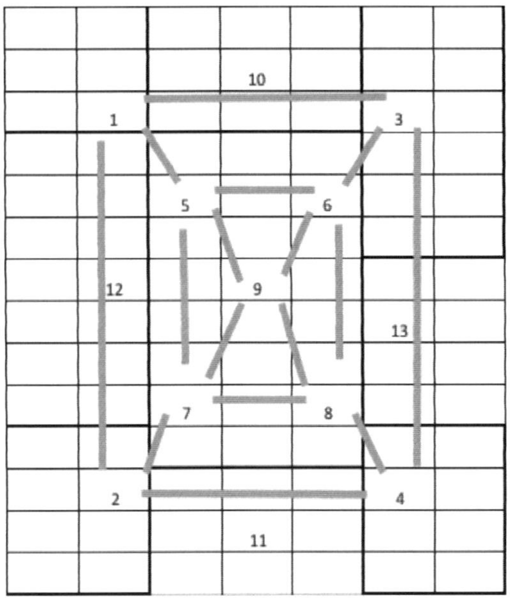

Figure 1. Spatial Structure grid to represent location and nodes for origin and destination of each trip

IOP Conf. Series: Materials Science and Engineering **277** (2017) 012049 doi:10.1088/1757-899X/277/1/012049

2.2. Service Variables

Based on previous research, mode selection of each trip is effected by previous experience with resulted with bad service quality rate which is included to be analyzed. This model is aim to maximize the performance value, which to represent the character of transportation user's decision making. Variable S or service from equation is representing the decision making of each user which will choose the comfortable level of each modes.

$$S_{ijl} = \frac{\mu_{ijl} \times \pi_{ijl}}{D_{ijl}} \quad (4)$$

Therefore, an on time and easy access variable is combined, then a dissatisfaction variable is included (D_{ijl}). Services variables are easy access which is translated from accessibility of services, Safety, Reliability as shown in table 1.

Table 1. Service Variables.

Variables	Details	Mean	Weighted Variables
Safety	Vehicle tracing	3,25	0.65
	Safety	4,5	0,9
Reliability	Well-timed arrival	5	1,00
	Precise and reliable information systems	4	0,8
Accessibility of services		4.25	0,85

2.3. Case Study

In this case study four mode of transportation are used, there are Automobile, Train, Bus, On-Demand transportation. Capacity of each modes are set to limit the users within each node. Automobile has high reliability score and accessibility, followed by On Demand transportation, and on last rank are buses and trains. Availability of buses and train are limited which resulted with rank lower on user options. Weighted variables are used to determine which mode is chosen by user then it will be counted as a trip. Demand of trips are from supporting area such as satellite city or suburb as center of the city as the destination.

3. Result and Discussion

The result of trip's quantity as parameter of deciding factor on how user decides to choose transportation modes from their origin point to destination. With service variable attached to the model, it is shown that user will choose a transportation mode with better or high service variable. With capacity variable limits each transportation users then is analysed to measure the ability of current system to fulfil trips demand. Using sensitivity analysis method this model able to show which progress to be made. A case study is analysed to verified the model. It shows a service variable is needed to measure performance of the system. First scenario is to increase road capacity (link between nodes) of automobile, this resulted with less trips using buses and trains. Second scenario, to limit automobile user, buses and trains has more demand than normal scenario. Third scenario, by increase the service variables and reducing dissatisfaction helps to increase demand on buses and trains. These results have shown the relation within model that can be used to reflect user decision making and represent performance model of urban transportation.

To add value and service variables of train and buses, more linked nodes are needed as it will increase the accessibility of each buses and train. Nodes to be developed can be seen by demand on each node from origin to its destination.

IOP Conf. Series: Materials Science and Engineering **277** (2017) 012049 doi:10.1088/1757-899X/277/1/012049

4. Conclusion

This research has developed a mathematical model to measure an urban transportation system performance which includes user decision making process. In order to represent user decision making a service variable is used based on previous research. The model is solved and verified by using Lingo Software. The result shows this method of measuring urban transportation system is also able to be used to project the next step transportation system by changing its variable. To be able to represent each city and its support areas a further research are to be made, including variable related to service.

5. References

[1] Cuevas V, Estrada M and Salanova J M 2016 Management of on-demand transportation services in urban contexts Barcelona case study *Transportation Research Procedia* **13** 155-165

[2] Mees P 2014 TOD and multi – modal Public Transport *Planning Practice & Research* **29**(5) 461

[3] Manaugh K, Badami M G and El-Geneidy A M 2015 Integrating social equity into urban transportation planning: A critical evaluation of equity objectives and measures in transportation plans in North America *Transport Policy* **37** 167-176

[4] Matijosius J, Aidas V, Virgilija V and Zygintas K 2016 The assesment of importance factors that predetermine the quality of a service of transportation by road vehicles.

[5] Mouwen A 2015 Drivers of customer satisfaction with public transport services *Transportation Research* A **78** 1-20

[6] Joewono T B, Tarigan A K M and Susilo Y O 2016 Road-based public transportation in urban areas of Indonesia: what policies do users expect to improve the service quality? *Transport policy* **49** 114-124

Proposed algorithm to improve job shop production scheduling using ant colony optimization method

Eka KA Pakpahan[1], Sonna Kristina and Ari Setiawan

Department of Industrial Engineering, Harapan Bangsa Institute of Technology, Jln. Dipatiukur 80-84, Bandung, 40132, Indonesia
[1]eka@ithb.ac.id

Abstract. This paper deals with the determination of job shop production schedule on an automatic environment. On this particular environment, machines and material handling system are integrated and controlled by a computer center where schedule were created and then used to dictate the movement of parts and the operations at each machine. This setting is usually designed to have an unmanned production process for a specified interval time. We consider here parts with various operations requirement. Each operation requires specific cutting tools. These parts are to be scheduled on machines each having identical capability, meaning that each machine is equipped with a similar set of cutting tools therefore is capable of processing any operation. The availability of a particular machine to process a particular operation is determined by the remaining life time of its cutting tools. We proposed an algorithm based on the ant colony optimization method and embedded them on matlab software to generate production schedule which minimize the total processing time of the parts (makespan). We test the algorithm on data provided by real industry and the process shows a very short computation time. This contributes a lot to the flexibility and timelines targeted on an automatic environment.
Keywords: scheduling, ant colony optimization, makespan

1. Introduction

Customer increasing demand volume and variety has put manufacturer on the verge of fierce competition. The one that could meet customer specification, quantity and customization at the right time will be the one that wins the competition. In order to be the winner, manufacturer has to adopt changes on their manufacturing practice. Quality, flexibility and timeliness will be the key performance indicators for their manufacturing system. Automation has been recognized for answering these challenges.

We consider here a specific automatic manufacturing system where machines and material handling system are integrated and controlled by a computer center. On this computer, schedule were created and then used to dictate the movement of parts from their storage to the assigned machine and the operations sequence at each machine. Machines are identical, meaning they have the same capability on processing any operation as they each has the same set of cutting tools installed on them. Schedules are generated on daily basis to provide maximum 24 hours unmanned operations. Each day, operator generate a schedule, setup the raw material and installed new set of cutting tools and then leave the work station until the next morning.

Several researches have been conducted to provide an effective schedule for the system mentioned above. They particularly consider the scheduling of jobs consisting of more than two sequentially processes onto several machines. Research conducted by [1] offers genetic algorithm to generate an operable schedule, but the process is done manually therefore it was very inefficient. [2], [3] and [4] then improved the work of [1] by adopting three different heuristic and meta-heuristic approaches and found that ant colony optimization lead to a better result. Unfortunately this previous work was conducted on a simplified case, which is by assuming that the cutting tools installed at each machine has unlimited life time. To be more realistic, this research is aimed to improve the work of [2] by considering cutting tools life time limit.

2. Methods

2.1 Problem Formulation

Consider P jobs ($p=1,2,...P$) each having Q processes ($q=1,2,...,Q$) and each process is consisted of R operations ($r=1,2,...R$). The processes and operations have predetermined sequence that cannot be violated. The jobs are to be scheduled on M machines ($m=1,2,...,M$). Each machine is equipped with S types of cutting tools ($s=1,2,...,S$) with T unit of tools ($t=1,2,...,T$) for each type. One particular unit of cutting tools has L lifetime limit.

Each operation requires particular types of cutting tools. The whole operations on a certain process have to be completed on one machine to minimize the risk of product failure due to frequent movement. Our aim is to generate production schedule by considering the availability of cutting tools on a particular machine, targeted to minimize the total processing time.

2.2 Algorithm Development

The algorithm was developed based on ant colony optimization method. It works by simulating the behavior of ants on finding source of food. They would choose a path and deposit a pheromone as guidance for their colony to follow their path. When the path is too long pheromone will decay, therefore the probability of the colony to follow their path will be smaller than the one that took a shorter path. The series process of choosing path and evaluating them would be represented by iteration. Iteration would be repeated for a number of times until it converges to one solution. More information concerning this algorithm can be found on [5].

On the case we considered here, the path that the ant took represents schedule solutions, and the number of ants represents the number of alternative solutions. The path that each ant took will be evaluated based on a certain fitness function, on this case it would be the makespan of the schedule. The best path which has the shortest makespan will have larger pheromone deposit, therefore has larger probability to be re-chosen by the ant on the next iteration. The description of each step on the algorithm is presented below:

Step 1 : Input parameter related to ant colony algorithm, A as the number of ants ($a=1,2,...,A$), initial pheromone (τ_{ij}^{0}), pheromone decay factor (ρ), scaling parameter (ς), and B as the number of iteration ($b = 1,2, ... , B$).

Step 2 : Input data of jobs to be scheduled: job number (p), process number (q), operation number (r), operations processing time (t_{pqr}), and the type of the cutting tools needed for the operation (s_{pqr}).

Step 3 : Input the number of machine available (M), number of tool types on each machine (S), and the number of unit per tools type (T).

Step 4 : Divide jobs based on the process number (q). There would be Q groups of data ($q=1,2,...,Q$). Calculate the processing time of a job by summing all operations on a particular process (t_{pq}).

$$t_{pq} = \sum_{r=1}^{R} t_{pqr}, \qquad \text{for } q=1,2,...,Q \text{ and } p=1,2,..., P \tag{1}$$

Step 5 : Set $q=1$.

Step 6 : Define the path (nodes and layers) based on the number of jobs (P) to be scheduled. Node represents the job number to be processed and layer represents the sequence. We will use i index, $i=1,2,...P$ for the node and $j=1,2,...P$ for the layer.

Step 7 : Begin ant colony optimization for jobs on grup q,

Step 7.1 : Set $b=1$, (iteration number)

Step 7.2 : Calculate the probability of node i at layer j (P_{ij}^{b}) getting selected on the bth iteration by the ant. Use equation (1) and (2).

$$P_{ij}^{b} = \frac{\tau_{ij}^{b-1}}{N_j^{b-1}} \qquad (2)$$

where

$$N_j^{b} = \sum_{i=1}^{P} \tau_{ij}^{b},\ for\ j = 1,2, \ldots P \qquad (3)$$

and

P_{ij}^{b} : Probability of node i at layer j (P_{ij}^{b}) getting selected on the bth iteration

τ_{ij}^{b} : Pheromone value on i^{th} node at j^{th} layer at the bth iteration

N_j^{b} : The sum of pheromone at the jth layer at the bth iteration

Step 7.3 : Create probability cumulative range (PCR_{ij}^{b}) for each node on a layer. The range stretch between a particular minimum ($PCR_{ij}^{b^-}$) and maximum value ($PCR_{ij}^{b^+}$) as described on equation (4), (5) and (6).

$$PCR_{ij}^{b} = \left\{PCR_{ij}^{b^-}, PCR_{ij}^{b^+}\right\} \qquad (4)$$

where

$$PCR_{ij}^{b^-} = \textstyle\sum_{n=1}^{i} P_{(n-1)j}^{b},\ \text{for}\ j = 1,2, \ldots P \qquad (5)$$

$$PCR_{ij}^{b^+} = \textstyle\sum_{n=1}^{i} P_{(n)j}^{b},\ \text{for}\ j = 1,2, \ldots P \qquad (6)$$

$$P_{(0)j}^{b} = 0,\ \text{for}\ j = 1,2, \ldots P\ \text{and}\ b = 1,2, \ldots B \qquad (7)$$

and

PCR_{ij}^{b} : Probability cumulative range for node i on layer j

$PCR_{ij}^{b^-}$: Minimum value of PCR_{ij}^{b}

$PCR_{ij}^{b^+}$: Maximum value of PCR_{ij}^{b}

Step 7.4 : Set $a=1$, (ant number)

Step 7.5 : Set $j=1$, (layer number)

Step 7.6 : Generate random number (0,1), determine where the random number fall on the probability cumulative range (see step 7.3) and choose the corresponding node (i).

Step 7.7 : Check whether the chosen node (i) have been selected on the previous layers ($j=1$ to $j=j-1$). If yes, repeat step 7.6. If no, go to step 7.8.

Step 7.8 : Assign the chosen node *(i)* onto machine by following these rules:

Step 7.8.1 : Find the corresponding processing time (t_{pq}) for the chosen node *(i)*, see equation (1).

Step 7.8.2 : Choose machine with the lowest utility. Machine utility is defined as the total time of scheduled job on a particular machine. If the utility is equal among machines, choose arbitrarily. If there is no more machine to be selected, place the job on unscheduled list.

Step 7.8.3 : On the selected machine, check on the cutting tools availability by comparing the tools remaining life time and the total operation time. Do so by considering all *T* units of cutting tools installed on the machine. If the tools remaining life time falls below the total operation time, erase selected machine from the list and then repeat step 7.8.2. If not, go to step 7.8.4

Step 7.8.4 : Allocate job i onto the selected machine, calculate utility of the machine after job i is scheduled by adding t_{pq}.

Step 7.9 : Set *j=j+1*. If *j=P* then continue to step 7.10. If no, then repeat step 7.6.

Step 7.10: Calculate the makespan of the ant. Makespan is defined as completion time of all jobs.

Step 7.11: Set *a=a+1*. If *a=A*, continue to step 7.12. If no, then repeat step 7.5.

Step 7.12: Among the ants ($a = 1,2, \dots A$), select the best and worst ant. The best ant refers to the ant that provides the shortest makespan and the worst ant refers to the ant that provides the longest makespan. Calculate the pheromone deposit according to equation (8).

$$\Delta\tau^b = \frac{\varsigma . fbest^b}{fworst^b} \tag{8}$$

where

$\Delta\tau^b$: pheromone deposit on iteration *b*

ς : scaling parameter

$fbest^b$: shortest makespan on iteration *b*

$fworst^b$: longest makespan on iteration *b*

Step 7.13: For the best ant, check their path. Denote the nodes and layer by asterisk and update the pheromone on each node on the path by considering the pheromone deposit. See equation (8) and (9).

$$\tau^b_{i^* j^*} = \tau^b_{i^* j^*} + \Delta\tau^b \tag{9}$$

Step 7.14: For all nodes and layers, update the pheromone by considering the pheromone decay factor. See equation (10).

$$\tau^b_{ij} = (1-\rho)\tau^b_{ij} \tag{10}$$

Step 7.15: Set *b=b+1*, If *b=B*, continue to step 7.p. If not, repeat step 7.2.

Step 7.16: Set *q=q+1*. If *q=Q*, then continue to step 8. If no, update the nodes and layers by erasing node i and layer j corresponding to unscheduled jobs resulted from *q=q-1,* then go to step 6.

Step 8 : Display the gantt chart of the schedule from *Q* groups of data. Ensure that the starting time of the process on stage *q* do not violate the completion time of the process on group *q-1*.

2.3 Software Development

Software was developed using Matlab software. To simplify the data input process GUI (graphical user interface) was developed. The algorithm (as explained on the previous sub point), were executed.

2.4 Data Testing

The algorithm was tested on data taken from real world industry. Complete data are given on table 1. There are 15 jobs, each has one to two processes and each process consist of several operations. On this particular data, process resembles work on a different surface of the part. Processes are sequence dependent and so were the operations. Operations on each process are preferred to be conducted only on one machine to prevent defect due to excess material handling.

IOP Conf. Series: Materials Science and Engineering **277** (2017) 012050 doi:10.1088/1757-899X/277/1/012050

Table 1. List of Jobs to be Scheduled.

Part #	*Stage #*	*Tools Requirement (Types Number)*											
		1	*2*	*3*	*4*	*5*	*6*	*7*	*8*	*9*	*10*	*11*	*12*
1	1	10 (1)[a]	10 (2)				15 (3)	10 (4)	10 (5)	25 (6)			
2	1	35 (2)	40 (1)	45 (3)	30 (5)	25 (4)	30 (6)	35 (7)					
	2	45 (1)	30 (2)						35 (6)	40 (3)	50 (5)	20 (4)	20 (7)
3	1	50 (1)	50 (2)	60 (3)		35 (4)	35 (5)			50 (6)			
	2	30 (1)	55 (2)					25 (3)	40 (5)	50 (4)	30 (6)		50 (7)
4	1	35 (1)	20 (2)		50 (3)	40 (4)	50 (5)			35 (7)	10 (6)		
	2		20 (1)						15 (2)	20 (3)	30 (4)	15 (5)	20 (6)
5	1	10 (2)	15 (1)		20 (3)	10 (4)					20 (6)		15 (5)
6	1	20 (1)	15 (2)							15 (4)	20 (5)	30 (3)	
7	1	45 (1)	30 (2)	25 (5)	30 (3)	25 (4)	35 (6)				50 (7)		
	2	35 (1)	50 (2)					35 (4)	30 (3)	20 (5)	30 (6)		
8	1	40 (1)	20 (2)	50 (3)	40 (4)	40 (5)				50 (6)			
	2		30 (1)				20 (2)		15 (3)	25 (4)		25 (5)	5 (6)
9	1			30 (1)	40 (2)	40 (4)	30 (3)	50 (5)		50 (7)	40 (6)		
	2	60 (1)	50 (2)						40 (3)	45 (4)	40 (5)		45 (6)
10	1	40 (1)	50 (2)	60 (3)			40 (4)	50 (5)					
	2	50 (1)	60 (2)						45 (3)	45 (4)		40 (5)	
11	1	40 (1)	50 (2)		30 (4)	20 (3)					20 (5)		
	2	40 (1)						50 (3)	60 (4)	40 (2)		20 (5)	30 (6)
12	1	50 (1)	40 (2)			60 (3)		60 (4)	50 (5)			60 (6)	
	2	45 (1)					45 (2)		30 (3)		30 (4)	50 (5)	
13	1	30 (1)	40 (2)	30 (3)	40 (5)	25 (4)		25 (6)			50 (7)		
	2	45 (1)							45 (2)	30 (3)		40 (4)	40 (5)
14	1		10 (1)		25 (2)		20 (3)	15 (4)			20 (5)		
15	1		15 (1)					15 (4)	10 (3)	20 (2)	20 (5)		

[a]Denotes the operation sequence number.

3. Result and Discussions

We create 6 scenarios. These scenarios can be divided into two parts. The first part is to analyze the effect of ant and iteration number onto the optimality of the solution and the computation length. The second part is conducted to create comparison against [2]. Table 2 and table 3 summarize the scenarios.

Table 2. Scenarios for Parameter Analysis (Part 1).

No.	**Number of Ants**	**Initial pheromone**	**Pheromone decay factor**	**Scaling parameter**	**Number of Iterations**
1	5	1	0.1	2	100
2	5	1	0.1	2	500
3	10	1	0.1	2	100
4	10	1	0.1	2	500

Table 3. Scenarios for Comparison Analysis (Part 2).

No.	**Number of Ants**	***Initial pheromone***	***Pheromone decay factor***	***Scaling* parameter**	**Number of Iterations**
1	5	1	0.05	2	200
2	5	1	0.05	2	300

The final result of the algorithm is a gantt chart showing assignment of part numbers onto each machine. The description is given on figure 1.

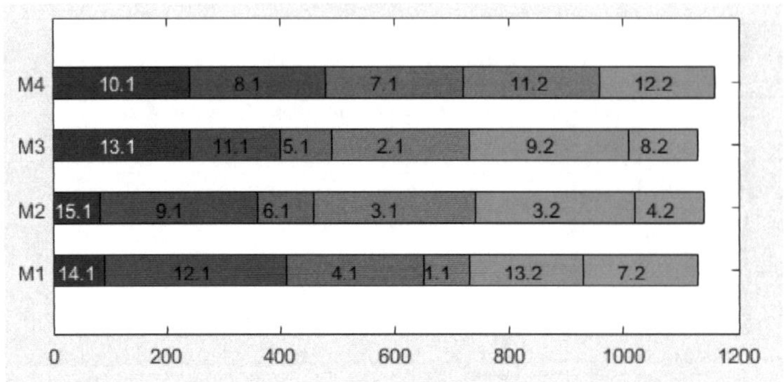

Figure 1. Gantt Chart of the Allocated Job on Machines.

Each scenario mentioned earlier was run 3 times. The result of the first part is displayed on table 4, table 5, table 6, and table 7.

Table 4. Result for Scenario 1, Part 1.

Running #	Makespan (minutes)	Number of Unscheduled Job	Computation Time (minutes)
1	1200	2	0.29
2	1210	1	0.29
3	1240	1	0.32

Table 5. Result for Scenario 2, Part 1

Running #	Makespan (minutes)	Number of Unscheduled Job	Computation Time (minutes)
1	1200	1	0.54
2	1200	1	0.63
3	1200	1	0.51

Table 6. Result for Scenario 3, Part 1.

Running #	Makespan (minutes)	Number of Unscheduled Job	Computation Time (minutes)
1	1180	2	0.33
2	1210	1	0.31
3	1220	1	0.32

Table 7. Result for Scenario 4, Part 1.

Running #	Makespan (minutes)	Number of Unscheduled Job	Computation Time (minutes)
1	1160	2	0.71
2	1200	1	0.69
3	1200	1	0.75

We can see few highlights from the tables above. The effect of the number of ants are obvious, the larger the number of the ants the better the result (in terms of makespan), it applies also for larger number of iteration. The best result is achieved when the parameter input are as follows: the ant number is 10, the pheromone decay factor are 0.05, scalling parameter 2 and iterations number 500.

The result of the scenarios for the second part is displayed on Table 8 and Table 10. To get comparison with [2], we displayed the result of [2] on Table 10 and Table 11.

Table 8. Result for Scenario 1, Part 2.

Running #	Makespan (minutes)	Number of Unscheduled Job	Computation Time (minutes)
1	1220	1	0.31
2	1180	2	0.30
3	1210	1	0.28

Table 9. Result for Scenario 2, Part 2.

Running #	Makespan (minutes)	Number of Unscheduled Job	Computation Time (minutes)
1	1210	1	0.35
2	1210	1	0.33
3	1200	1	0.33

Table 10. Result of [2], Scenario 1.

Running #	Makespan (minutes)	Computation Time (minutes)
1	1410	8.77
2	1420	6.50

Table 11. Result of [2], Scenario 2.

Running #	Makespan (minutes)	Computation Time (minutes)
1	1370	272.08
2	1440	267.80

Compared to the result of [2] we manage to improve the output performance especially in terms of computation time. When the number of iteration is set on 200, the computation time of [2] spans between 6 to 8 minutes (See table 10) and it worsen when the iteration is increased to 300. In terms of makespan, our result leads also to a better value. But the comparison cannot be done directly due to additional constraint considered on our research, which is the consideration of cutting tools availability. This constraint leads to the possibility that there are jobs that cannot be scheduled due to the unavailability of cutting tools, whereas on [2] all jobs is always scheduled.

4. Conclusions

We have managed to improve the performance of the algorithm developed by [2], in terms of computation time and makespan value. Schedule solution can be generated on a very short computation time and the result is more realistic due to the consideration of tools availability. Implementing the software on the computer center should be the next step of this research.

Acknowledgement
This research is funded by the Ministry of Research, Technology, and Higher Education of the Republic of Indonesia through the scheme of "Penelitian Dosen Pemula" Year of 2017.

5. References

[1] Setiawan A, Susan and Pakpahan E 2014 Penjadwalan Job Shop pada Empat Mesin Identik dengan Menggunakan Metode Shortest Processing Time dan Genetic Algorithm *Jurnal Telematika Institut Teknologi Harapan Bangsa* **Vol 9**, No 1.
[2] Surya T 2016 Penjadwalan Job Dua Stage pada Flexible Manufacturing System Menggunakan Metode Ant Colony Optimization *Laporan Tugas Akhir Teknik Industri Institut Teknologi Harapan Bangsa* (Bandung).
[3] Wahyudi K 2016 Algoritma Penjadwalan Job Dua Stage dengan Memperhatikan Preventive Maintenance Mesin Perkakas pada Flexible Manufacturing System *Laporan Tugas Akhir Teknik Industri Institut Teknologi Harapan Bangsa* (Bandung).
[4] Chandra A 2016 Penjadwalan Job Dua Stage pada Flexible Manufacturing System Menggunakan Metode Genetic Algorithm *Laporan Tugas Akhir Teknik Industri Institut Teknologi Harapan Bangsa* (Bandung).
[5] Rao SS 2009 *Engineering Optimization: Theory and Practice* (New Jersey: John Wiley & Sons, Inc).

IOP Conf. Series: Materials Science and Engineering **277** (2017) 012051 doi:10.1088/1757-899X/277/1/012051

Design and Simulation Plant Layout Using Systematic Layout Planning

D Suhardini[1], W Septiani, and S Fauziah

Department of Industrial Engineering, Faculty of Industrial Technology,
Trisakti University, Jakarta, 11440, Indonesia
didien1117@gmail.com

Abstract. This research aims to design the factory layout of PT. Gunaprima Budiwijaya in order to increase production capacity. The problem faced by this company is inappropriate layout causes cross traffic on the production floor. The re-layout procedure consist of these three steps: analysing the existing layout, designing plant layout based on SLP and evaluation and selection of alternative layout using Simulation Pro model version 6. Systematic layout planning is used to re-layout not based on the initial layout. This SLP produces four layout alternatives, and each alternative will be evaluated based on two criteria, namely cost of material handling using Material Handling Evaluation Sheet (MHES) and processing time by simulation. The results showed that production capacity is increasing as much as 37.5% with the addition of the machine and the operator, while material handling cost was reduced by improvement of the layout. The use of systematic layout planning method reduces material handling cost of 10,98% from initial layout or amounting to Rp1.229.813,34.
Keywords: Plant Layout, SLP, Cross Traffic, Simulation

1. Introduction

In line with the growth of residential development then there arose a tight competition between the industries of construction materials, so that companies that produce construction materials should continue to improve its performance so that it continues to grow and meet the expectations of consumers. The one thing that supports the improvement of efficiency and effectiveness of the factory is setting the right factory layout that encourages the achievement of the desired production capacity.

Table 1. Status of Demand of PT Gunaprima Budiwijaya 2016

Number	Month	Demand (Kg)	Capacity (Kg)	Deviation (Kg)	Status
1	January	10971.24	8000.00	-2971.24	Not Fulfilled
2	February	5745.00	8000.00	2255.00	Fulfilled
3	March	10495.00	8000.00	-2495.00	Not Fulfilled
4	April	6320.00	8000.00	1680.00	Fulfilled
5	May	9190.00	8000.00	-1190.00	Not Fulfilled
6	June	7928.60	8000.00	71.40	Fulfilled
7	July	4080.00	8000.00	3920.00	Fulfilled
8	August	8794.80	8000.00	-794.80	Not Fulfilled
9	September	3955.00	8000.00	4045.00	Fulfilled
10	October	10115.00	8000.00	-2115.00	Not Fulfilled
11	November	4837.00	8000.00	3163.00	Fulfilled
12	December	3900.00	8000.00	4100.00	Fulfilled

The company could not meet the demand for production in 2016 of 41.67% or 9566.04 Kg (Table 1). The cause of the problem is the layout which is less good on the production floor. The layout isn't good cause cross traffic in production floor. This research aims to design the layout uses a systematic layout planning (SLP). The SLP method applied to optimize the layout of existing machines and reduce the reject [1]; [2] this method provides the layout of the new machines that can improve the flow of materials and helps to improve the utilization of space in the plant [3]. Planning an effective facility layout can significantly reduce the operational costs of the company amounting to 10-30% [4]. Evaluation and design the layout needs to be done to reduce the cross traffic and optimize the use of floor area of production.

2. Methods

Procedure to relayout the production floor consists of three steps, such as analyzing the existing layout, design the plant layout based on the SLP and the evaluation and selection of alternative layouts using simulation program Pro model [5]. Simulation program used as analytical tools to help systematically designed the layout and to support appropriate decision-making to eliminate some of the problems [6] [7] [8].

Steps in the relayout and explanation can be seen here:

a. Analyzing the existing layout plant to identify the problem on flow of material and operation.
b. Designing plant layout based on SLP approach to increase production capacity and reduce material handling cost.
c. Evaluation and selection of alternatives using Simulation Pro model based on two kinds of criteria; cost of material handling and production time.

2.1 Evaluation Existing Layout

The initial plant layout was analyzed using three methods [9]:

1. Plant Layout Checklist
   The purpose of the plant layout Checklist is to identify the cause of the movement of the material not needed which can increase production time and cost of materials handling.
2. Material Handling Checklist
   The purpose of the material handling checklist is to analyze of material handling equipment utilization and other criteria.
3. Material Handling Evaluation Sheet
   The initial layout PT. Gunaprima Budiwijaya analyzed using material handling evaluation sheet. The results obtained from the MHES is total cost of handling materials, and then used as inputs in a systematic layout planning. The data needed in the calculation of the MHES is: transfer of material, the distance between the Department, the frequency of the transfer, material handling equipment used and the cost of the transfer.

2.2 Systematic Layout Planning

The systematic layout planning (SLP) is a procedure used to set the layout of workplace in a plant notice to the logical relationship between workplace with high frequency are placed close to each other. SLP technique applied to optimize the existing layout. The application is expected to make the fastest material flow with the lowest cost and least amount of material handling [10]; [11]. Systematic layout planning consists of four stages as follows:

Stage I: Determine the location where the facility will be built
Stage II: Make overall facility design
Stage III: Determine the design of facility layout in detail (to be done in this paper)
Stage IV: Preparation and installation of design results

The input data required by Systematic Layout Planning are divided into five categories:

P (Product) : The type of product (goods/service) produced.

Q (Quantity) : Volume of each type of goods/ components produced.
R (Route) : The order of operation for each product
S (Service) : Support service, such as locker rooms, monitoring stations, etc.
T (Timing) : In what time the type of component of the product was produced, what machine is used to produce it at that time.

It is a step-by-step planning procedure allowing users to identify, visualize, and rate the various activities, relationships, and alternatives involved in a layout project based on input data, the flow of materials, activity of relationships and relationship diagrams. The framework of SLP is shown in Figure 1

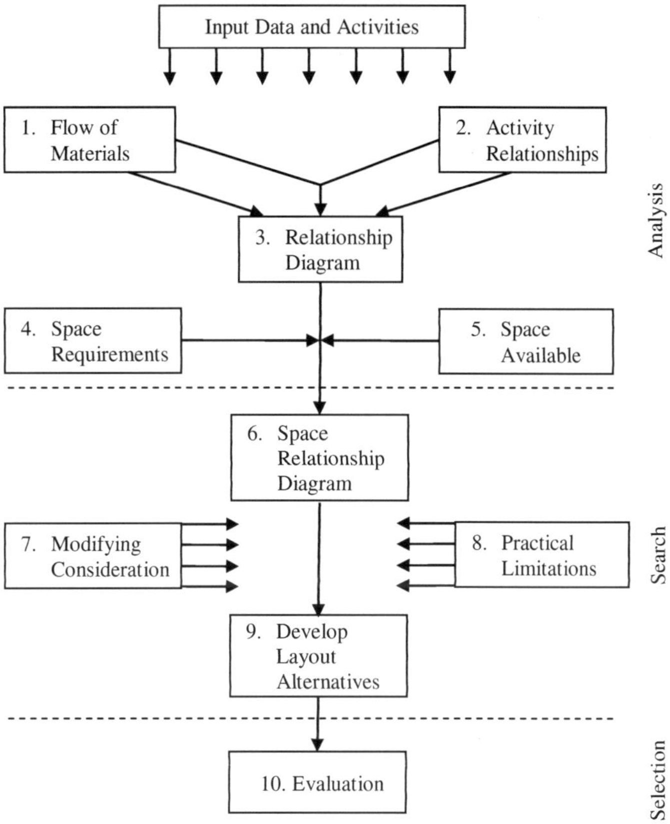

Figure 1. Systematic Layout Planning Procedure [12].

2.3. Evaluation and selection

Evaluation of the layout using two criteria, namely material handling cost and production time. To evaluate the total of material handling costs using Material Handling Evaluation Sheet and production processing time is using simulation using "ProModel version 6". Production time is the duration that is used to manufacture products including time transfering the material. After evaluating the result of SLP, then selection stage can be done using the best score of each criteria.

3. Result and Discussion

3.1 Evaluation Existing Layout

The result of the use of plant layout criteria checklist for initial layout of 38% answered "Yes". That means, 62% of the criteria is not met and material handling criteria checklist produces 49% answer "Yes" and the cost needed of handling materials on initial layout Rp 11,346, 590.68

3.2. Designing plant layout based on Systematic Layout Planning

3.2.1. Calculate the amount of machines used Routing Sheet

Capacity used to re-layout is the highest of the year 2016 i.e. request in January. the number of machines need to be calculated to produce the product capacity using the routing sheet. After obtained the number of machines needed, compared with the number of machines that already exist in the production floor. The result is shown in Table 2.

Table 2. The Comparison Of The Machine Quantity

No	Machine	Quantity of machine needed	Quantity of machine already exist	Machine shortage
1	*Mixer* 20 Kg	1	2	+1
2	*Mixer* 200 Kg	1	1	0
3	*Mixer* 500 Kg	1	1	0
4	Dempul	1	1	0
5	*Sandmill*	1	2	+1
6	Atritor Cie	2	1	-1
7	Timbangan	2	1	-1
8	*Packing*	10	1	-9
9	Manual	13	5	-8
10	Saringan	8	5	-3

The results of routing sheet calculation are used as the number of machines on an alternative layout.

3.2.2 Analyze The Material Movement Used From To Chart (FTC)

From To Chart (FTC) is the map that is used to analyze the movement of material that occurs in the production floor. In this study, the type FTC used are the FTC distance and FTC cost. The FTC distance is the chart showing distance between machines or departments in the production floor. Range data retrieved from a distance using the formula calculation of rectilinear. FTC cost are the result of the frequency multiplied by the costs/meter with the distance between the departments. Frequencies obtained from the MHES. Transfer cost for handling manual is Rp 156.25/meters and cost for handling with trolly is Rp1736,11/meter.

3.2.3 Activity Relationship Diagram

The data obtained in the FTC processed further by calculating the Inflow and outflow's coefficent (quantitative) is then converted into a form of qualitative called activity relationship chart (ARC). The ARC was changed to List the relationship closeness to see a priority in drafting the layout using Relationship Diagramming algorithm [12]; [13]. The Activity Relationship Diagram of production floor layout has seen in Figure 2.

3.2.4 Space Requirement

The space requirement for each workspace is a result from the required space compared with the space that is available. If there is a discrepancy, then it will use the larger area of the workspace for calculating the area of AAD, because of the inclusive allowance. The space requirement for each department show in Table 3.

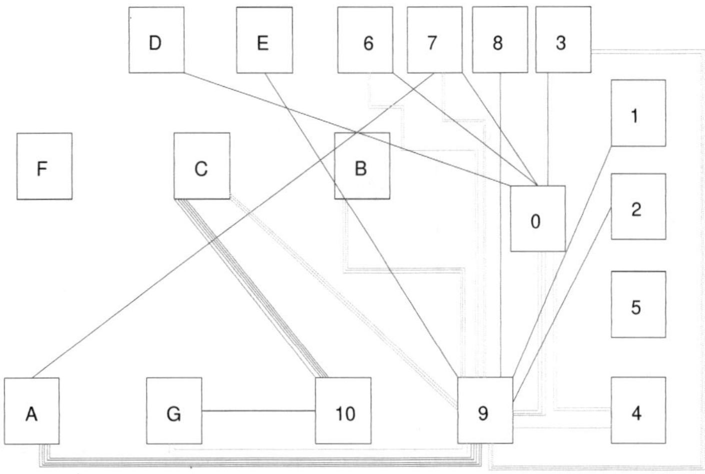

Figure 2. Activity Relationship Diagram of PT. Gunaprima Budiwijaya

Table 3. Space Requirement Each Department

Dept.	Production Area	Space Area (m^2)	Require Space
1	Mixer Machine 20 Kg	0.55	5.38
2	Mixer Machine 20 Kg	0.5	5.38
3	Mixer Machine 200 Kg	1.25	6.5
4	Mixer Machine 500 Kg	1.68	6.92
5	Putty Machine	1.1	5.62
6	Sand mill Machine	1.38	6.12
7	Sand mill Machine	1.38	6.12
8	"Atritor Cie" Machine	0.39	12.32
9	Weighing Area	ei	15
10	Packing Area	20	20
A	Raw Material Warehouse	30	55.52
B	Resin Warehouse	40	23.55
C	Storage	40	0
D	Pail Warehouse	10	8.608
E	Cans Warehouse	10	5.7875
F	Used Drum Warehouse	45	58.5
G	Temporary Storage	10	10
0	Empty Area	5	5
TOTAL AREA		233.23	256.3255

3.2.5. Space Relationship Diagram

Space relationship diagram is similar to Activity Relationship Diagram (ARD) the difference is just on the vast of the space. Activity relationship diagram only describing the relationship of each department. The space relationship diagram using the real space for each departmentin the layout.

3.2.6. Practical Limitation

Designing the layout based on Area Allocation Diagram and there will be limitations in placing departments on the production floor:

1. Sand mill machine requires 50 cm thick step.
2. Placement of machines should be close to the wall to facilitate electrical installation.
3. Placement of warehouse cultivated with rolling door.

IOP Conf. Series: Materials Science and Engineering **277** (2017) 012051 doi:10.1088/1757-899X/277/1/012051

3.2.7 Layout of Systematic Layout Planning

Systematic Layout Planning produced 4 alternatives layout that shown in figure 3, figure 4, figure 5, and figure 6 as follow

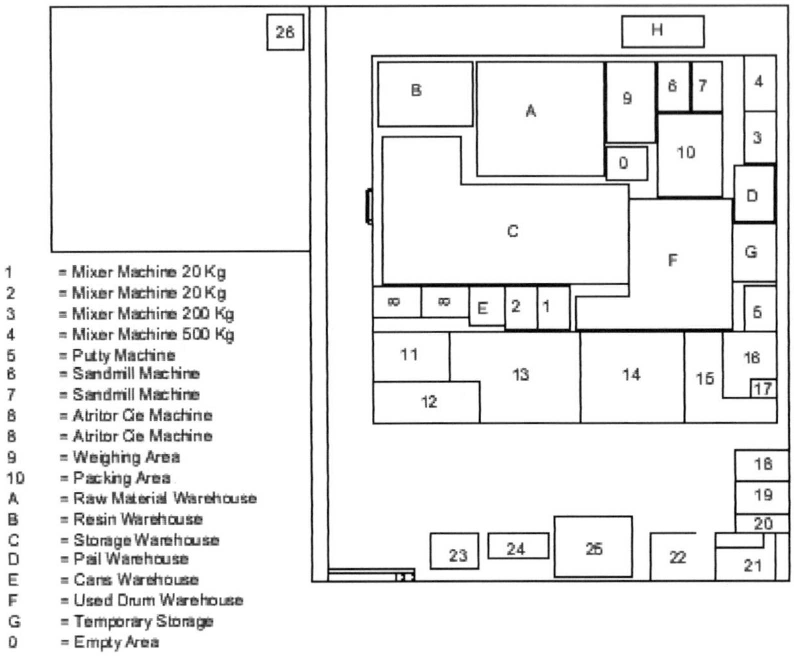

Figure 3. Layout of Systematic Layout Planning Alternatives 1

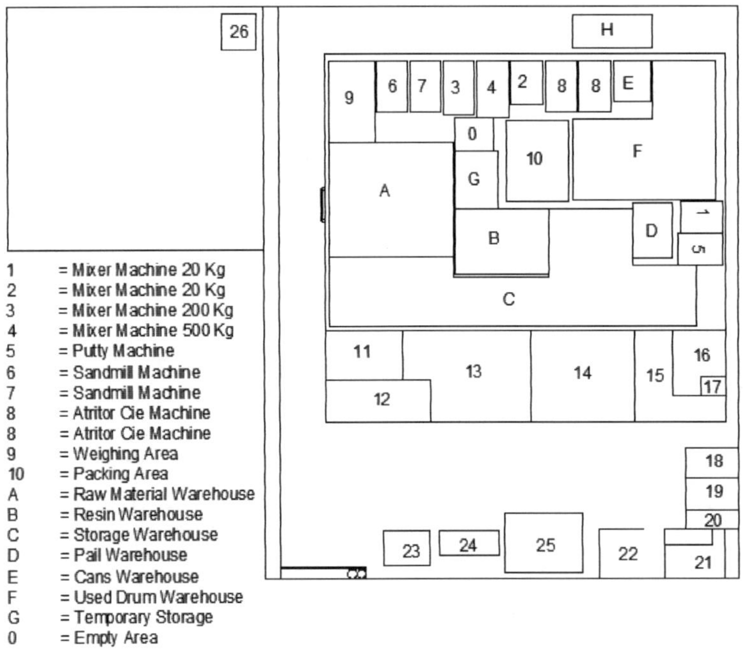

Figure 4. Layout of Systematic Layout Planning Alternatives 2

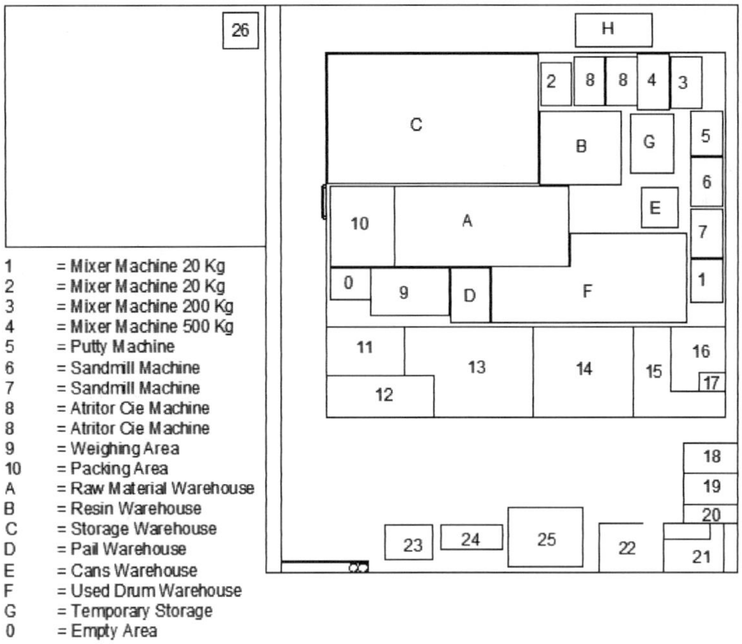

Figure 5. Layout of Systematic Layout Planning Alternatives 3

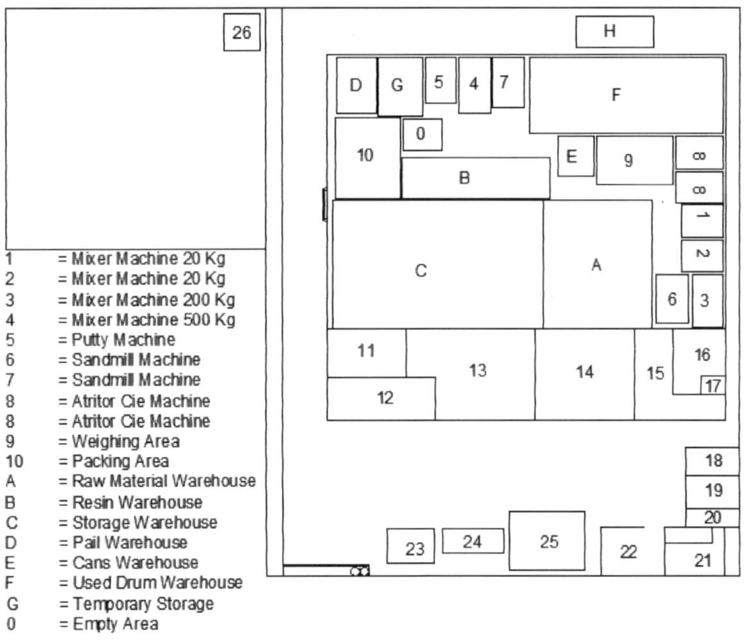

Figure 6. Layout of Systematic Layout Planning Alternatives 4

3.3. Evaluation and Selection

The result of reckoning material handling costs and production process time can be seen in Table 4. Each layout alternative results in different material handling costs. Alternatives 1, 2, and 4 results in lower material handling costs compared to the initial layout material handling material cost, but for alternative 3 material handling costs generated over the cost of the initial layout material handling.

Table 4. Result of Material Handling Costs and Production Process Time

Alternative	MHES Cost	The percentage decline or and increase in cost	Production Process Time (Hour)
Initial	Rp 11,292,201.80	-	-
1	Rp 10,119,807.57	10.38	148.5
2	Rp 10,062,388.46	10.89	142.53
3	Rp 15,241,479.21	-34.97	177.5
4	Rp 11,270,681.90	0.19	152.6

4. Conclusion

The initial layout analysis yielded that, the company's initial layout can only fulfill 38% on the checklist of layout criteria and 49% on material handling checklist. The cost of material handling in the initial layout is Rp 11.292.201.80. The search results using systematic layout planning can increase production capacity from 8,000 kg to 11,000 kg or 37.5% and can reduce material handling costs by 10.98% compared to the initial layout or approximately Rp1,229,813.34. Production time of simulation result is 142,53 hours. Production time can be fulfilled by the company within one month because available time is 192 hours. This propose layout must be improve using another improvement method like pairwise exchange method.

5. References

[1] Sutari O and Rao S 2014 Development of Plant Layout using Systematic Layout Planning (SLP) to Maximize Production – A Case Study. *International Journal of Mechanical And Production Engineering* **2**(8), 63-6.

[2] Wiyaratn W, Watanapa A and Kajondecha P 2013 Improvement Plant Layout Based on Systematic Layout Planning. *IACSIT International Journal of Engineering and Technology* **5**(1), 76-9.

[3] Shewale PP, Shete MS, and Sane SM 2012 Improvement in Plant Layout using Systematic Layout Planning (SLP) for Increased Productivity. *International Journal of Advanced Engineering Research and Studies* **1**(3), 259-61.

[4] Barnwal S and Dharmadhikari P 2016 Optimization of Plant Layout Using SLP Method. *International Journal of Innovative Research in Science, Engineering and Technology* **5**(3), 3008-15.

[5] Law AK 2000 *Simulation Modeling and Analysis*.(Singapore: McGraw Hill Higher Education).

[6] Guimarães AMC and Diallo M 2012 Using Simulation in an Industry of Gasket Products: A Case St. *Engineering Management Research* **1**(2), 153-62.

[7] Tearwattanarattikal P, Namphacharoen S and Chamrasporn C 2008 Using ProModel as a simulation tools to assist plant layout design and planning: Case study plastic packaging factory. *Songklanakarin J Sci Technol* **30**(1), 117-23.

[8] Akpınar ME, Yıldızel SA, Karabulut Y and Doğan E 2017 Simulation Optimization for Transportation System: A Real Case Application. *TEM Journal* **6**(1), 97-102.

[9] Apple JM 1990 *Plant Layout and Material Handling*.(USA: John Wiley & Sons, Inc).

[10] Muther R 1973 *Systematic Layout Planning*.(Boston: CBI Publishing Company).

[11] Jain S and Yadav TK 2017 Systematic Layout Planning: A Review of Improvement in Approach to Pulse Processing Mills. *International Research Journal of Engineering and Technology (IRJET)* **4**(5), 503-7.

[12] Tompkins JA 2010 *Facilities Planning*. fourth edition ed.(California: John Willey & Sons).

[13] Sugandhi SD and Bharule A 2016 Fine Blanking Plant Layout Improvement Using Systematic Layout Planning. *International Journal of Scientific Engineering and Research (IJSER)* **4**(4), 1-4.

IOP Conf. Series: Materials Science and Engineering **277** (2017) 012052 doi:10.1088/1757-899X/277/1/012052

Synchronization of production and sales target through pull inventory management policy

Roland Silitonga, Eka KA Pakpahan[1], Marrietta Anggraeni

Department of Industrial Engineering, Harapan Bangsa Institute of Technology, Jln. Dipatiukur 80-84, Bandung, 40132, Indonesia.
[1]eka@ithb.ac.id

Abstract. The aim of this research is to resolve high inventory problem at RBI Cileungsi warehouse due to the imbalance flow between production outputs compared to sales realization. As an attempt to synchronize these flows we will analyse sales realization data as a representation of customer demand and determine the optimal inventory policy to meet them. Analysis will be conducted to three SKUs which had high investment value. P inventory models will be implemented and then evaluated against current condition to determine their financial benefit. Evaluation suggested that the proposed policy would give total savings of Rp. 780,667,909.97 for the first SKU, Rp. 35,361,072.53 for the second SKU and Rp. 136,729,725.19 for the third SKU. These results lead to the conclusions that the proposed policy should be further considered to be applied to other SKUs.
Keywords: inventory, pull system, P model

1. Introduction

Reckitt Benckiser (RB) is one of the leaders among health and hygiene consumer product manufacturer. Their business spans across 60 countries and one of them is located in Indonesia. To support their business, RB uses 22.002 m^2 warehouses. The warehouses are bar soap warehouse and house hold/personal care warehouse each of them serves bar soap and house hold/personal care production department. At each warehouse, the area is divided in to two, the raw material and packaging material area and the finished goods area where they keep their final product.

Observation shows that during several last periods, inventory stockpile accumulates to a number that is significantly high (see fig 1). Figure 1 tells us that for almost half month of operation, on hand inventory is always above the warehouse capacity (the inbound is always higher than the outbound) and the next half month the situation is reversed, and in general on hand inventory is always high throughout the month. A lot of inventories are then placed along the warehouse aisle (see fig 2).

The inventory stockpile accumulation indicates a high capital cost. In addition, this accumulation also has a direct impact on the operational activities in the warehouse, such as the difficulty to move material handling equipment to enter the aisles during pickup process. Sometimes the implementation of a simple picking rule such, as FIFO (First in First Out) or FEFO (First Expired First Out) is waived, because picker will tend to simplify their job by picking the inventory stocked up on the aisle. If this case continues, the company will be facing a great risk of having expired products.

Interviews revealed an in line information with the inventory profile in figure 1, i.e unbalanced inbound and outbound flows. Production department states that they get production plan straight from marketing target. This target is set by optimistic scenario that the market would absorb the production output. Warehouse data shows that this target cannot be met. The monthly sales are still lower than the production output. If this continues, then potential warehouse problems as described previously will emerge.

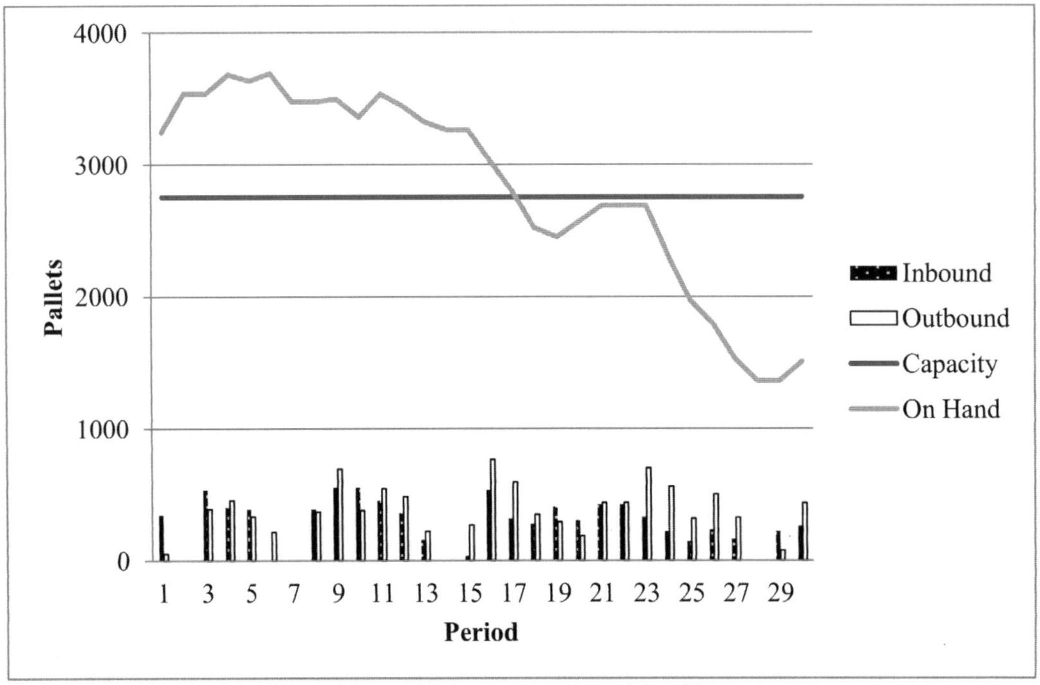

Figure 1. Inventory on Hand Profile.

Figure 2. Inventory stockpile located on aisle.

The description above shows that there is gap between production target and the sales realization. Sales tend to be lower than the production output, thus resulting high level of inventory. Theoretically, this problem can be solved through a proper sales and operation planning (S&OP), where all the parties involved such as the production function, marketing and warehouse get together and discuss the problem dynamic faced by each then try to solved them systemically [1]. A production plan that is based on a particular target (push system) is good as long as the market is able to absorb them, but if that is not the case then there will be risks of inventory stockpile.

This study recommends gradual synchronization between production plan and sales targets. The first attempt to gradually synchronize them would be the determination of inventory policies based on real demand. These policies would then be proposed as basis for production planning.

There are several models of inventory policies. For pull system conditions [2] states that there is a model perpetual review (Q model) and periodic review (P model). What distinguishes between the two models is that in the Q model, the inventory level becomes the trigger (signal) on when inventory needs to be replenished, while in P model, inventory replenishment is done periodically, each time interval (T*) is reached.

We will propose P model as a model of inventory management policy. Using this model, adjustments to production planning that take place periodically will be easier to do. The study was conducted by considering 3 variants of product occupying the A class inventory. Classification is based on the frequency of use and value of the product, so it can be concluded that the products that exist in A class are product with high investment value and the most absorbed by the market. The three product variants are Vanish Pouch 450 ml, Dettol Liquid 100 ml NUTN and Dettol 450 ml Lasting F COR.

2. Methods

The design of inventory policy will be done in three stages. The first stage is data collection and processing, the second is determining the policy decision variables and the last stage is comparative and managerial impact analysis on the proposed policy. These three stages are described more detail in the following paragraphs.

Stage I, is the stage of data collection and processing. Based on P model, data requirements are the product demand, production length and the costs associated with the production setup and planning process, holding cost and shortage cost. The flow of data collection and processing are depicted in figure 3.

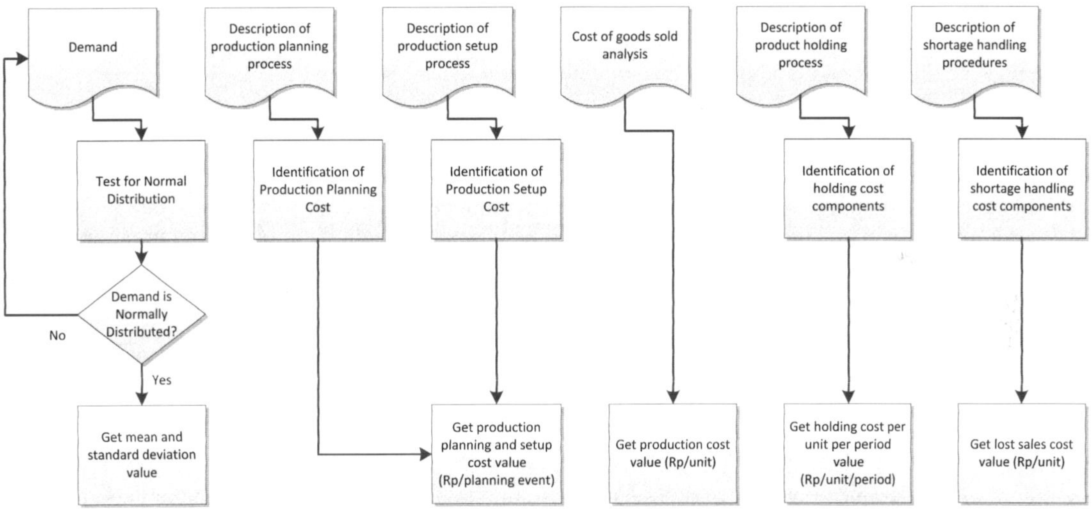

Figure 3. Data Collection and Processing Flow.

The data we gathered are summarized on table 1. To better understand the data, we give the notation below:

D	:	Product demand (unit/year)
S	:	Demand standard deviation (unit/year)
L	:	Production lead time (year)
A	:	Production planning and setup cost (Rp/setup)
p	:	Production cost (Rp/unit)
h	:	Holding cost (Rp/unit/year)
c_u	:	Lost sales cost (Rp/unit)

Table 1. Demand and Inventory Cost Data.

Variable	Vanish Pouch 450 ml (unit)	Dettol Liquid 100 ml (unit)	Dettol 450 ml (unit)
D	106,369,00	31,062.00	15,546.00
S	7,885.00	5,776.00	6,129.00
L	0.03	0.03	0.03
A	325,245.38	325,245.38	325,245.38
p	45,315.24	125,382.50	248,590.52
h	49,608.57	134,479.87	265,080.36
c_u	107,096.76	372,806.50	340,718.48

These variables are then used to determine the decision variables on P model. The goal is to minimize the total inventory cost as denoted on equation (1).

$$O_T = O_p + O_{ps} + O_s + O_k \quad (1)$$

where,

O_p : Total production cost (Rp)
O_{ps} : Total production planning and setup cost (Rp)
O_s : Total holding cost (Rp)
O_k : Total shortage cost (Rp)

Each of the total cost above will be calculated using equation (2), (3), (4), dan (5). [3].

$$O_p = Dp \quad (2)$$

$$O_{ps} = \frac{A}{T} \quad (3)$$

$$O_s = h\left(R - D_L - \frac{TD}{2} + N\right) \quad (4)$$

$$O_k = \frac{c_u N}{T} \quad (5)$$

where,

$$N = S\sqrt{T+L}[f(z_\propto) - z_\propto \psi(z_\propto)] \quad (6)$$

and,

T : Inventory review period (year)
R : Maximum inventory covering for T and L period (unit)
$f(z_\propto)$: Normal distribution ordinate value for a certain α
$\psi(z_\propto)$: Normal distribution partial expectation for a certain α

The second stage is the determination of decision variables. The decision variables are the review period (T) and the maximum inventory level allowed on the warehouse to enable the company to provide customer demand (R). We use equation (7) and (8) to get these values.

$$T^* = \sqrt{\frac{2A}{Dh}} \quad (7)$$

$$R = D(T+L) + z_\propto S\sqrt{T+L} \quad (8)$$

On the third stage we compare the total inventory cost for the proposed and actual system, along with their service level. On this stage we also conducted an analysis on the manajerial impact of the proposed method implementation.

3. Results and Discussion

We present inventory decision variables based on the P model and the expected inventory cost associated with it on table 1. The real cost occur on existing system is given on table 2.

Tabel 2. Inventory decision variables and total inventory cost expectation of the proposed policy.

	Vanish Pouch 450 ml	Dettol Liquid 100 ml	Dettol 450 ml
T	0.011	0.012	0.013
α	0.999	0.999	0.999
Z_α	3.990	3.990	3.990
R	10.750.453	6.069.406	5.706.950
O_p	4.820.136.763.560	3.894.631.215.000	3.864.588.223.920
O_{ps}	29.293.850.625	26.063.570.979	25.887.414.519
O_S	345.721.072.686	664.842.626.091	1.363.304.972.370
O_k	926.731.283	2.137.443.677	2.060.903.869
O_T	5.196.078.418.154	4.587.674.855.747	5.255.841.514.678

Table 3. Inventory cost on existing system.

Cost Component	Vanish Pouch 450 ml	Dettol Liquid 100 ml	Dettol 450 ml
O_p	5.403.615.793,80	4.136.494.057,50	4.583.512.007,76
O_{ps}	26.344.875,78	13.009.815,20	8.781.625,26
O_S	546.785.658,54	473.532.055,58	800.277.606,85
O_k	0	0	0
O_T	5.976.746.328,12	4.623.035.928,28	5.392.571.239,87

By comparing the cost value on table 2 and 3, we know that the implementation of the proposed method will bring cost saving to the company. The savings will be much more significant if the proposed method is applied to other products sitting on A class of inventory classification.

The implication of this proposed policy is that the company must shift from push to pull production. The determination of product replenishment is conducted on periodic base at T (see table 2) and production target are formulated to brought back inventory to the maximum level (R). These should be conducted as the first attempt to synchronize production output and the sales pull to minimize inventory stockpile.

4. Conclusions

The proposed policy has proven to give financial benefit, the savings are Rp. 780,667,909.97 for Vanish Pouch 450 ml, Rp. 35,361,072.53 for Dettol Liquid 100 ml and Rp. 136,729,725.19 for Dettol 450 ml. The savings are approximately 10% compared to the current total cost. The review period differ between the three product, but not too significant (see table 2 for the T value). To ease the implementation process, the company can consider conducting the inventory review on the same time by taking the smallest T among the three product. Demand data record should be conducted continuosly as input for policy evaluation. The main idea of this proposed solution is to see the problem at hand through systematic approach.

5. References

[1] Palmatier GE and Crum C 2003 *Enterprise Sales and operation Planning: Synchronizing Demand, Supply and Resource for Peak Performance* J. Ross Publishing.
[2] Bowersox D, Closs and Cooper 2002 *Supply Chain Logistic Management* McGraw-Hill New York.
[3] Nur Bahagia S 2002 *Sistem Inventori* 2nd Edition Penerbit ITB.

The recommendation of line-balancing improvement on MCM product line 1 using genetics algorithm and moodie young at XYZ Company, Co.

I K Sriwana[1], I A Marie[2] and D Mangala[3]

[1]Industrial Engineering, Esa Unggul University, Jakarta, Indonesia
[2]Industrial Engineering, Trisakti University, Jakarta, Indonesia
[3]Industrial Engineering, Tarumanagara University, Jakarta, Indonesia
[1]iphov.kumala@esaunggul.ac.id

Abstract. Kencana Gemilang, Co. is one electronics industry engaging in the manufacture sector. This company manufactures and assembles household electronic products, such as rice cooker, fan, iron, blender, etc. The company deals with an issue of underachievement of an established production target on MCM products line 1. This study aimed to calculate line efficiencies, delay times, and initial line smoothness indexes. The research was carried out by means of depicting a precedence diagram and gathering time data of each work element followed by examination and calculation of standard time as well as line balancing using methods of Moodie Young and Generics Algorithm. Based on results of calculation, better line balancing than the existing initial conditions, i.e. improvement in the line efficiency by 18.39%, deterioration in balanced delay by 28.39%, and deterioration of a smoothness index by 23.85% was obtained.
Keywords: Line Balancing, Moodie Young, Genetics Algorithm, Balanced Delay

1. Introduction

XYZ, Co. is one of electronics industries engaging in the manufacture sector. This company produces and assembles household electronic products such as, rice cooker, fan, iron, blender, etc. Line 1 is an initial line production at XYZ, Co., however production target on this line is less efficient which leads to underachievement of a production target established by the company. In order to overcome the problem, an analysis on line balancing to improve the efficiency is required, because according to [2], line balancing can solve combinatorial optimization problems and [6], Line balancing can be used to minimize the number of workstations, balancing the workload between work-stations. [4]said, that line balancing is a balancing assignment of task elements from an assembly line to work stations. The goal is to minimize the number of work stations and idle time on all work stations

MCM product line 1 has 34 work stations equipped with 35 operators. It is known that there is not optimum achievement of the established production target on this line. The target achieved by a production section is only approximately 275 units per hour as compared with 300 units per hour or 85% of its maximum number. Accordingly, this study was conducted aiming at identifying causes of the unachieved production target, line efficiencies, balanced delays, and smoothness indexes of initial line conditions followed by an analysis of line balancing to be able to improve efficiency and output of the line. Regarding employment of methods in the line balancing, According to [1], Moodie Young is considered appropriate for a precedence diagram starting from one or a couple of separated operations yet being united in one operational element and ending with one operational element. Thus the study was conducted using methods of Young and Genetics algorithm. The reason used GA because, according to [5], Genetic algorithms are optimal for complex problem solving. The most appropriate result of the analysis was taken as a recommendation of the line improvement.

2. Methods

Research methodology employed in the present study is as follows :

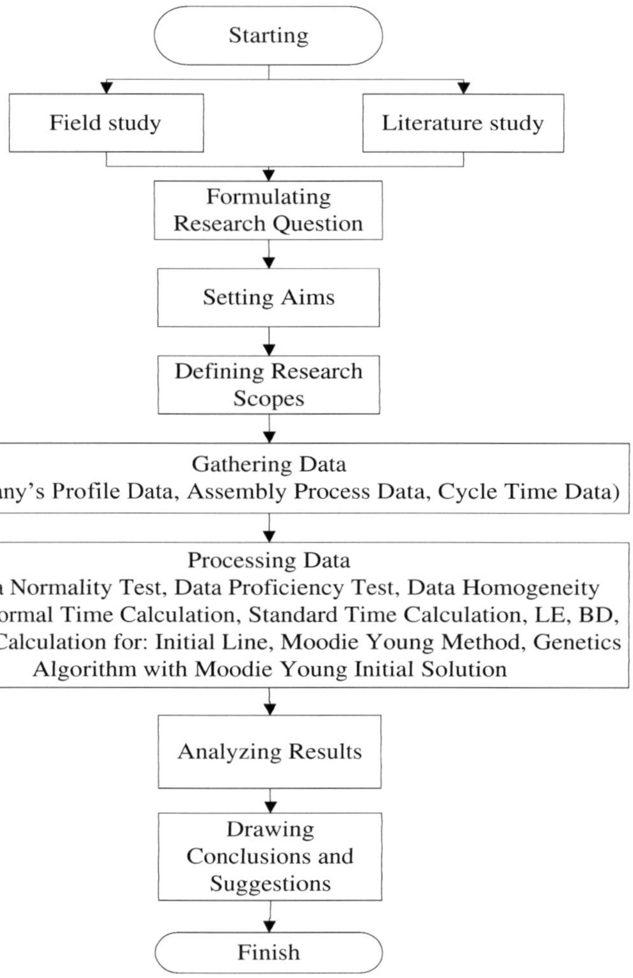

Figure 1. Research Flowchart

3. Results and discussion

3.1. Standard Time Calculation

Data gained was subject to a couple of tests before being used, i.e. normality, homogeneity, and proficiency tests. In this research, a confidence level of 95% and significance level of 5% were used. According to results of data tests, the gathered data was valid and able to be incorporated in the study.

In order to obtain a standard time, a normal time was calculated formerly. The normal time was gained from an average cycle time multiplied by an adjustment factor. An adjustment factor included was derived from the Westinghouse method. Standard time was obtained by multiplying the normal time by a leeway percentage then adding by the normal time itself. The leeway factor was given for three things, i.e. private needs, fatigue elimination, and inevitable obstacles.

3.2. Initial Line Conditions

In the initial line conditions, there were 34 work stations of which arrangement is showed in Table 1.

Table 1. Initial Line Conditions

Stat	Work Elements	Standard Time (sec)	Number of Workers (people)	Work station (sec)
1	1	7.80	1	9.83
	10	2.03		
2	2	9.48	1	12.56
	8	3.08		
3	3	9.96	1	9.96
4	4	6.54	1	11.01
	5	3.30		
	6	1.17		
5	11	2.30	1	9.05
	12	4.36		
	13	2.39		
6	52	3.92	1	10.26
	53	4.01		
	54	2.33		
7	51	3.30	1	3.30
8	55	3.64	1	7.44
	56	3.80		
9	59	2.13	1	12.04
	60	9.91		
10	50	7.67	1	7.67
11	48	3.70	1	6.47
	49	2.77		
12	15	9.85	1	11.68
13	19	1.75	1	6.27
	21	4.52		
14	20	3.23	1	10.73
	23	7.509		
15	18	2.15	1	8.34
	24	6.19		
16	22	11.81	1	11.81
17	14	2.04	1	8.27
	25	4.57		
	63	1.66		
18	26	11.11	2	9.16
19	31	6.13	1	6.13
20	32	2.19	1	4.90
	33	2.72		
21	37	7.29	1	7.29
22	42	5.86	1	12.93
	43	7.08		
23	17	5.28	1	5.28
24	7	5.01	1	8.30
	9	3.29		
25	47	14.38	1	14.38
26	57	6.55	1	10.52
27	61	9.41	1	9.41
28	62	9.65	1	9.65
29	28	2.13	1	6.88
	29	2.81		
	30	1.95		
30	34	1.99	1	7.35
	35	5.36		
31	38	7.81	1	10.52
	39	2.71		
32	40	2.36	1	9.77
	41	7.41		
33	44	4.65		10.85
	45	6.20		
34	46	5.91	1	5.91

A precedence diagram of line production at XYZ, Co. is presented in Figure 2.

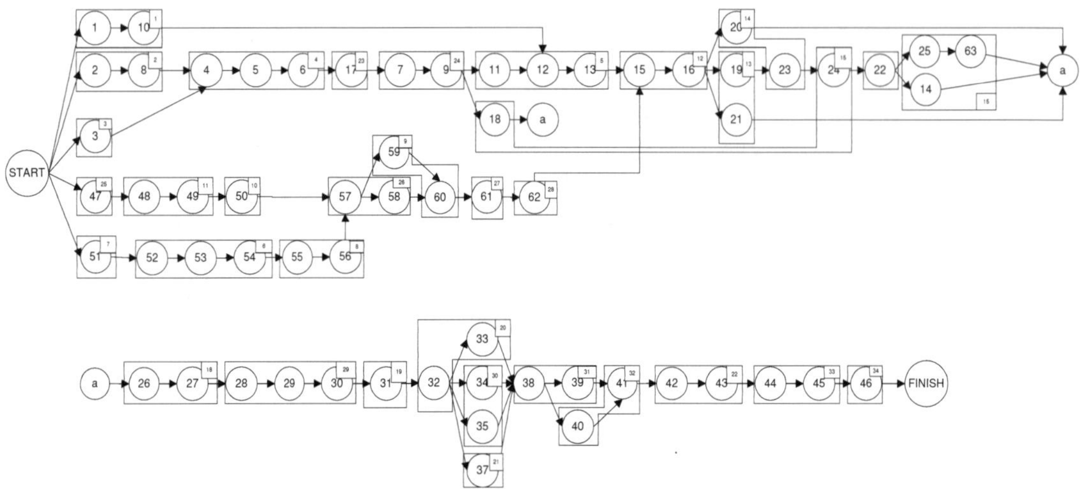

Figure 2. Precedence Diagram of XYZ, Co.'s Line 1

Based on the above initial line conditions, a line efficiency of 62.57, a balanced delay of 37.43%, and a smoothness index of 40.12% were obtained as shown in the following calculation.

3.3. Moodie Young Method-Based Improvement in Line Design
According to a Moodie Young method-based line design, 63 work elements were divided into 31 work stations (WS) as we can see in Table 2.

Table 2. Moodie Young method-based line design

WS	Work Elements	Standard Time (sec)	Number of Workers (people)	Time in WS (sec)	WS	Work Elements	Standard Time (sec)	Number of Workers (people)	Time in WS (sec)
1	3	9.96	1	9.96		21	4.52		
2	2	9.48	1	9.48	16	20	3.24	1	9.50
3	1	7.80	1	11.10		19	1..4		
	51	3.30			17	23	7.50	1	7.50
4	47	14.38	2	11.70	18	24	6.19	1	6.19
	52	3.92			19	22	11.81	1	11.81
	8	3.08				25	457		
5	4	6.54	1	10.55	20	14	2.04	1	8.27
	53	4.01				63	1.66		
	48	3.70			21	26	11.11	1	11.11
6	5	3.30	1	10.94	22	27	7.21	1	9.34
	49	2.77				28	2.13		
7	6	1.17	2	11.47		29	2.81		
	50	15.33			23	30	1.95	1	10.88
	17	5.28				31	6.13		
	54	2.33			24	32	2.19	1	9.48
						37	7.29		

Table 2. Moodie Young method-based line design (continued)

Stat	Work Elements	Standard Time (sec)	Number of Workers (people)	Time in WS (sec)	Stat	Work Elements	Standard Time (sec)	Number of Workers (people)	Time in WS (sec)
	7	5.01			15	15	9.85	1	11.68
8	55	3.64	1	11.94		16	1.84		
	9	3.29				33	2.72		
9	56	3.80	1	10.35	25	34	1.99	1	10.07
	57	6.55				35	5.36		
	58	3.97			26	38	7.81	1	10.52
10	11	2.30	1	10.64		39	2.71		
	12	4.36			27	40	2.36	1	9.77
	13	2.39				41	7.41		
11	18	2.15	1	6.67	28	42	5.85	1	5.85
	59	2.13			29	43	7.08	1	11.72
12	60	9.91	1	9.91		44	4.65		
13	61	9.41	1	9.41	30	45	6.20	1	6.20
14	62	9.65	1	9.65	31	46	5.91	1	5.91

An initial line efficiency of 80.96% with a balanced delay of 19.04% and a smoothness index of 16.27% were obtained from the calculation

3.4. Genetics Algorithm Calculation

Genetics Algorithm recognizes terms of gen, chromosome, and population. In this case, gen constituted the existing work elements. There are several ways to identify the arrangement of chromosomes, one of which is by using a random generator on Microsoft Excel software. Combination of each chromosome makes up a population. After the population was identified, fitness of each chromosome was calculated. The fitness value comprised a line efficiency of each chromosome.

Once the fitness of each chromosome was obtained, two best chromosomes from each population were selected as a pair of parents for crossover. The crossover was performed uniformly, in which every point can be an intersection. Random numbers between 0 and 1 were drawn in every chromosome, of which ≤ 0.5 was rounded down to 0 and > 0.5 was round up to 1. Random numbers were able to be drawn on Microsoft Excel. Numbers of 0 and 1 functioned to determine the selected chromosomes. If the number was 0, chromosome 1 was selected while if the number was 1, chromosome 2 was selected.

Mutated chromosomes obtained were followed by gain of a fitness value of 79.55%. The subsequent step was elitism in which a chromosome with the best fitness would be selected to be passed down to the next generation. The selected chromosome was chromosome 1 in population 1.

3.5. Genetics Algorithm-Based Moodie Young and Line Design Improvement

According to [3], When there are numerous alterations to calculate, Evolver 5.5 software can be employed to facilitate the calculation. There are up to 1000 iteration performed on the Evolver. The calculation was continuously carried out until reaching defined limits. It was found from the software calculation of an initial solution using the young moodie method that the line efficiency was 80.96, the balanced delay was 19.04%, and the smoothness index was 16.27%, while from the calculation of an initial line solution showed the line efficiency of 79.44%, the balanced delay of 20.56%, and the smoothness index of 17.37%.

IOP Conf. Series: Materials Science and Engineering **277** (2017) 012053 doi:10.1088/1757-899X/277/1/012053

3.6. Calculation results

Calculation results of comparison of the initial line between Young and Genetics Algorithm methods are presented in Table 3

Table 3. Comparison of the initial line between Young and Genetics Algorithm methods

	The first	Moodie Young	Genetic Algorith with Moodie Young
Line eficiency (%)	62.57	80.96	80.96
Balance Delay (%)	37.43	19.04	19.04
Smoothing Index	40.12	16.27	16.27
Amount of workstations	34	31	31
Output/Hour (Unit)	251	300	300

Table 3 shows that lines recommended with Moodie Young and Genetics Algorithm with Moodie Young initial solutions demonstrate the best results. These same results were due to limited movement of work elements based on a precedence diagram. Grouping of workstations based on the Moodie Young method and Genetic Algorithm with Moodie Young initial can be seen in Table 4.

Table 4. Proposed grouping of workstations

Stat	Work Elements	Standard Time (sec)	Number of Workers (people)	Time in WS (sec)	Stat	Work Elements	Standard Time (sec)	Number of Workers (people)	Time in WS (sec)
1	3	9.96	1	9.96	9	56	3.80	1	10.35
2	2	9.48	1	9.48		57	6.55		
3	1	7.80	1	11.10	10	58	3.97	1	10.64
	51	3.30				11	2.30		
4	47	14.38	2	11.70		12	4.36		
	52	3.92			11	13	2.39	1	6.67
	8	3.08				18	2.15		
	10	2.03				59	2.13		
5	4	6.54	1	10.55	12	60	9.91	1	9.91
	53	4.01			13	61	9.41	1	9.41
6	48	3.70	1	10.94	14	62	9.65	1	9.65
	5	3.30			15	15	9.85	1	11.68
	49	2.77				16	1.84		
	6	1.17			16	21	4.52	1	9.50
7	50	15.33	2	11.47		20	3.24		
	17	5.28				19	1.74		
	54	2.33			17	23	7.50	1	7.50
8	7	5.01	1	11.94	18	24	6.19	1	6.19
	55	3.64			19	22	11.81	1	11.81
	9	3.29							

Table 4. Proposed grouping of workstations (continued)

Stat	Work Elements	Standard Time (sec)	Number of Workers (people)	Time in WS (sec)	Stat	Work Elements	Standard Time (sec)	Number of Workers (people)	Time in WS (sec)
	25	4.57			25	33	2.72		
20	14	2.04	1	8.27		34	1.99	1	10.07
	63	1.66				35	5.36		
21	26	11.11	1	11.11	26	38	7.81	1	10.52
	27	7.21				39	2.71		
22	28	2.13	1	9.34	27	40	2.36	1	9.77
	29	2.81				41	7.41		
23	30	1.95	1	10.88	28	42	5.85	1	5.85
	31	6.13			29	43	7.08	1	11.72
24	32	2.19	1	9.48		44	4.65		
	37	7.29			30	45	6.20	1	6.20
					31	46	5.91	1	5.91

According to analysis from Table 4, it is shown that Moodie Young and Genetics Algorithm methods equipped with Moodie Young initial solutions are better than the existing initial lines therefore recommendation of Moodie Young and Genetics Algorithm with Moodie Young initial solutions can be implemented in order to improve the line productivity.

Grouping of work elements in the precedence diagram based on Moodie Young and Genetics Algorithm methods are shown in Figure 3.

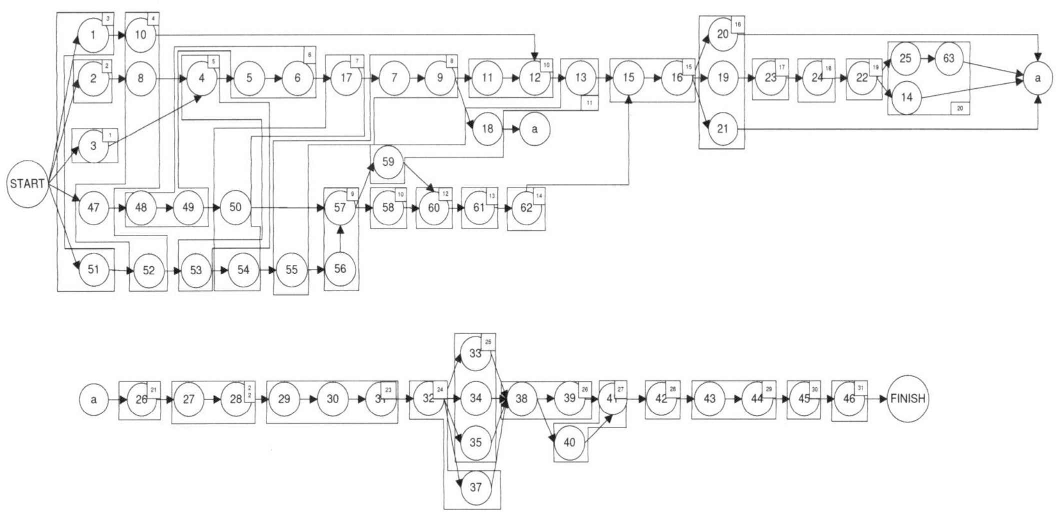

Figure 3. Grouping of Proposed Work Elements

Analysis from Figure 3 indicates that Moodie Young and Genetics Algorithm methods with Moodie Young initial solutions show same values of efficiency, balanced delay, and smoothness index. This was due to limited movement of work elements following the precedence diagram. This grouping of work elements was able to occur as every activity in each work element does not require high skills enabling the work element to be aggregated one another.

4. Conclusions

Based on the conducted present study, following conclusions are drawn.

a. Results of calculation show that initial line of the company indicated a line efficiency of 62.57%, balanced delay of 37.43%, and smoothness index of 40.12% with an output of 251 units per hour.
b. Results of calculation using Moodie Young and Genetics Algorithm with Moodie Young initial solutions show a line efficiency of 80.96%, balanced delay of 19.04%, and smoothness index of 16.27% with an output of 300 units per hour.
c. Results of calculation using Genetics Algorithm with initial line solutions show a line efficiency of 79.44%, balanced delay of 20.56%, and smoothness index of 17.37 with an output of 300 units per hour.
d. The best method recommends an implementation of line balancing based on Moodie Young and Genetics Algorithm methods with Moodie Young initial solutions to the company.

5. References

[1] Baroto T 2006 Simulasi perbandingan Algoritma Regionapproach, Positional Weight, Dan Modie-Young dalam efisiensi dan keseimbangan lini produksi, *Gamma*, **vol. II**, no. I, 49–54.
[2] Battaïa A D O 2013 A taxonomy of line balancing problems and their solution approaches, *Int. J. Prod. Econ.*, **142**, 259–277.
[3] Bernhard M 2011 Analisa Keseimbangan Lini Pada Departemen Chassis PT Toyota Manufacturing Indonesia Dengan Algoritma Ant Colony, Rank Positional Weight dan Algoritma Genetika. *Thesis*. Universitas Tarumanagara.
[4] Gaspersz V 2000 *Manajemen Produktivitas Total*. Jakarta: Gramedia.
[5] Sihombing R S 2014 Pemanfaatan algoritma genetik pada aplikasi penempatan buku untuk perpustakaan sekolah, *Pelita Informatika Budi Darma,* **vol. 6**, no. 2, 113–118.
[6] Shahabudeen P and Sivasankaran P 2014 Literature review of assembly line balancing problems, *Int J Adv Manuf Technol* **73**, 1665–1694.

IOP Conf. Series: Materials Science and Engineering **277** (2017) 012054 doi:10.1088/1757-899X/277/1/012054

The model for estimation production cost of embroidery handicraft

Nofierni[1], IK Sriwana[1] and Y Septriani[2]

[1]Industrial Engineering, Esa Unggul University, Jakarta, Indonesia
[2]Accounting, Polytechnic State Padang, West Sumatera , Indonesia
[1]nofi.erni@esaunggul.ac.id

Abstract. Embroidery industry is one of type of micro industry that produce embroidery handicraft. These industries are emerging in some rural areas of Indonesia. Embroidery clothing are produce as scarf and clothes that show cultural value of certain region. The owner of an enterprise must calculate the cost of production before make decision how many product are received from the customer. A calculation approach to production cost analysis is needed to consider the feasibility of each order coming. This study is proposed to design the expert system (ES) in order to production management of embroidery industry. The model will design used Fuzzy inference system as model to estimate of production cost. Research conduct based on survey and knowledge acquisitions from stakeholder of supply chain embroidery handicraft industry at Bukittinggi, West Sumatera, Indonesia. This paper are use fuzzy input were the quality, the complexity of the design and the working hours required and the result of the model are useful to manage production cost on embroidery production.
Keywords :fuzzy inference system, embroidery handicraft, production cost

I. Introduction

Embroidery is a handicraft product that is widely used to increase the value added of clothing especially for aesthetic value. Indonesia has many variations of embroidery craft products. One of the famous cities with embroidered craft is Bukittinggi. Embroidered sales of embroidered scarves are often used at various custom events. One of the famous is the embroidered scarf known as kapalosamek. Kapalosamek is hand embroidery was having produce without using machine (handmade).

Handicraft industry had been used skill to make embroidery which is applied on many kind clothing that are dispersed widely as microenterprise in Indonesia. The design and production through the creative and innovative process have been used of thread with various types and colors as the main raw material.

Many embroidery handicraft have been exported to many countries and local, regional market regularly. Rapid changes in clothing trend and variations of customer trends requires an approach on production management, especially with regard to calculate of production cost on enterprise unit. These research developed software with expert management system approach. The model used in this application refers to a method of managing production in modern industry, especially to solve a lot of problem in estimating the need for raw materials to make embroidery handicraft. Production planning at handicraft industry is a complex concept that refers to make proper decisions under uncertainty and lack information about the product quality and the availability of working hours.

Research on utilization of information technology in the handicraft industry has been done to improve the competitiveness of handicraft carpet industry in India. Related to knowledge management fuzzy expert system approach are used in order manage information and knowledge to achieve the company's goals [2]. The model based on fuzzy rules to simulate the behavior of the firms, is presented under the assumption of determined input parameters previously detected and an algorithm is developed to achieve the minimal structure of the model. Decision support system are used of fuzzy logic to consider inventory carrying, shortage and ordering cost as well as transportation cost [6]. Fuzzy Inference System are particularly suited for modeling the relationship between variables in complex environment because they introduce a process of decision making which is more human-like [2]. These system are based on fuzzy logic modeling approach, and allow reaching solution based on linguistic variables which makes reference to expert knowledge. They are useful in the cases were human knowledge is available and there is not enough information as quantitative value. FIS need parameters as inputs and outputs. The fuzzy numbers are quantified using fuzzy logic method using membership function. This research proposed to develop an expert system to make decision about production cost on embroidery handicraft. The embroidery handicraft enterprise has a representative to find the production cost.

2. Methods

Research methodology employed in the present study are used fuzzy inference system. The data have been collected from embroidery handicraft industry especially for scarf and clothing for women as cultural Minangkabau symbolic. In order to collect data, information about handicraft industry, the research have done at Bukittingi and the area of the surrounding that serves as part of the value chain embroidery bussiness. Data and information are collect from owners of business, craftman of embroidery, retailer where as selling the embroidery handicraft.

The lack of accuracy information about production cost on embroidery industry can be approached using fuzzy logic. This research are conducting with Fuzzy Rule-Based Inference System (FIS) to estimate production cost at handicraft embroidery. Despite the complexity of such decision making, FIS use linguistic value to define the input and output (figure1). FIS involves three important concepts : membership function, inference rules and fuzzy set operation. Membership functions represent the fuzzy sets of input and output variables, fuzzy set operations are main operations among fuzzy sets and inference rules are linguistic fuzzy rules in the form of " IF-THEN". In this study input fuzzy have been used to estimation of production cost on embroidery enterprise.

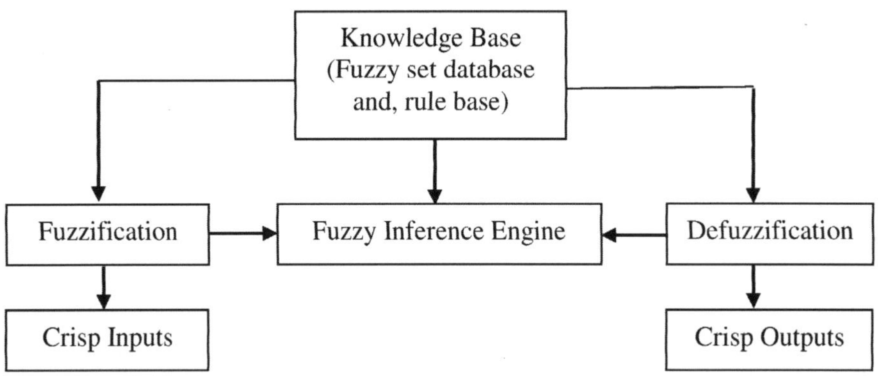

Figure 1. The inference process from fuzzification through defuzzification of FIS

3. Results and discussion

3.1. Value Chain and Production of Embroidery Handicraft

The expertise to make embroidery are usually as hereditary of rural people in Indonesia village. The process is not too complicated but require several days to obtain and patience in the create of embroidery. The main material were used are sheet of fabrics and yarn to be embroidered. Type and various yarn are supplied from retailer at regions. The motif will design by order of costumer or used the motif are available. Many motif embroidery was produced has a good appearance in design style, construction and colours. The more intricate embroidery making and more colourfull the design will be increase the price. Some of handicraft embroidery product are shown in Figure 2.

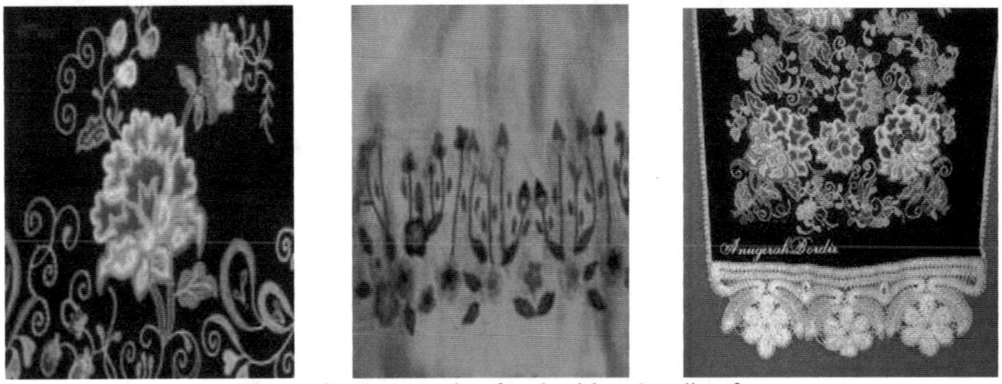

Figure 2. The sample of embroidery handicraft

The business activities of embroidery handicraft industry, based on observations and interviews with experts and respondents establish an order of the supply chain from upstream to downstream. Based on the viewpoint of the process sequence of the decision-making process and the implementation of the flow of products, information and funds, supply chain network embroidery industry shown in the Figure 3.

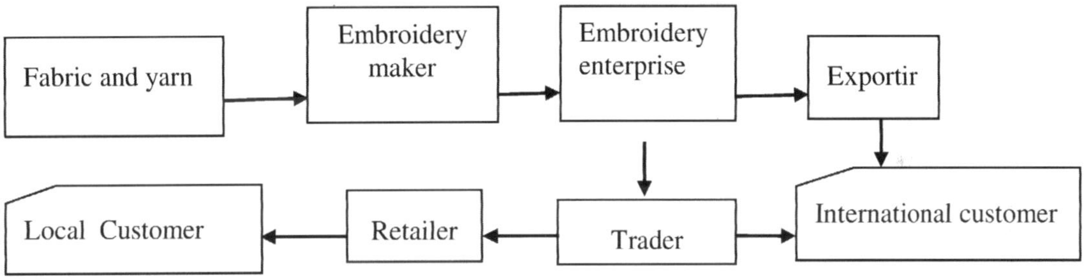

Figure 3. The structure of value chain embroidery handicraft industry

3.2. Developing Expert System

Knowledge management is an fundamental approach to collect and structuring the correct information transfer and dissemination within the organization. Knowledge is intellectual capital, it is critically important for organization. Knowledge is information that is contextual, relevant, and actionable. There are different meaning about data and knowledge. Where as data are collection of fact, measurement and statistic. Information is defined as organized or processed that are timely and accurate. While data, information and knowledge can all be viewed as asset of organization, knowledge provide a higher level of meaning about data and information. Knowledge encompasess the implicit and explicit restriction. There are conceptual difference and distinguished knowledge between an organization tacit knowledge and explicit knowledge. Tacit knowledge is usually in the domain of subjective, cognitive, and experimental learning, where as the explicit knowledge is the procedural guides, software, document.

There are many researchers who interest in how to acquisition knowledge in various research type. In order to construct knowledge stored in the minds of experts, are known methods of the expert

IOP Conf. Series: Materials Science and Engineering **277** (2017) 012054 doi:10.1088/1757-899X/277/1/012054

system. This method is capable of structuring knowledge and deposited into the machine to be used as a substitute for an expert in making decisions. Knowledge of embroidery handicraft businesses related to the production cost has not been compiled into a systematic knowledge and easy to navigate. Some of consideration in order calculate production cost of embroidery has not been calculated so as to ensure profit taking. The process for acquiring knowledge begins with the process of socialization is done through observation, focus group discussion and brainstorming.

3.3. Defining the criteria (inputs and outputs)

There are three fuzzy inputs used in preparing the FIS to calculate production cost of embroidery products. The significant consideration are the quality, the complexity of the design and the working hours required. The inputs have criteria such as :

1. Quality inputs are grouped into two categories such as high and low.
   a. High if the quality of embroidery product are use good material, tight embroidery, looks smooth.
   b. Low if the quality of embroidery product are used ordinary material, less tight embroidery, looks rough
2. The complexity of design determined by the complexity of embroidery maker based on decorative design and the usage of yarn. This input determined by variations in the color used. Criteria relating to the complexity design are grouped into three, namely
   a. Difficult if the embroidery product looks beautiful decoration design with embroidery surface area more than 50% of fabric and use yarn more than 8 colours
   b. Moderate if the embroidery product have ordinary decorative design with embroidery surface area 30 - 50% of fabric and use of yarn as many as5-7 colours
   c. Easy if the embroidery product have ordinary decorative design with embroidery surface area 10% to30% of fabric and use of yarn less than 4colours
3. Working hours is related to the required of working hours to make of embroidery handicrafts. The embroidery makers generally make embroidery as a side job to fill the spare time. Based on the required time are grouped into:
   a. High if work hours required as much 32 - 60 hours / (9 - 20 days)
   b. Moderate if work hours required as much 12 - 32 hours / (4 - 8 days)
   c. Low if work hours required as much 4 - 10 hours / (2-3 days)

In order to calculate the embroidery production costs are structured 3 category of production cost. There are the output criteria :

1. High if the production cost such as IDR 1.500.000 to 2.000.000
2. Moderate if the production cost such as IDR 700.000 to 1.499.000
3. Low if the production cost such as IDR 100.000 to 699.000.

3.4. Fuzzy Inference System

A fuzzy rules based system is developed by human operators with the aid of practical experience to face complex situations, with only a set imprecise linguistic if then rules and an imprecise system state. This system incorporate fuzzy inference and rule based expert system resembling what human do daily. Inputs and outputs are two basic elements in the system using handling approaches. The input constitutes some ambiguous verbal semantic or unclear concept for specific event. Following the fuzzy inference mechanism, the output can be fuzzy set or precise set certain features. Therefore, defuzzification is necessary to convert the output result into crisp number. Fuzzy inference infers the results from the existing rule-based system [4].

Expert systems are defined as consulting systems that simulate the reasoning behavior of human expert. The most important components of expert systems are the knowledge base and the inference engine. The main part of the FIS model is the rules. The behavior of a fuzzy system is characterized by a set linguistic rules which constitutes a rule base. The fuzzy "if-then" rules are defined on the basis of experts knowledge in each area. In Mamdani approach the premises and the consequences of the if-then are linguistic variables associated with fuzzy concept. Every rules has a weight as the number between 0 and 1 which assign the importance of each rule. A fuzzy rule can be written " if x_1 is a , and x_2 is b, where x_1 and $x_2$are variables, y is solution variable, and a, b, and c are fuzzy linguistic terms. The linguistic rules are extracted based on FIS approach. Table 1.shown the fuzzy interpretation of some parameters based on fuzzy linguistic. The extracted rules are entered in to the rule editor of software developed.

Table 1. Rule of alternative for calculation of production cost

Rules	Fuzzy Input Quality	 Design complexity	 Work hours	Fuzzy Output Production cost
1	Low	Difficult	Moderate	Moderate
2	Low	Difficult	Low	Low
3	Low	Difficult	High	Low
4	Low	Moderate	Moderate	High
5	Low	Moderate	Low	Moderate
6	Low	Moderate	High	Low
7	Low	Easy	Moderate	High
8	Low	Easy	Low	High
9	High	Easy	High	High
10	High	Difficult	Moderate	High
11	High	Difficult	Low	Moderate
12	High		High	Low
13	High		Moderate	High
14	High	Difficult	Low	High
15	High	Moderate	High	Moderate
16	High	Easy	Moderate	High
17	High	Easy	Low	Moderate
18	High	Easy	High	Low

The next step is fuzzification which is the process of converting precise or imprecise data into fuzzy data by assigning membership function. In this research are used triangular membership function is exploited due to its prevalent. The criteria linguistic are assigned as three categorize. The experts are involved in the formulation of criteria and their input factors are embroidery maker, the owner of enterprise who produce handicrafts, wholesalers and store that sent the embroidery products to end customers. Fuzzy set input and output model are entered to Matlab software. This rules will be construct in order develop application of expert system. Figure 4 shown relationship between fuzzy input data with fuzzy output data. Every input and ouput data based on rules are connecting as a rules under fuzzy value. An assessment of calculation production cost of embroidery handicraft in kind "kapalosamek" scarf using Matlab software, which is inputs data quality, complexity of design, work hours and output data as production cost has shown several fuzzy values.

IOP Conf. Series: Materials Science and Engineering **277** (2017) 012054 doi:10.1088/1757-899X/277/1/012054

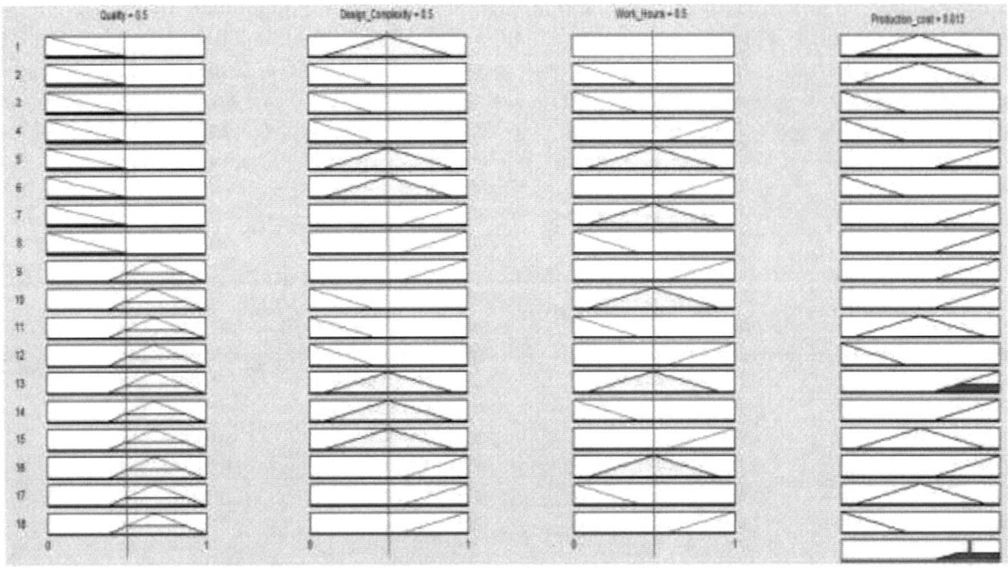

Figure 4. Relationship fuzzy input and output data

In order to calculate production cost using FIS system, each input data and output data were calculated using center of area under curved. An application to calculate production cost and result given shown at Figure 5.

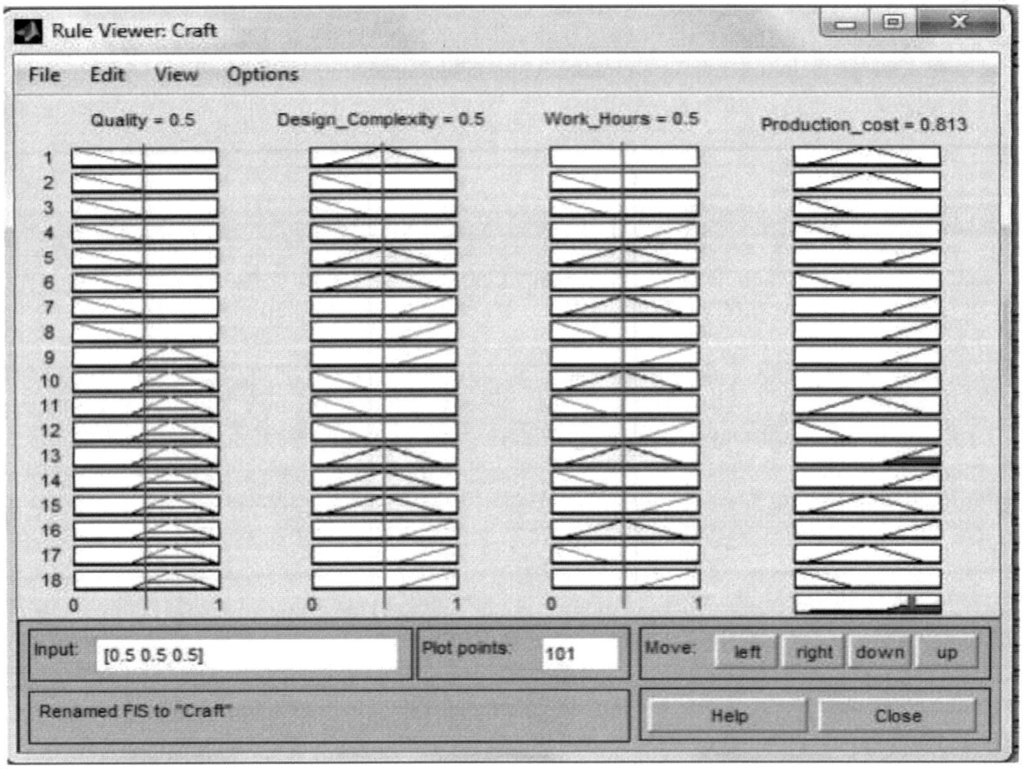

Figure 5. The sample of estimation cost using FIS on Matlab software

4. Conclusions

Based on the conducted present study, following conclusions are drawn.

a. A fuzzy inference system is proposed to solve the the problem for calculate cost of production of handicraft embroidery.

b. The production cost consider about quality, complexity design, availability working hours. The nature of the problem is complex due to the lack of accurate information as well as the need for knowledge of experts. The experts are involved in the formulation of criteria and their input factors are embroidery maker who make embroidery, embroidery business owner, local government, retailer, and customers.

c. Based on using FIS based on the expert opinions, the developing rules FIS can be helpful to develop the expert system. A fuzzy rule-based system with linguistic variables an some of set 'if-then" rules are applied to solve the production cost.

d. The output of this study to design software of expert system in order estimate embroidery production cost on dealing with quality, complexity design and working hours requirement.

5. References

[1] Agrawal, Sharma, Kumar 2010 IT Based KM Framework for improving Research and Methodology Process in Technical Education *Journal of Engineering, Science and Management Education,* **Vol.2**

[2] Arias A D, Castro J L, Navarro M, Sanchez J M and Zurita J M 2010 A fuzzy expert system for business management. *Journal of Expert Systems with Applications*, **Vol. 37**

[3] Azeem M F 2012 *Fuzzy Inference System: Theory and Application.* (London: In Tech Publications John Howkins)

[4] Juang Y S, Lin S S and Kao H P 2007 Design and implemetation of a fuzzy inference system for supporting customer requirement. *Journal of Expert Systems with Applications.*

[5] Wankee P, Alvarenga H, Hadi-Vencheh A and Azad M A K 2017 Fuzzy Inference Systems and inventory allocation decisions: Exploring the impact of priority rules on total costs and service levels. *Journal of Expert Systems with Application.*

[6] Yoshiyuki Oshita 2009 An analysis of Creative Industries Policy: Creativity and Design in the Context of Policies. *Quarterly Journal of Public Policy & Management.* **Vol.3**

IOP Conf. Series: Materials Science and Engineering 277 (2017) 012055 doi:10.1088/1757-899X/277/1/012055

Enhancing the management of the noise level using six sigma method: a case study on the machining industry

D Rimantho[1,2] and M W Hanantya [1]

[1] Department of Industrial Engineering, Pancasila University, Jagakarsa, Jakarta 12640, Indonesia
[2] dino.rimantho@univpancasila.ac.id

Abstract. The hearing disorder is caused by noises in the workplace, it has been a concern for numerous researchers. This study aims to improve the performance of the management of the noise level by applying the six sigma method. Data collection is done directly by using a sound level meter. In addition, several key informants also used in order to gather information related to the problem. The results showed the values of Cp and Cpk on the entire department are still below the recommended value. Moreover, the results also showed the potential failure (DPMO) approximately 115,260.6 and equivalent to the Sigma value is approximately 2.70. Furthermore, the highest value of the RPN in FMEA analysis is the workers not wear factor of earplug which is about 56. Thus, it is necessary to make enhancements to the process of managing the noise level in the framework of continuous improvement.
Keywords: Noise, six-sigma, FMEA, Fishbone, DPMO

1. Introduction

Noise can be defined as unwanted sound that can be found in the environment and the workplace [1]. Furthermore, noise is considered to be one factor contributing to occupational hazards in the workplace [2], [3], [4]. Several studies show that noise pollution is a major concern in many countries [5], [6], [7]. In addition, there are approximately 600 million workers exposed to noise in the workstation worldwide [8]. Furthermore, the WHO reports that workers in developed countries about 15% exposure to noise and potentially harmful to hearing [4].

The relationship between occupational noise and an increase in accidents and injury risk significantly has been the subject of study by several researchers [9], [10]. Several potential consequences arising from noise exposure that contribute to accidents and injury, such as disrupt the concentration [10], increased fatigue and extend the reaction time [9] reduces the ability to hear warning signals [11], increased stress [12], sleeping disorders [13], hypertension [14], and heart disease [15].

Furthermore, research conducted by Moll van Charante and Mulder emphasizes the increased risk of accidents triggered by the presence of hearing loss as an accumulation of exposure to noise [16]. Moreover, the lack of use of hearing protection devices (HPD) is suspected as one of the factors responsible for the increased risk of injury [17], [18], [19]. As a result of the lack of attention of workers in the use of HPD led to various problems such as a decreased ability to hear, interpret the message and a warning signal in the workplace and in order to analyze the ability of votes required in the monitoring machinery and work equipment [11]. In order to solve the problem of minimizing the potential risk to use various approaches. Several considerations are used in applying these approaches such minimal cost and on time. For example, U.S. Department of Health and Human Services has made a practical guide in order to prevent job loss of hearing [20].

One method that may improve the performance in reducing noise is to apply the Six Sigma method. There were challenges and exciting opportunities in every procedure that may be given by the Six Sigma method. Thus, six sigma method is applied as an approach method in this study to minimize the risk of noise. Six Sigma method has been applied in several areas of research, such as manufacturing [21], [22], health and safety [23], [24], environmental management systems [25], pharmaceutical industry [26], [27].

Six Sigma which applies a variety basic statistical methods and tools of traditional quality management has a huge impact on the quality management approach [28]. Through the application of basic statistics into innovation in quality as an initiative to make Six Sigma business strategy to achieve goals such as increased profitability, increased the market, increase customer satisfaction, reduce material wastage, the waiting time and the risk of accidents. Furthermore, the enhancing company performance is able to achieve by implementing the business strategy using six sigma method [29]. Thus, this study aims to improve the management of health and safety due to the risk of noise by applying six sigma methods.

2. Material and method

The study was conducted in the manufacturing industry, especially in the machinery and automotive divisions. This study analyzed the noise level management to review the factors that most influence on the management of the noise level. Furthermore, the study begins by analyzing and identifying the factors that lead to high levels of noise which exceeds the noise level standard quality standard defined by the Ministry of Labour in Indonesia. In order to address this problem, six sigma method is used to define, measure, analyze, improve and control. Six Sigma methods adopted and modified from US EPA Environmental Professional's Guide to Lean and Six Sigma [30].

In addition, several methods also used in analyzing the causes of problems such as Fishbone diagram, Failure Mode Effect Analysis (FMEA) and 5W1H. Furthermore, this study used observational data and noise level measurements conducted directly by using sound level in the department of construction, assembly and engineering. In addition, the study also conducted brainstorming on several key informants to obtain information related to the causative factors. In addition, the questionnaire used in order to explore the information related to factors that could potentially lead to failure of the noise management process. The questionnaire was distributed to five key people in the company. Primary data obtained from measurements of the noise level will be processed with the aid Minitab software.

In addition, by knowing the value of process capability, then the sigma value will also be determined. Moreover, Fishbone Diagram also used to identify the cause of the noise. Based on the calculation process, then performed an analysis of the factors that cause the problem by applying Fishbone diagram and determine the factors that have the highest number of potential risks by implementing FMEA. Thus, the proposed improvements to the process can be provided based on the ranking of risk factors.

3. Result and discussion

3.1. Define

Determination of the first stage in the current six sigma method is Define. This phase has four major step process that may be considered. For instance, investigation processes and work environment, SIPOC diagram preparation, gathering information and problem establishment. Clarify the issue a priority in the management of noise level. This phase uses SIPOC diagram instruments in order to obtain an overview the process in detail. SIPOC is an abbreviation of Supplier, Input, Process, Output, and Customer. The SIPOC diagram for the noise level management process is given in Table 1 below.

Table 1. SIPOC Diagram

SUPPLIER	INPUT	PROCCES	OUTPUT	CUSTOMER
Engineering *Assembly* *Construction*	• Material metal • standard work • Machinery and Equipment work • Personal protective equipment • The company's physical environment	Process design and assembly machines	The noise level more than 85 dB	• Employees feel uncomfortable • Employees get tired • Employees easily stressed • Employee potential against the risk of deafness.

3.2. Measure

In order to determine the size and nature of the problem to be solved is required initial data collection history. It is part of the determination of lean six sigma phase. According to the results of direct measurements of noise levels at three different location such as engineering, assembly and construction. Furthermore, the data shows that construction department has the extraordinary level of noise approximately 91.8 dB and the minimum level around 81.4 dB Moreover, at engineering department has the high level nearly 72.5 dB and the low level around 56.3. In addition, the assembly department provides the high level of noise roughly 86 dB and the small level around 67.1 dB. Therefore, overall it puts construction department with the highest noise levels compared with assembly and engineering departments. Based on this data, the normality test conducted to determine which the data have the normal distribution.

Based on normality test gives information that the value of Kolmogorov-Smirnov (KS) around 0.300 and p-value <0.010. With the hypothesis that the value of H0 is accepted if the p-value> 0.05, it can be stated that the level of noise on engineering department is a normal distribution. Furthermore, normality test of the data on the assembly department below provides information that the Kolmogorov-Smirnov (KS) is approximately 0.344 and a p-value of about 0.459. With the hypothesis that H0 is accepted if the p-value> 0.05. Thus, can be expressed that the level of noise on the assembly department is a normal distribution. The normality test at the construction department also provide the value of Kolmogorov-Smirnov (KS) around 0.090 and p-value approximately 0.150. With the similar hypothesis that the value of H0 will be accepted if the p-value>0.05. Therefore, it can represent that the level of noise on the construction department is a normal distribution.

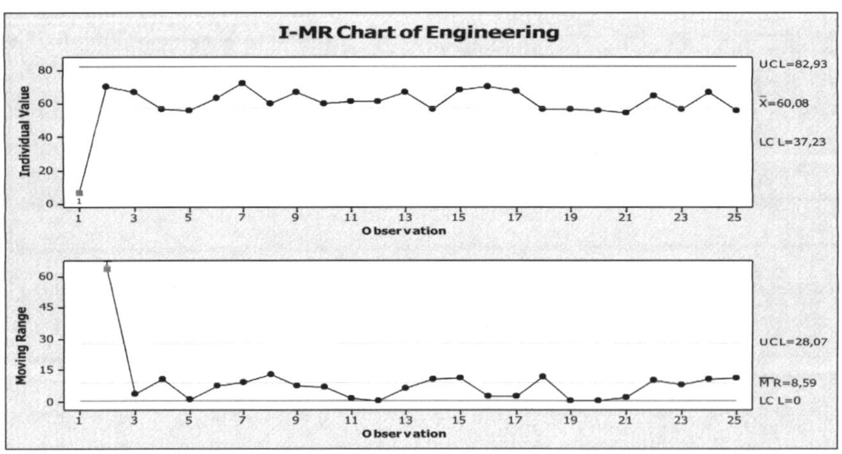

Figure 1. I-MR of noise control chart at The Engineering Department

From the normality test is known that all research data were normally distributed. Furthermore, individual moving range control chart (I-MR) were made on each of the test parameters. The control chart is useful to ascertain whether the performance of the process is currently under control (in control) in which the process is running with good stability without the influence of specific causes. The figures of control chart of noise at each engineering department are presented in Figure 1, 2 and 3.

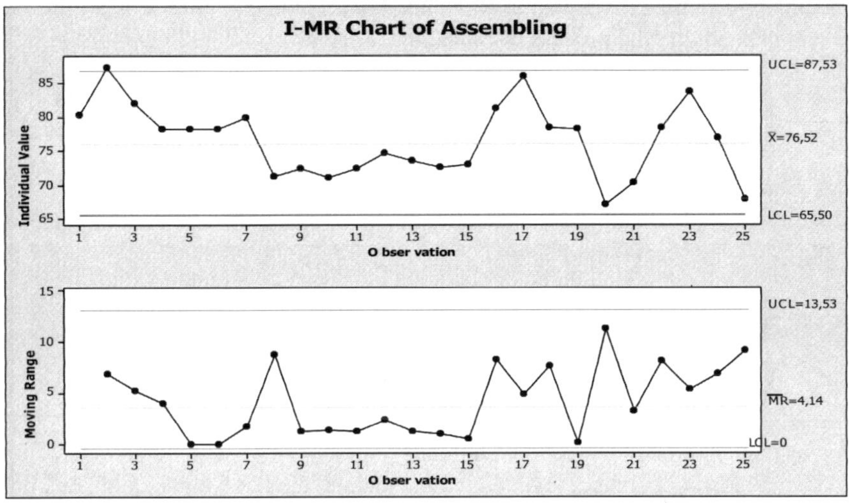

Figure 2. I-MR of noise control chart at The Assembly Department

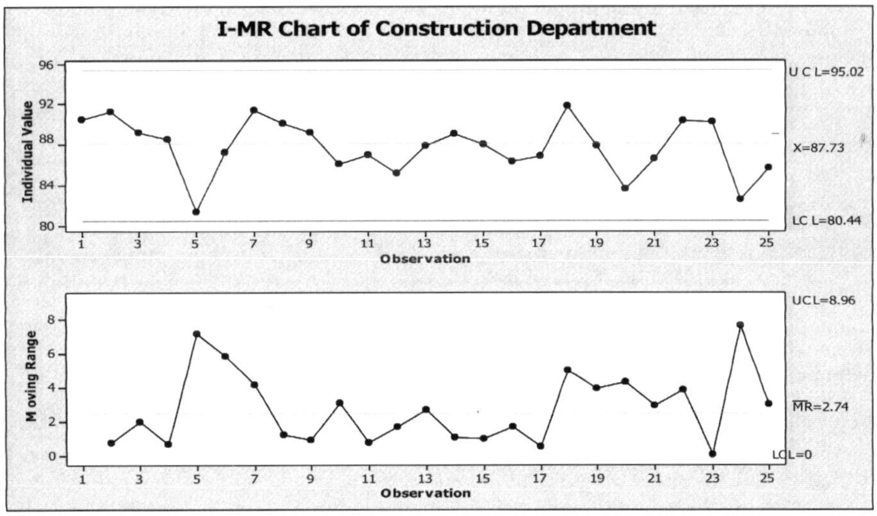

Figure 3. I-MR of noise control chart at The Construction Department

The next step in the measures phase is calculating process capability. The calculation of process capability is intended to determine the extent to which a management process to meet predetermined specifications. Process capability analysis conducted if all the data is in statistical control. The table 2 below represents the results of calculations to determine the value of the Process Capability Analysis of each department. Furthermore, the results show that the Cp values at the entire department were still under the prescribed specifications such as the engineering department around 0.39, the assembly department approximately 0.92 and the construction department roughly 0.71. Similarly, Cpk values are also under the prescribed specifications, such as the engineering department approximately 0.24, the assembly department around 0.85 and the construction department around 0.56. In addition, Cp and Cpk values on all three departments have a potential failure of the process of managing noise levels around 287,410.98 in the engineering department, the assembly department 6,823.09 and 51,547.65 on the construction department. Furthermore, based on the results of these calculations the average value of Cpk around 0.55 with the potential failure (DPMO) approximately 115,260.6. By using a conversion table DPMO to Sigma Value Based on Motorola Concept was obtained Sigma level around 2.70. Therefore, it can be explained that the process of managing noise levels should be improved.

Table 2. Summary of Process Capability Analysis

Department	Cp		Cpk		Possible process
	Result	Recomendation	Result	Recomendation	failure (DPMO)
Engineering	0.39	1.33	0.24	1	287,410.98
Assembly	0.92	1.33	0.85	1	6,823.09
Construction	0.71	1.33	0.56	1	51,547.65
Average			0.55		115,260.6

3.3. Analyze

Analyze is the third operational steps to improve the level of Six Sigma. In addition, the goal to be achieved in the Six Sigma is basically to bring the process under the conditions of stability and have a better capability and achieve zero failure rate. Analyses were performed by collecting the possible causes of the occurrence of out of specification by using the method of brainstorming and classify possible causes by using fishbone diagram.

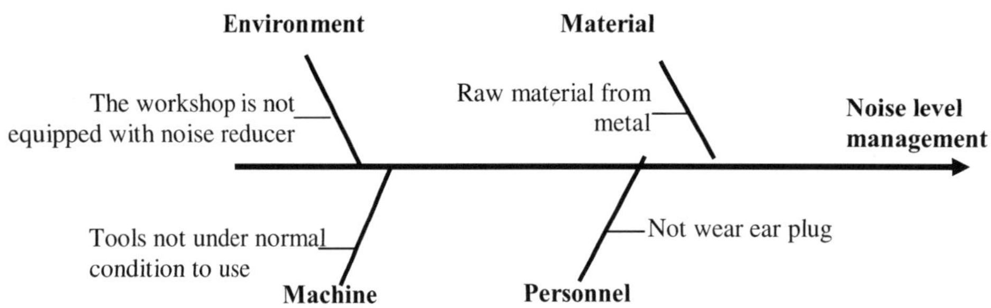

Figure 4. Fishbone diagram for noise level management

The next stage determines the root cause of the problem that caused the failure of the dominant noise level management. The method used is the Failure Modes and Effects Analysis (FMEA). Failure Mode and Effect Analysis (FMEA) is a systematic technique for identifying and analyzing and prevent failure. To be able to understand the type of failure, effects and how to make improvements so the next step is to process the results of brainstorming and fishbone diagram into FMEA worksheet. Furthermore, by applying the brainstorming the key informants provide an assessment of potential failure factors are based on the level of severity, occurrence, and detection. The scale of assessment for this calculation is 1-10. Assessment depends on the skills, knowledge, and experience of key informants in understanding the severity, occurrence, and the detection process. Then, calculate the risk priority number (RPN) and prioritize corrective action through the risk level.

Table 3 below shows that the factors that cause deviation dominant in the process of managing the noise level is the material used in the production process and also the rooms were not equipped with a silencer. This is demonstrated in the highest RPN value, such as the use of PPE (Earplug) around 56, the room was no noise reducer around 42 and the tool is not in a normal condition for use approximately 15.

Table 3. *Risk Priority Number* of noise management

ID Failure Modes	Potential Failure Modes	Risk Priority Number (RPN)
4.1	Wear PPE	56
2.1	Environment	42
3.1	Equipment	15
1.1	Material	10

3.4. Improve
The improve phases conducted after find out the root cause of noise levels that exceed the value of the noise standard. Thus, it is necessary to the determination of the corrective action plan in order to improve the quality of Six Sigma. At this stage formulated several actions aimed to increase or reduce the factors that can contribute significantly to the process of noise levels management. Therefore, in order to carry out the process of improving the noise level management activities may be carried out as follows:

a. Propose that material labeled as to the level of material hardness. This will provide insights into the working of the working methods of materials having different hardness levels. Thus, with this table will significantly reduce the noise level.
b. Provides a layer of noise reducer on the work area wall that has the potential to cause noise at a high level. Thus, it will be able to reduce the noise level significantly. In addition, employees who are in other rooms do not feel disturbed.
c. Proposed that companies implement workplace health and safety training in order to provide and enhance the knowledge of the workers in order to determine the potential risks arising from work activities. Through this training are expected to arise from employee behavior changes associated with the use of personal protective equipment, safe working methods and a safe working environment.
d. Companies should provide personal protective equipment appropriate for the work performed by employees. In addition, safety equipment must also consider the size of the body of each worker. This is done in conjunction with the comfort level of workers in the use of personal protective equipment. By using appropriate safety equipment will be able to influence the level of productivity of workers.
e. Perform periodically preventive maintenance on all types of equipment used in each production process. Through this activity will provide several advantages for companies and workers, such as reduced levels of machine breakdown, reducing the potential for accidents and improve engine performance. This maintenance activity can be made based on the period of daily, weekly and monthly.

3.5. Control
The final stage of the Six Sigma method is a process control. This activity aims to monitor and evaluates the overall implementation of activities to improve management of the noise level. This stage has a crucial role to Six Sigma approach. This is because the improvement process should be ensured that the repair process on the factors that caused the problems has diminished. In addition, this phase also provides assurance that the proposed changes have been made in order to achieve the targeted goals. Moreover, stakeholders may create a system that is standardized by the modification of policies, regulations, procedures and other management systems. It aims in order to deliver the certainty of continuous improvement. For example, use of statistical control charts to be able to control the noise

level minimization process Thus, it will be able to identify whether a statistically controlled process and meet the standard specifications or not. In addition, statistical process control can also observe the consistency of the process and make the process under control.

4. Conclusion

In conclusion, the Cp value at Engineering department around 0.39, Assembling department roughly 0.92 and Construction department approximately 0.71. In addition, the results also showed the average value of Cpk around 0.55 that generate potential failure (DPMO) approximately 115,260.6. This is equivalent to the sigma value is around 2.70. Furthermore, DMAIC method which involves of five phases of the Six Sigma method potentially be applied to the management of noise level. This method may bridge the various efforts in order to reduce the noise level. Moreover, Fishbone diagram shows several factors that cause problems. With the support of FMEA to give priority ranking of the factors that are considered the leading impact on the noise level. Thus, various alternative resolutions may be carried out as a reaction to the problem approach. Goal to condense the ratio of the noise level according to the target of the company will difficult to accomplish. This is because there is no systematic method that is able to support and resolve problems faced. Six Sigma method is a systematic method and is able to solve the problems of a reduction of noise level. The effectiveness and improvement can be achieved by integrating Six Sigma with other methods for the application of the noise level management plan. This study might a starting point in the implementation of six sigma method for designing an investigation into attempts of noise level management.

5. References

[1] Wang S, Qin Q, Liu L, Han L, and Chen Y 2013 A cross-sectional study on the effects of occupational noise exposure on hypertension or cardiovascular among workers from automobile manufacturing company of Chongqing, China *J. Biomedical Science and Engineering*, **6,** 1137-1142, http://dx.doi.org/10.4236/jbise.2013.612142.

[2] Doko-Jelinic J , Lukic J, Udovicic R, Zuskini E, Nola I A, and Zajec Z 2009 Workplace noise exposure after modernization of an aluminum processing complex. *Arh Hig Rada Toksikol*; **60,** 343-348.

[3] Pouryaghoub G., Mehrdad R., and Mohammadi S 2007 Interaction of smoking and occupational noise exposure on hearing loss: a cross-sectional study. *BMC Public Health*; **7:**137: 1-5

[4] Dias A., Cordeiro R 2008 Association between hearing loss level and degree of discomfort introduced by tinnitus in workers Exposed to noise. *Rev Bras Otorrinolaringol*; **74**(6):876-88

[5] Clark, C., Crombie, R., Head, J., van Kamp, I., van Kem-pen, E. and Ssfeldtan S.A 2012 Does traffic-related air pollution explain associations of aircraft and road traffic noise exposure on children's health and cognition? A secondary analysis of the United Kingdom sample from the RANCH project. *American Journal of Epidemiology*, **176,** 327-337. http://dx.doi.org/10.1093/aje/kws012

[6] Malakootian, M 2001 Noise pollution in Kerman-Iran. *Iranian Journal of Public Health*, **30,** 31-36.

[7] Ross, Z., Kheirbek, I., Clougherty, J.E., Ito, K., Matte, T., Markowitz, S., et al. 2011 Noise, air pollutants and traffic: Continuous measurement and correlation at a high- traffic location in New York City. *Environmental Research* **111**, 1054-1063. http://dx.doi.org/10.1016/j.envres.2011.09.004

[8] Attarchi MS., Labbafinejad Y, and Mohammadi S 2010 Contemporary exposure to cigarette smoke and noise of automobile manufacturing company workers. *J Public Health*, **18**(3):245-249

[9] Girard SA., Picard M., Davis AC., Simard M., Larocque R., Leroux T., et al. 2009 Multiple work-related accidents: tracing the role of hearing status and noise exposure. *Occup Environ Med.*; **66**(5):319–24. http://dx.doi.org/10.1136/oem.2007.037713. [PubMed: 19174422]

[10] Picard M, Girard SA, Simard M, Larocque R, Leroux T, and Turcotte F 2008 Association of work-related accidents with noise exposure in the workplace and noise-induced hearing loss based on the experience of some 240,000 person-years of observation. *Accid Anal Prev.* **40**(5):1644–52. http://dx.doi.org/10.1016/j.aap.2008.05.013. [PubMed: 18760091]
[11] Morata TC, Themann CL, Randolph RF, Verbsky BL, Byrne DC, and Reeves ER 2005 Working in noise with a hearing loss: perceptions from workers, supervisors, and hearing conservation program managers. *Ear Hear.* **26**(6):529–45. http://dx.doi.org/10.1097/01.aud.0000188148.97046.b8.[PubMed: 16377991]
[12] Leather P, Beale D, and Sullivan L 2003 Noise, Psychosocial stress and their interaction in the workplace. J *Environ Psychol.* **23**(2):213–22. http://dx.doi.org/10.1016/S0272-4944 (02)00082- 8.
[13] Gitanjali B and Ananth R 2003 Effect of acute exposure to loud occupational noise during daytime on the nocturnal sleep architecture, heart rate, and cortisol secretion in healthy volunteers. *J Occup Health.* **45**(3):146–52. http://dx.doi.org/10.1539/joh.45.146.
[14] Attarchi M., Dehghan F., Safakhah F., Nojomi M., and Mohammadi 2012 Effect of exposure to occupational noise and shift working on blood pressure in rubber manufacturing company workers. *Ind Health.* **50**(3):205–13. http://dx.doi.org/10.2486/indhealth.MS1321.
[15] Bortkiewicz A., Gadzicka E., Siedlecka J., Szyjkowska A., Viebig P., Wranicz JK., et al. 2010 Work-related risk factors of myocardial infarction. *Int. J Occup Med Environ Health.* **23**(3):255–65. http://dx.doi.org/10.2478/v10001-010-0030-7.
[16] Moll van Charante AW., Mulder PGH 1990 Perceptual Acuity and the Risk of Industrial Accidents. *Am J Epidemiol.* **131**(4):652–63.
[17] Rimantho D 2015 Identifikasi risiko kesehatan dan keselamatan kerja pada pekerja pengumpul sampah manual di Jakarta Selatan, *Optimasi Sistem Industri*, **14**(1) 1-5 (In Bahasa Indonesia), Available at: https://www.researchgate.net/publication/279861084_identifikasi_risiko_ kesehatan_dan_keselamatan_kerja_pada_pekerja_pengumpul_sampah_manual_di_jakarta_s el atan (Accessed: May, 21, 2016)
[18] Rimantho D., Cahyadi B 2015 Analisis Kebisingan Terhadap Karyawan Di Lingkungan Kerja Pada Beberapa Jenis Perusahaan, *Jurnal Teknologi.* **7**(1), 21-27. (In Bahasa Indonesia), Available at: https://www.researchgate.net/publication/283298400_analisis_kebisingan_terhadap_karyawa n_di_lingkungan_kerja_pada_beberapa_jenis_perusahaan (Accessed: May, 21, 2016)
[19] Choi S-W, Peek-Asa C, Sprince NL, Rautiainen RH, Donham KJ, Flamme GA, et al. 2005 Hearing loss as a risk factor for agricultural injuries. *Am J Ind. Med.* **48**(4):293–301. http://dx.doi.org/10.1002/ajim.20214
[20] NIOSH, *Preventing Occupational Hearing Loss - A Practical Guide* - U.S. Department Of Health And Human Services, Edited by John R. Franks, Mark R. Stephenson, and Carol J. Merry, 1996, available at: https://www.cdc.gov/niosh/docs/96-110/pdfs/96-110.pdf (accessed on April, 25, 2015)
[21] Linderman K, Schroeder RG, Zaheer S., and Choo AS 2003 Six Sigma: a goal-theoretic perspective. *Journal of Operations Management.* **21**:193–203.
[22] Schroeder RG, Linderman K, and Zhang D 2005 Evolution of quality: first fifty issues of production and operations management. *Production and Operations Management.* **14**(4):468–81.
[23] Ray S, Das P, Bhattacharya B.K 2011 Prevention of industrial accidents using Six Sigma approach, *International Journal of Lean Six Sigma.* **2**(3),196 – 21.
[24] Rimantho D., Cahyadi B 2016 Six Sigma Method Approach in The Prevention of Occupational Accidents on The Solid Waste Collector in South Jakarta, ARPN *Journal of Engineering and Applied Sciences.* **11**(16), 10014- 10022
[25] Calia RC., Guerrini F.M. and de Castro M 2009 The impact of Six Sigma in the performance of a Pollution Prevention Program, *Journal of Cleaner Production.* **17**, 1303–1310

[26] Rimantho D., Rahman T.A., Cahyadi B., and Hernawati T 2017 Application of six sigma and AHP in analysis of variable lead time calibration process instrumentation, *AIP Conference Proceedings*. **1813** (1), 040004
[27] Rimantho D., Hernadi D., Cahyadi B., Prasetyani R., and Kurniawan Y 2017 The Application of Six Sigma in Process Control of Raw Water Quality on Pharmaceutical Industry at Indonesia, *International Journal of Applied Engineering Research*. **12** (6) 848-860
[28] Goh TN, Xie M 2004 Improving on the Six Sigma paradigm, *TQM Magazine*, **16**(4): 235-240
[29] Park SH., 2002 Six Sigma for productivity improvement: Korean business corporations. *Productivity Journal*; **43**:173—183
[30] United States Environmental Protection Agency, 2009 *The Environmental Professional's Guide to Lean and Six Sigma*, Available at: www.epa.gov/lean (Accessed April, 25, 2015).

IOP Conf. Series: Materials Science and Engineering **277** (2017) 012056 doi:10.1088/1757-899X/277/1/012056

Determining a combination of factors on rubber end manufacturing using taguchi method at PT. Agronesia Bandung

Wahyukaton[1] **and Siti Qodriyah**[2]

Industrial Engineering Department, Engineering Faculty, Universitas Pasundan, Jl. Setiabudi 193 Bandung, West Java, Indonesia
[1]wahyukaton@unpas.ac.id, [2]sitiqodriyah22@gmail.com

Abstract. Rubber end is one of products that manufactured by PT. Agronesia Bandung. These products experienced defect and unconformity to the standard which is bubble trapped in rubber end products. This defect can be found from 3 factors – control factor, which are material thickness, temperature, and shift. Material thickness has 3 levels, which are 3 mm, 5.5 mm, and 7 mm; temperature has 3 levels, which are, 35^0 C - 40^0 C, 40^0 C - 45^0 C, and 45^0 C - 55^0 C; while shift has 3 levels, which are shift 1, shift 2, and shift 3. To solve the problem, Taguchi method is used to have the right combination of those 3 factors to produce rubber end. First step is selecting factors and determining level that will be a base for selecting orthogonal matrix, and then determining S/N ratio and analysis of variance. The result is factors combinations are material thickness at level 2 of 5.5 mm, temperature at level 3 of 45^0 C - 55^0 C, and shift at level 1 of shift I.
Keywords: taguchi method, analysis of variance, signal to ratio, defects

1. Introduction

PT. Agronesia is a rubber manufacture, and one of their product is rubber end. The company experience the problems on this rubber end that are defects on the products. These defects are bubble trapped, bare, raw, exceed dimension, and thickness, and these defects causing the products will be reworked or even rejected. Figure 1 shows that the most occurred defect is bubble which is 8 occurrences from 16 or 50%, so that this defect needs a special attention to reduce defects ratio and to improve the product quality.

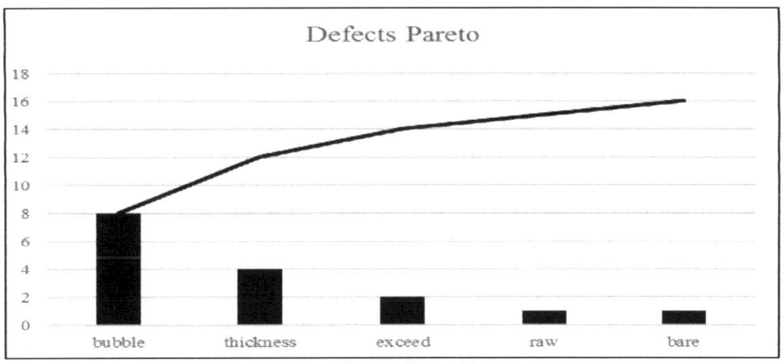

Figure 1. Defects Pareto

2. Method

In business, engineering and manufacturing, quality has a pragmatic interpretation as the *non-inferiority* or *superiority* of something; it's also defined as *fitness for purpose*. Quality is a perceptual, conditional, and somewhat subjective attribute and may be understood differently by different people. Consumers may focus on the specification quality of a product/service, or how it compares to competitors in the marketplace. Producers might measure the conformance quality, or degree to which the product/service was produced correctly [2].

2.1 Experimental design

Experimental design is the design of any task that aims to describe or explain the variation of information under conditions that are hypothesized to reflect the variation. The term is generally associated with true experiments in which the design introduces conditions that directly affect the variation, but may also refer to the design of quasi-experiments, in which natural conditions that influence the variation are selected for observation [4].

In its simplest form, an experiment aims at predicting the outcome by introducing a change of the preconditions, which is reflected in a variable called the predictor (independent). The change in the predictor is generally hypothesized to result in a change in the second variable, hence called the outcome (dependent) variable. Experimental design involves not only the selection of suitable predictors and outcomes, but planning the delivery of the experiment under statistically optimal conditions given the constraints of available resources.

2.2 Taguchi method

The Taguchi method of quality control is an approach to engineering that emphasizes the roles of research and development, product design and product development in reducing the occurrence of defects and failures in products. The Taguchi method considers design to be more important than the manufacturing process in quality control and tries to eliminate variances in production before they can occur [8].

Taguchi refers to experimental design as "off-line quality control" because it is a method of ensuring good performance in the design stage of products or processes. Some experimental designs, however, such as when used in evolutionary operation, can be used on-line while the process is running [7].

2.3 Factors, factors level

2.3.1 Factors

Based on defects pareto, it can be determined that the independent variable is bubble trapped due to the most occurrence defect.

For dependent variable (factors) can be determined from the early researches will be material thickness, temperature, and shift.

Table 1. Variables, Factors

Variables	Factors	Classification
A	Material thickness	Factor Control
B	Temperature	Factor Control
C	Shift	Factor Control

Uncontrollable factors (noises) in this research are humidity, workers performance is excluded in the matrix, so that the experiment only applied to controllable factors by doing replication on each experiment.

2.3.2 Factors level

Material thickness

The material thickness is significantly influence for bubble trapped in rubber end. The product thickness is 20 – 21 mm, so that material thickness will be set in 3 levels as follows,

- Level 1 = 3 mm
- Level 2 = 5.5 mm
- Level 3 = 7 mm

Temperature

There are 3 levels for this factor, as follows

- Level 1 = 35 - 40 mm
- Level 2 = 40 - 45 mm
- Level 3 = 45 - 55 mm

Shift

Shift is the interval work to produce rubber end.
There are 3 levels for this factor, as follows

- Level 1 = shift 1
- Level 2 = shift 2
- Level 3 = shift 3

By applying Taguchi method, the quality of manufactured goods, and engineering designs are developed by studying variations. The Composite Principle Component has been optimized by using Taguchi method. Analysis of variance (ANOVA) has been conducted for Composite Principle Component to find the optimal process parameters. Signal to Noise (S/N) Ratio has been found for PCA to find the optimal levels of the process parameters. Finally, a conformation test has been made for three different materials and the results have been plotted. [3]

The Taguchi method was used to determine optimum conditions for tire rubber in asphalt concrete with Marshall Test. The tire rubber in asphalt concrete was explored under different experimental parameters including tire rubber gradation (sieve #10–40), mixing temperature (155–175 1C), aggregate gradation (grad. 1–3), tire rubber ratio (0–10% by weight of asphalt), binder ratio (4–7% by weight of asphalt), compaction temperature (110–135 1C), and mixing time (5–30 min). The optimum conditions were obtained for tire rubber gradation (sieve #40), mixing temperature (155 1C), aggregate gradation (grad. 1), tire rubber ratio (10%), binder ratio (5.5%), compaction temperature (135 1C), mixing time (15 min). [5]

3. Result and Discussion

Before orthogonal matrix is constructed, first the degree of freedom (DoE) has to be determined. DoE is meant to calculate the number of minimum experiments to inquire the observed factors.

Table 2. Degree of Freedom

Factor	Degree of Freedom	Total
A	3 – 1	2
B	3 – 1	2
C	3 – 1	2
	Total DoF	6

After the experiments were conducted, the orthogonal matrix L_9 (3^4) for bubble trapped as follows,

Table 3. Orthogonal Matrix L_9 (3^4)

Experiments	A	B	C	Error	Result			Sum	Mean
1	1	1	1	1	3	4	0	7	2.33
2	1	2	2	2	2	2	3	7	2.33
3	1	3	3	3	3	3	0	6	2.00
4	2	1	2	3	3	2	2	7	2.33
5	2	2	3	1	0	3	3	6	2.00
6	2	3	1	2	1	3	1	5	1.67
7	3	1	3	2	0	3	2	5	1.67
8	3	2	1	3	4	0	3	7	2.33
9	3	3	2	1	2	0	3	5	1.67
							Total	55	18.33
							Mean	6.11	2.04

In this experiment, orthogonal matrix L_9 (3^4) is chosen, so that a linear graphic standard which is necessary to the experiments as follows,

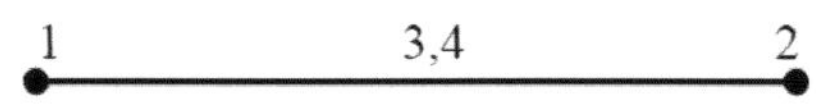

Figure 2. Linear Graphic of L_9 (3^4)

From the linear graphic in figure 2., it can be seen that factor A will be placed in column 1, factor B will be placed in column 2, factor C will be placed in column 3, and Error will be placed in column 4. The number of 1, 2, 3 are randomize process. Level on the orthogonal matrix column was made in number 1, 2, 3, etc. Or it can be done by using own method to describe the level, for example 0 and X, (+) and (–), L and H, to replace the number of 1, 2, 3 [6].

To find the factors that influence to bubble trapped, it needs to analyze by signal to noise (S/N) ratio in the analysis of variance. When the mean of responds was found, then the next step is to determine a robust design which is a design that will not affected by noise factor. The S/N which is been used is a quality characteristic, the smaller the better [6]. The quality of bubble trapped is good when the smallest bubble trapped is the better.

Table 4. Signal to Ratio (S/N)

Experiments	A	B	C	Error	Result			S/N
1	1	1	1	1	3	4	0	-9.21
2	1	2	2	2	2	2	3	-7.53
3	1	3	3	3	3	3	0	-7.78
4	2	1	2	3	3	2	2	-7.53
5	2	2	3	1	0	3	3	-7.78
6	2	3	1	2	1	3	1	-5.64
7	3	1	3	2	0	3	2	-6.37
8	3	2	1	3	4	0	3	-9.21
9	3	3	2	1	2	0	3	-6.37
							Total	-67.43
							Mean	-7.49

The mean of factor effect based on S/N ratio is an average of S/N respond from each particular levels and factors, the calculation was done based on the respond of experiments.

Table 5. Mean of S/N Ratio for Effect Factors

	A	B	C
Level 1	-8.17	-7.70	-7.07
Level 2	-6.99	-8.17	-8.09
Level 3	-7.31	-6.60	-7.31
Difference	1.19	1.11	1.02
Ranking	1	2	3
	A1	B3	C1

Table 5 shows that the setting of effected factors is factor C level 1, factor B level 3, and factor A level 2.

As the influence factors to the mean of bubble trapped were found, then to find a significant influence factors to S/N ratio can be done by construct a two-way analysis of variance model.

Table 6. Analysis of Variance for Bubble Trapped

Source	DoE	SS	MS	*f*
A	2	2.26	1.13	5.38
B	2	3.93	1.97	9.38
C	2	1.70	0.85	4.05
Error	20	4.11	0.21	
Total	26	12.01		

Test of hypothesis was done for all factors for 90% confidence level with $DoE_1 = 2$ and $DoE_2 = 20$, so that the $f_{(0.1;2,20)} = 2.59$.

1. Factor A
H_0: there is no effect of factor A to bubble trapped
H_1: there is an effect of factor A to bubble trapped
$f = 5.38 > f_{(0.1;2,20)} = 2.59$, reject H_0, there is an effect of factor A to bubble trapped
2. Factor B
H_0: there is no effect of factor A to bubble trapped
H_1: there is an effect of factor A to bubble trapped
$f = 9.38 > f_{(0.1;2,20)} = 2.59$, reject H_0, there is an effect of factor A to bubble trapped
3. Factor C
H_0: there is no effect of factor A to bubble trapped
H_1: there is an effect of factor A to bubble trapped
$f = 4.05 > f_{(0.1;2,20)} = 2.59$, reject H_0, there is an effect of factor A to bubble trapped

Table 7. Value and Percentage of Contribution for Each Factor

Factors	Value for Contribution (SS')	Contribution Percentage (ρ)
A	1.54	15.32
B	3.51	29.23
C	1.28	10.66

Based on the all calculations, a setting level design to reduce bubble trapped on rubber end product was found as follows,

IOP Conf. Series: Materials Science and Engineering **277** (2017) 012056 doi:10.1088/1757-899X/277/1/012056

Table 8. Setting Level Design

Factor	Level Chosen	Specification
Material Thickness	2	5.5 mm
Temperature	3	45° C – 55° C
Shift	1	I

4. Conclusion

1. The factors that causing bubble trapped defect on rubber end are,
   - Material thickness
   - Temperature
   - Shift
2. The setting level design to reduce defects are,
   - Factor A, material thickness is set to level 2 (5.5 mm.)
   - Factor B, temperature is set to level 3 (45° C – 55° C)
   - Factor C, shift is set to level 1 (shift I)

5. References

[1] Adil A and Moutawakil A 2012 The Quality Cost Reduction in Hollow Glass Manufacturing by Taguchi Method, *J. Sci. Res.* **4**(1), pp.155-172

[2] Hicks CR 1993 *Fundamental Concept in the Design of Experiments*, 4th Ed., New York: Holt, Reinehart, and Winstone Inc.

[3] Madhavi S K, Sreeramulu D and Venkatesh M 2017 Evaluation of Optimum Turning Process of Process Parameters Using DOE and PCA Taguchi Method, *Materials Today: Proceedings* **4**, pp.1937-1946

[4] Montgomery D C 2009 *Introduction to Statistical Quality Control*, 6th Ed., Singapore: John Wiley and Sons (Asia) Pte. Ltd.

[5] Sadr P, Kolahdooz A, and Eftekhari S A 2017 The Effect of Electrical Discharge Machining Parameters on Alloy Din 1.2080 Using the Taguchi Method and Determinant of Optimal Design of Experiments, *J. of Naval Architecture and Marine Engineering*, pp.47-64.

[6] Taguchi G and Phadke M S 1984 Quality Engineering Through Design Optimization, *Global Telecommunications Conference, GLOBECOM '84*, Atlanta, pp.1106-1113

[7] Tanabe I, Sakamoto H and Miyamoto K 2014 Development of Innovative Tool Using Taguchi-methods, *AASRI Procedia* **9**, pp.107 – 113

[8] Tortum A, Celik C and Aydin A C, 2005 Determination of the optimum conditions for tire rubber in asphalt concrete, *Building and Environment* **40**, pp. 1492–1504

IOP Conf. Series: Materials Science and Engineering **277** (2017) 012057 doi:10.1088/1757-899X/277/1/012057

Design on inspection system automation of curtain woven fabrics by image processing and the taguchi approach method at PT. Buana Intan Gemilang

M. Mirsyah[1] and T. Mulyana[2]

[1]Industrial Engineering Department, Telkom University, Indonesia, muhammadmirsyah@gmail.com
[2]Industrial Engineering Department, Telkom University, Indonesia, tatangmulyana@telkomuniversity. ac.id

Abstract. The textile industry is one of the industries that provide high export value by occupying the third position in Indonesia. The process of inspection on traditional textile enterprises by relying on human vision that takes an average scanning time of 19.87 seconds. Each roll of cloth should be inspected twice to avoid missed defects. This inspection process causes the build-up at the inspection station. This study proposes the automation of inspection systems using image processing and the Taguchi Approach Method to test the combination of factors that affecting the inspection process. The input for image processing and Taguchi Approach comes from the existing state and previous research as reference and trial and error process. The result of the Taguchi Approach is error rate and will be processed using S/N ratio and ANOVA, which will be determined which factor that has the most affecting the inspection process and determined the best combination (robust) for inspection process. The result showed combination of light intensity 1300lx, camera distance 16 cm, camera resolution 1280x960 pixels, grayscale value 10, threshold value 0.3, and edge detection operator sobel that has small error in inspection process.
Keywords: design, inspection, automation, image-processing, taguchi, woven-fabrics.

1. Introduction

Textile industry is one of the industrial sectors that ranks third largest exporter of export value that can increase foreign exchange. In order to produce products that conform to standards and quality based on consumer demand, it takes inspection process on the fabric. Inspection process at the company at this time is still done manually using the human eye with the help of lighting in the form of lights. After calculating the processing time to the manual inspection process obtained the average scanning time of 19.7 seconds. In addition, the limitations of human vision and human error affect greatly in the process of manual inspection, such as the defects that are missed when the fabric is inspected. Based on the company's production data, the volume of production and inspection volume are not balanced, since the total fabric produced is not entirely inspected on time. This has an impact on the unfulfilled consumer demand within the specified time. This research designs a proposal system in the form of an integrated automation system that aims to optimize the inspection process with the help of image processing. The purpose of this research is to be able to identify combination factors that has most influence in the inspection process and the combination of factors that robust using Taguchi Approach. Mathematical model for a process control plant is important because it provides key information as to the nature and characteristic of the system which is vital for the investigation and prediction of the system operation [1]. Through mathematical equation model, we can study about the dynamics of the process, stability of the system, design controller etc. The mathematical equation model is used to determine a performance of the system [2]. Automation is a technology in which a process or procedure can be accomplished without the aid of humans [3]. Currently automation system not only played a role that is very important in the world industry, but also in the management system electricity [4]. The modelization for the control system is made using Simulink graphical model. In the process, diagram block is used. This diagram block easing the understanding and modelling the system because it is stated in graphics. Some of the many of the main elements which is widely used for the modelization of the system using Simulink are source, system block diagram, and sinks. The input for the model depends on the value wanted and the form is fixed. The model used is based on control system model in which has been studied while output form is displayed in/not in graphical; graph and output value. The simulation is including continuous time system and discrete time system [5]. The Supervisory Control and Data Acquisition (SCADA) monitoring system is built based on Graphical User Interface (GUI) by using IGSS software [6]. Development of the SCADA application leverages the Wonderware Information Server with Active Factory and Generic Data Grid in order to obtain reporting system a system that can be accessed online [7]. Mathematical models are the foundation of most scientific and engineering methods. Models can be obtained by either a theoretical approach based on physical laws, or an experimental approach based on obtained measurements from the system. A system identification deal with the problem of building mathematical models of dynamic systems based on observed data from the system, and is thus an experimental modelling method [8]. Mathematical model for a process control plant is important because it provides key information as to the nature and characteristic of the system which is vital for the investigation and prediction of the system operation. The set of equations that make up that model is an approximation of the true process [9]. The system identification is the experimental approach to process modelling. It includes the following steps: experimental design, model structure, parameter estimation, and model validation [10]. A mathematical model to describe a plant can be created through derivation process of physical and chemical principles without experimental data. However, it is not always possible to develop a model theoretically. Therefore, system identification is another useful way to construct process models and estimate unknown model parameters. The goal of system identification is to determine system equations from a given input and output time history [11]. The System Identification problem is to estimate a model of a system based on observed input-output data. MATLAB software commonly used to help finding solution the identification problem [12]. Introducing a mathematical model to represent the actual system is very important especially in simulation and prediction purposes, particularly for designation of digital control and dynamic systems. Basically, there are two ways of constructing mathematical models i.e. mathematical modelling and system identification [13]. The analysis could be conducted to the result by using System Identification Toolbox in time domain response in order to know the characteristics of the models such as Step Response, Impulse Response, Bode Diagram and Poles Zeros [14].

2. Basic Theory and Methodology

There are three advantages of the Taguchi Method compared to other experimental design methods [15]:

1. The Taguchi method is more efficient because it can experiment with several factors but requires only a small amount of experiments.
2. The Taguchi method is capable of obtaining a process that produces a consistent and robust product against a factor that cannot be controlled (noise). This is because the strong design focuses on the influence of controllable factors and the uncontrolled factor average and the diversity of performance at the same time.
3. The Taguchi method gives conclusions to the factors that influence the experimental results and also the conclusions about the level of the factors that will yield optimal results.

IOP Conf. Series: Materials Science and Engineering **277** (2017) 012057 doi:10.1088/1757-899X/277/1/012057

The Taguchi method fits this research as it is in accordance with Taguchi's objective of quality, and also it has a high effectiveness compared to other methods. Besides this research has six parameters with three levels for each parameter, then there will be 729 combinations of experiments which will take a lot of time and process. By the Taguchi Method, the number of experimental combinations to be performed using orthogonal arrays was generated by only 27 experimental combinations to obtain accurate and efficient DOE results. Orthogonal Array is a factor and level matrix that is arranged so that the influence of a factor and level does not blend with other factors and levels. The matrix elements are arranged according to rows and columns. Rows are the state of a factor that can be changed in an experiment [16]. After the experiment was done 27 times then the result of the experiment that error will be input for signal-to-noise ratio and ANOVA. The S / N ratio is used to select the best factors or parameters that contribute to the reduction in the diversity of the experimental results. In addition, the S / N ratio is also implemented to minimize errors caused by factors. In the Taguchi ANOVA method is used to estimate the S/N ratio to identify the controllable parameters which will result in a performance that is resistant to parameters that cannot be controlled in other words "robust" so that we can know the optimal conditions [17]. In calculating the deviation between the experimental results and the desired value, the loss function is defined. The value of the current loss function will then be converted into signal-to-noise (S/N). The main purpose of the Analysis of Variance (ANOVA) is to find how much variation for each factor causes the relationship to the total variation shown in the final result, by knowing the variation of each factor we can know how the influence of a factor [15]. Significant influence can be read through the value of F Count (Fo) or P-value generated. If the value of Fcount is greater than Ftable with $\alpha = 5\%$ then the factor significantly has an effect on the experiment, where the greater the F value the greater the effect on the experiment. In addition, if the interaction test result is the interaction factor reject Ho or said significant interaction factor, it is necessary to do further test. The test is a Tukey Test, if the homogeneous variant and the number n of each variable are the same [19, 20, 21, 22]. Tukey test is done if the resulting ANOVA hypothesis the result is rejected then we will try to do the comparison every mean factor [5].

3. Result and Discussion

An experiment is performed on the device in Figure 1. Table 1 shows the results from image processing using MATLAB.

Table 1. Defect identification results

	Empty Feed Defect	Lacking Defective Defect	Oil Defect
RGB			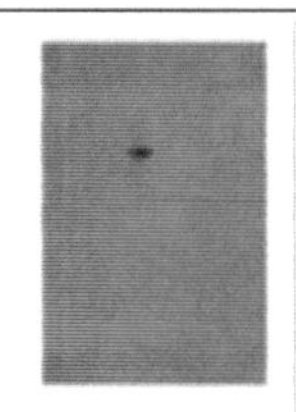
Defect Defected			

Table 2 showed that in the highest feed defect the highest error was found in the 9th experiment with a percentage of 98.33285842%, the combination in the 9th experiment was the light intensity of 1300 lx, the distance of the camera 20 cm, the camera resolution 1280 x 960 pixels, the grayscale value 10, the edge detection operator used is canny, and the threshold value is 0.3. The lowest percentage error is in the 18th experiment with the percentage error of 3.613085284%, the combination in the 18th experiment is the light intensity of 2220 lx, the distance of the camera is 20 cm, the camera resolution is 640 x 480 pixels, the grayscale value 5, the Canny edge detection operator, and a threshold value of 0.1.

Table 2. Error level empty feed defect

Exp.	Wide Fabric (px)	Defect Area (px)	Wide Fabric (cm^2)	Defect Area (cm^2)	Error (%)
1	301044.5	439.25	208	0.15	50.57498
2	301044.5	382.375	208	0.15	43.223433
3	301044.5	388	208	0.15	44.046547
4	301046.13	195.625	208	0.15	10.977928
5	301046.13	145.25	208	0.15	49.46683
6	300994.13	140.75	208	0.15	54.218878
7	273533.88	105	208	0.15	87.866672
8	273009.13	100.625	208	0.15	95.658714
9	272615.13	99.125	208	0.15	98.332858
10	301134.88	141.5	208	0.15	53.473197
11	301150	138.75	208	0.15	56.522869
12	301134.88	194.75	208	0.15	11.509409
13	300872.88	333.25	208	0.15	34.891033
14	300872.88	258.625	208	0.15	16.104154
15	300872.88	338.5	208	0.15	35.900848
16	272626.75	147.875	208	0.15	32.954069
17	272626.75	101.5	208	0.15	93.700324
18	272626.75	189.75	208	0.15	3.6130853
19	301150	137.625	208	0.15	57.802348
20	301150	125.125	208	0.15	73.566818
21	301150	121.875	208	0.15	78.195266
22	301150	137.25	208	0.15	58.233501
23	301150	145.125	208	0.15	49.647187
24	301150	155.375	208	0.15	39.775048
25	301150	115.125	208	0.15	88.643197
26	301150	133.75	208	0.15	62.374191
27	301150	182	208	0.15	19.327187

Table 3 showed that in the highest level lacking defective defect error in the 8th experiment with the percentage of 91.58502024%, the combination in the 8th experiment was the light intensity of 1300 lx, the distance of the camera 20 cm, the camera resolution 1280 x 960 pixels, the grayscale value 10, the operator the edge detection used is Sobel, and the threshold value is 0.2. Then the lowest error percentage in 21st experiment with error percentage 2,642,774,128%, combination of 21 light intensity experiment 2900 lx, camera distance 16 cm, camera resolution 1280 x 960 pixel, grayscale value 5, Canny edge detection operator, and Threshold value 0.2. Table 4 showed that the highest error in oil errors obtained in the 3rd experiment with a percentage of 72.73366565%, the combination in the third experiment is the light intensity of 1300 lx, the distance of the camera 16 cm, the resolution of the camera 640 x 480 pixels, the value of grayscale 0, the operator the edge detection used is Canny, and the threshold value is 0.3. The next lowest percentage error is in the 25th experiment with the error percentage of 13.28982429%, the combination in the 25th experiment is the light intensity of 2900 lx, the camera distance is 20 cm, and the camera resolution is 960 x 720 pixels, the grayscale value 0, the Prewitt edge detection operator, and a threshold value of 0.3.

Table 3. Error level lacking defective defect

Exp.	Wide Fabric (px)	Defect Area (px)	Wide Fabric (cm^2)	Defect Area (cm^2)	Error (%)
1	301046.13	573.375	208	0.22	44.466681
2	301046.13	469	208	0.22	32.107853
3	301046.13	378.5	208	0.22	15.874724
4	301094.75	349.75	208	0.22	8.9447889
5	301094.75	244.125	208	0.22	30.451859
6	301080.25	201.25	208	0.22	58.236156
7	283929	207.5	208	0.22	44.727479
8	283929	156.75	208	0.22	91.58502
9	283466.38	158.75	208	0.22	88.863121
10	301086.38	189.625	208	0.22	67.940273
11	301078.75	170.875	208	0.22	86.363528
12	301100.13	201.375	208	0.22	58.148373
13	301099.5	295.5	208	0.22	7.7734772
14	300872.88	258.625	208	0.15	16.104154
15	301099.5	388.125	208	0.22	17.946377
16	283360	380.625	208	0.22	21.259063
17	283360	228.875	208	0.22	30.9482
18	283360	440.125	208	0.22	31.903961
19	301079.25	193	208	0.22	64.999589
20	301079.25	193	208	0.22	64.999589
21	301079.25	310.25	208	0.22	2.6427741
22	301103.38	167.5	208	0.22	90.134163
23	301105.88	192	208	0.22	65.873629
24	301105.88	228.25	208	0.22	39.530063
25	283485	223.5	208	0.22	34.156557
26	283485	319.125	208	0.22	6.0431167
27	283485	349.5	208	0.22	14.208897

Figure 2 is the result of signal-to-noise ratio with Taguchi calculation using Minitab 16 software. The data used to get the result is by using error data obtained from the deviation calculation on three types of defects found on Table 2, Table 3, and Table 4. The result of the S / N ratio is converted to a line diagram. From Figure 3 and Figure 4 can be concluded for the selected light intensity factor is the low level of 1300 lux, then for the selected camera distance factor is the low level of 16 cm, for the selected camera resolution is 1280 x 960 pixels, then for the selected grayscale value is 10, then for the selected edge detection operator is Sobel, and for the threshold value is 0.3.

Table 4. Error level oil defect

Exp.	Wide Fabric (px)	Defect Area (px)	Wide Fabric (cm^2)	Defect Area (cm^2)	Error (%)
1	300300.5	536.75	208	0.2	46.203916
2	300300.5	506.25	208	0.2	42.962868
3	300300.5	1059	208	0.2	72.733666
4	300547.88	393	208	0.2	26.466071
5	300547.88	371.375	208	0.2	22.184223
6	300547.88	803.625	208	0.2	64.039404
7	286597	343.75	208	0.2	19.833007
8	286597	329.875	208	0.2	16.461072
9	286597	584.375	208	0.2	52.842945
10	300494.88	882.875	208	0.2	67.273127
11	300494.88	384.375	208	0.2	24.8293
12	300494.88	867	208	0.2	66.673889
13	300547.88	409.625	208	0.2	29.450512
14	300547.88	377.25	208	0.2	23.396066
15	300547.88	803.25	208	0.2	64.022615
16	285665.5	603.875	208	0.2	54.514036
17	285665.5	320.25	208	0.2	14.230019
18	285665.5	321.5	208	0.2	14.563494
19	300430	422.375	208	0.2	31.606984
20	300430	449.625	208	0.2	35.752016
21	300430	432.875	208	0.2	33.265954
22	300624.25	350	208	0.2	17.41092
23	300624.25	714.75	208	0.2	59.557639
24	300624.25	721.25	208	0.2	59.92211
25	284738.88	315.75	208	0.2	13.289824
26	284738.88	343.75	208	0.2	20.352762
27	284738.88	318.125	208	0.2	13.937169

Figure 1. Defect identification device.

Response Table for Signal to Noise Ratios
Smaller is better

Level	Light Intensity	Camera Distance	Camera Resolution	Grayscale	Edge Detection	Threshold
1	-33.56	-34.38	-34.47	-31.65	-32.48	-32.15
2	-32.67	-32.34	-33.05	-32.05	-32.80	-32.07
3	-33.27	-32.78	-31.98	-35.80	-34.21	-35.29
Delta	0.89	2.04	2.48	4.14	1.73	3.22
Rank	6	4	3	1	5	2

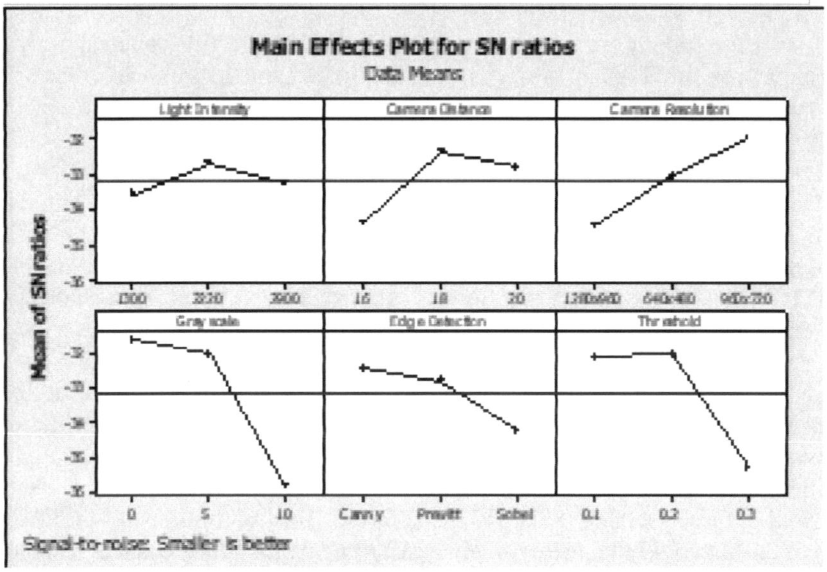

Figure 2. Signal-to-Noise ratio result.

Analysis of Variance for SN ratios

Source	DF	Seq SS	Adj SS	Adj MS	F	P
Light Intensity	2	3.694	3.694	1.847	0.23	0.798
Camera Distance	2	20.784	20.784	10.392	1.29	0.305
Camera Resolution	2	27.934	27.934	13.967	1.74	0.212
Grayscale	2	94.191	94.191	47.096	5.86	0.014
Edge Detection	2	15.290	15.290	7.645	0.95	0.410
Threshold	2	60.730	60.730	30.365	3.78	0.049
Residual Error	14	112.463	112.463	8.033		
Total	26	335.086				

Figure 3. Analysis of variance result.

Grayscale = 0 subtracted from:

Grayscale	Difference of Means	SE of Difference	T-Value	Adjusted P-Value
5	3.484	5.398	0.6455	0.7990
10	22.132	5.398	4.1003	0.0055

Grayscale = 5 subtracted from:

Grayscale	Difference of Means	SE of Difference	T-Value	Adjusted P-Value
10	18.65	5.398	3.455	0.0155

Figure 4. Defect identification device.

4. Conclusion

An experimental design with the Taguchi method was performed to obtain an optimum level factor for detecting fabric defects with image processing. 27 experiments were conducted based on orthogonal arrays obtained based on the number of factors and the number of levels for each factor. The defects detected in this experiment are empty feed defects, lacking defective defect, and oil defects, in order to obtain the accuracy or percentage error by comparing the area of defects and the width of the fabric in pixels with the width of defects and the width of the fabric in centimetres. Factors affecting error in the detection of defects through the S / N ratio are low for light intensity (1300lx), low level for camera range (16 cm), high level for camera resolution (1280x960 pixels), high level for grayscale value (10), intermediate level for edge detection operator (Sobel), and high level for threshold value (0.3). The quality measurements for Taguchi analysis that are implemented are smaller is better, because the smaller the error the better the results. The result of the error percentage measurement of 27 experimental combinations for the three types of defects can be concluded that for the empty feed defect the lowest error rate is in the 18th experiment with the error percentage of 3.613085284%, the combination in the 18th experiment is the light intensity of 2220 lx, the distance 20cm camera, 640 x 480pixels camera resolution, grayscale value 5, Canny edge detection operator, and threshold value 0.1. For the Lacking defective defect, the lowest error rate is in the 21st experiment with the percentage error of 2.642774128%, the combination in the 21st experiment is the light intensity of 2900 lx, the distance of the camera is 16 cm, the camera resolution is 1280 x 960 pixels, the grayscale value 5, the operator Canny edge detection, and threshold value 0.2. The lowest error rate for Oil Defect is in the 25th experiment with the error percentage of 13.28982429%, the combination in the 25th experiment is the light intensity of 2900 lx, the camera distance is 20 cm, the camera resolution is 960 x 720 pixels, the value of grayscale 0, the detection operator Prewitt edge, and threshold value 0.3. There is a significant influence of the grayscale value factor and the threshold value on the percentage of the error rate of the width of the defect and the width of the fabric. The order of the most influential factors to which the least influence is the value of grayscale, threshold value, camera resolution, camera range, edge detection, and light intensity.

5. References

[1] Tatang M, Mohamad Nor M.T, Dirman Hanafi B and Afrodi A 2010 Digital control design for the boiler drum BDT921 *Proceeding of Conf. on Industrial and Appl. Math., Indonesia* p 257-260

[2] Tatang M, Mohamaad Nor M.T and Noor Adzmira M 2011 Identification of heat exchanger QAD Model BDT 921 based on hammerstein-wiener model *Proceedings of the International Seminar on the Application of Science & Mathematics 2011 (ISASM 2011)*

[3] Rino Andias A and Tatang M 2015 Monitoring and Controlling of EMS-SCADA via SMS Gateway *Proceedings of the 3rd International Conference on Information and Communication Technology 2015 (ICoICT 2015)* p 615-619.

[4] Haris R, Rino Andias A and Tatang M 2015 EMS-SCADA Design of AC Usage on a Building *Proceeding 8th International Seminar on Industrial Engineering and Management 2015 (ISIEM 2015)* PS45-49.

[5] Tatang M, Mohamad Nor M.T, Dirman H, and Muhaimin 2009 Simulation of various PI controllers setting for boiler drums T11 QAD Model BDT 921 *IndoMS International Conference on Mathematics and its Applications 2009 (IICMA 2009)* (UGM, Yogyakarta – Indonesia)

[6] Fauzi F.D.M., Tatang M, Rizman Z.I, Miskon M.T, Chek W.A.K.W. and Jusoh M.H 2016 Supervisory fertigation system using interactive graphical supervisory control and data acquisition system *International Journal on Advanced Science, Engineering and Information Technology* **Vol. 6** Issue No. 4 p 489-494

[7] Haris R and Tatang M 2015 Website Design of EMS-SCADA for AC Usage on a Building *Proceedings of the 3rd International Conference on Information and Communication Technology 2015 (ICoICT 2015)* p 17-22

[8] Tatang M, Mohamad Nor M.T and Dirman H 2010 *A discrete time model of four heat exchanger types HE158C Proceedings of the 7th International Conference on Heat Transfer, Fluid Mechanics and Thermodynamics (HEFAT2010)* (Antalya, Turkey)

[9] Tatang M, Mohamad Nor M.T and Dirman H 2009 A Discrete Time Model of Boiler Drum and Heat Exchanger QAD Model BDT 921 *Proceedings of the International Conference on Instrumentation, Control & Automation (ICA2009)* (Bandung, Indonesia) p 154-159

[10] Tatang M, Farah Naja S, Dirman H and Mohamad Nor M.T 2011 ARX Model of Four Types Heat Exchanger Identification *Proceedings of the MUiCET 2011*

[11] Tatang M 2017 NNARX model structure for the purposes of controller design and optimization of heat exchanger process control training system operation *AIP Conference Proceedings* 1831.

[12] Tatang M and Mohamad Nor M.T 2010 Identification of Co Current Plate Heat Exchanger Model HE158C *Proceedings of 2010 IEEE Student Conference on Research and Development (SCOReD 2010)* (Putrajaya, Malaysia)

[13] Tatang M 2015 A nonparametric system identification based on transient analysis with plant process of heat exchanger as study case *International Journal of Innovation in Mechanical Engineering & Advanced Materials (IJIMEAM)* **Vol.1** (No.1) p 19-26

[14] Tatang M, Judi A and Ekki K 2016 Data Analysis using System Identification Toolbox of Heat Exchanger Process Control Training System *Proceedings of the 2016 Fourth International Conference on Information and Communication Technologies (ICoICT 2016)* p 466-471

[15] Roy R.K 2001 *Design of Experiments Using The Taguchi Approach* (New York: John Wiley & Sons, Inc)

[16] Nur Iriawan S.P.A 2006 *Mengelola Data Statistik dengan Mudah Menggunakan Minitab 14* (Yogyakarta: ANDI)

[17] Nur Firstiawan 2012 *Optimasi Parameter Proses Pemesinan CNC Milling Terhadap Kekasaran Permukaan Kayu Jati Dengan Metode Taguchi*

[18] Nalbant 2007 Application of Taguchi Method in THe Optimization of Cutting Parameters For Surface Roughness In Turning *Material & Design.*

[19] Montgomery D.C 2013 *Design and Analysis of Experiments* (New York: John Wesley and Sons)

IOP Conf. Series: Materials Science and Engineering **277** (2017) 012057 doi:10.1088/1757-899X/277/1/012057

[20] Groover M.P 2001 *Otomasi, Sistem Produksi, dan Computer-Integrated Manufacturing* (New Jersey: Pearson)
[21] Damajanti D.D, Rachmat H and Atmaja D.S 2013 Design of Integrated Scheduling and Automatic Controlling for Surface Treatment Process Using SCADA *IEEE International Conference of Industrial Engineering and Engineering Management* p1577-1581
[22] Geatari E.U, Rachmat H and Astuti M.D 2016 Design of User Requirements Specification to Automize Orthodox Black Tea Production System *Jurnal Rekayasa Sistem dan Industri 1* p43-48

Weighting analysis of pellet quality attributes using Multi Response Signal to Noise (MRSN) method

A Parkhan[1], I D Widodo[2] and F N Amin[3]

[1,2,3] Industrial Engineering Department, Universitas Islam Indonesia, Yogyakarta, Indonesia
[1]Aliparkhan@uii.ac.id, [2]Imamdjati@uii.ac.id, [3]10522321@uii.ac.id

Abstract. Quality of pellets, one form of animal feed, is not only measured by the nutritional content but also by its physical form. The physical strength of the pellet is determined from crushing and not easily moldy. Both quality characteristics are measured by reliability (pellet durability index) and resistance (water content percentage). In order to improve the quality of pellet, this study applied Multi Response Signal to Noise (MRSN) method. The weight of product quality attributes used will influence the method in determining the selected alternatives. To accommodate the weighting of dynamic product quality attributes, this study also ran weighting sensitivity analysis of product quality attributes. The results showed that the combination of factor level that produced the optimal pellet is A2, B1, C1, D1, E1, F1, G2 or combination of production process run with vapor pressure 1.9 bar, temperature conditioner 80 ° C, 3.5mm pellet diameter mold, cooler temperature 30 ° C, time in cooler 2 minutes, roller distance 1.5 cm, mixing time 175 seconds. This optimum combination can increase PDI percentage by 2.132% and decrease difference to target of water content by 0.234%. The optimum factor level combination will change if the weight for % PDI rises to be more than 0.77228 or decreases to be less than 0.00561, or in other words, the optimum combination will not change if the weight for % PDI is in the range 0.00561 - 0.77228.
Keywords: Quality attributes, Taguchi, MRSN, sensitivity Analysis

1. Introduction

Animal feed products consist of various forms, such as pellet (granular), concentrate, mash (flour), and crumble (fine grains). Besides depending on the nutritional content, the quality of poultry feed in the form of pellet is also influenced by the physical form such as not easily destroyed and not easily moldy. These depend on reliability measured by level of pellet durability index (PDI) and pellet resistance measured by their water content. The pellet-making process that influences the physical quality of the pellet is the vapor pressure, the temperature of the conditioner, the diameter of the pellet mold, the cooler temperature, the time in the cooler, the diameter of the roller and the mixing time [1]. Currently, the company produces pellets by PDI percentage is 86.794% and water content percentage is 9.738%. In an effort to improve the quality of products, this study conducted experiment designed based Taguchi method to develop robust design product. The Taguchi design experiment was an improvement of conventional experimental design. This method was a design process to produce products not sensitive to interference factors or have robust performance [2]. Taguchi based studies had been applied in some different area such as clothing [3][4][5], chemical study [6][7], food [8] and tool development [9]

This research used Multi Response Signal to Noise (MRSN) to optimize the different combination among product quality attributes or response variables. This method was done based on weight of response variables. Sensitivity analysis was done to guide researchers to be objective and to accommodate the dynamic product quality attributes weighting.

2. Literature review

2.1 Quality concept

Quality is a very important aspect in the progress of the company. One of the company success factors is determined by its final product quality. Many experts define quality in different words, however, all of them refer quality as meeting the customer requirements. Qualified products and services are products and services that match what their customers need [10]. Therefore, companies need to know their customers and their needs and desires. Quality has many different definitions, and varies from conventional to more strategic. The conventional definition of quality usually describes the direct characteristics of a product such as performance, reliability, easy to use, esthetics, perception and so on. Moreover, the strategic definition of quality is meeting the needs of customers [11].

2.2 Quality control

Quality control is a measurement process undertaken during the design of a product or process. Quality control activities include in every phase of product research and development, production process design, and customer satisfaction. Quality control is aimed to achieve all targets of continuous improvement, accelerated inventions, quick problem solving, and cost-effectiveness in improving product quality. Quality control can be divided into 2 parts, namely off-line quality control and on-line quality control [12]

2.2.1 Off-Line Quality control

Offline quality (QC) control is related to activities of QC during product development and process design. Activities undertaken include: identification of customers and their expectations, designing products in accordance with consumer expectations, designing products consistently and economically profitable and developing clearly and fairly specific standards, procedures, and production equipment. Off-line quality control is divided into 3 stages [13][14][15].

Phase I: Concept Design, relating to generating ideas in product design and development activities, in which the idea is identified from the needs of consumers

Phase II: Parameter Design, to optimize the level of controlling factors to the effects caused by other factors so that the resulting product can be robust to noise. Therefore, parameter design is often referred to as Robust Design

Stage III: Tolerance Design, is the last step by making orthogonal matrix, loss function, and Analysis of Variance (ANOVA) to balance cost and quality of product.

2.2.2 On-Line Quality control

On-line quality control is QC activities to observe and control the quality in every production process.

2.3 Experimental design

The experimental design is a design of any task that aims to describe or explain the variation of information under conditions that are hypothesized to reflect the variation. Each step/action of the experiment is completely defined so that related or required information for the issue studied can be collected [16]. Two experimental designs are known, namely the conventional experimental design and the Taguchi experimental design. The Taguchi method was first introduced by Dr. Genichi Taguchi. The purpose of the Taguchi experiment is to design how to minimize the deviation of quality characteristics from its target value. In the Taguchi method, a matrix (called orthogonal array) is used to determine the minimum number of experiments that can provide as much information as possible of

all factors affecting the parameters. The most important step of using orthogonal array lies in the selection of the level combinations of the input variables for each experiment [14].
Taguchi's philosophy of quality consists of three concepts [13][17][18]:

1. Quality should be designed into the product and not just check it. The best quality is achieved by minimizing the deviation from the target.
2. The product must be designed so that it is robust against environmental factors that cannot be controlled.
3. The cost of quality should be measured as a deviation function of a certain standard and losses must be measured across the entire system.

Based on the data responses, Taguchi can be divided into two, namely Taguchi single response and Taguchi multi response. Taguchi single response has only one response variable so the result can be directly obtained optimal combination of the response variable. Taguchi multi responses has more than one response variables (at least two response variables), and if each response variable has different factor levels combination then multi attributes decision making process is required to obtain an optimal combination of factors to improve the quality of each response variable. One of methods that can be used to solve multi-response Taguchi problems is MRSN (Multi Response Signal to Noise) method [19]. The steps of using Multi Response Signal to Noise Ratio (MRSN) are as follows [20]:

1. Calculate quality loss (Lij) for each trial. For quality characteristics:

Bigger the better — *Nominal the best* — *Smaller the best*

$$L_{ij} = k\frac{1}{n_i}\sum_{k=1}^{n_i}\frac{1}{y^2{}_{ijk}} \qquad L_{ij} = k\frac{1}{n_i}\sum_{k=1}^{n_i}\left(y_{ijk} - m\right)^2 \qquad L_{ij} = k\frac{1}{n_i}\sum_{k=1}^{n_i} y^2{}_{ijk}$$

Where: Y_{ijk} = data for i-th response, j-th trial, k-replication; m = target value

n_i = replication for the i-th response; k = coefficient of quality loss

Lij = loss function for the i-th response in the j-th trial.

2. Determine Multi Response Signal to Noise Ratio (MRSN).
   a. Specifies the maximum quality loss for each response.

   $$L_i^* = \max\left\{L_{i1}, L_{i2}, \ldots, L_{ij}\right\}$$

   b. Normalize quality loss (Cij) per experiment.

   $$C_{ij} = \frac{L_{ij}}{L_i^*}$$

   c. Calculate the total normalized quality loss (TNQL) per experiment:

   $$TNQL_i = \sum_{i=1}^{m} w_i C_{ij}$$

   Where: w_i = the weight of the normalization of the i-th response

   d. Calculates the MRSN ratio of each experiment.

   $$MRSN_j = -10\log(TNQL_j)$$

3. Determine the optimal combination of factor levels based on the largest MRSN value.

3. Research methods

The research was conducted on a factory that produced pellet. In general, this study was aimed to find out the factors that significantly affected the quality of pellets and the combination of factor levels that would produce the best quality pellets. Two response variables are selected: (1) the pellet reliability level measured by pellet durability index and (2) the pellet resistance measured by its moisture content. A good quality pellet is shown by the higher pellet durability index and water content of 10%.

IOP Conf. Series: Materials Science and Engineering **277** (2017) 012058 doi:10.1088/1757-899X/277/1/012058

4. Results and discussion

4.1 Experimental Planning

Some steps of the experimental planning phase include [5]:

1. Selection of product quality characteristics
   This stage is to determine the product quality characteristics and the objective function optimized. In this study, the dependent variable is the pellet durability index (% PDI) with the larger the better (LTB) objective function and the pellet resistance measured by its water content, with the nominal-the-better (NTB) objective function.
2. Identification and selection of factors that may affect the quality of the resulting product. Factors involved in this experiment are vapor pressure, conditioner temperature, pellet diameter, cooler temperature, processing time in the cooler, roller spacing, mixing time, and labor age. The factor level used and the reasons for choosing the level are as follows [1]:
   a. Vapor pressure: 1.6 bars and 1.9 bars. The levels are selected because if the vapor pressures are less than 1.6 bars then the output is the water not the steam that can cause the engine jam and pellet easily crumbled and moldy. If the vapor pressure is higher than 1.9 bars then the steam will be too hot. It causes the pellet too dry so that it will be easily destroyed.
   b. Conditioner temperature: 80 ° C and 85 ° C, the level is selected because if the conditioner temperature is lower than 80 ° C, the pellet will not be ripe so that the pellet will be easily crumbled and moldy. If the conditioner temperature is higher than 85 ° C, the protein will be broken and the pellet will be easily destroyed.
   c. Pellet mold diameter: 3.5 mm and 4 mm, the levels are selected because if the diameter of pellet mold is less that 3.5 mm then the shape of pellet will be too small and it will not fit to the standard. If diameter is bigger than 4 mm then pellet will be easily destroyed because of too many the air cavities inside.
   d. Cooler temperature: 30 ° C and 35 ° C, The levels are selected because if the cooler temperature lower than 30 ° C then the pellet will be easily destroyed when split using a roller and if it is higher than 35 ° C then the pellet feed will be too hot so that it cannot be broken by the roller into crumble.
   e. Time in the cooler: 2 minutes and 3 minutes, the levels are selected because if the time in the cooler is longer than 2 minutes then the water content in the pellet will be too much. It will be easily moldy. If it is longer than 3 minutes then the pellet water content will be too small. It will cause the weight of the pellet become light so that the company will lose benefit.
   f. The distance between the rollers: 1.5 cm and 2 cm, The levels are selected because if the distance between rollers is less than 1.5 cm then the pellet will be shaped like a crumble (if granules) and if it is bigger than 2 cm, the pellet size well be too big/long. This cannot fit the standard and it will be easily broken.
   g. Mixing time: 180 seconds and 175 seconds, the levels are selected because if the mixing time is faster than 175 seconds, the mixing will be uneven and the pellet will be easily destroyed. If it is longer than 180 seconds the raw material will be too hot so that it can clot.
   h. Labor age: <40 years and> 40 years, the age difference of the workforce will affect the pellets produced due to different experiences and concentrations.

   Based on initial evaluation, it is estimated that there is no interaction among factors.
3. Determination of Control Factor (Inner Array) and Noise Factor (Outer Array).
   Factors that allegedly affect the quality characteristics of the resulting product are divided into two factors, namely controllable factors and uncontrollable factors (noise). They can be seen in Table 1 and Table 2.

Table 1. Controllable factor

No	Controlable Factor	Code	Level 1	Level 2
1	Vapor pressure	A	1.6 bar	1.9 bar
2	Conditioner Temperature	B	80°C	85°C
3	Pellet Diameter Mold	C	3.5 mm	4 mm
4	Cooler Temperature	D	30°C	35°C
5	Cooling Duration	E	2 minutes	3 minutes
6	Distance **B**etween Rollers	F	1.5 cm	2 cm
7	Mixing Duration	G	180 seconds	175 seconds

Table 2. Noise factor

No	Noise Factor	Code	Level 1	Level 2
1	Labor ages	H	< 40 years	> 40 years

4.2. Experiment Implementation

To identify pellet reliability and pellet resistance, this experiment uses L8 for inner arrays and L4 for outer arrays. The experiment result data are shown by Table 3 and Table 4.

4.3. Data Processing

To determine the combination of factor levels for getting the optimum pellet quality, there are some assumption tests as follows:

4.3.1 Normality Test

The normality test indicates that data (both PDI and Water content data) follow normal distribution. For PDI data, $\chi^2_{\text{calculation}}$ (5.61) is less than χ^2_{table} (7.81), and for water content data, $\chi^2_{\text{calculation}}$ (6.33) is less than χ^2_{table} (9.49). These mean that Ho is accepted.

4.3.2 Homogeneity Test

Barlett test shows that both data (PDI and water content data) are homogeneous. The tests accept Ho because $\chi^2_{\text{calculation}}$ is less than χ^2_{tabel}. For PDI, $\chi^2_{\text{calculation}}$ (5.56) is less than χ^2_{tabel} (11.07) dan for water content, $\chi^2_{\text{calculation}}$ (8.18) is less than χ^2_{tabel} (11.07).

Table 3: Pellet Durability Index (PDI) Data

								Noise Factor (H)							
	Controllable Factor							1		1		2		2	
	A	B	C	D	E	F	G	Experiment Result Data				Experiment Result Data			
Trial	*Column Number*							Replication				Replication			
	1	2	3	4	5	6	7	1	2	3	4	1	2	3	4
1	1	1	1	1	1	1	1	88.4	88.9	87.1	87.0	85.4	88.2	89.6	88.7

Table 3: Pellet Durability Index (PDI) Data (Cont.)

								Noise Factor (H)							
	Controllable Factor							1		1		2		2	
	A	B	C	D	E	F	G	Experiment Result Data				Experiment Result Data			
Trial	*Column Number*							Replication				Replication			
	1	2	3	4	5	6	7	1	2	3	4	1	2	3	4
2	1	1	1	2	2	2	2	87.0	86.6	86.5	85.2	86.2	87.7	85.4	87.7
3	1	2	2	1	1	2	2	86.6	87.1	86.2	88.3	87.2	90.0	87.8	85.5
4	1	2	2	2	2	1	1	86.5	85.2	87.9	86.6	85.9	86.0	85.0	86.9
5	2	1	2	1	2	1	2	87.1	86.2	89.1	86.2	87.9	85.2	87.2	88.9
6	2	1	2	2	1	2	1	85.9	87.7	86.0	87.9	90.4	86.8	86.1	85.7
7	2	2	1	1	2	2	1	90.9	89.9	90.5	88.9	89.9	89.1	90.2	88.4
8	2	2	1	2	1	1	2	89.7	89.6	89.5	88.4	90.9	88.1	88.4	89.2

Table 4. Pallet water content data

								Noise Factor (H)							
	Controllable Factor							1		1		2		2	
	A	B	C	D	E	F	G	Experiment result Data				Experiment result Data			
Trial	*Column Number*							Replication				Replication			
	1	2	3	4	5	6	7	1	2	3	4	1	2	3	4
1	1	1	1	1	1	1	1	9.6	9.8	10.0	9.5	9.7	9.8	9.8	9.6
2	1	1	1	2	2	2	2	10.6	9.9	10.4	10.3	11.0	10.0	10.2	10.4
3	1	2	2	1	1	2	2	10.0	9.7	9.5	10.0	9.7	9.8	9.6	9.9
4	1	2	2	2	2	1	1	10.7	10.2	10.6	11.0	10.8	11.0	10.9	10.6
5	2	1	2	1	2	1	2	10.4	10.5	10.3	10.2	10.2	10.4	10.0	9.4
6	2	1	2	2	1	2	1	10.3	9.9	10.1	10.8	10.4	10.6	10.6	10.4
7	2	2	1	1	2	2	1	9.9	10.1	9.6	9.7	10.0	9.5	9.8	10.0
8	2	2	1	2	1	1	2	10.8	10.8	11.0	11.0	10.4	10.6	10.7	10.6

4.4 Analysis of Variance (ANOVA)

ANOVA results (Table 5 and 6) indicates that the effect of factors A, B, C and D is significantly different. For reliability (PDI data), $F_{calculation}$ of factors A, B, C and D (respectively 18.30; 9.84; 20.69 and 6.64) is higher than F_{table} (4.01). This means that Ho for factor A, B, C and D are rejected. For resistance (% water content data), $F_{calculation}$ of pellets for factors A, D and F (respectively 4.38, 105.44 and 15.20) are bigger than F_{table} (4.01). So, Ho for factor A, D and F must be rejected.

4.5. Signal to Noise Ratio (SNR) Analysis and Each Factor Effects

Signal to Noise Ratio (SNR) analysis is applied to identify the effect of each factor. This analysis is done based some significantly influent factors which are found in the previous ANOVA, namely A, B, C, and D for pellet reliability and A, D, and E for pellet resistance. For the factors that are not significantly influent, cost consideration is used to select the level factor for the combination. Because factor E, F and G do not have significant effect, so based on cost consideration (to select that is cheaper), the level factor E1, F1, and G2 are selected. Based on ANOVA and cost consideration, the best level factor combination for pellet reliability response variable is A2, B2, C1, D1, E1, F1, and G2. With the same analysis, the best level factor combination for resistance response variable is A1, B1, C1, D1, E1, F1, and G2.

Table 5. ANOVA table for PDI

Factor	Df	SS	MS	Fcal
A	1	27.0530	27.0530	18.30
B	1	14.5447	14.5447	9.84
C	1	30.5947	30.5947	20.69
D	1	9.8204	9.8204	6.64
E	1	3.1996	3.1996	3.16
F	1	0.0041	0.0041	0
G	1	0.3496	0.3496	0.24
Residual	56	81.3169	1.4785	
Total				

Table 6. ANOVA table for water content

Factor	df	SS	MS	Fcal
A	1	0.3025	0.3025	4.38
B	1	0.1806	0.1806	2.61
C	1	0.1806	0.1806	2.61
D	1	7.290	7.290	105.44
E	1	0.2025	0.2025	2.93
F	1	1.0506	1.0506	15.20
G	1	0.1406	0.1406	2.03
Residual	56	3.8025	0.0691	
Total				

4.6 Determination of Optimal Condition Factor Level Using MRSN

Multi Responses Signal to Noise analysis is done because the optimal combination of factor level of each response variable is different. The reliability of pellets is relatively more important than pellet resistance and the linguistic term "High" and "Medium" is chosen. The relative importance level is shown by the "linguistic term table" and then it is converted into fuzzy numbers. The score of each response variable is determined based on the crips scores of fuzzy number table. The result shows that the score of pellet reliability is 0.750 and pellet resistance is 0.583. Based on the scores, the weight of each response is calculated, they are

Pellet reliability weight (W1) = 0.750 /(0.750 +0.583) = 0.562641

Pellet resilience weight (W2) = 0.583 / (0.750 +0.583) = 0.437359.

The loss function (k) for every %PDI = Rp. 13,366, -; while the loss function (k) for each % water content = Rp. 22.050, -. The best formulation is obtained from the selection of the greatest MRSN. The best combination is A2, B1, C1, D1, E1, F1, G2. This combination is able to increase 2.132% of % PDI (from 86.794% to 88.926%). This combination also can reduce the difference to the target of water content from 0.262% to 0.028% (or 0.234%).

4.7 Sensitivity Analysis

The optimum factor levels combination was obtained on the weight of reliability attribute of 0.562641 and the weight of the pellet resistance attribute 0.437359. The analysis of attribute weight change shows that the optimum factor-level combination is sensitive to weighting changes of both responses (pellet reliability and pellet resistance). The optimum combination of factor level will change if the pellet reliability weight rises to be more than 0.77228 or decreases to be less than 0.00561. In other words, if the weight change for pellet reliability is in between 0.00561 - 0.77228, the change will not be sensitive to the original decision.

5. Conclusion

Based on the analysis above, some conclusions of this study are:

1. The optimum combination of factor levels is A2, B1, C1, D1, E1, F1, G2 or the production process is run with a vapor pressure of 1.9 bar; temperature conditioner 80 ° C; 3.5mm pellet molded diameter; cooler temperature 30 ° C; time in cooler 2 minutes, roller distance 1.5 cm, mixing time 175 seconds.

2. The optimum combination can increase 2.132% of % PDI (from 86.794% to 88.926%). This result can reduce the difference to the target of water content from 0.262% to 0,028% (or 0.234%).
3. The optimum factor level combination will change if the weighting for pellet reliability rises to be more than 0.77228 or decreases to be less than 0.00561, or in other words if the weighting for pellet reliability is in the range 0.00561 - 0.77228, it will not be sensitive to the original decision

6. References

[1] Amin F N 2014 *Desain Kualitas Pellet Dengan Metode Taguchi Multi Respon (Studi Kasus di PT.Charoen Pokphand Indonesia, Sidoarjo)* (Yogyakarta: FTI-UII)
[2] Belavendram N 1995 *Quality By Design: Taguchi Techniques for Industrial Experimentation* (New Jersey: Prentice Hall International)
[3] Latifah U 2008 *Setting Parameter Optimal Untuk Meningkatkan Kekuatan Tekan dan Meminimalkan Susut Kering Badan Keramik Dengan Menggunakan Metode Taguchi Multi Respon* (Yogyakarta: FTI UII)
[4] Zulfah L and Fajar M 2013 Penerapan Metode Taguchi Untuk Meningkatkan Kualitas Kain Tenun Pada Sentra Industri Kain Tenun Kabupaten Pemalang. *Jurnal Teknik Industri Universitas Pancasakti Tegal* **3** (3)13-18.
[5] Munawaroh S 2015. *Desain Kualitas Batik Tulis Dengan Metode TOPSIS (Technique For Order Preference By Similarity To Ideal Solution) (Studi Kasus Di Rifky Batik Ngasem, Yogyakarta),* (Yogyakarta: FTI-UII)
[6] Diwekar U M and Rubin E S 1994. Parameter design methodology for chemical processes using a simulator *Industrial and Engineering Chemical Research* **33** 292–298
[7] Hartono M 2012 Meningkatkan Mutu Produk Plastik Dengan Metode Taguchi *Jurnal Teknik Industri Politeknik Negeri Malang* **13** (1) 93-100.
[8] Setiawan A, Hayati N, and Yohanes A 2012 Setting Parameter Proses Pemasangan Lid Cup Air Minum Dalam Kemasan (AMDK) Dengan Metode Taguchi *Dinamika Teknik* (2) 27-34
[9] O'Hara A and Bogy D B 1995 Robust design optimization techniques for ultra-low flying sliders *IEEE Transactions on Magnetics* **31** 2955–2957
[10] Dorothea A 2003 *Manajemen Kualitas Pendekatan Sisi Kualitatif* (Jakarta: Ghalia Indonesia)
[11] Gasperz V 2001 *Total Quality Management* (Jakarta: PT. Gramedia Pustaka Utama)
[12] Soejanto I 2009 *Desain Eksperimen Dengan Metode Taguchi* (Yogyakarta: Graha Ilmu)
[13] Taguchi G 1987 *Introduction to Quality Engineering Course Manual* (American Supplier Institute)
[14] Peace G 1993 *Taguchi Method: A Hands-On Approach* (Addison–Wesley. Publishing Company).
[15] Ross J P 1988 *Taguchi Technique For Quality Engineering* (New York: Mc.Graw-Hill, Inc.)
[16] Sudjana 1991 *Desain dan Analisis Eksperimen* (Bandung: Tarsito)
[17] Taguchi G 1987 *System of Experimental Design* Vol. 1 & 2. (New York: Unipud Kraus International Publications)
[18] Montgomery D 2001 *Introduction to Statistical Quality Control 4th Edition* (New York. John Wiley & Sons, Inc.)
[19] Tong A and Su C T 1996 Robust design for the nominal-the-best performance characteristic *International Journal of Industrial Engineering* **3** 183–193
[20] Lestari D P 2009 *Peningkatan Kualitas Aspal Emulsi Menggunakan Metode MRSN* (*Multi Respon Signal to Noise*) (Yogyakarta: FTI-UII)

Service quality that improves customer satisfaction in a university: a case study in Institut Teknologi Indonesia

L Theresia[1] and R Bangun[2]

[1]Department of Industrial Engineering, Institut Teknologi Indonesia, Serpong, Indonesia, tarlind@yahoo.com
[2]Faculty of Administrative Science, Universitas Indonesia, Indonesia, bangun66@yahoo.com

Abstract. Universities should provide better service quality to get more customers. The purpose of this study was to find service quality which has impact on the increasing of customer satisfaction in a university. This study is a case study in Institut Teknologi Indonesia (ITI). The result of the study will help ITI to improve its service quality to increase customer satisfaction. This study employs path analysis technique. The data were collected through questionnaires developed from the literature. Questionnaires have two parts namely 1) service quality and 2) customer satisfaction. Service quality is measured through 22 questions with five service quality dimensions: 1) Tangibles, 2) Reliability, 3) Responsiveness, 4) Assurance and 5) Empathy. Customer Satisfaction is measured through 4 questions. The data then are processed by SPSS. The results showed that the students preferred to tangible quality improvement than intangible aspect.
Keywords: University, Service Quality, Customer Satisfaction

1. Introduction

The concept of quality is much easier to define in a manufacturing setting than in service sector. In the service sector, generally referred to as Service-Quality [1], the concept is much more difficult and complex to define. The quality in the service sector has become progressively more important. Service quality is referred to as the gap between the expectation of the customers on the service and real delivery of a service [19]. Measurement of a quality of service is a concept that is not uniform in various industries and there are differences in every type of service sector [8]. The service quality is also the strategy of the organization and it is called that there is integrity in service when the organization is doing what it is supposed to do. The service integrity helps the organization to measure the strategy of the organization and its implementation when it interacts with the customers [17]. In order to achieve its goals, the company must be able to give satisfaction to its customers. Measurement of service quality is important to find out the achievement of customer satisfaction. If there is a difference between the real quality of the service and the expected, then it indicates that customer satisfaction is not yet achieved. In this case, continuous service improvement is required to be successful in business development, particularly in today's highly competitive, dynamic and complex environments [14]. In order to survive the company must continuously improve its ability to serve its customers and have the advantage of competing better than existing competitors. Therefore, it is important to find the dimension of the service quality which has significant impact on the increasing of customer satisfaction.

ITI is a university located in Serpong and established in 1984. In operating its organization, ITI faces various problems, especially in the aspect of customer satisfaction. As a result, the number of enrolled students decreased continuously. It decreased more than 30% in the period from 1984 to 2004. Therefore, it is urgent for ITI to identify factors which would be able to increase organization's performance. Mmeasurement of service quality is important to find out the achievement of its performance.

2. Theoretical Background

Customer satisfaction is a principal element in achieving the goals in an organization and is a basic standard of excellence performance by the service organization [17]. Customer satisfaction is a person's feeling of pleasure or disappointment resulting from comparing a products perceived performance in relation to his or her expectations [14]. The satisfaction of customers is a collective response about the delivery of the service by the organization [29].Customers will be more satisfied with the high service quality. So the satisfaction of customers depends on the expectation of the customer about the service quality and perception of the customers about the service quality [3], [26].

The concept of service quality has been extended to a higher education context. Higher education plays an important role in the development of a country, as it promotes social, cultural and economic development and promotes active citizenship while instilling ethical values for citizens. The quality of services will affect student satisfaction. Student satisfaction as the favorability of the students' subjective evaluation on the various outcomes they experience associated with education. Student satisfaction is being shaped continually by repeated experiences in campus life [9].

The quality of services in the education sector has been the subject of numerous studies, because this matter is important for the sustainability of the education sector. A number of studies have been conducted, which indicate the importance of measuring the quality of services in higher education [21], [24]. However, there is no consensus on the dimensions and antecedents of the quality measurement of services perceived by students [27]. Consequently, there is no consensus on the best way to determine and measure the quality of service [5]. Parasuraman et al. has developed a SERVQUAL approach to find out the satisfaction of the services. The SERVQUAL model is based on analysis of gap between the expectations of the customers and perceived service quality. There were five dimensions of service quality as mentioned in Table 1 by various researchers.

Table 1. Dimension of Service Quality [20], [25]

Dimension	Definition
Tangibles	Physical facilities, equipment and appearance of personal
Reliability	Ability to perform the promised service dependably and accurately
Responsiveness	Willingness to help customers and provide prompt service
Assurance	Knowledge and courtesy of employees and their ability to inspire trust and confidence
Empathy	Caring, individualized attention, the firm provides it costumers

The SERVQUAL approach has been criticized by many researchers, include Brown, Churchill, and Peter [2], [22]. They argue that different score approach causes the issues related to the variance restriction related with the score of the components will produce poor reliability. Another critique was put forward by Teas [28]. He argued that SERVQUAL scale uses various types of expectations and the other researchers may not be able to distinguish between various expectations types while doing their evaluation. The alternative method to the SERVQUAL is the SERVPERF and was developed by Cronin and Taylor [6],[7]. They argued that this model is better. Although SERVQUAL model reaps lot of criticism, the experts, who used that instrument, concluded that the SERVQUAL provides better diagnostic result for quality of service [19]. The five dimensions of SERVQUAL are suitable for mass service and also banking industry as well [29]. The perception of customers about the service quality is based on various factors, for example social referrals, communications between sales persons, credibility of the consumers about the organization offering service and the different type of data collection [11].

3. Research Method

Based on the purpose of research, the type of research used is explanatory research. This research is a quantitative research and use path analysis technique. The data were analysed using SPSS version 16 program. The research model as shown in Figure 1 consists of two parts namely: Service Quality and Customer Satisfaction. Service Quality consists of five dimensions: (1) tangible, (2) reliability, (3) responsiveness, (4) assurance and (5) empathy. While Customer Satisfaction includes the performance of an organization's product and services compared to expectation of a customer.

To measure the level of student satisfaction, the authors used a questionnaire developed by Zeithaml [30]. In addition, the questionnaire in the constructs of the literature refers to the study conducted by Leonga et al. [16] and Ilyas et al. [13]. The questionnaire consists of two parts: service quality and customer satisfaction. The questionnaire for service quality contains 22 questions for the five dimensions of service quality; while for customer satisfaction consists of 4 questions. The questionnaire uses a Likert five scale that is 1 = Strongly Disagree to 5 = Strongly Agree.

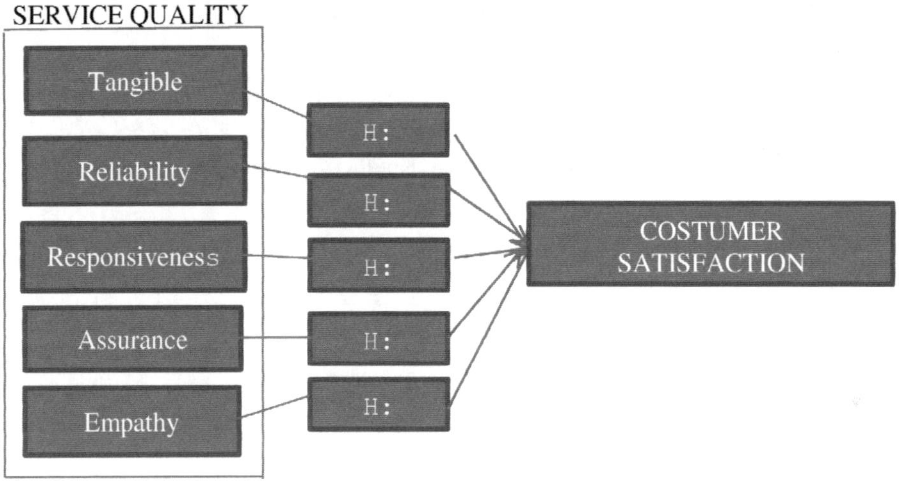

Figure 1. Conceptual Framework [13]

Based on research objectives and theoretical reviews, the research hypothesis is:

a. There is influence of tangible aspect to customer satisfaction
b. There is influence of reliability aspect to customer satisfaction
c. There is influence of responsiveness aspect to customer satisfaction
d. There is influence of assurance aspect to customer satisfaction
e. There is influence of empathy aspect to customer satisfaction

The population in this study is all regular students in ITI. Questionnaires were distributed to 1300 regular ITI students. The number of valid questionnaires is 1006. Table 2 shows the question in the constructs of the literature which was also mentioned in [8] and [11].

Table 2. Sources of Construct

Variable	No	Indicator
I. Tangibility	1	ITI Location
	2	Availability of parking space
	3	Appearance of ITI office
	4	Comfort and cleanliness of the office

Table 2. Sources of Construct (cont.)

Variable	No	Indicator
II. Reliability	5	Services such as canteens, mosques, sporting events, student activities, religious activities, security promised by ITI
	6	Physical infrastructure such as lecture hall, library room, student room, sports facility, worship room
	7	Internet facility
	8	Library
III. Responsiveness	9	Request letter service recommendation (practical work letter, final project, lab work)
	10	Handling lecture room problems, laboratory space, library
	11	Service provision during lectures, exams, seminars, task presentations, discussions
	12	Handling issues of security, cleanliness and order in the campus environment
	13	Handling the administrative problems of tuition fees, KRS and exam cards
IV. Assurance	14	Expertise and knowledge of faculty / staff
	15	Hospitality of employees in providing services
	16	Speed of employees in providing services
	17	Capability and knowledge of employees in providing services
	18	Timeliness of lecturers in teaching
	19	Hospitality of lecturers at teaching time
	20	Ability and knowledge of lecturers in teaching
V. Empathy	21	Ease in contacting lecturers/employees
	22	Ease in contacting lecturers/employees
VI. Customer Satisfaction	1	I am satisfied with the service provided ITI employees
	2	The decision to study at ITI is good
	3	I feel the study experience at ITI is fun
	4	I am easy to obtain services in ITI

4. Result and Discussion

Reliability is calculated by the value of Cronbach Alpha (a). In this study the value of (a) is 0.839 for all items that fit the criteria set by Nunnally [18]. Thus in this study the data collected reliable, which is also confirmed by the correlation relationship between variables. The resulting correlation matrix is shown in Table 3. From the correlation matrix shows that all variables of tangibles, reliability, responsiveness, assurance, empathy correlate with the dependent variable, customer satisfaction and no multicolinearity between variables, because there is no coefficient correlation greater than 0.8 [12]. The relationship of tangibles, responsiveness, assurance and reliability with customer satisfaction is very strong, while the empathy relationship with customer satisfaction is weak according to the criteria set by Field [10].

Table 3. Correlation Matrix

Correlation		Tangibility	Respon-siveness	Assur-ance	Reli-ability	Empathy	Customer Satisfaction
Tangibility	Pearson Correlation	1					
Respon-siveness	Pearson Correlation	.557**	1				
Assurance	Pearson Correlation	.583**	0.699**	1			
Reliability	Pearson Correlation	.606**	0.527**	0.512**	1		
Empathy	Pearson Correlation	.166**	.056*	.164*	.247**	1	
Customer Satisfaction	Pearson Correlation	.548**	.517**	.617**	.570**	.227**	1

**. Correlation is significant at the 0.01 level (2 tailed).
*. Correlation is significant at the 0.05 level (2 tailed).

Table 4 shows the variation in dependent variable due to independent variable. As the value of adjusted R square was 0.713 which showed that 71.3% of variation in customer satisfaction was being explained by all independent variables. While the rest of 28.7% influenced by other variables that are not included in this model.

Table 4. Regression Analysis

Model Summary				
Model	*R*	*R Square*	*Adjusted R Square*	*Std Error of Estimate*
1	.816*	0.764	0.713	0.2951

* Predictors: (Constant), Tangible, Reliability, Responsiveness, Assurance, Empathy

Anova calculation is then performed in the regression analysis, to see the significance of variation. From result of regression analysis, p value is 0.000. If p less than 0.05 which further predicted that overall the model was statistically significant. With a p-value probability level of 0.000, H0 is therefore rejected and H1 accepted. It can be concluded that the null (H0) hypothesis and the alternative hypothesis (H1) this proves service quality, consist of tangibility, reliability, responsiveness, assurance, empathy together significantly influence customer satisfaction.

Table 5. Regression Analysis – Coefficient

Coefficient[a] *Model*	Unstandardized	Coefficients	Standardized Coefficients	T	Sig
	B	Std Error	Beta		
1 (Constant)	0.342	0.144		2.050	0.032
Tangibles	0.333	0.044	0.039	0.647	0.011
Responsiveness	0.001	0.033	0.444	8.666	0.201
Assurance	0.199	0.031	0.599	11.200	0.120
Empathy	0.045	0.041	0.076	1.345	0.114
Reliability	0.321	0.039	0.099	2.667	0.004

a. Dependent Variable: Costumer Satisfaction

The next step is to see the influence of the independent variable to the dependent variable. From Table 4, it can be seen the effect of independent variables on the dependent variable. Only tangibles and reliability variables have statistically significant effects because they have a p value less than 0.05. While responsive, assurance and empathy are not statistically significant. The value of the regression coefficient for tangibles is 0.333 indicating that each increase of one unit in tangibles will result in an increase in customer satisfaction of 0.333 by keeping all other variables constant. Similarly, an increase of one unit in reliability will result in a 0.321 increase in customer satisfaction.

5. Conclusion

The results of the research show that the tangibles and reliability variables have a positive effect on customer satisfaction. This shows that students are concerned with tangible and reliability aspects. Therefore, ITI needs to improve the aspects related to buildings, such as office building, comfort and cleanliness of the building. Services, physical infrastructure, internet and libraries are also should be improved.

The aspects of responsive, assurance and empathy have no impact on student satisfaction because students are concerned with tangible aspect instead of physical factors. Similarly, the aspect of empathy, where students generally feel the nature of empathy is a personality that should be owned by an educator.

6. References

[1] Annamdevule S and Bellamkonda R S 2014 Effect of student perceived service quality on student satisfaction, loyalty and motivation in Indian universities Development of HiEduQual *Journal of Modelling in Management*, **Vol. 11** No. 2

[2] Brown T J, G A Churchill and J P Peter 1993 Improving the measurement of service quality *Journal of Retailing* **69**, no. 1 pp 127-139.

[3] Christou E and Sigala M 2002 Conceptualizing the measurement of service quality and TQM performance for hotels: the HOSTQUAL model. *Azta Touristica*, pp 140-169

[4] Clemes M D, Gan C, Kao T H and Choong M 2008 An empirical analysis of customer satisfaction in international air travel *Innovative Marketing*, pp 49-62

[5] Clewes D 2003 A student-centered conceptual model of service quality in higher education, *Quality in Higher Education*, **Vol. 9** No. 1, pp. 69-85

[6] Cronin J J and Taylor S A 1992 Measuring service quality: A re-examination and extension *Journal of Marketing* **Vol. 56** No. 3 pp 55-68.

[7] Cronin J and Taylor S A 1994 SERVPERF versus SERVQUAL: Reconciling Performance-based and perceptions minus expectations measurement of service quality *Journal of Marketing* **Vol. 58** No. 1 pp 125-131.

[8] Cullen et al. 2003 Quality in higher education: from monitoring to management *Quality Assurance in Education,* **Vol. 11**, pp. 30-34.

[9] Elliott K M and Shin D 2002 Student satisfaction: an alternative approach to assessing this important concept *Journal of Higher Education Policy and Management*, **Vol. 24** No. 2 pp 197 - 209

[10] Field A 2009 *Discovering Statistics Using SPSS* 3rd ed. (Los Angeles: Sage).

[11] Gounaris S P, Stathakopoulos V and Athanassopoulos A D 2003 Antecedents to perceived service quality: an exploratory study in the banking industry. *International Journal of Bank Marketing*, **21** (4), pp168-190.

[12] Gujarati D N 2003 *Basic Econometrics,* 4th ed. (Singapore: McGraw-Hill)

[13] Ilyas A, Arshed N and Hussain T 2014 Service quality and customer satisfaction: a study on Islamic Banks of Pakistan, *Journal of Business Strategies,* **Vol.10** No.1, pp 73–88

[14] Kotler P 2000 *Marketing Management.* (New Jersey: Prentice-Hall), p. 36

[15] Kuo Y F, Wu C M and Deng W J 2009 The relationships among service quality, perceived value, customer satisfaction, and post-purchase intention in mobile value-added services *Computers in Human Behaviour*, **25** (4),pp 887–896

[16] Leonga L Y, Hewb T S, Leea V H and Ooic K B 2015 An SEM–artificial neural- network analysis of the relationships between SERVPERF, customer satisfaction and loyalty among low-cost and full-service airline *Expert Systems with Applications*, pp 6620–34
[17] Munusamy J and Chelliah S 2011 An investigation of impact of service strategy on customers‟ satisfaction in the budget airline industry in Malaysia: a case study of air Asia. *Contemporary Marketing Review,* **1**(1), pp 1-13.
[18] Nunnally J C 1978 *Psychometric Theory* 2nd ed. (New York: McGraw-Hill).
[19] Parasuraman A, Zeithaml V A and Berry L L 1985 A conceptual model of service quality and its implications for Future Research *Journal of Marketing*, **49** (4), pp 41-50
[20] Parasuraman A, Zeithaml V and Berry L 1988 SERVQUAL: a multiple-item scale for measuring consumer perceptions of service quality *Journal of Retailing*, **64** (1), pp 12-40.
[21] Parri J 2006 Quality in higher education *Vadyba Management*, **Vol. 2** No. 11, pp. 107 – 111
[22] Peter J P, Churchill G A and Brown T J 1993 Caution in the use of difference scores in consumer research *Journal of Consumer Research* **19**, No. 1 pp 655-662.
[23] Prayetno S and Hapzi Ali 2017 Analysis Advocates Organizational Commitment and Advocates Work Motivation to Advocates Performance and Its Impact on Performance Advocates Office *International Journal of Economic Research*, **Vol 14** No. 13
[24] Quinn A, Lemay G, Larsen P and Johnson D M 2009 Service quality in higher education *Total Quality Management*, **Vol. 20** No. 2, pp. 139 - 152 .
[25] Shahin A, Mehrparvar E and Shirouyehzad H 2013 Prioritisation of departments based on service quality dimensions in Isfahan Steel Company: a multiple criteria decision making approach *International Journal of Productivity and Quality Management*, pp 116-130.
[26] Sigala M 2004 Investigating the factors determining e-learning effectiveness in tourism and hospitality education *Journal of Hospitality & Tourism Education*, **16** (2) pp 11-21.
[27] Sultan P and Wong H Y 2012 Service quality in a higher education context: an integrated model, *Asia Pacific Journal of Marketing and Logistics*, **Vol. 24** No. 5, pp. 755 -784.
[28] Teas R K 1993 Expectations, performance evaluation and consumer's perception of quality. *Journal of Marketing* **57**, no. 4, pp 18-34.
[29] Uddin M B and Akhter B 2012 Determinants of customer satisfaction of banking industry in Bangladesh Pakistan *Journal of Commerce and Social Sciences*, **6** (2), pp 242-256.
[30] Zeithaml V A 1987 *Defining and Relating Price, Perceived Quality and Perceived Valu*e (Cambridge, MA: Market. Sci. Inst).

IOP Conf. Series: Materials Science and Engineering **277** (2017) 012060 doi:10.1088/1757-899X/277/1/012060

Rice distribution planning for "the poor people" in Bandung, West Java, Indonesia.

Sutarman[1] and Endang Hidayat[2]

Industrial Engineering Department, Engineering Faculty, Universitas Pasundan
Jl. Dr. Setiabudi No. 193 Bandung – 40154 Indonesia
[1]sutarmanelank4@gmail.com, [2]endanghidayat@gmail.com

Abstract: Bandung is one of the big and developed cities in Indonesia. The high economic growth in this city has an impact on increasing the prosperity of its citizens. However, the prosperity has not spread evenly. In fact, many residents of the city are still lack of rice. The party in charge of distributing rice to the poor is a state company called Bulog. Bulog has several warehouses with limited capacity to supply the poor. Three of them are *Cisaranten Kidul,* Main Warehouse and *Citeureup* warehouse. *Bulog* supplies rice to poor villages spread over 30 districts in Bandung. The allocation of rice for all districts needs to be improved so that the needs of the poor can be fulfilled with minimum allocation cost. Distribution planning conducted in this research was used Linear Programming method. Through this method, the optimization of rice allocation for the poor in Bandung can be obtained. Based on the finding, Linear programming method can be recommended as rice distribution pattern for poor people in Bandung with minimal logistics cost.
Keyword: Linear programming, warehousing, inventory, distribution

1. Introduction

Rice is the main food for the people of Indonesia. By 2016, Indonesia's population was 255 million people with a yearly consumption of rice per capita of 98 kg. Thus, the need for rice reaches 25 million tons in a year. Indonesia is able to produce 48 million tons of rice per year, there is an excess supply of 23 million tons. Although the production of rice in Indonesia is excessive, but if the distribution is unplanned and well-controlled, it will lead to instability in food security in Indonesia where people do not get good quality rice at the right time and quantity.

Indonesia has a State-Owned Enterprise called "Bulog" in charge of maintaining price stability and the balance of rice supply and demand for the community, through a planned procurement and distribution process, so as to meet the needs of the poor or in Indonesian commonly referred to as '*Rumah Tangga Penerima Manfaat*'. The process of rice procurement is done in 2 ways, namely domestic and import. Domestic procurement is the cooperation of Bulog and its pre-established partners with various requirements. To support rice distribution activities, Bulog has several warehouses with different locations and capacities. Meanwhile, the demand for rice varies. In order to meet the demand in the right amount and time, Bulog should plan the distribution of rice by determining allocations for each destination and warehouse sources with minimum distribution costs.

Bandung is one of the big cities in Indonesia and the capital of West Java Province consisting of 30 districts wherein each district has a number of poor with varying numbers. Rice supplied by the Blog Bandung Regional Division, comes from 3 warehouses with different locations and capacities, they are: (1) Main Warehouse; (2) *Cisaranten Kidul* Warehouse, and (3) *Citeureup* warehouse.

The problems in this study are: (1) how to determine the allocation of rice for each district so that demand can be met with minimum distribution costs?, (2) how much rice should be stored in each warehouse for minimum storage cost?. Based on the two issues, the total minimum distribution cost will be earned.

2. Methods

A supply chain consists of all parties involved, directly or indirectly, in fulfilling a customer request. Supply chain includes not only the manufacturer, but also transporters, retailers, and even customers themselves [2]. The above issues require the support of relevant theories, to obtain the optimum amount of rice allocated from each warehouse to each district in Bandung, with minimum distribution cost. The next step is to determine the cost of the warehouse, because the cost of logistics consists of distribution costs and the storage cost.

2.1 The determination of allocation

The problem faced in this research is how to allocate the rice from several warehouses with different locations and capacities to several districts in Bandung with different locations and the number of requests. This issue is included in the distribution problem. For optimal allocation, the linear programming method is used. The problem can be illustrated as in Figure 1 below:

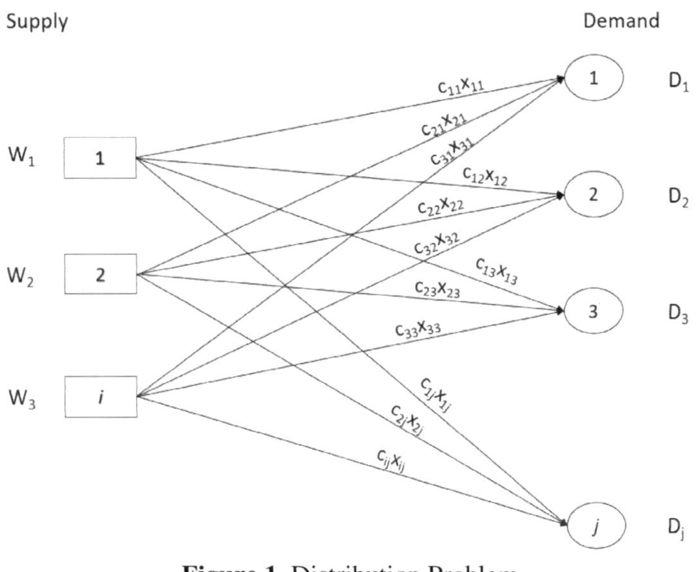

Figure 1. Distribution Problem

The problem described above can be modeled with a mathematical formulation, with an objective function to minimize the costs, as in equation 1 below:

$$Z = C_{11}X_{11} + C_{12}X_{12} + \cdots + C_{ij}X_{ij} \tag{1}$$

To complete the objective function, there are 2 functional constraints, namely (1) the effective warehouse capacity constraint, and (2) the rice demand constraint for each district.

2.1.1. The effective warehouse capacity constraint. The effective capacity is the capacity of each warehouse to meet the rice needs of the poor within a year. This capacity is designed to meet the demand for rice in Bandung and surrounding districts. The mathematical function of this constraint is illustrated in the following equation 2:

$$\sum_{j=1}^{n} a_{ij} X_{ij} \leq W_i \tag{2}$$

2.1.2. The rice demand constraint for each district The rice demand of the poor in every district in Bandung has been determined by the Indonesian government by "based on name, based on address" in accordance with the 2015 policy. Thus, the mathematical function of this constraint, as in the following equation 3:

$$\sum_{j=1}^{n} a_{ij} X_{ij} = D_j \tag{3}$$

The third constraint is to explain that the amount of rice allocated to each district in Bandung must be positives, thus the mathematical function as in the following equation 4:

$$X_{11} \geq 0, X_{12} \geq 0, \dots, X_{ij} \geq 0 \tag{4}$$

Where:
Z: Objective function with minimized distribution cost
C_{ij}: Distribution cost per kg from warehouse i to district j
X_{ij}: Decision variables for the quantity of rice allocated from warehouse i to district j
a_{ij}: The number of warehouses i that used by each district i
D_j: The quantity of rice demand in each district j
W_i : The effective capacity of warehouse i

2.2 *The total of distribution cost*
The key to achieving economic transportation is summarized in two basic principles. The first, often called the quantity principle, is that individual shipments must be as large as the involved carrier that can legally transport the equipment in use. The second, often called tapered principle, is that the delivery should be transported as long as possible [3].

The determination of total transportation cost is done after the allocation of each warehouse to each sub-district, with linear program formulation. Optimal solution assisted with Win QSB software. The optimal allocation amount and minimum distribution cost are obtained because it consider the objectives and functional constraints that have been formulated strictly in the formulation of linear programming.

2.3 *The warehousing cost*
A warehouse is traditionally viewed as a store or inventory. However, in contemporary logistics systems, the function of a warehouse is more appropriately seen as mixing and inventory modification to meet customer needs [4]

Warehousing costs consist of two types, namely (1) the fixed costs, i.e. the costs that do not depend on the volume of stored rice, and every warehouse has fixed costs differently, including depreciation costs, employees, warehouse management, maintenance, insurance and property taxes, (2) the variable costs, the amount of the cost depends on the volume of stored rice, and the carrying cost per ton of rice has been determined by the company.

Thus the warehousing cost is a fixed and a variable cost function, and based on the result of the study, the total warehousing cost is the fixed cost of the warehouse with the square root of stored rice volume, as in the following equation 5:

$$C_i = K_i \sqrt{Z_{wi}} \tag{5}$$

Where:
C_i:The warehousing cost of warehouse i
K_i: The fixed cost for each warehouse i
Z_{wi}: The volume of store rice at warehouse i

IOP Conf. Series: Materials Science and Engineering **277** (2017) 012060 doi:10.1088/1757-899X/277/1/012060

2.4 Determining the logistics cost

Another term of logistics is the "move and store" activity, the "move" is the movement of goods from the origin place to the destination, while the "store" is the storage of goods at a storage facility. Moving goods from the origin place to the destination is called transportation activities, and it requires transportation costs. Meanwhile, if the goods have arrived at the destination, the goods undergo the storage process to wait for the next process. The storage process requires a fee called warehousing costs. Thus the logistics cost is equal to the amount between transportation cost and warehousing cost.

3. Results and Discussions

The implementation of this research consists of 4 stages, namely: (1) determining the optimum allocation for each warehouse, and obtaining the minimum total transportation costs;, (2) determining the warehousing costs of the stored rice; (3) determining the logistics cost (the logistics cost is the total between transportation and warehousing cost), and; (4) determining the efficient distribution patterns. The steps mentioned can be explained as follows:

First, the illustration of the transportation model to determine the allocation of rice from each warehouse to each district in Bandung can be seen in figure 2:

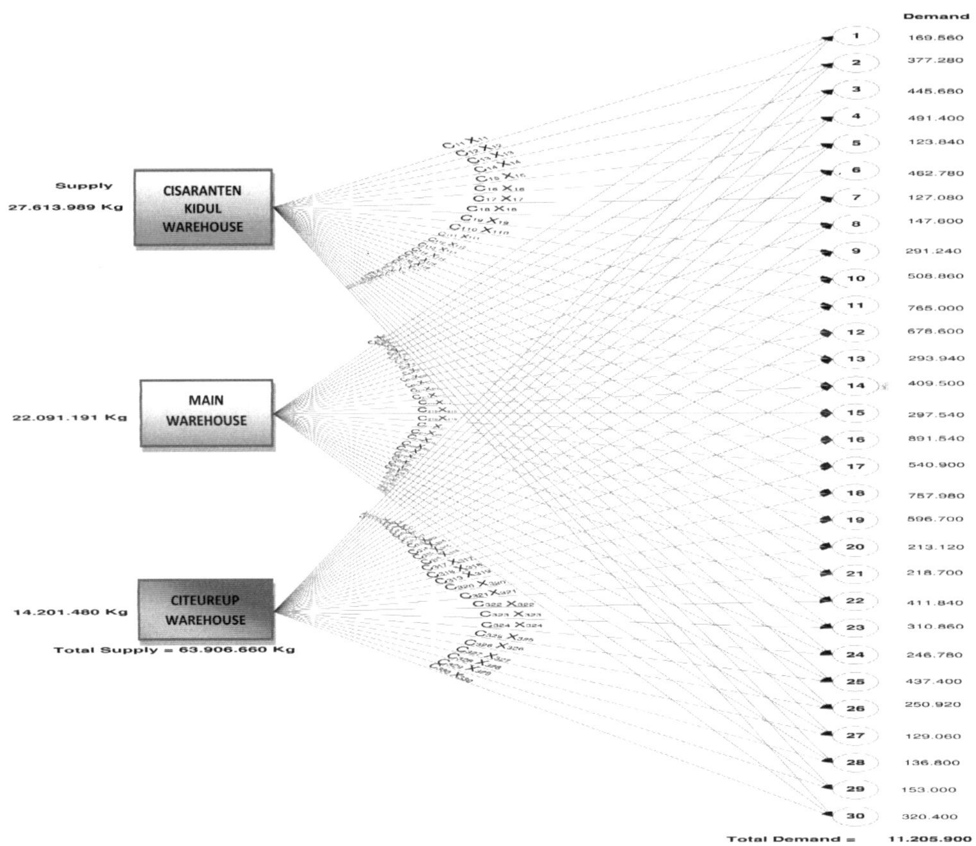

Figure 2. The Transportation Model Illustration

Based on figure 2 above, there are 3 warehouse locations with different capacity. They must supply 30 districts in Bandung with the number of different requests.

With the Linear programming formulation which consists of an objective function, consideration of warehouse capacity, the demand and the positive constraints, an optimum solution can be obtained, as in table1.

Minimum transportation cost:

$$Z = 58{,}20\,X_{11} + 57{,}25X_{12} + 56{,}42X_{13} + \cdots + C_{ij}X_{ij} + 0S_1 + 0S_2 + 0S_3 \quad (6)$$

a.The warehouse capacity constraint:

$X_{11} + X_{12} + X_{13} + X_{14} + X_{15} + X_{16} + \cdots + X_{130} + S_1$ =27.613.989
$X_{21} + X_{22} + X_{23} + X_{24} + X_{25} + X_{26} + \cdots + X_{230} + S_2$ =22.091.191
$X_{31} + X_{32} + X_{33} + X_{34} + X_{35} + X_{36} + \cdots + X_{330} + S_3$ =14.201.480

b.The districts demand constraint:

(1) $X_{11} + X_{21} + X_{31}$ =169,560
(2) $X_{12} + X_{22} + X_{32}$ =377,280
(3) ... +... +... = 445,680
(4) ... +... +... = 491,400
(5) ... +... +... =123,840
... ... +... +... = ...
... ... +... +... = ...
(30) $X_{130} + X_{230} + X_{330}$ =320.400

c.The positives constraint

$X_{11} \geq 0, X_{12} \geq 0, \ldots, X_{330} \geq 0$
$S_1 \geq 0, S_2 \geq 0, S_3 \geq 0$

Where:

Z = objective function with the minimum transportation cost
C_{ij} = transportation cost from warehouse i to district j
X_{ij} = Decision variable for qunatity of rice to be allocated from warehouse i to district j
S_i = The number of unused resources-1es for allocation in Bandung City.

The optimum solution for the above formulation is shown in table1.

Table 1. The Rice Allocation for 30 Districts

No	Districts	Allocation (tons)		
		Cisaranten Kidul Warehouse (X_1)	Gudang Utama (X_2)	Citeureup Warehouse (X_3)
1	Sukasari	-	169.56	-
2	Sukajadi	-	377.28	-
3	Cicendo	-	445.68	-
4	Andir	-	491.40	-
5	Cidadap	-	123.84	-
6	Coblong	-	462.78	-
7	Bandungwetan	-	127.08	-
8	Sumur Bandung	-	-	147.60
9	Cibeunying Kaler	291.24	-	-
10	Cibeunying Kidul	508.86	-	-
11	Kiaracondong	765.00	-	-
12	Batununggal	-	-	678.60
13	Lengkong	-	-	293.94
14	Regol	-	-	409.50
15	Astana Anyar	-	-	297.54
16	Bojongloa Kaler	-	-	891.54
17	Bojongloa Kidul	-	-	540.90

Table 1. The Rice Allocation for 30 Districts (Cont.)

18	Babakan Ciparay	-	-	757.98
19	Bandung Kulon	-	596.70	-
20	Antapani	213.12	-	-
21	Arcamanik	218.70	-	-
22	Ujungberung	411.84	-	-
23	Cibiru	310.86	-	-
24	Rancasari	246.78	-	-
25	Buahbatu	437.40	-	-
26	Bandung Kidul	-		250.92
27	Gedebage	129.06	-	-
28	Panyileukan	136.80	-	-
29	Cinambo	153.00	-	-
30	Mandala Jati	320.40	-	-
31	S1	23,470.93	-	-
32	S2	-	19,296.87	-
33	S3	-	-	9,932.96
	Total	27,613.99	22,091.19	14,201.48

The allocation of rice at each warehouse for each district is the total allocation minus the unused amount (slack variable). The allocation of rice from *Cisaranten Kidul* warehouse to fulfill the demand of Bandung is 4,143.06 tons. Meanwhile, the allocation of rice from Main Warehouse as much as 2,794.32 tons, and from *Citeureup* warehouse is 4,268.52 tons. The surplus of 52,700.76 tons of rice supply is allocated to other areas outside Bandung. After the optimum allocation is obtained, substitute the allocation quantity as a decision variable to the equation in the objective function. With assisted by Win QSB software, the total transportation cost is IDR 602,797,400, equivalent to US $ 46,370.

Second, as explained above that the warehousing cost is the total between fixed cost and variable cost where the fixed cost of each warehouse is determined by the company. While the variable cost is the square root of the amount of rice allocated to each warehouse. The warehousing cost is the total of the three warehouse costs, namely: *Cisaranten Kidul*, *Gedebage* (Main Warehouse) and *Citeureup*, as can be seen in Table 2.

Table 2. The Total Warehousing Cost

i	Warehouses	Fixed Cost (K_i)	Amount allocated (Z_i)	$C_i = K_i\sqrt{Z_{wi}}$
1	Cisanaten Kidul	53,264	4,143.06	3,428,431
2	Main Warehouse	71,598	2,794.32	3,784,785
3	Citeureup	77,463	4,268.52	5,060,948
			11,205.90	12,274,165

Third, the logistics cost is obtained by totaling the transportation cost and the warehousing cost. Based on the above calculation, the transportation cost is IDR 602,797,400 and the warehousing cost is IDR 12,274,165, thus the logistics cost is IDR 615,071,565, equivalent to US $ 47,313.20.

Fourth, in accordance with the above description, an efficient distribution pattern can be determined, because: (1) rice allocation can be optimized with minimum transportation costs, (2) rice allocation can be optimized with minimum warehousing costs. Although transportation costs and warehousing costs are minimum, but all the rice demand for the whole district are fulfilled. For that reason, the design of rice distribution patterns for the poor in Bandung can be obtained, as can be seen in the figure 3.

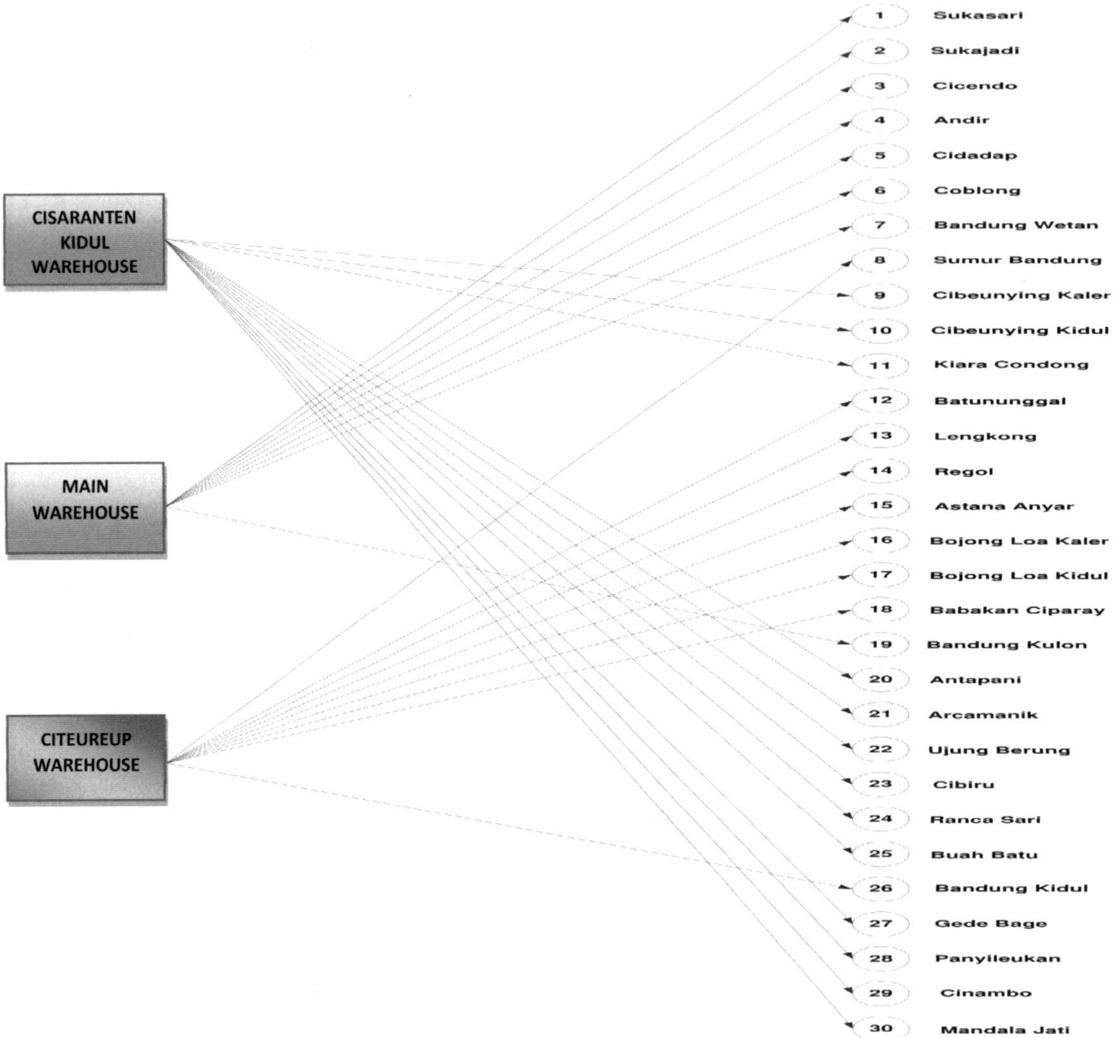

Figure 3. Rice distribution pattern

From the above distribution pattern, the rice demand for 13 districts can be supplied from *Cisaranten Kidul* Warehouse totaling 4,143,06 tons, while 8 districts can be processed by the Main Warehouses of 2,794.32 tons, and 9 districts of *Citeureup* warehouse totaling 4,268.52 tons.

4. Conclusion

Based on the use of theories, the steps and calculations that have been done, the conclusions of this study are as follows: (1) rice allocation for the poor in 13 districts in Bandung can be fulfilled from *Cisaranten Kidul* Warehouse as 4,144.06 tons, while 8 districts can be supplied by Main Warehouse as much as 2,794.32 ton while the remaining 9 districts can be supplied by *Citeureup* Warehouse as much as 4,268.52, with the transportation cost Rp 602,797,400; (2) the warehousing cost as optimal allocation result of *Cisaranten Kidul* warehouse is Rp 3,428,431, while the warehouse costs in the Main Warehouse amounting to Rp 3,784,785 and the warehousing costs in *Citeureup* Warehouse is Rp 5,060,948. Thus, the total cost of logistics if it is using the design of the distribution pattern is Rp 615,071,565.

IOP Conf. Series: Materials Science and Engineering **277** (2017) 012060 doi:10.1088/1757-899X/277/1/012060

5. References

[1] Bahagia Senator Nur 2006 *Sistem Inventori* (Bandung: ITB Press)
[2] Ballou Ronald H 2004 *Business Logistics/Supply Chain Management* (New Jersey: (Pearson Prentice Hall)
[3] Bowersox Donald J David J Closs M Bixby Cooper and John C. Bowersox 2013 *Supply Chain Logistics Management* (New York: McGraw Hill)
[4] Chopra S and P Meindl 2013 *Supply Chain Management: Strategy, Planning and Operation* (Pearson Global Edition)
[5] Frazelle E.H 2002 *World Class Warehousing and Material Handling*(New York: McGraw-Hill)
[6] Gaspersz Vincent 2005 *Total Quality Management* (Jakarta: PT Gramedia Pustaka Utama)
[7] Kisperska, Danuta Moron 1999 *Warehousing Conditions for Holding Inventory in Polish Supply Chain* (Elsevier) p123
[8] Lambert, D M and J R Stock 1993 *Strategic Logistics Management* (USA:Homewood)
[9] Maede Laura 1998 *Strategic Analysis of Logistics and Supply Chain Management Systems Using the Analytical Network Process* Vol 34 (Elsevier) p 201
[10] Li Dong and Christopher O Brien 1999 *Integrated Decision Modelling of Supply Chain Effisiency* (Elsevier) p147
[11] Pujawan I Nyoman 2010 *Supply Chain Management* (Surabaya: Guna Widya)

Risk mapping and mitigation design of small and medium enterprises clothing products using supply chain risk management

Muhammad Nurman Helmi[1] and M N Masri[2]

[1] Department of Industrial Engineering, Universitas Pasundan, Bandung, 40154, West Java, Indonesia, helmi@unpas.ac.id
[2] Department of Industrial Engineering, Universitas Kebangsaan, Bandung, 40261, West Java, Indonesia, masrimn@gmail.com

Abstracts. To ensure the supply of clothing product from SMEs as the object of research in this article, efforts are made to anticipate the failure of supply chain activities so that future losses can be avoided. All causes of risk must be identified and measured so that risk mitigation can be developed. A House Of Risk approach (HOR) is used to identify risks at each stage of the Supply Chain Operation Reference (SCOR). Further identified the possibility of occurrence of risk (occurance) and its impact (severity) with Failure Mode Effect Analysis (FMEA) model. Based on the level of risk can then be mapped the risk that ultimately can be determined risk mitigation. The identification process of HOR phase 1 found 24 risk events and 24 risk agents. The implementation of the second phase of HOR resulted in 14 mitigation actions to anticipate the occurrence of risks that would disrupt the supply chain activities of SMEs clothing products.
Keywords : Supply Chain Risk Management, Risk Event and Risk Agent, House Of Risk, Mitigation

1. Introduction

Micro and Small Medium Enterprises (SMEs) is one solution of economic problems in Indonesia, which until now has not been stable. SMEs are expected to reduce the number of unemployed in Indonesia. In terms of turnover generated contribution of SMEs apparel/fashion sector to the overall turnover of SMEs sales amounted to 30.8%, while the other 5 sectors only contributed an average of 14.0% such as food and beverage sector, services, production, and handicraft/craft [4].

There are some problems that the clothing industry of SMEs must face today. Although the government has tried to do the empowerment programs to the SMEs, especially on the issue of managing the supply chain and distribution network, as well as efforts to reduce the risks that may be faced by many SMEs. The role of SMEs clothing industry is very important in absorbing the weaving and knitting industry output. If the clothing market is disturbed, it will also disrupt the upstream industry of weaving and knitting, even fiber producers [6].

Various challenges must be faced by SMEs clothing industry on the one hand as well as the increasing number of clothing companies on the other hand, resulting in intense competition in the industry [6]. The performance of the clothing industry ultimately determines whether SMEs within the industry are well managed.

Thus, the research is interested to conduct mapping and mitigation the supply chain of SMEs clothing products. All the causes of risks that potentially disrupt the supply chain of SMEs can be mapped and mitigated by proactive action so that the cause of risk can be reduced or eliminated.

2. Literature Review

2.1 Supply Chain Management
Supply Chain Management in [3], [5], [16] states that the term distribution, logistics or supply chain management has the same meaning, that is the process by which companies move materials, components and products to customers.

Supply chain management builds upon this framework and seeks to achieve linkage and co-ordination between the *processes* of other entities in the pipeline, i.e. suppliers and customers, and the organisation itself. Thus, for example, one goal of supply chain management might be to reduce or eliminate the buffers of inventory that exist between organisations in a chain through the sharing of information on demand and current stock levels [16].

2.2 Supply Chain Risk Management
For other matters relating to risk, as states in [3], [8], [10] risk is an event that results in a loss when it occurs during a certain period. While likelihood is a qualitative explanation of probability and frequency [1]. International journal on Physical Distribution and Logistics Management, states that *Supply chain risk management (SCRM)* is the implementation of strategies to manage both everyday and exceptional risk along the supply chain based on continuous risk assessment with the objective of reducing vulnerability and ensuring continuity [16].

Supply chain risk management is collaboration with partners in supply chains to implement risk management processes to address the risks and uncertainties caused by logistic and supply chain activities [15], [17]. Meanwhile according to Hillman M and Keltz H [8] a common risk management processes of a company consists of four main activities : risk identification; risk assessment; decision making and implementation and risk supervision.

2.3 Supply Chain Operation Reference Model
Supply Chain Operation Reference Model (SCOR) is a reference model of supply chain operations. SCOR is basically a process-based model. This model integrates 3 main elements in management ie business process reengineering, benchmarking, and process measurement into the cross-functional framework in the supply chain [12]. The SCOR-model has been developed to describe the business activities associated with all phases of satisfying a customer's demand. The Model itself contains several sections and is organized around the five primary management processes of Plan, Source, Make, Deliver, and Return [12].

2.4 Failure Modes and Effects Analysis
To help in identifying where the priority should be placed in supply chain risk management a useful tool is failure mode and effect analysis (FMEA). According to [12] FMEA is a qualitative analysis of risk identification and can be universally applied to various industries. FMEA is a tool that can be used by management in managing risk, especially for execution of the analysis phase, ie risk identification, risk assessment and measurement, and risk prioritization. Furthermore state by [12] risks in FMEA are evaluated in three components : severity, occurrence, and detection. These three components then are used in calculating the Risk Priority Number (RPN). FMEA uses the RPN to determine the risk priorities of failure modes [14].

3. Methodology

3.1 Desain of Risk Mitigation Strategy
Standard framework of risk management is a modification of existing standards with the main reference standard AS/NZ 3460 [1]. While for the process of strategy design, it is done by developing

the House of Quality matrix (HOQ) to develop mitigation actions in dealing with potential risks arising from the supply chain. Risk Priority Number (RPN) calculation by FMEA method is to conduct risk Assessment in the House of Risk (HOR).

3.2 Risk Identification

The data input stage into the first phase HOR model is as follows :

1. Identification of business process/supply chain activity based on SCOR model.
2. Identify risk events for each business process identified in the previous stage.
3. Identify the level of impact (severity) of an occurrence of risk to the company's business process, the scale used is 1 – 10.
4. Identify the consequences (potential causes) of an event risk to the company's business.
5. Identification of risk-causing agents, ie what factors are causing the occurrence of identified risk events.
6. Identify the correlation between a risk event and the risk-causing agent. This correlation value is denoted by Rij, where Rij ∈ {0,1}, for Rij = 1 then there is a correlation between the occurrence of risk to the agent risk i with risk agent j and Rij = 0 then there is no correlation. This correlation value has a weight (w), where the greater the correlation between a risk agent and the risk event will be marked by a larger value scale. This weight represents how much a risk agent causes the occurrence of risk events. The scale used is 9 (the strong correlation), 3 (the correlation moderate), 1 (the correlation is weak) and 0 (when there is no correlation).
7. Identify opportunities for the occurrence of a risk agent. Opportunities for the occurrence of a risk agent state the magnitude of the probable value of an agent of risk which may cause disruption to a business process with a certain degree of impact. Scale 1 - 10 is used to indicate the probability of occurrence of an agent of risk.
8. Calculation of risk priority index value (Pj). This priority index will be used as a consideration for determining which risk agent priorities need to be designed for mitigation strategies. Determining the value of the risk priority index (Pj) from the risk agent using the following formula [6] :

$$Pj = Oj\sum_{i=1}^{n} Si\ x\ (Rij\ x\ wij)\quad \forall j \tag{1}$$

where :

j = 1, 2, ... m; i = 1, 2, ... n
Rij = correlation between the risk agent j and the risk event i;
Rij = ∈ {0,1}: is a binary function for Rij = 1 if there is a correlation between the risk agent j and the risk event i.
Wij = the correlation weight between the risk agent j and the risk event i.
Oj = Occurrence level of risk

$$Oj = \sqrt[k]{Oj1\ x\ Oj2\ x\ ...Ojk}\quad \forall j \tag{2}$$

j = 1, 2, ...m, k = assessment of person to k.
Si = Level of impact of a risk (severity level of risk).

$$Si = \sqrt[k]{Si1\ x\ Si2\ x\ ...x\ Sik}\quad \forall i \tag{3}$$

where:

i = 1, 2, ... n, k = assessment of person to k.
Ei = Risk Events (Risk Events) where i = 1, 2, ..., n.
Ci = Potential causes of risk;
Aj = Risk agents (Risk agents) where j = 1, 2, ..., m

9. Aggregate calculation of potential risk agent j (ARPj).

Since a risk agent can induce or cause a number of risk events, it is necessary to quantify potential risk agents, or Aggregate Risk Potential (ARP). If Oj is the probability of the occurrence of the risk agent *j*, *Si* is the severity of the impact if the risk event *i* occurs, and *Rij* is the correlation between the risk agent *j* and the risk event *i* (which is defined as how likely the risk agent induces/causes the risk agent *i*, ARP*j* (potential aggregate risk agent *j*) can be calculated as follows:

$$ARPj = Oj\sum_i Si.Rij \quad (4)$$

3.3 Second Phase HOR

Stage of second phase HOR model process (HOR-2) of risk handling is as follows [6] :

1. Select a number of high priority ranking risk agents using Pareto analysis of ARP*j*.
2. Identify actions deemed relevant to prevent risk agents or Preventive Action (PA).
3. Determining the relationship between each precaution and each risk agent, *Ejk*. The values can be {0, 1, 3, 9} representing each, 0 = none, 1 = low, 3 = medium, and 9 = high relation between action *k* and agent *j*. This relationship (*Ejk*) can be considered as the level of effectiveness of action in reducing the likelihood of risk agent *j*.
4. Calculate the total effectiveness (TE*k*) of each action as follows [6] :

$$TEk = \sum_i ARPj.Ejk \qquad \forall k \quad (5)$$

5. Assess the difficulty level (D*k*) in performing each action, and place the values in a row below the total effectiveness. The degree of difficulty, which can be represented by a scale (such as Likert or other scale), must reflect the funds and other resources needed to take action.
6. Calculate the ratio of total effectiveness to difficulties (ETD*k*) [6], ie :

$$ETDk = TEk/Dk \quad (6)$$

7. Establish a priority ranking for each action (R*k*) in which the first rank is awarded for the highest ETD*k* action.

3.4 Risk Mapping and Assessment

A Sample Risk Matrix (Risk Map) introduced by the Airport Handling Manual in its 33rd edition in January 2013 shows an overview of Likelihood or Probability of an Accident or Damage on the impact of Severity or Scope of Damage. Risk Map describe the Likelihood or Probability of an Accident or Damage given levels such as; often, occasionally, possible, unlikely and practically impossible, whereas on Severity or Scope of Damage is given a level like; insignificant, minor, moderate, critical, and catastrophic.

After risk mapping is done, the next step is to analyze or determine the level of risk identified. Efforts taken to define the risk level are as follows:

1. Determine the possibility of risk occurs with the following standards:
   - The likelihood is enormous: it is certain that it will be possible to affect a strategic objective with a potential risk value ranging above 80%.
   - Most likely: most likely to affect a strategic goal with a potential risk value occurring in the range of 60% to 80%.
   - Moderate possibility: the likelihood of risk occurring to affect a strategic goal with a potential risk value occurring in the range of 40% to 60%.
   - Possibilities are small: a small risk may occur to affect a strategic objective with a potential risk value occurring in the range of 20% to 40%.
   - The likelihood is very small: the possibility of very little risk can occur to affect a strategic goal with a probable risk value of less than 20%.
2. Determine the impact of risk with the following benchmarks:
   - The impact is insignificant: the risk has very little effect on a strategic goal, but strategic objectives can still be achieved.
   - Minor impact: risk has little influence on a strategic objective and requires little handling.
   - Medium impact: risk has a moderate effect on a strategic objective and requires considerable effort to deal with it.

- Large impact: risk has a major impact on a strategic goal and requires serious handling
- Disaster impact: risk has the effect of not meeting a strategic goal and requires very serious effort to deal with it.

4. Result And Discusion

In the design of supply chain mitigation of SMEs Clothing, HOR tool model applied for strategic planning which is divided into 2 (two) process : the first phase of HOR-1 is to conduct a risk identification and Assessment process and the second phase HOR (HOR-2) is to mitigate/manage risk.

4.1 Risk Events (risk agent)

The results of interviews and questionnaires to identify risks were obtained 24 (twenty four) risk events consisting of (see Table 1) :

- In the Plan of SCOR business process : 3 risk events
- In the Source of SCOR business process : 5 risk events
- In Make of SCOR business process : 7 risk events
- In the Deliver of SCOR business process : 5 risk events, and
- In business process Return of SCOR : 4 risk events.

4.2 Impact of Risk Events (Severity)

Each risk event can be identified by estimating the magnitude of the disturbance it generates, assessing the impact of a risk event using a Likert scale of 1 to 10 and converted into the unit of costs/losses incurred when the risk event occurs, and the determination the value of the impact of these risk events is done on the entire business process. The result can be seen at Table 1.

4.3 Correlation (Rij) Risk Occurrence (Ei) and risk agents (Aj)

Furthermore, it is still in the process of identifying the HOR-1 model on the relationship/correlation between 24 risk events and 24 causes of risk. To simplify and accelerate the calculation of all correlation values of risk agents and risk events, a matrix was designed as mentioned before. The values listed in the matrix use a scale of 0, 1, 3, and 9 which states that there is no correlation, weak correlation, moderate correlation, and strong correlation. The results are shown in Table 1.

4.4 Opportunity and frequency of Occurrence

The appearance of each risk agency was identified on 20 SMEs that were sampled. The chance of occurrence of a risk agent on a single SME can vary in frequency or how many times to other SMEs. Let's say the risk agent A1 can appear twice on SME1, whereas in SME3 it can appear three times. This is due to many factors, whether caused by internal environmental factors or external environment SMEs Clothing products themselves. An example, A1 risk agent (demand forecast error) may result in a risk of error on E1 (yarn purchase plan, fabric, and accessories) for clothing.

Extrernally, it may be that market research is inaccurate, and internally SMEs make purchasing planning materials excessive or lack of needs. There is no plan of ordering/purchasing the optimal materials. Such risk agents will certainly lead to other risk events such as excessive material inventory, increased cost of ownership and material storage costs that will eventually increase production costs. While the cost of production exceeds the optimum limit, only losses earned by SMEs business of clothing products.

4.5 Priority Index Assessment

This priority index is calculated in aggregate so it is denoted by ARPj (Agregate Risk Potential) performed on each risk agent throughout the SME business process of the clothing product. The correlation of Ei and Aj as well as the frequency of occurrence of a previously identified risk agent will complement the calculation of the ARPj value indicating which risk agency priority index will be mitigated in order to reduce the risk of SME supply chain of clothing products, Table 2.

Table 1. Matrix Model HOR-1 Risk Identification

SCOR CODE		Causes/Risk Agent (A*j*)																								Severity of Risk Event i (S*i*)	Si x Rij
		A1	A2	A3	A4	A5	A6	A7	A8	A9	A10	A11	A12	A13	A14	A15	A16	A17	A18	A19	A20	A21	A22	A23	A24		
Plan	*E1*	9	9	0	1	3	3	3	9	0	0	0	0	1	1	9	1	9	0	0	3	1	0	1	3	3	198
	E2	9	9	0	3	1	3	9	9	0	0	0	0	0	1	9	1	3	0	0	3	3	1	9	3	6	456
	E3	0	0	9	0	1	1	1	3	9	3	9	9	3	0	0	0	1	1	1	3	1	1	1	1	4	232
Source	*E4*	3	9	1	9	9	3	3	9	1	1	0	0	1	9	1	3	0	1	1	3	1	3	3	9	6	498
	E5	1	3	1	3	9	9	9	1	0	0	3	0	0	1	3	3	3	1	9	3	1	3	1	3	6	420
	E6	1	3	1	3	9	9	3	9	3	1	9	1	1	9	1	3	1	3	0	9	3	3	1	9	3	285
	E7	3	3	1	3	9	9	9	9	1	1	1	1	1	1	1	1	9	3	9	3	1	3	9	1	5	460
	E8	3	9	1	9	9	3	9	9	0	0	3	1	0	0	9	1	3	1	1	3	9	1	3	1	6	528
Make	*E9*	1	1	1	1	1	1	1	1	9	1	3	1	0	0	3	1	1	9	0	9	1	3	9	9	3	201
	E10	0	0	9	0	0	1	0	0	3	9	9	3	3	1	3	1	0	3	0	9	3	0	0	0	2	114
	E11	1	1	3	0	0	0	0	0	0	1	9	0	1	9	1	3	0	9	1	3	1	1	0	0	2	88
	E12	0	0	1	0	0	0	0	0	3	1	3	9	9	3	0	0	0	3	0	0	0	0	0	0	3	96
	E13	1	1	3	1	1	0	1	1	0	0	1	3	9	3	0	0	1	3	1	1	1	0	1	3	8	288
	E14	1	1	3	1	1	1	1	1	3	3	9	3	3	9	3	9	1	9	1	3	1	1	3	3	4	296
	E15	3	1	1	3	3	1	3	3	1	1	3	1	1	1	9	1	3	3	1	3	1	1	1	1	3	150
Deliver	*E16*	1	1	3	1	3	3	3	3	3	3	9	3	3	3	3	9	1	3	1	3	1	9	9	9	5	450
	E17	3	3	0	3	9	9	9	9	1	1	3	1	3	1	1	1	9	3	9	3	1	3	0	0	2	170
	E18	9	3	1	3	3	3	9	9	0	1	3	1	3	1	3	3	3	3	3	3	3	9	0	3	6	492
	E19	1	1	3	1	1	1	1	1	3	3	9	1	1	0	0	3	0	3	9	3	1	9	0	0	2	110
	E20	3	3	0	1	3	3	3	3	0	0	3	0	1	0	0	0	3	3	1	9	3	3	1	1	2	94
Return	*E21*	1	3	0	1	1	3	3	3	0	1	3	1	3	1	9	3	3	3	1	3	9	9	3	9	6	456
	E22	1	3	1	3	3	9	3	3	3	3	3	0	1	3	1	3	3	3	1	3	1	9	9	9	3	243
	E23	9	9	1	1	1	1	1	1	0	3	3	3	9	1	3	3	3	3	3	1	3	3	9	9	3	249
	E24	3	3	1	3	1	9	3	1	0	1	1	1	3	3	1	3	1	3	3	3	1	9	9	9	3	225
(Oj)		3	3	2	3	3	4	4	4	3	4	4	4	3	3	3	4	4	4	3	4	3	4	3	3		
(ARPj)		594	1368	464	1494	1260	1140	1840	2112	603	456	352	384	864	888	450	1800	680	1968	330	376	1368	972	747	675		
Rank		17	6	18	5	8	9	3	1	16	10	23	21	12	11	20	4	14	2	24	22	7	10	13	15		

4.6 Risk Mitigation Process

Risks to be mitigated by applying this HOR-2 model are the high-value, pre-calculated risk agents ARP*j* in the HOR-1 process. The results of the calculation of 24 ARP*j* items that have been ranked in rank from 1 to rank 24, starting from the largest ARP*j* value as the 1st rank and the smallest ARP*j* as the 24th rank.

In order to suppress the risk agent that incurs a risk event, it is necessary to design a Preventive Action step for each selected risk agent (14 risk agents) and determine how strong the correlation is with the designed precautions, then using the 0,1, 3, and 9 correlation values are obtained.

A matrix of HOR-2 model is built so as to make it easier to calculate the total effectiveness of each action using the sum total formula of multiplication between the ARP*j* value and the correlation value E*jk*. Similarly, the assessment of difficulty level (D*k*) applying preventive action using Likert scale. From both the total value of effectiveness and the level of witness of the implementation of preventive action can be calculated the ratio of total effectiveness of ETD*k*.

Then rank the priorities of each action (R*k*), the action with the highest ETD is the first rank in this case where the 9th Preventive Action or PA9 (choose qualified suppliers) to the first rank with the value ETD*k* = 20520 and so on for other PA. Furthermore, this risk mapping is intended for the purpose of identifying various risks that could occur and could potentially threaten for the purpose of the realization of SMEs vision and mission of the apparel product itself.

To prepare mitigation actions in handling potential risks in the supply chain, design process is carried out using the second phase of the House Of Risk matrix (HOR-2). Selecting several risk agents from HOR-1 with high values to be handled is using the Pareto Diagram for ARP*j* to be followed up on HOR-2. A*j* risk agents with ARP*j* values are each ranked from the largest value to the smallest value, then calculated the percentage value of each ARP*j* to the total ARP*j* and then the percentage is cumulatively up to 100%. The Pareto Diagram can be constructed as seen in Figure 1 and Matrix Model HOR-2 Risk Mitigation rank, Table 2.

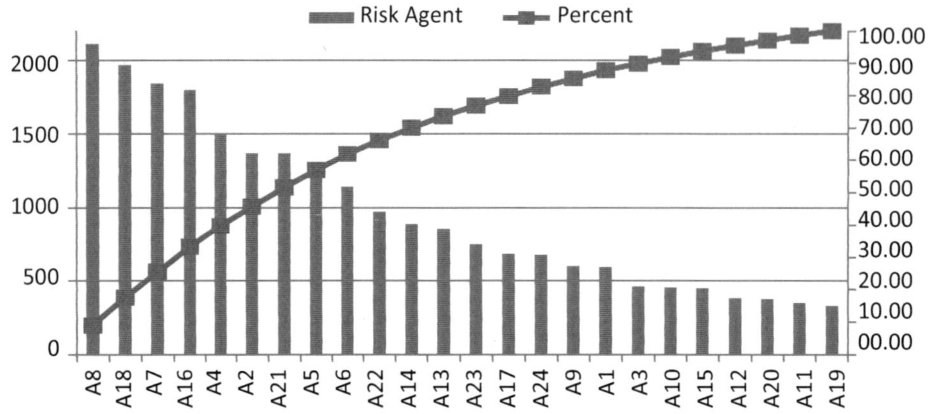

Figure 1. Pareto ARPj Diagram of All Risk Agents

Table 2. Matrix Model HOR-2 Risk Mitigation Rank

Code	Cause of Risk (A *j*) *to be treated*	PA 1	PA 2	PA 3	PA 4	PA 5	PA 6	PA 7	PA 8	PA 9	PA 10	PA 11	PA 12	PA 13	PA 14	ARP*j*
A8	Supplier stock is reduced	9	0	9	0	3	9	3	3	3	1	0	0	1	3	2112
A18	Worker's inaccuracy	1	9	0	3	0	1	0	0	0	3	0	3	1	1	1968
A7	Supplier capacity fluctuates	3	0	9	0	9	9	1	3	3	0	0	0	1	3	1840
A16	Non-standard packaging	0	3	0	9	1	0	1	0	1	3	1	1	0	0	1800
A4	Capacity of domestic suppliers	3	0	3	0	9	3	1	3	3	0	1	0	1	3	1494
A2	Rising material prices	9	0	3	0	1	9	1	1	3	1	0	0	3	3	1368
A21	Information delays	1	0	1	1	1	1	9	0	1	0	1	1	3	3	1368
A5	Dependence on one supplier	9	0	1	0	3	9	1	9	9	3	0	0	1	3	1260
A6	Supplier not qualified	3	0	3	0	3	9	1	3	9	1	0	0	1	3	1140
A22	Delivery process error	1	3	1	1	1	3	0	0	3	9	1	0	0	1	972
A14	Warehouse condition is not feasible	1	0	0	1	1	3	0	0	0	0	9	3	0	1	888
A13	Fire hazard	3	9	1	1	0	3	3	1	1	1	1	9	0	0	854
A23	Behind the fashion	1	1	0	0	0	3	9	0	3	3	0	0	9	1	747
A17	Late ordering of ingredients	9	1	1	0	1	3	1	3	3	3	0	0	0	9	680
	TE*k*	92928	43296	75440	36000	44820	46512	31464	60480	41040	23328	16872	28182	22410	23120	
	D*k*	7	3	8	5	9	7	4	3	2	2	5	4	6	2	
	ETD*k*	13275	14432	9430	7200	4980	6644,6	7866	20160	20520	11664	3374,4	7045,5	3735	11560	
	R*k*	4	3	7	9	12	11	8	2	1	5	14	10	13	6	

In order to map the risk agent to the extent of the level of impact it can cause when the risk agent occurs, it can be seen the result showed at Table 3 as follows :

a. which is included in the extreme impact level are : A8 (supplier stock is reduced) and A16 (non-standard packaging), which is marked in red.
b. which is included in the high impact level are : A22 (delivery process error), A18 (Employee's inaccuracy), A6 (Supplier not qualified), A7 (Fluctuating Supplier Capacity), A4 (Capacity of Domestic Supplier), A21 (Delay Information), and A5 (Dependency on one supplier), marked with yellow.
c. which is included in the small impact level are : A2 (high material prices), A13 (Fire hazard), A17 (Late ordering of ingredient), marked in blue.
d. which is included in the low impact level are : A14 (Warehouse condition is not feasible), A23 (Behind the fashion), marked in blue.

Table 3. Risk Mapping

Likelihood level	Level of impact				
	1 Not Significant	2 Minor	3 Medium	4 Mayor	5 Disaster
(E) Very Large	A22		A8, A16		
(D) Large		A18	A6		
(C) Medium		A2, A13, A17	A7, A4		
(B) Small		A14, A23		A21, A5	
(A) Very Small					

One of the objectives of this research is to prepare risk mitigation at each risk agency. Risk mitigation is prepared especially in the group of risk agents included in the extreme and high impact levels. Both of these impact levels require risk mitigation prepared to avoid risk occurrence by any risk agent. The recommended series of risk mitigation in the red, yellow, moderate and low groups can be seen in Table 4.

Table 4. Risk Mitigation Plan

Risk	Level of Risk	Risk Mitigation Plan and Recommendations
Supplier stock is reduced	Extreme	Provision of sufficient stock
Worker's inaccuracy	High	Train workers regularly
Supplier capacity fluctuates	High	Ensuring supplier capacity
Non-standard packaging	Extreme	Application of packaging standardization
Capacity of domestic suppliers	High	Increase supplier capacity
Rising material prices	Moderate	Prepare optimal inventory
Information delays	High	Improved information system
Dependence on one supplier	High	Increase the number of suppliers
Supplier not qualified	High	Selecting qualified suppliers
Delivery process error	High	Implement SOP submission
Warehouse condition is not feasible	Low	Warehouse layout setup
Fire hazard	Moderate	Safety training
Behind the fashion	Low	Market research periodically
Late ordering of ingredients	Moderate	Calculates ROP materials

5. Conclusion

After assessing and mapping the risks of SMEs clothing products, the risk mitigation plan is prepared by taking into account the risk agents and their risk levels. The process and outcome of risk mitigation is then discussed with the SME players in the apparel product to verify and validate. Verification and validation of the risk / reward mitigation plan is particularly important if it is to be implemented. SME players in clothing products can see whether the recommendations can reduce or even eliminate the causes of risk in the supply chain.

6. References

[1] AS / NZS 2004 *Risk Management Guidelines* Standard Australia, Australia

[2] Ariyanti Fransisca Dini and Aditya Andika 2016 Supply Chain Risk Management in the Indonesian Flavor Industry : Case Study from a Multinational Flavor Company in Indonesia *Proceedings of the International Conference on Industrial Engineering and Operations Management Kuala Lumpur*, Malaysia

[3] Badr Y and Stepen J 2007 Security and Risk Management in Supply Chains *Journal of Information Assurance and Security* Vol 2 p 288-296 National Institute of Applied Sciences Lyon, France

[4] Bappeda 2016 *Statistik Dalam Angka* BPS, Bandung

[5] Cameron I and Raman R 2005 *Process Systems Risk Management* 1st Edition Vol 6 Academia Press, Sydney, Australia

[6] Catharina Tan Lian Soei, Ria Satyarini and Ivan Prasetya 2015 *Identifikasi Key Success Factor Pada Industri Clothing di Kota Bandung* Pusat Penelitian dan Pengabdian pada Masyarakat Universitas Katolik Parahyangan Bandung

[7] Douglas M. Lambert. Martha C. Cooper, and Janus D. Pugh 1998 Supply Chain Management : More than a New Name for Logistics *The International Journal of Logistics Management* No.2 .p.4.

[8] Hillman M and Keltz H 2007 *Managing Risk in The Supply Chain – A Quantitative Studi* (AMR Research , Inc)

[9] Laudine, H. Geraldin, Pujawan, I.N. and Dyah S. D. 2007 Manajemen Risiko dan Aksi Mitigasi Untuk Menciptakan Rantai Pasok Yang Robust *Jurnal Jurusan Teknik Industri ITS* Surabaya. p 53-64

[10] Mc Beath Bill 2013 *Supply Chain Risk Solutions : A Market Overview* A Publication of Chainlink Research.

[11] Nina Bucalo and Borut Jereb 2017 Risk Management in the Pharmaceutical Industry in Slovenian Companies *The Journal of Logistics & Sustainable Transport* Vol. 8, No. 1, May 2017, 42–49

[12] SCOR 2010 *Supply Chain Operations Reference Models Version 10.0* Supply Chain Council, Inc. America

[13] Sharma S. and R. Pratap 2013 A Case Study of Risks Prioritization using FMEA Method *International Journal of Scientific and Research Publications*, Vol. 3(10), pp. 1-4.

[14] Tang Christoper S 2005 *Perspectives in Supply Chain Risk Management : A Review*. UCLA Underson School, 110 Westwood Plaza, UCLA, Los Angles, CA 90095, USA.

[15] Waters, Donald. 2007. *Supply Chain Risk Management – Vulnerability and Resilience in Logistics*, Kogan Page, United Kingdom.

[16] Wieland A and Wallenburg CM 2012 Dealing With Supply Chain Risks Linking Risk Management Practices and Strategies to Performance *International Journal On Physical Distribution And Logistics Management* Vol 42 Issue : 10 p 887-905

[17] William Ho, Tian Zheng, Hakan Yildiz and Srinivas Talluri 2015 Supply Chain Risk Management : A Literature Review *International Journal of Production Research*, 53:16, 5031-5069

Stakeholder engagement in quattro helix model for mobile phone reverse logistics in Indonesia: a conceptual framework

H Maheswari[1,2,3], G Yudoko[1], and A Adhiutama[1]

[1]School of Business and Management Institut Teknologi Bandung, Bandung - Indonesia
[2]Mercu Buana University, Jakarta - Indonesia
[3]hesti.maheswari@sbm-itb.ac.id

Abstract. The number of e-waste from mobile phone industry is still dominating until now. This is happened because there is no mutual commitment from all of parties i.e. businesses, government, and societies to reduce the use of mobile phone that has the shortest product life cycle. There are many researches study about firms' motivation and government's role, other discuss about actions of communities in supporting reverse logistics implementation. Unfortunately, research about engagement mechanism that involving all parties is still rare. Therefore, it is important to find the engagement model through this conceptual paper and it is expected useful to build the novel model. Through literature review, the results of this research are establishing the Quattro helix model as the appropriate structure to build the robust team by exploring stakeholder theories; mapping the engagement model either in form of collaboration or participation that consider stakeholders' role and motivation and finding six types of engagement that consider their interest; and determining the novel model of engagement through Quattro helix model for implementing reverse logistics in handling e-waste by describing the linkage and the gaps among existing model.
Keywords: Engagement, Quattro Helix, Reverse Logistics, Stakeholder,

1. Introduction

The number of e-waste from mobile phone industry is still dominating until 2025 [1]. Besides that, e-waste generation globally in 2015 is 22.5kg/inhabitant and rise to 40kg/inhabitant in 2018 [2]. The main problem of e-waste is they contain hazardous and toxic materials besides precious material and metal. For separating their content, required the high accuracy and sophisticated equipment if not it will pollute the environment.

Therefore, one of key issues in dealing with the global e-waste crises is technology for repair, refurbishment, and remanufacturing as well as recycling [3] through conducting reverse logistics (RL). Actually, some countries have tried to implement sustainable reverse logistics (SRL) for handling this problem but this activity needs specific and integrated treatment from the businesses, academics, and government as well as society involvement. Most researches study about firms' motivation and government's role, while others discuss about actions of communities in supporting SRL implementation. Moreover, they are useless if their research separates the role of all stakeholders because waste reduction initiatives will be feasible if strong political commitment from the governments and funds for conducting public educations are available [4] and other parties such as businesses and community support this program.

The need of societies to mobile phone increase significantly since some factors, such as: 1) economic development that improve the people's purchasing power that exacerbated by consumptive consumers' behaviour and affluent urban population that urge the societies to have mobile phone more than one for communicating with others; 2) the industrial growth in term of information and technology so the firms can find many features to upgrade their electronic product easily for persuading the consumers replace their mobile phone faster; 3) the poor e-waste management and doing nothing of Government as if there is no urgent problem and he is still expecting the tax revenue from mobile phone industries.

Table 1. Communities' behaviour to mobile product and recycling program

Questions/Statement	Answers (%)					
Demografy Factors						
Age (year)	16-20 6.6	21-25 25.8	26-30 27.6	31-40 29.3	41-45 4.5	46-50 6.2
Income/month, Pocket money (million)	1-5 37.3	5-6 23	6-10 28.4	10-15 7.5	15< 3.8	
Education	Senior High School 38.7			Bachelor 47.7		Master 13.6
Communities' behavior						
What are the communication tools are often used?		Mobile 96.9	Tablet 1.1	Laptop/PC 2.3		
How many mobile phones do you have? (unit)		1 52.1	2 41.4	3 4.6	3< 1.9	
How long is the average service life your mobile phone? (year)		1 12.1	2 24.6	3 20.5	4 15.9	End of Life 26.9
What is the reason for replacing your mobile phone?		Outdate style 3.3	Outdate function 35.2	Bored 8.5	Damage 53	
How do you treat your used mobile phone?		Keep at home 38.6	Give to others 27.7	Thrown in a bin 9.7	Sell to the market 25.8	
Why do you not participate in waste mobile phone recycling?		Would rather give the phone to others than recycling it at low price Do not know where to send the mobile phone Mobile phone as data storage equipment Being afraid of disclosure of privacy				32.1 40.9 15.7 11.3
What is the reason for the low recycling rate?		Absence of a sound recycling system Low of environmental awareness Lack of government regulation Absence of laws and regulation				62.2 25.2 10.3 2.3
What is your preference pattern of recycling model?		Door-to-door collecting service Collecting side in community built by government Service centre and maintenance centre The reasonable of reimbursement cost				22.2 26.8 21.5 29.5
Do you know that waste mobile phone contains toxic & hazardous substance?		Know 31.8	Have no idea 35.1	Knowing very little 21.4		Very familiar 11.7
Do you agree that consumers are the ultimate beneficiaries of product and should they afford a part of charge for the recycling of their mobile phone?		Disagree Agree, Pattern: deposit system Agree, Pattern: embedded in product price Agree, Pattern: Paying when the product is replaced It is up to government policy				23.8 26 6.8 24.1 19.3
What percentage of recycling cost can you accept? (%)		0 18.3	1-5 47.3	6-10 34.4	10 < 0	

Source: Preliminary Research in 2016

From my preliminary research found the average mobile phone replacement in Indonesia is less than three years (2.57 years) with the three-most reasons were damage, outdate function, and bored that illustrated in table 1. The short service life would lead to a large number of waste mobile phones, which could be potentially harm to the environment. From 363 respondents, most of them choose mobile phone as communication tool (96.9%) and have only one mobile phone (52.1%) although almost a half of them have two mobile phones (41.4%). From the questionnaire, the main reasons listed for why many waste mobile phones were not recycled were that most consumers did not know where to send the phones for recycling (40.9%) or they would rather give it to others, families or friends than recycle it at low price (32.1%). Moreover, some respondents prefer keep it at home (38.6%) because they are hard to find someone else who really want to use it. The survey revealed that the recycling rate of waste mobile phone is notably low. Asked what cause the low rate, 62.2% of respondents indicated the absence of a sound recycling system besides low of communities' environmental awareness (25.2%), lack of government regulation (10.3%), and absence of laws (2.3%) besides that do not know that waste mobile phone contain toxic substance. The investigation of respondents' pattern of waste mobile phone recycling indicates that 29.5% of them are willing to send their phone if they get the reasonable of reimbursement cost and the second answer is there need collecting side in community built by government so easy for them to participate in the program. Finally, related to charge for the recycling of their mobile phone, 26% agree with deposit system as the payment pattern, 24.1% agree if they pay when they replace it, and 23.8% disagree if they have to bear the cost. Thus, 4,17% of respondents are willing to pay the recycling cost for 1-5% of the product price.

Many researchers recognized the successful of reverse logistics program depend on the number of stakeholders involved and the type of engagement among them either empowerment, collaboration, or involvement. The appropriate collaborations and/or participations form is needed so the SRL program will run well. There is no research that explain the collaboration and participation mechanism inter-stakeholders at once that involving all parties i.e. manufactures, takeback operators (suppliers, distributors, retailers, service centres), consumers, and government clearly. Most of SRL collaboration researches focused on manufacture and takeback operator collaboration [5], [6]. Some papers discussed about the need of collaboration between manufacture and consumer [7]. There are a few researches concerned with collaboration among manufacturer, operator, and consumer [6], [8]. Only one papers discussed four stakeholders' involvement i.e. research of reference [12]. Therefore, by exploring the linkage and the gaps among theories and researches, this research wants to find potential alternatives a model of collaboration and/or participation in conducting RL which describes the further developments of those theories that consider all stakeholders cooperation that very relevant with current situations in Indonesia but has not emerged in the research publication.

2. Research Method

According to reference [10] for conducting systematic literature review, this research should include two steps: 1) Category selection: define and refine a topic, design a search, and locate research reports. There are three main topics of this research namely: stakeholder engagement, Quattro helix, and reverse logistics, therefore the researcher explores many papers refers to these research topics without limiting the year of publication paper. After that, they are reviewed and classified by differing their characteristics and distinguished and recorded in a prepared spreadsheet to be analysed holistically. 2) Material evaluation: validity test, crosschecked to ensure enrichment of the study such as SCOPUS indexed and journal accredited, or research quality such as it answers all research questions.

3. Results and Discussion

Intent of this section is to discuss the Quattro helix model as the appropriate structure to build the robust team by exploring stakeholder theories; the engagement model either in form of collaboration or participation that consider stakeholders' role and motivation and potential six types of engagement that consider their interest; and the novel model of engagement through Quattro helix model for implementing reverse logistics in handling e-waste by describing the linkage and the gaps among existing model.

3.1. Stakeholders Concepts through Engagement in The Quattro Helix Model

For handling waste problems, the firms need at least the three groups of stakeholders, they are government [4], [5], [9], [11], consumer [6], [7], [12], [13], [14], [15], and retailers, distributors, wholesaler, service centre, e-waste collector, transporter, and recycler [8], [12], [13], [14], [16], [17]. The last groups will be united in term of takeback operator in this research and also used in some research [5], [18], [19]. Reference [20] divided the stakeholders into three groups, namely: the first, second, and third batch stakeholders. The first batch stakeholders are *manufacturers* that must responsible of their product once they become obsolete; *assemblers* that provide electrical and electronic product are much cheaper than the branded product; *retailers* that dominated by individual proprietary shop with exchange offers and take-back value which attract the customers to a great extent; and *raw material producers* such as smelters for various metals who buy scrap from trader and recyclers of e-waste. The second batch stakeholder is *consumers* both domestic or individual household and official consumers. The third batch stakeholders are responsible for collection, segregation, dismantling, treatment and disposal e-waste. They comprise *scrap dealers* that collect the used product from consumers; *recyclers* that dominated by informal sector; and *disposers* who responsible to the incineration and landfilling process of e-waste for secure and safe disposal. Actually, the group of stakeholders is not the most important thing, because the engagement of all parties believed will improve the businesses' SRL performance. For getting the high-quality stakeholder engagement, we have to consider nature of relationship, level of engagement, and types of communication.

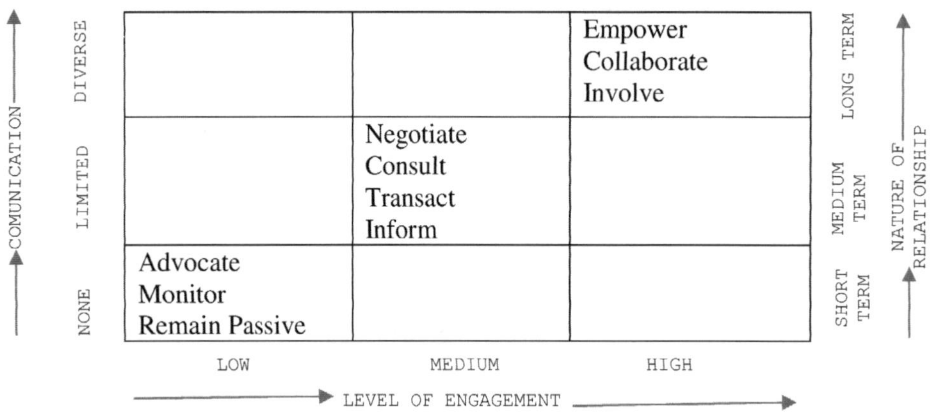

Figure 1. Different Level and Approaches to Engagement [27]

In determining level of engagement, it is needed to define the nature of the relationship. Besides that, the method of engagement should be selected that appropriate with the needs, capacity and expectation of the relevant stakeholders. Figure 1 describe the different level and approaches to engagement based on the types of communication and the nature of relationship. The owners of engagement should involve stakeholders in the design and implementation of the engagement purpose and scope (Table 2). These issues will be a guidance in determining how they are engaged. The engagement will be made as an alternative conceptualization of the Quattro helix model. The four networks might thus be described as organizing a Quattro Helix instead of a Triple Helix. The four helices in this research are occupied by private companies (manufacturers), public authorities (government), intermediaries (takeback operators), and civil societies (consumers) that established as the robust team to implement SRL.

3.2. Stakeholder Engagement Model

For implementing SRL, this study has established the four parties' engagement, namely: manufacturers, government, takeback operators, and consumers. Because the nature of relationship based on network theory is long term trust, level of engagement is high and it will use diverse communication, so the appropriate types of engagement i.e.: involvement, collaboration and empowerment.

Table 2. Level and Methods of Engagement [27]

No.	Level of Engagement	Methods of Engagement
1	Remain Passive (No active communication)	• Stakeholders concern expressed through protest • Letters • Media • Website
2	Monitor (one-way communication: stakeholders to organization)	• Media and internal tracking • Second-hand report from other stakeholders possibly via targeted interviews
3	Advocate (one-way communication organization to stakeholders)	• Pressure and regulatory bodies • Other advocacy efforts through social media • Lobbying efforts
4	Inform (one-way communication organization to stakeholders, there is no invitation to reply)	• Bulletins and letters • Brochures • Reports and Websites • Speech, conference, and public presentation
5	Transact (limited two-way engagement: setting and monitoring performance according to terms of contract)	• Public private partnership • Private finance initiative • Grant-making • Cause-related marketing
6	Consult (limited two-way engagement: organization ask question, stakeholders answer)	• Survey • Focus group • Meeting with selected stakeholders • Public meeting • Workshop
7	Negotiate (limited two-way engagement: discuss a specific issue or range of issues with the objective of reaching consensus)	Collective bargaining with workers through their trade unions
8	Involve (two-way or multi-way engagement: learning on all sides but stakeholders and organization act independently)	• Multi-stakeholder forum • Advisory panels • Consensus building processes • Participatory decision-making processes • Focus groups • Online engagement tool
9	Collaborate (two-way or multi-way engagement: joint learning, decision making and actions)	• Joint projects • Joint ventures • Partnership • Multi-stakeholder initiative • Online collaborative platform
10	Empower (new forms of accountability: decision delegated to stakeholders; stakeholders play a role in shaping organizational agenda)	Integration of stakeholders into governance, strategy and operation of the organization

3.2.1. *Involvement.* "Participation is defined as mental and emotional involvement of a person in group situation in lich encourages him to contribute to group" [27]. Social (civic or citizen) participation has been defined as ''a process in which individuals take part in decision making in the institutions, programs, and environments that affect them" [25]. The type of participation will be adjusted with involvement, empowerment, and partnership characteristics. Reference [26] identified five levels of participation with relates to the stance an organization promoting participation, namely: *information*, merely telling people what is planed; *consultation*, offering some options, listening to feedback, but now allowing new ideas; *deciding together*, encouraging additional options and ideas, and providing opportunities for joint decision making; *acting together*, not only do different interest by deciding together, they form a partnership to carry it out; and supporting independent community interest, local groups or organizations are offered funds, advice or other support to develop their own agendas.

3.2.2. Collaboration. Collaboration model is used to simplify the complexity of relation and interorganizational interaction [21]. Collaboration between organizations is supported by many factors, among others; confidence (trust) between the parties, an effective method of communication, interpersonal relations, and leadership. There are at least six issues in processing collaboration between organizations [22], namely: managing aims; compromise; communication; language; democracy and equality; power and trust; and determination, commitment, and stamina. In collaboration, all parties work together and build a consensus to reach a decision that results the benefits to all of parties. The relationship between the parties is ongoing and therefore collaboration is dynamic and interdependence. As a dynamic concept, collaboration is an incremental process through several stages [23], namely (1) The floating vision of collaboration that explains the common interest, (2) Approaches to visioning in the form of equalization understanding and collective experience that documented in the form of operating principles as a reference of how stakeholders work, (3) Appreciative inquiry, which is a tool to find a better way, effective and constructive covering (4): discovery (finding the best), dream (vision wish generated), design (visible proposal), deliver (application design).

Reference [21] explained the characteristic of jurisdiction-based model is active and opportunistic collaborative. All actors assume that the activities as part of their own job but the activities follow pattern from outside. Therefore, bargaining and negotiating is the important instrument in this model. Bargaining generate unilateral concession and mutually beneficial solution. The horizontal dimension explains the process creating policy and governance. Nobody has a power to determine strategy other actor because each actor has strategy and operationalization severally. While, The Top-down model emphasizes the central government control vertically to regional and local government. From this model appear a dilemma how to realize the national government program through local government that legally independent. Therefore, the most important aspect in this model is local government voluntary in conducting central government program. Reference [21] stated that the successful collaboration if it impresses cross function integration [24]. It comprises of bureaucracy function, across of policy arena, and government level that member (societies and NGO) participate in problem solving and implementation process. The three models are: vertical collaboration reflects on relations among inter-government institutional and societies, non-governmental organizations in top down relations; horizontal collaboration shows relations among interorganizational in the same level; while, partnership linkage is depicted as blending vertical-horizontal relation for generating effective arrangement.

Reference [17] developed a series of procedures for implementing contracts between manufacturers and retailers to maximize individual and total profits through the reverse supply chain i.e. revenue sharing, collect payment support, and transportation cost support. The contract is analyzed according to: 1) the manufacturer proposes a set of contracts that can be considered by the retailer; 2) the retailer evaluates the proposed contracts and chooses the one that is expected to maximize profit; 3) they adjust the parameters of the best contract to achieve the mutual goal of the supply chain.

3.2.3. Empowerment. The key issues of empowerment are new form of accountability, decisions delegated to stakeholders, and stakeholders play a role in shaping organizational agenda [27]. The methods of engagement are integration of stakeholders into governance, strategy and operations of the organization. Reference [28] stated that Chinese government has made effort to tackle problems associated with the recycling and disposal waste mobile phone. By engaging the communities through empowerment program such as green box program or recycled old-for-new activity as well as the willingness to pay, will regulate consumers' economic responsibility and awareness. Not only treat communities for engaged in handling waste mobile phone but also businesses must have an active role to seek a breakthrough for reducing e-waste by providing storage point, disposal center, environmental friendly design [12], [14], [18].

3.3. The Proposed Model of Stakeholder Engagement in Sustainable Reverse Logistics
The collaboration and participation model from previous study can inspire to develop a new stakeholder engagement model for implementing reverse logistics. Figure 2 endeavor to create the comprehensive model of stakeholder engagement in SRL that involve the all of main parties i.e. manufacturers, government, takeback operator, and consumers for filling the research gap or potential future research in SRL.As mention above, the proposed model involves only four parties as the main actors in conducting RL activities. Each party must play its role for reducing the number of e-waste. Society must reduce the desire to replace their mobile phone. The businesses must have a method to maximize the use of electronic product by recycling, refurbishing, reusing, and repairing. Government must control the businesses' behavior in gaining the benefit and provide the societies to handle their own e-waste. However, government needs support from the two parties, societies and businesses to jointly overcome the e-waste problem because government is powerless if must handle it alone.

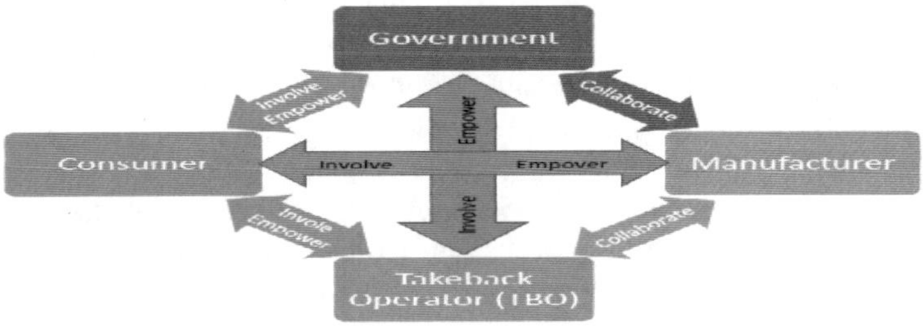

Figure 2 Stakeholder Engagement Model in Quattro Helix Model for RL Implementation

4. Conclusions and Future Work
The research gap shows there is a big opportunity to study about stakeholder engagement model in RL for handling e-waste problem. The Quattro helix model (manufactures, government, takeback operators, and societies) is the appropriate structure to build the robust team to implement the SRL program and keep all parties reach their goals. The new model, stakeholder engagement in SRL must be made clearly so this research must find six types of collaborations and/or participation i.e. government to manufacturer, takeback operator, and consumer; Manufacturer to takeback operator and consumer; and takeback operator to consumer. For future research, the importer, NGO, and e-waste agency will be considered as a part of takeback operator because they often influence either positive or negative to the successful the SRL implementation. The researcher also should aim to test the new model, to ensure whether the new model appropriate to be implemented in developing countries.

5. Acknowledgements
This study would not possible without support material and non-material from LPDP - The Ministry of Finance, The Ministry of Research Technology and Higher Education, and Mercu Buana University. The author also would like to thank to my research partner for helping me to explore the papers.

6. References
[1] Andarani, P and Goto, N., 2014, Potential E-Waste Generated from Household in Indonesia Using Material Flow Analysis. *J Mater Cycles Waste Management*, **Vol. 16**:306-320

[2] Balde, C.P., Kuehr, R., Blumentha, I.K., Gill, S.F., Kern, M., Micheli, P., Magpantay, E., and Huisman, J. 2015, J. 2015. E-Waste Statistics: Guidelines on Classification, Reporting, and Indicators. UNU-IAS

[3] Nnorom. I.C. and Osibanjo, O., 2010. Overview of Prospects in Adopting Remanufacturing of End-of-Life Electronic Products in the Developing Countries. *International Journal of Innovation, Management and Technology*, **Vol. 1**, No. 3,

[4] Yudoko G, 2007, Integrated Municipal Solid Waste Planning and Management (IMSWPM) in Developing Countries: A Conceptual Framework, *Jurnal Manajemen Teknologi*

[5] Bernon, M. and Cullen, J., 2007, An Integrated Approach to Managing Reverse Logistics, *International Journal of Logistics*. Research and Application **Vol 10** p. 41-56

[6] Bernon M and Rossi S., 2011, Retail Reverse Logistics: A Call and Grounding Framework for Research, *International Journal of Physical Distribution and Logistics Management* **Vol 41**, No 5, pp. 484-510
[7] Sharma S and Singh G., 2013, Reverse Logistics: Design Implication on The Basis of Product Types Sharing Identical Supply Chain Member Motivation, Uncertain Supply Chain Management I, pp 33-44
[8] Agrawal, S., Singh, R.K., and Murtaza, 2016. Prioritizing Critical Success Factors for Reverse Logistics Implementation Using Fuzzy-TOPSIS methodology. *J Ind Eng Int* **Vol. 12**: 15-27
[9] Sheuh, Jiuh-Biing, 2014. Green Supply Chain Collaboraion for Fashionable Consumer Electronic Products under Third-Party Power Intervention – A Resource Dependence Perspective. *Sustainability*, **Vol. 6**, 28832-2875
[10] Neuman, W.L., 2006. *Social Research Methods: Qualitative and Quantitative Approaches.* Pearson International Edition, Sixth Edition, United State
[11] Santos R.F. and Marins F.A.S., 2015, Integrated Model for Reverse Logistics Management of Electronic Products and Components. *Information Technology and Quantitative Management, Procedia Computer Science* pp. 575 – 585.
[12] Vereecke A and Muylle S, 2006. Performance Improvement Through Supply Chain Collaboration in Europe. *International Journal and Production Management* **Vol. 26** No. 11 pp 1176-1198
[13] Gloria M and Talavera V, 2008, Supply Chain Collaboration in The Phillipines. *Journal of International Business Research*, **Volume 7**, Special Isuue 2
[14] Budijati, S.M 2016. Model of Reverse Logistics Management Closed Loop and Opened Loop as well as Consumer Behavior Accomodation, UGM
[15] Nnorom I.C., Ohkwe J., and Osibanjo O, 2013, Survey of Willingness of Residents to participate in Electronic Waste Recycling in Neigeria – A Case Study of Mobile Phone Recycling, *Journal of Cleaner Production* 17, pp 1629 – 1637
[16] Migliano J.E.B., Demajorovic J., and Xavier L.H., 2014, Shared Responsibility and Reverse Logistics Systems for E-Waste in Brazil, *Journal of Operation and Supply Chain Management* **Vol 7**., No. 2., p91-109
[17] Yoon S.W and Jeong S.K, 2016, Implementing Coordinative Contracts Between Manufacturer and Retailer in a Reverse Supply Chain, *Sustainability Journal*, **Vol. 8**, 913
[18] Aitken J and Harrison A, 2012 Supply Governance Structures for Reverse Logistics Systems, International *Journal of Operations and Production Management*, **Vol 33**., No. 6, pp 745-754
[19] Sauza R.G., Climaco, J.C.N., and Sant'Anna, A.P., (2016), Sustainability Assesment nd Prioritisation of E-Waste Management Option in Brazil, *Waste Management* 57, pp. 46-56
[20] Borthakur, A and Sinha, K 2013, Electronic Waste Management in India: A Stakeholders Perspective. *Electronic Green Journal*, 1 (36)
[21] Agranoff and Mc Guire, 2003. Collaborative Public Management: New Strategy for Local Government. P 35
[22] Huxham, Chris and Vangen Siv, 1996, Key Themes in The Management of Relationship Between Public and Non-Profit Organization, *The International Journal of Public Sector Management* **Vol 9**, No 7
[23] Munt, Richard, 2003. Building Collaboration, Stronger Families Learning Exchange, Bulletin No. 3 p6-8
[24] Cullen J., 2010, Tools to Manage Reverse Logistics, Cranefield University
[25] Heller, K., Price, R. H., Reinhartz, S., Riger, S., Wandersman, A., and D'Aunno, T. A., 1984. Psychology and community change: Challenges of the future. Monterey: Brooks/Cole
[26] Wilcox D, 1994, Community Participation and Empowerment: putting theory into practice, *RRA Notes Issue* 21, pp 78-82, IIED London
[27] AA1000 SES, 2015. Accountability 1000 Stakeholder Engagement Standard
[28] Yin, J., Gao, Y, and Xu, H., 2013. Survey and Analysis of Consumers' Behaviour of Waste Mobile Phone Recycling in China.

Supplier selection in supply chain management using analytical network process for Indonesian cement industry

A H Ismail[1,2] and R Z Z Mahardika[1]

[1] Department of Industrial Engineering, Pancasila University, Jagakarsa, Jakarta 12640, Indonesia
[2] asrul@univpancasila.ac.id

Abstract. Supply chain management has increased more significance with the impact of globalization. In the present worldwide market, well-managed supply chain is a standout amongst the most vital requirement to be more competitive in the market. For any organization incorporate cement industry, the most critical decision in initial process of supply chain management is to buy products, materials or services from suppliers. So the role of suppliers is irrefutable important in the global aggressive markets. Appropriate decision of supplier selection can lead to reducing cost in supply chain management. However, it is becoming more complex because of existing various criteria and involving the suitable experts in the company to make valid decision in accordance with its criteria. In this study, the supplier selection of an Indonesia's leading cement company is analyzed by using one of the popular multi-criteria decision making method, Saaty's analytical network process (ANP). It is employed for the selection of the best alternative among three suppliers of pasted bag. Supplier with the highest rank comes from several major steps from building the relationship between various criteria to rating the alternatives with the help of experts from the company. The results show that, Communication capability, Flexible payment terms, Ability to meet delivery quantities are the most important criteria in the pasted bag supplier selection in Indonesian cement industry with 0.155, 0.110 and 0.1 ANP coefficient respectively. And based on the ANP coefficient values in limit supermatrix, the A2 or supplier 2 had the highest score with 64.7% or 0.13 ANP coefficient.
Keywords: ANP, supply chain management, supplier selection, multi criteria decision making

1. Introduction

The implementation of the concept of the Supply Chain Management (SCM) is very important in affecting the performance of the company. Supply Chain Management to become a competitive strategies to bridge the company with suppliers and distributors in interorganizational system [1]. There are three decision levels in supply chain management, which are strategic, tactical and operational. Strategic level decisions involve long-term issues such as demand planning, strategic alliances, outsourcing, supplier selection and pricing [8]. Supplier selection is a critical strategic decision process in the supply chain, which affects the quality of product directly. Supplier selection problem aims to determine the right suppliers for certain product or service. Since some qualitative and quantitative factors are decisive in the selection process, the decision maker has to determine these suitable factors at the first step [9].

Different decision-making methods are designed to solve supplier selection problem, such as the analytical hierarchy process (AHP), data envelopment analysis (DEA), analytic network process (ANP), case-based reasoning (CBR), mathematical programming, and artificial intelligence (AI) techniques. The traditional cost-based solution methodologies are not as efficient and effective as multi-criteria decision-making approaches since they are not capable of considering qualitative supplier selection criteria such as flexibility, reliability, culture, and service capability [4]. ANP is one of the suitable methods to cope supplier selection problems because it considers the interdependencies between criteria and sub-criteria.

The main goal of the procurement department is to select the proper source to minimize cost and maximize quality, customer satisfaction and market share. However, supplier selection is a multidimensional problem, which is complex to solve because of its intangible and tangible factors [2]. This problem has been considered in many studies for years, and a variety of criteria and solution methods are generated to solve this problem efficiently. In the literature, some of the researchers define some important criteria in the selection process. Dickson (1969) addresses 23 selection criteria as critical factors in the supplier selection problem. So he identifies quality, price and delivery as the most critical factors among these factors. Following to Dickson's work, Weber, Current, & Benton, (1991) review 76 articles published between 1966 and 1991 and they deduce that the net price of the product, quality and delivery time of the product are the most used criteria in the literature shows that a significant change could be observed in the relative importance of various critical success factors with the increasing competition and globalization after 1990's [3]. Furthermore, they state that reliability, flexibility, consistency, and long-term relationship are four new critical criteria considered in the supplier selection process.

The priority is ranked by [4] using supplier selection criteria as follows: quality, delivery, price/cost, manufacturing capability, service, management, technology, research and development, finance, flexibility, reputation, relationship, risk and safety and environment. They point out that quality and delivery time have become more important than price and the traditional evaluation approach, which is a selection of a supplier based on only the lowest price, is not effective in today's competitive supply chain management. Beside the classical supplier evaluation criteria (price, quality and delivery time), there are other critical criteria such as operational performance, service quality, educational status of the personnel, technology, financial capacity, process control capability, after-sales service, and sustainability that may help to select competent suppliers.

Some of the latest studies about supplier selection in cement industry are as follows: Sharma [7] gave an examination of an AHP model for supplier selection in cement industry using addition green activities criteria. Mukherjee [5] developed an integrated model of fuzzy-AHP-VIKOR-MOGA for supplier selection in cement industry in India. Although still rare the study about ANP in this field but Mukherjee in 2016 have a comprehensive study about supplier selection method using MCDM or the integrated model.

In this study, criteria is determined directly by four experts in company with their expertise. Although in the prior interview we put all criteria based on literature review but only 10 criteria we had selected since it is the most significant to the problem based on our experts in company. The criteria seem too small in number if compared with other literatures, but we convinced that it will increasing during the increment of expertise in company.

2. Analytical Network Process (ANP)

ANP is a Multi Criteria Decision Making (MCDM) method which was introduced by Saaty. It is the extension of Analytical Hierarchy Process (AHP) as a MCDM method that describes the interrelationships and feedback between elements and alternatives. The ANP method is able to improve the weakness of the AHP in the form of the ability to accommodate the relation between the criteria or alternative. The ANP is comprise of three major steps such as determined the criteria and the relations between them, made pairwise comparisons and rated alternatives. These steps need the involvement of experts from the company that appropriate to answer the question related to the criteria. The relevance of the ANP method there are 2 types of namely relevance in a set of elements (inner dependence) and the interrelationship between different elements (outer dependence). The relevance is causing the ANP method is more complex than AHP method [6].

The ANP is an approach in the decision making process that provides the general framework in treating decisions without making assumptions about the independency of the elements on a higher level of elements on a lower level and about the independency of the elements in a levels. In fact the ANP network using without having to specify the level as on the hierarchy used in the Analytic Hierarchy Process (AHP), which is the starting point of ANP. The concept of home in the ANP is the influence of 'influence', while the concept of home in AHP is preference 'preferences'. AHP with assumptions dependencies on cluster and the element is a special case of ANP (see Fig.1 below).

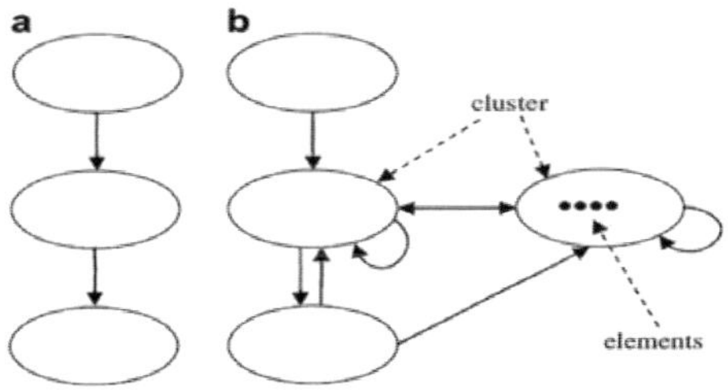

Figure 1. Difference of the structure of the hierarchy and Network [6]

Similar with AHP, every relevance in the network will generate pairwise comparisons and the results of these comparisons form a super matrix. And then the super matrix transforms into a weighted super matrix which sum of the weights in each column is equal to one. The general form of the supermatrix can be described as follows:

$$
W = \begin{array}{cc} & \begin{array}{ccc} C_1 & C_2 & C_m \\ e_{11}\ e_{12} \cdots e_{1n_1} & e_{21}\ e_{22} \cdots e_{2n_2} \cdots & e_{m1}\ e_{m2} \cdots e_{mn_m} \end{array} \\ \begin{array}{c} C_1 \begin{array}{c} e_{11} \\ e_{12} \\ \vdots \\ e_{1n_1} \end{array} \\ C_2 \begin{array}{c} e_{21} \\ e_{22} \\ \vdots \\ e_{2n_2} \end{array} \\ \vdots \quad \vdots \\ C_m \begin{array}{c} e_{m1} \\ e_{m2} \\ \vdots \\ e_{mn_m} \end{array} \end{array} & \begin{bmatrix} W_{11} & W_{12} & \cdots & W_{1m} \\ W_{21} & W_{22} & \cdots & W_{2m} \\ \vdots & \vdots & \ddots & \vdots \\ W_{m1} & W_{m2} & \cdots & W_{mm} \end{bmatrix} \end{array} \quad (1)
$$

Where C_m denotes the m cluster, e_{mn} denotes the nth element in the mth cluster and W_{ij} is the principal eigenvector of the influence of the elements compared in jth cluster to the ith cluster. If the jth cluster has no influence on the ith cluster, then $W_{ij} = 0$ [8]. Finally, the weighted super matrix raise to a larger power until generating the limit matrix, the matrix in which the values in each column are equal. The priorities of alternatives can be seen in that limit matrix [9].

In the methodology of the ANP, data that is used is the primary data obtained from the results of the interview (in-depth interview) with experts in company, who have understanding about the issues discussed. Continue with charging the questionnaire on the second meeting with respondents. The data is ready to prepare in the ANP is the variables assessment of the respondents against the problem of the research object in the numerical scale.

3. Result and Discussion

In this study, the supplier selection problem is analyzed in a manufacturing supply chain concept and the proposed model of pasted bag supplier selection is applied in Indonesian leading cement manufacturing (PT ITP). PT ITP is one of the largest cement companies in Indonesia. In 2016 PT ITP using paper bag of pasted bag majority votes in the process of the distribution of cement 5,034,290 units.

PT ITP is supported by some divisions SCM that support the cement production activities to meet the needs of the customer daily. One of the divisions that its existence has an important role in the activities of the distribution of the results of the production of cement, namely is Paper Bag Division that deal with directly creating paper bag of cement. The main function of Paper Bag Division that is concentrated on the production of cement paper bag. The selection pasted bag supplier problem will be analysed in this study. This model were worked with experts professional in the company and defined four criteria and ten sub criteria with 3 suppliers (see Table 1 below).

Table 1. Clusters and Criteria of Supplier Selection Problem in Cement Industry.

Clusters	Nodes
A: Alternatives	A1: Supplier 1 A2: Supplier 2 A3: Supplier 3
B: Technical	B1: ISO certification (9001, etc.) B2: Communication capability (responsiveness) B3: Ability to meet delivery quantities
C: Experience	C1: Working years in this sector >= 10 years C2: Working years in this sector < 10 years
D: Price-Cost	D1: Net price of the final product (including Transportation) D2: Flexible payment terms D3: Flexible contract conditions
E: Quality	E1: Process control capability E2: Ability to meet delivery due dates

Interviews were performed with the procurement department managers in order to identify weights of the criteria and comparisons. Past experience and the background of the experts are utilized in the determination of the criteria and 4 important criteria (clusters) to be used for supplier selection are established. These 4 clusters are as follows: "Technical" (B), "Experience" (C), "Price-Cost" (D), and "Quality" (E). The list of clusters and nodes of the problem is shown in Table 1 above. These specialists likewise examine the connection between criteria appeared in Figure 2 beneath.

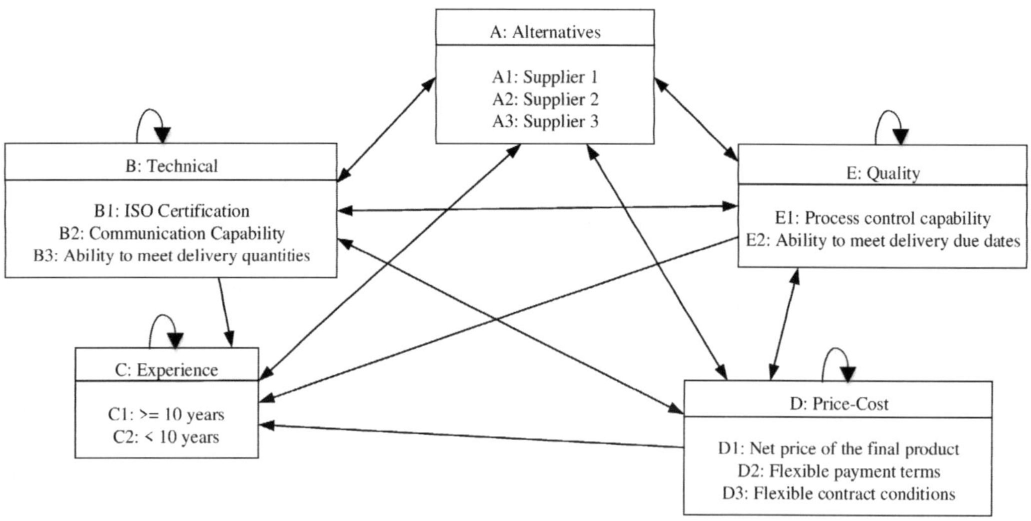

Figure 2. Conceptual ANP Model of Supplier Selection Problem in Cement Industry

This interdependency will generate comparisons between element and those comparisons form the unweighted supermatrix. However the comparisons have to assess the consistency ratio (CR < 0.1) before generating the unweighted supermatrix. Unweighted Supermatriks was obtained from eigenvector derived from every paired comparison matrix and then reform as single integrated matrix or supermatrix by inserting the comparisons as sub column of the prior matrix. Some entries can be worth 0 in accordance with the elements/clusters that do not have the influence or impact.

Weighted Supermatrix can be generated by multiplying cells of cluster matrix and each corresponding cells of Unweighted Supermatrix. Weighted Supermatrix is raised to powers for obtaining the Limit Supermatrix. Priorities for all the criteria and alternatives can be read from the Limit Supermatrix. All columns of the Limit Supermatrix have the same value (constant). Therefore the first column is considered as solution with more decimal numbers. Unweighted Supermatrix, Weighted Supermatrix and Limit Supermatrix are all shown in the appendix part of this paper in Appendix A, Appendix B and Appendix C.

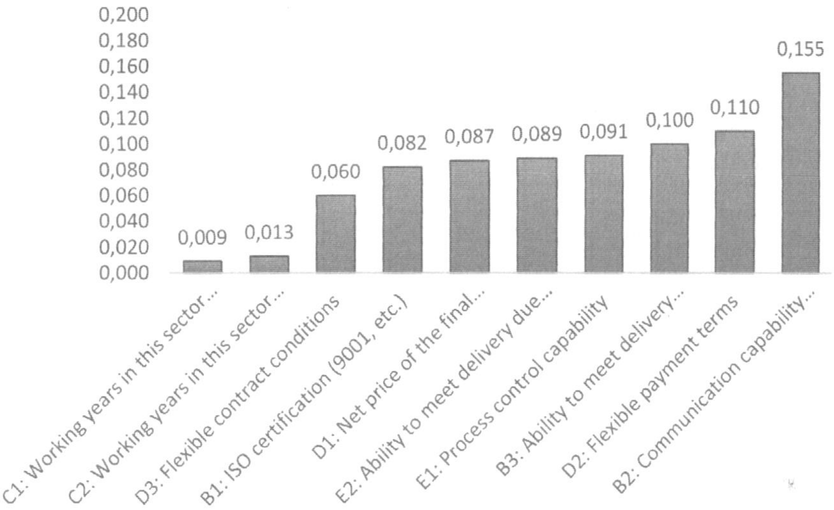

Figure 3. The rank of sub-criteria of supplier selection problem in Cement Industry.

In this study "B2: Communication capability" (0.155), "D2: Flexible payment terms" (0.110), and "B3: Ability to meet delivery quantities" (0.100) are determined as the three most important sub-criteria in the supplier selection problem using ANP. And the 4th to 7th position are so close (the range about 0.8-0.9). Interestingly, experiences criteria (C1 and C2) are the criteria that less important compare to others. It means as long as the suppliers could fulfil the non-experience criteria in a good manner, the chance of being selected is big although the newbie supplier. Consistency ratios of the all pair-wise comparison matrix are calculated less than 0.1. So the weights are shown to be consistent and they are used in the supplier selection process.

Table 2. The Ranking of Suppliers

Rank	Supplier	ANP Coefficient	Score
1	A2: Supplier 2	0.13	64.7%
2	A3: Supplier 3	0.05	26.0%
3	A1: Supplier 1	0.02	9.3%

Relies upon the ANP coefficient values in limit supermatrix (Appendix C), the ranking of the three alternatives from top to bottom order are A2, A3 and A1 as shown in Table 2. Proposed model results show that A2 is the best supplier alternative with 64.7%. Depends on the analysis the least suitable supplier is A1.

4. Conclusion

A successful supply chain management is a basic factor to gain a competitive advantage in the present worldwide condition. The principle idea of supply chain management is generation of good relationships between individual chain members to serve customers precisely. For sure, definitive operational levels between supply chain members determine the quality of final product or service in a supply chain. In this way, supplier selection process in the supply chain management has a significant impact on competitive advantage. Choosing the appropriate supplier is more than finding the supplier who offers the best price or cost, especially in a leading cement company. Experiences, on-time delivery, the quality of the product, and order accuracy are usually as important as the price.

In the chain relationship, basically long-term relation can be preferable to the short-term savings. In order to gain long-term savings, companies establish a satisfactory relationship with their suppliers. Hence, the supplier selection criteria list may include other factors such as communication and relations, reliability, service quality, financial capability, and sustainability that ensure to develop a long-term supplier relationship. And to manage the risks in this fierce competition multi-suppliers could be the best solution using MCDM like ANP rather single-supplier.

In this study, supplier selection problem is considered as a multi criteria decision problem and a model is produced by utilizing ANP. The assessment criteria are identified according to the company's objectives and the proposed model is applied to a real life case study. Three main supplier evaluation criteria (quality, cost, and delivery) are determined as the main criteria. Additionally to the classical supplier selection criteria, relatively new criteria such as communication and relations and experience are also been considered in this supplier selection problem. The results show that, Communication capability, Flexible payment terms, Ability to meet delivery quantities are the most important criteria with 0.155, 0.110 and 0.1 ANP coefficient respectively, in the pasted bag supplier selection in Indonesian cement industry. (See Fig. 3 above)

Then again, the experience criteria are considered as the minimum essential supplier selection criteria by the procurement division of PT ITP. It exhibits that experience could be the less need satisfied by suppliers. Supplier's consciousness of QCD, quality cost delivery, in supplier selection process should be enhanced later on to pick up a competitive advantage. Also to this, the proposed technique is practicable for assessing potential supplier as far as their exactness as for multiple interdependence criteria.

From three suppliers that ranked based on the ANP coefficient values in limit supermatrix (see Appendix C), the A2 or supplier 2 had the highest score with 64.7% or 0.13 ANP coefficient. Followed by A3 and A1 with 26% or 0.05 ANP coefficient and 9.3% or 0.02 ANP coefficient respectively.

Appendix A. Unweighted Supermatrix for Supplier Selection in Cement Industry.

		B			**C**		**D**			**E**		**A**		
		B1	B2	B3	C1	C2	D1	D2	D3	E1	E2	A1	A2	A3
B	B1	1	0	0	0	0	0.161	0.162	0.171	0.349	0.455	0.251	0.334	0.226
	B2	0	1	0	0	0	0.38	0.419	0.462	0.438	0.316	0.375	1,043	0.294
	B3	0	0	1	0	0	0.456	0.419	0.354	0.212	0.219	0.375	0.316	0.39
C	C1	0	0	0	1	0	0	0	0	0	0	0.833	0.227	0.599
	C2	0	0	0	0	1	0	0	0	0	0	0.167	0.773	0.401
D	D1	0.476	0.441	0.512	0.469	0.543	1	0	0	0.349	0.455	0.376	0.286	0.354
	D2	0.362	0.384	0.295	0.341	0.276	0	1	0	0.438	0.316	0.448	0.39	0.463
	D3	0.14	0.169	0.171	0.189	0.159	0	0	1	0.212	0.219	0.16	0.313	0.178
E	E1	0.5	0.5	0.5	0.5	0.568	0.5	0.5	0.5	1	0	0.5	0.568	0.432
	E2	0.5	0.5	0.5	0.5	0.432	0.5	0.5	0.5	0	1	0.5	0.432	0.568
A	A1	0.088	0.089	0.104	0.208	0.094	0.087	0.087	0.084	0.092	0.093	1	0	0
	A2	0.671	0.64	0.584	0.472	0.634	0.639	0.667	0.657	0.647	0.615	0	1	0
	A3	0.214	0.238	0.305	0.314	0.261	0.252	0.241	0.252	0.245	0.272	0	0	1

Appendix B. Weighted Supermatrix for Supplier Selection in Cement Industry.

		B			C		D			E		A		
		B1	B2	B3	C1	C2	D1	D2	D3	E1	E2	A1	A2	A3
B	B1	0.342	0	0	0	0	0.038	0.038	0.04	0.099	0.129	0.093	0.124	0.084
	B2	0	0.342	0	0	0	0.089	0.098	0.108	0.124	0.089	0.139	0.388	0.109
	B3	0	0	0.342	0	0	0.107	0.098	0.083	0.06	0.062	0.139	0.118	0.145
C	C1	0	0	0	0.058	0	0	0	0	0	0	0.095	0.026	0.068
	C2	0	0	0	0	0.058	0	0	0	0	0	0.019	0.088	0.046
D	D1	0.108	0.1	0.116	0.125	0.145	0.058	0	0	0.099	0.129	0.11	0.084	0.104
	D2	0.082	0.087	0.067	0.091	0.074	0	0.33	0	0.124	0.089	0.132	0.114	0.136
	D3	0.032	0.038	0.039	0.05	0.042	0	0	0.33	0.06	0.062	0.047	0.092	0.052
E	E1	0.075	0.075	0.075	0.101	0.115	0.089	0.089	0.089	0.175	0	0.097	0.11	0.084
	E2	0.075	0.075	0.075	0.101	0.087	0.089	0.089	0.089	0	0.175	0.097	0.084	0.11
A	A1	0.021	0.022	0.025	0.045	0.02	0.02	0.02	0.02	0.023	0.023	0	0	0
	A2	0.163	0.155	0.142	0.102	0.137	0.149	0.156	0.154	0.16	0.152	0	0	0
	A3	0.052	0.058	0.074	0.068	0.056	0.059	0.056	0.059	0.06	0.067	0	0	0

Appendix C. Limit Supermatrix for Supplier Selection in Cement Industry.

		B			C		D			E		A		
		B1	B2	B3	C1	C2	D1	D2	D3	E1	E2	A1	A2	A3
B	B1	**0.082**	0.082	0.082	0.082	0.082	0.082	0.082	0.082	0.082	0.082	0.082	0.082	0.082
	B2	**0.155**	0.155	0.155	0.155	0.155	0.155	0.155	0.155	0.155	0.155	0.155	0.155	0.155
	B3	**0.1**	0.1	0.1	0.1	0.1	0.1	0.1	0.1	0.1	0.1	0.1	0.1	0.1
C	C1	**0.009**	0.009	0.009	0.009	0.009	0.009	0.009	0.009	0.009	0.009	0.009	0.009	0.009
	C2	**0.013**	0.013	0.013	0.013	0.013	0.013	0.013	0.013	0.013	0.013	0.013	0.013	0.013
D	D1	**0.087**	0.087	0.087	0.087	0.087	0.087	0.087	0.087	0.087	0.087	0.087	0.087	0.087
	D2	**0.11**	0.11	0.11	0.11	0.11	0.11	0.11	0.11	0.11	0.11	0.11	0.11	0.11
	D3	**0.06**	0.06	0.06	0.06	0.06	0.06	0.06	0.06	0.06	0.06	0.06	0.06	0.06
E	E1	**0.091**	0.091	0.091	0.091	0.091	0.091	0.091	0.091	0.091	0.091	0.091	0.091	0.091
	E2	**0.089**	0.089	0.089	0.089	0.089	0.089	0.089	0.089	0.089	0.089	0.089	0.089	0.089
A	A1	**0.019**	0.019	0.019	0.019	0.019	0.019	0.019	0.019	0.019	0.019	0.019	0.019	0.019
	A2	**0.132**	0.132	0.132	0.132	0.132	0.132	0.132	0.132	0.132	0.132	0.132	0.132	0.132
	A3	**0.053**	0.053	0.053	0.053	0.053	0.053	0.053	0.053	0.053	0.053	0.053	0.053	0.053

5. References

[1] Bozarth C C 2008 *Introduction to Operations and Supply Chain Management* (Pearson Education, Inc.,Upper Saddle River, New Jersey)

[2] Chan F T S, Chan H K, Ip R W L, Lau H C W 2007 A decision support system for supplier selection in the airline industry *Proc.of The Institution of Mechanical Engineers Part B- J.of Engineering Manufacture*. **221(4)** 741-58

[3] Cheraghi S H, Dadashzadeh M, Subramanian M 2011 Critical success factors for supplier selection: an update. *J.of Applied Business Research.* **20(2)** 91-108

[4] Ho W, Xu X, Dey P K 2010 Multi-criteria decision making approaches for supplier evaluation and selection: A literature review. *Euro.J.of Operational Research.* **202(1)** 16-24

[5] Mukherjee K, Sarkar B, Bhattacharyya A 2013 Supplier selection by F-compromise method: a case study of cement industry of NE India *Int.J.Computational System Engineering.* **1(3)** 162-74

[6] Saaty T L 2004 Fundamentals of the analytic network process - Dependence and feedback in decision-making with a single network. *J.of System Science and System Engineering.* **13(2)** 129

[7] Sharma D G and Rawani A M 2016 Green Supplier Selection for Indian Cement Industry: AHP based approach. *Int.Research.J.of Engineering and Technology*. **03(04)** 2368–73.

[8] Hwang, Ching-Lai, Yoon, Kwangsun 1981 *Multiple attribute decision making* (Springer-Verlag)

[9] Yilmaz O, Gülsün B, Güneri A F, Özgürler Ş 2011 Supplier Selection of A Textile Company with ANP *Int.Research.Conf.Trends in Development of Machinery and Associated Technology.* **15** 257–60.

IOP Conf. Series: Materials Science and Engineering **277** (2017) 012064 doi:10.1088/1757-899X/277/1/012064

Identifying research advancements in supply chain risk management for Agri-food Industries: Literature review

W Septiani [1] and P Astuti [2]

[1]Magister of Industrial Engineering, Faculty of Industrial Technology, Trisakti University, Jakarta, 11440, Indonesia
[1]winnie_septiani@yahoo.com
[2]pudji@trisakti.ac.id

Abstract. Agri-food supply chain has different characteristics related to the raw materials it uses. Food supply chain has a high risk of damage, thus drawing a lot of attention from researchers in supply chain management. This research aimed to investigate the development of supply chain risk management research on agri-food industries. These reviews were arranged in steps systematically, ranging from searching related to the review of SCRM paper, reviewing the general framework of SCRM and the framework of agri-food SCRM. Selection of literature review papers in the period 2005-2017, and obtained 45 papers. The results of the identification research were illustrated in a supply chain risk management framework model. This provided insight toward future research directions and needs.
Keywords: supply chain risk management, literature review, framework model, agri-food

1. Introduction

Logistic and supply chain management play a crucial role in food industries. Supply chain in food industries related with safety, healthy and nutritious begin from primary production to final preparation and consumption [1]. Refer to [2], the assessment of food supply chain was integrated between quality, safety, sustainability and logistic efficiency. In the context of work carried out by United Nations agencies and international organizations, "food safety is the assurance that food will not cause harm to the customer when it is prepared and/or eaten according to its intended use [3].

Supply chain risk management refers to risk management with a focus on the supply chain and its risk, which have an impact on the company [4]. A typical of agri-food supply chain may consist of a number of entities linked from "a farm to fork", such as farmers, input suppliers, co-operatives, pack-houses, transporters, exporters, importers, wholesalers, retailers, and finally costumer [5]. All actors of the supply chain (from supplier to customers, from employees to top management) are involved in risk management activities [4]. Supply chain activities include procurement, inventory management, product design and new product development, manufacturing (planning), order processing, transport (distribution), sales, demand management, customer service [5].

According to [6] defined risk management as an effort to reduce the risk and minimize the loss due to risk uncertainty. [7] in his research states that the implementation of supply chain risk management can increase the amount and quality of knowledge and can reduce the chances of risk and risk impact. According to [8] there is a strong influence of the implementation of supply chain risk management on the improvement of sustainable process in supply chain. While [4] in their research explain that it requires a thorough and comprehensive analysis to explore supply chain risk management.

This research aimed to identifyng research advancements in supply chain risk management for agri-food industries. The results of identification are written in a framework model supply chain risk management for agrifood industries. This model is built based on the results of literature review from several studies related to supply chain management, supply chain risk management, food supply chain assessment, food safety, food quality and agri-food industries. The scope of agri-food in this research referred to the agri-food product grouping by [9], horticultural product (e.g. fruit general, pineapple, tomato, fresh cut salads, flowers general, grapewine), meat product (e.g. meat general, beef), dairy product (e.g. milk, cheese), fishery product (e.g. seafood general, fish, lobster, smoked seafood), bakery product (e.g. bakery general, bread), beverage (e.g. beverage general, beer, wine) and other food product (e.g. egg, sushi, pasta, cofee, olive oil).

2. Methods

A literature survey has been undertaken of quantitative and qualitative empirical studies published in several leading journal international : supply chain management journals, risk management journals, production economic journals, journal of food composition and analysis, journal of food engineering, food bioprocess technology journal, journal of food microbiology and others journal. Paper traced using the publisher websites: Springerlink, Willey interscience, Elsevier ScienceDirect, Emerald Insight. Publication were analyzed for the period between 2005 and 2017. Paper traced by using a few keywords such "supply chain", "risk management", "risk assesment", "food industry", "food supply chain" and "agri-food industries".

The overall sample contain forty-two papers in totals (status papers published up to the end of 2017). Out of this sample only 45 papers specially discuss about food supply chain. The number of used publication papers for each step can be seen in table 1.

Table 1. Distribution of Journals of reviewed, 2005-2017

Distribution of journal review	Count
Journal Review	
Supply Chain Risk Management : Review	3
Agri-food Supply Chain : Review	5
Supply Chain Management (SCM) : Review	3
Supply Chain Management integrated with Knowledge Management : Review	1
Sustainable Supply Chain Management (SSCM) : Review	3
Framework SCRM Food Industry	7
Framework SCRM	3
Journal Agri-food Supply Chain Risk Management	20

The number of publications on each years is indicated in Figure 1.

3. Result and discussion

3.1 Supply chain risk management

Consequently, we follow the SCRM definition [10] "the management supply chain risks through coordination or collaboration among the supply chain partners so as the ensure profitability and continuity". [6] defines global SCRM as "the identification and evaluation of risks and consequent losses in the global supply chain, and implementation of appropriate strategies through a coordinated approach among supply chain members". Furthermore [6] defines risk management as an effort to reduce the risks and minimize losses due to the uncertainty of risk.

Risk cannot be eliminated completely but the influence and impact on the company can be reduced, with proper risk management. SCRM aims to identify potential sources of risk in the supply chain, through the coordination of all supply chain members, to reduce the overall supply chain vulnerability. Supply chain activities include : procurement, inventory management, product design and new product development, manufacturing (planning), order processing, transportation/distribution, sales, demand management and customer service [5].

[11] states that sources of uncertainty in the supply chain is divided into two internal source and external source. Internal source uncertainty include available capacity, custom regulation, information delays, internal organization. Whereas external source uncertainty meliputi competitor action, manufacturing yield, political environment, price fluctuations, stochastic cost, supplier quality.

[10] four main approaches in the management of SCR is supply management, product management, demand management, information management. [12] explain drivers of supply chain consists of risk globalization, product variants, outsourcing, reduction of suppliers, focus on efficiency, central distribution, centralized production.

According [13], states there are three main elements of SCRM definition, they are risk identification and evaluation/assessment, gobal supply chain and coordinated risk management strategies. [14] defines risk into three dimensions, (1) Likelihood of occurrence of a particular event or outcome, (2) Consequences of the particular event or outcome occurring; and (3) Causal pathway leading to the event. [15] apply matrix loss exposure for assessment of risk factors. Risk factors are categorized as operational risk, financial risk and physical risk. [6], presents the supply chain risk categorize consists of supply risk, operational risk, demand risk, security risk, macroeconomic risk, competitive risk and resource risk. [16] suggest the supply chain risk categorize includes the supply chain risk demand, the damage risk, inventory risk, the risk of breakdown process, the risk of physical capacity of the plant, supply risk, system risk, the risk of power/authority and transportation risk. While [4], suggests the supply chain risk categorize in the three groups, the risk in the company, the risks outside the company and in the supply chain and risks beyond the supply chain, but it has an influence on the supply chain. [6] propose a framework for the integrated global supply chain risk management by using the relation of table risks. [16] develop a Supply Chain Risk Management Process (SCRMP) as the development of Risk Management Process (RMP) to identify, assess and manage risks of supply chain. Risk evaluation is based on the total value of the index Hazard Totem Pole (HTP) of the integration of the three dimensions, namely the risk of severity, probability and cost. [8], develops models of supply chain risks that connecting the process from upstream risk identification, risk assessment, risk mitigation and performance risk to the continuous improvement process. [13] SCRM framework was developed based on taxonomy defined for risk events and risk management approach.

3.2 Paper Review Related Supply Chain Risk Management for Agri-food

[17] identified risks in supply chain. The supply chain management discussed includes material flow risk, financial flow risk, and information flow risk. [10] represents four main approaches in supply chain risk management: supply management, product management, demand management, and information management. [6] created an integration framework for global supply chain risk management with the help of risk table linkage. [14] risk and performance risk [4] developed a framework for future analysis and exploration of supply chain risk management. The supply chain environment factors considered include external threats and resource requirements to improve the economy. [18], developed an integration framework of marketing strategy and supply chain.

Table 2. List of paper review related with SCRM for Agri-food

Article	Subject area	Main objective of the paper
[19]	Supply chain risk management	SCRM literature review combined Systematic Literature Review (SLR) with Citation Network Analysis (CNA) through dynamics perspective approach
[17]	Supply chain risk management	Described about the supply chain operation management includes material flow risk, financial flow risk, information flow risk

Table 2. List of paper review related with SCRM for Agri-food

Article	Subject area	Main objective of the paper
[4]	Supply chain risk management	Built Risk-related terminology, drivers of supply chain risk, Deriving 17 principles of supply chain risk management
[9]	Agri-food supply chain	Analyzed the development of RFID technology in the agri-food sector
[20]	Agri-food Supply Chain	Conducted an assessment of the state of the art food supply chain planning models from the different components
[21]	Agri-food supply chain	Developed hierarchical decision-making framework and critical taxonomy from agrifood supply chain
[22]	Risk assessment Agri-food	Analyzing the risk assessment of Campylobacter result in broiler meat
[23]	Agri-food supply chain	The design of food transportation system model with notices the temperature parameter, bacterial growth and others in the food transport
[24]	Supply network design	Review about: (1) decisions addressed in the model, (2) performance metrics, (3) the degree to which the model supports integrated decision processes, and (4) globalization considerations.
[25]	Supply chain management	Described about SCM from knowledge philosophy perspective and future SCM research
[26]	Supply chain management : RPID Technology	Developed a state-of-the-art on RFID technology deployments in supply chains
[27]	Supply chain knowledge management	Evaluated the relationship between knowledge management and SCM with analysis of the existing theoretical and empirical works
[28]	Sustainable Supply Chain	Evaluated quantitative model sustainable supply chain
[29]	Sustainable Supply Chain	Explained about design and management sustainable supply chain and future research opportunity
[30]	Sustainable Supply Chain	Literatur review sustainable Supply chain.

3.3 Framework Model Supply Chain Risk Management for Agri-food Industries

[16] developed Supply Chain Risk Management Process (SCRMP) as the development of Risk Management Process (RMP) to identify, assess, and manage supply chain risk. The risk evaluation was performed based on the total value of a Hazard Totem Pole (HTP) index of the integration of three dimensions, namely, severity, probability, and cost. However, this research has not been equipped with data management system to store risk relationship information and risk update as. [8] developed an upstream supply chain risk model that links the processes of risk identification, risk assessment, risk mitigation, and risk performance to continuous process improvement. [13] conceptualization and planning of supply chain risk management process.

[31] conducted a study on risk assessment in agricultural supply chain in developing countries. The major risks assessed include weather-related risks, natural disaster, biological, and environmental risks, market-related risks, logistical and infrastructure risks, management and operating risks, public and institutional policy risks, and other risk magnitudes. While [32], arranged risk management in New Zealand. [5] drew up a conceptual framework of a food supply chain development collaboration. [2] developed a food supply and logistics supply chain assessment framework by integrating four elements, namely, quality, security, sustainability, and logistics efficiency. In addition, transportation

problems in the supply chain were also considered. [33] developed a knowledge-based supply chain risk framework model for dairy agroindustry. The risks were identified from the breeders, the cooperatives, to the dairy processing industries.

Table 3. List of paper review related with framework model of SCRM for Agri-food

Articles	The approach used	Main objective of the paper
[34]	Mathematical model, numerical experiment	Decision of pricing strategy, shelf space allocation, and replenishment policy in a single-item food supply chain setting.
[35]	Qualitative analysis, expert judgment	Identified the major risks that Japanese food companies : Production risk, price purchase risk, demand risk and procurement risk.
[36]	Extensive literature review	Developed an integrated framework for food supply chain resilience. A conceptual framework that highlights the importance of logistics, collaboration, sourcing, and knowledge management in achieving supply chain resilience.
[37]	Theory of fuzzy sets	Identified RFs in PSCs, based on the distance between fuzzy numbers. The development of some mathematical areas, such as probability theory, theory of fuzzy sets, and rough theory, enables that uncertainty and imprecision of any kind be sufficiently well described quantitatively.
[33]	Expert system and fuzzy logic	Framework model of sustainable supply chain risk management for dairy agroindustry based on knowledge Base.
[2]	Integration of multiple decisions affecting the performing of the system	Conceptual framework for the assessment of food supply chain (FSC) and logistics of food products in agreement with a multidisciplinary and integrated view.
[8]	Empirical analysis	Empirical testing to see the relationship between risk identification, risk assessment, risk mitigation and risk performance with continuous improvement processes
[16]	The total value of the index Hazard Totem Pole (HTP)	Develop a systematic structure and approach to incorporate supply chain risk and assess severity and likelihood so that risk mitigation plans can be developed.
[31]	Supply chain situation analysis	framework for undertaking assessments of risk and risk management capability within agricultural supply chains in developing countries
[5]	In depth interviews in the two companies	A conceptual framework of collaboration development in the agri food supply chain

[38] describes a conceptual ABM that is based on a regional food system in western Iowa and eastern Nebraska. Agents in the model include: Food Hub Manager Agent, Producer agent and Customer Agents. [39] the study showed that external drivers of the organizations affect internal drivers, and compared to external drivers, internal drivers have a more important role in creating sustainable orientation within an organization. [40] the aim of this paper is to highlight the determinants of competitiveness and complementarity of different organizational models of SFSCs, led by actors who play different roles along the supply chain, in urban and peri-urban areas across Europe. [41] the overall aim of the current research is contribute to a better scoping practice of Life Cycle Assessment (LCAs) and Life Cycle Coasting (LCCs) of side flows in a food waste context. [42] proposed a method based on data from the testing of food-waste feed with comprehensive evaluation of its product safety by integrating fuzzy mathematics effectively and the model of AHP. [43], the model suggest that the institutional factors are the most influential factors to food supply chain risk management, followed by the basic characteristic of enterprise and finally the characteristic of employees. The summary from the identification research supply chain risk management for agri-food industries displayed in figure 1.

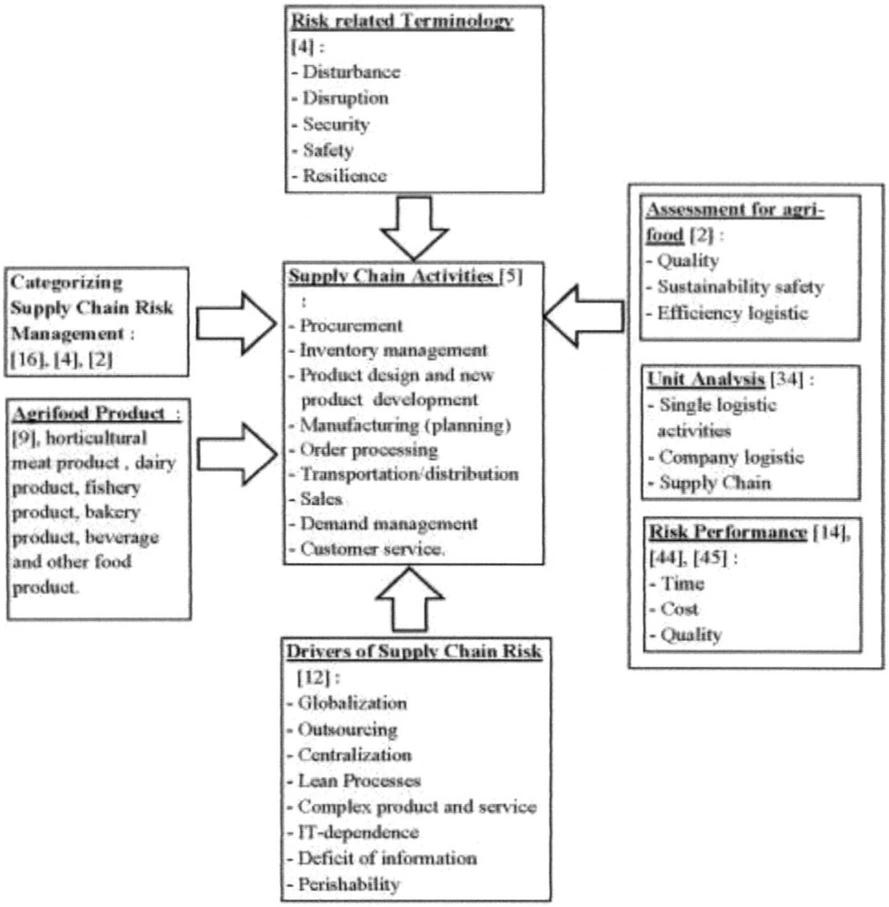

Figure 1. Framework model of Supply Chain Risk Management for Agrifood Industries : Literature Review

4. Conclusion

- Research on supply chain risk management for agri-food industries highly developed in accordance with needs for food products. Risk that occurs on supply chain food had the chance larger than a manufactured product. The unique features of the food industry such as the perishable nature of food products differentiate it from other sectors of the economy, and have intensified the need for an efficient supply chain.
- risk assessment that occurs in supply chain agri-food in general have to take attention into aspects of quality, safety, sustainability and logistic efficiency
- Research has been widely implemented in supply chain risk measurement, but few was discussed about risk mitigation plans in the agri-food supply chain so that it gives a great opportunity to do such research.

5. Reference

[1] Burlingame B and Pineiro M 2007 The essential balance: Risks and benefits in food safety and quality. *J Food Compost Anal* **20**, 139-46.

[2] Manzini R and Accorsi R 2013 The new conceptual framework for food supply chain assessment. *J Food Eng* **115**, 251-63.

[3] FAO and WHO 2001 Codex Alimentarius Commission Geneva, Swtzerland Twenty-fourth Session.

[4] Pfohl HC, Kohler H and Thomas D 2010 State of the art in supply chain risk management research: empirical and conceptual findings and a roadmap for the implementation in practice. *Logistic Research* **2**, 33-44.
[5] Matopoulos A, Vlachopoulou M and Manthou V 2007 A conceptual framework for supply chain collaboration: empirical evidence from the agri-food industry. *Supply Chain Management: An Int. J.* **12**(3), 177-86.
[6] Manuj I and Mentzer JT 2008 Global supply chain risk management strategies. *International Journal of Physical Distribution & Logistics Management* **38**(3), 192-223.
[7] Ellegaard C 2008 Supply risk management in a small company perspective. *Supply Chain Management: An Int. J.* **13**(6), 425-34.
[8] Kern D, Moser R, Hartmann E and Moder M 2012 Supply risk management: model development and empirical analysis. *Int. J. of Physical Distribution & Logistics Management* **42**(1), 60-82.
[9] Costa C, Antonucci F, Pallottino F, Aguzzi J, Sarriám D and Menesatti P 2013 A Review on Agri-food Supply Chain Traceability by Means of RFID Technology. *Food Bioprocess Technol* **6**, 353-66.
[10] Tang CS 2006 Review Perspectives in supply chain risk management. *Int. J. Product Econ* **103**, 451-88.
[11] Cucchiella F and Gastaldi M 2006 Risk management in supply chain: a real option approach. *J. of Manufacturing Technology Management* **17**(6), 700-30.
[12] Thun JH and Hoenig D 2011 An empirical analysis of supply chain risk management in the German automotive industry. *Int. J. Production Economics* **131**, 242-9.
[13] Bandaly D, Satir A, Kahyaoglu Y and Shanker L 2012 Supply chain risk management – I: Conceptualization, framework and planning process. *Risk Manage* **14**(4), 249-71.
[14] Ritchie B and Rindley C 2007 Supply chain risk management and performance A guiding framework for future development. *International Journal of Operations & Production Management* **27**(3), 303-22.
[15] Yang YC 2011 Risk management of Taiwan's maritime supply chain security. *Saf Sci* **49**, 382-93.
[16] Tummala R and Schoenherr T 2011 Assessing and managing risks using the Supply Chain Risk Management Process (SCRMP). *Supply Chain Management: An International Journal* **16**(4), 474-83.
[17] Tang O and Musa N 2010 Identifying risk issues and research advancements in supply chain risk management. *Int. J. Production Economics* **133**, 25-34.
[18] Jüttner U, Christopher M and Godsell J 2010 A strategic framework for integrating marketing and supply chain strategies. *The Int. J. of Logistics Management* **21**(1), 104-26.
[19] Colicchia C and Strozzi F 2012 Supply chain risk management: a new methodology for a systematic literature review. *Supply Chain Management: An Int. J.* **17**(4), 403-18.
[20] Ahumada O and Villalobos JR 2009 Application of planning models in the agri-food supply chain: A review. *European Journal of Operational Research* **195**, 1-20.
[21] Tsolakis NK, Keramydas CA, Toka AK, Aidonis DA and Iakovou ET 2014 Agrifood supply chain management: A comprehensive hierarchical decision-making framework and a critical taxonomy. *Biosys Eng* **120**, 47-64.
[22] Nauta M, Hill A, Rosenquist H, Brynestad S, Fetsch A, van der Logt P, Fazil A, Christensen B, Katsma E, Borck B and Havelaar A 2009 A comparison of risk assessments on Campylobacter in broiler meat. *Int. J. Food Microbiol* **129**, 107-23.
[23] James SJ, James C and Evans JA 2006 Modelling of food transportation systems - a review. *Int. J. of Refrigeration* **29**, 947-57.
[24] Meixell MJ and Gargeya VB 2005 Global supply chain design: A literature review and critique. *Transportation Research Part E* **41**, 531-50.

[25] Burgess K, Singh PJ and Koroglu R 2006 Supply chain management: a structured literature review and implications for future research. *Int. J. of Operations & Prod. Mgmnt* **26**(7), 703-29.
[26] Sarac A, Absi N and Pérès SD 2010 A literature review on the impact of RFID technologies on supply chain management. *Intern Journal of Production Economics* **1**, 77-95.
[27] Marra M, Ho W and Edwards JS 2012 Supply chain knowledge management: A literature review. *Expert Systems with Applications* **39**, 6103-10.
[28] Seuring S 2012 A review of modeling approaches for sustainable supply chain management. *Decision Support Systems*, 1-8.
[29] Gupta S and Desai ODP 2011 Sustainable supply chain management: Review and research opportunities. *IIMB Management Review*, 1-12.
[30] Seuring S, Muller M. 2008 From a literature review to a conceptual framework for sustainable supply chain management. *J. of Cleaner Production* **16**, 1699–710.
[31] Jaffee S, Siegel P and Andrews C 2010 Rapid Agricultural Supply Chain Risk Assessment: A Conceptual Framework. *Agriculture and Rural Development Discussion Paper* **47**.
[32] Melyukhina O 2011 Risk Management in Agriculture in New Zealand. *OECD Food, Agriculture and Fisheries Working Papers* **42**.
[33] Septiani W, Marimin, Herdiyeni Y and Haditjaroko L 2014 Framework Model of Sustainable Supply Chain Risk for Dairy Agroindustry Based on Knowledge Base. *ICACSIS 2014*, 255-60.
[34] Yang S, Xiao Y, Kuo Y-H. 2017 The Supply Chain Design for Perishable Food with Stochastic Demand. *Sustainability* **9**(7).
[35] Radzi RM, Saidon IM and Ghani NA 2017 Risk Mitigating Strategies in The Food Supply Chain by Japanese Food Companies in Malaysia. *International Journal of Management and Applied Science* **3**(1), 89-93.
[36] Umar M, Wilson M and Heyl J 2017 Food Network Resilience Against Natural Disasters: A Conceptual Framework. *SAGE Open* **7**(3).
[37] Vujović A, Đorđević A, Gojković R and Borota M 2017 ABC Classification of Risk Factors in Production Supply Chains with Uncertain Data. *Mathematical Problems in Engineering*.
[38] J. Craven T and Krejci C 2017 An Agent-Based Model of Regional Food Supply Chain Disintermediation. *The Agent-Directed Simulation Symposium*; Virginia Beach, USA: Society for Modeling & Simulation International (SCS).
[39] Emamisaleh K and Rahmani K 2017 Sustainable supply chain in food industries: Drivers and strategic sustainability orientation. *Cogent Business & Management* **4**(1).
[40] Grando S, Carey J, Hegger E, Jahri I and Ortolani L 2017 Short Food Supply Chains in Urban Areas: Who Takes the Lead? Evidence from Three Cities across Europe. *Urban Agriculture & Regional Food Systems* **2**.
[41] Östergren K, Davis J, Menna FD, Vittuari M, Unger N and Loubiere M 2017 Food supply chain side flows management through Life Cycle Assessment and Life Cycle Costing: a practitioner's perspective. *System Dynamics and Innovation in Food Networks 2017: Int. J. on Food System Dynamics*.
[42] Xiaoping W 2016 Food Supply Chain Safety Risk Evaluation Based on AHP Fuzzy Integrated Evaluation Method. *Int. J. of Security and Its Applications* **10**(3), 233-44.
[43] Yongsheng L, Xuan W. 2016 Food Supply Chain Risk Management Situation Evaluation Model Based on Factor Analysis. *Int. Business and Management* **12**(2), 40-6.
[44] Septiani W, Marimin, Herdiyeni Y, Haditjaroko L. 2016 Risk dependency chain model of dairy agroindustry supply chain using fuzzy logic approach. *Supply Chain Forum: An Int. J.* **17**(4), 218-30.
[45] Septiani W and Djatna T 2015 Rancangan Model Performansi Risiko Rantai Pasok Agroindustri Susu dengan Menggunakan Pendekatan Logika Fuzzy. *Agritech* **35**(1), 88-97.

IOP Conf. Series: Materials Science and Engineering **277** (2017) 012065 doi:10.1088/1757-899X/277/1/012065

Land transportation model for supply chain manufacturing industries

Fajar Kurniawan

Saint Mary's University of Hong Kong
fajar.kurniawan.hk@gmail.com

Abstract. Supply chain is a system that integrates production, inventory, distribution and information processes for increasing productivity and minimize costs. Transportation is an important part of the supply chain system, especially for supporting the material distribution process, work in process products and final products. In fact, Jakarta as the distribution center of manufacturing industries for the industrial area. Transportation system has a large influences on the implementation of supply chain process efficiency. The main problem faced in Jakarta is traffic jam that will affect on the time of distribution. Based on the system dynamic model, there are several scenarios that can provide solutions to minimize timing of distribution that will effect on the cost such as the construction of ports approaching industrial areas other than Tanjung Priok, widening road facilities, development of railways system, and the development of distribution center.
Keywords: Land Transportation, Supply Chain, material distribution process

1. Background

Jakarta is the capital city of Indonesia and the distribution center of various manufacturing industries area, such as Cikarang, Cikampek and Tangerang. Tanjung Priok is the entry point for material supply for industries, especially at Port Tanjung Priok, where almost 60% of export import activities are done through this Port.

[11] published in Logistics Performance Index (LPI) has reviewed the national logistics force, position of Indonesia is ranked 53rd from out of 160 countries surveyed. Compared to other ASEAN countries, Indonesia's performance has been relatively low in terms of the competitiveness of the country's logistics system. Indonesia's logistics competitiveness ranking is still below neighboring countries, such as Singapore (5), Malaysia (25), Thailand (35), and Vietnam (48). Studies conducted by Asia Foundation and The Institute for Economic and Social Research, University of Indonesia (2008), attribute this to the following: (1) the cost of logistics in Indonesia is very high, averaging 14.08% of total sales; (2) problems include not only high of average logistics costs, but also the potential loss of revenue due to the lack of the role of a national shipping Company in overseas transport. Logistics costs per 55 Kilometers in Indonesia is 550 USD; whereas in Malaysia it costs only 300 USD. Several types of export products, account total cost before shipping and land transport in the country to reach more than 40% of the total cost of logistics and transport costs [12].

It has been noted that Transportation is the most expensive logistics activity, representing over 40 percent of most corporate logistics expenses and over $400 billion in annual expenses in the United States alone. The strategy of substituting transportation and information for inventory has yielded big productivity dividends [5].

All the problems that arise in relation to the performance of Indonesian Logistics, originated from the low performance of the transport handling and distribution activities. Transportation for distribution in Indonesia mostly uses road as the main facility for distribution. Road transport of goods is expected to reach 91.25% according to Directorate General of Land Transportation of the Republic of Indonesia (2006).

Jakarta as the entry point of manufacturing industries distribution continuous to experience problems with its transportation system. Jakarta is the most congested traffic city in the world. The Start-Stop Index estimates the average number of starts and stops per vehicle in 78 cities around the world. It was noted that Jakarta drivers had 33,240 starts and stops annually according to the survey [2]. Congestion problem gave considerable influence on the transport and distribution system, aggravated by the fact that the port of Tanjung Priok as a distribution center for manufacturing industries is located in the northern part of the Jakarta.

The manufacturing industries acts as consumer that should be served by the system government design. Customer satisfaction is the main focus that will support the viability of a business system especially in manufacturing system. It is necessary to conduct in-depth study of the physical condition to the current distribution process located in the areas of around Jakarta, such as: Cikampek, Cikarang and Tangerang. Mapping of current situation will give a figure of the main cause inefficiency for distribution process.

A number of questions arise, such as:

a. "What is the influential variable that causes the transport and distribution problem in Jakarta, especially for manufacturing industries ?"
b. "What are the related cost figures of the recent transport and distribution system in Jakarta ?"; and
c. "What kind of model, if any, can possible minimize the potential loss of revenue caused by the low performance of transportation and distribution system in Jakarta ?"

Distribution process from Tanjung Priok Port to the industrial area in a significant variable of supply chain in manufacturing industries. Supply chain is a network of facilities and distribution options that performs the following functions; the procurement of materials, transformation of these materials into intermediate and finished products; distribution of these finished products to customers. Supply chain management is a strategy through of these different functions can be achieved [6]. The objective of supply chain management (SCM) is to integrate the different function of procurement, transformation and distribution for improving the performance of organizations to achieve efficient fulfillment and customer satisfaction. The organization of supply chain is the manufacturing industries.

Model for land transportation will use System Dynamics approach. According to [3] System Dynamics associated with time which depend on how the behavior of the system describes system for reaching the main goals using qualitative and quantitative models, how information feedback will effect the behavior of the system, and design for the information feedback structure and controlling policies through simulation and optimization.

Therefore, the intended research will focus on the most efficient design of transportation and distribution model in relation to the expected reduction in delays, considering the characteristics of the region, hand in hand with the needed involvement of various stakeholders and their expected involvement and contribution to the implementation of supply chain management.

2. Problem Statement

There is a tremendous need of optimal model for land transportation in manufacturing industries around Jakarta. Model will develop based on the problem that arise by current situation of distribution. There are 2 obstacles regarding the distribution process complaint by the manufacturing industries such as: inefficiency of the distribution process and long delivery time.

The manufacturing industries as the consumer that should be served by the system government design. Customer satisfaction is the main focus that will support the viability of a business system especially in manufacturing system. It is necessary to conduct in-depth study of the physical condition to the current distribution process located in the areas of around Jakarta, such as: Cikampek, Cikarang and Tangerang.

3. Objectives

The followings are the perceived objectives of the study :

- To obtain full perspective of the present condition of transportation and distribution system for manufacturing industries in Jakarta.
- To determine the main variables that become the causes of delay in transportation and distribution activities.
- Designing model for transportation and distribution by implementing system dynamics to reduce the level of delay, optimizing strategy that can minimize potential loss of revenue caused by the mentioned low performance system.
- Developing and validating improvements in the transportation and distribution model.
- Supporting the strategic decision making aspects of the government resulting strategic policy in transportation and distribution control system which can provide positive effect on Logistic Performance Index of Indonesia.

4. Methodology

The methodology used in this activity is a system approach. This approach was chosen, because the development of a region is influenced by many interrelated elements are dynamic [7]. The system approach is the right choice, because this approach is able to solve a problem that has characteristics [7] as follows :

a. Complexity, which is quite a complicated interaction between all the elements.
b. Dynamic, in the sense that there are factors that change every time and forecasting in the future.
c. Probabilistic, need for opportunities in the inference functions conclusions and recommendations.

One thing that stands out from the system approach is the searching of key factors in the assessment of the problem to obtain an appropriate settlement, with the use of quantitative models for assist in the rational decision-making process. System approach can solve the problems of high complexity.

Region is made based on two sub-systems, as follows :

a. Infrastructure
b. Information Technology
c. Regulation

Sub-systems can be studied and analyzed to produce the most optimal model of land transportation system for manufacturing industries.

System approach in policy model, performed using soft system and a hard system. Model Policy called Soft System, because the model is designed based on recent quantitative data, as well as data in the past time, such as Gross Domestic Product and Human Development Index. The model also entered into the category of soft systems, because after the simulation process, the results are used to make qualitative policy scenarios.

In an effort to achieve the desired objectives of this activity, there are several approaches to implement, as follows:

1. Identification of potential Infrastructure sub-system consisting of several factors such as : Road, Railway system, Distribution Center, Port, which can be an indicator support of the added value to maximize transportation model.
2. Identification of information technology and regulation, which is a reflection of the improvement of supporting distribution model. The results of this analysis can be reflected the extent of the model can improve to minimize delivery time.

3. Identification of prime factors based on criterias that can describe the advantages of these factors, the sub-systems of production, as follows :
   - How far the factors can reduce delivery time, logistic cost and inefficiency ?
   - Analysis of the government's commitment to the improvement of infrastructure.
   - Identification of information and technology for the development of Land transportation system.
4. Analysis of linkages between sub-system.
5. Policy Analysis of infrastructure, modes of transport, roads, public facilities, facilities, and social, which can describe the existence of infrastructure, so that it appears the advantages and disadvantages of the road transportation system.
6. Identify physical potential of port development, to analyze the potential and challenges of natural resource development, support and institutional factors.
7. Analysis of the relationship between four main sub-systems, carried out by using a system dynamic.

Correlation between each sub-system and the influence of the sub-systems to the minimize the delivery time, implemented as a policy option that needs to be tested in a system linkages. Results of these analyzes are formulated in the form of alternative development policies. The sequencing of policy priorities using multiple criteria set by the experts and all stakeholders associated with the development policy of regional integration. Visually, the conception of the study can be shown in Figure 1.

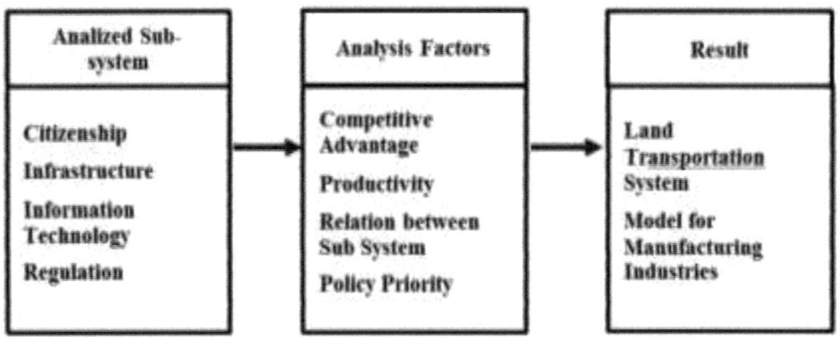

Figure 1. Conception Activity

In order to provide supporting data of the study and examine the main questions raised, the following methodology will be employed. Research methodology used as follows:

a. Participant Observation
   Participant observation with a group of manufacturing industries logistics manager will be carried out. The aim is to collect more qualitative information about the perceived impact, critical point and the cause of delay due to supply chain agent point of view as the users of transportation and distribution system in daily activity.
b. Individual Interviews
   There will be an attempt to interview stake holders and experts connected to the logistic system in Jakarta. Individual interviews by a way of structured questionnaire will also be used to collect information about the alternative solution for developing policy in order to develop improvements in transport and distribution system for manufacturing industries in Jakarta . The solution will be based on the perception of the environment. The interview is especially important for examining questions like "What kind of logistic system policy can be encouraged to make the rational choice' for the delay and loss revenue?". Participanta in the individual interviews based on Purposive Sampling Methods.
c. Analysis of Official Statistic

IOP Conf. Series: Materials Science and Engineering **277** (2017) 012065 doi:10.1088/1757-899X/277/1/012065

In order to gain a fuller picture of recent transportation and distribution condition in Jakarta and fullfil conpliance data in system Dynamics, a comprehensive analysis of the official statistics will also be undertaken aside from the official data, other sources of statistical data produced by agencies such as Badan Pusat Statistik (Statistic Central Board), Direktorat Lalulintas Polda Metro Jaya (Traffic Directorate Police Force of Jakarta), PT. Pelindo, and Pemda DKI Jakarta (Government of Province DKI Jakarta). Special attention to information about all variables that have correlation with the performance of logistic system in Jakarta will be given due consideration.

d. Data Collection
Quantitative data will largely be gathered through existing database and datasets in the field of transportation and distribution. Datasets publicly available in official Indonesian Statistics Board as well as logistics database of manufacturing industries which have distribution chain in Jakarta.

Figure 2. Research Methodology

5. Model Development

Systems dynamic approach has an excellent prospects, as a tool to answer a change of paradigm in the development model for transportation of manufacturing industries. Through this approach the expected prediction of the Implementation various policy scenarios regional development, both spatial and non-spatial, can be done [8]. In other words, this approach can serve as an early warning system of the implementation of a land transportation model, so can choose the most optimal policy scenarios. If there are certain consequences as a result of the implementation of these policies, can be prepared measures to anticipate as soon as possible. Stages for model development shown in figure 2.

Forrester is known as the founder of system dynamics, which deals with the simulation of interactions between objects in dynamic systems. System Dynamics will study problems with the system point of view, in which the elements of the system interact in a feedback relationship that will produce a particular behavior. Interactions in this structure is translated into the mathematical models were subsequently using computer simulation to obtain historical behavior. System Dynamics factors is the concept of feedback information from the behavior system, a mathematical model of dynamic interaction, and computer simulation, will a conduct a series of controlled experiments concerning the state of the system in a "laboratory" [4]. In Industrial Dynamics, [4] presented a methodology for the simulation of dynamics models, which is the origin of systems dynamics. The main objective of system dynamics is to understand the structural causes that bring about the behavior of a system [10].

System Dynamic that will be done to make a model of regional integration policies, consists of several stages (Figure 2), as follows :

1. Mapping of the Real System
   Current system which components consist of various problem mapping, so it would appear the existence of such a system overview.
2. Formulate a mental model. The relationship between sub-systems, causal, and behavior between each sub-system, and the development of model.
3. Making Causal Loop Diagrams
   This diagram will express about the causal relationship between subsystem models of land transportation and the factors within each sub-system, into the language of a particular image. This diagram is used to form the structure of the model policy which represent in the form of the circumference of a causal diagram

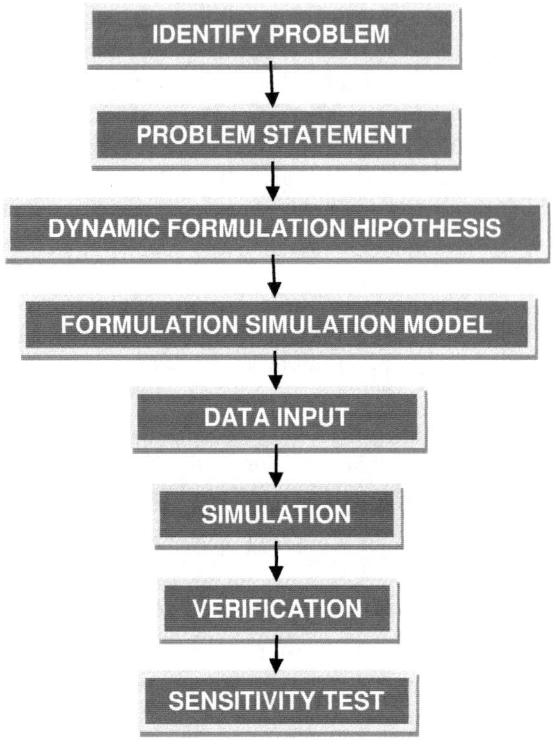

Figure 3. Stage for Model Development

4. Preparation of the model, which is done by building Soft Floor Diagram (SFD)
   Model of land transportation system is the basis of the investigation experimental relatively inexpensive and time saving than if you held a direct experiment on a real system. Modelling Process is done in making the policy model, as follows :
   a. Identify the problem (determination of limit)
      At this stage will be selected on the policy model based on need, which is accelerating the growth of the port area, and determine variables that affect the growth of the industrial area. This stage will define the problems encountered in the growth of the region in the form of dynamic models.
   b. Dynamic formulation hypothesis
      Done by sorting the initial hypothesis and mapping (limit diagram models, subsystem diagrams, causal diagram, mapping stock and flow, policy structure diagram).
   c. Formulation of Simulation model
      The relationship between the sub-systems model of transportation integration policies can be described in the form of the formulation, wherein the formulation will be used as input to the simulation models. The formulation can be obtained from :
      - Bibliographic from sub-systems and factors used in the model of Land transportation policy.
      - Justification of relevant experts
      - Trends of the data sub-systems and factors of the data in the past, in order to obtain a pattern that can be formulated to predict the model of land transportation today and in the future.

      At this stage also made estimates of parameters that affect the purposes of the study, and the relationship of behavior and the initial conditions of parameters.

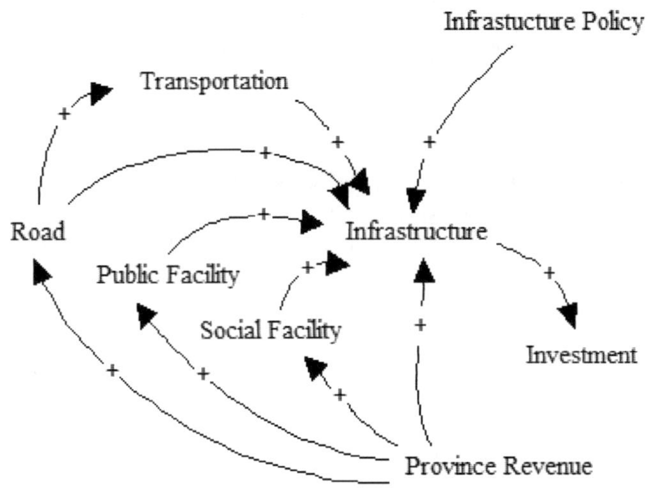

Figure 4. Causal loop subsystem infrastructure

Figure 5. Causal Loop Subsystem Information Technology

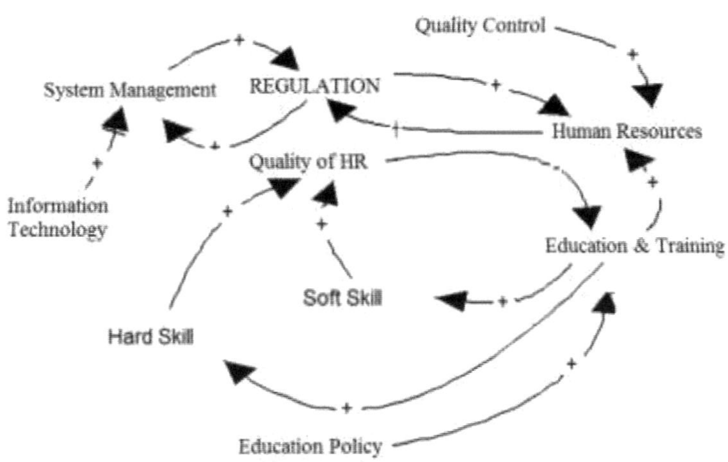

Figure 6. Causal Loop Subsystem Regulation

d. Assessment
Policy model which has been formed to be tested by comparing the results of model calculations with the real conditions which should be obtained through a real phenomenon, as well as references from relevant experts.
e. Data Input
This activity will be done by filling the data and defining all relations between the sub-systems with other subsystems, as well as the relationship between the factors with other factors, both of which are in a sub-system, and which are beyond the sub-system.
f. Simulation
Simulation is a dynamic process behavior models done by running models, and will show in resulting graphs and data tables behavior models.
g. Verification
A test of consistency and behavior models, through testing and evaluation activities, if the result is not verified, then should do the implementation of CLD, if the results are valid will continue on the next process. Policy model will be tested through the simulation process and the results will be compared with the conditions that should logically be obtained through a real phenomenon, as well as references from relevant experts. Verification of this activity is also carried out by the workshop, so the assessment of the workshop will appear on the simulation results, if the results are considered relevant, the process continues,
h. Sensitivity Test
Goal of the test is gaining leverage point and the output is in the form of the dominant factors that influence the development of each sub-system A model of regional integration policies.

Based on the observation, land transportation categorized in two facility, such as: Road and Rail. Road transport offers "door to door" transport flexibility through Full Truck Loads (FTL) and Less Truck Load (LTL) options. Road Tonnage per kilometers has increased greatly in recent times and is still increasing around the globe as India and China become wealthier [9]. In today's activity, truck transport system has the best position and opportunity in Jakarta. Truck businessman has good relationship with the stakeholders in supply chain system including government. This situation will stop the development of other transportation facilities, such as railway system. Rail is excellent for high density products over long distance and suitable for low value, non time sensitive deliveries. The downside to rail transport is that transport routes are limited to fixed-track and terminal facilities, which may differ in gauge size when railing internationally [9]. Very difficult to build the railway system into the Tanjung Priok Port, because because a group of truck businessman will lose of demand if the port of Tanjung Priok have those facilities.

6. Conclusion

Simulation shows the result that the main problem of land transportation facility in servicing the manufacturing industry as follows:

a. Capacity of the road is too small compared to the number of road transportation mode.
b. Railway system is difficult to develop, because the monopolize of trucks business activities and their relation with stakeholders in supply chain of manufacturing industries.
c. Lack of Information Technology support and inspection system for human resources activity in regulation implementation.

Based on the model development, stake holders of supply chain in manufacturing industries should implement the following activities :

a. Planning for distribution infrastructure should be in long term base, to compete with the increasing of the transportation mode.
b. Government should develop railway system into the Port Tanjung Priok.
c. Government should build distribution center.
d. Implementation of Information Technology System for decreasing delivery time.

e. Develop open service system to public, and inspection for human resources activities.

7. References

[1] Aminullah E 2004 *Berpikir Sistemik: Untuk Pembuatan Kebijakan Publik, Bisnis dan Ekonomi.* (Jakarta: PPM).

[2] Cox W 2015 Is Jakarta The World's Most Congested City? *New Geography Report.*

[3] Coyle R G 1996 *System Dynamics Modelling: A Practical Approach.* (United Kingdom: Chapman & Hall).

[4] Forrester J W 1961 *Industrial Dynamics.* (New York: MIT Press & Wiley).

[5] Frazelle E H 2002 *Supply Chain Strategy.* (USA: McGraw-Hill).

[6] Shapiro J F 2000 *Modelling the Supply Chain.* Duxbury Press. Pacific Groove.

[7] Marimin 2004 *Teknik dan Aplikasi Pengambilan Keputusan Kriteria Majemuk.* Grasindo

[8] Muhammadi, Aminullah E, and Soesilo B 2001 *Analisis Sistem Dinamis : Lingkungan Hidup, Sosial, Ekonomi, Manajemen.* (Jakarta: UMJ Press).

[9] Scott C, Lundgreen H, Thompson P 2011 *Guide to Supply Chain Management.* (London: Springer Heidelberg Dordrecht)

[10] Sterman J D 2000 *Business Dynamics: Systems Thinking and Modelling for a Complex World.* (New York: McGraw-Hill Higher Education).

[11] World Bank 2014 Connected to Compete Trade Logistics in Global Economics. *The World Bank Report.*

[12] The Asia Foundation and LPEM-UI 2008 *Biaya Transportasi Barang Angkutan, Regulasi dan Pungutan Jalan di Indonesia.* (Jakarta: The Asia Foundation)

IOP Conf. Series: Materials Science and Engineering **277** (2017) 012066 doi:10.1088/1757-899X/277/1/012066

Business model in marketplace industry using business model canvas approach: An e-commerce case study

Yana Erlyana[1,2] and Henny Hartono[1,3]

[1]Bunda Mulia University, Jakarta, Indonesia
[2]yerlyana@bundamulia.ac.id, [3]hhartono@bundamulia.ac.id

Abstract. The advancement of technology has huge impact on commerce world, especially in the marketplace that has shifted from brick-and-mortar to digital/online marketplace. Grasping the opportunity, ABC joined venture with DEF to create a new online venture namely XYZ Online Shop – an e-commerce website that has large segmentations. The objective of this research is to analyze the business model conducted by XYZ Online Shop by utilizing Business Model Canvas Framework and SWOT analysis. The results obtained from the research are that the business model conducted by XYZ Online Shop excels in customer relationship block and still needs to improve key partner and key activities blocks. Business Model Canvas along with SWOT analysis describes how XYZ Online Shop creates, delivers, and captures value based on its internal and external environments.
Keywords: business model canvas, swot, e-commerce

1. Introduction

E-commerce is not a foreign word in the business world nowadays. E-commerce is a new system or paradigm in the business world, which shifts the paradigm of traditional commerce into electronic commerce by utilizing ICT (Information and Communication Technology), or in other words internet technology. The survey conducted by Mars Indonesia-a research company-revealed that 90.5 million Indonesians are connected to Internet. Therefore Indonesia has the highest e-commerce potential growth in the world.

In recent years, more and more business doers, both large companies and retailers, have shifted or expanded into digital business. The number of e-commerce players continues to grow, reinforced by the surveys conducted by domestic and foreign information technology research institutes.

Electronic commerce is a powerful concept and process that has fundamentally changed the current of human life. Electronic commerce is one of the main criteria of revolution of Information Technology and communication in the field of economy [1]. E-Commerce is emerging as a new way of helping business enterprises to compete in the market and thus contributing to economic success. Basically e-commerce is the buying and selling of goods and services on the Internet and on different online networks especially World Wide Web. The development of e-commerce has led to companies moving much of their business efforts to online environments [5]. E-commerce has been noted as a strong tool for sustainable organizational growth.

Aulia E. Marinto, the Chairman of the Indonesian E-Commerce Association (idEA), through a recent press release quoted from Economic Census 2016 data released by the Central Bureau of Statistics (BPS) that e-Commerce industry in the last ten years grew about seventeen percent with the total number of e-commerce business reached 26.2 million units [7].

Meanwhile, global research from Bloomberg stated that in 2020 more than half of Indonesia's population would be involved in e-Commerce activities. McKinsey also stated in a report titled

'Unlocking Indonesia's Digital Opportunity' that the switch to the digital realm would boost economic growth to $ 150 billion by 2025. The report also added that 73 percent of internet users in Indonesia access the internet via mobile devices [4]. This figure is expected to continue to grow in the next five years.

Aulia also supported this by stating "The massive use of smart phones is also one of the supporting factors for the growth of e-Commerce industry in Indonesia."

This phenomenon is what makes one of the big mobile service provider companies in Indonesia, PT ABC to join venture with foreign company PT DEY to form an open marketplace. In this research hereinafter referred to as XYZ online store. XYZ website is an e-commerce website that presents a virtual online shop to potential buyers as well as sellers in one place exactly like the concept in a shopping mall. The prospective sellers can advertise as many products as they can and they can expand their market share while the potential buyers can freely purchase any product of rich variety and enjoy the benefits such as discounts provided by the sellers. Therefore, XYZ Online shop has the concept of C2C and B2C marketplaces.

The development of XYZ Online shop continues to grow well from year to year. Based on the data recorded by XYZ Online shop until June 2017, XYZ Online shop already has more than 63,000 sellers and more than 8 million products that divided into 8 categories: gadgets/computers, electronics, fashion, beauty & health, children and infants, home/garden, hobby, and service/food. Additionally, in 2016, XYZ Online Shop was the only e-commerce company from Indonesia that participated in the Korean program Sale FESTA; Korea's largest shopping festival held in cooperation with the Ministry of Commerce, Industry and Energy, with the Ministry of Culture, Sports and Tourism of South Korea.
In order to run the business successfully in the middle of intense business competitions, XYZ Online shop determines to have the right business model in accordance with the goals and objectives set by the XYZ Online shop management. The business model will describe the rationale of how an organization creates, delivers and captures value [3][8][9].

In reference to the issues faced by XYZ Online shop, the objectives of this study are to identify the business model applied using Business Model Canvas approach and to formulate business development strategy alternatives from the study of SWOT.
SWOT method is utilized to analyze the internal and external factors affecting the company and results of the SWOT matrix are used to formulate an alternative strategy of business development [13][6][11]. While Business Model Canvas approach is used to obtain the right business model as a refinement and improvement of the current business model aiming to achieve the company's business development objectives [2] [3] [10] [14].

Business Model is a hypothesis about how a company makes money in the long run: what the company will sell, and to whom, how the company will collect revenue, what technology will be used, when the company will rely on its business partners and about the cost management [3].
There are nine basic building blocks on a business model that shows how to think about how a company makes money- the combined ninth block is called Business Model Canvas (BMC). The nine blocks cover four key areas of business: customers, supply, infrastructure, and financial viability. The division consists of 9 interlinked business blocks, namely: (1) Customer segment; (2) Value Proposition; (3) Channel; (4) Customer Relationship; (5) Revenue Stream; (6) Key Partners; (7) Key Activities; (8) Key Resource; and (9) Cost Structure [9].

2. Methods

The study was conducted at XYZ Online Shop Company located in Jakarta from May 2017 to June 2017. XYZ Online shop was selected intentionally (purposive) on the basis that the company is currently developing its business in e-commerce. The data utilized in this study were primary and secondary data. The primary data were the main sources of information required to reveal the research objectives that were obtained from observations, focused-group discussions, and direct interviews with both internal and external respondents. The information of secondary data was obtained from the company documents related to the issues along with the supporting data from the literature, Internet, scientific journals and results of previous studies.

This research is a descriptive case study. A case study aims to make a more detailed, in-depth and thorough study of a specific object, which is usually relatively small in a certain period of time [14]. A descriptive research method is used to obtain the overview of the information, explanations and conditions related to the object of study in a factual, accurate and systematic sense.

This study utilized a Business Model Canvas approach. SWOT analysis and pairwise comparison were also utilized on the nine elements of Business Model Canvas to formulate and determine strategy priorities on each business development of XYZ Online shop [12].

Some interviews were also conducted to collect the data. The respondents in this study were classified into internal and external respondents. Internal respondents were selected based on the respondents' knowledge on the condition XYZ Online shop specifically and their full authority in determining policies in XYZ Online shop.

To identify the XYZ Online shop business model that currently applied, the business model was identified through the nine elements of the Business Model Canvas i.e. customer segments, value proposition, channels, customer relationship, revenue streams, key resources, key activities, key partners, and cost structure. Therefore, the next stage is to perform a SWOT analysis on the nine elements of the Business Model Canvas, in which they were a mixture of XYZ Online shop company's internal and external conditions. The research framework can be seen in figure 1.

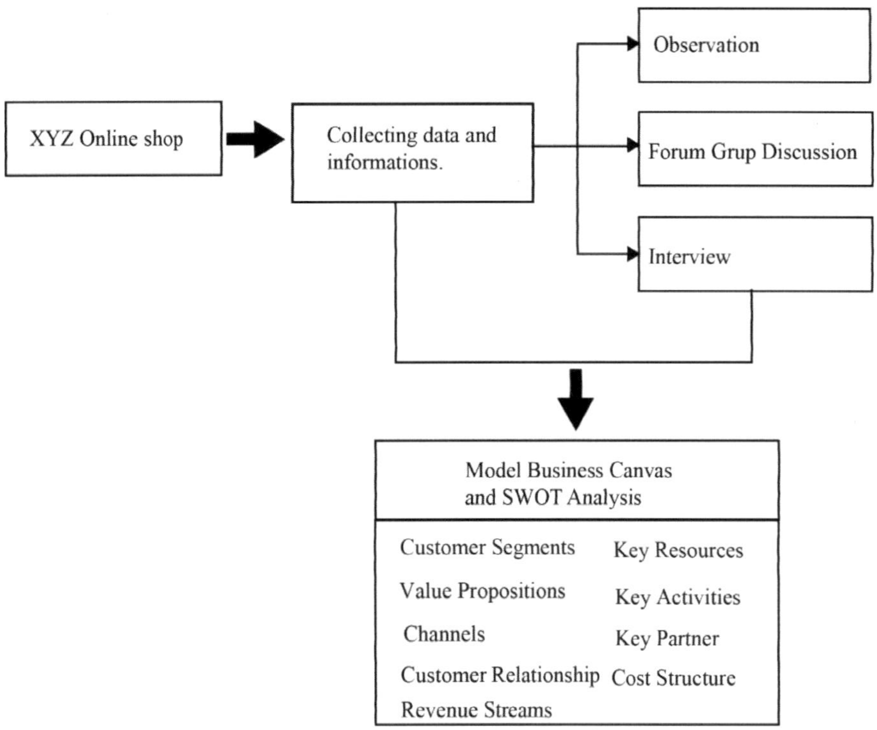

Figure 1. Research framework

3. Results And Discussion

3.1 Business Model and Business Type of E-Commerce XYZ Online Shop

In June 2017, XYZ Online shop has successfully added sellers to more than 63,000 sellers and a live listing of 8 million products. Within a year, XYZ Online shop also managed to reach 2 Million users and send over 400,000 thousand products. In addition, they are also able to achieve revenue of 360 billion rupiah in 2016.

IOP Conf. Series: Materials Science and Engineering **277** (2017) 012066 doi:10.1088/1757-899X/277/1/012066

XYZ Online shop is one of the sites that use marketplace business model. Marketplace is a business model where the website not only helps to promote merchandise, but also facilitate online transactions.

3.2 Identification on Elements of XYZ Online Shop Business Model Canvas

Result of identification on the nine elements of the XYZ Online shop Business Model Canvas were obtained by conducting observations and Focus-Group Discussion on the internal respondents. The nine elements are interrelated and provide an overview of the state of XYZ Online shop.

1. Value Proposition: XYZ Online shop offers value propositions as a facilitator site that offers different concepts from other sites. The other sites only facilitate advertising space and then submit a deal agreement on the parties' sellers and buyers, in contrast to the concept offered by XYZ. XYZ Online shop facilitates the transaction processes between sellers and buyers. Sellers and buyers do not need to deal directly or conduct a COD (Cash on Delivery), but the security and product purchase warranty will be facilitated by XYZ. XYZ Online shop site takes the open marketplace platform, bringing together sellers and buyers to transact online.
2. Customer Segments: The customers segment served by XYZ Online shop are people who live in big city (urban community) with an active lifestyle. With age ranging from young age category (infants, toddlers and children), adolescents to adults with age group of 17 - 45 years with an estimated monthly income of 500.000 IDR to 10,000,000 IDR. Products categories are available for each age group. The customers profiles of XYZ Online Shop are from middle to upper class segments.
3. Channel: Since XYZ Online shop is a business with a major electronic business base, the XYZ Online shop only has an online channel using: website as the media, mobile websites and mobile/smartphone apps.
4. Customer Relationship: There are two main activities conducted by XYZ Online shop to establish and build relationships with its customers, namely activities to establish a relationship with seller and activities to establish a relationship with buyer.
   In establishing a relationship with seller, the activities undertaken are as follows: XYZ presents Seller Zone service, a physical facility service dedicated to the sellers who want to learn about e-commerce; procurement of workshops to assist sellers in optimizing the use of XYZ website on a regular basis; and award for seller with a good performance within the period of time specified by XYZ.
   In establishing a relationship with buyer, the activities undertaken are as follows: XYZ regularly organize events/promotional activities for consumers or potential customers; provide customer service for 24 hours; and provide warranty in the payment system.
5. Key Partners: In conducting business activities, XYZ currently has several partners that include sellers as suppliers. There are two types of sellers in XYZ Online shop: individuals and businesses. This site has also opened registration for sellers located outside of Indonesia. The seller at XYZ Online shop must scan and upload some official documents such as ID cards and passbooks to register. Another key partners are the payment gateway (FAST PAY), freight forwarders (JNE, TIKI, RPX, FL and GOJEK), and media partners as services provision of advertisements, and buzzers, as well as public figures as ambassadors.
6. Key Activities: Key activities or activities undertaken by XYZ Online shop include several main points, first is the development, fixing and maintenance of the website by the platform department in XYZ company. Second is the marketing activity of the site itself and the planning of promotional strategies undertaken by the marketing department. Third is the activity of maintaining the relationship directly between the XYZ as a provider of facility service for sellers and buyers made by the customer service.
7. Key Resources: The key resources of XYZ Online Shop to run the business model or everyday business activities include the capabilities and expertise possessed by the partners, capabilities and

expertise of XYZ Online shop employees, the number of human resources and physical assets in the form of buildings and office equipment.

8. Cost Structure: Costs incurred by XYZ Online Shop in creating the value proposition on the product through several series of key activities include employee salary, marketing costs, operational costs, and building rental fees. Among the costs incurred by XYZ Online shop every month, the cost of employee salaries takes the highest percentage up to 40%, followed by 30% for marketing cost, 20% for operational cost, and 10% for building rental fee.
9. Revenue Streams: XYZ Online shop's revenue stream comes from a seller sales commission.

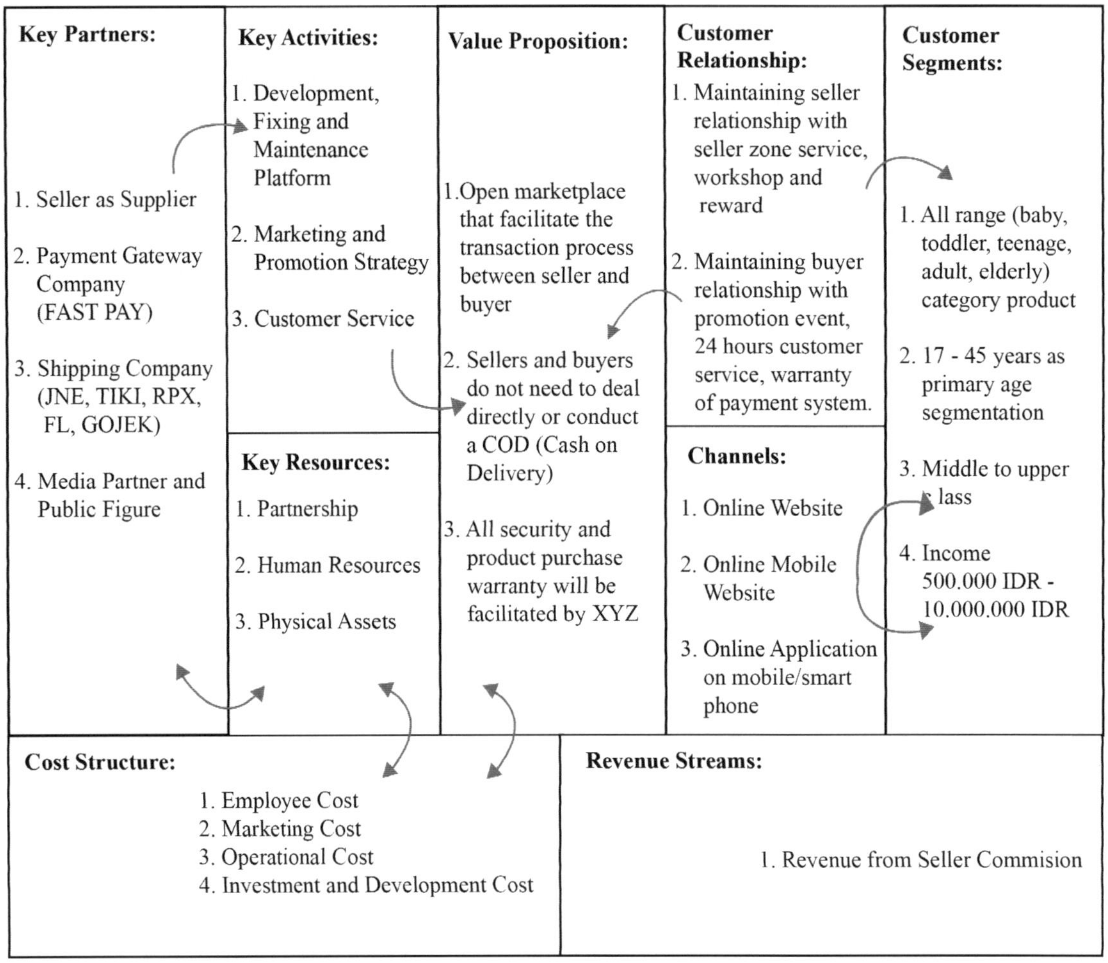

Figure 2. Business Model of XYZ Online shop

3.3 SWOT Analysis

The following stage after identifying XYZ Online shop business model was to conduct a SWOT analysis on the nine elements of Business Model Canvas obtained from the observations, focus-group discussions with the internal respondents, and interviews with the external respondents. SWOT analysis on the nine elements of XYZ Online shop Business Model Canvas can be seen in Figure 3.

Element	Strength (S)	Weakness (W)	Opportunity (O)	Threat (T)
Customer Segments	Product variations Customer network	Buyers often complain about shipping costs	No scale of segmentation	Competitor with same segmentation
Value Propositions	Topten market leader No COD (Cash On Delivary) Security warranty Payment process warranty	Some menu on site is confusing	Platform innovation New technology	Inovation from another marketplace site New government rules
Channels	Three different media (website, mobile web, smartphone apps)	Online service only	Worldwide	Hackers Virus
Customer Relationship	Seller Zone Service Workshop for seller Award for seller Promotion event for buyer 24 hr customer service payment waranty	Customer Relationship budgets	New business opportunity and access to stakeholders	Unhealthy competition Unsynchronized government policies
Revenue Streams	Marketing works Strong shareholders	Limited working capital Inaccurate administration	Large capital foreign investors with large capital	Payment system is not in accordance with the schedule and commitment
Key Resources	Have lot of seller as supplier Partnership with third party for payment system Experienced human resources	Limitation of MOU	Foreign Supplier MEA Projects in Indonesia	High interest of loan Fluctuation of foreign currency values
Key Activities	Stable platform Three main department business (IT, Marketing, Customer service)	Limitation of employees with appropriate expertise or knowledge required	Supply of manpower from the high education institutions	Low responses from new supplier
Key Partner	Strategic partner network (suppliers, payment gateway, shipping company, media partner, public figure)	Selection and assessment on suppliers are not maximal	New seller in local home indutry	Consultants and customers have direct access to the suppliers Decrease customer trust level
Cost Structure	Funding sources from banks	Payment policy to ABC and DEF group	Funding from international finance agency	Increase of regional minimum wages

Figure 3 SWOT Analysis On The Nine Elements Of Business Model Canvas Of XYZ Online Shop

4. Conclusion

The research resulted a clear and detailed description of a picture of the business model used by XYZ Online shop, mapped in nine blocks of business model canvas. The proposition value offered by XYZ Online shop is a marketplace that becomes facilitator between seller and buyer that comes with a guaranteed payment system. The next block discusses the customer segments that are divided by products categories and it can be concluded that the customers' segmentation of XYZ Online shop based on age and income are 17-45 years old with middle to upper class segmentation. On channel block XYZ Online shop only has one type of channel that is online. However XYZ Online shop offers several media that are user friendly ie website, mobile web and smart phone apps. XYZ Online shop has several ways in terms of building a good relationship with its customers that includes sellers and buyers such as activities to provide seller zone service and workshop that enrich the knowledge of seller to e-commerce as well as to hold many promotional programs to increase interest in buying products for buyers.

Key activities conducted by XYZ Online shop include three main steps: first is the development, fixing and maintenance of the website by the platform department in XYZ company; second is the marketing activity of the site itself and the planning of promotional strategies undertaken by the marketing department, and the last is the activity of maintaining the relationship between XYZ as a provider of facility service for sellers and buyers made by the customer service. XYZ Online shop also cooperates with various parties to improve its services such as cooperation with payment Gateway Company and shipping company. Every partner also becomes one of XYZ Online shop's main resource. The other major resources of XYZ Online shop also include the amount of human resources they own and also the company's assets, such as buildings, and office equipment. The addition of employees, suppliers, and marketing management improvements can increase XYZ Online shop revenue.

Based on the results of the conclusions described above, researchers can provide some suggestions that would be useful for the XYZ Online shop as follows:

1. Increase the number of consumers on the segmentation outside the main segmentation by adding local or non-local seller.
2. Continue to innovate in its platform both in the system information and interface design to improve the user experience.
3. Doing more innovative promotional events from competitors to avoid consumers or prospective customers moving to other marketplaces.
4. Increase the number of experienced human resources in the marketing department to improve targeted promotional strategies.
5. The expansion of cooperation with some parties, especially the shipping parties to yield more beneficial services for both parties.

5. References

[1] Niranjanamurthy M, Kavyashree N, Jagannath S and Chahar D 2013 Analysis of e-commerce and m-commerce: advantages, limitations and security issues *Int. J. of Advanced Research in Computer and Communication Engineering* **2(6)** pp 2360-2370

[2] Bonazzi F L Z and Zilber M A 2014 Innovation and business model: a case study about integration of innovation funnel and business model canvas *Revista Brasileira de Gestao de Negocios* **16(53)** pp 616-637

[3] Eisenmann T 2001 *Internet Business Models: Text and Cases* (New York: McGraw-Hill/Irwin)

[4] Das K, Gryseels M, Sudhir P and Tan K T 2016 *Unlocking Indonesia's digital opportunity* (Jakarta: McKinsey Indonesia Office)

[5] Ahmed A A, Dalbir S and Ibrahim M 2011 Potential e-commerce adoption strategies for libyan organization *Int. J. of Information and Communication Technology Research* **1(7)** pp 321-328

[6] Awais M and Samin T 2012 Advanced swot analysis of e-commerce *Int. J. of Computer Science* **9(2)** pp 569-574

[7] Abdurrahman M S 2017 Pertumbuhan e-commerce indonesia tertinggi di dunia *Liputan6* Retrieved from https://kepulauanseribukab.bps.go.id/backend/fileMenu/Kliping-Berita-Online-Edisi-17-21-Mei-2017.pdf

[8] Odeniyi, Ayodeji O, Lawal, Tunde N, Kareem, and Adebowale A E 2015 An appraisal of groupon e-business model *Int. J. of Scientific & Technology Research* **4(1)** pp 291-297

[9] Osterwalder A and Pigneur Y 2012 *Business model generation* (Jakarta: Elex Media Komputindo)

[10] Permadi B, Nurmalina R and Kirbrandoko 2015 Analisis pengembangan model bisnis kanvas cv kandura keramik bandung *J. Aplikasi Manajemen* **14(1)** pp 88–97
[11] Rochman N, Sa'id E G, Daryanto A and Nuryantono N 2011 Analysis of indonesian agroindustry competitivness in nanotechnology development perspective using SWOT-AHP Method *Int. J. of Business and Management* **6(8)** pp 235–244
[12] Zikmund W G, Babin B J, Carr J C and Griffin M 2013 *Business research methods 9th edition* (USA: Cengage Learning)
[13] Kahraman C, Nihan C D and Taufan D 2007 Prioritization of e-government strategies using swot-ahp analysis: the case of turkey *European J. of Information System* **16(8)** pp 284–298
[14] Mazhaly N, Umar D D, and Moengin P 2014 Penerapan blue ocean strategy di pt. x dalam menghadapi persaingan penjualan automatic tank gauging di indonesia *J. Teknik Industri* **3(2)** pp 241– 254

IOP Conf. Series: Materials Science and Engineering 277 (2017) 012067 doi:10.1088/1757-899X/277/1/012067

Discrete-event system simulation on small and medium enterprises productivity improvement

J Sulistio[1,2] and N A Hidayah[1]

[1]Departement of Industrial Engineering, Universitas Islam Indonesia, Jl. Kaliurang, DI Yogyakarta, Indonesia
[2]jokosulistio@uii.ac.id

Abstract. Small and medium industries in Indonesia is currently developing. The problem faced by SMEs is the difficulty of meeting growing demand coming into the company. Therefore, SME need an analysis and evaluation on its production process in order to meet all orders. The purpose of this research is to increase the productivity of SMEs production floor by applying discrete-event system simulation. This method preferred because it can solve complex problems die to the dynamic and stochastic nature of the system. To increase the credibility of the simulation, model validated by cooperating the average of two trials, two trials of variance and chi square test. Afterwards, Benferroni method applied to development several alternatives. The article concludes that, the productivity of SMEs production floor increased up to 50% by adding the capacity of dyeing and drying machines.
Keywords: productivity improvement, SME, discrete-event system simulation

1. Introduction

The economic and trade openness resulted created challenges and opportunities among countries. Open trade between one country and another extend market access of domestic products in the international market as well as challenges to the competitiveness of the domestic industry not to mention small and medium enterprises (SME). SME is define as a company that has maximum asset of 10 billion IDR, not more than 50 billion IDR revenue, and employed less than 100 workers [1]. Based on data from the ministry of cooperatives and small and medium enterprises in mind that over the years the number of small and medium enterprises always increases, in the year 2013 there were 57,895,721 units of small and medium enterprises in Indonesia [2]. In order to compete, SME should improve the quality and service available. These are the obstacles often faced by small and medium enterprises, (a) poor system used, and (b) the lack of service. Other issues faced by SME include trade liberalization, such as the ASEAN-China Free Trade Area (ACFTA). On the other hand, Indonesian Government has agreed on an ACFTA cooperation agreement or other agreement, without considering the readiness of SMEs to compete [1]. Successful SME is the one who have the ability to utilize its available resource, and leverage the low economic value (including waste) to higher economic value commodities [3].

This article analyzes the production system improvements in on one of the SMEs engaged in craft bag in Yogyakarta as a case study. SME marketing program's success in taking chances causes increased production target to 1,200 units of bags for 20 working days. With the existing conditions, the target is difficult to achieve by the company. As an attempt of meeting the target, the company practice overtime. By doing so, the company reach the target. Nonetheless, it had to incur additional costs, which reduce its profit.

2. Methodology

Efforts to increase productivity on the production floor already investigate by some previous researchers. Productivity achieved by using a goal programming production planning in order to meet market demand. Goal programming method production companies can minimize costs while maximizing the available resources [4]. Another attempt to increase the productivity of SME's production floor exhibit by applying nonlinear optimization on its supply chain planning. However, the nonlinear optimization method has several disadvantage. One of them is it requires quite a lot of scenarios that result in the longer computing time [5]. In addition, productivity improved by increasing the quantity and quality of the product. Interviews and direct observations produced some recommendations, namely: (1) the supply of electricity fabric cutting machines (2) procurement cutting tablemat, and (3) procurement walking foot machine. The recommendations implemented by companies with good results [6]. However, this method has a risk of failure if applied directly in the real system. Other efforts made through the analysis of customer fulfillment management using system dynamics modeling. From this analysis, we found the factors that influence the production process of SMEs. Thus, a series of improvements made to increase productivity [7]. Another method to improve SME's productivity is discrete-event system simulation. Simulation is capable to solve the problems occur in companies productivity through schedule planning. With discrete systems, modelers can specify standard time subsequent impact on the productivity of the production floor 6. The previous studies have produced findings that are beneficial for the company. Nevertheless, discrete-event system simulations carried out previously intended to test the results of a specific method of analysis. This article applies discrete-event system simulation with the aim of seeking alternative and determine the best alternative in the face of complex system problems.

Discrete-event system simulation is widely used in manufacturing industry, business, logistics and supply chain management. Simulation applied because the problem cannot solved by the complex dynamic equations of analytical methods [8]. It is powerful to solve problems in a complex system, which incorporated numbers of interrelated variables [9–12]. It is suitable for capturing the potential improvement regarding wide area of logistics by cooperating the analysis of complex and stochastic process [10,13]. Modeler can use simulation to review alternatives in process design, scheduling, resource allocation, and management change [14]. Other use of simulation is to assess the effects on productivity of a supply chain [15].

The execution steps of discrete-event system simulation are vary. A study in input modeling consider it involve eight level of reiterative steps. They are identify problem, state objectives, collect/ prepare data, formulate model, modify/ refine model, verify/ validate model, experiment/ interpret results, and conclude/ implement [16]. The framework to conduct discrete-event system simulations developed to serve its unique purpose. Nonetheless, they share major similarities [9,17,18]. This article runs a resembling framework as shown in Figure 1.

In order to build a robust and credible model, they should verified. Verification is the step to make sure that the program is actually works. In addition, model should validated. It means to make sure that the model is a representation of the real system. Thus, modeler could run one or more experiment toward it [9]. Validation also means to compare both model and actual system, and then improve the model credibility based on its gap. This activity is iterative until the modeler consider it accurate enough [19]. This paper used mean test and Chi-square goodness of fit test for validation. Once a robust model developed, alternatives designed by applying what-if analysis. Afterwards, alternatives analyzed in order to determine which one is the best to implement. The methods to perform such task is Bonferroni test. It is useful in model alternative design runs with both independent sampling or with common random numbers. The Bonferroni approach recommended only when the research have a small number of comparisons. The practical upper limit is 20 comparison [17].

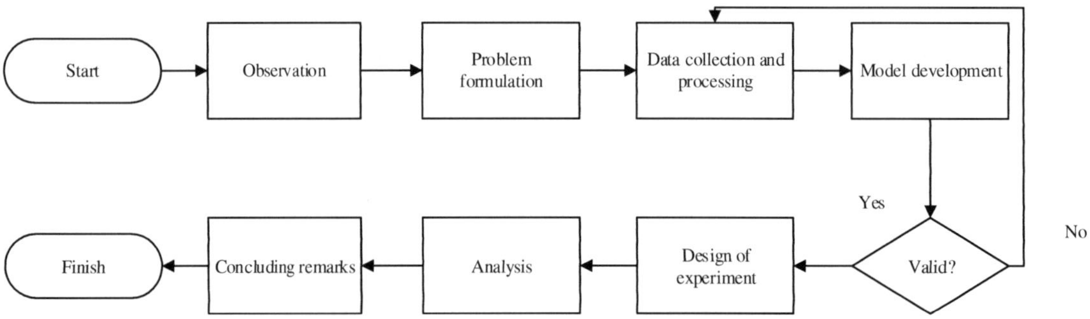

Figure 1. Discrete-event system simulation steps

3. Discussion

Flexsim is cooperated to build the model and after the necessary runs, it provide a rigorous report. An evaluation made accordingly. The data collected from 30 workday's observation. In order to build the model the following data are required. Total output, which is the overall units produced in each day. In this case, the outputs average is 44.83 units per day as shown in Table 1. Processing time, it is the amount of time needed to process a single work in every process. There are five processing times recorded, they are layer sewing, zipper sewing, accessories sewing, handle sewing, and cutting. Location capacity include warehouse capacity1,000 unit, cutting capacity 25 unit/roll, coloring capacity 15 unit/batch, machine capacity 1 unit. Arrival events, the arrival in this research are fabrics, work in process, zipper, handle, and accessories. Those entities arrive once in every day. The fabric arrival are 20 units per day. Work in process, zipper, and accessories are each arrive in 100 unit per day. As for handle, it arrive in 150 unit per day. Resources, the resource in this model are two operators assigned to inspect finished goods prior to store in warehouse.

Every input analysis required to establish probability distribution. This research use Expert Fit to complete the task. The model constructed, consist of three source, nine queue, three processor, eight combiner, one separator, and two operator. Modeling in Flexsim start with defining the source. There is only one arrival in fabric arrival and work in process. Therefore, arrival sequence was chose in arrival style. Nine queues made to hold entities from source. This SME has warehouse to storage fabric, work in process, accessories, and finish goods. All storage that has queue can hold up to 1,000 units. Three machines in the production line modeled using processor, one coloring and two sewing machine with the processing time johnsonbounded probability distribution. Eight machines modeled using combiner because it involve join activity. The processing time in these machines, follow inversegaussian probability distribution. The cutting process is executed bay one machine. It is modeled using separator because from one roll of fabrics can be divided in twenty-five units of materials. The process time follow johnsonbounded probability distribution. The workhour in this company is 7.5 hours. To model the workhour, timetable is used.

Table 1. Real system and simulation output in units

	Real system	Simulation result
Mean	44.83	43.5
SD (v)	3.69	1.04
n	30	30

Series of output validation test performed to ensure that the model represents the real system. Based on those test, the simulation model constructed has the same behavior of the real system. The tests are:

- Mean test. This research use level of significance 0.05. Since $-Z_{0.025} < Z < Z_{0.025}$ which is $-1.96<1.90< 1.96$, therefore Ho is accepted. It means that statistically the real system's mean is not different with the simulation's mean.
- Chi-square goodness of fit test. This test determines if a sample data matches a population. The results shows that the calculated chi-square value (5.76) is less than the critical value from a chi-square table (42.55). Thus, the simulation distribution fits the real system's distribution.

Analysis and evaluation in article this is based on the summary report and the state of model the simulation for one working day (27,000 seconds). The simulation report shows high value of idle at the body sewing machine number one (59%) and two (50%), as well as in mounting accessories station number one (60%) and two (49%). Both machines placed after the handle sewing machine because it waits the entities from staining and drying machines. The condition worse by the 83% block on the mantle sewing machine number one and two. It means, the process could not continue to staining machine. In addition, the processing percentage on staining machine is already quite high (85.71%). Therefore, the problems occur on floor shop, is inefficiency due to long-range of process time in coloring and drying machines.

Developing alternatives are essential to address the problem. This article proposed two alternatives. Figure 2 and figure 3 shows the comparison of those alternatives.

- First, increase the capacity of dyeing and drying machines. This scenario proposed because dyeing and drying machines are critical for the overall manufacturing times due to their long process time. By doing so, sewing machine idle dropped to 35% - 55% and 31% - 49%. In addition, the total output generated is increasing to 49-50 units per day.
- Increase the capacity on the coloring and drying machine whilst reducing the number of sewing machine. This scenario proposed as a reaction to the firs scenario. Even though decreased, the idle still considered high. This scenario managed to reduce idle time and increase the output value of the company up to 50% of the initial conditions.

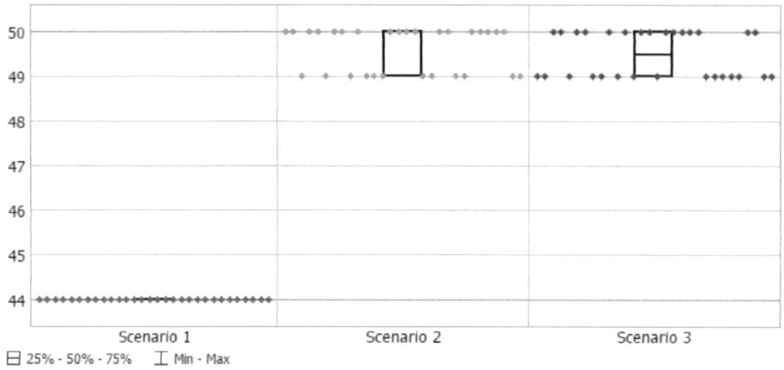

Figure 2. Alternatives output

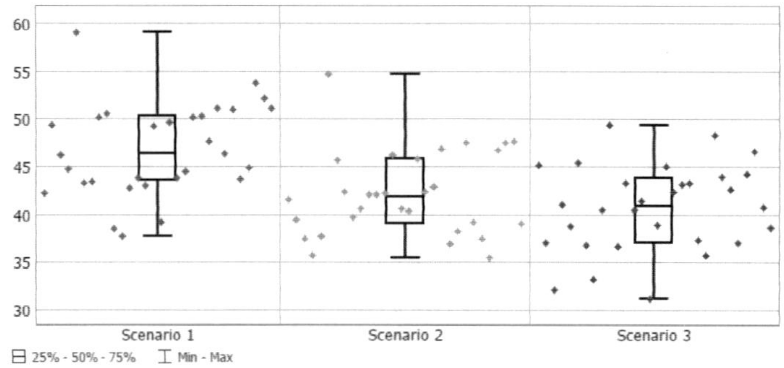

Figure 3. Idle in sewing machines

In order to determine which alternative is the most suitable, Bonferroni test is perform. Scenario 1 is the existing model, scenario 2 is the first alternative, and scenario 3 is the second alternative. Bonferroni test shows that scenario 1 is inferior to scenario 2 nor scenario 3. Nonetheless, scenario 2 and 3 are not significantly different. The most suitable alternative is scenario 2. The reason is, adding capacity means buying new equipment.

4. Conclusion

Based on the results, discrete-system system simulation method is suitable to increase SME's productivity. Therefor it satisfy SME's consumer demand. The chosen alternative is adding capacity on staining and drying machines. Therefore, the staining machine can work more quickly. Parallel to that, SME can reduce the number of sewing machine due to its high idle time. It can lead to high maintenance costs, which will be a burden to the company if the output generated from the engine is quite low.

5. References

[1] Sudaryanto R and Wijayanti R R 2013 Strategi pemberdayaan UMKM menghadapi pasar bebas Asean *Pus. Kebijak. Ekon. Makro. Badan Kebijak. Fiskal. Kementeri. Keuangan, Jakarta*

[2] Depkop 2012 Perkembangan Data Usaha Mikro , Kecil , Menengah (UMKM) Dan Usaha Besar (UB) Perkembangan Data Usaha Mikro , Kecil , Menengah (UMKM) Dan Usaha Besar (UB) *www.depkop.go.id* 2011–2

[3] Budiretnowati R G 2015 Kajian Tentang Profil UKM Sukses *J. Pengkaj. KUKM* **3**

[4] Nafisah L, Sutrisno S and Hutagaol Y E H 2016 Perencanaan Produksi Menggunakan Goal Programming (Studi Kasus di Bakpia Pathuk 75 Yogyakarta) *SPEKTRUM Ind.* **14**
[5] Sitompul C and Hariandja J 2012 Analisis Metode Optimasi Nonlinear untuk Perencanaan Rantai Pasok yang Berbentuk Umum *Res. Report-Engineering Sci.* **1**
[6] Andayani S, Murti I and Sajiyo S 2016 Peningkatan Kuantitas Dan Kualitas Produk Perajin Handicraft Perca dan Bordir KSM "Kesra" Kelurahan Kedurus, Kota Surabaya *JPM17* **1**
[7] Rahmanto N, Hartono B and Tontowi A E 2016 Analisis manajemen pemenuhan pelanggan pada industri kecil menengah pembuatan tas dan bordir menggunakan pemodelan system dynamics (Studi kasus pada CV. Kurnia Jaya--Yogyakarta) *J. Manaj. dan Organ.* **2** 104–15
[8] Rabelo L, Sarmiento A T and Jones A 2011 Stability of the supply chain using system dynamics simulation and the accumulated deviations from equilibrium *Model. Simul. Eng.* **2011** 7
[9] Sowlati T and Opacic L 2016 Applications of Discrete-Event Simulation in the Forest Products Sector: a Review *For. Prod. J.*
[10] Stoldt J, Schlegel A and Putz M 2016 Enhanced integration of energy-related considerations in discrete event simulation for manufacturing applications *J. Simul.* **10** 113–22
[11] Gosavi A, Cudney E A, Murray S L and Masek C M 2016 Analysis of Clinic Layouts and Patient-Centered Procedural Innovations Using Discrete-Event Simulation *Eng. Manag. J.* **28** 134–44
[12] Mauricio T B, Montevechi J A B, Leal F, de Carvalho Miranda R and Lombardi F 2015 Using discrete event simulation to change from a functional layout to a cellular layout in an auto parts industry *Acta Sci. Technol.* **37** 371
[13] Suh E S 2015 Cross-docking assessment and optimization using multi-agent co-simulation: a case study *Flex. Serv. Manuf. J.* **27** 115–33
[14] Brailsford S, Churilov L and Dangerfield B 2014 *Discrete-event simulation and system dynamics for management decision making* (John Wiley & Sons)
[15] Windisch J, Väätäinen K, Anttila P, Nivala M, Laitila J, Asikainen A and Sikanen L 2015 Discrete-event simulation of an information-based raw material allocation process for increasing the efficiency of an energy wood supply chain *Appl. Energy* **149** 315–25
[16] Minadeo-Cook M and others 2007 Filling the Educational Void in Input Modeling: Non-Parametric Tests for Independence *IIE Annual Conference. Proceedings* p 926
[17] Banks J, Carson J S, Nelson B L and Nicol D M 2013 *Discrete-Event System Simulation: Pearson New International Edition* (Pearson Education Limited)
[18] Law A M 2013 *Simulation Modeling and Analysis* (McGraw Hill Education India Pvt Ltd)
[19] Sharma P 2015 Discrete-event simulation *Int. J. Sci. Technol. Res.* **4** 136–40

IOP Conf. Series: Materials Science and Engineering **277** (2017) 012068 doi:10.1088/1757-899X/277/1/012068

Control excess stock and calculating damaged products as the effort to increase revenue (case study of SME FBS)

N Nurhasanah[1], D A Mardhika[2], W Tanjung[3], A M Gayatri[4], Q A Suri[5], Jingga[6], R Safitri[7] and A Supriyanto[8]

[1,2,3,4,5,6]Industrial Engineering Department, Faculty of Science and Technology, Universitas Al Azhar Indonesia, Jakarta 12110, Indonesia
[7,8]Informatic Engineering Department, Faculty of Science and Technology, Universitas Al Azhar Indonesia, Jakarta 12110, Indonesia
[1]nunungnurhasanah@uai.ac.id

Abstract. Of small and medium scale (SME) is a business engaged in production. The growth product innovation of each year to year made competitiveness every SME very tight, and the sales must be high that avoid goods the product last year will be tough sold in the following year. Forecasting demand is needed so that no its production. In production process, besides products should also be considered about damaged products, resulting in a loss. In this study, researchers conducted a observations on SME FBS producing pants, shirts and shirts. SME FBS not having planning previous production, also in any period of production there always products be damaged. This study attempts to increase their SME FBS by controlling waste products, and those damaged products. According to the research conducted other products in some excess pants 1609 unit , and the shirts 187911 unit, and increase the income through control over the excess product obtained by 1% to the pants, and 52% to the shirts. For damaged product on period last year and future, increase 0.07% if the damaged on shirts can be sold, and 0.29% on pants if the broken sold.
Keywords: Fuzzy Averaging For Forecasting, Excess Stock, Interpolation, Reject Product

1. Introduction

Small and medium-sized enterprise (SME) is one of the driving force of the Indonesian economy today. This has proven to be able to survive during economic crisis and become the dynamist of the economic growth after the crisis. Home based businesses as part of SME has become the backbone of the Indonesian economy. There are many business advantages in the house; besides it does not cost for renting a place, this business could grow into not only a merely side business, and the transportation costs can also be minimized [6]. SMEs are required to make innovations in their various products to survive in the market.

Forecasting is an attempt to predict any future circumstances through testing the past condition. Forecasting sales means to determine the approximate amount of sales volume, even to determine the potential sales and market area that is controlled in the future. The usefulness of forecasting is to make the right decisions based on the past events [4].

A company should be able to predict the size of customer demand for its products. This kind of forecasting is also called sales forecasting. Demand forecasting is an attempt to know whether the number or group of the future products is in a certain constraint or condition and to reduce the risks or uncertainty faced. It also deals with production control activities, capacity, and scheduling systems and becomes input for financial, marketing, and human resources planning. The maximum result of a forecasting activity is to minimize the uncertainty that may occur in the future. To be able to reach the optimal decision, it requires ways/methods of accurate, systematic and accountable forecasting.

This study will calculate to increase the SME FBS profits and sales. The SME FBS does not have production planning; it relies on market demand that refers to market conditions at the time. Since there is no systematic production planning, there are many unsold products because the production is not in accordance with demand. The large number of unsold products drives us to know the production planning in the next year to minimize the product that accumulates in the warehouse.

In this research, the production planning will be conducted using forecasting demand method. The result is to anticipate when the production will experience a surge in demand, and the warehouse will bear for excessive goods. Through this way, SME FBS can be more anticipate and maximize its sales, so that the product can be sold entirely. As for the unsold products, the researcher uses excess stock method to find out the number of the residual products and the minimum price of the outdated products, and to determine the price of the damaged products, so that the products can come out quickly and can increase the profit.

2. Methods

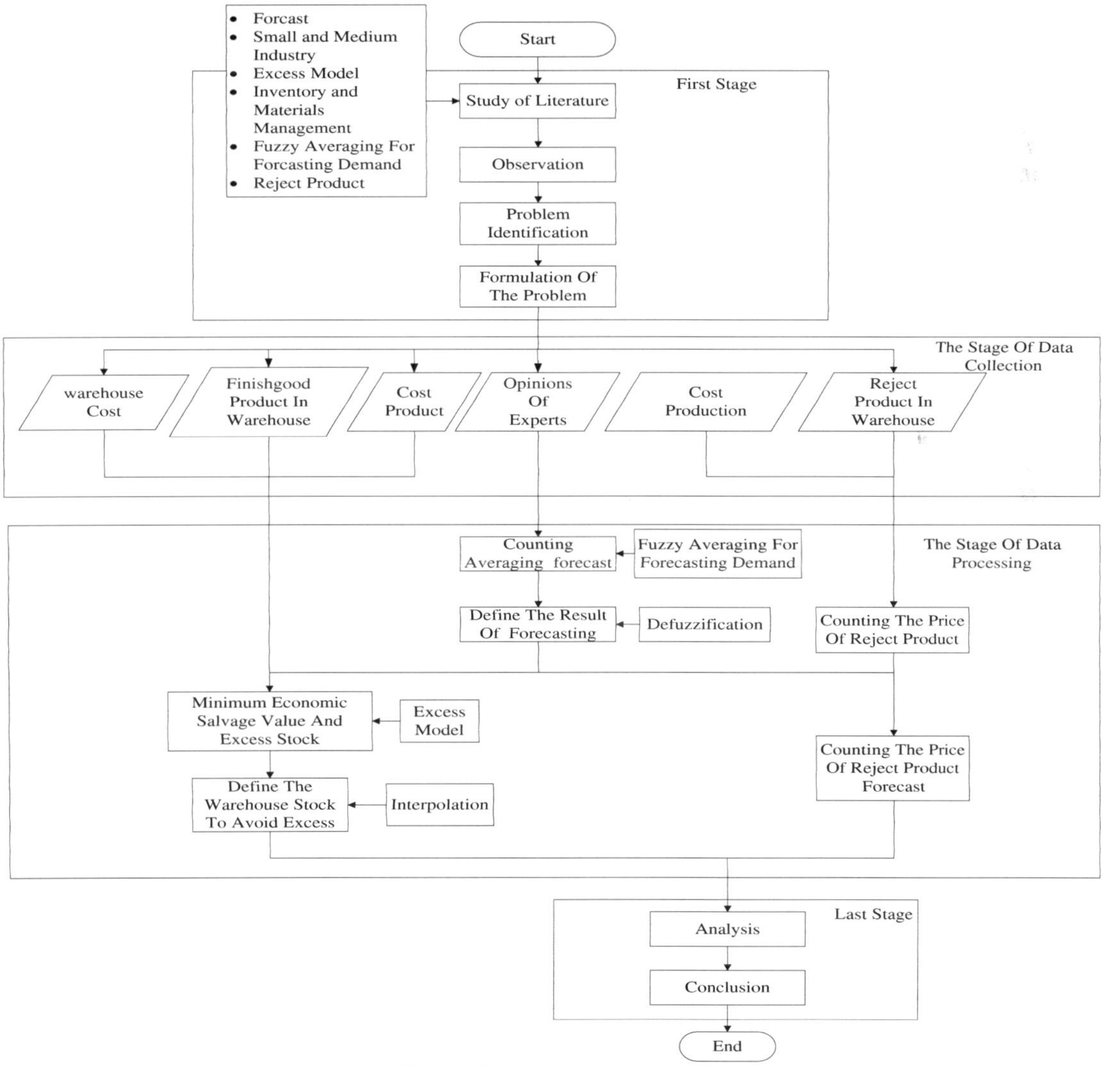

Figure 1. Flowchart Research

This study consists of 4 stages. The first stage is to conduct a review of the previous research and study of scientific articles related to this research topic. The second stage is to observe and collect data related to expert opinion, inventory cost, production quantity, number of products stored in warehouse, number of reject goods stored in warehouse, and production cost.

The third stage is the data processing, which consists of the calculation of product demand planning for 2018. The method used for demand planning is fuzzy averaging for forecasting method [1]. This method is processed by the expert opinion, calculating the excess model to specify the excess stock, and fixing the reject product price. The last stage is to do the analysis based on the calculations that have been done, then validate them with the company. Figure 1 presents the research flow diagram.

3. Result and discussion

3.1 Forecasting Demand

The way to forecast with the fuzzy averaging for forecasting demand method from interview with SME FBS as an expert (Ei) is as follows:

- Calculate A_{ave}=(m_1 , m_M, m_2)

$$\mathbf{A}_{ave} = (m_1, m_M, m_2) = \left(\frac{1}{n}\sum_{i=1}^{n} a_1^{(i)}, \frac{1}{n}\sum_{i=1}^{n} a_1^{(i)}, \frac{1}{n}\sum_{i=1}^{n} a_2^{(i)}\right). \quad (1)$$

  Explanation:
  m_1 : the smallest production value
  m_M : medium production value
  m_2 : the largest production value

- Specifying Defuzzification
  Explanation:
  Maximum value of forecast A_{ave}: m_M

$$Xmax = m_M \quad (2)$$

Here are the results of calculations from the forecast demand for pants, shirts and T-shirts presented in tables 1, 2, and table 3:

Table 1. Calculation Result of Demand Forecasting for Product Type of Pants

		Pants		
Ei	Ai	smallest production a1(i)	medium production aM(i)	largest production a2(i)
E1	A1	3300	3600	3750
E2	A2	3000	4000	5000
E3	A3	3100	3500	4200
Total		9400	11100	12950
Average		3133	3700	4317

From the table above, we get the result A_{ave}= (3133, 3700, 4317) based on the calculation using formula (1). By using the second (2) formula, the forecast value for pants is 3700 dozen.

Table 2. Calculation Result of Demand Forecasting for Product Type of Shirts

		Shirts		
Ei	Ai	smallest production a1(i)	medium production aM(i)	largest production a2(i)
E1	A1	16500	18000	18750
E2	A2	16500	17000	18750
E3	A3	16600	17600	18500
Total		49600	52600	56000
Average		16533	17533	18667

From the table above, we get the result A_{ave}= (16533, 17533, 18667) based on the calculation using formula (1). By using the second (2) formula, the forecast value for shirt is 17533 dozen.

Table 3. Calculation Result of Demand Forecasting for Product Type of T-Shirt

		T-shirts		
Ei	Ai	smallest production a1(i)	medium production aM(i)	largest production a2(i)
E1	A1	27500	30000	31250
E2	A2	27500	29000	31250
E3	A3	28000	30000	31000
Total		83000	89000	93500
Average		27667	29667	31167

From the table above, we get the result A_{ave}= (27667, 29667, 31167) based on the calculation using formula (1). By using the second (2) formula, the forecast value for T-shirt is 29667 dozen.

3.2 Excess Stock

After calculating the forecast for each product, the researcher can calculate the product surplus in the coming period. The excess stock can be calculated using the following formula:

- calculating actual time supply (t)

  $$t = {}^{M}/_{R} \tag{3}$$

  Explanation :
  M :available stock in unit
  R :annual demand in unit
- calculating economic time supply (t*)

  $$t^* = \frac{P - Ps + {}^{C}/_{Q}}{PF} + \frac{Q}{2R} \tag{4}$$

  Explanation :
  P :unit cost or market value of the item
  Ps :unit resale or salvage value of the item
  C :order cost per order
  Q : lot size per unit
  F : annual holding cost fraction

After the t and t* is known, the next is to compare both results; if t> t*, the calculation can be proceed, but if the opposite happens, no further action is required because the supply is sufficient.

- specifying excess excess (q)

  $$q = (t - t^*) \text{ x } R \tag{5}$$
- Calculating Net Benefit

  $$\textit{Net benefit} = \textit{salvage revenue} + \textit{holding cot saving} - \textit{repurchase cost} - \textit{reorder cost} \tag{6}$$

  $$\textit{salvage revenue} = qP_s = P_s(M - tR) = P_s M - P_s Rt \tag{7}$$

  $$\textit{holding cot saving} = (M^2 PF / 2R) - (M - q)^2 PF / 2R) - (QqPF / 2R) = (M^2 PF / 2R) - (RPFt^2/2) - (MQPF/ 2R) + (QqPF/2) \tag{8}$$

  $$\textit{repurchase cost} = Pq = PM - PRt \tag{9}$$

  $$\textit{reorder cost} = (Cq / Q) = (CM / Q) - (CRt / Q) \tag{10}$$

If the net benefit is positive (+), the product can be sold at the price of Ps; but if not, the process is continued to find the minimum economic salvage value (Ps*).

IOP Conf. Series: Materials Science and Engineering **277** (2017) 012068 doi:10.1088/1757-899X/277/1/012068

- minimum economic salvage value (Ps*)

$$Ps^* = P + \frac{C}{Q} - \frac{PF(M - Q/_2)}{R} \tag{11}$$

After the value of Ps* is known, the product can be sold more than Ps* or its equivalent.

Here are the results of the calculations of pants, shirts and T-shirts using the excess stock, which is presented in tables 4. 5 and 6.

Table 4. Excess Stock Calculation Result on Pant Products

Output			
Symbol	Description	Denomination	Value
t	Actual Time Supply	Year	1
t*	Economic Time Supply	Year	0.97
q	Excess Stock	Unit	1522
Net Benefit	Profit	Rupiah	89864
Ps*	Minimum Economic Salvage Value		49882

Table 5. Excess Stock Calculation Result on Shirt Products

Output			
Symbol	Description	Denomination	Value
t	Actual Time Supply	Tahun	1.0
t*	Economic Time Supply	Tahun	2.1

In the calculation of shirt product, it is obtained t < t*, there is no further action required because the stock is sufficient.

Table 6. Excess Stock Calculation Result on T-Shirt Products

Output			
Symbol	Description	Denomination	Value
t	Actual Time Supply	Year	1
t*	Economic Time Supply	Year	0.47
q	Excess Stock	Unit	187761
Net Benefit	Profit	Rupiah	98946825
Ps*	Minimum Economic Salvage Value		28946

- Interpolation

  To determine the result 0 on q, which means there is no excess, the calculation using interpolation is done with the following formula:

$$\text{Interpolation (xe)} = xs0 + \frac{xs1 - xs0}{xe1 - xe0}(x - xe0) \tag{12}$$

  Explanation:
  Xe : The sought/desired excess value which in this case is 0
  Xso : stock value before
  Xs1 : stock value after
  X0 : excess value before
  Xe1 : excess value after

Here are the results of interpolation calculations on pants and t-shirts, presented in table 7, and for increase profit from excess to zero excess on pants and t-shirts are presented in table 8 and 9:

Table 7. Interpolation Calculation Result to Eliminate Excess on Pants and T-shirts

Available Stock in Unit (M)		
Product	Excess	No Excess
Pants	44400	42878
T-shirts	356000	168239

For stock on pants it will be excess if the warehouse have 44400 unit, but no excess is 42878 unit, and for T-shirt if the warehouse have stock 356000 it will be excess, for no excess is 168239 unit in warehouse.

Table 8. Profit Calculation Result From Excess to Zero Excess and The Percentage of Profit Increase on Pants

Pants	
IncomeWith Excess	Rp 2,329,576,078.70
Income Without Excess	Rp 2,352,053,211.43
Increase	1%

Table 9. Profit Calculation Result From Excess to Zero Excess and The Percentage of Profit Increase on T-Shirt

T-shirts	
IncomeWith Excess	Rp 5,271,236,167.05
Income Without Excess	Rp 10,944,757,200.00
Increase	52%

3.3 Damaged Product

To determine the price of damaged products is as follows:

- Calculating the percentage of damaged products at once production
  Percentage T = (number of damaged products/amount of production) *100% (13)
- Calculating the Damaged Product Price Based Cost Production
  The Price of Damaged Products = total production/number of the produced goods (14)
- Calculating the profits derived from damaged products
  Profit of Damaged Product (PDP) = (total production/number of produced goods) *number of damaged products (15)
- Calculating the percentage of additional profit if the damaged products are sold
  Percentage P = (PDP/sales profit)*100% (16)

The following this is the result the determination of the minimum price products damaged, and the reckoning increase in revenue if products broken can be sold a whole by the selling price minimum specified shown in table 10 until 13:

Table 10. Percentage calculation result of the damaged products at once production

Product	Production	Damaged Products	Precentage
Shirts	187906	3579	2%
Pants	40178	205	0.51%

Table 11. Calculation Result of Damaged Products Cost in Unit

Product	Production	The Production Cost	Cost Of Damaged Product in Unit
Shirt	210,400	Rp 8,828,384,000.00	Rp 6,250.00
Pants	44,400	Rp 1,960,082,400.00	Rp 6,250.00

Table 12. Profit calculation result if the damaged products on shirts are sold and the percentage of profit increase

Shirt	Production	Profit Increase	Precentage
Good Product	184327	Rp 7,734,360,920.00	0.29%
Damaged Products	3579	Rp 22,368,750.00	
Total		Rp 7,756,729,670.00	

Table 13. Profit calculation result if the damaged products on Pants are sold and the percentage of profit increase

Pants	Production	Profit Increase	Precentage
Good Product	39973	Rp 1,764,662,773.33	0.07%
Damaged Products	205	Rp 1,279,166.67	
Total		Rp 1,765,941,940.00	

For performing calculations products broken in the period next calculation products they would damaged old as has been in count before, that to the product of a shirt had damaged of 2 % and for products pants namely 0,51 %.
The following the calculation on damage will happen to shirts and pants , and calculation increase in revenue for them shown in table 14 until 16:

Table 14. Percentage calculation result of the damaged forecast products at once production

Product	Production	Damaged Products	Precentage
Shirts	44400	0.51%	226
Pants	210400	2%	4007

Table 15. Profit calculation result if the damaged products on shirts are sold and the percentage of profit increase

Shirt	Production	Profit Increase	Precentage
Good Product	206393	Rp 8,660,231,911.53	
Damaged Products	4007	Rp 25,046,486.01	0.29%
Total		Rp 8,685,278,397.54	

Table 16. Profit calculation result if the damaged products on Pants are sold and the percentage of profit increase

Pants	Production	Profit Increase	Precentage
Good Product	44174	Rp 1,950,097,743.44	
Damaged Products	226	Rp 1,413,584.55	0.07%
Total		Rp 1,951,511,327.99	

To product price damaged same as product price damaged last year, Rp.6.250 in unit, because the value of the debt is considered as minimum from the sale of products based on the production costs on a shirt, pants, and t-shirts.

4. Conclusions

According to the data processing done, these conclusions as follows:

1. Using fuzzy averaging for forecasting demand, and produced planning production to the product of 3700 dozen pants, shirts 17533 dozen, and t shirt 29667 dozen, converted into units be 44400 unit for products pants, 210400 unit for shirts , and 356000 unit for shirts.
2. Based on the calculation of uses the excess stock producing shirts and pants will experience in excess of products 1522 unit, and t shirt about 187761 unit. To avoid excess products to the pants and shirts, then came the determination of the capacity the warehouse with reducing the stock be 42878 unit for trousers with the time of a cycle of 0.96 years or 352 day, and 168239 unit for shirts with the time of a cycle of 0.47 years or 172 days
3. By calculation both method of excess stock products on pants be obtained minimum price is Rp.49.882 per unit, and the price they shirts the minimum for Rp.28.946 per unit. Accounting for damaged products, obtained the minimum price sales for the damaged product Rp.6.250 per unit based on the production costs.
4. Excess control stock has been done get to the income rose by 1 % pants, and for producing shirts of 52%, the calculation damaged products, obtained income rose for products pants of 0.07% and for shirts 0.29 %

Acknowledgements
Thanks to the Directorate of Research and Community Service of The Ministry of Research and Technology of The Republic of Indonesia which funded this research activity in 2017 with an Applied Product Research scheme.

5. References

[1] Bojadziev G and Bojadziev Maria 2007 *Fuzzy Logic for Business, Finance, and Management*. USA: Clearance Center, Inc.

IOP Conf. Series: Materials Science and Engineering **277** (2017) 012068 doi:10.1088/1757-899X/277/1/012068

[2] Dewi H 2012 *Tinjauan Atas Perlakuan Akuntansi Untuk Produk Cacat Dan Produk Rusak Pada PT Indo Pacific*. Proceedings ISSN 225-3936, Universitas Widyatama.
[3] Kusuma H 1999 *Manajemen Produksi Perencanaan dan Pengendalian Produksi*. Yogyakarta: Andi Offset.
[4] Markidakis 2005 *Metode dan Aplikasi Peramalan*. Jakarta: Penerbit Binarupa Aksara.
[5] Hidayat S, Nurhasanah N and Zulkifli A S 2013 Perencanaan dan Penjadwalan Distribusi pakaian jadi dengan metode. *Jurnal Optimasi Sistem Industri,* Vol. **12** No.2, pp 343-351
[6] Soekartawi 2000 *Pengantar Agroindustri*. Jakarta: Rajagrafindo Pustaka.
[7] Tersine R J 1994 *Principles of Inventory and Materials Management Fourth Edition*. New Jersey: Prentice Hall Inc.
[8] Wayan N 2011 Aplikasi Perhitungan Interpolasi Newton Dengan Borland Delphi 5.0.0, *Jurnal Ilmiah Faktor Eksacta*, vol **4** no 1, Universitas Indraprasta PGRI.

Cheapest insertion heuristics algorithm to optimize WIP product distributions in FBS Fashion Industry

WN Tanjung[1], N Nurhasanah[2], QA Suri[3], Jingga[4], B Aribowo[5], DA Mardhika[6], AM Gayatri[7], R Safitri[8] and A Supriyanto[9]

[1,2,3,4,5,6,7]Industrial Engineering Department, Faculty of Science and Technology, Universitas Al Azhar Indonesia, Jakarta 12110, Indonesia
[8,9]Informatic Engineering Department, Faculty of Science and Technology, Universitas Al Azhar Indonesia, Jakarta 12110, Indonesia
[1]widya@uai.ac.id

Abstract. The textile industry is one of the 10 commodities of industrial products which are still survives in Indonesia due to the crisis in the year 2009 until 2016. Drawback happened in 2017 by increased the number of demand by approximate 3% compares with previous year. In this case, the research conducted in Small Medium Enterprise (SME) called FBS. SME is a business group that is able to absorb a lot of labor and a source of income for society. SME FBS producing clothing boys and domiciled in Jakarta. To complete FBS product, the WIP products are sent to CMT or depot in Sukabumi. In this study, aims to do the shortest route in the determination of the distribution of WIP product to 10 CMT scattered in the area of Sukabumi. After optimization hapened, the route must be started from the Depot SME FBS Sukabumi-Shell Sand Village – village of Sukamaju Village – Margaluyu Village – Narogong Cicurug, – the village of Parakanlima, Cuguha – Padabeunghar Village – Sagaranten Village – village of Ciherang, Ciguyang, Sagaranten – the village of Bojong Waru, Pasirsalam Village, Purabaya – Students – return to Depot SME FBS Sukabumi with mileage in a single trip of 403.6 kilometers. It spents 10 hours 09 minutes and cost distribution issued amounting to IDR 296,928.52. The route length was optimized 47% from 759.1 become 403.6 kilometers.

Keywords: Distribution, Travelling Salesman Problem, Cheapest Insertion Heuristics Algorithm, Routes, Costs

1. Introduction

The textile industry is one of the 10 commodities of industrial products which are in Indonesia because as one of the biggest labour absorber in Indonesia (more than 1.3 million people directly) of the workforce, more than half (600 thousand people) worked in the textile industry which is also a labour-intensive industry and textile product contribution to GDP (gross domestic product) is pretty significant, although National had come down because of the crisis in 2009 to 2016 [1]. According to the Kemenperin in the period January-March 2017, export of textile industry reached USD 2 billion, this figure is experiencing a rise of three percent compared to the same period in the previous year [4]. Not only the textile industry of textiles products that may develop well, but the IKM (small and medium Industry) are also able to compete in developing domestic products. Small and medium industry (IKM) business group is capable of absorbing a lot of labour and a source of Community income [3]. With the concomitant development of the textile industry sector in the efforts in particular on SME then faced increasing competition, each want to master the market widely.

SME FBS is the small and medium industries which produce clothes of boys. Problems faced by SME FBS namely in the distribution of WIP product from Jakarta to Sukabumi in the Village Office which became Purabaya gatherers or depot to IKM FBS. In which the Depot SME FBS Sukabumi funneling WIP product to 13 CMT (Cut Make and Trim or Tailor) spread in Sukabumi areas. In order for the WIP product can be distributed quickly then needed an efficient distribution channel from the village to Purabaya respectively 13 CMT. Therefore the shortest path or route and the time chosen to optimize the distribution of WIP product, because they have to consider the time that will be spent in the way by looking at the area road conditions. View of the problem in the shortest route search can be used the Travelling Salesman Problem (TSP). TSP is a problem to optimizing and finding travel (tour) the shortest length [6]. TSP is a problem to determine the sequence of all the city has to be visited by the salesman and the city should only be visited exactly one route. The problem of how the salesmen can arrange a trip route, so long he is the optimum route i.e. minimum distance is best [5].Constraints that occur in the field showed the need for the determination of the optimal route with minimum mileage which pay attention to the capacity of the vehicle used by one of the model approach of the Travelling Salesman Problem (TSP) Cheapest Insertion Heuristics Algorithm to deliver the optimal solution to the problem of TSP, including on the selection of the route of travel, costs, and time.

2. Methods

2.1. Cheapest Insertion Heuristic Algorithm

Completed undirected edge weighted graph as G = (V, E), whereas V = {v1,...,vn}, and V = {1,...,n}. c_{ij} is represent the length of (i, j) ∈ E. A tour is a cycle in G with n vertices. A partial tour is a pair of parallel edges in G with at most n – 1 vertices. For a partial tour T and a vertex k / ∈ T we define insert (T, k) to be the cycle obtained by deleting an edge (i, j) ∈ T and inserting instead the two edges (i, k) and (k, j).We say that k is inserted into T. We define by c (T, k) the length increase caused by the insertion of k into T, that is c(T, k) = c(insert(T, k)) – c(T). The Cheapest Insertion algorithm is useful in traveling salesman problem and known as heuristic methods. In our version, the initial partial tour is a minimum cost 2-edge cycle. Alternatively, it could start with an arbitrary vertex as the initial partial tour. For a partial tour T and a vertex k ∈ T we define delete (T, k) to be the cycle obtained by deleting from T the two edges incident with k, say (i, k) and (k, j), and replacing them by (i, j). We say that k is deleted from T and mark the resulting subtour by delete (T, k). Thus, deletion is the inverse operation of insertion. We define by c-(T, k) the length reduction caused by the deletion of k from T, that is c-(T, k) = c(T) – c(delete(T, k)).

```
Input
A weighted graph G= (V,E).
Begin
k, l: arg, min {cij + cji: I, j Є V, I ≠ j},
T: = (k,l, k).
while T is a partial tour
        v: = arg min {c(T,k): k ¢ T}.
        T: = insert (T,v).
end while
end Cheapest Insertion
```

The early stages are done in this research is conducting a study of literature by making reference books and journals that have a relation to the Travelling Salesman Problem (TSP) Cheapest Insertion Heuristics Algorithm. And referring to previous research that is "Enhancing Competitiveness of Ready Made Garment Small-Medium Enterprises Through Logistics Performance Measurement Using SCOR Method" compiled by [2].

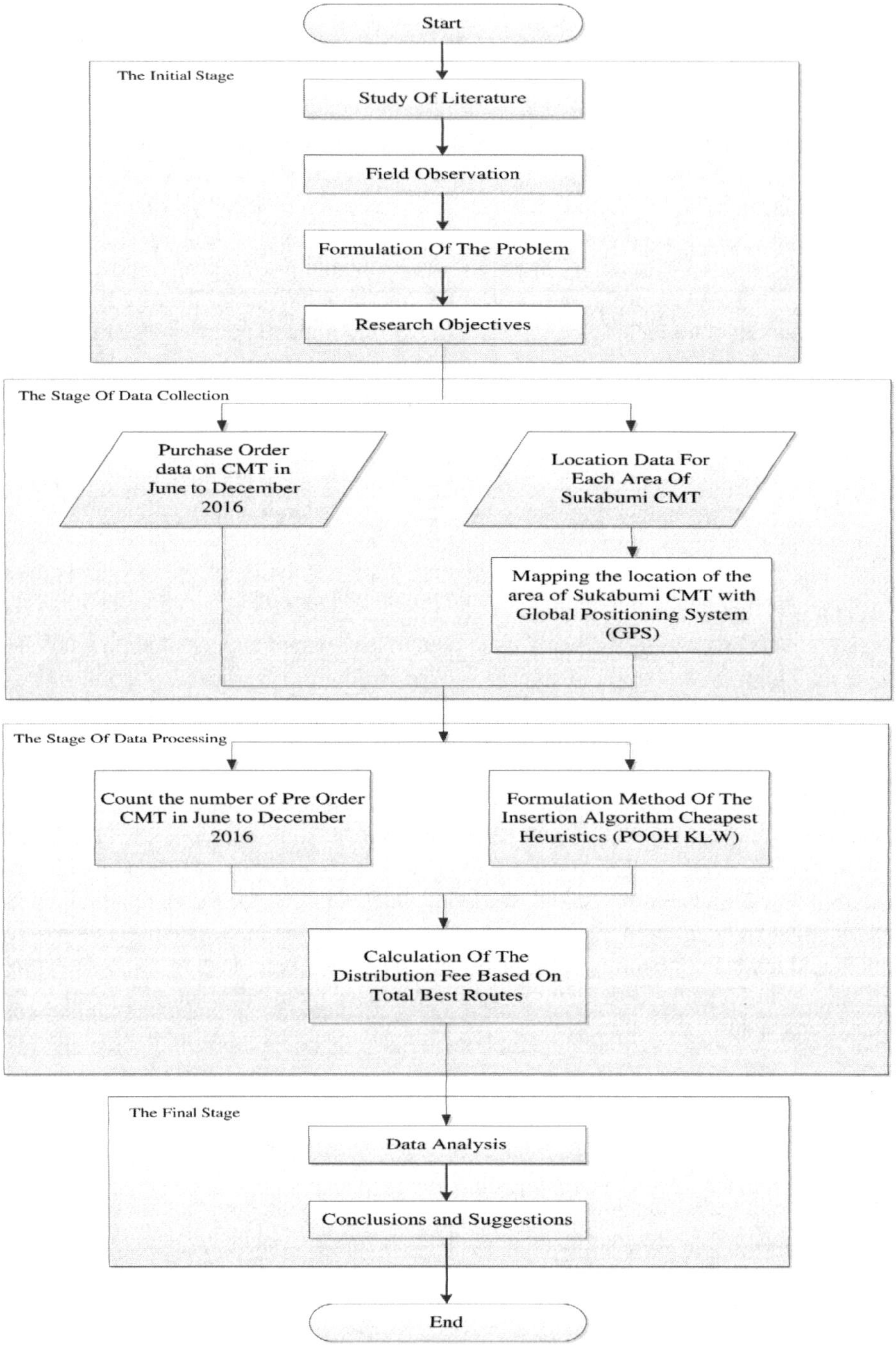

Figure 1. Flowchart Research

The second step is the observation field which aims to select the appropriate location to conduct research. This research was conducted at SME FBS Depot located in the village of Purabaya, Sukabumi-Jawa Barat and CMT are scattered in 13 villages. The third step is to choose a problem researchers review distribution of intermediate goods SME FBS to figure out the optimal distribution channel. The fourth step is to assign the research objectives based on the basic theory of the concept of Cheapest Insertion Heuristics Algorithm. The next step is the collection of data, the data that is collected in the form of primary data and secondary data. Primary data is data obtained directly from observations of researchers such as company history, a list of operational data, purchase order distribution of intermediate goods in June to December 2016, each location data in CMT The region of Sukabumi, mapping the location of CMT in Sukabumi area with GPS. The next step is data processing that is calculate the number of purchase order cmt from June to December 2016, formulation of CIH Algorithm Method, and last step is analysis, conclusion, and suggestion.

3. Result and discussion

3.1 Operational Data of SME FBS Distribution

The distribution of raw materials in the form of semi-finished goods is done from the warehouse or the center of SME FBS uction located at Jl. Z number 1, South Sukabumi, South Jakarta, and the materials will be sent to Depot SME FBS located in Purabaya Village, Sukabumi. The operational activities of the distribution of SME FBS from Jakarta to Sukabumi are conducted every Friday, starting at 23.00 pm with travel time for ± 5 hours and 12 minutes until Purabaya Village, Sukabumi on Saturday at 04:15 pm. Currently the 13 CMT (tailors) take the semi-finished goods directly to the collecting office in Purabaya Village, Sukabumi. Figure 1 is the distribution route of semi-finished goods in Sukabumi area.

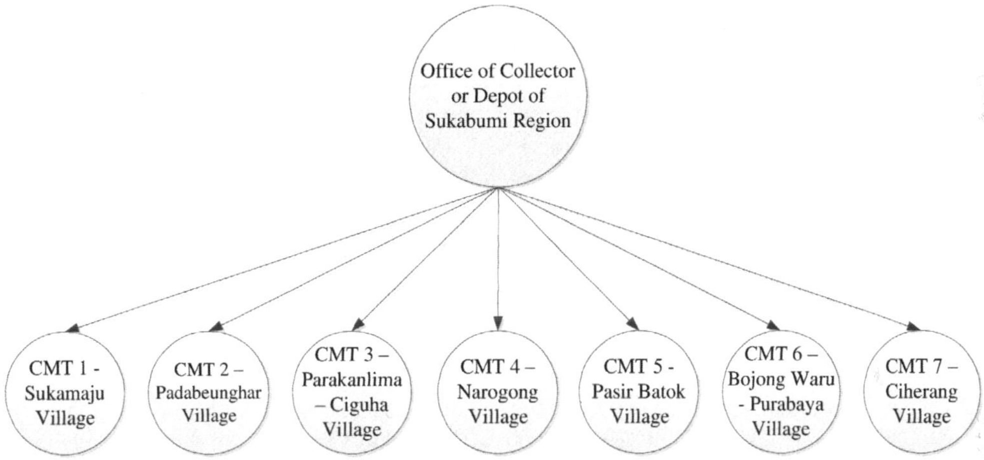

Figure 2. The Distribution Route of Semi-Finished Product in Sukabumi

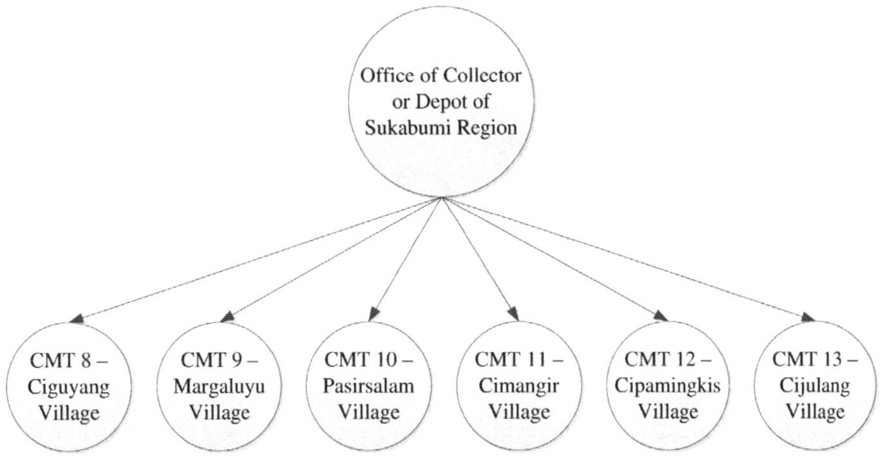

Figure 2. The Distribution Route of Semi-Finished Product in Sukabumi (continue)

It can be seen from Figure 2 that each CMT is scattered in different regions and distance. Table 1 data for the distance and the time taken by each of the CMT to the location of the Depot SME FBS Sukabumi.

Table 1. Number of PO in each CMT from June to December 2016 in Sukabumi

No.	Name CMT	The number of requests (pcs)							
		Jun	Jul	Agst	Sept	Okt	Nov	Des	Total
1	CMT 1	0	0	0	486	0	480	0	966
2	CMT 2	0	0	1476	2058	771	894	972	6171
3	CMT 3	0	0	672	732	2520	540	546	5010
4	CMT 4	0	0	3252	1668	4020	2478	4629	16047
5	CMT 5	0	0	564	1698	0	984	1260	4506
6	CMT 6	0	0	0	0	540	4668	6018	11226
7	CMT 7	0	0	0	0	0	480	1020	1500
8	CMT 8	0	0	384	480	972	3480	2157	7473
9	CMT 9	0	0	0	486	0	0	0	486
10	CMT 10	0	0	720	1128	1224	0	0	3072
11	CMT 11	0	0	0	0	0	0	0	0
12	CMT 12	0	0	0	0	0	0	0	0
13	CMT 13	0	0	0	0	0	0	0	0

In Table 1, it can be seen that each CMT can complete different POs every month from June to December 2016. And for the CMT that does not do the sewing or the total is zero (0), then the data and CMT is not used to process the next data.

Table 2. Data of Distance and Time of CMT to Depot SME FBS

No.	Name CMT	The Name of The Village	The Point Coordinates	Travel Activities	Travel Distance (km)	Total Trip Distance (km)	It takes 1 x Travel	Total Travel Time Travel
0	Depot IKM FBS Sukabumi		-7.100231, 106.890475	-	-	-	-	-
1	CMT 1	Sukamaju	-6.831008, 106.939914	2	52.2	104.4	2 hours 3 minutes	4 hours 6 minutes
2	CMT 2	Padabeunghar	-7.013959, 106.786794	2	24.6	49.2	1 hours 3 minutes	2 hours 6 minutes
3	CMT 3	Parakanlima - Ciguha	-6.996134, 106.826723	2	31.7	63.4	1 hours 26 minutes	2 hours 52 minutes
4	CMT 4	Narogong	-7.006560, 106.837744	2	34.2	68.4	1 hours 35 minutes	3 hours 10 minutes
5	CMT 5	Pasir Batok	-6.805060, 106.704736	2	69.6	139.2	2 hours 37 minutes	5 hours 14 minutes
6	CMT 6	Bojong Waru - Purabaya	-7.098825, 106.888350	2	0.25	0.5	3 minutes	6 minutes
7	CMT 7	Ciherang - Sagaranten	-7.179652, 106.860263	2	17	34	42 minutes	1 hours 24 minutes
8	CMT 8	Ciguyang - Sagaranten	-7.193913, 106.876841	2	14.5	29	33 minutes	1 hours 6 minutes
9	CMT 9	Margaluyu - Cicurug	-6.877141, 106.991827	2	47.9	95.8	1 hours 50 minutes	3 hours 40 minutes
10	CMT 10	Pasirsalam - Nyalindung	-7.346611, 106.661676	2	87.6	175.2	2 hours 57 minutes	5 hours 54 minutes
TOTAL					379.55	759.1		29 hours 38 minutes

Table 2 is the distance and travel time data of CMT to Sukabumi collecting office, in which each CMT has different travel distance and time. The coordinate point uses GPS (Global Positioning System) coordinates with the help of Google Maps application. The next is the required fuel data and costs incurred by each CMT.

In Table 3, it can be seen the total cost incurred to travel from each CMT area to the collecting office in Purabaya-Sukabumi as much as IDR 497.211.

Table 3. Fuel Required and Cost Borne by Each CMT

No.	Name CMT	The Name of The Village	Titik Koordinat	Travel Distance (km)	Fuel Type Premium Required (Litres)	Costs Incurred (1 Trip)	Total Costs Incurred (2 Trips)
0	Depot IKM FBS Sukabumi		-7.100231, 106.890475	-	-	-	-
1	CMT 1	Sukamaju	-6.831008, 106.939914	52.2	5.22	IDR 34,191	IDR 68,382
2	CMT 2	Padabeunghar	-7.013959, 106.786794	24.6	2.46	IDR 16,113	IDR 32,226
3	CMT 3	Parakanlima - Ciguha	-6.996134, 106.826723	31.7	3.17	IDR 20,764	IDR 41,527
4	CMT 4	Narogong	-7.006560, 106.837744	34.2	3.42	IDR 22,401	IDR 44,802
5	CMT 5	Pasir Batok	-6.805060, 106.704736	69.6	6.96	IDR 45,588	IDR 91,176
6	CMT 6	Bojong Waru - Purabaya	-7.098825, 106.888350	0.25	0.025	IDR 164	IDR 328
7	CMT 7	Ciherang - Sagaranten	-7.179652, 106.860263	17	1.7	IDR 11,135	IDR 22,270
8	CMT 8	Ciguyang - Sagaranten	-7.193913, 106.876841	14.5	1.45	IDR 9,498	IDR 18,995
9	CMT 9	Margaluyu - Cicurug	-6.877141, 106.991827	47.9	4.79	IDR 31,375	IDR 62,749
10	CMT 10	Pasirsalam - Nyalindung	-7.346611, 106.661676	87.6	8.76	IDR 57,378	IDR 114,756
			TOTAL				IDR 497,211

3.2. Delivery Calculation of Semi-Finished Goods

It is illustrated with 11x11 distance matrix table where the M_{ij} element is the distance from *I* to *j*. The table is described in Table 4 as follows:

Table 4. Distance Matrix

	Depot SME FBS Sukabumi	CMT 1 "Sukamaju"	CMT 2 "Padabeunghar"	CMT 3 "Parakanlima - Ciguha"	CMT 4 "Narogong"	CMT 5 "Pasir Batok"	CMT 6 "Bojong Waru - Purabaya"	CMT 7 "Ciherang - Sagaranten"	CMT 8 "Ciguyang - Sagaranten"	CMT 9 "Margaluyu - Cicurug"	CMT 10 "Pasirsalam - Nyalindung"
From / To	0	1	2	3	4	5	6	7	8	9	10
0		52.2	24.6	31.7	34.2	69.6	0.25	17	14.5	47.9	87.6
1	52.2		36.4	28.3	28.8	38.9	52.2	66.6	64.1	25.8	130
2	24.6	36.4		8.8	11.3	40.9	24.8	34.6	32.1	36.4	69.9
3	31.7	28.3	8.8		2.6	37.8	32	41.7	39.2	27	77
4	34.2	28.8	11.3	2.6		38.7	34.4	44.2	41.7	27.5	79.5
5	69.6	38.9	40.9	37.8	38.7		69.6	84	81.5	49.1	106
6	0.25	52.2	24.8	32	34.4	69.6		17.3	14.8	48.1	87.8
7	17	66.6	34.6	41.7	44.2	84	17.3		5.6	60	76.1
8	14.5	64.1	32.1	39.2	41.7	81.5	14.8	5.6		57.5	74.1
9	47.9	25.8	36.4	27	27.5	49.1	48.1	60	57.5		104
10	87.6	130	69.9	77	79.5	106	87.8	76.1	74.1	104	

3.3. Calculation of Optimal Route Using Cheapest Insertion Heuristics Algorithm Method for TSP

Step 1: Search is started from the first city linked to the last city.
The data processing in the first step is taken from area or region 0 to 10.

Step 2: Creating a subtour relationship between the two cities.
Data processing of the second step makes the subtour: (0,10) → (10,0)

Step 3: Creating a table that stores the city that can be inserted in the subtour along with the additional distance. The added point is the point that has never been passed, which is shown in Table 5

From Table 5, it can be seen the smallest temporary insert is 0.45 kilometers using the formula at (1):

$$C_{ik} + C_{kj} - C_{ij} = (0{,}25 + 87{,}8) - 87{,}6 \quad (1)$$
$$= 0{,}45$$

Table 5. Arc Addition of Subtour to 1

The Arc Will be Replaced	The Arc To be Added	The Extra Distance
(0,10)	(0,1) - (1,10)	94.6
(0,10)	(0,2) - (2,10)	6.9
(0,10)	(0,3) - (3,10)	21.1
(0,10)	(0,4) - (4,10)	26.1
(0,10)	(0,5) -(5,10)	88
(0,10)	(0,6) - (6,10)	0.45
(0,10)	(0,7) - (7,10)	5.5
(0,10)	(0,8) - (8,10)	1
(0,10)	(0,9) - (9,10)	64.3
(10,0)	(10,1) - (1,0)	94.6
(10,0)	(10,2) - (2,0)	6.9
(10,0)	(10,3) - (3,0)	21.1
(10,0)	(10,4) - (4,0)	26.1
(10,0)	(10,5) - (5,0)	88
(10,0)	(10,6) - (6,0)	0.45
(10,0)	(10,7) - (7,0)	5.5
(10,0)	(10,8) - (8,0)	1
(10,0)	(10,9) - (9,0)	64.3

Arc (0,10) is replaced with *Arc* (0,6) and *Arc* (6,10) or *Arc* (10,0) is replaced with *Arc* (10,6) and *Arc* (6,0). If there are 2 routes that have the same total distance, one can be choosen. For example the first possibility is chosen, then the new subtour becomes: (0,6) → (6,10) → (10,0).

Step 4: Next is a table that stores the biased area or region is inserted in subtour along with the additional distance
The smallest additional distance is 29.05 kilometers by using the formula at (1) which replaces *Arc* (0,6) with *Arc* (0,8) and *Arc* (8,6), so the new subtour produced is: (0,8) → (8,6) → (6,10) → (10,0).

Step 5: Because there are villages that still have not yet been logged in, it is necessary to create a table storing cities that can be inserted in the subtour along with the additional distance
The smallest additional distance is 8.1 kilometers by using the formula at (1) which replaces *Arc* (0,8) with *Arc* (0,7) and *Arc* (7,8), so the new subtour produced is: (0,7) → (7,8) → (8,6) → (6,10) → (10,0).

Step 6: Because there are villages that still have not been entered, then a table is created to store villages that can be inserted in the subtour along with the additional distance
The smallest additional distance is 42.2 kilometers by using the formula at (1) which replaces *Arc* (0,7) with *Arc* (0,2) and *Arc* (2,7), so the new subtour produced is: (0,2) → (2,7) → (7,8) → (8,6) → (6,10) → (10,0).

Step 7: Because there are villages that still have not yet entered, then the table will be created to store villages that can be inserted in the subtour along with the additional distance
The smallest additional distance is 15.9 kilometers by using the formula at (1) which replaces *Arc* (0,2) with *Arc* (0,3) and *Arc* (3,2), so the new subtour produced is: (0,3) → (3,2) → (2,7) → (7,8) → (8,6) → (6,10) → (10,0).

Step 8: By looking at villages that have not yet entered the subtour, then a table is created to store the village that can be inserted in the subtour along with the additional distance
The smallest additional distance is 5.1 kilometers by using the formula at (1) which replaces *Arc* (0,3) with *Arc* (0,4) and *Arc* (4,3), so the new subtour produced is: (0,4) → (4,3) → (3,2) → (2,7) → (7,8) → (8,6) → (6,10) → (10,0).

Step 9: Since there are villages that still have not yet been logged in, the table is then rebuilt to store the village that can be inserted in the subtour along with the additional distance
The smallest additional distance is 41.2 kilometers by using the formula at (1) which replaces *Arc* (0,4) with *Arc* (0,9) and *Arc* (9,4), so the new subtour produced is: (0,9) → (9,4) → (4,3)→ (3,2) → (2,7) → (7,8) → (8,6) → (6,10) → (10,0).

Step 10: Because there are villages that still have not been entered, then a table is created to store the village that can be inserted in the subtour along with the additional distance
The smallest additional distance is 30.1 kilometers by using the formula at (1) which replaces *Ar* (0,9) with *Arc* (0,1) and *Arc* (1,9), so the new subtour produced is: (0,1) → (1,9) → (9,4) → (4,3)→ (3,2) → (2,7) → (7,8) → (8,6) → (6,10) → (10,0).

Step 11: Then the table is rebuilt to hold the village that can be inserted in the subtour along with the additional distance
The smallest additional distance obtained is 56,3 kilometers using the formula at (1) which replaces *Arc* (0,1) with *Arc* (0,5) and *Arc* (5,1), so the new subtour is: (0,5) → (5,1) → (1,9) → (9,4) → (4,3)→ (3,2) → (2,7) → (7,8) → (8,6) → (6,10) → (10,0).

It can be concluded that from all the above steps and the additional subtour, the optimal route calculation which can be passed in the distribution of semi-finished product in SME FBS which is the shortest route for 10 villages is (0,5) → (5,1) → (1,9) → (9,4) → (4,3)→ (3,2) → (2,7) → (7,8) → (8,6) → (6,10) → (10,0) that can be seen in figure 2.

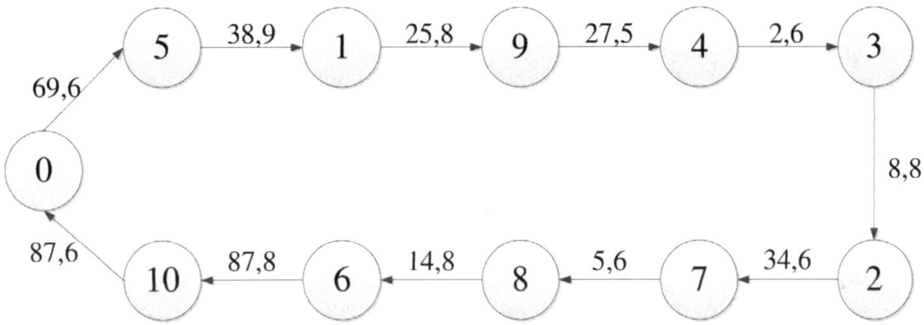

Figure 3. Trajectory or Shortest Route among 10 Villages at Each CMT of SME FBS in Sukabumi Region

Figure 3 is resulted from the trajectory or the shortest route for the distribution of semi-finished materials that will be distributed to each of the 10 CMT on SME FBS in Sukabumi Region. This trajectory or route is acquired the mileage which is $C_{0,5} + C_{5,1} + C_{1,9} + C_{9,4} + C_{4,3} + C_{3,2} + C_{2,7} + C_{7,8} + C_{8,6} + C_{6,10} + C_{10,0}$ = 403,6 km starting from 0-5-1-9-4-3-2-7-8-6-10-0. with the following explanation Depot SME FBS Sukabumi - Pasir Batok Village – Sukamaju Village –Margaluyu, Cicurug Village – Narogong Village – Parakanlima, Cuguha Village – Padabeunghar Village – Ciherang, Sagaranten Village – Ciguyang, Sagaranten Village – Bojong Waru, Purabaya Village – Pasirsalam, Nyalindung Village – then back again to the Depot SME FBS Sukabumi.

Time that must be traveled to do the distribution of intermediate goods starting from Depot SME FBS Sukabumi and back again to the Depot SME FBS Sukabumi using a mode of transportation owned SME FBS "Mitsubishi Colt L300" and Trucks Engkel Box "110ps Diesel Mitsubishi Colt" with a speed of 40 km/hours due to view the road conditions in the area of Sukabumi, here is the time needed to perform the pendisribusian intermediate goods:

$$t = \frac{s}{v} = \frac{403{,}6\ km}{40\ km/hours} = 10\ hours\ 09\ minutes \tag{2}$$

IOP Conf. Series: Materials Science and Engineering **277** (2017) 012069 doi:10.1088/1757-899X/277/1/012069

Obtained results it takes to do the distributing intermediate goods for 10 hours 9 minutes.

Obtained information that in previous distribution where CMT come straight to Depot SME FBS Sukabumi takes time for 29 hours 38 minutes with a total distance traveled 759.1 km, while route optimization is done only after the takes 10 hours 09 minutes total distance traveled 403.6 kilometers. Using the distribution route optimization done every once a week be exact on Saturday by sending just one time only. The results can be seen from the comparison between the old distribution system with the best route proposal by using Cheapest Insertion Heuristics Algorithm Method retrieved the level of efficiency of approximately 47%.

3.4. Cost Calculation Based on Best Total Route

Mileage in route delivery of transportation costs. In this case the cost of transportation efficiency calculation is done using the efficiency of distance traveled, which can reach a distance of 403.6 km correspond to the optimal route calculation.

SME FBS has 2 types of vehicles that is "Mitsubishi Colt L300" with 2477 cc engine with a maximum capacity of 2.5 tons, while for the type of Engkel Box Truck Box "Colt Diesel 110ps Mitsubishi" with 3908 cc engine with a maximum capacity of 2 tons, from 2 types the vehicle is a fuel in use is diesel in which 1 litre of gasoline type diesel can travel ± 7 km [3]. And the cost of fuel for the type of diesel costs Rp 5.150/litre [3]. And see the road condition of the region sukabumi car can run with speed 40 km/hour. From the following calculation results obtained to do the distribution to 10 CMT region as follows:

1 Litre of Solar Fuel = Rp. 5.150
1 Litre of Solar Fuel can travel = ± 7 km
Then,
We get

$$1\ km = \frac{IDR\ 5.150}{7\ km} = IDR\ 735{,}7\ /km$$

$$BFuel\ needed\ = \frac{403{,}6\ km}{7\ km} = 57{,}66\ Litre$$

So obtained

$$C_{the\ efficiency\ of\ the} = (The\ cost\ of\ Gasoline\ for\ one\ kilometer)\ x\ (d_{e\ the\ efficiency\ of\ the}) \quad (3)$$

= IDR 735,7 /km x 403,6 km = IDR 296.928,52 /once delivered

Judging from the results obtained to distribute goods to half so to 10 CMT in Sukabumi Region by using vehicle types "Mitsubishi Colt L300 Engkel Box Truck" and "Colt Mitsubishi 110ps Diesel" takes over 10 hours, 10 minutes, in addition the materials needed for ingredients amounted to 57.66 litres and costs that must be incurred to buy fuel at IDR 296,928.52 that distinguish only capacity.

It can be seen that the cost that must be issued before the route optimization of IDR 497,210.50 while the cost that must be issued after the route optimization of IDR 296,928.52 can efficient a fee of 40% in 1 trip. Moreover, it can be seen from these two types of vehicles have similar results to do the distribution of intermediate goods to 10 CMT in Sukabumi Region in both the time taken, the fuel that is needed, and the cost to purchase the materials burn that sets it apart is the maximum capacity of each type of vehicle.

4. Conclusion

Based on the calculation concerning the shortest proposed route to distribute the semi-finished products in SME FBS to the 10 villages in Sukabumi by using Cheapest Insertion Heuristics (CIH) algorithm for the completion of the Traveling Salesman Problem (TSP) and cost efficiency inreference to the best route, it can be summarized as follows:

1. The shortest route from the calculation is 1-6-2-10-5-4-3-8-9-7-11 with the distance of 403,6 kilometers, with the following explanation: Depot SME FBS Sukabumi - Pasir Batok Village –

Sukamaju Village –Margaluyu, Cicurug Village – Narogong Village – Parakanlima, Cuguha Village – Padabeunghar Village – Ciherang, Sagaranten Village – Ciguyang, Sagaranten Village – Bojong Waru, Purabaya Village – Pasirsalam, Nyalindung Village – then back again to the Depot SME FBS Sukabumi with the distance efficiency level of 47% in 1 time travel and done every once a week on Saturday.

2. The distribution of Costs issued by SME FBS after route optimization done IDR 296,928.52 whereas before the optimization of IDR 497,211 to the level of cost efficiency of 40% in 1 trip.

5. References

[1] MP3EI 2016. *Road Map Industri Tekstil dan Produk Tekstil.* (Jakarta: Asosiasi Pertekstilan Indonesia).

[2] Nurhasanah N, et.al. 2016 Enhancing Competitiveness of Ready Made Garment Small-Medium Enterprises Through Logistics Performance Measurement Using SCOR Method. *Proceeding 2nd International Conference of Industrial, Mechanical, Electrical, Chemical Engineering (ICIMECE)*, p:E-24 - E-27.

[3] Nuryati S 2017 Penggunaan Bahan Bakar Minyak Terhadap Kecepatan Kendaraan Dan Nilai Waktu Perjalanan Di Wilayah Kota Bekasi. *Jurnal Bentang*. **Vol. 5** No.1, 45-61.

[4] Ridwan 2017 Ekspor Industri Tekstil Januari-Februari 2017 Mencapai USD 2 Miliar. http://industry.co.id/read/7342/ekspor-industri-tekstil-januari-februari-2017-mencapai-usd-2-miliar (accessed on July 28, 2017)

[5] Hassin R and Keinan A 2008 Greedy heuristics with regret, with application to the cheapest insertion algorithm for the TSP. *Operations Research Letters* **36** (2008), 243 – 246.

[6] Tanjung, N W and Sopiah 2014 Implementing Ant Colony Optimization (ACO) In Traveling Salesman Problem (Case Study at Distribution Store PB). *Proceeding the 7th International Seminar on Industrial Engineering and Management (7th ISIEM)*, p.OR18-30.

Productivity analysis to overcome the limited availability of production time in SME FBS

N Nurhasanah[1], Jingga[1], B. Aribowo[1], AM Gayatri[1], DA Mardhika[1], WN Tanjung[1], QA Suri[1], R Safitri[2], and A Supriyanto[2]

[1]Industrial engineering department, Faculty of science and technology, Universitas Al Azhar Indonesia, Sisingamangaraja Road, Jakarta 12110, Indonesia
[2]Informatic engineering department, Faculty of science and technology, Universitas Al Azhar Indonesia, Sisingamangaraja Road, Jakarta 12110, Indonesia
[1]nunungnurhasanah@uai.ac.id

Abstract. Good industrial development should pay attention to the human factor as the main driver. Condition of work procedures, work area, and environment can affect the production result because if not optimal, the production will run slowly. If the work system is less than optimal, the productivity will do so, the operator will work uncomfortably and be easy to undergo work fatigue, even it can cause work accidents. Thus, the optimal and ergonomic arrangement of the the overall work system mechanism and work environment design is required for workers to work well, regularly, safely and comfortably with the aim of improving work productivity. This research measures the performance in textile SME (Small and Medium Enterprise) located in Sukabumi which is SME FBS which produces children's clothing. This performance measurement is aimed at improving the competitiveness of the textile IKM so that it has the equal competitiveness with other SMEs or with textile industries that already have their name in market. Based on the method of hour standard time and TOC calculation at 2 FBS CMT (Cut-Make-Trim) in Sukabumi, which are the CMT Margaluyu Village and CMT Purabaya Village, the result is that the standard time of shirt work on CMT Margaluyu Village is less than that of CMT Desa Purabaya. It can be seen that more effective in SME FBS production is by process method.
Keywords: Standard Time, Theory of Constraint, Productivity.

1. Introduction

In this modern era, technological progress is growing. The development is in accordance with the increase of the various needs of human beings, both basic and additional needs. As consumptive beings, humans are never satisfied with what has been obtained, so they are constantly looking for something new to meet their needs. In the field of industry, it becomes a major factor to continue to innovate in order to meet the needs of consumers. With the help of better technological developments, the process can be done quickly and easily so as to maintain the trust of consumers.

In Indonesia itself, the level of intense competition occurs in various fields of industry, for example in the textile industry. The textile industry is one of the prioritized industries to be developed because it has a strategic role in the national economy that is as a contributor to the country's foreign exchange, absorbing labor in large quantities, and as a reliable industry. As one of the basic needs that must be used, the needs of clothing increase every year as the Indonesian population increases. In addition, the modern era is associated with the growing of the need in clothing. Therefore, textile companies in Indonesia are required to make innovations and improvements to meet those needs.

Time is the element to be considered in designing and improving the work system. Generally the determination of the systems and methods used in an industry depends heavily on its purpose, for example in a manufacturing industry is producing goods, a bank serves transactions with customers, sales of cow's milk from farms, and so on. Then, the industry needs approaches to increase its productivity with problem solving and so forth. The measurement of working time is intended to obtain the standard time. In the manufacturing system, this standard time is used as a basis for, among others, production scheduling, planning, financing, and productivity evaluation.

The previous research that used as reference is performance measurement at SME (Small and Medium Enterprise) textile industry located in Jakarta, that is SME VSH that produce women moslem clothes. Measurement of production scheduling performance is done by SCOR (Supply Chain Operation Reference) method with five indicators: reliability, responsiveness, agility, cost, and asset management. The problems that occur in the SME VSH are the emergence of defective products that affect the value of existing product assets in the warehouse. Judging from the worker factor, the problems that happened were the ability and experience of the less trained workers resulting in the resulting product not in accordance with the wishes and needs of consumers. Then look in terms of work environment, the emergence of defective products occur due to poor working environment, whether from noise levels, air temperature, working positions are not ergonomic and so forth. However, despite it all after analyzed using SCOR method performance measurement in production scheduling at SME VSH is 99% which means that the performance of SME VSH is good in production scheduling.[4]

CMT Margaluyu Village in the production process of the child shirt with each worker station has workers who do their respective duties (the sequent working process), so that the workers are not moving around. Meanwhile, in CMT Purabaya Village the workers are required to produce by method 1 person must complete 1 shirt, so that the worker should run from one work station to another work station until the shirt is finished. The reason is that CMT Margaluyu Village only has 4 workers and 2 sewing machines and 1 ham machine while CMT Purabaya Village has 8 workers with 4 sewing machines, 1 ham machine and 1 serger. The research team wanted to see what methods are suitable for the production of child shirts in SME FBS, whether by process method or by method 1 worker has to finish 1 shirt from start to finish.

2. Methods

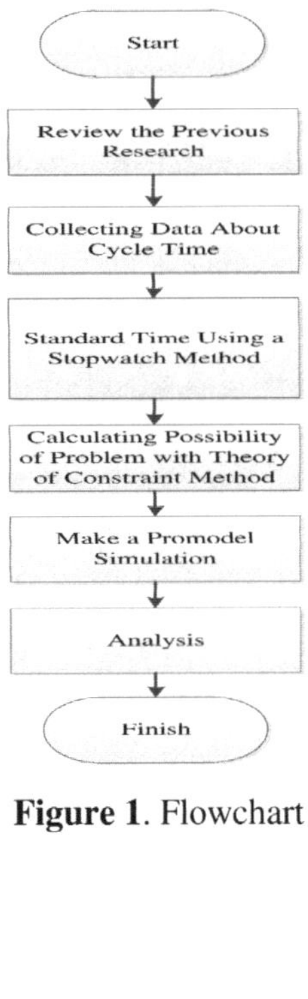

Figure 1. Flowchart

The first step taken in this research is to review the previous research with the similar theme. The research team conducted a review of the research by a graduate of Industrial Engineering Department of UAI who conducted research on SME VSH in 2016, that produce Women Moslem Clothes. Measurement of production scheduling performance is done by SCOR (Supply Chain Operation Reference) method with five indicators: reliability, responsiveness, agility, cost, and asset management [4]. After reviewing the previous research, the researcher conducts data collection on work stations in 2 CMT, CMT Margaluyu and CMT Purabaya which are both located in Sukabumi. Furthermore, the researcher calculates the cycle time with 30 times-observation of the workmanship using stopwatch then calculates the normal time and standard time of each work station. After obtaining each standard time from work stations in the 2 CMT location, the researcher calculates the Theory of Constraint, to find and see which work station has a constraint (in this case the constraint of working time on a work station against its available time). Therefore, the researcher analyzes what kind of method that is suitable for the child shirt production model in SME FBS in order to increase its production even with limited time, labour and machinery.

3. Result and Discussion

3.1 Working Methods

CMT Margaluyu Village in the production process of the child shirt with each worker station has workers who do their respective duties (the sequent working process), so that the workers are not moving around. Meanwhile, in CMT Purabaya Village the workers are required to produce by method 1 person must complete 1 shirt, so that the worker should run from one work station to another work station until the shirt is finished. The reason is that CMT Margaluyu Village only has 4 workers and 2 sewing machines and 1 ham machine while CMT Purabaya Village has 8 workers with 4 sewing machines, 1 ham machine and 1 serger. We need to know which method that has more effectiveness in the production of sewing shirts in SME FBS.

3.2 Cycle Time

The way to get raw data from the collected data is as follows[2]:

$$Ws = \frac{\sum x}{N} \tag{1}$$

Explanation:
Ws : Cycle Time
$\sum$x : Sub-Group Average Price
N : Number of Performed Observations.
We didn't used the "data adequacy test" after we get cycle time in each activities in 2 CMT. Because according to [1] it said that the use of the calculation model is used for population N which has been determined whose value must be greater than the value of the sample n sought from the calculation. This means that the data taken should be always enough. Therefore, we recommend that the term 'test adequacy data' be avoided and replaced by the term 'the number of samples should be' or any other meaningful terms.

3.2.1 Normal Time Calculation. This is the formulation to get the normal time[2]:

$$Wn = Ws\ x\ p \tag{2}$$

Explanation:
P : adjustment factor of 1 normal average cycle, p1 for too slowly working or vice versa.

3.2.2 Standard Time Calculation. And this is how we get the standard time for each process[2]:

$$Wb = Wn + (Wn\ x\ \%\ Allowance) \tag{3}$$

Allowance is the factors given to the worker to complete his work besides the normal time.

Table 1. Cycle Time and Calculation of Normal Time in CMT Margaluyu Village

Number	Work Elements	Skill			Effort			Consistency			Condition			Total	Rating	Cycle Time	Normal Time
		Explanation		Score	Explanation		Score	Explanation		Score	Explanation		Score		Factor	(Second)	(Second)
1	Reach Fabric	Excellent	B1	0,11	Excellent	B2	0,08	Excellent	B	0,03	Good	C	0,02	0,24	1,24	5,48	6,79
2	Assembly Pocket	Good	C2	0,03	Good	C2	0,02	Excellent	B	0,03	Good	C	0,02	0,1	1,1	151,15	166,27
3	Sew the Body Shirt	Good	C1	0,06	Good	C1	0,05	Average	D	0	Good	C	0,02	0,13	1,13	126,33	142,75
4	Sew using Serger	Excellent	B1	0,11	Good	C1	0,05	Good	C	0,01	Good	C	0,02	0,19	1,19	101,01	120,21
5	Trim	Good	C1	0,06	Good	C1	0,05	Good	C	0,01	Good	C	0,02	0,14	1,14	75,70	86,30
6	Sew Shirt Sleeve	Good	C2	0,03	Average	D	0	Good	C	0,01	Good	C	0,02	0,06	1,06	311,99	330,71
7	Sew Cuff	Good	C2	0,03	Good	C1	0,05	Excellent	B	0,03	Good	C	0,02	0,13	1,13	355,82	402,08
8	Sew using Serger	Good	C1	0,06	Good	C2	0,02	Excellent	B	0,03	Good	C	0,02	0,13	1,13	79,17	89,46
9	Trim	Good	C1	0,06	Good	C1	0,05	Good	C	0,01	Good	C	0,02	0,14	1,14	64,23	73,22
10	Assembly Shirt Sleeve to Body Shirt	Good	C1	0,06	Excellent	B2	0,08	Good	C	0,01	Good	C	0,02	0,17	1,17	129,52	151,54
11	Assembly Collar	Good	C2	0,03	Good	C1	0,05	Good	C	0,01	Good	C	0,02	0,11	1,11	146,90	163,06
12	Assembly Label	Excellent	B1	0,11	Good	C2	0,02	Good	C	0,01	Good	C	0,02	0,16	1,16	23,07	26,76
13	Sum	Good	C1	0,06	Good	C1	0,05	Good	C	0,01	Good	C	0,02	0,14	1,14	129,97	148,17
14	Put the Finished Shirt	Excellent	B1	0,11	Excellent	B2	0,08	Excellent	B	0,03	Good	C	0,02	0,24	1,24	8,13	10,09

Table 2. Cycle Time and Calculation of Normal Time in CMT Purabaya Village

Number	Work Elements	Skill			Effort			Consistency			Condition			Total	Rating	Cycle Time	Normal Time
		Explanation		Score	Explanation		Score	Explanation		Score	Explanation		Score		Factor	(Second)	(Second)
1	Reach Fabric	Excellent	B1	0,11	Excellent	B2	0,08	Excellent	B	0,03	Good	C	0,02	0,24	1,24	4,48	5,55
2	Sew Shirt Sleeve	Good	C1	0,06	Good	C1	0,05	Good	C	0,01	Good	C	0,02	0,14	1,14	372,54	424,70
3	Sew Cuff	Good	C1	0,06	Excellent	B1	0,1	Good	C	0,01	Good	C	0,02	0,19	1,19	343,30	408,53
4	Sew the Body Shirt	Good	C1	0,06	Good	C2	0,02	Good	C	0,01	Good	C	0,02	0,11	1,11	205,57	228,18
5	Assembly Pocket	Excellent	B1	0,11	Excellent	B1	0,1	Good	C	0,01	Good	C	0,02	0,24	1,24	183,74	227,84
6	Sew using Serger	Excellent	B2	0,08	Good	C1	0,05	Good	C	0,01	Good	C	0,02	0,16	1,16	136,57	158,42
7	Trim	Good	C2	0,03	Good	C1	0,05	Good	C	0,01	Good	C	0,02	0,11	1,11	104,44	115,93
8	Assembly Collar	Excellent	B2	0,08	Good	C2	0,02	Good	C	0,01	Good	C	0,02	0,13	1,13	106,17	119,97
9	Assembly Label	Average	D	0	Good	C2	0,02	Average	D	0	Good	C	0,02	0,04	1,04	23,11	24,03
10	Assembly Shirt Sleeve to Body Shirt	Good	C1	0,06	Good	C1	0,05	Good	C	0,01	Good	C	0,02	0,14	1,14	112,77	128,56
11	Sew using Serger	Good	C1	0,06	Good	C1	0,05	Good	C	0,01	Good	C	0,02	0,14	1,14	163,50	186,39
12	Trim	Good	C2	0,03	Good	C2	0,02	Average	D	0	Good	C	0,02	0,07	1,07	113,29	121,22
13	Sum	Good	C1	0,06	Excellent	B1	0,1	Good	C	0,01	Good	C	0,02	0,19	1,19	193,19	229,89
14	Put the Finished Shirt	Excellent	B1	0,11	Excellent	B2	0,08	Excellent	B	0,03	Good	C	0,02	0,24	1,24	6,09	7,56

Before performing the standard time calculation, the calculation of normal time is done in advance by considering several factors in accordance with the theory of Westinghouse Rating System such as skill or the ability of workers, consistency of workers, performance of workers and working conditions.

Table 3. Calculation of Standard Time in CMT Margaluyu Village

Number	Work Elements	Unvoidable Delay Score (%)	Personal Needs Score (%)	Effort Score (%)	Work Attitude Score (%)	Work Movement Score (%)	Eye Strain Score (%)	Workplace Temperature Score (%)	Atmosphere Situation Score (%)	Workplace Environment Score (%)	Allowance	Normal Time (Second)	Standard Time (Second)
1	Reach Fabric	1	1	1	0,1	0	1	3	1	1	0,09	6,79	7,4
2	Assembly Pocket	2	1	1	0,2	0	5	3	1	1	0,14	166,27	189,9
3	Sew the Body Shirt	2	1	2	0,5	0	6	3	1	1	0,17	142,75	166,3
4	Sew using Serger	3	1	1	0,5	0,5	7	3	1	1	0,18	120,21	141,8
5	Trim	3	1	1	0,5	0,5	7	3	1	1	0,18	86,30	101,8
6	Sew Shirt Sleeve	2	1,5	2	0,7	1	7	3	1	1	0,19	330,71	394,2
7	Sew Cuff	3	2	1	0,7	1	7	3	1	1	0,20	402,08	481,3
8	Sew using Serger	3	1	1	0,5	0,5	7	3	1	1	0,18	89,46	105,6
9	Trim	3	1	1	0,5	0,5	7	3	1	1	0,18	73,22	86,4
10	Assembly Shirt Sleeve to Body Shirt	3	1,5	1	0,2	0,4	3	3	1	1	0,14	151,54	172,9
11	Assembly Collar	2	1	1	0,2	0	5	3	1	1	0,14	163,06	186,2
12	Assembly Label	1	1	1	0,1	0	5	3	1	1	0,13	26,76	30,3
13	Sum	3	1	1	0,2	0	5	3	1	1	0,15	148,17	170,7
14	Put the Finished Shirt	1	1	1	0,1	0	1	3	1	1	0,09	10,09	11,0

Table 4. Calculation of Standard Time in CMT Purabaya Village

Number	Work Elements	Unvoidable Delay Score (%)	Personal Needs Score (%)	Effort Score (%)	Work Attitude Score (%)	Work Movement Score (%)	Eye Strain Score (%)	Workplace Temperature Score (%)	Atmosphere Situation Score (%)	Workplace Environment Score (%)	Allowance	Normal Time (Second)	Standard Time (Second)
1	Reach Fabric	1	1	1	0,1	0	1	3	1	1	0,09	5,6	6,1
2	Sew Shirt Sleeve	2	1,5	2	0,7	1	7	3	1	1	0,19	424,7	506,2
3	Sew Cuff	3	2	1	0,7	1	7	3	1	1	0,20	408,5	489,0
4	Sew the Body Shirt	2	1	2	0,5	0	6	3	1	1	0,17	228,2	265,8
5	Assembly Pocket	2	1	1	0,2	0	5	3	1	1	0,14	227,8	260,2
6	Sew using Serger	3	1	1	0,5	0,5	7	3	1	1	0,18	158,4	186,9
7	Trim	3	1	1	0,5	0,5	7	3	1	1	0,18	115,9	136,8
8	Assembly Collar	2	1	1	0,2	0	5	3	1	1	0,14	120,0	137,0
9	Assembly Label	1	1	1	0,1	0	5	3	1	1	0,13	24,0	27,2
10	Assembly Shirt Sleeve to Body Shirt	3	1,5	1	0,2	0,4	3	3	1	1	0,14	128,6	146,7
11	Sew using Serger	3	1	1	0,5	0,5	7	3	1	1	0,18	186,4	219,9
12	Trim	3	1	1	0,5	0,5	7	3	1	1	0,18	121,2	143,0
13	Sum	3	1	1	0,2	0	5	3	1	1	0,15	229,9	264,8
14	Put the Finished Shirt	1	1	1	0,1	0	1	3	1	1	0,09	7,6	8,2

3.3 Theory of Constraint (TOC)

After obtaining the standard time of each workstation, the Theory of Constraint is performed to see which work station becomes the constraint (in the form of processing time constraints exceeding the time capacity available in producing the shirt).

The formulation to calculate the TOC is:
First, we calculate the total amount of time for process work on each sewing machine.We need to know how much the cost of materials and the price of each shirt.And we calculate the market potential per week with this formula[3]:

$$Market\ Potential\ each\ Week = \frac{Target\ Amount\ of\ Shirt\ Production\ each\ Year}{Total\ Weeks\ per\ Year} \quad (4)$$

Then, calculate the Gross Profit[3]:

$$Gross\ Profit = Price\ of\ Shirt - Cost\ of\ Materials \quad (5)$$

We get the number of capacity requirement of each machine by use this formula[3]:

$$Capacity\ Requirement = Market\ Potential\ each\ Week\ x\ Total\ Amount\ of\ Time\ on\ Each\ Machine \quad (6)$$

Here are the calculation of time for each work station:

Table 5. Processing Time in each CMT Work Station of Margaluyu Village

Product	Price	Market Potential per Week (dozen)	Processing Time per Unit (minutes) Sewing Machine 1	Sewing Machine 2	Sewing Machine 3	Ham Machine	Cost of Materials	Gross Profit
Forboys Shirt	70000	288,5	12,5	2,8	14,6	7,3	6250	63750

Table 6. Calculation of Theory of Constraint in CMT Work Station of Margaluyu Village

			Capacity Requirement			
Product	Price	Market Potential per Week (dozen)	Sewing Machine 1	Sewing Machine 2	Sewing Machine 3	Ham Machine
Forboys Shirt	70000	288,5	3605,6	799,5	4209,1	2094,4

Based on the calculation in the table above, it shows that Sewing Machine 1 and Sewing Machine 3 become the constraint due to over time (in 2640 minutes for 4 working days, in one day work for 11 hours).

Table 7. Processing Time in Each CMT Work Station of Purabaya Village

			Processing Time per Unit (minutes)				
Product	Price	Market Potential per Week (dozen)	Sewing Machine	Obras Machine	Ham Machine	Cost of Materials	Gross Profit
Kid Dream Shirt	70000	288,5	34,9	6,8	4,7	6250	63750

Table 8. Calculation of Theory of Constraint in CMT Work Station of Purabaya Village

			Capacity Requirement		
Product	Price	Market Potential per Week (dozen)	Sewing Machine	Serger	Ham Machine
Kid Dream Shirt	70000	288,5	10081,6	1956,1	1345,4

From the calculation of TOC at CMT work station in Purabaya Village as the table above, it is seen that the Sewing Machine is also a constraint due to the workmanship time exceeds the available work time. Therefore, at the CMT work station both in Margaluyu and Purabaya Village, there are several possible solutions that can be done so that the process of working time does not exceed the working time previously set.

1. The division of work on the work station must be more structured, because the previous data show that the sewing machine 2 only does 1 job that is the front and rear assembly, while the sewing machine 1 works on 5 types of work and sewing machine 3 does 2 kinds of work. If the problem lies in a less competent workforce in sewing, it should be necessary to train the worker.
2. To add the work station (in this case sewing machine) if possible.

Then we use some formulas to calculate the capacity utilizations :
First, we calculate the gross profit per constraint resource minute[3]:

$$Gross\ Profit\ per\ Constraint\ Resource\ Minute = \frac{Gross\ Profit\ per\ Unit}{Processing\ Time\ per\ Unit\ in\ a\ Machine} \tag{7}$$

Then, we calculate the net profit [3]:

$$Net\ Profit = Throughput - Expenses \tag{8}$$

Table 9. Capacity Utilization at CMT Work Station of Margaluyu Village

Work Center	Product	Process Time (minute)	Available Capacity (minute)	Capacity Utilization %
Sewing Machine 1	Forboys Shirt	77.25(12.5) = 965.625	2640	36,58
Sewing Machine 2	Forboys Shirt	288.5(2.8) = 807.8	2640	30,60
Sewing Machine 3	Forboys Shirt	107.54(14.6) = 1570	2640	59,47
Ham Machine	Forboys Shirt	288.5(7.3) = 2106.3	2640	79,78

IOP Conf. Series: Materials Science and Engineering **277** (2017) 012070 doi:10.1088/1757-899X/277/1/012070

Table 10. Capacity Utilization at CMT Work Station of Purabaya Village

Work Center	Product	Process Time (minute)	Available Capacity (minute)	Capacity Utilization %
Sewing Machine	Kid Dream Shirt	213.2(34.9)= 7440.68	2640	100,00
Serger	Kid Dream Shirt	288.5(6.8)= 1961.8	2640	74,31
Ham Machine	Kid Dream Shirt	288.5(4.7)=1355.95	2640	51,36

Based on the calculation in table 9 and 10, it is seen that the machine with the largest capacity utility is on the Ham Machine at CMT work station Margaluyu Village that is 79,78%, while at CMT Purabaya Village the Sewing Machine has the largest capacity utility that is 100%.

3.4 Promodel Simulation

After getting the TOC calculation result, the simulation of shirt production process at CMT of Margaluyu and Purabaya Village is done, with the following result:

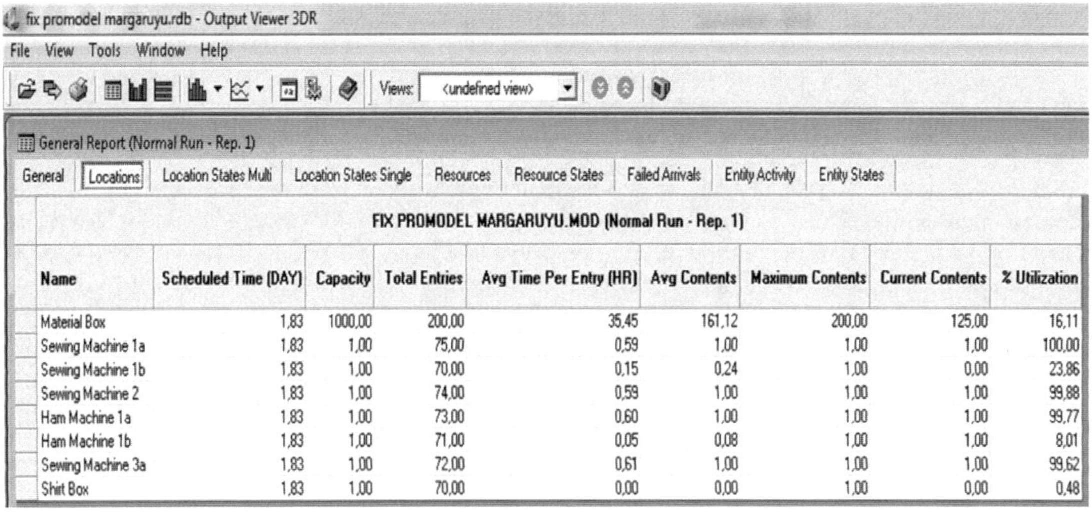

FIX PROMODEL MARGARUYU.MOD (Normal Run - Rep. 1)

Name	Scheduled Time (DAY)	Capacity	Total Entries	Avg Time Per Entry (HR)	Avg Contents	Maximum Contents	Current Contents	% Utilization
Material Box	1,83	1000,00	200,00	35,45	161,12	200,00	125,00	16,11
Sewing Machine 1a	1,83	1,00	75,00	0,59	1,00	1,00	1,00	100,00
Sewing Machine 1b	1,83	1,00	70,00	0,15	0,24	1,00	0,00	23,86
Sewing Machine 2	1,83	1,00	74,00	0,59	1,00	1,00	1,00	99,88
Ham Machine 1a	1,83	1,00	73,00	0,60	1,00	1,00	1,00	99,77
Ham Machine 1b	1,83	1,00	71,00	0,05	0,08	1,00	1,00	8,01
Sewing Machine 3a	1,83	1,00	72,00	0,61	1,00	1,00	1,00	99,62
Shirt Box	1,83	1,00	70,00	0,00	0,00	1,00	0,00	0,48

Figure 2. Promodel Results at CMT Margaluyu Village

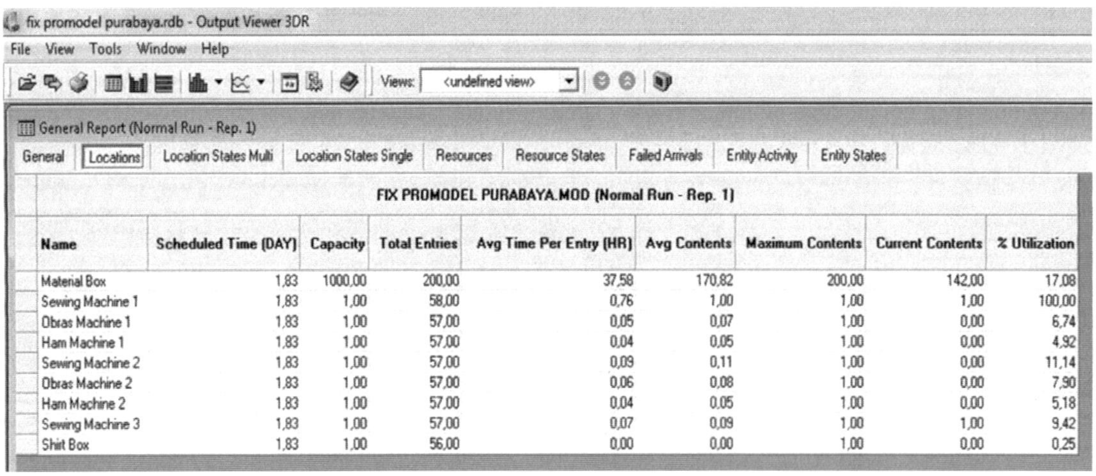

FIX PROMODEL PURABAYA.MOD (Normal Run - Rep. 1)

Name	Scheduled Time (DAY)	Capacity	Total Entries	Avg Time Per Entry (HR)	Avg Contents	Maximum Contents	Current Contents	% Utilization
Material Box	1,83	1000,00	200,00	37,58	170,82	200,00	142,00	17,08
Sewing Machine 1	1,83	1,00	58,00	0,76	1,00	1,00	1,00	100,00
Obras Machine 1	1,83	1,00	57,00	0,05	0,07	1,00	0,00	6,74
Ham Machine 1	1,83	1,00	57,00	0,04	0,05	1,00	0,00	4,92
Sewing Machine 2	1,83	1,00	57,00	0,09	0,11	1,00	0,00	11,14
Obras Machine 2	1,83	1,00	57,00	0,06	0,08	1,00	0,00	7,90
Ham Machine 2	1,83	1,00	57,00	0,04	0,05	1,00	0,00	5,18
Sewing Machine 3	1,83	1,00	57,00	0,07	0,09	1,00	1,00	9,42
Shirt Box	1,83	1,00	56,00	0,00	0,00	1,00	0,00	0,25

Figure 3. Promodel Results at CMT Purabaya Village

It can be seen that at CMT Purabaya Village in 44 hours the workers can produce as many as 56 pieces of shirts, while in CMT Margaluyu Village with the same time the number of shirts produced is 70 pieces. We use the promodel to see that if the simulation is same with the real situation and the calculation we had. Based on the result, CMT Margaluyu Village has more produced the shirt rather than CMT Purabaya Village, with the same time. Eventhough CMT Margaluyu Village only has 4 workers and 2 sewing machines and 1 ham machine while CMT Purabaya Village has 8 workers with 4 sewing machines, 1 ham machine and 1 serger, it seems like the process method that CMT Margaluyu Village (the sequent working process) is more effective than the process method that the worker should run from one work station to another work station until the shirt is finished.

4. Conclusion

Based on the result of the research on the standard time of child shirt production in SME FBS located in Sukabumi, with 2 CMT locations that is CMT Margaluyu and CMT Purabaya Village, the result is that the standard time of shirt work at CMT Margaluyu Village is less than that of at CMT Purabaya Village. It appears that the more effective in the production of sewing shirts in SME FBS is by process method, each person has his own type of work, because it is feared that if with the method 1 worker must make 1 shirt from the beginning until the shirt is finished, it will create cummulation at a work station and increase the time because the worker must walk to one work station to another work station.
Based on the calculation of TOC, the result is that the constraint at 2 CMT is the work station on the sewing machine.

5. Acknowledgement

Thanks to Directorate of Research and Community Service of the Ministry of Research and Technology of the Republic of Indonesia which funded this research activity in 2017 with an Applied Product Research scheme.

6. References

[1] Budi A 2007 Studi Kritis Atas Uji Kecukupan Data. *Industrial and Systems Engineering Assessment Journal Bina Nusantara University* pp 82 - 87
[2] Barnes and Ralph M 1980 *Motion and Time Study* (Singapore: John Wiley & Son)
[3] Richard J T 1994 *Principles of Inventory and Materials Management* 4th Edition. (New Jersey: Prentice Hall) pp 401-443
[4] Mariyatul Q 2016 Enchancing Competitiveness of Textile and Clothing Small - Medium Industries Through Performance Measure of Material Plan Using Scor Method *9th Int. Conf. on Industrial Engineering and Management Journal* pp 77 -81

Analysis of information quality attribute for SME towards adoption of research result

E Febriani[1], W S Dewobroto[2] and R D Anggraini[1]

[1] Department of Industrial Engineering, Faculty of Industrial Technology, Trisakti University, Jakarta Barat, Indonesia
[2] Podomoro University Centre of Entrepreneursial Leadership, Jakarta Barat, Indonesia
[1] elfira.febriani@trisakti.ac.id, rannydwianggraini@gmail.com
[2] wisnu.s@gmail.com

Abstract. Small and Medium Enterprises (SME) holds significant role in fostering Indonesian economy. However, the research that is supposed to support the development of SMEs business has not yet fully adopted or utilized. Information attributes may be used as the benchmark to find the intention of SMEs from a research result and develop the strategy of quality information for all organizations both SMEs and the researcher. Therefore, because of the importance of information quality attribute required by SMEs, the research aims to analyses the information quality required by SMEs to clarify the information quality into the dimension of information quality. The research was started by distributing online questionnaire to SMEs. The questionnaire result showed that the content dimension is the most aspect required by SMEs, followed by time and form dimension, respectively. Quality information attribute required by SMEs from a research is that the result may be applied to the business.
Keywords: SMEs, quality information, dimension of information quality

1. Introduction

Small Medium Enterprises (SMEs) is an individual business or business with legal entity that runs the activity in economy and operated in a simple manner with the purpose to gain profit with certain limitations [1]. In developed countries, the number of employees working in one enterprise or establishment tends to be one of the main criteria used in size categorization of SMEs [2]. SMEs as one of the alternatives of new job opportunity also contributes in fostering the economic growth during post-monetary crisis in 1997 when the large enterprises experienced difficulties in developing their business. Additionally, SMEs also has significant contribution to the regional income and state income of Indonesia.

However currently, the adoption or absorption of research result (both from Universities and research agencies) has unfortunately not yet fully utilized by the community, especially the SMEs. SMEs especially in the manufacturing sector make a significant contribution to economic growth, yet most of the research sector has focused on large organizations [3]. Therefore, it is necessary to have a link or media to facilitate the SME and researchers so that the requirements of SMEs may be fulfilled and the research result may be fully utilized. In order to realize this matter, it is necessary to understand what information attributes are required by SMEs from a researcher. Information attributes may be used as the benchmark to improve the effectiveness of information system and to develop the strategy of quality information for all organizations both SMEs and the researcher [4].

Quality information shows that information is presented according to the expectation and requirements of the users based on the dimensions of information quality. Useful information must be supported by three pillars of (1) people-targeted (accurate); (2) relevant, that is the information must confirm the data and target of the data or information so that the information will be beneficial to the users; and (3) on time [5].

Quality dimension is important to see the condition of quality information, which may be considered from several aspects. According to James A. O'Brien [6] , three dimensions of information quality are time, content and form. The time dimension describes the time period that the information deals with and the frequency at which the information is received, content dimension describes the scope and contents of the information, and the form dimension describes how the information is presented to the recipient.

By understanding what kind of information is required and desired by SME, it is expected that in the future, the research result may be fully adopted by SMEs. Therefore, because of the importance of information quality attribute required by SMEs, the research aims to analyze the information quality required by SME to clarify the information quality into the dimension of information quality.

2. Methods

The method used in this research is descriptive analysis method. Descriptive analysis is a method that aims to describe or provide an overview of a research object under study through samples or data that have been collected and make conclusions generally accepted [7]. This research is included into descriptive research to give clear and detailed description on the information quality required by SME so that they may adopt the research result.

Data collection was conducted using interview, questionnaire distribution, and observation. The selected respondents were the owner of SMEs in Jakarta, Bogor, Depok, Tangerang and Bekasi from various business types. The number of respondents is 124 respondents.

Data collection was assisted by online. Questionnaire is a research instrument consisting of a series of questions (or other types of prompts) for the purpose of gathering information from respondents. The questions in the questionnaire were related to the facts and opinion of the respondents, and the questionnaire is closed type where respondents were requested to answer the questions by choosing the provided alternatives. Characteristics questionnaire of SME actors were multiple choices which aimed to find out the characteristics of markets and target customers. Attribute questionnaire was used to find the factors that mostly influenced the customers (SME) in terms of research to support the business.

In preparing the questionnaire, the research utilized Likert scale. Likert scale is used to measure the attitude, opinion, and perception of individual or a group of individuals regarding the social phenomenon. Data measurement used in this research was Likert scale to measure the research variables such as the attitude, opinion, and perception of individual or a group of individuals [7].

3. Result and discussion

3.1. Description of respondents

The questionnaire result (Figure 1) shows that the respondents who run a business or have a business were dominated by male of 57% and female of 43%. The questionnaire result did not represent the overall SME actors in because Indonesia is home by millions of businesswomen.

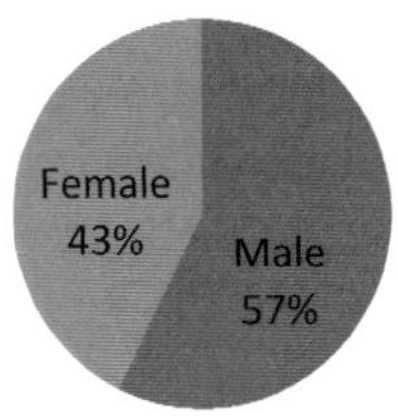

Figure 1. Sex of respondents

From the questionnaire result in figure 2, it shows 75% of the respondents were 20-39 years old, 23% of 40-40 years old, 1% of 50-59 years old, and 1% of < 20 years old. The result shows that many business persons were between 20-29 years old. Majority of SME actors were in the productive age range, with the population of 57.5%.

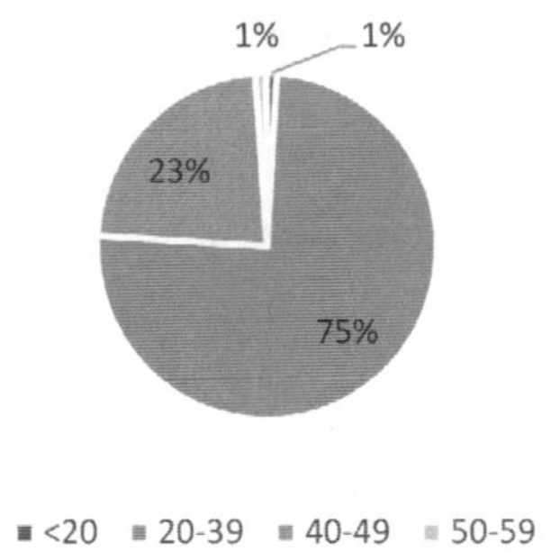

Figure 2. Age of respondents

In figure 3, there are 31% respondents who has been in the business for 3-5 years. Then 27% respondents have been in the business for 1-3 years, and 12% were less than one year and 11% has been in the business for more than 9 years. The percentage of length of business above 5 years is decreasing because many businesses fails in the first years. Many factors have caused the business failed to survive such as the competition, low commitment in doing the business as well as financial problems. Not only in Indonesia, about 20% of small businesses fail in their first year, and 50% of small businesses fail in their fifth year in the US [8]. Lu and Beamish [9] observed similar failure rates in Australia, the United Kingdom, Japan, Taiwan, and Hong Kong. Wheelen and Hunger [10] found the high failure rate to be largely due to informal strategic planning processes and a lack of systems to keep track of the SMEs' performance.

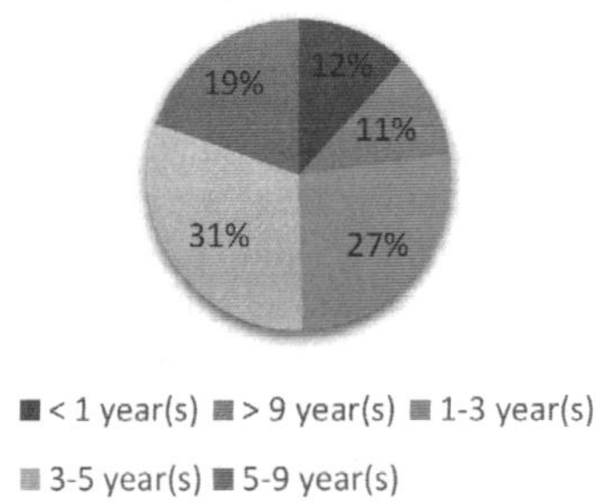

Figure 3. Length of respondents' SME business

The questionnaire results on the turnover or income (Figure 4) of the respondents shows that mostly less than 300 million IDR, which accounts for 56%. Then 29% respondents have the turnover of 301 million to 2.5 billion and 13% respondents have annual profit of 2.5 billion to 50 billion. This shows that the majority types of business are micro business.

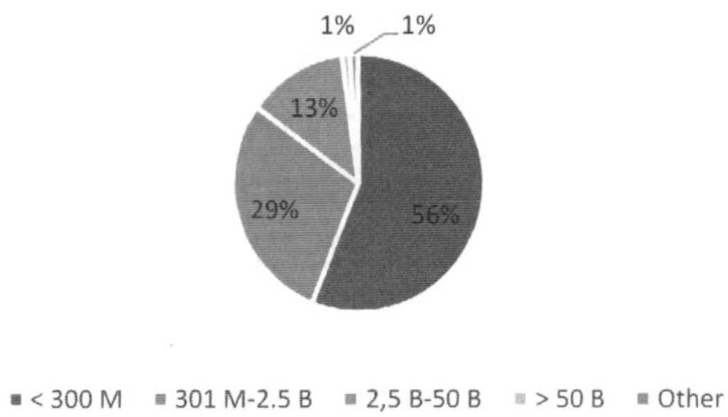

Figure 4. Average annual business income of the respondents

Micro business in Indonesia emerged largely due to the limited formal business opportunities, that may people switched to other business filed and started small business. This development shows the

IOP Conf. Series: Materials Science and Engineering **277** (2017) 012071 doi:10.1088/1757-899X/277/1/012071

growth of low quality economy during post-monetary crisis in 1998 where mostly developed sectors were informal sectors

In Figure 5., about 57% respondents said to unable to do their own research for the business and 43% of the respondents said to have their own research. This shows that the actors of SME indeed required other parties to do the research, especially in devising strategy and decision making for the business in the future.

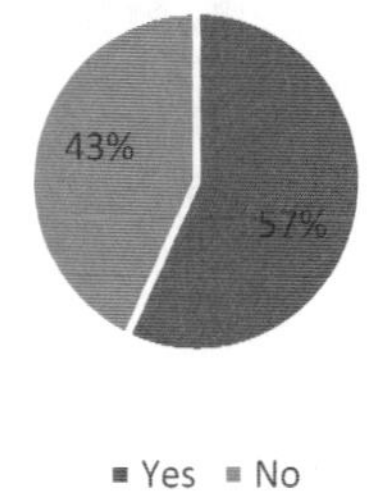

Figure 5. Capability of respondents to do the research

Then, in Figure 6, in terms of research costs, majority of respondents of 38% answered to be able to pay IDR. 2,000,000. The second highest of 30% respondents were able to pay research service of IDR. 5,000,000. Only few (8%) who were able to pay research above IDR. 15,000,000. These may be related to the previous questionnaire result that showed that most business of the respondents were in micro category. Therefore, capability of research was insignificant.

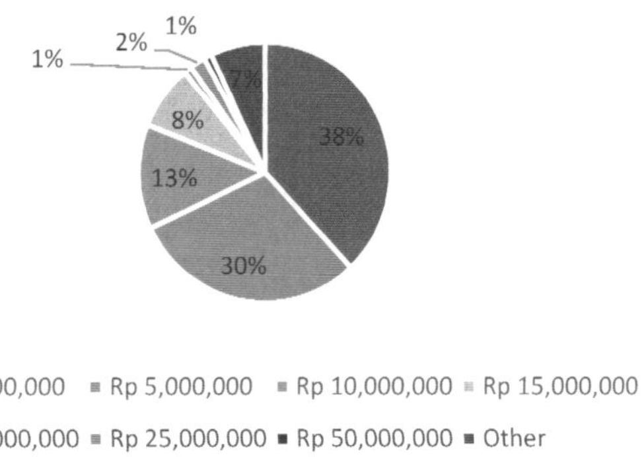

Figure 6. Capability of SME to fund the research

Business fields based on the respondents result in this research were mostly trading of 42 respondents as shown in Figure 7. The second highest business fields were culinary of 34 respondents. Trading business may include small trading (retailer), wholesale, agency, and also export and import. The questionnaire result gives the information related to the previous result that showed that the respondents in trading field require research as the effort to better develop their business.

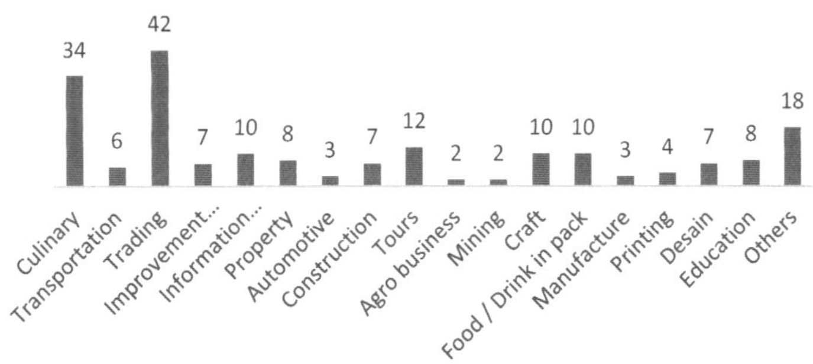

Figure 7. Business field of the respondents

When respondents were asked the importance of research result to their business, as much 98% answered that the research is the most important matter, as shown in Figure 8. Although the market research does not guarantee 100 percent successful business, this step may reduce the business risk and expand the potential of business success.

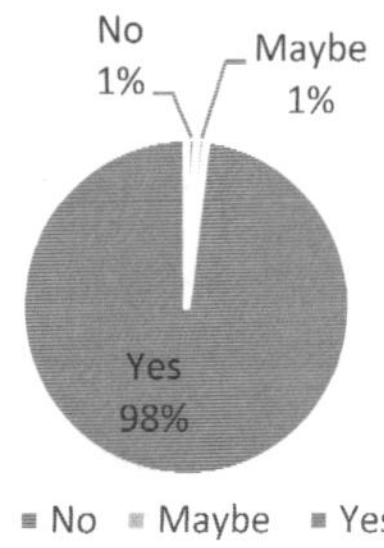

Figure 8. Percentage of research interest for SME

Based on this research SMEs urgently require a research for their business. Even from a business beginner to the already doing business in a long time, they need it. SMEs need improvement as well as large organizations whatever their field business.

3.2. Information Quality Attribute

Research is a process of steps used to collect and analyse information to increase understanding of a topic or issue [11]. The research should be feasible for adoption by community or SMEs from the beginning to help developing their business.

To find out what kind of information is required by SMEs, the questionnaire was distributed online to the respondents (SMEs). The questionnaire is a measurement tool to find out the importance information attribute from a research result for SMEs. The followings are the Information Quality Attribute [6] which are adopted in form of questions to be distributed to the SME owners.

Table 1. Information quality attribute for questionnaire

No.	Attribute
1	Information from the research result must be up to date
2	Information from the research result must be available at any time necessary
3	Research result must be able to present information as frequent as necessary
4	Research result must be able to present past, present and future information
5	Information must be error-free
6	Research result must be able to present information relevant to the required information
7	Information must be complete/comprehensive
8	Research result must be applicable to the business
9	Research result must be able to present required (accurate) information
10	Information has broad or narrow scope according to the requirement
11	Information must be able to present work performance (progress, completed job, current achievement, etc.)
12	Information presentation must be easy to understood
13	Information must be presented in detail
14	Information may be adjusted for future situation
15	Information is presented in visual (graphic, info graph, etc.)
16	Information is presented in various media (paper, video, etc.)

Quality dimension is said to be the condition of quality information, which may be considered from several aspects. According to O'Brien [6], three dimensions of information quality are time, content and form. Time dimension emphasizes that the information should be provided when needed, be up-to-date when it is provided, should be provided as often as needed and having information about the past, present, and future time periods.

Content dimension emphasizes that information should be free from errors, should be related to the information needs of a specific recipient, all the information needed must be provided, only the needed information should be provided, information can have a broad or narrow scope, or an internal or external focus, and information can reveal performance by measuring activities accomplished and progress made. Form dimension emphasizes that that information must be attractive and easy to understand and use. [6]

Quality dimension is the respondent (SMEs) requirements to be accepted when the research is conducted to their business. Figure 9 is the result of classification of information attribute to the quality dimension. Content dimension consists of the types of information required by SMEs such as error-free information, and the research result must be able to present information relevant to the required information, present complete/comprehensive information, the research result must be applicable to the business, the research result must be able to present the required information (accurate), information has broad or narrow scope according to the requirement, and information is able to present work achievement (progress, completed job, current achievement, and so forth). Percentage of content dimension is higher than other dimensions.

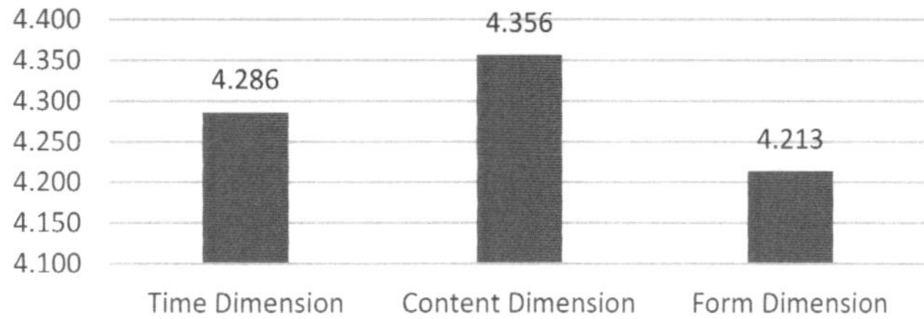

Figure 9. Quality dimension of SME information

Time dimension has the second largest percentage, that is the information from the research result must be up to date, information from the research result must be available at any time necessary, research result must be able to present information as frequent as necessary, and research result must be able to present past, present and future information. The third dimension is form dimension, which describes the what form of research result is required by SMEs such as the presentation must be easy to understood, information must be presented in detail, information may be adjusted for future situation, information must be presented in visual (graphic, info graph, etc.), and information must be presented in visual (graphic, info graph, etc.).

Based on the previous questionnaire result, it was find out that the content dimension has the highest percentage required by SMEs if the research is to be used for their business. The graphic of information quality attribute shows in detail the respondent requirement for the research result (Figure 10). The three attributes of information quality required by the respondents (SMEs) if the research is to be done then the result should be applicable to the business, the result must be able to present the required information (accurate) and information from the research result must be up to date.

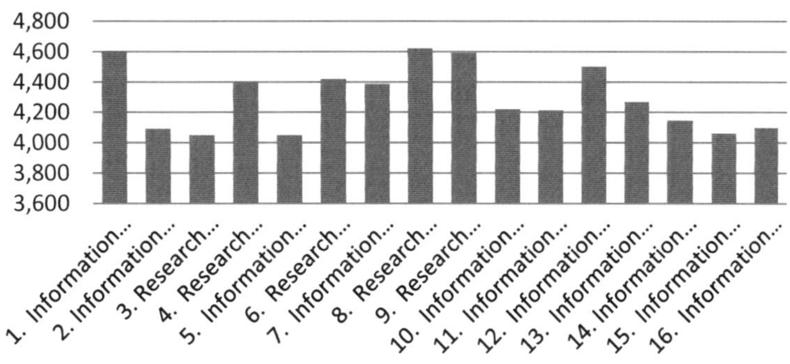

Figure 10. Information quality attribute of SME

4. Conclusions

Based on the research, it was found that respondents with length of business from one year is unable to do their own research for the business. They claimed that can pay research fees according to the income they earn. Because research is important for them, information quality needed. Dimension of information quality which important based on the research is the content dimension has the highest percentage required by SMEs, then followed by time dimension and form dimension. The graphic of information quality attribute shows in detail the respondent requirement for the research result. The most important information quality attribute required by the respondents (SMEs) if the research is to be done then the result should be applicable to the business, the result must be able to present the required information (accurate) and information from the research result must be up to date. The three attributes are the most important and required by SMEs if a research is to be done to their business.

5. References

[1] Anoraga P and Sudantoko H D 2002 *Koperasi, Kewirausahaan dan Usaha Kecil* (Jakarta: Penyalur Tunggal, Rineka Cipta)

[2] Jamil C M and Mohamed R 2012 Performance Measurement System (PMS) In Small Medium Enterprises (SMES): A Practical Modified Framework *W.J. Soc. Sci* **1** 200

[3] Terziovski M 2010 *J. Strat. Manag* **31** 892

[4] Maniah and Hamidin 2017 *Analisis dan Perancangan Sistem Informasi: Pembahasan Secara Praktis dengan Contoh Kasus* (Yogyakarta: Deepublish)

[5] Offset J 2009 *Sistem Teknologi Informasi* (Ed. III) (Yogyakarta: Andi Offset)

[6] J. A. O'Brien 2010 *Introduction to Information Systems: Business and Managerial Perspectives. 12th edition* (McGraw-Hill Education)

[7] Sugiyono 2009 *Metode penelitian kuantitatif, kualitatif dan R&D* (Bandung: Alfabeta)

[8] J McIntyre G 2017 What Percentage of Small Businesses Fail? [Online]. Available: https://www.fundera.com/blog/what-percentage-of-small-businesses-fail

[9] Lu J W and Beamish P W 2001 *J. Strat. Manag* **22** 565

[10] Wheelen T L and Hunger J D 1999 *Strategic Management and Business Policy* (Addison-Wesley)

[11] Creswell J W 2008 *Educational Research: Planning, Conducting, and Evaluating Quantitative and Qualitative Research* (3rd ed) (Upper Saddle River: Pearson)

IOP Conf. Series: Materials Science and Engineering **277** (2017) 012072 doi:10.1088/1757-899X/277/1/012072

Cost of unreliability method to estimate loss of revenue based on unreliability data: Case study of Printing Company

Judi alhilman

Industrial and System Engineering School Telkom University
Jl.Telekomunikasi Terusan Buah Batu, Bandung, West Java, 40257, Indonesia
alhilman@telkomuniversity.ac.id

Abstract. In the production line process of the printing office, the reliability of the printing machine plays a very important role, if the machine fail it can disrupt production target so that the company will suffer huge financial loss. One method to calculate the financial loss cause by machine failure is use the Cost of Unreliability(COUR) method. COUR method works based on down time machine and costs associated with unreliability data. Based on the calculation of COUR method, so the sum of cost due to unreliability printing machine during active repair time and downtime is $ 1003,747.00.
Keywords: Unreliability, COUR, Financial loss.

1. Introduction

In a production business, a lot of factors that affect the profits and losses of the company, one of them is due to the unreliability of machines in the production line. If the reliability of the machine decreases then the production will stop and will interfere to the production target that has been determined, this will certainly impact on loss revenue that should be obtained. To estimate the potential losses caused by unreliability machine it will be calculated by COUR method. This method is chosen because the calculation based on downtime data history from printing machine and based on costs that relation to down time printing machine, so the results of this calculation is expected close to real conditions. All manufacture industries must ensure that their production is accordance with demands, one of them is that the machine must reliable enough. [1]. The manufacturing industry strives to find ways to reduce the cost of production and materials in order to always be competitive and sustainable in the business world. It is very difficult to produce with good quality, low production cost regardless of engine reliability. As long as machine condition becomes a good element and source for the company in reducing cost, evaluation and selection of machine conditions become an important component in the production line. [1]. There have been many studies of potential loss calculations using the COUR method but it is still rare to examine to the printing press. In the printing industry, the problem also still occurs where the machine suddenly breakdown while the production running. If the printing machine fail to operate, it will disrupt the newspaper production activity which will impact to unstable demand fulfilment [2]. Based on previous research using COUR method in industry area, the research on printing machine using COUR method is very useful because from the results of the research we can know the loss of revenue caused by unreliability machine.

2. Methods

2.1. COUR Model

Calculation by COUR method will get the result of loss potential revenue due to the inability of a machine in production line. Unreliable systems will show a high COUR value [2]. Here is a model that explains the costs used in COUR calculations.

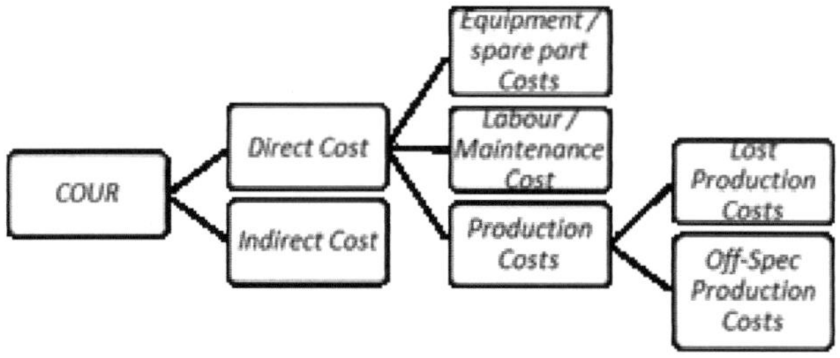

Figure 1. Cost of Unreliability Model

$$COUR = DC + IC \quad (1)$$

DC: Direct Cost
IC: Indirect Cost

2.2. Direct Cost

Direct cost is all costs that have a direct cause and effect relationship with reliability events. Direct cost is influenced by following aspects:

- Equipment / spare parts costs (EC): costs used for replacement of machinery or engine components, such as spare part purchases.
- Labor Maintenance cost (LC): the cost used to pay for labor in maintenance and repair activities.
- Production cost (PC): costs incurred due to loss of production, consisting of:
  - Loss Production costs (LPC): revenue that should be earned if production activity is not disrupted by production failure.
  - Off-Spec Productions costs (OPC): costs incurred because the resulting product is incompatible with product specifications, which may mean that it cannot be sold because of reject.

$$DC = EC + LC + PC \quad (2)$$

2.3. Indirect Cost

According to (Daley, 2009), indirect cost of COUR is a cost charge that does not have a direct causal relationship with reliability events. Costs included in the indirect cost are as follows.

- The Cost of being a reactive organization: organization or company conducting reactive maintenance activities, such as predictive maintenance and preventive maintenance.
- The Cost of Sloppiness: poor reliability is usually associated with other elements of production, such as health & safety, quality and environmental performance.
- The Cost of Loss Business: poor reliability can affect production targets and production quantities, so customer needs are not met, resulting in dissatisfaction and loss of customers that can result in loss of profits that should be obtained.

3. Result

3.1. Data Collection and Processing

Data collection is taken from a large printing company in Bandung that has distribution of its printing products throughout West Java, data used in 2016. Data collected are damage duration time, time to repair, wage and cost data related to unreliability machine. Maintenance activities on the printing machine is done by maintenance crews with details as follows: team maintenance consists of 4 people crews, each get a wage of IDR. 5000,000,00 / month. The total wage for 4 maintenance crews is IDR. 240,000,000,00 / year. Loss profit (LP) calculation is: LP = total production / hour x production price = 40,000.00 exp. x IDR 3000.00 = IDR. 120,000,000.00/hour, it is mean that if the printing machine down then the company will loss revenue of IDR. 120,000,000.00 / hour for one machine.

3.2. Determination of Representative Distribution

To find out the most representative distribution of Time to Failure (TTF) and Time to Repair (TTR) data for each machine, a data distribution match against the Weibull, Exponential and Normal distributions . Determination the distribution representing TTF and TTR distribution of printing machine can be seen in figure 2.

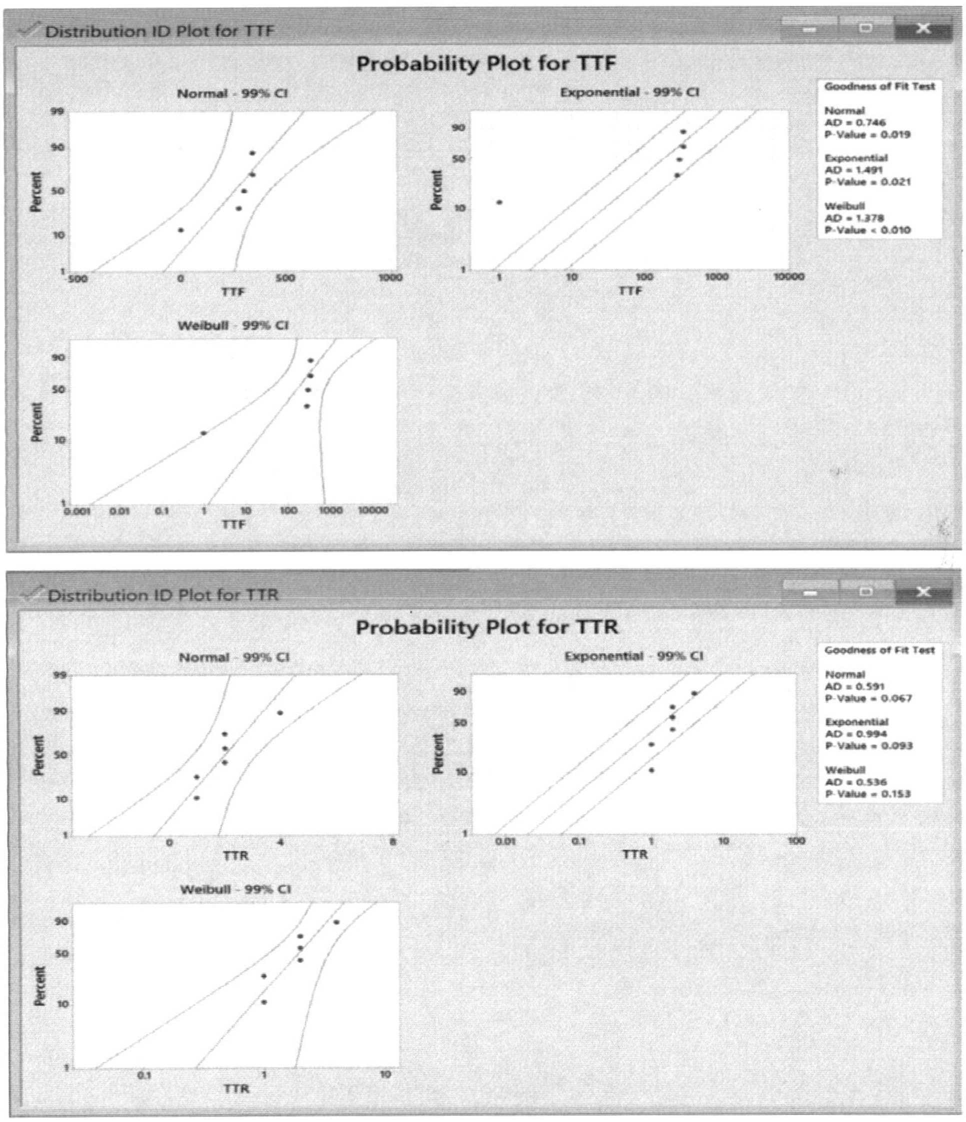

Figure 2. Distribution of TTF and TTR data

The selected distribution is Normal distribution. From each distribution it is found that Mean Time To Failure (MTTF) and Mean Time To Repair (MTTR) for normal distribution with parameter value equal to 192.85 and 2.075 respectively

3.3. Cost of Unreliability Calculation

Failure Rate Calculation, the data used for COUR calculations are machine failure data from January to December 2016. The first stage of COUR calculation is to calculate the failure rate of the unit.

$$\text{Failure rate} = 1/\text{MTTF} \tag{3}$$

In this study, interval study was determined during the observation time as long as 8640 hours. Details of failures number during the observation period, study interval and calculation of the failure rate can be seen in Table 1.

Table 1. Failure Rate Calculation Results

	MACHINE
Study Interval (hours)	8640
Number of Failures	27
MTTF	192.85
Failure Rate	0.0051854

3.4. Corrective Loss Time and Downtime Loss

The value of loss time is obtained from Corrective Loss Time (CLT) and Downtime Loss (DTL) calculations, as follow.

$$\text{CLT / Years} = \text{Corrective Time / Failure x S} \tag{4}$$

$$\text{DTL Hours / Years} = \text{Downtime Time x S} \tag{5}$$

Details results of Loss Time calculations can be seen in Table 2.

Table 2. Corrective Loss Time and Downtime Loss Calculations

	MACHINE
Failure Rate	0.0051854
Number of Failures	27
Correvtive Time/Failure (MTTR)	2.075907
Downtime Time Hours	2.5244
Corrective Loss Time Hours/Years	56.049489
DT Time Hours/Years	68.1588

3.5. Money Loss Calculation

The next stage of COUR calculation is to calculate the value of money loss. To obtain the calculation of money loss required data downtime loss, loss time, loss production cost, equipment / spare parts cost, and labour maintenance cost.

$$\text{LPC} = \text{loss time x Production Loss / hour} \quad (6)$$

$$\text{EC} = \text{loss time x maintenance cost / hour} \quad (7)$$

$$\text{LC} = \text{loss time x LC / hour} \quad (8)$$

$$\text{Value of corrective COUR} = \text{LPC} + \text{EC} + \text{LC} \quad (9)$$

Assumption loss profit / hour IDR. 120,000,000.00, maintenance cost IDR. 600,000.00/hour, labour maintenance cost is IDR. 20,000,000.00/month. Details of COUR calculations from one machine can be seen in Tables 3 and 4.

Table 3. Corrective COUR Calculation

	MACHINE
Corrective Loss Time Hours/Years	56.049489
Loss Product Cost	IDR. 6,725,938,680
Spare Part Cost	IDR. 33,629,693.4
Labour Maintenance Cost	IDR. 11,209,897.8
Corrective COUR	IDR. 6,770,778,271 = $451,385.00

Table 4. Downtime COUR Calculation

	MACHINE
DT Loss Time Hours/Years	68.1588
Loss Production Cost	IDR. 8,230,560,000
Spare Part Cost	IDR. 41,152,800
Labor Maintenance Cost	IDR. 13,717,600
DT COUR	IDR. 8,285,430,400 = $552,362.00

The formula to calculate Downtime COUR is use the same formula with corrective COUR but different in LPC calculation, for corrective COUR use Corrective loss time hours/year but for downtime (DT) COUR use DT lost time hours/year. The calculated costs converted into US Dollar ($) units, with exchange rate with rupiah is IDR 15,000/ 1 $.

4. Discussion

4.1. Failure Rate Analysis

The printing machine has a failure rate is 0.52% with MTTF of 192.85 hours. This indicates that the possible functional failure will occur in the unit is not too large because the value of failure rate close to zero, so the company is sufficient to apply preventive maintenance policies, such as inspection, or on-condition maintenance to prevent downtime that may occur due to machine failure.

4.2. Money Loss Analysis

Money loss calculations are used to help identify problems regarding loss income or expenditure incurred due to reliability issues. The analysis of money loss based on corrective time and downtime can be seen in Table 3 and 4 above. It is shows the relationship between cost elements that arise due to reliability issues. Between loss time and the cost are interconnected, the higher the loss hour per year, the cost of the company will be greater. Based on the money loss, which causes the money loss this printing machine unit is high due to high labour maintenance cost and high corrective loss time / year. This shows that the reliability problem will have a very bad impact on the cost that must be issued by the company. The high value of money loss in the printing machine is caused by the amount of loss time that resulting the production process will stop, so that income is not earned. Can be concluded that the amount of time loss is very influential on the amount of money loss that occurred.

5. Conclusion

The reliability of a machine is very influential to earn amount of company revenue, based on the cost calculation with Cost of Unreliability method, the cost caused by the system's inactivity based on active repair time is $ 451,385.00, and $ 552,362.00 based on downtime, so the total is $ 1003,747.0, its mean the company will suffer a loss of revenue due to the unreliability machine per year is $ 1,003,747.00.

6. References

[1] G Kanagaraj S G Ponnambalam and N Jawahar 2016 Reliability-based total cost of ownership approach for supplier selection using cuckoo-inspired hybrid algorithm *Int. J. Adv (Manuf. Technol* vol. **84**) no. **5– 8** pp 801–816

[2] F Vicente 2012. Assessing the cost of unreliability in gas plant to have a sustainable operation (Europe: Petroleum and Chemical Industry Conference Europe *Conference Proceedings PCIC* EUROPE)

[3] Daley 2009 The cost of unreliability: a case study *(Journal of Quality in Maintenance Engineering)* volume **4** issue **3**

[4] Ebeling C E 1997. *An Introduction to Reliability and Maintability Engineering* the McGraw-Hill Companies

[5] Martinez L B 2008 RAM Analysis of Mining Process: A Case Study of a Copper Smelting Process in the Field of Mining (Chile: *IFAC Conference*)

[6] Moubray J 1997 *Reliability Centered Maintenance Second Edition* Industrial Press Inc.

[7] A Kleyner and P Sandborn 2006 Forecasting the Cost of Unreliability for Products with Two Dimensional Warranties. *Safety and Reliability for Managing Risk* – Guedes Soares & Zio (eds) Taylor & Francis Group, London, ISBN 0-415-41620-5

[8] P V Suresh and K Sudhakar 2013 Life cycle cost assessment of solar-wind-biomass hybrid energy system for energy centre MANIT (Bhopal: *Green Computing Communication and Conservation of Energy* (ICGCE)) pp 635-639

[9] Blanchard B S and Fabrycky W J 2006 *System Engineering and Analysis* Prentice Hall International Series in Industrial & Systems Engineering.

[10] Alhilman Judi, Fransiskus Tatas Dwi Atmaji and Nurdinintya Athari 2017 Software Application for Maintenance System: A Combination of Maintenance Methods in Printing Industry. *ICOICT Conference Proceedings*

[11] A Pride 2013 *Reliability-Centered Maintenance (RCM)* Wbdg vol. **15** pp 1-15

Implementation of virtual corporation: the challenge of future business in Indonesia

Dadan Umar Daihani

Doctoral Program in Strategy, Universitas Trisakti, Jakarta, Indonesia
Magister Program in Industrial Engineering, Universitas Trisakti, Jakarta. Indonesia
dadan@trisakti.ac.id

Abstract This paper presents a descriptive analysis of the development of virtual organizations in the business. Recently, many commercials activities were carried virtually. Online shops are growing everywhere; furthermore hotel reservations, ticket booking, as well as purchasing process are all done through the cyberspace. In the government, administrative process has migrated to electronic systems such as E-budgeting, E-KTP, E-procurement and online visa application. This era is not excessive if it is said as the era of E-anything. To address the issue, we employed survey based study on APJII survey and other related sources. Furthermore, the data is analysed based on the theory of development of virtual organizations and virtual corporations by using the six social sciences paradigm. This study showed that the Internet users in Indonesia have increased progressively from year to year. Therefore, the e-commerce market in Indonesia is growing rapidly as well. By looking at the growth of Internet users and the reasons for their use, it can not be denied that the internet-based business in Indonesia is a very promising (from the electronic shop, logistics and online transportation (known as online taxi) such as, Gojek, Grab, UBER, etc). One thing that should be prepared is the wise and proper government regulation. The electronic technology-based business activity of course is a necessity, and can't be inhibited its development. However, if the infrastructure such as the regulation and culture of the community are not ready yet, then it can lead to the social conflict such as a conflict between conventional taxi and online taxi players (Gojek, Grab, Uber etc). We hope this paper will open the horizon and our mind of looking at the emergence of new patterns in business.
Keywords: Virtual Corporation, Virtual Company, Virtual Organization, E-commerce

1. Introduction

1.1 Background
Nowadays, the computer has been considered not as new thing. In almost every field, all activities require a computer. An expert says that he is not afraid of the computer but he is more terrified to be left behind by the development of computers. We are at the stage where all problems can't be solved without computers. I do not fear computer. I fear the lack of them (Issac Asimov).

Computers was created originally to only act as a tool for calculating, then turned into a data processing tool, then developed further as an information processing tools, even now has functioned as a knowledge processing tool. The database has changed into a knowledge base. Especially after the discovery of a networking system, capable of connecting various computers in the worldwide. With that said, computer seemed to change its meaning from the calculator into the entrance of a virtual world. Especially after the creation of internet technology.

The virtual world has become a relatively new world for various human activities, as well as in business. As we know, Amazon.com has long been known as an electronics store. The founder of this electronics store was Jeff Bezos in 1994. Initially this store was named Cadabra and its website began to be launched under the name amazon.com in 1995. The company was named by the name of the river; Amazon, one of the largest rivers in the world. Amazon.com started as an online bookstore, then became a seller of DVDs, CDs, MP3s, software, video games, electronic goods to furniture, food, toys and jewellery [3].

Now various similar sites have mushroomed and virtual companies have sprung up everywhere. As stated by Ulrich Franke of Cranfield University England and his colleague Bernd Hickmann, of RWTH Aachen, Germany that [17]: " Due to the accelerating development of information and communication technology (ICT) our business world has been changed. It is a double sided coin. On one side, sophisticated ICT generates higher customer expectations and intense competition (market side). On the other side of the coin, ICT enables economic actors to reorganize their value adding processes (organization side) in response to the market side ".

Widespread and very rapid influence of information technology advances in the business, then almost all business activity, has now migrated from conventional media to electronic media. Now known as e-education, e-gouverment, e-budgeting, e-procurement, e-commerse etc.

1.2 Problem Definition

Besides giving positive impact, of course, internet-based business development also emerge a negative impact in some manner, especially related to people's readiness, government regulation and the availability of supporting facilities. without doubt, through the internet, the reach of the market has increased widespread. A product that was previously only known locally, or regionally, can be recognized globally. So from the aspect of marketing is became easier, but on the other hand, the level of competition will be more complicated as well. The products offered in West Java (for example), formerly only compete in West Java region, but now have to compete also with products coming from abroad, because people already know it through global trade sites. This is what needs to be explored and studied, so that Indonesian entrepreneurs are able to take advantage of the development of the internet and not otherwise become victims of the progress of the internet.

Again it should be emphasized that business activity based on electronic technology is a necessity, and can not be inhibited its development. However, if the infrastructure and culture of the community is not ready, it can lead to social conflict. As recently, We were surprised by the conflict between conventional taxi and online taxi players (go-jeg, uber etc).

Based on the general description as mentioned above there are four major issues of this study, namely: 1). How far the development of internet usage in Indonesia 2). What are the characteristics of a virtual corporation? 3). What needs to be prepared to adapt to this new form? 4). What government policies need to be formulated to support the realization of virtual corporation.

2. Theoretical Bacground

2.1 The Concept of Virtual Coporation

An interesting question to contemplate in this digital age is "what is exactly the meaning of workplace?". Simply, workplace can be said to be just a place where people work. But if examined more closely, often the workplace has a fairly complex meaning, and have a varied understanding, but mainly physical. In this digital age, the workplace is not always interpreted strictly as a physical space or linked to a geographical location. Because with the discovery of World Wide Web technology then the space or workplace can now be digital. Currently we are doing a variety of real-life activities in digital space. Everyday we enter in to the internet and penetrate various aisles to locate the workspaces and also communicate in both directions as where it should be done in physical space [2]. This is all thanks to the growing development of internet-based technologies and applications.

The Internet itself was originally developed for military purposes in 1957. Three years later the

American military developed an increasingly sophisticated communication technology called Arpanet. This technology is basically the embryo of the birth of the internet. Having discovered TCP / IP technology, the use of computer networks is increasingly widespread and university researchers continue to work to develop this technology. This technology is then used not only for military purposes but business actors have already utilized it. Furthermore, after developed the World Wide Web application (WWW), then the use of internet widespread even has created a new world of social media (medsos).

The Internet has changed the way we communicate. Now, online media, such as email and chat rooms allow individuals to overcome time and space barriers. Everyone can interact directly with people who are far away. Through the internet, we can share information easily and quickly. We can send messages via email to many people (group) in one time.

Internet technology allows humans to interact widely. [2] With the discovery of the internet, the limits of work space become more lax. In the digital space, everything can be done, digital space is a modern workspace and not just a site, an office, a house, or a college course. This place is not an illusion. It is as real as the physical space we are now living.

Digital space or can be called also with the virtual world is very effective and efficient to serve as a business space. As we know that the main challenge facing today's companies is the increasingly complex competition. One alternative to achieving competitiveness is to adopt the principles of Virtual Corporation (VC). The concept of VC is basically like a cooperative capable of integrating various sources and competencies without having to be physically. For example a virtual travel agency offering services ranging from hotels around the world, to booking airline tickets and other tourism facilities. If carefully considered, in reality all entities involved in travel.com it does not have one company. They are only joined in a network and tied through a particular application. For consumers travelbiro.com can be considered a real company, in which incorporated various companies in a virtual corporation. Virtualness can be defined as the ability to offer a complete product or service to the customer, where the company itself has only some of its competencies [18], other competencies are supported by other companies through cooperation in a network.

2.2 Characteristics of Virtual Corporation

There are several notions about virtual which include [16] [17]:

Virtual means 'unreal, looking real'.

Virtual Reality' is a good example for this sub-concept of 'Virtual'. It simply means, that a 'Virtual Corporation' has the appearance of a real (traditional) company for externals, but in reality this company does not exist, it is only a conglomerate of independent network partners.

Virtual means 'immaterial, supported by information and communication technology (ICT)'.

This sub-concept of 'Virtual' means, that something does not physically exist, it is only created by data. For example, the 'Virtual Shopping Mall' does only exist on the Internet. The 'Virtual Office' does not exist in physical terms; employees work from home and are connected to each other by ICT.

Virtual means 'existing, but changing' The dynamic network follows this meaning of 'Virtual'.

The organizational unit is existing, but the composition of partners is temporary. This kind of organization reconfigures itself permanently; it is dynamic and progressive. On the company level such temporary networks are called 'Virtual Corporations', on the worker level such temporary networks are called 'Virtual Teams'.

In conclusion, the Virtualness is the ability of the organization to consistently obtain and co-ordinate critical competencies through its design of value-adding business processes and governance mechanisms involving external and internal constituencies to deliver differential, superior value in the market place.

Based on various virtual concepts as mentioned above, there are four basic characters of Virtual Organization, which are: 1) Differentiation, 2) Configuration, 3) Integration and 4) Technology. Quoted by Ulrich J. Franke from Bultje R and van Wijk J [18] [7].

Differentiation: Modularity and Heterogeneity.

IOP Conf. Series: Materials Science and Engineering **277** (2017) 012073 doi:10.1088/1757-899X/277/1/012073

To meet complex customer needs, a variety of different satisfiers (i.e. modules) with specific competencies and strengths are needed. Each satisfier develops its specific core competence to optimally meet a specific requirement. The flexible and dynamic combination of core competencies to unique value chains gives firms the opportunity to achieve competitive advantages by virtually increasing their resource and know how endowment

Configuration: Temporary and Loose Coupled Network.

Although the customer perceives the VC as one entity, it consists of independent companies, which are configured dynamically in a network according to the prevailing customer requirements. The overall structure of the VC is therefore determined by the configuration of the network and changes with customer requirements.

Integration : Trust as Coordination Mechanism.

Heterogeneity and the dynamical reconfiguration of independent core competencies re-quire clear criteria and rules for the assignment of tasks. Shared organizational goals and the co-destiny of participating satisfiers provide an important integration function. Beyond that, trust between the cooperating partners has a fundamental impact on the success of VCs.

2.3 Virtual Corporation Development

In line with the development of information and communication technology, the behaviour of business and media to do business is growing. In the past, business or trade was done by physical agents in physical space through physical processes. and traded products were physically shaped as well. Examples are traditional markets and conventional stores that trade goods directly. Buyers and sellers transact directly. In line with the discovery of a means of communication then, people transact through means of communication, such as telephone, telegram, telefax, etc. Similarly, the goods traded have begun to take the form of digital like e-books, digital music and movies in digital format.

Currently people transact using digital money digitally and goods in digital format. So it is foreseen that in the future will form a digital commerce community in cyberspace where the process, its players and also its products are all digital (see figure 1).

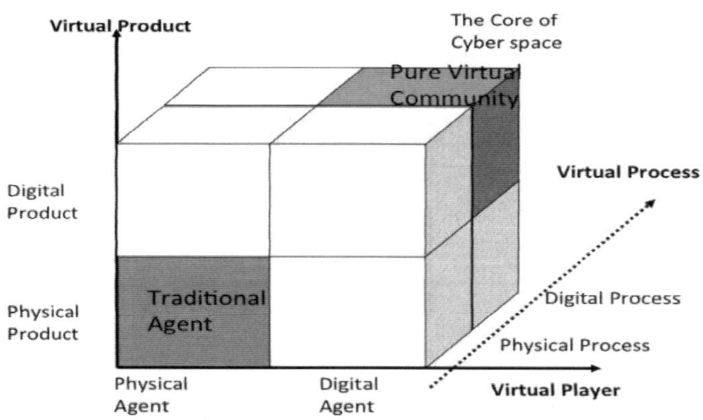

Figure 1. Form of business in the modern era

3. Research Method

3.1 The six Social Sciences Paradigm

The development of the social life of the community can be said as the transformation of a paradigm to the other paradigm. An American philosopher named Thomas Kuhn (1922-1996) was the first to define the concept of paradigms. In this study the approach will be based on The six social science paradigms are considered as the most important set of paradigms to understand and explain business activities. They represent the different categories of creation of knowledge about and in a given reality [3].

The six principles of social paradigms are as follows [3].

1. **Reality as concrete and conformable to law from a structure dependent of the observer** .i.e. an external, tangible and objective reality. The relations between the components are easily predictable and measurable because of their deterministic characteristic.
2. **Reality as the process of determining the concrete world.** That is a process that develops organically, is concrete but changes in detail. The relationship between components, considered "social facts", affects and is influenced by the environment.
3. **Reality as an interrelated field of information**. That is a constantly changing reality depending on the exchange of information. The relationship is dynamic, probabilistic and relative to the "relational network" component, because the individual reacts to and interprets the information received. The research method is based on gathering information about the results achieved, or "contextual dependency".
4. **Reality as a world of symbolic discourse.** It is an inter-subjective interpretation, based on dynamic exchange and negotiation that leads to mutual understanding and social rules. Relationships are based on culture, symbols and language, making the results of relative and specific methods of study for the immediate context.
5. **Reality as social construction.** We have no concrete status, the fact evolves through a continuous process, because of the subjective but subjective and subjective perception of reality. The research method is based on a "hermeneutic spiral"
6. **Reality as a manifestation of human intentionality**. That is, individuals create reality in their own consciousness or "transcendental consciousness". They are seen as deliberate beings, whose intentions are not conscious while creating knowledge is the main determinant for understanding.

So basically to understand a paradigm is a learning cycle process. Starting from the pre-understanding stage, then enter the stage of understanding, then to the post understanding stage. After that the process repeats, according to the context and purpose of the study. (See figure 2)

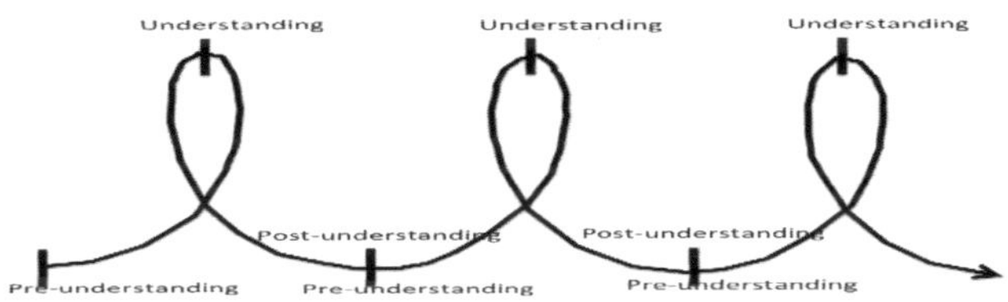

Figure 2: The hermeneutical spiral created by [11], inspired by [4]

3.2 Source of Information.

In accordance with the nature of the study is to examine the phenomenon that occurs in cyberspace, the approach used in this study is a descriptive method of analysis. The database and information used are secondary data sourced from various research documents and the results of previous studies. Further validation and valid test conducted by conducting a series of FGDs with various parties, both from academics, ministries and industry players.

4. Research Finding

4.1 The Growth of internet usage.

One of the driving force of the increasingly widespread electronic-based trading is the increasing number of people who use the internet. According to the survey conducted by the Association of Internet Service Providers Indonesia (APJII), internet service users in Indonesia from year to year increased rapidly. In 2016, internet service usage in Indonesia has reached 132.7 million people from 256, 2 million population of Indonesia or about 51.8%. Most of the internet users are the population of

Java island, which is about 86.3 million people or 65%. Whereas in 2015 the number of usage is still 88.1 million people. Thus every year there is a huge increase, as can be seen in the picture below. [1] In 2005 international users were still around 16 million, then increased sharply in 2006 to reach 20 million people, and then, ten years later had reached 132.7 million people or nearly seven times.

What is interesting to note is the reason for using the internet. If you see the reason, it is still the most related to finding information and update information related to work, entertainment. Meanwhile, the use of the internet to do business is increasing. This means that more and more people of Indonesia who use the Internet as a workspace or activity. (see table 1 below). Furthermore, the results of this survey revealed that commercial content is accessed by 123.5 million people, or about 93.1% of Internet users have accessed commercial content. Almost all internet users know that the Internet other than used for scientific problems is also used as a medium for the place of sale and purchase of goods and services. From this survey, it was revealed that there are about 130 million internet users in Indonesia who know that the internet is also used as a commercial medium, and 84, 2 million of whom have transacted on line.

Other interesting user characters to note is the age group. The most common age group of Internet users is a productive age group from the age of 25 to 55 years old. This shows that the use of the Internet as a workspace is very dominant. In detail can be seen in table 2 below

Table 1. Reasons for Internet Usage [1]

No	Reasons	Number	%
1	Updating Information	31,3	25,3
2	Related with Jobs	27,6	20,8
3	Leisure	17,9	13,5
4	Socialization	13,6	10,3
5	Education	12,2	9,2
6	Entertainment	11,7	8,8
7	Business, Trade and searching of goods/product	10,4	8,5

Table 2. Internet users viewed from the Age group [1]

No	Age Group (years)	Number	%
1	35 - 44	38,7	29,2
2	25 - 34	32,3	24,4
3	10 - 24	24,4	18,4
4	45 - 54	23,8	18
5	55 up	13,2	10

4.2 The development of E_Commerce.

By looking at the growth of internet users and the reasons for their use, it can not be denied that the internet-based business sector is a very promising business sector. Therefore, the e-commerce market is growing rapidly as well. With the number of internet users who reach 132 million people more or about 51.2% of the population in Indonesia, then the e-commerce market becomes a potential gold mine. Therefore, it is not surprising that the number of online stores from day to day is increasing. According to the Ministry of Communications and Information, in 2013, e-commerce transactions have reached Rp130 trillion . With reference to data from ICD, Rhein Mahatma predicts that e-commerce market in Indonesia in the period 2012-2015 will grow by 42%. This growth rate is estimated to be larger than other ASEAN countries such as Malaysia (14%), Thailand (22%), and Philippines (28%). This large growth rate has attracted many big investors both from within and abroad. Even VC (Venture Capital) such as Rocket Internet, Cyber Agent, East Ventures and Ideo Source have invested in several e-commerce companies based in Indonesia, such as LAZADA, ZALORA BERRYBENKA, TOKOPEDIA, BILNA, SAQINA, VIP PLAZA, RALI DLL. The world's e-commerce players are able to take advantage of a fairly large Indonesian market of 132.7 million internet usage.

The future of internet-based business is very promising, especially when associated with the number of middle class in Indonesia whose number is increasing from year to year. In 2015, the number of middle class Indonesia has reached 170 million people. This amount is a huge domestic market potential. The middle class is defined as a group with a daily expenditure of about US $ 2 - 20. This middle class, in Indonesia grew very rapidly from 45 million (25% of the population) in the early 2000s, then 10 years later reached 134 million people and now In 2015 reached 170 million people or 70% of the population.

Besides the domestic market, with the opening of AEC (ASEAN Economic Community) then e_commerce market will be more widespread. Currently it is estimated that the market potential in AEC reaches the number of 600 million consumers. What a huge market. If in the conventional era, ASEAN market must be reached physically by opening branch or cooperating with destination country, now just design electronic store, then the whole world will get to know national product.

4.3 The development of logistic and transport services based on Internet

The impact of the growing e_comerce, has now opened a variety of new jobs, including web designers. At the past, to open the store required physical land and strategic location, but now enough to design the site in cyberspace and then uploaded to the Internet network, then realized the electronic store. Besides business web designer, also developing transportation business and now also has penetrated the field of logistics (delivery of goods etc.). The delivery of goods business is growing not only to deliver goods from the electronic store only but has evolved into the purchase of goods from conventional stores as well. This business activity is now evolving from the application-based shuttle business, such as Uber, GoJek, Grab, etc. In Indonesia, the presence of transportation services creates a conflict between business-based transport apps (online) with conventional business people.

By using the application, online taxi players are able to provide many services and benefits for the customers. Some of the benefits that customers get are very easy way of booking, pick up direct from the customer's place of domicile, travel time is measurable well and most importantly the cost is relatively cheaper than conventional taxi.

To understand the development of this business, let us first note the early development of the emergence of online taxi named Uber. This field of business began with the establishment of 'Uber Cab' by Travis Kalanick and Garrett Camp in 2009 [8]. Travis stated that the idea of Uber's establishment emerged from his own experience. As he was going to attend LeWeb conference in Paris, Travis had trouble getting a taxi. Based on this experience, Travis then think it would be nice if there is a Taxi company that can be contacted anytime and anywhere.

From his experience, Travis founded UberCab in San Francisco. Apparently the company is growing rapidly and began to be established in various other cities (in 2011) including in New York City, Chicago and Washington, DC. Then developed also in Paris as the city of the birth of the idea. Uber Paris is the first city outside America and opened ahead of the Le Web International Conference in December 2011. Further expanded to Canada in 2012. Then In July 2012, the company launched its application in London in order to welcome the Olympics. At the beginning of its establishment there are already 90 partners who register themselves into drivers.

UberCab continues to expand almost everywhere in the world such as Australia, Singapore, South Korea, India, South Africa, China, Mexico, Poland, Denmark, Germany, Thailand and Nigeria. In its development, Indonesia began to bring up UberTaxi in 2014. In August 2015 this service has been available in 59 countries and more than 200 cities around the world.

As a new business, Uber is very successful worldwide, and is very helpful to Taxi users. Likewise investors are trying to invest in this business. Uber itself has absorbed about $ 49 million in 2011 business funds [7].

In Indonesia this online-based taxi has also developed rapidly. According to KOMPAS.com [11], just one year stands, Uber claims to have experienced a significant growth of 30 times, as asserted by Head of Communication Uber Southeast Asia Karun Arya. This "30-fold" growth figure refers to the number of Uber app downloads, though the data details are hard to tell. The number of partners (drivers) who joined Uber Indonesia has now reached 12 thousand. Besides, the area of operation has been spread to various major cities such as Bandung and Bali.

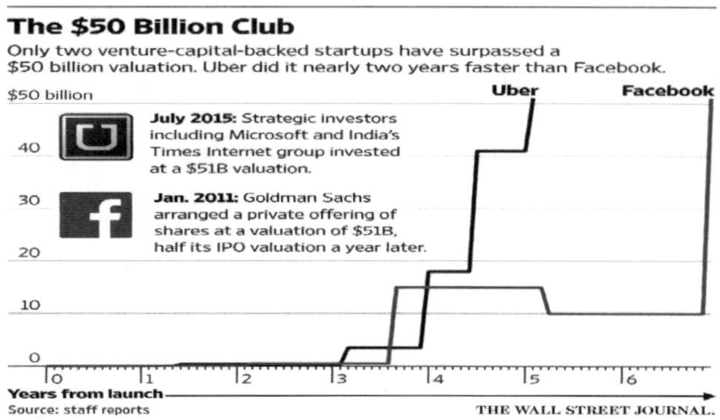

Figure 3 : Comparative value of Investment between Uber and Facebook [7]

There are several obstacles which are now faced by online-based businesses, such as regulatory readiness to adopt this technology. Until now for example, Uber itself is still operating "illegally" in the city of Jakarta, because it has not met the four conditions proposed by the Provincial Government of DKI Jakarta. The four requirements are: having legal existence in the form of Foreign Investment (PMA), paying tax according to the regulation, having adequate insurance, and ensuring that the rental vehicle that become partner follow the motor vehicle eligibility test or KIR.

In the era of ICT, it is difficult not to adopt the power of crowd share technology. Various logistic activities are perceived to be more efficient and effective as well as productive when run by utilizing the various features and conveniences offered by this technology. Even in the field of health now can be served through the facility on line doctor. For example known service "Heal", namely - a service in Los Angeles and San Francisco that provides the facility to ask a trained doctor visiting a user's home or wherever they are needed immediately [7]. This service charges a flat rate of $ 99 per house per call. Through this service open a new business opportunity, no need to establish a building as a consultation place for doctors, because the doctor directly to visit the patient. This service program has attracted many investors to join it. Similarly, the application on Uber, the patient will be notified when the doctor would arrive at a place with a map and application contacts.

5. Conclusion

From various findings of this study there are some interesting conclusions are as follow :

1. In the past, business or trade was done by physical agents in physical space through physical processes and traded products were physically shaped as well. Due to the development of ICT, nowadays business has adopted the principles of Virtual Corporation (VC). The concept of VC is basically like a cooperative capable of integrating various sources and competencies without having to be physically. The are four main characters of Virtual Corporation that are : 1) Differentiation, 2) Configuration, 3) Integration and 4) Technology. Some forms of virtual corporation among them are E-Business, E-commerce, E-government.
2. E-Business, E-Commerce, has been highly developed in the world. The use of ICT is not only limited to administrative processes and product sales (E-Mall), but also in the service industry (Taxi online, logistic services, food delivery). By using internet, the reach of the market is increasingly widespread. A product that was previously only known locally or regionally now can be recognized globally.
3. The internet users in Indonesia are increased progressively from year to year. In 2016, internet service usage in Indonesia has reached 132.7 million people from 256, 2 million population of Indonesia or about 51.8%. The real sector of trade in goods and services industry has adopted many principles of virtual corporation in term of E-business or E-commerce. Therefore, the e-commerce

IOP Conf. Series: Materials Science and Engineering **277** (2017) 012073 doi:10.1088/1757-899X/277/1/012073

in Indonesia is growing rapidly as well, but is often hampered because the available regulations have not been able to touch this world. Information technology can't be dammed, therefore the capabilities and convenience offered by ICT technology should be utilized.

4. In order to develop business based on ICT we have to prepare a certain prerequisite related to people's readiness, government regulation and the availability of supporting facilities.

6. References

[1] APJII 2016 *Penetrasi dan Perilaku Pengguna Internet Indonesia 2016.*

[2] Amant K and Pavel Z 2005 *Internet-Based Workplace Communications: Industry & Academic Application*, Information Science Publishing, Hershey • London • Melbourne • Singapore.

[3] Ann Byers 2006 Jeff Bezos: The Founder of Amazon.com (*Internet Career Biographies*)

[4] Arbnor I and Bjerke B 1997 *Methodology for creating business knowledge*, 2nd edition, SAGE Publications.

[5] Bultje R and van Wijk J 1998 Taxonomy of Virtual Organisations, based on definitions, characteristics and typology. *VoNet: The Netwsletter @ http://www.virtual-organization.net*, **2**(3), 7-20

[6] Bremer C, Walz M, Molina A and Eversheim W 1999 Global Virtual Business - A Systematic Approach For Exploiting Business Opportunities in Dynamic Markets. *International Journal of Agile Manufacturing*. Vol. 2, Issue 1.

[7] David Jinks MILT 2016 The uberfication of Deliveries: Why Uber's move into logistic will transform the delivery market and encourage the Sharing Economy Revolution, parcelhero.com.

[8] http://thenextweb.com/uk/2012/06/29/uber-officially-launches-its-private-driver-for-hire-service-in-london/

[9] http://vator.tv/news/2011-12-07-uber-pulls-in-another-32m-for-app-based-car-service

[10] http://tekno.kompas.com/read/2016/10/24/15064727/2016.pengguna.internet.di.indonesia.capai.132.juta.

[11] Kim Eriksen, Claus Hemmingsen 2008 " Online Marketing New models of advertising?, *Thesis Master Degree*, Aalborg University Cand.Merc.Ibe Fibigerstræde 2 DK 9220 Aalborg Øst.

[12] Mertens P and Faisst W 1995 Virtuelle Unternehmen - eine Organisationsstruktur für die Zukunft, *http://www-bior.sozwi.uni-kl.de/tum/tum:2:95/295mer.htm; Stand 22.09.97*,.

[13] Klüber R, Alt R and Österle H Emerging Electronic Services for Virtual Organizations - Concepts and Framework 1999 Institute for Information Management, University of St. Gallen, Switzerland.{roland.klueber; rainer.alt; hubert.oesterle}@unisg.ch

[14] S. Balint and Kourouklis A 1998 The Management of Organisational Core Competencies in Organizational Virtualness, P. Sieber and J. Griese, Eds. Bern: Simowa Verlag, pp. 165-172.

[15] Schräder 1996 *Managment virtueller Unternehmungen: organisatorische Konzeption und informations-technische Unterstützung flexibler Allianzen.* New York: Campus,.

[16] Ishaya T and Macaulay L 1999 The Role of Trust in Virtual Teams , Department of Computation, University of Manchester Institute of Science and Technology, England. {mcaigti3, lindam}@co.umist.ac.uk

[17] Ulrich Franke, (Cranfield University, England), Bernd Hickmann, (RWTH Aachen, Germany) 1999 Organizational Virtualness and Electronic Commerce *Proceedings of the 2nd International VoNet - Workshop Zurich, Is the Net-Broker an Entrepreneur?* What Role does the Net-Broker play in Virtual Webs and Virtual Corporations? September 23-24, Simowa Verlag Bern.

[18] Ulrich J and Franke BA 1999 Organizational Virtualness and Electronic Commerce *Proceedings of the 2nd International, VoNet - Workshop Zurich The Concept of Virtual Web Organizations of Virtual Web* September 23-24, Simowa Verlag Bern

IOP Conf. Series: Materials Science and Engineering **277** (2017) 012074 doi:10.1088/1757-899X/277/1/012074

Feasibility model study for Blumbangreksa product model based on lean startup method

A K Pakpahan[1],W S Dewobroto[2] and R Y Pratama[3]

Organization and Business Development Laboratory, Industrial Engineering, Faculty of Industrial Engineering, Trisakti University, Jakarta, Indonesia
[1]arnolt.pakpahan@trisakti.ac.id

Abstract. Based on the data from Ministry of Maritime Affairs and Fisheries in 2015, the productivity of shrimp farmers in Indonesia is still below China, India and Thailand, because of the low survival rate of shrimp seeds were planted in Indonesia. Water quality factors become a significant factor that increasesthe survival rate of shrimp seeds plantation, therefore team of PT. Atnic EkoteknoWicaksana create a tool called Blumbangreksa that able to monitor water quality of shrimp farms, measure temperature, salinity, pH, DO (dissolved oxygen), TDS (total dissolve solid) in water and moist air over the surface of the water and GSM -based and Internet of things. Based on the research results, unique value proposition of Blumbangreksa products is the measurement result of water quality are accurate, real-time measurements, based on Internet of things and have the ability measurements at once. Based on the feasibility study using the opportunity assessment of Marty Cagan, it can be seen that the product has fulfilled ten elements of assessment opportunity, so Blumbangreksa products are considered feasible. Initial investment fund of Blumbangreksa products is Rp 1,369,856,574, with profitability index of 1:51 and average breakeven products each year as many as 18 products are sold, and the payback period for 4 years and 2 months, therefore the business of Blumbangreksa product is feasible.
Keywords: Lean Canvas, Experiment Board, Opportunity Assessment, Blumbangreksa

1. Introduction

1.1. Background

The current area of shrimp in Indonesia is around 344.759 ha or around 39.78% from the potential lands roommates are available around 866.550 ha in Indonesia [10]. Based on the area of cultivated, 80% of the main problem of this farming business is low productivity of shrimps with 267 kg/ha, considered as low number if compared with China (538 kg/ha), India (750 kg/ha) and Thailand (2444 kg/ha).

This problem caused by low survival rate of shrimp seeds planted. Water quality factor become significant aspect of shrimp seed survival rate. Therefore, team of PT. Atnic EkoteknoWicaksana creates a water quality gauge called Blumbangreksa whose monitoring water quality of shrimps, measuring temperature, salinity, pH, DO (dissolved oxygen), TDS (total dissolve solid) in water and humidity of water surface based on GSM and Internet of things.

IOP Conf. Series: Materials Science and Engineering **277** (2017) 012074 doi:10.1088/1757-899X/277/1/012074

1.2. Scope of Issue

Based on the background, Blumbangreksa did not have a valid product design business model for customer needs, so it can be at risk of failing when it hits the market. [5] stated that innovation is feasible if 10 opportunity assesment is reached.

1.3. Research Purposes

The purpose of this research includes:

1. To design value proportion from Blumbangreksa product.
2. To design business model for Blumbangreksa product.
3. To create a business model based on the feasibility Blumbangreksa business model.

2. Theoretical Background

2.1. Lean Startup

The basis of the lean startup is minimizing waste to improve the frequency of contact with the customer, so that the trial could be done and to avoid falsemarket assumption as early as possible. Lean Startup method could be done with creatingbusiness model using several tools, which are questionnaire, empathy map, experiment board, and lean canvas. Application in new business is to start with WHO hypotheses (candidate) customer, what are the demand, how much and etc. Next, the customer's hypotheses are validated towardcustomer to determine whether it fulfilled costumer's demand [1].

2.2. Eight Quality Dimensions

This dimension is divided into 8 (eight) categories, those are performance, performance which related tobasic operation characteristics of a product. Features is product characteristic which designed to improve function or increase costumer's interests. Durability is the length of a product's life before that product must be replaced. Conformance is how far the characteristic of a product meets certain specificaion from costumer. Reliability is possibility of a product to work smoothly within specific period of time. Serviceability is ease, speed, and ability to repair the product. Aesthetics is size, design, smell, and color of the product. Perceived Qualities is a customer's perspective about acquired product quality.

2.3. Empathy Map

Empathy Map was created because customer profile will determine the making of a better value proposition, a way of approach with the customer to be more comfortable and ways to engage with costumer to be more appropriate.Where in the end will provide a deeper understanding of how customers get satisfaction on the value of products / services when compared with the price paid [8]. Here are four aspects of making an empathy map: Say (What the respondent says), Think (What respondents think), Feel (What respondents feel), and Do (What respondents do).

2.4. Experiment Board

Experiment Board is a tool used to measure the validity of a measurement against a research or survey based on assumptions so that customer desires can be translated well [9]. Experiment board is used to validate customer needs and once validated will result in MVP (Minimum Viable Product).

2.5. Lean Canvas

Lean canvas is a template business model that is widely used for businesses that are just starting or usually called startup business. This tool cannot be used for business that has been developed and running. While in the business model canvas widely used to identify businesses that has been running because the method of business tools canvas model is very applicable to the business that has been running [6].

3. Research Method

Research methodology is a systematic thinking process consisting of research stages that must be done in order to find solutions to solving a problem. Research methodology is used as a guide during research to be more directed towards a solution and conclusion. The following are the steps in data processing:

3.1. Questionnaire

The research questionnaires were distributed to 100 respondents in 5 (five) regions: DKI Jakarta, D.I Yogyakarta, Karawang, Lampung and Pemalang to know the characteristics of potential customers and the importance of Blumbangreksa product attributes.

3.2. Interview using Empathy Map

Interviews using empathy maps were conducted to identify the problems of potential customers of Blumbangreksa, interviews were conducted on 10 prospective customers in 5 (five) areas: DKI Jakarta, D.I Yogyakarta, Karawang, Lampung and Pemalang.

3.3. Validation using Experiment Board

Validation using experiment board is done to test the results obtained from emphaty map. At this stage, the experiment board validation was conducted on 10 (ten) prospective consumers of Blumbangreksa product in 5 (five) areas: DKI Jakarta, D.I Yogyakarta, Karawang, Lampung and Pemalang.

3.4. The minimum viable product (MVP)

This stage is done to determine the interest and opinion or feedback of potential customers of Blumbangreksa product that have a unique value proposition. Testing MVP conducted to 10 (ten) prospective consumers of Blumbangreksa products.

3.5. LeanCanvas

Lean canvas will explain the business model for Blumbangreksa product, in this business model there are 9 (nine) parts, namely: problem, solution, unique value proposition, customer segment, key metrics, channels, unfair advantage, cost structure and revenue stream. This lean canvas result is obtained through focus group discussion to Blumbangreksa development team and prospective customer of Blumbangreksa product.

3.6. Business Feasibility Study using Opportunity Assessment

Opportunity Assessment is a method used to test the feasibility of a lean canvas business model. The criteria worthy or not worth a product is based on the fulfillment of 10 (ten) questions that should be answered by the business idea, following 10 (ten) questions opportunity assessment:

1. What is the problem that Blumbangreksa product will solve? (Problem, Solution and Unique Value Proposition)
2. Who is the target market of Blumbangreksa product? (Customers Segment)
3. How big is the Blumbangreksa product opportunity? (Market Size)
4. What are the alternative product of Blumbangreksa? (Competitor)
5. What is the differentiator of Blumbangreksa product? (Unfair Advantage)
6. Why Now? (Window Market)
7. How to introduce Blumbangreksa product to market? (Go-to-Market Strategy, Channels)
8. How does the success of Blumbangreksa product make money? (Revenue Stream)
9. What are the key factors that measure the success of Blumbangreksa products? (Key Metrics)
10. Based on the 9 (nine) statements above, is Blumbangreksa product feasible as a business? (go-or-no go)

4. Result and Discussion

4.1. Questionnaire

4.1.1. Introduction of Questionnaire. Questionnaire preliminary questions about water quality measuring device attributes of Blumbangreksa, where each attribute is classified based on 8 (eight) quality dimensions.

Table 4.1. Blumbangreksa Questionnaire Attribute

No.	Attributes	Dimension
1	Gauge water quality improving living shrimp life.	Performance
2	Water quality measuring device can be used various types of shrimp.	Features
3	Gauge water quality indicators that are easy to understand.	
4	Gauge water quality has an affordable price.	Conformance
5	Water quality measuring device is easy to operate.	
6	Gauges are easy to obtain water quality in the market.	
7	Gauge water quality to provide accurate information.	Reliability
8	Gauge water qualitybased on in real timethe Internet.	
9	Gauge water quality has good sales.	Serviceability
10	Gauge water quality come with manuals.	
11	Gauge water quality can be used long term.	Durability
12	Gauge water quality has materials that are not easily corroded	
13	Gauge water quality has a bright color for easy viewing or found	Aesthetics
14	The exterior gauge water quality with functional instructions use a tool (eg, the hint button operation)	
15	Gauge own water quality certificate ISO	Perceived quality
16	Gauge water quality has been licensed by the Ministry of Marine and Fisheries.	

4.1.2. Questionnaire Research and Behavioral Characteristics of Respondents. Based on questionnaire contains characteristics and user behavior gauge water quality can be concluded that:

a. Works respondent is a collector ornamental shrimp 11%, 79% farmers, self-employed 9%, fisheries extension 1%.
b. All respondents (100%) ever gauges water quality.
c. Older respondents in the shrimp business is less than 1 year 15%, 1-2 years 30% 2-3 years 20%, and more than 5 years by 25%.
d. Based on the age of the respondents that respondents aged under 20 years as much as 1%, aged 30-39 years were 48%, aged 50 years or more as much as 4%.
e. For the ideal price, selected respondents showed 4% of respondents chose under IDR. 1,000,000. For the ideal price Rp 1,000,000 - Rp 9,000,000 selected by 21% of respondents, for the ideal price of USD 10,000,000 - USD 19,000,000 was chosen by 17% of respondents, for the ideal price of USD 20,000,000 - USD 29,000,000 was chosen by 49% of respondents, for the ideal price of Rp 30,000,000 - Rp 39,000,000 selected by 7% of respondents, and for the ideal price above Rp 40,000,000 selected by 2% of respondents.
f. On the question questionnaire about the frequency change of the measuring instrument water quality, for respondents who have never made the turn as much as 1%, to turn the tool needed 55% of respondents, for the turn of the tool every month to do 10% of respondents, to turn the tool every 3 months carried out by 1% of the respondents, to turn the tool annually performed annually carried by 28% of respondents, for a turnover of measuring instruments over 3% a year, for respondents who did over 2 years as much as 2%.
g. At the frequency of use of measuring instruments water quality of 15% of respondents take measurements 2 times a day, 30% of respondents take measurements once a day, 6% of

respondents take measurements once a week, and 46% of respondents take measurements of water every time,

4.1.3. Blumbangreksa Questionnaire Research Importance Attributes. Based on questionnaires attributes of users gauge water quality obtained the following results:

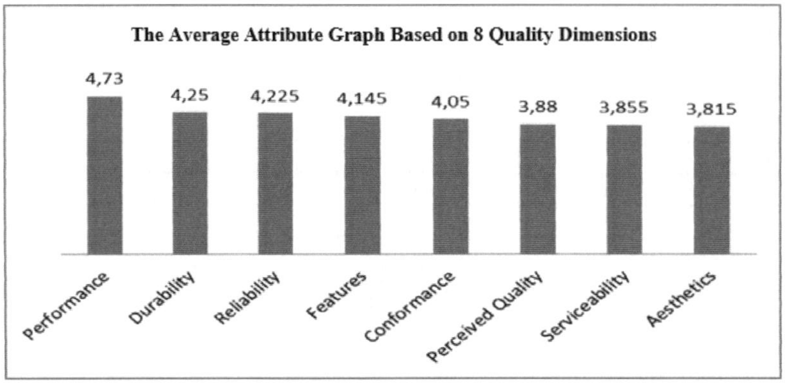

Figure 4.1 Questionnaire Graph Based On 8 (eight) Quality Dimensions

From these results can be seen that the dimensions of performance, durability, and reliability into dimensions that have the greatest average value of this is because the factors in these 3 (three) dimensions are considered respondents directly affect the shrimp seedling survival rate so that affect the productivity of the harvest. In the questionaire results can be seen the average value of the performance dimension is 4.73. Furthermore, the durability dimension has the second highest average value with a value of 4.25. Next dimension reliability has the third highest average value with a value of 2.225.

While the dimensions of features, conformance, and perceived quality are considered important enough for the respondents because the factors in these three dimensions are considered to have an effect on the main function of the Blumbangreksa product. Furthermore, the features dimension has the fourth highest average value with a value of 4.145, the tool factor can be used on various types of shrimp and has an easily understandable indicator considered quite important on the water quality gauge. The conformance dimension has the fifth highest average value with a 4.05 value, the tool factor has an affordable price and is easily obtainable in the market. The perceived quality dimension has the sixth highest average value of 3.88.

In the aesthetics and serviceability dimensions these two dimensions have the lowest average value because they do not have a direct impact on shrimp seedling survival associated with shrimp harvest productivity. Dimensions of aesthetics with an average grade of 3.815 and serviceability with an average grade of 3.855.

4.2. Empathy Map

Empathy map obtained by an interview to 10 respondents who had used the water quality measuring device. There are four important aspects of Empathy folder that Say, Do, Think and Feel. These factors will support the results of the questionnaire as the map empathy mind of the consumer side can be known.

4.3. Experiment Board

Experiment board was made to validate the result of empathy folder following results of the experiment board prospective users gauge the quality of water (figure 4.3).

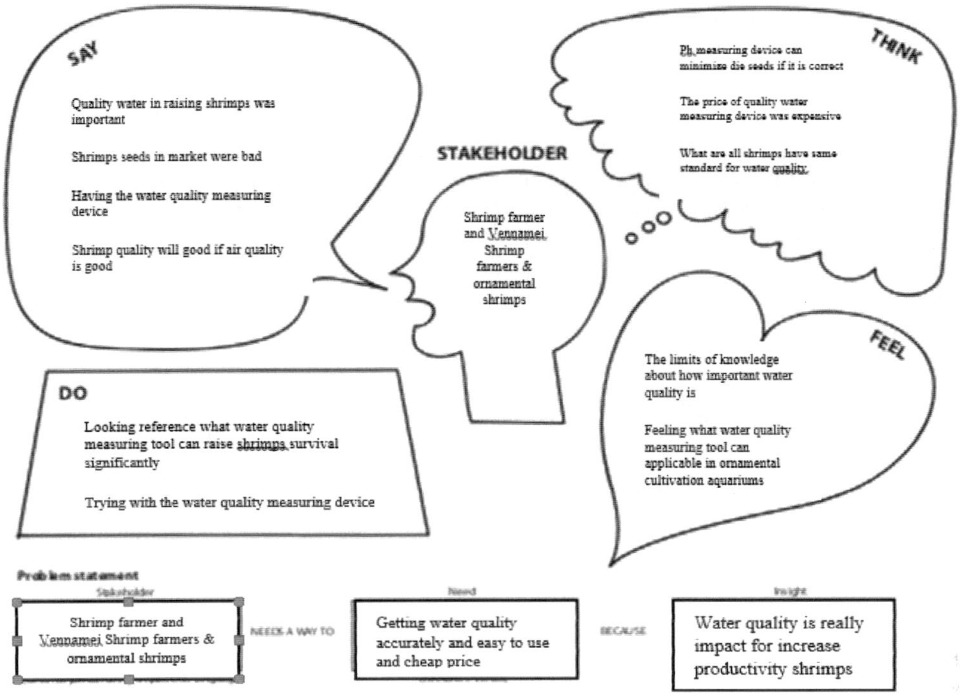

Figure 4.2. Empathy Map

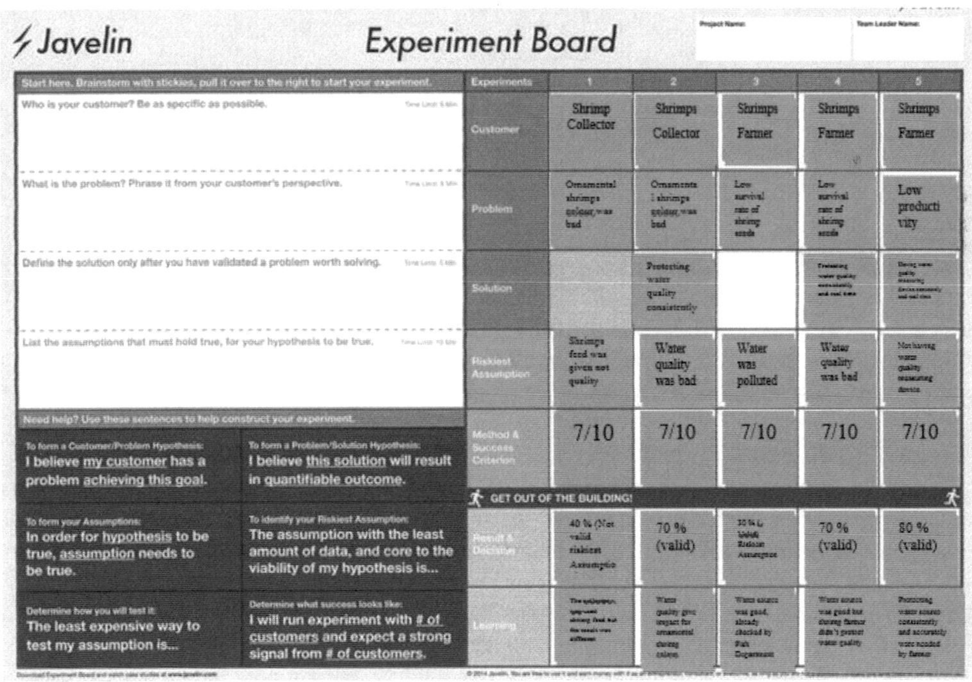

Figure 4.3. Experiment Board

From the experiment board result, the collector ornamental shrimp and the shrimp farmers obtain valid results, where as valid problem is the color of ornamental shrimp is not maximum, low survival rate shrimps, and low productivity because low survival rate of shrimp seeds. Valid solution must protect water quality consistently and real time with the valid riskiest assumption is water quality is not maintained properly and the user does not have a measuring device water quality.

4.4. Minimum viable Product

MVP (minimum viable product) is a product concept with a minimum feature into the market for market testing [9]. MVP for Blumbangreksa product is gauge water quality with a high accuracy in measurement because it uses a sensor Dissolved Oxygen Sensors, Low Cost Epoxy Body Conductivity Sensor, Low Cost Submersion and In-Line pH Electrodes, EZO ™ Dissolved Oxygen Circuit, EZO ™ Conductivity Circuit, EZO ™ pH Circuit, PWR-ISO, 1m Digital Thermal Sensor DS18B20 are all imported from the United States. In the video MVP Blumbangreksa product has few things to offer such as:

- Has the accuracy of height measuring kuatas water.
- Based Internet of things so as to facilitate access to data.
- Based on GSM to areas not reached the internet yet.
- Have the ability to measure simultaneously.

MVP tried to be introduced to the 10 (ten) potential customers in regions : DI Yogyakarta Pemalang, Karawang and Lampung. 9 (nine) out of 10 (ten) people interested in the product potential customers of Blumbangreksa.

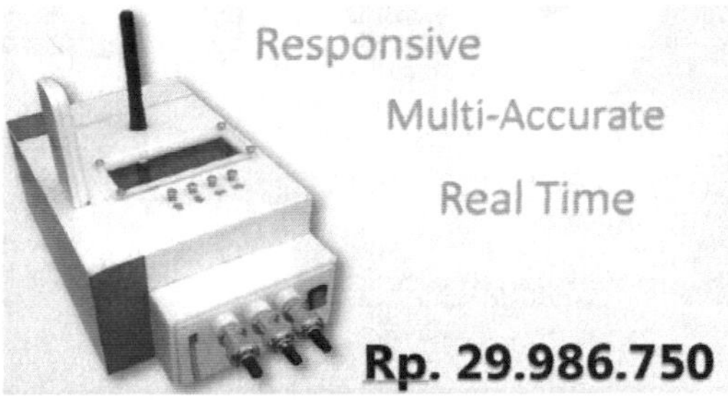

Figure 4.4. Figure Illustration of MPV Blumbangreksa

4.5. Lean Canvas

Leancanvas was created to describe a business model Blumbangreksa product and has an unique value proposition to attract customers (figure 4.5).

Lean canvas will used as a reference for the feasibility study of Blumbangreksa product by 9 (nine) blocks that customer segment, unique value proposition, channels, unfair advantage, key metrics, problem, solution, cost structure and revenue streams.

4.6. Business Feasibility Study Using Opportunity Assessment

In a feasibility study the business model product Blumbangreksa using the criteria of opportunity assessment by [5]. There are ten (10) questions or aspects that must be answered to decide whether the business model is mainly can be said to be proper or improper. In the following Blumbangreksa products opportunity assessment conducted based on the 9 (nine) statements whether the product Blumbangreksa feasible as a business? (Go-or-no go). Based on 9 (nine) answers from questions for the opportunity assesment of Blumbangreksa products criteria, this is conclusion from opportunity assessment of Blumbangreksa product (table 4.2)

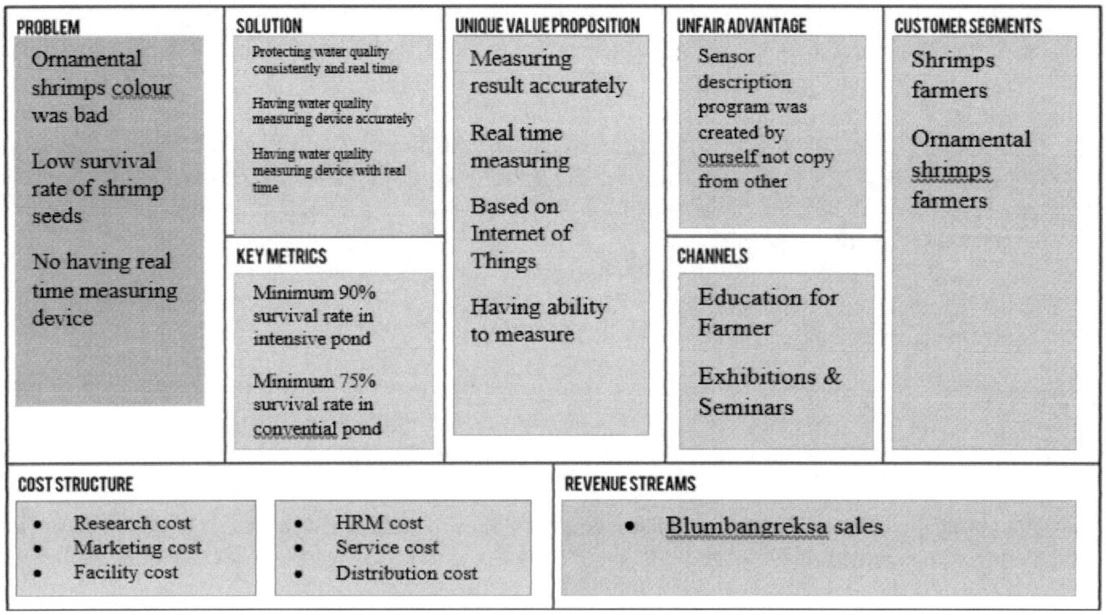

Figure 4.5. Lean Canvas Blumbangreksa Product

Table 4.2. Table Conclusions Opportunity Assessment

No.	Statement	Go or No Go?
1	Unique Value Proposition	Has 4 unique value
2	Customer Segments	Shrimp Farmers and Ornamental Shrimp Farmers
3	Market Size	866.550 ha
4	Competitor	Sensor less and manual measurements
5	Unfair Advantages	Program sensor
6	Market Window	It takes
7	Go to Market Strategy	Education, seminars and pilot
8	Cost Structure & Revenue Stream	Worthy
9	Key Metrics	90% survival rate (intensive), 75% (traditional) and 37% increase in sales each year
10	Business Model feasible Go or No Go?	Go / Eligible

5. Conclusion

In Blumbangreksa product research concluded 3 (three) conclusions as follows:

1. The unique value proposition of Blumbangreksa product is the result of water quality measurement was accurate, water quality measurement with real time, water quality measurements based on internet of things, the measurement can be done once at the same time.
2. The lean canvas of Blumbangreksa product has the customer segmentsare shrimp farmers and collector ornamental shrimps. This product has a unique value proposition that has a water quality measurement results are accurate, real-time measurement, Internet-based of things, have the ability to measure time. Blumbangreksa product had key metrics in the form of survival rate a minimum of shrimp seeds to 90% for shrimp farms with intensive systems, survival rate the shrimp seedling of at least be 75% for shrimp farms the traditional system, and increased

IOP Conf. Series: Materials Science and Engineering **277** (2017) 012074 doi:10.1088/1757-899X/277/1/012074

product sales by 37% per year. The cost structure of Blumbangreksa product is was used include initial investment costs, raw material costs and fixed costs of products, so the initial capital investment for Blumbangreksa product business is Rp 1,369,856,574. Revenue stream company's revenue through sales of products Internal Rate of Return of 33.29% and Profitability Index itsof 1.51, which means decent. So the Payback Period of Blumbangreksa product business is 4 years 2 months and break event point in the first year is 15 units, the sales of second year is 16 units, the sales of third year is 18 units, the sales of fourth year is 19 units and the sales of fifth year is 21 units.

3. In a feasibility study for a Blumbangreksa product business model has opportunity assessment criteria was stated in the Blumbangreksa product business model are feasible base on 10 (ten) assessment opportunity criteria's.

6. References

[1] Blank S and Dorf B 2012 *The Startup Owner's Manual.* California: K&S Ranch, Inc.

[2] Brown T and Wyatt J 2010 *Design Thinking for Social Innovation.* California: Stanford Social Innovation Review.

[3] Dewobroto W S 2014 *Journal of Business Model Generation and Lean Startup Method as The Basis for Business Development Feasibility Study, Case Study of PO. Gajah Mungkur Sejahtera.* Jakarta: Fakultas Teknologi Industri, UniversitasTrisakti.

[4] Kasmirand Jakfar 2003 *Studi Kelayakan Bisnis*. Jakarta: Kencana.

[5] Marty Cagan 2008 *How to Create Products Customers Love*, SVPG Press.

[6] Maurya A 2012 *Running Lean, Second Edition.* United States: O'Reilly Media, Inc.

[7] Mueller R and Thoring K 2012 *Design Thinking vs. Lean Startup: A comparison of two user-driven innovation strategies.* Berlin School of Economics and Law, Anhalt University of Applied Sciences.

[8] Osterwalder A and Pigneur Y 2010 *Business Model Generation: A Handbook for Visionaries, Game Changers, and Challengers.* New Jersey: John Wiley & Sons, Inc.

[9] Soekanto S 2000 *Sosiologi Suatu Pengantar.* Jakarta: King Grafindo Persada.

[10] Ministry of Marine Affairs & Fisheries 2015 *http://statistik.kkp.go.id/sidatik-dev/Publikasi/src/kpda2015.pdf*

Feasibility study analysis for multi-function dual energy oven (case study: tapioca crackers small medium enterprise)

N W Soraya[1], R M El Hadi[2], E Chumaidiyah[3], and W Tripiawan[4]

Industrial Engineering Program, Telkom University, Bandung, Indonesia
[1]nendenwidha@gmail.com, [2]rosadmeh2014@gmail.com,
[3]endangchumaidiyah@yahoo.co.id, [4]wawan.tripiawan@gmail.com

Abstract. Conventional drying process is constrained by weather (cloudy / rainy), and requires wide drying area, and provides low-quality product. Multi-function dual energy oven is the appropriate technology to solve these problems. The oven uses solar thermal or gas heat for drying various type of products, including tapioca crackers. Investment analysis in technical, operational, and financial aspects show that the multi-function dual energy oven is feasible to be implemented for small medium enterprise (SME) processing tapioca crackers.
Keyword: Feasibility Study Analysis, Multi-Function Dual Energy Oven, Incremental Rate Of Return (Incremental ROR), Tapioca Crackers.

1. Introduction

Until today, most of industries is still implementing conventional process in produce their product. One of these are utilize sunlight as the only one source of heat energy for drying the product. The process is done by storing the product outdoor to be able to exposed the sun's heat directly. Then, there are several challenges to be faced by small medium enterprise (SME).

First challenge is the need for a wide area as a place of drying. Spacious of drying land owned by small medium enterprise (SME) can also affect the amount of product that can be produced. If small medium enterprise (SME) have a wide area as a place of drying, more products can be dried and produced, and vice versa. So, when the market demand for the product has increased, small medium enterprise (SME) need to consider wide area of drying, the amount of product to be dried, and the length of time for drying the product until it is completely dry. It would be very unfortunate if a large market opportunity cannot be fulfilled by the small medium enterprise (SME) because the limited area of drying.

Next challenge faced by small medium enterprise (SME) are still utilizing the sunlight as the only one source of heat energy for drying the product. When cloudy or rainy, then the product being dried outdoor must be secured by the workers. But it would be more worried if the wind blew hard and damaging the product. small medium enterprise (SME) would worry because the weather can change suddenly. When the product is secured, then the drying process can be continued for the next day. But, if the product is damaged by the wind, rain, or fungus (temperatures are too humid) then it can cause great harm for industries.

Other challenge that need to be considered by small medium enterprise (SME) when drying the product conventionally is the contamination of germs or dirt that comes from the wind. small medium enterprise (SME) needs to pay attention to the equipment used, clean environment, away from garbage or pollution. Considering that the product will be directly contact with customers, especially food products that will enter into the human body.

Oven multi-function dual energy is one of the appropriate technology that designed based on the various challenges faced by small medium enterprise (SME) as described previously. As the name implies, the machine has two types of energy or heat source that can be used to dry the product. The heat energy can be derived from sunlight and also gas. One such energy can be selected for use by weather conditions or by the needs of the small medium enterprise (SME).

Sunlight or solar thermal can be used when the weather is bright. The roof of oven multi-function dual energy can be opened so that the sun's heat can get into the oven space. Whereas when the weather is cloudy, rainy, or when small medium enterprise (SME) want to do the production process in the evenings (overtime), then the drying process can use the heat energy generated from the gas fuel. Oven space must be covered by a roof when the drying process using heat energy generated from the gas fuel. This technology able to dried various type of products (multi-function), one of them is tapioca crackers.

2. Methods

2.1 Forecasting

A seasonal pattern is a repetitive increase and decrease in demand. There are several methods for reflecting seasonal patterns in a time series forecast. We will describe one of the simpler methods using a seasonal factor. A seasonal factor is a numerical value that is multiplied by the normal forecast to get a seasonally adjusted forecast. One method for developing a demand for seasonal factors is to divide the demand for each seasonal period by total annual demand, according to the following formula [1]:

$$Si = \frac{Di}{\sum D} \qquad (1)$$

where
Di : Historical data

2.2 Depreciation

Straight line method assumes that the loss in value is directly proportional to the age of the asset [1].

$$dk = \frac{B-S}{N} \qquad (2)$$

where
dk : annual depreciation deduction in the k-th year
B : unadjusted cost basis
S : salvege value at the end of the depreciable life of the asset
N : depreciatable life of the asset in year

2.3 Feasibility Study

Feasibility study is to reduce the level of risk of loss and ensure that the investment to be made is profitable [2][3].

2.3.1. Technical and Operational Aspect Analysis

The material will be covered in the technical and operational aspects, such as product design, raw material and product specifications, production processes, the location of the company (factory), factory layout, equipment and machines used, the layout engine, as well as quality control [2].

2.3.2. *Financial Aspect Analysis*

Financial aspects discuss about capital requirements and investment needed in the establishment / development effort is planned, then summarize it in the form of financial statements (balance sheet, income sheet, and cash flow), and analyze them to determine feasibility of the business [2].

A. Net Present Value (NPV)

Net Present Value (NPV) is the present value (present value) of the difference between the benefit (benefit) at cost (cost) at a certain discount rate. NPV is an excess benefit (benefit) compared to the cost / expense.

$$NPV = \sum_{t=0}^{t=n} \frac{B_t - C_t}{(1+i)^t} \tag{3}$$

where

Bt : Benefit / cash inflow in year t

Ct : Cost / cash outflow in year t

i : interest rate

B. Internal Rate of Return (IRR)

Internal Rate of Return (IRR), investor's method, discounted cash flow method, or profitability index solve for the interest rate that equivalent worth of an alternative's cash flow (receipts or savings) to the equivalent worth of cash out flows (expenditures, including investments).

$$\sum_{k=0}^{N} Rk(\tfrac{P}{F}, i\%, k) = \sum_{k=0}^{N} Ek(\tfrac{P}{F}, i\%, k) \tag{4}$$

*i is often used in place of i to mean the interest rate that is to be determined

where

Rk : net receipts or saving for the k-th year

Ek : net expenditure including investments for the k-th year

N : project life (or study period)

Once i' has been calculated, it is then compared with the MARR to assess whether the alternative in question is acceptable. If i' ≥ MARR the alternative is acceptable; otherwise, it is not.

C. Incremental Rate of Return (incremental ROR)

This policy can be reduced to three principles when applying rate of return methods:

a) Each increment of capital must justify itself by producing a sufficient rate of return on that increment;
b) Compare a higher investment alternative against a lower investment alternative only when the latter is acceptable;
c) Select the feasible alternative that requires the largest investment of capital as long as the incremental investment is justified by saving that earn at least the MARR.

D. Sensitivity Analysis

Sensitivity analysis means the relative magnitude of change in the measure of merit (such as present worth) caused by one or more changes in estimated study parameters. Sometimes sensitivity is more specifically defined to mean the relative magnitude of the change in one or more factors that will reserve a decision among alternatives.

IOP Conf. Series: Materials Science and Engineering **277** (2017) 012075 doi:10.1088/1757-899X/277/1/012075

3. Research Methodology

3.1. Initial Research

3.1.1. Background of Problem

Background of problem is obtained from the field studies and literature review. Where, small medium enterprise (SME) still utilize the sunlight to drying their product, Small medium enterprise (SME) requires a wide area for drying, need a long duration for drying the products when the weather is cloudy / rainy, and the level of products cleanliness is low.

3.1.2. Statement of the Problem

Statement of the problem is a problem of analysis feasibility study for investment of oven multi-function dual energy based on the background described previously.

3.1.3. Goal Setting

The purpose of this feasibility study is to determine the appropriate solutions for small medium enterprises (SME) in improving the effectiveness and efficiency of the drying process of the product to make investments of the oven multi-function dual energy. This feasibility study will be reviewed by the technical, operational, and financial aspects.

3.2. Data Collection Stage

3.2.1. Data Collection of Technical and Operational Aspect

Technical and operational data collected in this feasibility study comes from interviews, observation, experiment, and also related institutions include the District Office, the Central Statistics Agency (BPS) and the Meteorology, Climatology, and Geophysics (BMKG).

3.2.2. Data Collection of Sales and Production Capacity

Sales and production capacity data will represent the amount of income earned for small medium enterprise (SME) that processing tapioca crackers.

3.2.3. Data Collection of Financial Aspect

The data collection of financial aspects includes the investment and operational costs by data collection in technical and operational aspects. In addition, there is also a data entry from the sales and production capacity as a revenue/income.

3.3. Data Processing

3.3.1. Data Processing of Technical and Operational Aspect

The data processed includes oven's capacity, location, the need for tools and materials, until building process of the multi-function oven dual energy. From technical and operational aspects will also be acquired investment costs, operating costs, up to the cost of treatment of oven multi-function dual energy which would be used on data processing of financial aspects.

3.3.2. Data Processing of Sales and Production Capacity

Sales and production capacity which useful as revenue or income which would be used on data processing of financial aspects.

3.3.3. Data Processing of Financial Aspect

IOP Conf. Series: Materials Science and Engineering **277** (2017) 012075 doi:10.1088/1757-899X/277/1/012075

Financial aspects include the investment costs, operating costs, sources of funds, then the revenue will be calculated on the financial statements include an income statement, cash flow, and balance sheet. In addition, the processing of financial data also includes the calculation of investment feasibility criteria, incremental ROR, and sensitivity analysis.

3.4. Analysis Stage

3.4.1. Analysis of Investment Feasibility

The feasibility analysis of investment aims to answer questions about the feasibility of the investment of oven multi-function dual energy when it is implemented in small medium enterprise (SME) that processing tapioca crackers.

3.4.2. Sensitivity Analysis

A sensitivity analysis was conducted to determine the effect of variables. By looking at the changes in these variables that will be known how much sensitivity of the feasibility.

4. Analysis

4.1. Technical and Operational Aspect

Here are the specifications of the oven multi-function dual energy.

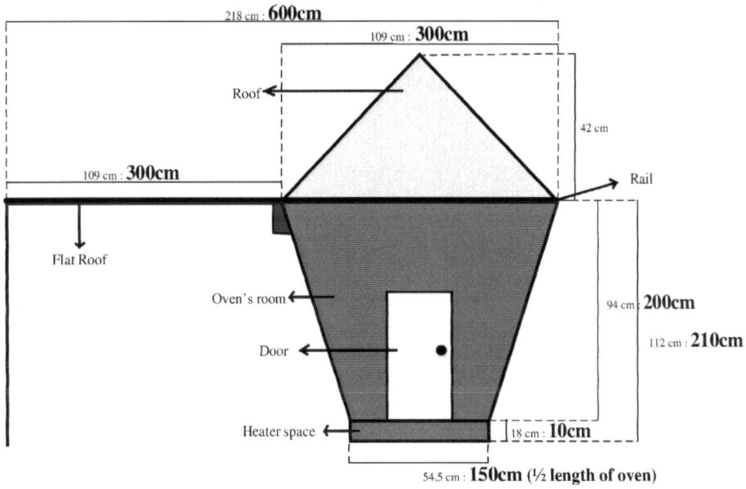

Figure 1. oven multi-function dual energy

When the small and medium enterprise (SME) is not using oven multi-function dual energy, they need to rent a land for drying tapioca crackers about 60 x 10 meters.

Oven multi-function dual energy has a capacity adjusted production capacity of tapioca cracker / day in small medium enterprise (SME). The details of prototype compare with the actual oven multi-function dual energy are:

1) Rack size is 60 x 20 cm (prototype) and 400 x 50 cm (oven);
2) Crackers size is about 3 x 5 cm;
3) Capacity per racks is 60 crackers (prototype) and 1000 crackers (oven);
4) Number of rack is 8 racks (prototype) and 32 racks (oven).

The production capacity of tapioca crackers is 31200 pieces / day on the small medium enterprise (SME). So, the oven's capacity to dry the tapioca crackers is enough.

4.1.1. Result Of The Experiment

Here is the result of experiment (drying tapioca crackers) using prototype oven multi-function dual energy with solar thermal energy

1) Location : Tapioca crackers small medium enterprise (SME) owned by H. Sumarno ST. Rt. 01 Rw. 02;
2) Date : Friday, Mei 27th 2016;
3) Capacity : 480 piece;
4) Result : Temperature 35 ^{o}C; duration 10 hours; and the result is normal.

Then the result of experiment (drying tapioca crackers) using prototype oven multi-function dual energy with gas fuel energy.

1) Location : Tapioca crackers small medium enterprise (SME) owned by H. Sumarno St. *Rt.* 01 *Rw.* 02
2) Date : Saturday, Mei 28th 2016 – Tuesday, Mei 31st 2016
3) Capacity : 480 piece;
4) Result:
   - Temperature 60 ^{o}C; duration 3 hours; gas used 0,8 kg; and the result is normal without roasted;
   - Temperature 70 ^{o}C; duration 2,5 hours; gas used 0,4 kg; and the result is normal without roasted;
   - Temperature 80 ^{o}C; duration 2 hours; gas used 0,6 kg; and the result is normal without roasted.

The result of experiment (drying tapioca crackers) using prototype oven multi-function dual energy with gas fuel shows that the optimum temperature is 70°C with a time of 2.5 hours and spend 0.4 kg (give the lowest cost).

4.2. Sales and Production Capacity

Here's a sales and production capacity of tapioca crackers in 2017, 2018, and 2019 before - after using the oven based on the result of forecasting (seasonal method).

1) In 2017 the sales and production capacity of tapioca crackers is about 317 kg before using the oven and 472 kg after using the oven;
2) In 2018 the sales and production capacity of tapioca crackers is about 330 kg before using the oven and 472 kg after using the oven;
3) In 2019 the sales and production capacity of tapioca crackers is about 340 kg before using the oven and 472 kg after using the oven.

4.3. Financial Aspects

4.3.1. Income Statement

Income statement is calculated based on the estimated profits and losses derived from the production process tapioca crackers in small medium enterprise (SME) processing tapioca crackers.

1) Sales revenue before and after using the oven has a same number;
2) Direct material cost before and after using the oven has a same number;

3) Direct labor cost before using the oven is bigger than after using the oven. It's because the number of roasting labor before using the oven is more than after using the oven;
4) Overhead cost before using the oven is bigger than after using the oven. It's because before using the oven, small medium enterprise needs to pay the rent cost of field/drying area;
5) Depreciation cost before using the oven is cheaper than after using the oven. It's because after using the oven, small medium enterprise has more investment (oven multi-function dual energy);
6) There is no interest because the capital of investment only comes from the owner of small medium enterprise;
7) Based on the government regulation, small medium enterprise needs to pay 1% of tax from their revenue / year;

4.3.2. Feasibility Criteria
Feasibility criterias of investment oven multi-function dual energy include NPV, IRR, and incremental ROR with MARR (Minimum Attractive Rate of Return) is 17%.

Table 1 Feasibility Criteria

	BEFORE USING THE OVEN	**AFTER USING THE OVEN**
NPV	Rp84,089,575	Rp231,741,027
IRR	45.04%	29.29%
INCREMENTAL ROR	46.01%	

The result of calculation above said about investment of oven multi-function dual energy is the best alternative to be implemented in the small medium enterprise (SME) that produce tapioca crackers.

4.3.3. Sensitivity Analysis
Sensitivity analysis was conducted to determine the changes that occur in the variable investment oven multi-function dual energy.

Table 2 Sensitivity Analysis

Increasing the Investment Cost		
Condition	Incremental ROR	MARR
Initial	51.06%	
10%	41.95%	
20%	34.26%	
30%	27.65%	17%
40%	21.89%	
49%	17.31%	
50%	16.83%	

When the cost of investment oven multi-function dual energy is increase until 50%, the number of incremental ROR is 16,83%. So, the best alternative is using conventional drying process again.

5. Conclusion

a. Sales and production capacity in produce tapioca crackers in small medium enterprise (SME) can continue production capacity 1,3 kw tapioca flour/days however the weather is rainy or cloudy using the oven multi-function dual energy (gas heat). The oven multi-function dual energy can dry 32000 pieces of tapioca crackers in one time with temperature 70°C, spend 2,5 hours, and only using 0,8 kg (based on the experiment using prototype of oven multi-function dual energy). It means the investment of oven multi-function dual energy in technical and operational aspects is feasible;

b. The results of the feasibility analysis in financial aspect shows the value of NPV is Rp165,788,070 before using the oven and Rp231,741,027 after using the oven; value of IRR is 22.34% before using the oven and 29.29% after using the oven; as well as the value of incremental ROR is 50.06%. This indicate that the investment of oven multi-function dual energy is the best alternative to be implemented on a small industrial processing tapioca cracker;

c. The limit of investment costs oven multi-function dual energy on small medium enterprise processing tapoka cracker is 49% with an incremental ROR of 17.31% > MARR (17%).
Therefore, when the investment cost of the oven multi-function dual energy increase is higher than 49%, the alternative can use conventional drying process again.

6. References:

[1] DeGarmo E P 1989 *Engineering Economy*. Eight edition. (New York: MacMillan Publishing Company)
[2] Ahmad S 2008 *Studi Kalayakan*. Edisi Kedua. (Jakarta: PT Elex Media Komputindo)
[3] Overton R 2007 *Feasibility Study Made Simple*. Series l. (Martin Books Pty Ltd)